COST-EFFECTIVE SPACE MISSION OPERATIONS
SECOND EDITION

SPACE TECHNOLOGY SERIES

This book is published as part of the Space Technology Series, a cooperative activity of the United States Department of Defense and the National Aeronautics and Space Administration.

Wiley J. Larson
Managing Editor

From Kluwer and Microcosm Publishers:
Space Mission Analysis and Design - Third Edition by Larson and Wertz.
Spacecraft Structures and Mechanisms: From Concept to Launch by Sarafin.
Reducing Space Mission Cost by Wertz and Larson.
Fundamentals of Astrodynamics and Applications by Vallado.

From McGraw-Hill:
Understanding Space: An Introduction to Astronautics - Third Edition by Sellers.
Space Propulsion Analysis and Design by Humble, Henry and Larson.
Cost-Effective Space Mission Operations-Second Edition by Squibb, Boden and Larson.
Modeling and Simulation: An Integrated Approach to Development and Operation by Cloud and Rainey.
Modeling and Simulation for Space Systems by Rainey.
Human Space Mission Analysis and Design by Larson and Pranke with Connolly and Giffen.

Future Books in the Series:
Space Launch and Transportation Systems: Design and Operations by Larson, Kirkpatrick, Ryan and Weyers.
Applied Space System Engineering by Larson, Kirkpatrick, Thomas and Verma.
Applied Project Management for Space Systems by Chesley, Larson and Menrad.

COST-EFFECTIVE SPACE MISSION OPERATIONS
SECOND EDITION

Edited by

Gael F. Squibb
Gael Squibb Consulting

Daryl G. Boden
United States Naval Academy

Wiley J. Larson
United States Air Force Academy

*This book is published as part of the
Space Technology Series, a cooperative activity
of the United States Department of Defense and the
National Aeronautics and Space Administration.*

 **Custom Publishing**

Boston Burr Ridge, IL Dubuque, IA Madison, WI New York San Francisco St. Louis
Bangkok Bogotá Caracas Lisbon London Madrid
Mexico City Milan New Delhi Seoul Singapore Sydney Taipei Toronto

The McGraw·Hill Companies

COST-EFFECTIVE SPACE MISSION OPERATIONS

Copyright © 2006, 1996 by The McGraw-Hill Companies, Inc. All rights reserved. Printed in the United States of America. Except as permitted under the United States Copyright Act of 1976, no part of this publication may be reproduced or distributed in any form or by any means, or stored in a data base retrieval system, without prior written permission of the publisher.

This work was created in the performance of a Cooperative Research and Development Agreement with the Department of the Air Force. The Government of the United States has certain rights to use this work.

1 2 3 4 5 6 7 8 9 0 DOC DOC 0 9 8 7 6

ISBN-13: 978-0-07-331321-4
ISBN-10: 0-07-331321-1

McGraw-Hill Editor: Judith Wetherington
Cover Design: Dale Gay
Text Design: Anita Shute
Technical Editors: Perry D. Luckett and Marilyn McQuade
Printer/Binder: RR Donnelley—Crawfordsville

Table of Contents

Chapter			Page
	List of Authors and Editors		xi
	Preface		xv
1	**Space Mission Operations**		**1**
	1.1	The Space Mission Lifecycle	4
	1.2	Elements of a Space Mission	8
	1.3	Cost-Effective Strategies for Mission Operations	11
2	**Designing Space Mission Operations**		**15**
	Step 1.	Determine Mission Objectives, Requirements, Constraints, and Type of Mission	17
	Step 2.	Develop Alternative Mission Concepts that Support Step 1	20
	Step 3.	Identify and Do Key Trades	21
	Step 4.	Characterize Acceptable Mission Concepts and Associated Space Mission Architectures	22
	Step 5.	Assess Items in Step 4 and Select a Baseline Mission Concept with Supporting Space Mission Architecture for Future Development	22
	Step 6.	Develop Alternative Mission Operations Concepts to Support the Mission Concept	23
	Step 7.	Do Key Trades Within Mission Operations	26
	Step 8.	Allocate Resources to Functions for Each Mission Phase	27
	Step 9.	Assess Mission Utility, Lifecycle Cost, Relative Complexity, and Cost of Operations	28
	Step 10.	Iterate and Document Reasons for Choices	29
	FireSat		30
3	**Mission Operations Functions**		**31**
	3.1	Mission Planning	37
	3.2	Activity Planning and Development	42
	3.3	Mission Control	49
	3.4	Data Transport and Delivery	54
	3.5	Navigation Planning and Analysis	59
	3.6	Spacecraft Planning and Analysis	62
	3.7	Payload Planning and Analysis	66
	3.8	Payload Data Processing	70
	3.9	Archiving and Maintaining the Mission Database	75
	3.10	Systems Engineering, Integration, and Test	78
	3.11	Computers and Communications Support	80

Chapter			Page
	3.12	Developing and Maintaining Software	82
	3.13	Managing Mission Operations	85
4	**Developing a Mission Operations Concept**		**91**
	4.1	Process for Developing a Mission Operations Concept	92
	4.2	Steps for Developing the Mission Operations Concept	96
	4.3	Mission Operations Concept, Bi-spectral Infrared Detection (BIRD)	114
5	**Operations Cost Estimations**		**129**
	5.1	Assessing Operations Complexity	130
	5.2	Operations Cost Modeling	148
6	**Defining and Developing the Mission Operations System**		**193**
	6.1	Definition and Development Process	195
	6.2	Monitoring Definition and Development	214
	6.3	Scenarios and Subsystems to Monitor	235
7	**Activity Planning**		**243**
	7.1	Evolution of Activity Planning	245
	7.2	Understanding the Dimensions of the Planning Components	249
	7.3	Analyzing Components of the Activity Plan	254
8	**Conducting Space Mission Operations**		**267**
	8.1	A Day in the Life of Operators	270
	8.2	Uplink Process	281
	8.3	Downlink Process	290
	8.4	Planning and Analysis Functions	292
9	**Launch and Early-Orbit (L&EO) Operations**		**301**
	9.1	Handling the Demands of the L&EO Environment	301
	9.2	Developing L&EO Operations	312
	9.3	Conducting L&EO Operations	334
	9.4	Reducing the Cost of L&EO Operations	336
	9.5	An Update for Today's Environment	338

Table of Contents

Chapter			Page
10	**Space Navigation and Maneuvering**		**343**
	10.1	Keplerian Orbits	344
	10.2	Orbit Perturbations	355
	10.3	Orbit Maneuvering	360
	10.4	Launch Windows	370
	10.5	Orbit Maintenance	372
	10.6	Interplanetary Trajectories	375
	10.7	Mission Geometry	384
11	**Communications Architecture**		**389**
	11.1	Defining and Evaluating Requirements	390
	11.2	Evaluating Design Options	408
	11.3	Forming the Architecture	429
12	**Ground Systems**		**451**
	12.1	Defining the Ground System	452
	12.2	Hardware, Software, and Staffing Requirements	465
	12.3	New Systems and Existing Networks	475
	12.4	Operational Concerns	489
13	**Processing Data and Generating Science Data Products**		**495**
	13.1	Characteristics of Science Data Products	496
	13.2	Instrument Development and Impact on Data System Design	498
	13.3	Relationship Between Uplink System and Science Data Processing System	499
	13.4	Downlink Data Processing Flow	501
	13.5	Design Considerations for Science Data Processing Systems	506
	13.6	Science Data Processing System Development and Test	509
	13.7	Science Data Processing System Architecture	512
	13.8	Science Data Processing System Implementation	516
	13.9	Science Data Archive Considerations	519
	13.10	Examples	521
14	**Assessing Payload Operations**		**529**
	14.1	Developing the Payload Operations Concept	532

Chapter			Page
	14.2	Assessing Requirements for Payload-Data Systems	541
	14.3	Assessing Drivers for Payload and Platform Interfaces	547
	14.4	Characteristics of a Good Payload Operations Plan	552
15	Spacecraft Performance and Analysis		557
	15.1	How Typical Spacecraft Subsystems Work	559
	15.2	How We Carry Out Typical Spacecraft Operations	572
16	Spacecraft Failures and Anomalies		577
	16.1	Spacecraft Failures and Anomalies	577
	16.2	Defining Anomalies	595
	16.3	Resolving Anomalies	599
	16.4	Planning for Anomalies	605
	16.5	Case Studies	608
17	Interplanetary Space Mission Operations		613
	17.1	Elements of Interplanetary Mission Operations	614
	17.2	Differences Between Interplanetary and Earth-Orbiting Missions	623
	17.3	Operations Activities Throughout the Mission Lifecycle	634
18	Software Engineering		643
	18.1	Software Engineering Process and Software Standards—CMM® and CMMI®	644
	18.2	The Eyes and Ears of Management—Project Monitoring and Software Quality Assurance	647
	18.3	Sizing and Pricing Software	652
	18.4	Staffing Software Projects—Organization and the Right Mix of Skills	664
	18.5	Software Lifecycles	667
	18.6	Focusing on What is Needed for Launch	675
19	Microsatellite Mission Operations		677
	19.1	Microsatellite Teams	679
	19.2	Exploring the Mission Concept	685
	19.3	Detailed Development	688
	19.4	Ground Element	693

Chapter			Page
	19.5	Pre-Launch Operations	697
	19.6	Post-Launch Operations	698
20	**Human Space Flight Operations**		707
	20.1	Rationale for Human Missions	707
	20.2	Implications of Humans in Space	709
	20.3	Planning and Analyzing Space Shuttle Missions	712
	20.4	International Space Station	717
	20.5	Human Lunar Missions	723
	20.6	Human Interplanetary Missions	726
21	**FireSat**		735
	21.1	Mission Information and Requirements (Inputs to an Operations Concept)	735
	21.2	Clarification and Validation of Requirements	739
	21.3	Operations Functions Needed and Selected Trades (Section 4.2.2)	740
	21.4	Mission Operations Concept (Chapter 4)	745
	21.5	Cost Estimate for 2 Satellites Versus 1 Satellite	779
	21.6	Conclusions	780
App. A	**Communications Frequency Bands**		781
App. B	**Mission Summaries**		783
Index			813

List of Authors and Editors

Carolyn Blacknall. Operations Integration Officer, NASA Johnson Space Center, Houston, Texas. M.S. (Aerospace Engineering) University of Texas at Austin; B.S. (Astronomy) University of Texas at Austin. Chapter 20—*Human Space Flight Operations*.

Daryl G. Boden. Professor, United States Naval Academy, Annapolis, Maryland. Ph.D. (Aeronautical and Astronautical Engineering) University of Illinois; M.S. (Astronautical Engineering) Air Force Institute of Technology; B.S. (Aerospace Engineering) University of Colorado. *Co-Editor*, Chapter 1—*Space Mission Operations*; Chapter 2—*Designing Space Mission Operations*; Chapter 10—*Space Navigation and Maneuvering*.

Klaus Briess. German Aerospace Center, Institute of Space Sensor Technology and Planetary Exploration. Chapter 4.3—*Mission Operations Concept, Bi-spectral Infrared Detection (BIRD)*.

John Carraway. Senior Member of the Technical Staff, Jet Propulsion Laboratory, California Institute of Technology, Pasadena, California. B.S. (Electrical Engineering) Massachusetts Institute of Technology. Chapter 5—*Operations Cost Estimations*.

Juan Ceva. Department Manager, Navigation and GPS Applications Department, Raytheon, Pasadena, California. Eng. and M.S. (Aeronautics and Astronauts) Stanford University, Stanford, California. Chapter 18—*Software Engineering*.

Kim Chacon. Senior Principal Software Engineer, Software Engineering Center, Raytheon—Space and Airborne Systems (SAS), El Segundo, California. B.S. (Computer Science) California State Polytechnic University. Chapter 18—*Software Engineering*.

Gary M. Comparetto. Senior Principal Engineer, The MITRE Corporation, McLean, Virginia. M.S. (Nuclear Engineering) Pennsylvania State University; M.S. (Electrical Engineering) Georgia Institute of Technology; B.S. (Electrical Engineering) Bucknell University. Chapter 11—*Communications Architecture*.

Richard S. Davies. Technical Director, Stanford Telecommunications, Inc., Sunnyvale, California. Engineer, Stanford University. M.S. and B.S. (Electrical Engineering) University of Pennsylvania. Chapter 11—*Communications Architecture*.

Eileen Dukes. Consultant, Interplanetary Horizons, Pine, Colorado. B.S. (Aeronautics and Astronautics) Massachusetts Institute of Technology. Chapter 8—*Conducting Space Mission Operations*.

Michael Fatig. Vice President, EMS Technologies Defense and Space, Norcross, Georgia. M.S. (International Business); B.S. (Technical Management) University of Maryland. Chapter 9—*Launch and Early-Orbit (L&EO) Operations*.

List of Authors and Editors

Neal Gaborno. Senior Manager, Image Processing, Guidance, and Control Systems Department, Software Engineering Center, Raytheon Company, El Segundo, California. B.S. (Electrical Engineering) University of Southern California. *Chapter 18—Software Engineering.*

Felix Godwin. Retired, Teledyne Brown Engineering, Huntsville, Alabama. B.S. (Physics) University of London. *Chapter 20—Human Space Flight Operations.*

Paul Graziani. President and Chief Executive Officer, Analytical Graphics, Inc. Exton, Pennsylvania. B.S. (Biology and Computer Science) LaSalle University. *Section 10.7—Mission Geometry.*

William B. Green. Retired, California Institute of Technology, Pasadena, California. M.S. (Nuclear Engineering) and B.S. (Physics) University of California Los Angeles. *Chapter 13—Processing Data and Generating Science Data Products.*

Mark K. Jacobs. Senior Systems Engineer, Science Applications International Corporation, Schaumburg, Illinois. B.S. (Metallurgical Engineering) University of Wisconsin. *Section 5.2—Operations Cost Modeling.*

David E. Kaslow. Chief Engineer, Lockheed Martin Corporation, King of Prussia, Pennsylvania. Ph.D. (Physics) University of Michigan; M.S. (Physics) Indiana University; B.A. (Mathematics) Indiana University. *Chapter 6—Defining and Developing the Mission Operations System; Chapter 7—Activity Planning; Chapter 16—Spacecraft Failures and Anomalies; Chapter 21—FireSat.*

Wiley J. Larson. Space Technology Series Editor for the U.S. Air Force Academy's Space Mission Analysis and Design Project, President, CEI, Colorado Springs, Colorado. D.E. (Spacecraft Design) Texas A&M University; M.S. and B.S. (Electrical Engineering), University of Michigan. *Co-Editor, Chapter 1—Space Mission Operations; Chapter 2—Designing Space Mission Operations.*

Matthew J. Lord. Manager, Communications Engineering, Globalstar LLC, Loral Space and Range Systems, Milpitas, California. M.S. (Engineering Management) Santa Clara University; B.S. (Electrical Engineering) University of California at Santa Barbara. *Chapter 12—Ground Systems.*

Perry D. Luckett. Director and Chief Consultant, Executive Writing Associates, Colorado Springs, Colorado. Ph.D. (American Studies) University of North Carolina at Chapel Hill; M.A. and B.A. (English) Florida State University. *Technical Editor.*

Edward B. Luers. Future Missions Planning Manager, Jet Propulsion Laboratory, California Institute of Technology, Pasadena, California. A.S. (Electrical Engineering), University of Cincinnati. *Chapter 12—Ground Systems.*

Marilyn McQuade. M.S. (Nuclear Engineering) Massachusetts Institute of Technology; B.S. (Nuclear Engineering) Massachusetts Institute of Technology. *Technical Editor.*

Richard B. Miller. Retired, Jet Propulsion Laboratory, California Institute of Technology, Pasadena, California. M.S. (Physics) University of Oregon; B.A. (Physics) Occidental College. Chapter 12—*Ground Systems*

Ray B. Morris. Multi-mission Ground Systems and Services Operations Manager, Jet Propulsion Laboratory, Pasadena, California. B.S. (Biology) Caltech. Chapter 17—*Interplanetary Space Mission Operations.*

Mac Morrison. Retired. B.S. (Physics) Ohio University. Chapter 15—*Spacecraft Performance and Analysis.*

David W. Murrow. Member of Technical Staff, Systems Division, Jet Propulsion Laboratory, California Institute of Technology, Pasadena, California. M.S. (Aerospace Engineering) University of Texas; B.S. (Aerospace Engineering) University of Colorado. Chapter 17—*Interplanetary Space Mission Operations.*

Paul Ondrus. Earth Science Mission Operations Project Manager, NASA Goddard Space Flight Center, Greenbelt, Maryland. M.S. (General Administration) University of Maryland University College; B.S. (Electrical Engineering) University of Akron. Chapter 14—*Assessing Payload Operations.*

Emery Reeves. Aerospace Consultant, Palos Verdes Estate, California. M.S. (Electrical Engineering) Massachusetts Institute of Technology; B.E. (Electrical Engineering) Yale University. Chapter 16—*Spacecraft Failures and Anomalies.*

Jeffrey K. Shupp. Principal Systems Engineer, Systems Integration Programs, Lockheed Martin, King of Prussia, Pennsylvania. M.S. (Nuclear Engineering) University of California at Berkeley; B.S. (Nuclear Engineering) Pennsylvania State University. Chapter 6—*Defining and Developing the Mission Operations System;* Chapter 7—*Activity Planning.*

Aaron Silver. Senior Scientist, Raytheon Intelligence & Information Systems, Aurora, Colorado. Ph.D. (Operations Research) Ohio State University; M.S.E. (Operations Research) The Johns Hopkins University; M.S.E.E. (Electrical Engineering) Columbia University; B.E.E. (Electrical Engineering) Pratt Institute. Chapter 18—*Software Engineering.*

Gael F. Squibb. Retired, Jet Propulsion Laboratory, California Institute of Technology, Pasadena, California. M.S. (Systems Management) University of Southern California; B.S. (Physics) Harvey Mudd College. Chapter 1—*Space Mission Operations;* Chapter 2—*Designing Space Mission Operations;* Chapter 3—*Mission Operations Functions;* Chapter 4—*Developing a Mission Operations Concept;* Chapter 21—*FireSat.*

Gary Thomas. Engineering Fellow, Software Engineering, Raytheon, Garland, Texas. M.S. University of Chicago; B.S. University of Missouri. Chapter 18—*Software Engineering.*

Craig I. Underwood. Reader in Spacecraft Engineering, Surrey Space Centre, University of Surrey, Guildford, Surrey, England. Ph.D. (Radiation Effects in Space Electronics) University of Surrey; P.G.C.E. University of York; B.S. (Physics with Computer Science) University of York. Chapter 19—*Microsatellite Mission Operations.*

Jeffrey F. Volosin. Honeywell International. Chapter 9—*Launch and Early-Orbit (L&EO) Operations.*

Jeffrey W. Ward. Managing Director, Surrey Satellite Technology Limited, Guildford, Surrey, England. Ph.D. University of Surrey; B.S. (Computer Engineering) University of Michigan. Chapter 19—*Microsatellite Mission Operations.*

Robert Zmarziak. Program Manager, Consolidated Operations and Anomaly Support Team, Northrop Grumman, Redondo Beach, California. B.S. (Math) Regents College; M.B.A. LaVerne University. Chapter 15—*Spacecraft Performance and Analysis.*

Preface

The book, *Cost-Effective Space Mission Operations*, is part of a collection of books in the Air Force Academy's Space Technology Series. The series provides practical approaches to designing and operating space systems; captures experiences, wisdom, and lessons learned by key people in our space programs; and helps educate space professionals and students in the "big picture" of space missions. Carefully selected experts from around the world write each book in the series.

The goal of this book is to provide processes, tools, and data that can help us do space mission operations better and more cost effectively. It should prove especially helpful to a mission operations manager, who is responsible for the early planning of mission operations, designing the mission operations system, and conducting daily operations. This book defines a process to help translate mission objectives and requirements into a viable mission operations concept. The mission operations manager must develop this operations concept early enough, during concept exploration, so the project manager can trade future cost of operations with current development costs.

The first two chapters describe how the mission operations element blends in with the other elements of a space mission. Chapters three through seven describe the thirteen mission operations functions, provide a process for developing a mission operations concept, present a model for evaluating the cost and complexity of conducting operations, and explain how we define and develop a mission operations system. Chapters eight through sixteen describe how we conduct routine, launch and early orbit, and special operations; plan and analyze the mission; and transport and process data. Finally, the last five chapters describe how we conduct interplanetary, micro-satellite, and crewed mission operations, the main differences between these missions and Earth-orbiting, uncrewed missions, and software engineering for space missions.

Leadership, funding, and support essential to developing the book came from the Air Force Space Missile Systems Center, Air Force Space Command, Naval Research Laboratory, Office of Naval Research, Headquarters-National Aeronautics and Space Administration, Goddard Space Flight Center, Johnson Space Flight Center, Lewis Research Center, the Jet Propulsion Laboratory, the Advanced Projects Research Agency, and the Department of Energy. The European Space Agency also supported this effort through their European Space Operations Center. Getting money to develop much-needed reference material is exceptionally difficult in the aerospace community. We're deeply indebted to the sponsoring organizations—particularly the Air Force's Phillips Laboratory—for their support and their recognizing the importance of projects such as this one.

This book is intended for professionals and students involved in space systems engineering and space operations, including program managers, mission operations managers, spacecraft engineers and designers, project scientists, and operators. It's suitable as a textbook for a course in space mission operations, a supplementary text in other courses, or as a professional reference.

Preface

Our thanks go to the many people who made this book possible—in particular,

- Robert Giffen and Michael DeLorenzo, Astronautics Department Heads at the Air Force Academy, who furnished the leadership and resources to complete this book.
- Perry Luckett and Marilyn McQuade, who made the text much more concise, readable, and grammatically correct.
- Anita Shute, who designed the book, edited and formatted the manuscript, interpreted and incorporated our numerous changes, and prepared drafts and camera-ready copy.
- Several people who spent many hours reviewing the book: Curt Heftman, Dave Linick (Jet Propulsion Laboratory), Felix Godwin (Teledyne Brown Engineering), Tim Gillespie (Onizuka Air Force Base), Mike Violet (Falcon Air Force Base), Ron Humble (United States Air Force Academy), and the students who used a draft version of this book in the space mission operations course at the University of Colorado at Colorado Springs.
- Our families for their support during this project.

Through much iteration of reviewing, editing, and revising, we've tried to eliminate all ambiguities and errors. If you find any, please contact us.

Daryl G. Boden *Department of Astronautics*
Wiley J. Larson *United States Air Force Academy*
Gael F. Squibb *USAF Academy, Colorado 80840*
Voice: 719-333-4110   FAX: 719-333-3723
Email: wileylarson@adelphia.net

Chapter 1

Space Mission Operations

Daryl G. Boden, *United States Naval Academy*
Wiley J. Larson, *United States Air Force Academy*
Gael F. Squibb, *Jet Propulsion Laboratory, California Institute of Technology (Retired)*

> 1.1 The Space Mission Lifecycle
> 1.2 Elements of a Space Mission
> 1.3 Cost-Effective Strategies for Mission Operations

So you ask, "Why would anyone write a book about space mission operations?" We've asked ourselves the same question—many times. Much of mission operations is subjective, with no absolute laws, like laws of physics, to guide the way. In many cases there are no concrete answers, only opinions. We often depend on a "seat of the pants" approach and rely solely on the "seat" of the person chosen to do the work. One of our main goals is to identify objective processes, tools, and data that can help us do space mission operations better and more cost-effectively. We want to document this material so people doing the work—perhaps, someone like you—can start from higher ground and benefit from previous experience. Remember, we stand on the shoulders of giants!

We have written this book for the hypothetical *mission operations manager* (MOM), who is responsible for early planning of mission operations, designing the mission operations system, and conducting day-to-day operations. Our MOM can represent the operations or project organization because many of the users are the same for both organizations. We take an aggressive approach to reducing the cost of mission operations, tempered with a perspective on lifecycle cost.

We advocate a concurrent approach to designing mission operations instead of the traditional approach. In the traditional approach, operations develops after design of the mission and spacecraft. Often this approach restricts cost-reduction strategies to mission operations, which overly constrains our ability to reduce the lifecycle cost. Table 1.1 shows a timeline for traditional mission operations planning. Note that operational planning begins well after the spacecraft is designed—too late to make constructive changes to the operations concept and spacecraft design. In contrast, the concurrent approach to designing mission operations forces planners to address key issues in these areas with other mission elements. The proposed concurrent approach considers many of the traditional issues but provides *more time* to make decisions that reduce operational complexity and, ultimately, lifecycle cost. Issues about operations and spacecraft design arise early, when changes are easier and less costly to implement. Figure 1.1 shows a timeline for the concurrent approach.

Table 1.1. Functional Timeline for Spacecraft Operations. Status of the spacecraft and ground system determines the level of detail of these activities. For new spacecraft or a new generation of spacecraft, test and integration become quite detailed. Integrating operations activities, a new ground system, or a major ground-system upgrade requires much lead time, typically driven by software development.

Prelaunch	Launch	Early Orbit Checkout	Normal Operations
		2 Days – 6 Months	Several Weeks – 30 Years

2 – 5 Years
• Spacecraft design
• Payload design
• Mission design
1 – 2 Years
• Develop flight plan
  - Spacecraft
  - Payload
  - Ground system
• Develop training plan
• Identify simulator requirements
• Integrate and test support systems
6 Months
• Assemble operations team
• Validate ground-system database
• Validate ground-system hardware
• Validate flight software
3 Months
• Start prelaunch training
• Rehearse launch
• Demonstrate communications protocol
1 Month
• Simulate launch operations
• Review readiness
	• Support launch team
	• Transfer spacecraft to initial orbit
		• Validate components
		• Validate subsystems
		• Validate subsystems interfaces
		• Validate systems
		• Detect and analyze anomalies
		• Calibrate instruments
		• Validate instrument processing
		• Validate protocol for external interfaces
		• Maneuver spacecraft to mission orbit
			• Perform real-time spacecraft operations
			• Process and distribute payload data
			• Translate requirements into operational activities
			• Resolve anomalies
			• Maintain ground-system database
			• Maintain flight software
			• Maintain ground software
			• Continue operator training
			• Recover and repair spacecraft
			• Dispose of non-operational spacecraft

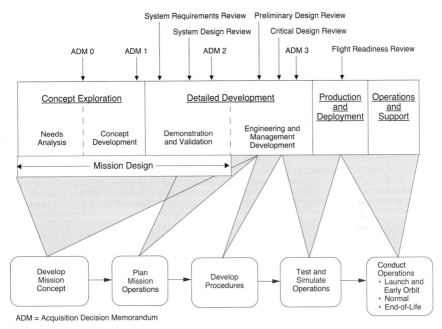

Fig. 1.1. Concurrent Approach to Space Mission Operations. We must design mission operations in parallel with all other elements of the space mission. Mission phases are defined in the next section.

Figures 1.1 and 1.2 show a similar concurrent approach to Space Mission Operations. The National Aeronautics and Space Administration (NASA) and Department of Defense (DOD) use different names for the phases but both involve experts in mission operations early in the lifecycle of the mission concept and mission development.

Over nearly 45 years in the United States, Europe, and the former Soviet Union, we've learned a lot of lessons—some the hard way—while doing commercial, military, and civil space missions. We hope to merge the lessons learned and wisdom gained over these years, analyze what we've done, identify what has worked well, and try to develop better ways of planning and conducting space mission operations.

Because we're addressing a diverse group of people, missions, and cultures, we begin by providing definitions for key terms used in this book. The terms may not be exactly what you use, but we define them so you can understand and apply them to your situation. We focus on Earth-orbiting, uncrewed missions in the US and discuss the differences between them and interplanetary missions. Finally, we describe additional operational requirements for a crewed mission. We hope to give you a broad perspective on mission operations and how different cultures do business in space.

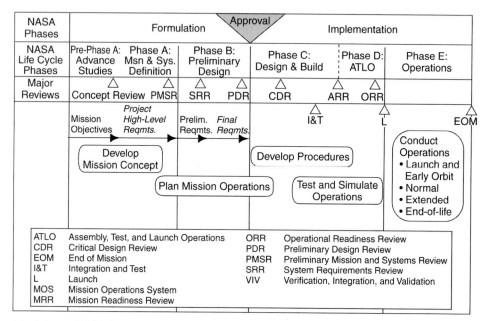

Fig. 1.2. NASA Concurrent Approach to Space Mission Operations. This diagram shows the approach that NASA uses in designing and planning for operations during all of the lifecycle phases of the project. NASA uses different names for the phases but the outcome is similar to that used by DOD.

1.1 The Space Mission Lifecycle*

Table 1.2 illustrates the lifecycle of a space mission, which typically progresses through four phases:

Concept exploration—the initial study phase, which results in a broad definition of the space mission and its components.

Detailed development—the formal design phase, during which we define the system components and, in larger programs, develop test hardware or software.

Production and deployment—including constructing the ground and flight hardware, writing software procedures, and launching the first full constellation of spacecraft.

---

* Adapted with permission from *Space Mission Analysis and Design*, Larson and Wertz [1999].

Operations and support—the day-to-day operation of the space mission, its maintenance and support, and finally its deorbit or recovery at the end of the mission.

Table 1.2. Development Phases for Space Systems. Every space program progresses through the top-level phases. Subphases may or may not be part of a given program. The time required to complete the process varies with the program's scope. Major programs take as long as 15 years from concept exploration to initial launch, whereas some programs may require only 12–18 months.

Phase	Concept Exploration		Detailed Development			
Subphase	Needs Analysis	Concept Development	Demonstration and Validation	Engineering and Management Development	Production and Deployment	Operations and Support
Typical DOD Milestones	ADM 0	ADM 1	SRR SDR ADM 2	PDR CDR ADM 3	Launch	Deorbit
Typical Products	Statement of needs; studies	Breadboards and studies	Advanced prototypes	Engineering prototypes; detailed design	Production hardware	Operational system
Time required in major program	Continuous	1 – 2 years	2 – 3 years	3 – 5 years	4 – 6 years	5 – 15 years

▲ Program Milestones
ADM - Acquisition Decision Memorandum

▼ Program Reviews
SRR - System Requirements Review
SDR - System Design Review
PDR - Preliminary Design Review
CDR - Critical Design Review

This book discusses concept exploration and detailed development with emphasis on preparing for operations and support. Mission operations costs make up 12%–50% of a space mission's lifecycle cost; the percentage varies with duration and complexity of the mission. In Chap. 5 we'll discuss how to assess mission operations complexity and cost.

These phases may be divided and named differently depending on whether the *sponsor* (the group which provides and controls the program budget) is the DOD, NASA, foreign agencies like the European Space Agency (ESA), or a commercial enterprise. The time required to progress from initial concept to deorbiting or end of the mission appears to be independent of the sponsor. Large, complex space missions typically require 10–15 years to develop, and they operate from 5–15 years, whereas small, relatively simple missions require as little as 12–18 months to develop and will operate for 1–6 months.

Procurement, operating procedures, and risk policies vary with sponsoring organizations, but the key players are the same: the space mission operator, end user, and developer. *Operators* control and maintain the space and ground assets, and are usually applied-engineering organizations. *End users* receive and use the mission's products. They include astronomers and physicists for science missions, meteorologists for weather missions, the general public for communication and navigation missions, geologists and agronomists for Earth resources missions, and the war fighter for offensive and defensive military space missions. The *developer* is the procuring agent, such as DOD, NASA, ESA, or a commercial enterprise, and includes the contractors, subcontractors, and government organizations that handle development and testing. The operators and users must generate technically and fiscally responsible requirements; the developer must provide the necessary product or capability on time and within the changing constraints of politics and funding.

Three basic activities occur during the Concept Exploration Phase (see Fig. 1.3). Users and operators develop and coordinate a set of broad needs and performance objectives based on an overall concept of operations. At the same time, developers generate alternative concepts to meet the perceived needs of the user and operating communities. The sponsor does long-range planning, develops program structure, budgets, and estimates available funding to meet the needs of the users, operators, and developers. To produce and deploy a cost-effective space capability successfully, the four key players must closely integrate their activities.

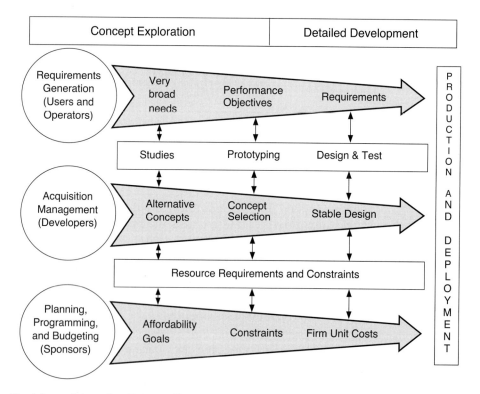

Fig. 1.3.   Interaction Between Users and Operators, Developers, and Sponsors. All players should be involved early and continually in the development of the mission.

The goal during concept exploration is to assess the need for a space mission and to develop affordable alternatives that meet the operators' and end users' requirements. The *needs analysis* continues until it culminates in a new program start. Operators and end users develop potential mission requirements based on the considerations shown in the left-hand column of Table 1.3. The process is different for each organization, but at some point a new program begins with a set of mission objectives, concept of operations, and desired schedule. In DOD, the *Mission Needs Statement* documents this information and becomes part of the planning, programming, and budgeting system [Defense Systems Management College, 1990]. If approved, the program receives funding and proceeds to concept development.

Table 1.3. Further Breakdown of the Concept-Exploration Phase. During concept exploration the operator and user define their requirements and pass them to the developing organization for concept development. The operator, user, and developer must work together to develop realistic, affordable mission objectives and requirements that meet the user's needs.

Concept Exploration and Definition	
Needs Analysis	**Concept Development**

Generate potential requirements based on
  Mission objectives
  Concept of operations
  Schedule
  Lifecycle cost and affordability
  Changing marketplace
  Research needs
  National space policy
  Long-range plan for space
  Changing threats to national defense
  Military doctrine
  New technology developments
	Reassess potential requirements generated during needs analysis
	
	Develop and assess alternative concepts for mission operations
	
	Develop and assess alternative architectures for the space mission
	
	Estimate
	  performance supportability
	  operational complexity produceability
	  schedule funding profiles
	  risk lifecycle cost

At the *Program Initiation* milestone (ADM 0), the funding organization commits to concept development. Scrutiny of the program will depend on its scope, political interest, and funding requirements. In DOD, *major programs* receive the utmost attention at the highest levels. Military components use distinct criteria to identify major programs [Defense Systems Management College, 1990]. A DOD program is "major" if it requires more than $200 million for research, development, test, and evaluation or more than $1 billion for production. Programs that require participation by more than one component of the armed forces or have congressional interest may also be classified as major programs.

During *concept development* the developer must generate alternative ways to meet the operator's and end user's needs. This includes developing and assessing different concepts and components for mission operations, and estimating the factors in the right-hand column of Table 1.3. The results become part of the system concept. High-level managers in the user, operator, and development communities decide if the concepts, initial mission objectives, and potential requirements meet the mission's intentions. If the program satisfies the need at a reasonable cost, it passes the *Requirements Validation* milestone (SRR) and proceeds into the Detailed Development Phase (see Fig 1.3).

This book provides the technical processes and information necessary to explore concepts for many space missions. One major problem that can undermine the entire process is that in many cases, users and operators analyze the needs and formulate mission requirements apart from the development community. Then they pass these requirements "over the wall" without negotiating. The developer often generates alternatives without the operators and users. These unilateral actions produce minimum performance at maximum cost.

> To explore a concept successfully, we must remove the walls between the sponsor, space operators, users, and developers and become a team.

A good team considers the mission's operations, objectives, and requirements as well as the available technology to develop the best possible mission concept at the lowest possible lifecycle cost.

1.2 Elements of a Space Mission*

All space missions consist of a set of *elements* or *components* as shown in Fig. 1.4. Arranging these elements forms a *space mission architecture*. Organizations and programs define their mission elements differently, although all elements shown in Fig. 1.4 are normally present in any space mission. The space mission architecture revolved around the mission concept, which is driven by the mission objectives, top-level requirements and a number of other factors (like politics) that we'll discuss later. Let's review each element of the space mission architecture. Most elements are shown in Fig. 1.4.

The *subject* of the mission is the phenomenon that interacts with or is sensed by the space payload, e.g., moisture content, atmospheric temperature or pressure for weather missions; types of vegetation, water, or geological formations for Earth-sensing missions; or a rocket or intercontinental ballistic missile for defense missions. We must decide what part of the electromagnetic spectrum to use to sense the subject; this determines the type of sensor as well as payload weight, size, and power. In many missions, we may trade off the subject. For example, if we are trying to track a missile during powered flight, the subject could be the rocket body, the exhaust plume, or both.

For communications and navigation missions, the subject is a set of equipment on the Earth or on another spacecraft, such as communication terminals, televisions, receiving equipment for navigation using the global positioning system, or other user-furnished equipment. The key parameters of this equipment characterize the subject for these types of missions.

---

* Adapted with permission from *Space Mission Analysis and Design*, Larson and Wertz [1999].

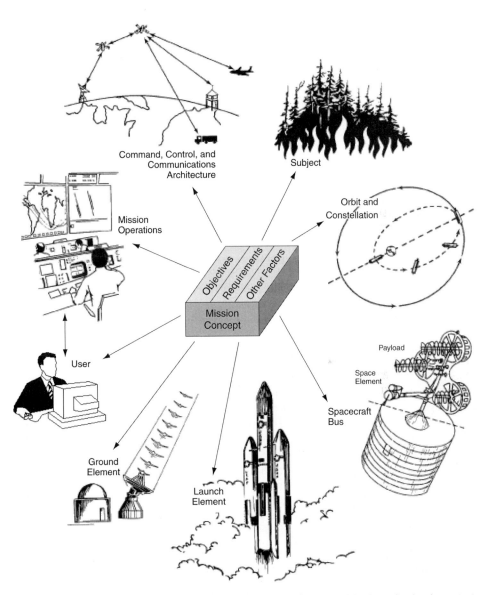

Fig. 1.4. **Space Mission Architecture.** Most space missions include these basic elements to some degree. See the text for definitions. Requirements for the system flow from the operator, end user, and developer and are allocated to the mission elements based on the mission concept. [Adapted from *Space Mission Analysis and Design*.]

The *payload* consists of the hardware and software that sense or interact with the subject. Typically, we trade off and combine several sensors and experiments to form the payload, which largely determines the mission's cost, complexity, and effectiveness. The subsystems of the *spacecraft bus* support the payload by providing orbit and attitude maintenance, power, command, telemetry and data handling, structure and rigidity, and temperature control. The payload and spacecraft bus together are called the *spacecraft, space element,* or *launch-vehicle payload*.

The *orbit* is the spacecraft's trajectory or path and typically is different for initial parking, transfer, and the final mission. There may also be an end-of-life or disposal orbit. The mission orbit significantly influences every mission element and provides many options for trades in the mission architecture.

The *launch element* includes the launch facility, launch vehicle, and any upper stage required to place the spacecraft in orbit, as well as interfaces, payload fairing, and ground-support equipment. The selected launch system constrains the spacecraft's size, shape, and mass.

The *command, control, and communications* (C^3) *architecture* is the arrangement of components that satisfy the mission's communication, command, and control requirements. It depends strongly on the amount and timing requirements of data to be transferred, and on the number, location, availability, and communicating ability of the space and ground assets.

The *ground element* consists of control centers and fixed and mobile ground stations around the globe, together with their data links. They allow us to command and track the spacecraft, receive and process telemetry and mission data, and distribute the information to the operators and users.

Finally, *mission operations* consists of the people occupying the ground and space assets, as well as the hardware, software, facilities, policies, and procedures that support the mission operations concept. A key consideration is the C^3 architecture, which connects the spacecraft, ground elements, and mission operations elements. The operators must be aware of and in most cases know the users, who they are, what they want, and what products are being generated and sent to the users. This will be a key point throughout this book—**know the users and what they need!** This interaction with the users starts early in the conceptual design of the mission and continues throughout the lifecycle of the mission.

In this book we'll focus on three pieces of the space mission architecture that form the basis of the mission operations concept: mission operations; ground systems; and C^3. We will propose alternative approaches for operating a mission and trade among all the mission elements.

We may combine the elements shown in Fig. 1.4 in different ways to form alternative concepts for a space mission. A mission concept describes how all elements work together to meet the mission's objectives. To develop a reasonable and complete mission concept, all players—users, operators, developers, and sponsors—must actively develop alternatives and select the baseline mission concept.

The *mission operations system (MOS)* is the collection of people, procedures, hardware, and software associated with elements shown in Fig. 1.4 for C^3; ground systems; and mission operations. The MOS also includes parts of the space element needed to do mission operations. Because the spacecraft is part of mission operations, we must consider the MOS early in the spacecraft and mission design. The *mission operations concept*, discussed in Chap. 4, describes how the MOS will carry out the mission according to the mission concept and supporting architecture.

1.3 Cost-Effective Strategies for Mission Operations

Much of the motivation for this book is based on today's trends of reduced funding for space systems and fewer projects. We believe these trends will continue well into the next decade. Furthermore, almost half of the world's new space systems and upgrades to existing systems will be designed and developed in the next three to five years. This means we have a wonderful chance to design more cost-effective operations for space missions. We will live with our successes and failures for the next several decades.

Speaking of decades, this country has emphasized the lifecycle cost of space systems for over a decade, yet very little has been done about accounting for mission operations costs in DOD or NASA. We tend to give the mission operations infrastructure a pot of money to support space missions without allocating operations costs to particular missions. The infrastructure has grown and serviced our missions well under the stewardship of its managers. Now, as the infrastructure is forced to downsize, we must look for ways to accurately assess its cost and the cost of operating space missions. The idea is to charge projects directly for their use of the operations infrastructure. This single change will profoundly affect the way we trade between spacecraft design and mission operations, ultimately helping us better predict and control the lifecycle cost of individual missions.

Funding for mission operations is decreasing while operations become more complex. As budgets shrink, NASA and DOD are looking for money to support new and existing programs. One way to get this money is to save it from operations.

We recommend strategies for developing cost-effective operations of space missions with emphasis on reducing the lifecycle cost of space systems. Table 1.4 lists these strategies and provides references.

The MOM must be actively engaged in developing the mission and mission operations concepts from the very beginning of the project. Many of the cost-effective approaches we're discussing stem from efficient concepts for the mission and mission operations. We can usually reuse existing or evolving resources and affect spacecraft designs only if we try early in conceptual design. Later, when concepts, hardware, and software have been determined, it becomes expensive to change systems to make their operation easier and more cost-effective.

We must train more people on all aspects of mission operations, so that early interaction among users, operators, developers, and sponsors is effective. Organizations must strive to develop more people with the perspective of a MOM.

Standardize functions for mission operations. Many spacecraft and mission designers don't understand what happens during operations. This problem isn't

Table 1.4. Strategies for Cost-Effective Operation of Space Missions. Select a strategy or combination of strategies you can use on your mission to reduce operations and lifecycle costs.

Cost-Effective Strategy	Key Features	Where Discussed
Get the mission operations manager actively engaged in developing the mission and mission operations concepts	• Focus development and operations efforts on key cost-effective approaches early in the process, while negotiation and change are practical • Do space/ground trade-offs early • Consider centralized vs. distributed data processing and control early	Chaps. 1, 2, 3, and 4
Standardize functions for mission operations	• Standardized functions allow direct comparison—this may facilitate, combine, or eliminate functions	Chap. 3
Standardize communication techniques and protocols	• One way to standardize interfaces	Chaps. 11, 12, and 13
Directly account for operations costs	• Assess actual cost of operations • Provides better information for trade-offs of lifecycle cost	Chaps. 1 and 5
Promote reuse of procedures, software, hardware, and people	• Reuse works well if planned from the beginning • Keep abreast of changes in the infrastructure because it evolves with time	Chaps. 5, 8, and 18
Make spacecraft operability a primary concern early in designing the mission and spacecraft	• Assess operability using key factors • Maintain margin in all spacecraft systems	Chaps. 5, 14, and 15
Combine technical and operational demonstrations to improve technical and operational performance	• Use operational demonstrations to validate and improve mission operations concepts and reduce operations cost	Chaps. 1 and 2
Carefully consider using automation and autonomy	• Remember that automation and autonomy have mixed results • Use autonomy and automation to improve technical and operational performance, reduce risk, and reduce lifecycle cost • Don't automate just to automate—automate to reduce lifecycle cost	Chap. 5 and 8

unique to space. For decades, designers, builders, and users have had difficulty communicating what they need and how to achieve the lowest cost and best performance. If our industry can focus on the key (most costly and risky) operations tasks, we can save money. Reviewing what operators do during operations can have significant benefits. Once functions are understood, we can combine them, reduce their scope, or even eliminate them all together—ultimately reducing operations costs.

Standardize communication techniques and protocols. Standard interfaces provide fertile ground for reducing cost. The communication link is the main interface between spacecraft and ground, users and operators, and data providers and data reducers. Standardizing these interfaces will enable us to streamline our missions. We may be able to use more cost-effective commercial capabilities to carry out daily communications.

Account directly for operations costs. We've been doing space operations since about 1960; yet, our industry doesn't know what operations cost. We do know, in total, how much funding goes to operations organizations, but we don't have a way to track how much a particular space mission spends for mission support. Thus, inefficient missions can hide among all the other operations.

If we're sincerely trying to reduce operating costs, we must account directly for mission support. This will allow us to trade costs intelligently between space and ground assets. Managers of development programs can also use this accounting to justify spending more or less money to develop the spacecraft.

One particularly effective approach is to use *zero-base operations*—starting a project with no operations people and having to justify adding people and capability on the ground. We believe a more acceptable approach is to begin accounting for operations costs and to use the information to make sensible decisions about how to operate.

Promote reuse of procedures, software, hardware, and people. The mission operations infrastructures within NASA and DOD have developed extraordinary capabilities over the years. In fact, they're struggling under the weight of these capabilities because we can no longer support the size of these infrastructures. The stewards of our infrastructures must trim down by identifying and maintaining their best procedures, software, hardware, and people.

Representatives like the MOM must interact with new projects early in the design stage of the mission and spacecraft. During these early discussions, developers can determine which procedures, software, and hardware might be appropriate, so we can re-use many capabilities and save the developers money. If this interaction doesn't take place early, developers will design systems using their own procedures, hardware, and software. Then, the operators will offer capabilities that are unacceptable because they've become changes that will drive up cost.

Make spacecraft operability a primary concern early in designing the mission and spacecraft. For centuries designers have created designs that looked great on paper but were very difficult and expensive to construct and operate. Our recent experiences in space suggest that things haven't changed. We still design spacecraft that are one of a kind and therefore difficult to build and operate. The Global Positioning System in DOD and Earth Observing System in NASA devoted about 11% and 51% of their 1994 budgets to operate their respective space and ground systems. That's an incredible amount of money!

We must assess spacecraft and mission operability before we release the spacecraft design to production. We can no longer afford to spend any money in operations that isn't absolutely justified.

Combine technical and operational demonstrations to improve technical and operational performance. Our industry tends to focus on technical demonstrations to show that technology is ready to fly in space and enhance technical performance. Today we need to focus on technological **and** operational approaches that reduce the cost of operations and, ultimately, lifecycle cost. The focus should be on technologies that can reduce the cost of operations. We can do this by reducing the number of ground contacts; minimizing (or eliminating) tracking requirements; and associating attitude, position, and time information with the mission data on the spacecraft. We must try new operational approaches, such as allowing the spacecraft computer to schedule ground contacts only when necessary.

Consider carefully using automation and autonomy. Ground-station automation and spacecraft autonomy are phrases we hear whenever we consider reducing the cost of mission operations. But, we must consider carefully the overall lifecycle costs before we move blindly forward with automation and autonomy. Automation and autonomy increase development costs. If we don't apply them correctly, they can also increase operational costs, especially if management adds procedures to either verify or prevent automatic or autonomous operations. Still, for long missions that require frequent, repetitive actions, automation and autonomy can save a lot of money.

We begin developing cost-effective mission operations by describing in Chap. 2 how to design a mission operations system. This process takes us from stating a mission objective to describing the operations hardware, software, people, and procedures needed to meet the mission's objectives while ensuring its safety. In Chap. 3 and 4, respectively, we describe integral parts of the mission operations system: how we do each of the mission operations functions and the mission operations concept.

The first five chapters are brought together in Chap. 21, where we discuss a mission called FireSat. This mission is developed from a design point of view in *Space Mission Analysis and Design*—Third Edition (SMAD III). However, the discussion of the operational aspects is relatively light. In Chap. 21 we go through the nine functions for this mission, discuss trades, develop a mission operations concept and do an initial cost estimate.

References

Defense Systems Management College. 1990. *System Engineering Management Guide*. Ft. Belvoir, VA:U.S. Government Printing Office.

Larson, Wiley J. and James R. Wertz. 1999. *Space Mission Analysis and Design*. Third Edition. Netherlands: Kluwer Publishing.

Rechtin, Eberhardt. 1991. *Systems Architecting*. Englewood Cliffs, NJ: Prentice Hall.

Chapter 2
Designing Space Mission Operations

Daryl G. Boden, *United States Naval Academy*
Wiley J. Larson, *United States Air Force Academy*
Gael F. Squibb, *Jet Propulsion Laboratory, California Institute of Technology (Retired)*

Space designers typically consider the mission operations concept after designing the mission and the spacecraft, so mission operations affect only slightly the design of other mission elements. Often the mission operations element must develop costly and complex approaches to work around inadequate mission and spacecraft designs. By including mission operations early in the mission design, we can evaluate how design decisions affect mission operations and then trade between operations and other elements of the space mission.

Concurrent design for mission operations sounds like a good idea[*], but how do you do it? This chapter defines the process that takes us from a statement of the mission objective to a mission operations concept. The mission objectives broadly state the goals the system must achieve to be productive. Space missions usually have several primary and secondary objectives. The mission operations concept describes how operations will work to meet these objectives.

Table 2.1 lists the steps for designing space mission operations and tells where to find more information in this book explaining each step. The key word in this table is *process*. In this chapter we walk through the process step by step, describing what needs to be done, how to do each step, the inputs and outputs of each step, and where we provide more detail. To develop a mission operations concept, we must continually update and refine the concept as the mission design matures and then review and revise decisions from earlier steps.

Some people think using concurrent design drives cost up, not down.

Table 2.1.   Developing a Space Mission Concept. This process takes us from a statement of the mission objective to a mission operations concept. The process is iterative and we must continually update the concept as the design matures.

Step	Issues	Where Discussed
1. Determine mission objectives, requirements, constraints, and type of mission.	State the mission objectives. Identify requirements and constraints for the mission, program, hardware, and software.	Chaps. 1, 2, and 10–12; Larson and Wertz [1999] Chap. 1–4
2. Develop alternative mission concepts that support step 1.	Describe the mission and any underlying philosophies, strategies, and tactics. The idea is to identify significantly different ways of doing the mission using a mix of space, ground, and air assets. Identify essential organizations and interfaces.	Chaps. 1 and 2; Larson and Wertz [1999] Chap. 2 and 22
3. Identify and do key trades: • Among mission elements for each of the selected concepts • Among organizations.	Inter-element trades are the biggest cost savers. Do trades early in the design to be most effective. Minimize the number of organizations and interfaces.	Chaps. 1–3, and 9–15; Larson and Wertz [1999] Chap. 2 and 3
4. Characterize acceptable mission concepts and associated space mission architectures. • Information-system characteristics • Payload characteristics • Spacecraft characteristics • Data products • Ground-system characteristics	Summarize needed payload, spacecraft bus, ground-system, and communications capabilities and characteristics. Describe products needed by users and operators for each mission concept.	Chap. 4, and 9–15
5. Assess items in step 4 and select a baseline mission concept with supporting space mission architecture for future development.	Use the baseline mission concept to develop the mission operations element, ground element, and architecture for command, control, communication, and computers.	Chaps. 1 and 2; Larson and Wertz [1999] Chap. 3
6. Develop alternative mission operations concepts to support the mission concept.	Step 4 lists inputs to developing a mission operations concept.	Chap. 4
7. Do key trades within mission operations.	This usually involves trading people tasks with hardware and software.	Chaps. 3–6, 8, 12, and 15
8. Allocate resources to functions for each mission phase.	Identify operationally difficult requirements and requirements derived from the mission operations concept.	Chaps. 3, 4, and 6–9
9. Assess mission utility, lifecycle cost, relative complexity, and cost of operations.	Is mission utility enhanced? Are lifecycle cost, complexity, and risks within acceptable limits?	Chap. 5; Larson and Wertz [1999] Chap. 3 and 20
10. Iterate and document reasons for choices.	More iterations generally improve the result.	--

Step 1. Determine Mission Objectives, Requirements, Constraints, and Type of Mission.

When developing a mission concept we must consider the end-user of the data, the mission objective, and the user's requirements. The user may be the military command and control system for early warning data, project scientists for remote sensing data, or the general public for communications and navigation data. A mission may have several different users with different requirements for a single payload or it may have multiple payloads with multiple users. The user's requirements define the data volume, timeliness, quality, and level of processing. For example, if the user is the project scientists on an exploratory or scientific mission, they may require level-zero (unprocessed) mission data plus engineering data from the spacecraft. But they may not need the data immediately. Conversely, a missile warning crew needs the data as soon as an event occurs, but they need only the processed data displaying the event. The mission operations manager (MOM) must develop an operations plan that will meet the user's needs within the mission constraints. The MOM must also be prepared to question a user's requirements, estimate their cost, and show how to spend less on mission operations. In the above-mentioned example of a scientific mission, one requirement may specify unprocessed and uncompressed data for scientific analysis. But this increases data volume and may strain the communications architecture selected for the mission.

In addition to mission objectives and user's requirements, we must also determine how mission constraints such as cost, schedule, politics, or the environment will affect our operations. Political constraints may require us to use an existing ground network or place our ground station at a particular location. To develop a workable operations plan, we balance constraints on the mission with mission objectives and the user's requirements.

We define space missions based on three categories: trajectory, type of payload, and payload complexity. Most trajectories fall into one of the following categories: low-Earth orbits (LEO), semi-synchronous orbits, geostationary orbits (GEO), and interplanetary trajectories. Most spacecraft are in the LEO category because LEO requires the least launch-vehicle energy to achieve the orbit. The altitudes for LEO usually range from 150 km to 1000 km, and inclinations vary from 0° (equatorial) to 180°. Some typical inclinations for LEO are 28°–55° for Space Shuttle missions, 90° (polar) for Earth mapping, and 95°–105° for sun-synchronous (see Chap. 10 for a definition of sun-synchronous orbits). LEO has several advantages over GEO and other higher orbits. Because it requires less energy for the launch vehicle per kilogram of payload, LEO is easier to achieve. So we can launch a heavier spacecraft using a given launch vehicle or we can use a smaller launch vehicle to launch our spacecraft. Also, the distance from the spacecraft to a point on the Earth's surface is much shorter for LEO than GEO (a few hundred km compared to 36,000 km). Shorter path lengths allow for better sensor resolution and lower space losses in the communications system. Finally, the natural radiation hazards are lower in LEO than in higher orbits.

However, LEO increases operational complexity, mostly because of reduced coverage and short pass times. A spacecraft in LEO may be visible to any one ground tracking station during only a small percentage of revolutions and it may be invisible to all ground stations for several orbits. We overcome this problem with relay satellites or multiple ground stations, both of which increase operations complexity. Also, even when a ground station can see the spacecraft, coverage time is extremely short and the angular rate of the spacecraft relative to the tracking station is high. For example, a spacecraft in a circular orbit at an altitude of 200 km has a maximum time in view of a ground station of slightly less than seven minutes and a maximum angular rate of two degrees per second [Larson and Wertz, 1999]. The operators must establish a communications link, determine the spacecraft's state of health, upload commands, and download the payload data—all in less than seven minutes. Again, we overcome this problem by using relay satellites or multiple ground stations. Finally, in LEO, atmospheric drag perturbs the orbit, making orbit prediction difficult and limiting the spacecraft's lifetime. Without periodic orbit maneuvers, spacecraft in LEO will eventually re-enter the atmosphere and burn up. At the very least, atmospheric drag will change the orbit parameters (semi-major axis and period), and we may need periodic maneuvers to maintain our mission orbit.

Semi-synchronous spacecraft orbit the Earth twice a day, yielding an orbit period of about twelve hours. Two classes of semi-synchronous orbits are *Global Positioning System (GPS) orbits*—circular orbits inclined at about 55°, and *Molniya orbits*—highly elliptical orbits inclined at 63.4°. We define both orbits in more detail in Chap. 10. Semi-synchronous orbits feature long viewing times, low relative angular velocities, and immunity to atmospheric drag. These characteristics make operations easier, but it requires more launch energy to achieve these orbits and the radiation environment is more severe than for LEO.

Geostationary (GEO) spacecraft orbit the Earth once per day, exactly matching the Earth's rotational velocity. The orbit plane is the Earth's equatorial plane. In this orbit, the spacecraft remains stationary relative to the Earth's surface. This position provides for continuous coverage of the Earth and continuous contact with the ground site, making scheduling less complicated. Tracking is easy because the relative velocity between the ground site and the spacecraft is zero. But GEO does have some disadvantages. First, it takes more launch energy than LEO or semi-synchronous to reach this orbit. Also, the path lengths between the spacecraft and the target or the ground station are long (36,000 km). Finally, perturbations caused by a non-spherical Earth, the Moon, and the Sun cause the spacecraft to drift from its assigned position. We correct for this drift by doing periodic station-keeping maneuvers, which increases operational complexity.

Some people argue that spacecraft in GEO orbit cost more to operate than spacecraft in LEO because GEO spacecraft are continuously visible and operators will interact with the spacecraft continuously, driving up operations costs. This is a management issue, not a design issue, and the operations concept must state how often the satellite will be monitored by operators. We must consider carefully how much time is needed to interact with the spacecraft based on the mission and spacecraft design.

The final category of trajectories is *interplanetary trajectories*, in which spacecraft escape Earth's orbit, traverse the solar system, and either fly by or rendezvous with other bodies in the solar system. Interplanetary trajectories have large launch energies, precise navigation requirements, extremely long path lengths for communications, and long flight times. All of these characteristics increase the complexity of mission operations.

The second category we use to define a mission is the type of payload on the spacecraft. Most common payloads are for

- Communications
- Navigation
- Remote Sensing
- Scientific
- Interplanetary Exploration
- Technology Demonstrations
- Research, Development, Testing, and Evaluation (RDT&E)

Communications spacecraft transfer information from one point to another. Most communication spacecraft are in GEO, taking advantage of the hemi-spherical coverage and continuous availability. A new class of communications spacecraft is being developed for LEO to provide continuous, global coverage for personal communications with cell phones. Although LEO reduces power and launch requirements, this system is much more operationally complex than a small number of spacecraft in GEO. Space communications systems give us high volumes of data and continuous coverage.

Navigation spacecraft transmit a signal that contains highly accurate spacecraft position, velocity, and time data, which allows us to determine the position and velocity of users anywhere on, or near, the Earth. Navigation systems such as GPS, a constellation of 24 satellites in semi-synchronous orbits, feature highly accurate orbit determination and prediction, as well as continuous, worldwide coverage. They are thus able to provide extremely accurate position and velocity information for ground, air, and space users.

Remote-sensing spacecraft normally carry sensors for observing the Earth or some other object from space. Some typical uses of remote-sensing spacecraft are weather observation, photo reconnaissance, missile detection, and Earth-resources mapping. Remote-sensing spacecraft in LEO take advantage of short path lengths for high resolution, whereas remote-sensing spacecraft in GEO take advantage of global coverage. Remote-sensing spacecraft are characterized by high data volume, precise orbit navigation, and frequent station-keeping maneuvers and payload calibrations.

Scientific spacecraft are similar to remote-sensing spacecraft and possess many of the same features. Additionally, scientific spacecraft like the Hubble Telescope may observe objects in space rather than on Earth. Besides the requirements for remote sensing, scientific missions may also require frequent slewing maneuvers, extremely accurate attitude determination, and precise pointing control.

Exploratory *interplanetary spacecraft* are actually remote-sensing (Mars Global Surveyor), scientific (Cassini), or even communications satellites (Mars Global Surveyor carries a communications package to communicate between Mars orbit and the surface of Mars, and there are plans for a series of Mars communications satellites). However, because of their trajectories and communications differences we treat them separately. Chapter 17 describes the operational differences between interplanetary and Earth-orbiting spacecraft.

Technology demonstration and *RDT&E spacecraft* are experimental and typically high-risk; they may place more stress and demands on operations than other missions do.

The final category that defines the type of mission is payload complexity. One factor we use to evaluate payload complexity is how many separate payloads the spacecraft carries. Other factors are the frequency and volume of payload commands, type and number of payload constraints, timeliness and volume of data, and criticality of payload operations. Chapter 14 discusses payload operability and relates the payload's design and operations concept to the complexity of mission operations.

Our plan for mission operations is incomplete until we've identified the hardware, software, and interfaces required to support the mission. Before we develop the operations concept to support a particular mission architecture, we must decide if we can use existing hardware, software, and ground facilities. Mission architectures that require daily eight-hour tracks using the Deep Space Network may not work, and developing new software and dedicated ground-support facilities will drive up costs.

Step 2. Develop Alternative Mission Concepts that Support Step 1.

A *mission concept* is a broad statement of how the mission will work in practice [Larson and Wertz, 1999]. While not assuming a particular approach, we should identify significantly different ways of carrying out the mission with a mix of ground, air, and space assets. For example, suppose our mission objective were to travel from Los Angeles to New York. Our alternative mission concepts may include traveling by car, bus, train, or airplane, or some combination of the four. We may also choose to hitch-hike or ride a bicycle across country.

We can eliminate several options for meeting the overall mission objective by applying such constraints as schedule, cost, politics, and technology. For example, on our trip from Los Angeles to New York, schedule constraints may keep us from hitch hiking or riding a bicycle.

Most space missions involve getting and transferring information from one place to another, so our mission concepts should focus on how we obtain and transfer data. Therefore, we must consider issues like data acquisition and processing, communications architecture, ground-systems tasking and scheduling, and mission timeline.

Step 3. Identify and Do Key Trades.

In this step the MOM identifies and offers ways of controlling those system drivers that increase the cost and complexity of mission operations. Doing trades with other mission elements can reduce the cost of mission operations without increasing lifecycle costs or mission risk. We must do these trades as early as possible to avoid costly re-designs and to keep from recommending changes to parts of the mission design that are substantially frozen. In this section we isolate trades between mission operations and other elements. In actual programs these trades may have more complex effects on other mission elements. For example, a trade between orbit altitude and operational complexity affects several elements. Orbit altitude is a major input to communications (propagation path length), space (distance to the target and space environment), and launch (launch-vehicle energy needed to attain orbit). Table 2.2 summarizes some typical trades between mission operations and other mission elements.

Table 2.2. Typical Trades Between Mission Operations and Other Elements of a Space Mission. We must do these trades early in the mission design before element designs are frozen.

Element	Trade	Issues
Orbit	LEO versus GEO	• LEO reduces launch costs and communications path length but increases operational complexity and reduces coverage compared to GEO
Spacecraft (payload and bus)	Resource margins versus development costs	• Positive resource margins reduce operational complexity but increase development cost and spacecraft mass
	Spacecraft autonomy versus development costs	• Autonomy can reduce operations cost but increases development costs
	Software re-use versus development cost	• Re-using software from mission to mission lowers development costs but reduces flexibility
	Real-time versus stored commands	• Real-time commands increase flexibility but also increase operational complexity
Ground Systems	Centralized versus distributed data processing	• Centralized data processing decreases interfaces but increases staffing costs
	Dedicated versus existing facilities	• Using existing facilities reduces development costs but leads to more complex scheduling and less flexibility
	Automation versus development cost	• Automated ground stations reduce operational cost (staffing) but increase development costs
	Software re-use versus development cost	• Re-using software from mission to mission lowers development costs but reduces flexibility

Table 2.2. Typical Trades Between Mission Operations and Other Elements of a Space Mission. (Continued) We must do these trades early in the mission design before element designs are frozen.

Element	Trade	Issues
Communications	Stored versus real-time data transfer	• Real-time data transfer increases timeliness of data and reduces onboard memory needs but increases operational complexity
	Data rate versus pass duration	• For a given volume of data, short passes require less ground-network support but also require higher data rates over the communication link

This trade between operational complexity and development costs has been ignored. Usually, designers have reduced development costs (current dollars) and driven up operational costs (future dollars). Of course, it's easier to spend fewer dollars now and more later than to spend more dollars now with the promise of reducing future costs. But if we are to reduce lifecycle costs, we must do informed trades between current development costs and future operational costs.

Step 4. Characterize Acceptable Mission Concepts and Associated Space Mission Architectures.

Describe operations for each element of the mission (see Sec. 1.2):

- Subject
- Launch and replenishment
- Orbit and coverage
- Spacecraft (payload and bus) operation and requirements
- Communications architecture and data flow
- Ground element and data flow
- Mission operations

If you are the MOM, work closely with the program manager during this phase, developing plans of operations for each element of the mission as the mission design matures. Be alert for potential operational conflicts among the mission elements and be prepared to identify operational system drivers and propose alternatives to make operations simpler.

Step 5. Assess Items in Step 4 and Select a Baseline Mission Concept with Supporting Space Mission Architecture for Future Development.

Justify and document the "best" approach. Base it on lifecycle costs and ability to meet mission objectives and top-level requirements. Evaluate operational costs and complexity for each mission concept and make sure cheaper operational alternatives don't drive up development costs. Evaluate the mission concepts and ensure mission operations can meet objectives.

Step 6.   Develop Alternative Mission Operations Concepts to Support the Mission Concept.

The *mission operations concept* explains how the mission will be flown through all phases. It combines the hardware, software, people, and procedures needed to meet mission objectives. At this step in the design process, we concern ourselves with operations concepts that support each mission architecture. If this is truly a concurrent design, we must develop the operations concept with other mission elements so the mission is operable. Chapter 4 describes how to develop a mission operations concept.

Figure 2.1 shows the inputs to the operations concept, which vary with each mission and change as the mission design matures. We begin by stating the mission objective and mission concept. Normally, the concept includes the mission type, user, and plan for fulfilling the mission objectives.

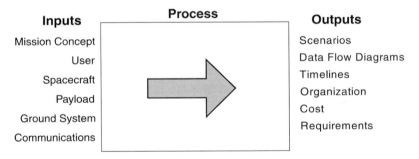

Fig. 2.1.   Developing an Operations Concept. Concept development begins with a mission concept and continues throughout the mission (see Chap. 4).

Payload operability, discussed in Chap. 14, is central to developing an operations concept. Factors include scheduling, commanding, data processing, data flow, spacecraft support, and physical constraints needed to complete the mission. The payload schedule can be either time-driven, event-driven, or adaptive. A time-driven schedule requires the payload to execute a given set of commands at specified times. An event-driven schedule requires the payload to execute a given set of commands following an event, such as acquisition of a target. An adaptive schedule is the most complex form of payload scheduling because it requires the payload (or operator) to determine which commands to execute based on observed data. Commanding the payload includes several issues. Simple payloads use repeated commands (several chances to execute) and don't require spacecraft support to carry them out. Complex payloads use commands that are time critical (only one chance to execute), require spacecraft support (maneuvers), are frequent, and don't repeat.

Data processing and data flow affect the communications architecture of the mission. Payloads with high data volume, multiple data formats, and short latency demand strong operational support. Payloads with low data volume, single data formats, and delayed transmission are relatively easier to operate. Finally, payloads with physical constraints such as temperature limits or pointing restrictions make operations more complex. Table 2.3 summarizes factors that influence payload operability.

Table 2.3. Payload Operability. Payload operability is central to developing an operations concept for a space mission (see Chap. 14).

Factor	Issues
Scheduling	Time-driven, event-driven, or adaptive
Commanding	Repetitive commands, time critical, spacecraft support required
Data Processing	Multiple formats, data volume, data latency, data storage
Payload constraints	Pointing accuracy, temperature limits, pointing restrictions

Several factors affect the spacecraft's operability, but perhaps the two most important are adequate resource margins and autonomy. Typical resources monitored on the spacecraft are communications link, power, thermal, pointing, and memory. If we have adequate margins for all resources, we only need to model spacecraft operations at the systems level. If any margins are negative, or nearly zero, we must model the spacecraft at the sub-system level to ensure safe operations. For example, if the spacecraft's peak power load exceeds the available power, some systems must draw power at different times. Thus, we have to model all spacecraft operations at the subsystem level to determine if a particular operation will exceed available power. Other issues affecting spacecraft operability are frequency and timing of maneuvers, autonomy of spacecraft operations, number of telemetry channels monitored, requirements for spacecraft commanding and validation, monitoring and analyzing the spacecraft's state of health, engineering calibrations, and software maintenance. We discuss spacecraft operability further in Chap. 15.

The final inputs to an operations concept are the communications architecture and ground element needed to support the mission. The communications architecture represents the data flow (uplink and downlink) between the spacecraft, the operator, and the user. The communications architecture depends on data volume, ground-station coverage, latency requirements, and data reliability. A spacecraft in LEO that must deliver near-real-time, processed data to the user will require extensive operational support. A spacecraft in GEO that can delay data transmission will need less support. Early in the mission we must decide whether to use an existing ground-support network, such as the Deep

Space Network or the Air Force Satellite Control Network, or to develop a new, dedicated network. Using an existing network may reduce development costs but may also make operations more complex because of potential scheduling conflicts with other missions and the need to match the network's interfaces. We describe existing ground-support networks in Chap. 12.

Table 2.4 lists the steps in developing an operations concept for a space mission. The results of developing an operations concept and the contents of a concept document are shown in Fig. 2.1 and described in Chap. 4.

Table 2.4. Developing an Operations Concept for a Space Mission. Chapter 4 describes this process in detail; it's an important part of the space mission concept.

Step	Key Items	Where Discussed
1. Identify the mission concept and supporting space mission architecture and gather information.	• Information system characteristics • Payload characteristics • Spacecraft bus characteristics • Definition of data product • Ground system	Chap. 2; Larson and Wertz [1999] Chap. 2
2. Determine what mission operations must do.	Functions usually vary for each mission concept and architecture. We can combine or eliminate some functions.	Chap. 3
3. Identify ways to carry out operations and determine whether capability exists or must be developed.	• Where conducted (space or ground) • Degree of automation on ground • Degree of autonomy on spacecraft • Software reuse (space and ground)	Chaps. 1–3, 8, and 18
4. Do trades for items identified in Step 3.	Options are often selected before development of operational scenarios. These trades take place within the operations element, which includes the flight software.	Chaps. 1–5, 11, 12, and 18 Table 3.1
5. Develop scenarios for operations determined in Step 2 and the options selected in Step 4.	Operational scenarios are step-by-step descriptions of how to do mission operations. Key issues and drivers on the operations system are identified during this step.	Chaps. 4, 6–8
6. Develop timelines for each scenario.	These timelines establish performance parameters for each function within mission operations. Identifies what, how fast, and when events occur.	Chaps. 4, 6–8
7. Determine the type of resources (hardware, software, or people) needed to do each step of each scenario.	Allocation to hardware, software, or people depends on what is done, how quickly it must be done, and how long the mission lasts.	Chap. 4

Table 2.4. Developing an Operations Concept for a Space Mission. (Continued) Chapter 4 describes this process in detail; it's an important part of the space mission concept.

Step	Key Items	Where Discussed
8. Develop data-flow diagrams.	These diagrams underpin the ground and flight data systems and the command, control, and communications architecture.	Chaps. 4 and 13
9. Characterize the organization and team responsibilities.	Identify type of organization, product responsibility, interfaces, and number of people. For cost-effective operations, minimize the number of organizations and interfaces	Chaps. 4 and 5
10. Assess mission utility and complexity and cost of mission operations.	The cost estimates include costs for development and operations. We refine them each time we update the mission operations concept.	Chap. 5; Larson and Wertz [1999] Chap. 20
11. Identify derived requirements and cost and complexity drivers and negotiate changes to mission concept.	Document derived requirements and ensure consistency with top-level requirements.	Chaps. 1, 2, and 4
12. Generate technology development plan.	Technology may or may not exist to support the mission concept	Chaps. 4 and 6
13. Iterate.	--	--

Step 7. Do Key Trades Within Mission Operations.

Trades within mission operations usually involve trading people tasks with hardware and software. Pertinent factors are development costs, operational complexity and cost, level of acceptable risk to the mission, and flexibility of operations. Table 2.5 lists typical trades and issues associated with mission operations and where to find more information in this book.

Table 2.5. Trades with Mission Operations. These trades represent areas that contribute most to the complexity of space mission operations.

Trade	Issues	Where Discussed
Spacecraft autonomy	Development cost, mission duration, modeling required, "trust" given to autonomous operations	Chaps. 3 and 15
Ground-station automation	Development cost, mission duration, amount of software	Chaps. 3 and 12
Ground-station staffing	Daytime operations or around the clock, acceptable level of risk, how much data latency acceptable	Chaps. 4, 5, 8, and 12

Table 2.5. **Trades with Mission Operations. (Continued)** These trades represent areas that contribute most to the complexity of space mission operations.

Anomaly response	Acceptable level of risk, how much down time acceptable, level of spacecraft safing required	Chaps. 4, 5, 15, and 16
State-of-health monitoring	Alarming and trending, acceptable level of risk	Chaps. 8 and 15
New or existing hardware	Development cost, flexibility, maintenance	Chaps. 4–6, and 8
New or existing software	Development cost, software maintenance, flexibility	Chaps. 4–6, 8, and 18

Step 8. Allocate Resources to Functions for Each Mission Phase.

Our mission operations concept must translate user requirements to a process that works—a system that safely and reliably provides the required data to the user.

Table 2.6 lists what we must do for all space missions. We describe each of the thirteen functions in Chap. 3 and explain further how to do them in the chapters listed in the table. We're not prescribing how to organize mission operations. For some missions, especially smaller ones, we can combine several functions into one activity. In other missions, we may divide the operations organization into uplink and downlink sections and operations functions into uplink and downlink components.

Table 2.6. **What Operations People Do Throughout a Space Mission.** These functions are typical of most space missions and are not intended to imply any organizational structure.

	Function	Where Discussed
1.	Mission planning	Chaps. 3, 6, and 8
2.	Activity planning and development	Chaps. 3, 4, and 7
3.	Mission control	Chaps. 3, 4, 8, and 12
4.	Data transport and delivery	Chaps. 3, and 11–13
5.	Navigation planning and analysis	Chaps. 3, 7, 8, and 10
6.	Spacecraft planning and analysis	Chaps. 3, 4, 7, and 15
7.	Payload planning and analysis	Chaps. 3, 4, 7, and 14
8.	Payload data processing	Chaps. 3, 13, and 14
9.	Archiving and maintaining the mission database	Chaps. 3, 13, and 14
10.	Systems engineering, integration, and support	Chaps. 3, 4, and 6
11.	Computers and communications support	Chaps. 3, 4, 8, and 11–13
12.	Developing and maintaining software	Chaps. 3, 4, and 18
13.	Managing mission operations	Chaps. 3 and 5–8

Operations tasking and demand for resources vary as the mission matures. We divide the mission into four phases:

- Launch and early-orbit (L&EO) operations
- Routine operations
- Anomaly operations
- Extended operations

We describe launch and early-orbit operations in detail in Chap. 9. This phase is high risk because the time to make decisions is short, the pace is fast, and decisions may determine whether the mission succeeds or fails. In addition, we're operating in a harsh and dynamic environment, where many activities take place for the first time in the mission.

As we transition to routine operations, our attention moves from the health and safety of the spacecraft to supporting the mission (see Chap. 8). Routine operations usually involve well established activities, static procedures, and fully trained operators.

If something unexpected occurs, we begin anomaly operations (see Chap. 16). The approach we follow to resolve the anomaly and return to normal operations depends on the nature of the anomaly, the type of mission, the user's requirements, and the operators' qualifications.

Finally, missions that continue beyond the design life of the mission enter into extended operations. Extended operations usually involve reduced staffing, limited onboard resources, and well understood operations. Even though operations are well understood, operator complacency and reduced resources can increase risk and potential for operator errors during extended operations.

Step 9. Assess Mission Utility, Lifecycle Cost, Relative Complexity, and Cost of Operations.

We use the cost model described in Chap. 5 and summarized in Table 2.7 to assess the complexity and cost of our mission. The model is divided into four factors: operability of the mission design, operability of the spacecraft and payload, risk policies, and inheritance and complexity of the ground system. Each of the four factors divides into approximately twenty sub-factors, and the sub-factors divide into low, medium, and high tasking levels. We determine the tasking level (high, medium, or low) using the criteria described in the cost model. To determine the values required for the sub-factors, we turn to the appropriate chapter in the book. For example, turn to Chap. 10 to determine the frequency and number of maneuvers our mission requires. The cost model we use in this book estimates only the cost of mission operations, not development costs. For example, this model predicts lower operational costs if we use a dedicated ground station because we don't have to adapt our mission to an existing system. It doesn't predict the cost of developing the dedicated ground station.

Table 2.7. Complexity and Cost Factors for Assessing the Mission Operations Concept. This model estimates the cost of operating a space mission based on top-level requirements (see Chap. 5).

Mission Operations Cost Model Factors	
Complexity Metrics for Mission Design and Planning Science Events Engineering Events Maneuvers Tracking Events	**Complexity Metrics for the Flight System** Commanding Monitoring Pointing Resources and Margins Automation
Complexity Metrics for Risk Avoidance Command and Control Data Return Performance Analysis Fault Recovery	**Complexity Metrics for the Ground Systems** Interfaces Complexity of Data Automation Organization and Staffing

After we evaluate all of the sub-factors in the cost model, we convert the tasking levels to numerical values, called scale factors, and average them for each of the four factors. Finally, we combine these four scaling factors and compare them to a similar mission in a look-up table. We multiply the number of full-time equivalents (FTEs) needed to operate the nominal mission by the scale factor to estimate the FTEs needed to operate the new mission.

The four factors in the cost model cover all aspects of mission operations and include the thirteen mission operations functions described in Chap. 3, but functions don't map directly into cost factors. For example, each of the four cost factors includes different aspects of navigation. After we've determined the number of FTEs needed to support our mission, we can estimate the number of FTEs necessary to support each function by comparing our mission to a similar, existing mission. Chapter 3 contains data for allocating a percentage of resources to operations functions.

Step 10. Iterate and Document Reasons for Choices.

We present two extremely important points in this step:

- The operations concept matures continuously over the mission's lifetime
- We must document and justify our decisions

The operations concept document grows and changes continuously. As the mission design matures and changes, we must evaluate the effects of these changes on mission operations. For example, someone may decide to save weight by placing fewer batteries on the spacecraft. But this decision may create a negative power margin for the spacecraft during eclipse periods. Whenever we reduce resource margins on the spacecraft, we increase operational complexity and cost.

Similarly, someone may decide to remove or replace one of the payloads on the spacecraft. Removing a payload may increase resource margins or reduce maneuver requirements, thus reducing operational complexity. Each time the mission design changes, we have to determine whether the operations concept must change.

It's hard to evaluate the need for these changes if we haven't documented and justified our previous choices and decisions. This step is especially important if a change to the mission design significantly increases operational complexity and cost. If we can't justify our estimate of operational costs, it will be difficult to trade against other mission elements in order to reduce lifecycle costs.

FireSat

One way to help you understand the key concepts in this book is to use a specific mission as an example and to indicate some of the key trades that can be made. We are going to use a mission called FireSat as an example mission. This mission is described in great detail in *Space Mission Analysis and Design* [Larson and Wertz, 1999] and we will take the requirements and mission concepts that are described there as a given and use them as inputs to our operational considerations. The concepts and lessons that are included in this book are applied to FireSat in Chap. 21.

References

American National Standard. 1993. *Guide for the Preparation of Operational Concept Documents*. Washington, DC: American Institute of Aeronautics and Astronautics.

Larson, Wiley J. and James R. Wertz, eds. 1999. *Space Mission Analysis and Design*. 3rd ed. Torrance, CA: Microcosm, Inc. and Dordrecht, Netherlands: Kluwer Academic Publishers.

Chapter 3
Mission Operations Functions

Gael F. Squibb, *Jet Propulsion Laboratory,
California Institute of Technology (Retired)*

3.1	Mission Planning
3.2	Activity Planning and Development
3.3	Mission Control
3.4	Data Transport and Delivery
3.5	Navigation Planning and Analysis
3.6	Spacecraft Planning and Analysis
3.7	Payload Planning and Analysis
3.8	Payload Data Processing
3.9	Archiving and Maintaining the Mission Database
3.10	Systems Engineering, Integration, and Test
3.11	Computers and Communications Support
3.12	Developing and Maintaining Software
3.13	Managing Mission Operations

We've discussed the big picture and how mission operations fits into the overall scheme. But you may still not know what space mission operators do. You're not alone—many people don't. Different countries, and indeed different organizations within a country, may organize differently to operate space missions. For example, in the United States, NASA and DOD organize their operations differently, and people within each organization differ on philosophy and implementation. In this book, *mission operations* includes actions needed to prepare for launch, activities that take place after launch, and activities required to maintain the infrastructure that supports space missions.

Recall that we must distinguish between the *mission concept*—how the elements of the mission fit together (Chap. 1) and the *mission operations concept*—

how we'll do mission operations to carry out the mission concept (Chap. 4). In this chapter we'll define the 13 functions crucial to mission operations everywhere and discuss how we can combine them to meet the mission operations concept. Hardware, software, and people work together to complete these 13 functions. The mission operations manager (MOM) must carefully trade automation against human operations because automating some of these functions can lead to lower operations costs and, in most cases, lower lifecycle costs. Organizations may group or name those functions differently, but we believe they embody the tasks essential to mission operations.

The MOM must decide *which functions* to do, as well as their *scope* and *how* to do them. Depending on the size of the mission, a manager may even have to add functions to our list. The MOM must also address organizational, hardware, and software interfaces between the functions. For example, he or she has to decide whether to use an existing mission operations infrastructure or to create one for this mission. In any event the decision will depend on what needs to be done and what is the most cost-effective way to do it.

The 13 functions don't normally correspond to operational teams. The number and size of the operations teams depend on the mission, its complexity, and the organization's philosophy. For example, if we examine a communications mission that uses a geostationary satellite,

- A commercial operation requires about 12 people per spacecraft to do everything
- A DOD operation requires about 27 people per spacecraft to do the same tasks
- A European Space Agency operation requires about 22 people per spacecraft

These examples illustrate a common mission with minor differences in complexity and tremendous differences in philosophy. We'll examine these philosophical differences throughout this book and, in Chap. 5, we'll discuss how to assess the complexity and cost of space mission operations.

Figure 3.1 overviews a mission operations system, which implements the mission concept and mission operations concept. It processes information from, and controls, the ground and space assets, so users and operators get needed information and services. Most missions today—science, communications, navigation, and remote sensing—focus on providing information to users or customers.

We divide mission operations into two main pieces:

- *Data* —the hardware and software on the ground and in the spacecraft that help us operate the mission
- *Operations Organization*—the people and procedures that carry out the mission operations concept

Mission Operations Functions

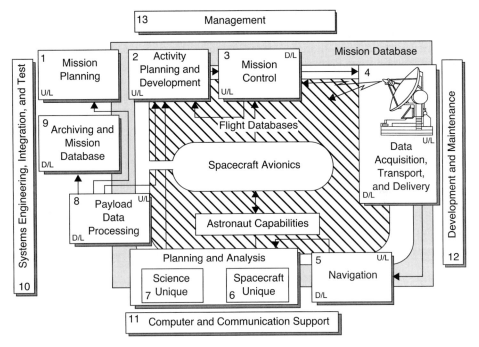

Fig. 3.1. **The 13 Functions of Mission Operations.** The figure shows how the functions interact and whether they're involved in downlink (D/L) or uplink (U/L). Functions in the shaded area share data within the mission database. The white area in the center containing spacecraft avionics and astronaut capabilities are located on the spacecraft or flight vehicle.

Mission operators should help develop the mission concept, but they often get involved too late. We intend to attack this problem aggressively in the future to reduce the lifecycle costs of space missions. Mission operators *do* generate the concept that leads to successful mission operations.

Figure 3.1 and Table 3.1 describe the 13 functions mission operators do for space missions. We define the functions so we can discuss them clearly. We can combine or eliminate functions to support the desired mission operations concept. The first nine functions, shown in the mission database box in Fig. 3.1, are basic to all space missions. We can tie them together in order to support the two essential processes for mission operations—uplink and downlink (U/L and D/L). Chapter 8 discusses uplink and downlink, which operate interactively for many Earth-orbiting spacecraft. However, for a planetary mission with long, one-way light times (>10 minutes), they operate separately. The nine functions share data through the mission database, the spacecraft's avionics, and for the Space Shuttle or the International Space Station, astronauts may carry out portions of these functions. These functions require extensive data processing on the ground or on the flight vehicle on most missions. The order from upper left (mission planning) clockwise to lower left reflects the usual emphasis in processing: from uplink to downlink analysis, and then to planning new uplink activities. The other four functions, which form an infrastructure for mission operations, are

- Managing mission operations
- Developing and maintaining software
- Computers and communications support
- Systems engineering, integration, and test

The arrows in Fig. 3.1 designate important interactions. For example, payload data processing provides data for archiving and is a major input to mission planning. The payload data can profoundly affect mission planning, especially if we're doing an *adaptive mission*, in which operators learn from the data they see and change the mission to get different information or do other activities.

The spacecraft and ground systems must work together on at least seven of these functions. Thus, mission operations people must take part early in decisions that divide responsibility for designing, developing, testing, and operating these systems. Indeed, we believe mission operations should share control of spacecraft avionics to meet mission needs and hold down lifecycle costs. Key space-to-ground trade-offs in these areas are integral to the mission and mission operations concepts.

Table 3.1 lists the 13 functions and illustrates the issues and trades we must consider for each. We'll describe them in detail in the following sections. We'll also discuss the information required to design them, recommend a way to carry them out, and describe what they should provide—inputs, process, and outputs. We must do some functions, or parts of them, before launch, after launch in real time, and after launch in non-real time. Timing often depends on the mission type. For example, planning and developing activities for Earth orbiters often occur before launch because we usually carry out these activities right after launch. But for interplanetary missions, we may develop many activities after launch—while the spacecraft is traveling to its target planet.

All 13 functions are important, and achieving mission objectives depends on parts of each. Some are also more expensive, so we must emphasize them during design to lower lifecycle costs. Table 3.2 highlights five functions, each of which consumes more than 12% of total human resources. Four of these five vary more than 6% from low to high percentages of cost. They therefore demand more of our attention to keep lifecycle costs down.

The following sections describe the information and steps needed to carry out all 13 functions, as well as the products they generate. We use background shading for post-launch activities and no shading for pre-launch activities in the tables in this chapter.

Mission Operations Functions 35

Table 3.1. Mission Operations Functions. This table shows key issues, trades, and where we discuss functions in more detail.

Function	Key Issues and Trades	Where Discussed
1. Mission Planning	• Provide positive spacecraft resource margins • Automate certain space and ground activities • Restrict the number of mission and flight rules • Focus on operability	Chaps. 7, 8, 14, and 15 Sec. 3.1
2. Activity Planning and Development	• Reduce the required number of command loads • Automate validation of command loads and keep approval at lowest possible level • Avoid late changes to plan	Chaps. 7 and 8 Sec. 3.2
3. Mission Control	• Consider sharing operations between missions or multi-tasking operators within a mission • Design ground and flight systems that can be easily upgraded • Automate analysis of ground system and spacecraft performance	Chaps. 8, 12, and 15 Sec. 3.3
4. Data Transport and Delivery	• Keep tracking requirements to a minimum • Design data structures and formats to use standard services • Use variable-length data packets	Chaps. 11–13 Sec. 3.4
5. Navigation Planning and Analysis	• Minimize number of maneuvers • Match mission requirements to accuracy of orbit determination method • Match mission requirements with standard services provided by tracking network	Chaps. 8 and 10 Sec. 3.5
6. Spacecraft Planning and Analysis	• Maintain positive resource margins • Minimize subsystem interactions • Automate spacecraft safe modes and analysis	Chaps. 15 and 16 Sec. 3.6
7. Payload Planning and Analysis	• Use existing planning tools • Automate payload data gathering and calibrations • Insist on payload operability • Minimize payload operations that require knowledge of previously acquired data	Chap. 14 Sec. 3.7
8. Payload Data Processing	• Use existing tools for data processing • Address data processing requirements early in mission design • Understand need for and availability of ancillary data	Chap. 13 Sec. 3.8
9. Archiving and Maintaining the Mission Database	• Consider data push vs. data pull approach • Develop plan early and ensure capabilities exist • Understand need for and availability of ancillary data	Chap. 13 Sec. 3.9
10. Systems Engineering, Integration, and Test	• Involve users, developers, and operators early in mission design • Do key trades on mission operations concept early • Maintain the big picture perspective • Use rapid development and prototyping processes	Chaps. 4 and 6 Sec. 3.10

Table 3.1. **Mission Operations Functions. (Continued)** This table shows key issues, trades, and where we discuss functions in more detail.

Function	Key Issues and Trades	Where Discussed
11. Computers and Communications Support	• Understand existing capabilities and define additional requirements • Consider maintenance and upgrades when defining system	Chaps. 11 and 12 Sec. 3.11
12. Developing and Maintaining Software	• Make software re-use a priority • Involve operators and users early in the design	Chaps. 6 and 18 Sec. 3.12
13. Managing Mission Operations	• Implement automation and autonomy carefully • Consider multi-tasking staff • Keep organization flat and reduce interfaces	Sec. 3.13

Table 3.2. **Relative Cost of Mission Operations Functions.** We've compiled data for four missions and expressed the cost for each function as a percentage of the annual mission operations cost. The missions included are Magellan, Mars Observer, Galileo, and TOPEX. The functions highlighted with gray background use the most resources. [CSP Associated, Inc., 1993]

Function	% Cost*		% FTE†	
	Low %	High %	Low %	High %
1. Mission planning	1	3	2	4
2. Activity planning and development	4	9	7	13
3. Mission control	6	9	10	11
4. Data transport and delivery	4	9	5	11
5. Navigation planning and analysis	3	6	4	7
6. Spacecraft planning and analysis	13	20	18	25
7. Payload planning and analysis	3	19	3	14
8. Payload data processing	7	25	6	13
9. Archiving and maintaining the mission database	0	3	0	3
10. Systems engineering, integration, and test	3	6	3	7
11. Computers and communications support	3	9	2	5
12. Developing and maintaining software	5	8	5	8
13. Managing mission operations	9	11	11	14

* Percent of average annual mission operations costs
† Percent of annual, average, full-time equivalents each function uses

3.1 Mission Planning

The *mission plan* is a top-level description which spans the mission's lifecycle. It describes how the mission will be flown, expresses objectives in operational terms, and sets in place major activities. Mission planning concentrates on uplink but takes into account the downlink abilities of the spacecraft and ground system. The mission plan is consistent with and generated after the mission and mission operations concepts. In this activity we develop mission objectives and plan how to meet them. Mission planners create a plan before launch and then change the plan as required. Before launch, mission planners play a major role in trades between mission functions. During operations, they respond to unforeseen events with updated mission plans and make sure operations meet objectives.

3.1.1 Information Required for Mission Planning

Table 3.3 lists information needed for mission planning—mostly during mission development. Early in development we may have only mission objectives, a mission concept, and a mission operations concept. We have to get the other inputs or assume them so we can begin mission planning. As a design matures, we learn enough about the mission requirements, spacecraft, and mission operations to update our plan.

Table 3.3. Information Required for Mission Planning. Before launch, we know the mission objectives, concept, requirements, and abilities of the spacecraft (bus and payload) and mission operations system. We then compare current status and mission plan throughout the mission. Background shading indicates a post-launch mission phase.

Information Required	Comments	Where Discussed
Mission objectives	• State in terms of science or payload return • Usually, state in qualitative terms	Chap. 1
Mission concept	• State how the mission elements will work together	Chap. 1
Mission operations concept	• Emphasize mission operations; ground systems; and command, control, and communications system • Focus on goals and vision of customer, program management, and users of payload data	Chap. 4
Mission requirements	• Make sure they are quantitative, measurable, and consistent with mission objectives	Chaps. 4 and 6
Spacecraft capabilities (Includes spacecraft bus and payload)	• Include capabilities of spacecraft bus • Focus on areas that are changeable to simplify operations	Chaps. 14 and 15

Table 3.3. Information Required for Mission Planning. (Continued) Before launch, we know the mission objectives, concept, requirements, and abilities of the spacecraft (bus and payload) and mission operations system. We then compare current status and mission plan throughout the mission. Background shading indicates a post-launch mission phase.

Information Required	Comments	Where Discussed
Mission operations system capabilities	• Identify current and planned abilities of the existing operations system, including tracking stations, ground stations, hardware, and software	Chaps. 11–13, and 18
Current status versus plan (ongoing activity)	• Obtain reports from other 12 functions • Identify deviations caused by changes in spacecraft performance, customer requirements, or data requirements	Chap. 3

3.1.2 How We Plan a Mission

Table 3.4. How We Plan a Mission. We develop a plan during mission development and change it after launch only when spacecraft capabilities or mission requirements change.

Steps	Where Discussed
1. Quantify mission objectives and goals so they are meaningful to operators	Chap. 1
2. Define orbit or trajectory and calculate launch windows, number and frequency of maneuvers, and viewing periods	Chap. 10; Chap. 7 [Larson and Wertz, 1999]
3. Describe the payload and define operational characteristics	Chap. 14; Chap. 9 [Larson and Wertz, 1999]
4. Describe spacecraft bus and define operational characteristics	Chap. 15; Chaps. 10 and 11 [Larson and Wertz, 1999]
5. Define mission phases, allocate activities to phases, and establish a mission timeline—basis for first high-level activity plan	Chaps. 7 and 8
6. Evaluate requirements for operations and identify those that are difficult to meet	Chaps. 6, 11, and 12
7. Decide whether to use existing mission operations system or develop a new one	Chap. 12
8. Identify mission rules not related to health and safety imposed by program office, users, or operators and express rules in quantitative terms that software can check	Chaps. 4 and 6 Sec. 3.1.3
9. Document and iterate as necessary	

In mission planning we prepare the baseline set of documents described in Sec. 3.1.3. The mission plan describes how the mission will be flown until in-flight events require it to change. The plan establishes operational criteria to meet objectives, as well as major activities and events. Although planning produces a document or set of documents, the object is to understand the mission and to begin trading among the mission functions to get the best return for the customer's money. Much of this work begins when we're developing the mission concept.

After launch, mission planners periodically compare achievements with the mission plan. We learn about these achievements from payload or spacecraft planning and analysis. We then change the mission plan, mission rules, and orbital plan based on the flight system's changing abilities or processed data from the payload.

3.1.3 Products of Mission Planning

Table 3.5. **Products of Mission Planning.** A document describing the mission plan, mission rules, and orbital phases. We complete the document during mission development and update it as needed after launch.

Products	Contains
Mission Plan	1. Mission objectives and goals 2. Orbit or trajectory description and orbital plan 3. Payload description and operation 4. Spacecraft bus description and operation 5. Mission phases 6. Description and techniques of mission operations 7. Mission rules and method of verification

The mission plan matures as launch approaches and continually reflects trades across the project's elements. The mission plan is usually placed under configuration control before the first review of project requirements. Configuration control allows changes only after everyone they affect has had a chance to review them and their costs. A proper authority decides whether or not to change the plan. Each plan contains:

Objectives and Goals. Quantify the objectives and goals of the mission so operators can understand them. For example, "discovering and understanding the relationship between newborn stars and cores of molecular clouds" is meaningful to a scientist, but "observing 1000 stars over two years with a repeat cycle of once every five months using each of the four payload instruments" is much more meaningful to an operator. Describing the objectives and goals in these terms requires experienced operators and payload users to interact. This interaction becomes very detailed during the development of the mission operations concept, as discussed in Chap. 4, but the plan must include top-level objectives and goals.

Trajectory or Orbit Requirements and Description. This section establishes a context for the mission by describing the number and frequency of maneuvers, as well as constraints on the view periods of tracking facilities and target, launch time, and launch windows. (See Chap. 10.)

Payload Requirements and Description. This section includes the number of instruments and how they'll gather the data needed for the mission. It also includes the payload's operational characteristics, which derive from the mission operations concept. (See Sec. 4.2.2 and Chap. 14.)

Spacecraft Bus Requirements and Description. Here we define performance characteristics; allocate to the spacecraft mass, power, and flight information—such as memory reserve as a function of time from launch. During mission design, we must be sure to show the spacecraft is operable. (See Sec. 4.2.2 and Chap. 15.)

Definition of Mission Phases. We define the mission phases associated with the orbit or trajectory while developing the mission concept; then, refine and modify them while generating the mission operations concept. Normally, we specify the duration and function of these activities:

- Orbit insertion
- Spacecraft checkout
- Payload checkout
- Cruise (interplanetary)
- Maneuvers
- Orbit operations

The orbital plan or mission-phase plan defines in more detail what takes place during each of the mission phases described in the mission plan. In describing the payload checkout phase, we may define the priority of checking out the payload instruments relative to calibrating the spacecraft's attitude-control sensors, as well as the order of checking out these instruments. The mission-phase plan reflects agreements between spacecraft and payload designers and users. It therefore defines, at a very high level, the operational activities that the spacecraft and ground system will carry out during the mission timeline.

Mission Operations System Requirements and Description. This describes and includes requirements for key functions. It includes allocations of performance parameters by end-to-end information system (EEIS) engineering. The EEIS distributes and defines the nature of bit-error rates between the spacecraft and the ground, as well as between the ground antenna and the mission database. Most NASA missions follow standards of the Consultative Committee for Space Data Systems (CCSDS) for protocols regarding data transmission. These protocols define the way packets of information are assembled for transport. Designers of the information system must understand how packets will react to bit-error distributions, especially if the information in the packets is compressed.

Mission Rules (Guidelines and Constraints). Developing mission rules is crucial to mission planning. *Mission rules* describe how to conduct the mission and are usually not related to the spacecraft's health and safety. The health and safety rules are called *flight rules*; we discuss them in Sec. 3.6, Spacecraft Planning and Analysis. Mission rules set policy for what should be done, not for how to do it. Examples of mission rules include:

- Use a specific payload sensor only over the ocean and turn it off ten miles before crossing any land

- Schedule spacecraft events that require real-time monitoring during prime shifts
- Record telemetry 15 minutes before and after critical activities
- Don't overwrite critical data stored on the flight recorder until you verify playback data

The program office may impose mission rules, but in general they represent agreements between the users and operators of the mission operations system. Thus, the rules tend to depend on the mission operation team's background and experiences.

Many mission rules are checked through processes while developing and generating *activities*—time-ordered set of contiguous events. Thus, you must quantify mission rules so operators can understand and translate them into software that can check spacecraft activities automatically for compliance with the mission rules.

3.1.4 Key Considerations

Depending on the complexity of the mission, the spacecraft, and the payload, integrating engineering and payload requests can be time consuming and difficult. Complexity is a function of the

- Number of payload instruments
- Operability of the spacecraft
- Adaptability required because of the data received from the payload instruments

Mission planners, working with designers, may be able to lower the cost of operations by providing positive margins, automating certain space or ground functions, restricting the number of mission or flight rules, and focusing on operability.

Positive Margins. Planners participating in trades across the mission elements during the design phase should make sure the spacecraft and instruments have positive margins. A healthy positive margin may mean we don't have to model, evaluate, and constantly monitor a spacecraft. A spacecraft designed with positive margins is easier and less expensive to operate.

Consider, for example, the power system. When the power available to the spacecraft is less than the power the spacecraft uses with all instruments and subsystems turned on, we must be very careful what we do to the spacecraft. Imagine you're at home, and you could have only half your lights, your clothes dryer **or** your stove, and your garbage disposal **or** your refrigerator on at any given time. You don't control your refrigerator, so you would have to model its performance to predict when it would be turning on and off. Then, you would have to plan to use your disposal only when the refrigerator is off, or turn off your refrigerator every time you turn on your disposal. You would also have to plan your life so you wouldn't need to dry your clothes for a dinner party at the same time you'd be cooking dinner for the party. With positive margins, we don't need to pay for modeling and the tools used in modeling the spacecraft.

Earth-orbiting spacecraft typically have positive margins. Providing positive margins is more difficult for planetary missions beyond Mars, but we can do it. Of course these margins often reduce the scientific return of the mission, and we must trade off operational complexity and cost against the additional scientific return.

Spacecraft Autonomy Versus Ground Generation of Appropriate Functions. Planners seldom trade well between what the spacecraft and ground units do. Decisions are often based on the cost of designing and building the spacecraft, and ground operations must later absorb the cost of accommodating the decisions. The trade-offs include the cost of making the spacecraft autonomous versus doing the same thing on the ground for the life of the mission. The mission duration is an important consideration in making this trade.

The Number of Mission and Flight Rules. We must be sure to keep the mission and flight rules to a minimum during the design of the mission and the spacecraft. Each rule that must be checked involves software, people, or both, and adds to the expense of operations.

Operability and Interaction Between Subsystems. During the design phase and during the element trades, we should make the spacecraft as operable as possible and keep interactions between spacecraft subsystems to a minimum.

3.2 Activity Planning and Development

In this function we convert each orbital-phase plan generated in mission planning into detailed commands that are ready for uplinking to the spacecraft bus as command mnemonics. Normally, we script the spacecraft's actions so each command takes place at a particular time. Many tools within the ground data system have been developed to generate these activities. Each of the NASA centers involved in flight operations has these tools. For example, JPL uses a software program called SEQGEN. Commercial packages are also available and are adequate for many missions. For the next generation of spacecraft, we are beginning to look at process control as a way to store rules on board and to have actions occur when events satisfy these rules. These rule-based actions have usually been limited to fault-protection rules related to the health and safety of the spacecraft and payload. As the technologies mature, putting process-control rules on board spacecraft should save operational costs and increase the payload return. Certain missions planned by NASA, such as returning asteroid samples, will require onboard decision making based on the "state" of the payload, spacecraft, and the environment. This type of onboard technology will make up in large part for our inability to plan the spacecraft events or control them from the ground due to long delays in transmission over great distances. (See Chap. 17.) NASA is still in the early R&D phase of developing computer operating systems and programs to enable spacecraft to rely on the "state" information as a means of activity planning.

3.2.1 Information Required to Plan and Develop Activities

Table 3.6 lists the information we need. Mission planning provides most of this information for the initial activities. After launch, the mission operators' activity requests become increasingly important to achieving the mission objectives.

We also plan activities before launch for the spacecraft—during its integration and test phase—so we can make sure it will work the way we expect it to work.

Table 3.6. Information Required to Plan and Develop Activities. The mission plan (Sec. 3.1) gives us most of this information. Shaded activities occur after launch.

Information Required	Comments
Mission Plan, including the Mission Rules and the Orbital Phase Plan	• Generated by mission planning • Changes occur during the mission as the mission deviates from the nominal plan
Requests for Test and Integration Activities	• Used to support spacecraft integration and test • Some of these activities may occur while verifying the spacecraft after launch • Comparing integration and test data to the in-flight data enables rapid understanding of the spacecraft's health
Requests for Post-Launch Activities	• Requests come mainly from spacecraft and mission control • The activity requests may have been generated before launch, but most are generated in detail after launch using building blocks, or macros of tested and validated groups of commands • The activity requests are the details which implement the higher-level plan for the orbital phase

3.2.2 How to Plan and Develop Activities

The steps for planning and developing activities, as listed in Table 3.7, provide a command load for uplink that is ready to transmit to the spacecraft. In the following pages, we'll amplify these steps. Chapter 7 provides even more detail.

Define Activities. Activity planning supports payload and spacecraft-bus engineers in their early planning of the flight system. The result is one or more activities on the spacecraft and the ground needed to return certain sets of data. Each activity relates directly to many spacecraft commands. Payload-planning tools are provided that allow the payload user to superimpose the field of view of the instrument on the target body of interest. These tools correct for spacecraft trajectory, target body rotation, instrument location, boresight offsets, allowable scan platform, and mirror motion. They also provide a realistic planning footprint. As a result, an investigator can look at different ways of gathering data and determine how to set up observation patterns so they cover the target fully. Later, mission operators will carry out these activities at the scheduled times.

Table 3.7. How to Plan and Develop Activities. This process converts the mission plan into activities, timelines, and commands. Shaded areas occur after launch.

Post-Launch Steps	Comments	Where Discussed
1. Define Activities	• Next level of detail from the orbital-phase plan • Planning usually considers a series of activities, each of which relates directly to spacecraft commands • Timelines of these activities are generated • Function works closely with spacecraft and payload engineers and scientists	Chaps. 7, 14, and 15 Sec. 3.2
2. Generate and Integrate Activities	• Next level of detail after activity planning • Integrates new requests with the activity-planning timeline • Ensures shared resources don't conflict • Automated tools are required to do this process efficiently	Chap. 7 Sec. 3.2
3. Check Mission and Flight Rules	• Final check of mission and flight rules • Software helps generate activities more efficiently	Chap. 7 Sec. 3.1 and 3.2
4. Generate Timelines	• The timeline displays the activities of the spacecraft and the ground	Chaps. 7 and 8 Sec. 3.2
5. Validate Activities	• Verifies the safe interaction of planned activities • Should be automated • Checks constraints for health and safety • Checks to ensure the activities do what is necessary to support the payload • Review and approval should be at the lowest possible level and add value to the process	Chaps. 7 and 8 Sec. 3.2
6. Translate Activities	• Converts the command mnemonics and associated parameters to a binary stream packaged as required by the network being used	Chap. 13 Sec. 3.2

Generate and Integrate Activities. Before launch we must define how long it takes to generate and complete an activity. We use increments of an orbit for Earth-orbiting spacecraft, or weeks to months for a planetary spacecraft. In any event the mission-phase plan lays out pre-launch priorities for the given period as the starting point for generating activities.

The Infrared Space Observatory (an ESA mission) generates a command load for seven orbits (seven days) at a time. The process begins 21 days before uplink and is frozen three days before uplink. Other missions have plans tailored to their mission attributes.

New requests typically come from four functions: mission control, navigation, planning and analysis for the spacecraft bus or planning and analysis for the payload. New and unplanned requests are a part of any mission for several reasons:

- We need an unplanned calibration (for example the star tracker, the attitude-control gyros, or the movement of a platform) to understand some unexpected data
- Unexpected changes in the availability of tracking facilities
- The spacecraft's abilities have changed, so all the standard activities for a certain function need changing before the next command period
- We need to send a new set of tables up to the spacecraft based on the last calibration

These new activities may be merely executing an activity at a different time. If they're truly new, we have to design and test them.

Using the mission-phase plan as the baseline, we take new requests from the other mission operations functions and integrate them into the command load. We remove conflicts based on rules and priorities or negotiation with the various functions. First, we integrate activity periods at a planning level and then expand these activity periods into groups of commands and individual commands. We keep integrating activities and resolving conflicts at increasing levels of detail so individual commands never conflict.

We must make sure shared resources don't conflict. The desire to maximize the use of the spacecraft often results in placing many activities close together. Late changes to activities or adding new activities can cause oversubscription of such spacecraft resources as power, time, data storage, command and telemetry links, and memory. The resource most often exceeded and hardest to resolve late in the process is time—an attempt to do too much, either on the ground or the spacecraft.

Usually, we fix allocations; that is, during a specific period, each payload instrument can use a pre-defined maximum amount of a given resource. As events begin to drive activities, we have to consider dynamic allocation of resources. Of course, we only need to allocate resources when a particular resource doesn't have enough margin. So to avoid some of these conflicts, it's important that resources have enough margins.

Then, we have to ensure the margins are suitably defined, applied, and consistently enforced. If we set margins too small or use them too quickly, the uplink will be overwhelmed with conflicts, requiring last-minute deletions or rework of activities. Spacecraft commanding is one of the few areas where the final deadline is firm. If the commands aren't ready when the spacecraft needs them, there is a BIG problem—probably calling for a complete rework of the command set while the spacecraft is idle.

Some integration occurs during mission design. We usually need to do more after the design is complete because certain information, such as final tracking schedules, may not be available earlier, or we must incorporate new activities. Many activities must happen at specific times, while others, such as various engineering activities, can fit in wherever possible. Except for very simple activities, we can improve integration by using automated tools that identify conflicts and missing dependencies. At this point, we have an integrated command load with no conflicts at the activity level. The commands are mnemonics with their required parameters.

Check Mission and Flight Rules. We can check many of the rules with the same tools used to generate the activities. But we can check other rules only after we've generated the final command load. We use software as much as possible to make activity development faster and therefore less expensive.

Generate Timelines. These timelines display at a high level—and usually in ground-received time—the activities of the spacecraft and the ground. This display allows the controllers and analysts to monitor the spacecraft's activities as the ground station receives data. The one way light time correction to show spacecraft and ground events in the same time domain is required for interplanetary missions.

Validate Activities. We verify the interactions of the planned activities and make sure the command load meets the intended goals of the activities while posing no risk to the spacecraft. We should automate this step as much as possible. Software tools can check to see that the command load doesn't harm the spacecraft and often go through the sequence to make sure it gathers the information requested. Early understanding of the need and method for validation is essential to minimizing operations costs.

Because we can achieve the same end in many ways, the risk for an activity or group of activities can be difficult to quantify. Each approach has good and bad, as well as unknown, aspects. Fear of the unknown pushes perceived risks higher and drives constant searching to find a perceived "safer" way. People who know the interactions on the ground and spacecraft can accurately judge the risk without complicated analysis, but the results must still be quantified. For these reasons, projects often use hardware or software to evaluate the activity interactions. Electronic or manual reviews, hardware or software simulations, or a combination of these techniques can carry out these valuations.

Operations are constrained for all flight and ground systems. Good spacecraft design and automation can make this task extremely easy and reduce the need for staff. Hardware constraints (known as *flight rules*) are those which, if violated, may damage or stress a piece of the flight system, for example, by exceeding thermal, electrical, or radiation limits. There are obvious constraints, such as Sun impingement on a sensor array designed for viewing deep space, and less obvious constraints on the abilities of power subsystems for spacecraft in certain radiation environments. Other constraints (known as *mission rules*) provide guidelines on how to operate the flight hardware and ground systems within acceptable bounds. Violating mission rules usually won't cause permanent damage but may exceed management guidelines on effort or expense, or compromise project goals. These rules may also set restrictions on instrument operating margins or limit operation to modes that have been tested and validated.

If we document constraints properly, we can verify compliance simply and automate constraint checking readily. Having to interpret the rules can delay the process, especially if different interpretations are supportable. For manual reviews, we should make statements clear, so reviewers don't need special knowledge to understand them. For electronic reviews, the constraints statement should be precise and detailed enough to convert it into code.

Because we must understand constraints early and try hard to validate them automatically, we often run command loads through simulators to make sure they're coherent and correct. We can make spacecraft simulators by setting up flight computers and flight hardware in a testbed and running the actual command load through the system. Hybrid simulators use software to simulate some of the components, such as the attitude-control subsystem. Of course, we can also build simulators entirely with software, such as modeling the flight computer in software and then running it on a workstation or other computer. But simulation costs money because we must build, operate, and check the output of the simulator. These steps require time and people.

The amount of simulation required varies tremendously within the NASA missions. We use less for Earth-orbiting spacecraft because they usually have power and telecommunication margins and round-trip transmission times of a fraction of a second. Planetary missions simulate many of their activities because they have negative margins and their data transmissions take tens of minutes to tens of hours.

Complexity of the command load and acceptable risk determine whether review and approval is long or short. Today, mission operators often risk the loss of some data and therefore shorten reviews to save money. Each review should require the fewest people, meet technical specifications, and add to the command load's integrity. If the review is being held merely to have a manager sign off on the command load, we should probably eliminate the review.

Translate Activities. At this point the *command load* is in the form of words and parameters—typically called a command mnemonic for the spacecraft and/or commands which are programs for onboard computers (i.e., payload specific computers), plus some key words for the tracking net. A *command dictionary* has translation tables to convert each command mnemonic to a binary bit pattern that is sent to the spacecraft. The computer programs are already in the form of the binary bit pattern to be sent. Translating a command load means converting activities into a format the ground transport and uplink system (such as the Deep Space Network or TDRSS) and the spacecraft can handle. For a simple flight system, this translation can convert the commands into a binary stream for ground transport and subsequent uplink. For a more sophisticated system, it can use a predicted state of the onboard memory and compile the computer program commands into a direct memory load with required memory management.

3.2.3 Products of Activity Planning and Development

Table 3.8 describes these products. We have converted the mission plan into a series of commands or instructions that the spacecraft will execute for the next activity period. We have checked the command load to ensure that it's safe and will return the desired payload data.

Table 3.8. Products of Activity Planning and Development. These three products enable the spacecraft and ground elements to operate for the next activity period. Shaded areas occur after launch.

Post-Launch Products	Comments
Detailed command loads in mnemonic form	• Spacecraft and payload analysts review the command loads in detail • The mnemonic form is in English, in that the commands are names and parameters that relate to the command
Timeline	• The timeline graphically represents the command load. The x-axis is time, with different spacecraft, payload, and ground actions plotted in parallel against this time. • Different timelines emphasize different aspects of the mission and display at varying degrees of resolution: at minutes, hours, days, or weeks • Chapter 8 shows an example of a timeline
Command load	• The command load, ready for transmission to the spacecraft, is a series of bits generated as described in the translation process

3.2.4 Key Considerations

The most difficult step in planning and developing activities is avoiding conflicts while integrating the mission plan with current requests. More complex spacecraft and planning tools, as well as many automatic checks of ground data, increase this difficulty.

Larger mission margins make activity planning easier because they mean we don't have to manage and simulate resources as accurately. Spacecraft autonomy is also important; in fact, a completely autonomous spacecraft would require no activity planning and development. Finally, a mission plan that depends heavily on receiving and analyzing data to determine what to do makes activity planning and development more difficult.

Following these guidelines will help to make your activity design more cost effective:

- Reduce the required number of command loads. Having many command loads and dense activities within each of them will add to the staff and therefore to your budget. Whenever possible, generate command loads one after the other instead of in parallel. We need parallel generation only when the time to generate a command load is longer than the time to the execute it.

- Validate command loads electronically and automate validation as much as possible.
- Make sure approval of command loads occurs at the lowest possible level while maintaining acceptable levels of risk. Always ask, how does this review and approval add value?
- Minimize changes to the command load once development starts. If mission objectives require late changes, define when they can occur and limit the amount of change at each time window. Document these constraints and make them part of the mission rules.

The speed, memory size (random access memory), and storage size (for example, hard disk space) of today's computers allow us to integrate tools for planning and developing activities. Integrating these tools will usually automate file interfaces, make the process faster, and reduce staffing needs.

3.3 Mission Control

Mission control runs the mission. By following a script, it directs a spacecraft's operation in real time, mainly to ensure that ground crews safely receive the required data. We use mission control whenever mission needs demand it and spacecraft tracking coverage allows it. Mission control includes carrying out the detailed activity plan and ensuring the spacecraft's health and safety during the station contact. It also means setting up and verifying the ground configuration, and then briefing everyone involved before we contact the spacecraft. Mission control then sends commands to the spacecraft, monitors its performance and that of the ground data system, and directs its recovery from any nonstandard conditions.

3.3.1 Information Required for Mission Control

We get needed information mainly from mission planning and activity planning and development. Other information comes from navigation and spacecraft planning and analysis. Table 3.9 summarizes this information.

Table 3.9. **Information Required for Mission Control.** Mission control receives information from mission planning, activity planning, spacecraft planning and analysis, and navigation planning and analysis. Shading indicates information required after launch.

Information Required	Where Discussed
Mission Plan and Mission Rules	Sec. 3.1
Command Loads and Command Files	Sec. 3.2
Real-time Telemetry	Chaps. 14 and 15; Sec. 3.6 and 3.7
Tracking Schedules and Data	Chap. 10; Sec. 3.5
Pass Plan	Chap. 8

3.3.2 How to Do Mission Control

Mission control uses the information in Table 3.9, along with procedures, to operate the mission in real time. We develop procedures for spacecraft contacts, make sure data transfers take place for all uplinks and downlinks during the contact, monitor all systems during the contact, and document all activities that occur during the contact. As Table 3.10 shows, a mission-control team

1. **Develops Procedures.** Controllers fly the mission using the baseline procedures. These procedures tell controllers what to do before, during, and after a contact with the spacecraft. They cover how to configure and control the ground system and the spacecraft.

2. **Supports Spacecraft Integration and Test.** By helping integrate and test the spacecraft bus and payload before launch, we see how the spacecraft and ground system work and get hands-on training.

    We also do end-to-end tests, during which the ground element is integrated with the spacecraft to verify uplink and downlink. These tests validate the operations system's ability to command, deliver telemetry, and monitor alarms in real time.

3. **Configures the Ground System to Support Passes.** Mission controllers configure and control ground-system tools and procedures. Like the conductor of an orchestra, we use this step to make sure everything happens at the right time.

4. **Transmits Commands to the Spacecraft.** During the pass, mission controllers transmit scheduled real-time commands and the spacecraft's command loads. These commands are nearly always pre-planned, but procedures or oral orders may authorize mission controllers to issue real-time commands under certain predefined conditions. These predefined conditions usually cover situations in which the spacecraft has violated limits or is in a non-standard state.

5. **Verifies the Spacecraft's Receipt of Commands.** A mission controller must use spacecraft telemetry and information on the ground system's status to verify that the spacecraft has correctly received its commands. Verification is usually automatic, and detected errors produce messages and alarms.

6. **Monitors the Spacecraft's Health and Safety.** We monitor telemetry measurements versus alarms, the onboard memory readout and verification, and the spacecraft's predicted and actual states. We use graphical displays and text which the ground system presents based on processed information from the spacecraft.

7. **Monitors the Ground System's Operations.** During a pass, we must monitor the ground system's performance just as we do that of the spacecraft, so the mission can meet its objectives.

8. **Coordinates Mission-Control Functions.** Before the pass we must coordinate some activities needed to support real-time contact between the ground and the spacecraft. For example, we may need to

- Schedule short-term tracking support
- Generate an integrated plan for flight and ground activities
- Schedule institutional support and make sure it's available
- Generate logs and reports
- Update databases and files

Efficient mission operations require coordination. It ensures that the ground system is able to transmit planned commands to the spacecraft and receive, process, and store the spacecraft's telemetry data.

Table 3.10. How to Do Mission Control. Mission control develops procedures and supports integration testing before launch and directs the real-time operation of the mission after launch. Mission control supports several planning and analysis functions throughout the mission. Shading indicates steps we do after launch.

Steps	Comments	Where Discussed
1. Develop Procedures for Controllers to Configure the Spacecraft and the Ground System	• Step-by-step instructions for pass support	Chaps. 7 and 8
2. Support Spacecraft Integration and Test	• Verifies uplink and downlink systems • Tests end-to-end compatibility	Chaps. 8, 9, and 12
3. Configure Ground System to Support Passes	• Real-time with some pre-launch planning	Chaps. 8 and 12
4. Transmit Commands to the Spacecraft	• Done in real time during pass • Real-time commands and command loads	Chap. 8 Sec. 3.4
5. Verify the Spacecraft's Receipt of Commands	• Done in real time during pass • Usually automated	Chaps. 7 and 8
6. Monitor the Spacecraft's Health and Safety	• Monitor telemetry in real time during the pass • Compare predicted and actual spacecraft states	Chaps. 7, 8, and 15
7. Monitor the Ground System Operations	• Monitor performance in real time	Chaps. 7, 8, and 12
8. Coordinate Mission-Control Functions	• Planning and scheduling before and after passes	Chaps. 7 and 8
9. Generate an Integrated Plan for "As-Flown" Activities	• Post-pass report • Document deviations from pass plan	Chap. 8
10. Support Activity Planning and Development	• Post-launch support for new activities	Chap. 7

Table 3.10. How to Do Mission Control. (Continued) Mission control develops procedures and supports integration testing before launch and directs the real-time operation of the mission after launch. Mission control supports several planning and analysis functions throughout the mission. Shading indicates steps we do after launch.

Steps	Comments	Where Discussed
11. Generate Operations Schedules and Plans for Future Passes	• Post-launch planning and scheduling for passes	Chaps. 7 and 8
12. Negotiate and Schedule Tracking Support	• Coordinate with activity planning and development to generate pass plans	Chap. 8 Sec. 3.5
13. Support Planning and Analysis Teams	• Help investigate anomalies	Chap. 16

9. **Generate the Integrated Plan of "As-Flown" Activities.** This activity plan is identical to the one prepared before the pass if everything went according to plan. But if we deviate from the plan, we must do an "as-flown" plan to record what was done and when. Typically, mission controllers use it to analyze anomalies.

10. **Support Activity Planning and Development.** Mission controllers help generate and approve the activities described earlier. They typically bring to the process the interface issues between the spacecraft and the ground.

11. **Generate Operations Schedules and Plans for Future Passes.** To do plans for future passes, we must get ready and check out all information and data needed for the next spacecraft contact.

12. **Negotiate and Schedule Facility and Tracking Support.** These schedules for tracking stations become part of the activity plans described earlier in this chapter.

13. **Support Planning and Analysis Teams.** We must help teams analyze anomalies in the spacecraft bus and payload. Experience and information from real-time controllers is invaluable to the analysts who understand the spacecraft and its subsystems but don't operate it.

3.3.3 Products of Mission Control

Pass plans, pass reports, database files, and alarm notices are the post-launch products of real-time operations. We list these products in Table 3.11 and describe them below.

Pass plans describe what will be done during each pass, whereas procedures describe how controllers will do it. A typical pass plan will describe the planned times for acquisition, as well as key events anticipated or required on the spacecraft and the ground. It is a level higher than a command load and is used to brief everyone in the ground system before the pass.

A pass report describes the activities that took place during a contact between the ground and the spacecraft. It typically includes the commands sent, the telemetry channels that were in alarm, and any unplanned action that procedures or on-duty analysts authorized.

Table 3.11. Products of Mission Control. All products are completed after launch and are related to real-time mission operations.

Products	Comments
Procedures	• Describe in detail what the operators do • Describe what commands are authorized to be sent and under what anomaly conditions
Pass Plans	• Describe what is scheduled to happen during pass
Pass Report	• Describes activities which took place during a pass • Notes deviations from the pass plan
Database Files	• Uplink-command and command-verification files • Downlink telemetry files • Ancillary data
Alarm Notices	• Alarm notices and responses occur in real time

Database files generated and validated by the mission control team are the command validation files and the files or updates to telemetry and tracking data received during the pass.

When alarms occur, mission control notifies the appropriate people as identified by the procedures. Although controllers have issued notices manually in the past, systems are now in place that automatically send an electronic page to someone who is on duty but not at the control center. The person can then use a computer to connect with the ground system. After proper identification and authorization, this person can receive and analyze the spacecraft telemetry on his or her computer. This technology keeps teams at the control center small during spacecraft contacts.

3.3.4 Key Considerations for Mission Control

Mission control is easier when we can easily modify ground and flight systems to react to unplanned changes or conditions. For example, workstations and networks are more robust than a central system. Graphical user interfaces that allow operators to change configurations by point and click are easier to use than those that take lines and lines of commands.

We can lower mission control costs in several ways. Possibilities include sharing operators between missions, multi-tasking operators within a given mission, designing flexible ground and flight systems, and increasing automation in the ground system.

When several missions have common features in flight at the same time, we can train people so they may work in more than one mission at a time, thus reducing the total staff required. For individual missions, multi-tasking lowers the total staff required and makes jobs more interesting. Many of the specialists typical of early space flight now do related, but new, tasks to reduce costs.

If we design the spacecraft to remain flexible to change, we'll need fewer resources to make these changes. By automating analysis of the performance of the ground system and the spacecraft, we can reduce staff and probably improve performance and reliability.

3.4 Data Transport and Delivery

This function transmits data to the spacecraft, receives data from the spacecraft, and processes tracking data. It accepts commands or command files from mission control and prepares the data for transmission, ultimately modulating the carrier signal and radiating the telecommands to the spacecraft. This function also receives the signal from the spacecraft and processes the spacecraft's engineering and mission data. Finally, it gathers and processes radiometric data used for planning and analyzing navigation.

3.4.1 Information Needed to Transport and Deliver Data

Mission control and navigation planning and analysis provide information before launch to validate the system's capabilities and after launch to support real-time transmission of data to and from the spacecraft. Table 3.12 summarizes this information.

Table 3.12. **Information Needed to Transport and Deliver Data.** Mission control and navigation planning and analysis provide this information before and after launch. Shaded areas occur after launch.

Information Required	Where Discussed
Antenna pointing predictions	Chap. 10; Sec. 3.5
Binary commands or command files	Chap. 8; Sec. 3.4
Ground activities	Chaps. 7, 8, and 12; Sec. 3.4
Radio frequency predictions	Sec. 3.5
Pre-track calibration data	Chaps. 8 and 12; Sec. 3.4

People in navigation planning and analysis issue antenna pointing predictions based on either pre-flight nominals or orbit determination after launch. These predictions help us point the ground-tracking or in-orbit-tracking antennas to the spacecraft during a scheduled pass. Also, if a spacecraft uses other spacecraft for orbital tracking and relay (such as NASA's TDRSS), it will have the ephemeris of these spacecraft on board and use it to point a high-gain antenna at the tracking spacecraft during the pass.

Navigation planning and analysis generates predictions of radio frequencies (rf) and frequency shifts expected during each pass. Activity planners generate commands for rf transmission. Mission control then transmits commands to staff that handle data delivery and transport. Mission control also configures the ground system by using activities that set up the ground system to support passes.

3.4.2 How to Transport and Deliver Data

We must test and validate our approach to handling data before launch so we can be sure the spacecraft and ground system are compatible. After launch, this function uplinks commands to the spacecraft and receives and delivers the data downlinked from the spacecraft. Table 3.13 lists the steps in this process.

Table 3.13. How to Transport and Deliver Data. This process establishes our communications link to the spacecraft.

Steps	Comments	Where Discussed
1. Validate Each Function's Abilities	• Done before launch • Validates uplink and downlink tasks • Part of end-to-end test of the system	Chap. 8 Sec. 3.8
2. Send Commands to Spacecraft	• Tested before launch and carried out in real time after launch • Commands uplinked to spacecraft	Chaps. 8 and 11 Sec. 3.3
3. Manage Data Flow	• Tested before launch and carried out in real time after launch • Data downlinked to ground station • Standard formats recommended	Chaps. 8 and 11–13
4. Determine Data Quality, Continuity, and Completeness	• Evaluates uplink and downlink data after activities are complete	Chaps. 11 and 13

1. **Validate Each Function's Abilities.** Do pre-launch steps similar to those of mission control. Test and validate facilities for transporting and delivering data. Participate in end-to-end tests of the spacecraft to validate the uplink and downlink functions. Generate procedures used for station's setup, calibration, and pass support.
2. **Send Commands to Spacecraft.** Telecommanding is the ground-to-spacecraft step used to instruct a spacecraft, its subsystems, or scientific instruments. To do telecommanding, we modulate telecommand data on an rf carrier. When received by the spacecraft's rf subsystem and distributed to the appropriate device, this data starts, changes, or ends an action.

Mission control supplies telecommands. Once the tracking network receives a telecommand message, we determine the telecommand's destination and route it to the appropriate tracking station (or spacecraft). We then check the telecommand's format to verify that it is acceptable. Often, destination codes are reversed and the telecommand message is returned to the control center for a bit-by-bit comparison with the original. Once we've found the telecommand to be acceptable, we handle it by one of these methods:

Thruput telecommanding. In this mode, we route the telecommand immediately to the transmitting station's modulator and transmit it to the spacecraft without delay.

Store-and-forward telecommanding. We route the telecommand to a specific tracking complex together with a release time. At the appointed time, the telecommand will be modulated on the rf carrier and transmitted to the spacecraft.

Telecommand. Once some or all of the above steps have taken place, we modulate the telecommand on a sine wave subcarrier; then, phase-modulate it on an rf carrier, amplify it, and transmit (radiate) it to the receiving spacecraft.

3. **Manage Data Flow.** This step returns information from a spacecraft to Earth-based users. Data can include the results of scientific measurements or information about the spacecraft or its subsystems. We package telemetry data to conform to one of the acceptable data formats recommended by the Consultative Committee for Space Data Systems (CCSDS). We then encode it (optional), phase-modulate it on an rf carrier, and transmit it to a receiving tracking station or relay satellite. At the tracking station, systems or people:

Capture data. Once a tracking station's receiver is locked to a spacecraft's rf carrier, we demodulate telemetry data and synchronize symbols.

Extract data. Depending on whether or not coding is used and its type, there may be several times as many symbol bits as there are original data bits. To extract data, we need to take several typical steps. For data with convolutional encoding, we use a maximum-likelihood algorithm to do convolutional decoding while detecting and correcting errors. We supply partially decoded data to a frame synchronizer, which continually tests for a unique, synchronization marker. When the synchronizer finds this sync word at successive periodic intervals, frames are synchronized. Finally, we decode the Reed-Solomon block code at the end of the transfer frame correct errors before archiving transfer frames. We discard any data the system couldn't correct or, if the project requests, pass it through to the project's facility for capturing data.

Time tag transfer frames. Once decoding is complete and the transfer frame is valid, we time tag the frame. For NASA, this time stamp is usually accurate to within 1.5 microseconds with respect to the time that a specific bit enters the antenna's feed horn. The time stamp appears in the secondary header of the standard formatted data unit (SFDU).

Add valued services to transfer frames. Plans predetermine how to read headers of decoded transfer frames and handle different virtual channels. For example, we can return a virtual channel containing spacecraft engineering data in real time. But we may want to record a channel containing imaging data at the stations and trickle the data back as line capacity permits or mail it back. Last, we can route a third virtual channel containing a specific instrument's data directly to the payload-processing center.

Deliver data. Finally, we must deliver the data to the project as agreed. This usually involves sending the telemetry data from the tracking facility to the project control center and populating a project or mission database. We may also send the data directly to a payload center.

Process spacecraft data. This processing is necessary if the spacecraft sends data to the ground faster than the tracking facility can send it to the control center.

4. **Determine Data Quality, Continuity, and Completeness.** If possible, we complete this set after the pass and collect missing data from archives in the tracking station. This task is more automated now, but in most operations centers before 1992, people had to do it.

3.4.3 Products of Data Transport and Delivery

Our main products are the telemetry and tracking data collected during a pass. We also produce reports on tracking passes and on the quantity, quality, and continuity (QQC) of stations. Table 3.14 lists these products.

Table 3.14. **Products of Data Transport and Delivery.** We generate these products in real time during a station pass and when time permits following a station pass.

Product	Comments
Telemetry data	• Generate in real time during station pass
Tracking data	• Generate in real time during station pass
Tracking pass report	• Post-pass, when time permits • Generate with mission control
Station QQC reports	• Post-pass, when time permits

We process the telemetry data as described in Chap. 13 and Step 3 above and then deliver it into the project or mission database.

Tracking data, also defined as radiometric data, serves to determine a spacecraft's position as well as for scientific investigations. These data quantify the relative position or motion of the spacecraft and the Earth. Several types of data are available:

- Doppler
- Ranging
- Angle
- Very long baseline interferometry (VLBI)
- Radio science measurements

Station pass reports typically identify the configuration of the tracking station or spacecraft during the pass and concentrate on deviations from normal conditions. We list these reports as a project deliverable and include them in the mission database.

Station QQC reports describe the quantity, quality, and continuity of the telemetry gathered for the project during the contact. They include the statistics on both the real-time data and the high-rate data received but not necessarily processed during the pass. Most networks include this information electronically, but we also deliver it to the project database.

3.4.4 Key Considerations

Network loading and out-of-date systems cause most of our problems in handling data. For example, sending data from the tracking stations to the control centers with protocols of the 1970s causes errors in the mission database. NASA developed these protocols before commercial standards were in place. Now, we need to follow CCSDS standards or those of the network tracking our spacecraft, or we'll have problems getting standard services.

We can reduce the cost of handling data by first keeping tracking coverage to a minimum. Tracking consumes resources from every part of mission operations. Overloaded tracking networks already have trouble allocating these resources to space projects.

Next, we can design structures and formats for flight data to match the abilities of our transport and delivery system. Using standard services is reliable and cost effective. Having an engineer design a new spacecraft characteristic to enhance a mission may result in spending a lot of money on developing and operating ground systems that must support this characteristic.

Finally, we should use variable-length packets instead of many unique, predetermined formats. Today's standards and technology allow us to identify a data packet's length in the header, so we don't need to use inefficient fixed-length packets.

3.5 Navigation Planning and Analysis

We must determine the spacecraft's position and predicted flight path, as well as how to correct that flight path to achieve mission objectives. To do so, we must

- Acquire radiometric, tracking, or optical measurements
- Determine the statistically best estimate of the trajectory based on these measurements
- Compute trajectory-correction maneuvers (TCMs) to achieve the desired targeting objectives or orbit changes
- Plan and analyze performance to ensure it meets mission and payload objectives

3.5.1 Information Needed to Plan and Analyze Navigation

Because determining and predicting spacecraft orbits is uncertain, we must continually update our estimate of the spacecraft orbits by analyzing tracking data. We also have to know the locations of other celestial bodies, such as the Sun, Moon, or Jupiter, to predict their effects on the spacecraft's trajectory. Finally, when orbital maneuvers are necessary, we must have an accurate model of the spacecraft propulsion system to plan the maneuver. Table 3.15 summarizes these requirements.

Table 3.15. **Information Needed to Plan and Analyze Navigation.** We need this information to determine and predict orbits, as well as to plan spacecraft maneuvers. Shading indicates information required after launch.

Information Required	Where Discussed
Ephemeris data of celestial bodies	Chap. 10
Planetary atmospheric and dynamic models	Chap. 10
Tracking data	Chap. 10; Sec. 3.4
Information on spacecraft propulsion	Chap. 15; Sec. 3.6

3.5.2 How to Plan and Analyze Navigation

We must convert raw tracking data to current and future estimates of the spacecraft's orbit or trajectory. We also calculate the maneuvers necessary to maintain our spacecraft orbit or change the orbit to meet mission objectives. We've listed steps in Table 3.16 and described them below.

1. **Support Pre-Launch Mission Planning.** We use predictive tools and other tests to show that the navigation equipment can do what the mission requires.

Table 3.16. How to Plan and Analyze Navigation. We combine tracking data statistically to estimate the current orbit and then use this estimate to generate trajectories and plan maneuvers.

Steps	Comments	Where Discussed
1. Support Pre-Launch Mission Planning	• Demonstrate the orbit propagator's abilities before launch	Chap. 10 Sec. 3.1
2. Determine Orbit(s)	• Statistically determine best estimate of orbit using tracking data, onboard data, or other sources	Chaps. 8 and 10
3. Design and Analyze Maneuvers	• Use to maintain and modify orbit	Chap. 10
4. Determine and Plan Spacecraft Attitude	• Done with spacecraft planning and analysis • Combine with position information to accurately determine pointing information	Chaps. 10 and 15
5. Generate and Regenerate Trajectories	• Use to generate future spacecraft location predictions	Chap. 10

2. **Determine Orbit.** We extract radio-tracking data (Doppler, range, range-rate, and VLBI) at the tracking stations, transmit it over high-speed data lines to the control center, and buffer it on computer-disk storage. Newly acquired data of different types from individual tracking stations for various spacecraft are automatically sorted and merged into a single time-ordered array for the spacecraft of interest. After removing data of poor quality, we have a data file ready for orbit determination.

   Chapter 10 explains this process in detail. Missions are now using Earth-orbiting spacecraft that can use onboard GPS receivers and reduce or eliminate the need for radiometric tracking data. Interplanetary missions are using onboard optical navigation (see Chap. 17) to determine trajectories during approach to planets. This is an example of moving the function completely inside the avionics part of the mission operations diagram (Fig. 3.1).

3. **Design and Analyze Maneuvers.** Propulsive maneuvers are required to maintain a spacecraft's orbit and to navigate a deep-space mission to its target. Before launch, we determine how to place these maneuvers to satisfy mission objectives and operational constraints while using as little fuel as possible. During the mission, we factor in the spacecraft's actual performance. To make future maneuvers more accurate, we analyze each maneuver to improve orbit determination and the spacecraft's performance. Chap. 10 discusses spacecraft maneuvers in more detail.

4. **Determine and Plan Spacecraft Attitude.** For missions that require the spacecraft to point at an object on Earth, in space, or on the surface of another planet or asteroid, we need a predictive tool to help generate commands. People handling the spacecraft or payload usually operate these tools, but we give them information on the spacecraft's attitude and the payload's pointing direction. By calculating these values, we give payload users inputs that are more accurate than the predictions used to generate the commands.
5. **Generate and Regenerate Trajectories.** Each time we update the orbit or trajectory, we must do new predictions, which are used by nearly everyone in mission operations. On board the spacecraft, they help us point the payload instruments, antennas, and communications.

3.5.3   Products of Navigation Planning and Analysis

Accurate, post-launch predictions of the spacecraft's trajectory, required maneuvers, and attitude history make the products listed in Table 3.17 important to other mission operations functions, especially in mission planning and data processing.

Table 3.17. Products of Navigation Planning and Analysis. Products are generated post-launch and used as inputs to other mission operations functions.

Product	Comments
Trajectories (past, present, and future)	• Input to data processing, mission planning, and mission control
Maneuver designs	• Input to spacecraft planning and analysis and mission planning
Attitude history	• Input to data processing

3.5.4   Key Considerations

Accuracy drives the cost of orbit determination. We have to make sure the mission needs certain levels of accuracy before spending money on navigation systems that must support them.

Having fewer propulsive maneuvers simplifies navigation because we must plan them and then determine and analyze new orbits or trajectories after each maneuver is complete.

After selecting a tracking network, we should try to match our mission requirements to their standard services for determining orbits and predicting trajectories. Sometimes, we can change our requirements so they're compatible with the network's services.

Because onboard computational abilities and supporting technologies are improving, we must understand the cost and accuracy trade between determining orbits and propagating trajectories on board versus doing the same things on the ground. GPS's capabilities to support onboard navigation may well be sufficient for many LEO missions.

3.6 Spacecraft Planning and Analysis

Here, planning and analysis make sure we maintain the spacecraft's health and safety and get back its mission and scientific data. Spacecraft engineers

- Prepare telemetry predictions
- Analyze the real-time and processed data
- Identify and resolve anomalies
- Design maneuvers
- Maintain attitude-control and flight software
- Develop, analyze, and test the engineering commands for uploads and real-time commanding

Spacecraft engineers also use the analysis software and hardware (testbed) on the ground to analyze data, prepare predictions, and simulate commands. Finally, they maintain and update various documents or databases (dictionaries, maps, procedures, flight rules and constraints, plans) and reports (consumables, trends, and in-flight performance) needed to complete a space mission.

3.6.1 Information Needed to Plan and Analyze a Spacecraft

To do this function, we get inputs from the five previously discussed mission operations functions: mission planning, activity planning and development, mission control, data transport and delivery, and navigation planning and analysis. The information describes how we expect the spacecraft bus to operate, the spacecraft bus's actual performance, and the spacecraft's upcoming maneuvers and planned activities. We need telemetry data, plans, and activities before launch to validate the spacecraft bus's capabilities and all five inputs after launch to plan the spacecraft bus's activities and analyze its performance. Table 3.18 summarizes this information.

Table 3.18. **Information Needed to Plan and Analyze a Spacecraft.** We get information needed to plan and analyze spacecraft-bus operations from previously described functions.

Information Required	Where Discussed
Plans	Sec. 3.1
Command files	Sec. 3.2
Alarms and pass reports	Sec. 3.3
Channelized telemetry data	Sec. 3.4
Maneuver designs	Sec. 3.5

3.6.2 How to Plan and Analyze Performance of a Spacecraft

In this function we compare the spacecraft bus's actual performance with the expected performance to make sure it meets mission objectives. Many of the steps listed in Table 3.19 begin before launch, and all but the first step, validation, continue until the end of the mission. We further describe these steps below.

Table 3.19. How to Plan and Analyze a Spacecraft. Steps are not necessarily sequential. We do them as needed before and after launch to maintain the spacecraft's health and safety and ensure we get back all mission data. Shaded areas occur after launch.

Steps	Comments	Where Discussed
1. Validate Processing Abilities of the Spacecraft and Ground System	• Done before launch during end-to-end tests of the system • Helps train operators	Chaps. 12, 13, and 15
2. Generate Reports and Maintain Database	• Reports describe the spacecraft bus and its operation • After launch databases document actual performance	Chaps. 13 and 15 Sec. 3.1 and 3.4
3. Plan Calibrations	• Pre-launch planning of calibrations • Post-launch updates based on spacecraft performance	Chaps. 7 and 15
4. Generate Reports on Consumables, Trends, and Performance	• Compare actual and expected performance of spacecraft hardware and resources	Chap. 15
5. Plan Spacecraft Bus Commands	• Generate all commands uplinked to spacecraft bus • Coordinate with mission planning and activity planning and development	Chap. 15 Sec. 3.1 and 3.2
6. Operate and Maintain the Flight Simulator	• Used to verify command loads before uplinking commands to spacecraft	Chap. 8 Sec. 3.3
7. Maintain the Spacecraft Bus's Flight Software	• Manage flight software • Generate software memory loads • Control software configuration	Chap. 15 Sec. 3.12
8. Analyze Engineering Data	• Determine health, safety, and performance of spacecraft • Investigate spacecraft anomalies	Chaps. 15 and 16

1. **Validate Processing Abilities of the Spacecraft and Ground System.** Before launch, test and validate the ground processing system and its compatibility with the spacecraft. Show compatibility mainly through end-to-end testing, with the spacecraft and ground station flowing data into the control center. You'll often do these tests while the spacecraft is in the thermal-vacuum chamber.

We can also participate in spacecraft system tests to ensure that the part of the ground system used for spacecraft planning and analysis is operating properly. This step also best trains people in mission operations on the spacecraft's characteristics before launch.

2. **Generate Reports and Databases.** Include:

   Flight rules and constraints. These list all in-flight operational limitations imposed by the spacecraft hardware and software.

   Decommutation maps. These list the telemetry channels contained in each of the different maps, which define the variable part of the telemetry commutator. Nearly all missions are now using CCSDS standards which define packets of information. Each packet is self identifying and, when used properly, replaces decommutation maps.

   Telemetry dictionary. This describes in detail the engineering and science telemetry measurements, as well as the operational limits and parameters needed to understand each measurement.

   Command dictionary. This describes and show bit patterns of the messages and commands that may be sent to the spacecraft; we cross-reference the flight rules and telemetry dictionary.

   Spacecraft idiosyncrasies. These include unusual or anomalous performance characteristics.

   Operating procedures. These define how to do spacecraft planning and analysis.

   Spacecraft contingency plan. This identifies potential anomalies that would require ground response and plan corrective actions.

3. **Plan Calibrations.** We show the strategies and activities for calibrating subsystems of the spacecraft that affect its performance. Subsystems we must usually calibrate after launch (and sometimes during the mission) include attitude control, star trackers, and moveable platforms.

4. **Generate Reports on Consumables, Trends, and Performance.** We identify uses, changes with time and operation, and periodically performance of the spacecraft. Although we do most of this after launch, we start reporting during the system test and then maintain and update reports during the mission.

5. **Plan Spacecraft Bus Commands.** We generate all command inputs and review all system commands for uplink to the spacecraft. Command requests (planned real-time and command loads) are to calibrate spacecraft subsystems and maintain the spacecraft's health. We review planned command uploads for correct engineering and completeness.

   We develop maneuver designs using inputs from navigation planning and analysis and determine appropriate commands to carry out the maneuver. In the future, the ground system may transmit a new state vector to the spacecraft, which would then compute and execute the maneuver. If so, detailed ground commands won't be necessary.

Spacecraft planning and analysis supports mission planning and activity planning and development as required during the mission.

6. **Operate and Maintain the Flight Simulator.** A flight simulator (such as a testbed, simulator, or testlab) is often required to verify commands before transmitting to the spacecraft.
7. **Maintain the Spacecraft Bus's Flight Software.** Maintain flight software, generate memory loads for this software, and control all changes to flight programs, including databases.
8. **Analyze Engineering Data.** We analyze the spacecraft's engineering data (real-time and non-real-time) and determine its health, safety, and performance. We develop performance models before launch. These models may be software programs or hardware models that predict the spacecraft's performance and analyze its data. We assess performance by analyzing the spacecraft's system data. Investigate spacecraft anomalies and correct them. We investigate payload anomalies as requested. We maintain trends and supply information to mission planning about the spacecraft's deviations from pre-launch assumptions of its abilities.

3.6.3 Products of Spacecraft Planning and Analysis

We use the products from this function to develop future spacecraft requests. We coordinate with mission control, mission planning, and activity planning and development to schedule necessary commands during future spacecraft passes. We also use the products listed in Table 3.20 to support payload planning and analysis and data processing.

Table 3.20. Products of Spacecraft Planning and Analysis. We complete these products after launch and input them to other mission operations functions.

Product	Comments
Spacecraft status versus plan	• Generate post-launch • Use to track performance and modify mission plan
Spacecraft limits	• Generate pre-launch and update post-launch based on actual performance
Spacecraft activity requests	• Generate pre-launch and update post-launch • Input to mission control
Processed spacecraft data	• Generate post-launch • Input to payload planning and analysis and data processing
Reports	• See Step 2 (p. 63) for a list of the most important reports

3.6.4 Key Considerations

The biggest driver of operational complexity is the level of resource margins available on the spacecraft. A large positive power margin means we don't need detailed analysis of planned activities. If there is a negative power margin with all spacecraft and payload sources on, analysis becomes more difficult and detailed. If subsystems interact under negative margins, spacecraft planning and analysis becomes even more difficult.

Other factors are level of spacecraft analysis required, number of interactions, spacecraft safing, the need for real-time engineering analysis, and the level of automation available. Proper design of the spacecraft and the associated telemetry measurements will make it possible to analyze the spacecraft as a system. There is still a tendency to design the "best" subsystems, integrate them into a spacecraft, and then attempt to figure out how to analyze the spacecraft in flight. This approach will usually mean analyzing each of the subsystems for proper operation and then integrating this information at the system level. Subsystem analysis of a spacecraft is labor intensive.

The spacecraft that minimizes interactions between spacecraft components, between spacecraft and payload components, and between payload components will be easier to operate and require fewer resources from planning and analysis.

A spacecraft that will go to a safe state when an error occurs is easier to operate than one that is fragile and needs a lot of monitoring.

Many of today's spacecraft can go unattended for a week or more at a time, but operators tend to want to observe and monitor the spacecraft in real time. Frequent real-time operations require more resources than real-time passes once a day or once a week.

Finally, using automated analysis tools to decrease staffing lowers operations costs. Artificial-intelligence techniques are now becoming useful for mission operations.

3.7 Payload Planning and Analysis

We need to identify and prioritize payload opportunities in order to design observations and activities and to carry out correctly the command load transmitted to the spacecraft. We analyze payload data to assess the payload's performance and to change planned observations when necessary to get better data.

3.7.1 Information Needed to Plan and Analyze Payload

We get information from mission planning, activity planning and development, mission control, and data transport and delivery. We use it before launch to make sure the payload is operable and the data system is compatible. After launch, the information helps us assess the payload's performance and plan observations. Table 3.21 lists the inputs for this function.

Table 3.21. Information Needed to Plan and Analyze a Payload. We need information before launch to validate payload operability and after launch (shaded rows) to plan observations and assess performance.

Information Required	Where Discussed
Plans	Sec. 3.1
Activities	Sec. 3.2
Alarms and pass reports	Sec. 3.3
Payload data	Sec. 3.4

3.7.2 How to Plan and Analyze a Payload

Table 3.22 lists the steps for payload planning and analysis. Before launch, we monitor the design and development phases to ensure we have a payload operations concept that will work and meet mission objectives. After launch, we monitor the payload performance, help plan future operations, and plan necessary payload calibrations.

Table 3.22. How to Plan and Analyze a Payload. These steps are similar to those done in spacecraft planning and analysis. They focus on the payload's proper and accurate operation. Shaded areas occur after launch.

Steps	Comments	Where Discussed
1. Validate Processing Abilities of the Payload and Ground System	• Conduct pre-launch during end-to-end tests of the spacecraft • Helps train operators	Chaps. 12 and 14
2. Generate Reports and Databases	• Write reports pre-launch with mission planning • Maintain databases post-launch	Chaps. 13 and 14 Sec. 3.1 and 3.4
3. Plan Payload Observations	• Identify opportunities • Assign priorities and resolve conflicts • Generate and validate command loads	Chaps. 7 and 14
4. Plan Payload Calibrations	• Use to interpret data correctly • Use for calibrations based on analysis of payload data	Chaps. 7 and 14
5. Analyze Payload Performance	• Quick-look analysis • Calibration analysis • Trend analysis	Chap. 14
6. Assess and Maintain Payload Flight Software	• Manage payload software • Generate software memory loads • Control configuration of payload software	Chap. 14 Sec. 3.12
7. Investigate Anomalies	• Do as required • Support spacecraft planning and analysis	Chap. 16

1. **Validate Processing Abilities of the Payload and Ground System.** Before launch, these activities are much like those for spacecraft planning and analysis. We usually participate in the same tests as those described for the spacecraft. One big difference is that some of the instruments may not be turned on, except during the thermal-vacuum testing with its simulations of space environments.
2. **Generate Reports and Databases.** Similar to those for spacecraft planning and analysis.
3. **Plan Payload Observations.** We identify payload opportunities by evaluating trajectory information and the ephemerides of the body being investigated, be it Earth or another planet. These opportunities represent times in the mission during which the payload's observations will achieve mission objectives. Once opportunities have been identified, we design, implement, and integrate these observations to form the command load and transmit it to the spacecraft. Depending on the payload, we may divide the observations into discipline groups, so that each discipline group identifies observations to meet the mission's objectives. If the payload has only one instrument, this process is considerably simpler than if it has a dozen instruments. Each discipline group then prioritizes its observations and activities.

   Once priorities are set, we combine into a single file the inputs from each group and the observations requested from the payload and spacecraft engineering disciplines. Priorities help resolve conflicts between the discipline groups and produce a conflict-free timeline of activities that will generate commands for the spacecraft and instruments. The degree to which this process can be automated varies with mission and spacecraft design. At NASA, mapping missions are highly automated because mapping strategies are automated and observations don't require decisions. The astrophysics community uses a very automated process to determine observations for each orbit. On the other hand, planetary missions tend to require a lot of interaction and many human decisions.

   Payload planning often begins months or even years before final command load goes to the spacecraft. Hence, observations may need changing to account for discoveries or new information about the observation. Thus, the ground system allocates resources (people, hardware, and spacecraft) to these changes. For essential observations, we can also modify the command load slightly on board the spacecraft.

   Payload planning ends when we validate internal commands for the spacecraft and instruments. We base these commands on scientific needs, the spacecraft's abilities, and mission guidelines and flight rules. We submit commands to activity planning and development, where people will expand, constraint-check, and compile them. Then, validate them again to ensure the final commands don't harm the spacecraft and do accurately reflect the initial observation requests.

4. **Plan Payload Calibrations.** We need calibrations to interpret the payload data correctly. First, analyze the payload telemetry during downlink and then design and do changes to the calibration plan, or do more calibrations, to ensure accurate processing and interpretation of the received data. Early definition of this process during payload design reduces the time we must spend on calibration.

5. **Analyze Payload Performance.** We operate and calibrate the payload hardware and analyze trends. This step ensures the instrument is within specifications and no trend is developing that would keep it from meeting future demands.

   To analyze the payload, we can do

   Quick-look analysis. We inquire into the health and quality of the instrument data.

   Calibration analysis. We analyze calibration observations and selected observations by the instrument. If necessary, we ask for new calibration observations or even a change to the basic calibration plan written before launch.

   Trend analysis. We analyze mainly the engineering measurements on the instrument that show the health and performance trends during the mission. We note carefully anything that suggests a need to change the instrument's operating plan.

6. **Assess and Maintain Payload Flight Software.** Many of the instruments flown today have processors as large as the spacecraft's processors. We maintain this software just as we would the spacecraft's software.

7. **Investigate Anomalies.** We do these as required and help the spacecraft group analyze spacecraft anomalies as appropriate.

3.7.3 Products of Payload Planning and Analysis

As with spacecraft planning and analysis, products from this function help mission planning and mission control plan and conduct future operations. Table 3.23 lists these products.

3.7.4 Key Considerations

The spacecraft and data-handling enable us to get payload data to the user for planning and analysis. When these systems introduce errors or don't meet specifications, we must reevaluate payload planning and often must change it to solve these problems. We must also do more payload planning when the mission

Table 3.23. Products of Payload Planning and Analysis. Products are generated after launch and are inputs to mission planning, mission control, and data processing.

Product	Comments
Payload status versus plan	• Input to mission planning • Use to monitor payload performance
Payload limits	• Input to mission planning and mission control
Payload activity requests	• Input to activity planning and development
Processed payload data	• Input to payload data processing

plan asks us to generate payload activities based on analysis of received data. When the ground system has to be more adaptable, mission operations costs go up.

To reduce these costs, we use existing planning tools, automate payload data-gathering, emphasize payload operability during design, and minimize required adaptivity of the payload in the operations concept.

Many planning tools exist, and others are "new" designs of present capabilities. We tend to re-invent capabilities instead of finding tools we can use "as is" or with slight changes.

Next, we should consider automating the payload data-gathering and calibration. Onboard automation of data-gathering reduces the resources required to operate the payload but usually demands resources up front for planning the automation.

We must continue to insist on payload operability during the design phase. Typical questions to instrument builders are: "How are the instrument commands going to be determined? What information is required, and how is this information converted into instrument commands?" We must make sure the team addresses the number and complexity of calibrations before completing the payload design.

Finally, we should minimize the amount of adaptivity required to achieve mission success. A mission that requires information from the payload **before** the next command load can be generated is more expensive to operate than one that collects data in a standard manner. Adaptivity is sometimes necessary but designers often use it to put off understanding how to establish observations or command loads.

3.8 Payload Data Processing

First, we must bring together instrument data packets, engineering data, and ancillary data (e.g., orbital/navigation data) into instrument data records. We may also need to do higher-order processing to support payload analysis and generate digital and photography products of archival quality. In fact, we may need to do any of the processing steps described below—at either the control center, a dedicated payload-processing facility, or on the spacecraft. Table 3.24 lists the types of data needed to process payload data.

Table 3.24. Information Needed to Process Payload Data. To process payload data, we need it plus other data regarding the spacecraft and ground system. We collect all information after launch.

Information Required	Comments	Where Discussed
Payload data	• Payload data in the form sent to the spacecraft avionics system (usually packets) is received by payload data processing • Instrument-level testing, payload integration testing, spacecraft integration testing, and flight data all make up the payload data	Sec. 3.4
Ancillary data	• Ancillary data is information about the spacecraft and ground system required to process and analyze the payload data • The specific data and the formats are specified before launch. These data include spacecraft and payload engineering information.	Sec. 3.4
Orbital/navigation/ attitude-predict data	• This data is sometimes included as ancillary data but is important enough to be called out separately • Most observations require information relative to where the payload is pointing to be able to interpret the data • Sometimes, as information about the orbit or trajectory improves, it may appear in several versions, which we must be able to identify separately	Sec. 3.5

The software for processing payload data is often unique (not commercially available) because some scientific processing requires specialized routines that have few commercial customers. Examples include

- Cartographic projections for bodies other than the Earth
- Radiometric reconstruction of color imagery from multiple images acquired through spectral filters
- Image registration to less than one pixel accuracy

After this—and perhaps more specialized—processing, the data is ready to be archived for use by other scientists (in the case of NASA missions). Chapter 13 further discusses processing of payload data.

As mentioned earlier, processed payload data may change the mission plan, but at this processing level, we usually pick up very subtle but important instrument characteristics.

3.8.1 Information Needed to Process Payload Data

3.8.2 How to Process Payload Data

We take the data received from transport and delivery, sometimes through the mission database, and construct the sensor information, which may be an image, spectra, or other meaningful product. This process transforms the data into a product usable by a discipline expert, as opposed to a sensor expert. We start processing payload data before launch and continue throughout the mission and often well past the end of the mission.

Payload data processing often occurs in a facility separate from the mission control center and under the control of scientists and specialists. Depending on the mission, organizations separate from mission operations may do some of this processing. Table 3.25 shows the steps for processing payload data. We discuss it in more detail in Chap. 13.

Table 3.25. Steps for Processing Payload Data. The steps will vary depending on the payload, but these are typical for scientific missions.

Steps	Comments	Where Discussed
1. Validate Payload-Processing System	• Process data before launch from one or all of these sources: – Bench-level testing – System-level testing – Simulated data or real ground based data in a format similar to the flight instrument Note: Participation in system-level tests and end-to-end tests is helpful, but usually there are so many constraints from the non-space environment that the tests don't fully check out the processing capabilities. The "real" test often doesn't occur until after launch, when the instrument is in space.	Chaps. 13 and 14
2. Generate Payload Data Records	• Data transport and delivery provide the payload data—either to the mission database or directly to the payload-processing facility—in the form of packets separated by instrument type • This data is then aggregated into payload data records—typically called level 0 processing • No value is added to the data at this point. It is the best set of data from the instrument that data transport and delivery can produce.	Chaps. 13 and 14 Sec. 3.4
3. Process Sensor-Specific Data	• Process instrument engineering data • Decompress data • Remove sensor signature • Apply calibration information to the raw data	Chap. 13
4. Correlate Ancillary Data with Sensor Data	• Add to the sensor-specific data the ancillary data needed to process it into meaningful payload products	Chap. 13
5. Generate Products	• Generate a data record that contains the instrument data and its ancillary data with proper identification of the processing completed	Chap. 13

Table 3.25. Steps for Processing Payload Data. (Continued) The steps will vary depending on the payload, but these are typical for scientific missions.

Steps	Comments	Where Discussed
6. Analyze Products	• Product analysis is usually separate from, but associated with, product generation • Before the data goes to the user, the operations staff makes sure it's meaningful and doesn't contain incomplete information or processing errors. This is especially true for scientific missions, in which the spacecraft is an observatory used by hundreds of astronomers.	Chap. 13
7. Manage Data	• Deliver products to the payload user or store them in a database for further processing and product generation • We need to manage data so we can locate for analysis products of the same region but taken at different times • The archiving function also uses data management	Chap. 13 Sec. 3.9

3.8.3 Products of Payload Data Processing

Although these products vary from mission to mission, Table 3.26 lists the most common ones. The objective of any mission is to obtain the sensor output. As we process data, we may detect deviations from the mission plan even when all other mission indications are normal. The Viking project reached Mars and had to wait several weeks for a dust storm to diminish before mapping the landing sites and then landing on the surface. All engineering aspects of the mission were normal. Only the payload processing showed that the mission plan had to change.

Table 3.26. Products of Payload Data Processing. These are typical post-launch products for a scientific mission. Commercial and defense missions will have similar products, except for the press releases.

Product	Comments
Deviations from plan	• The mission plan describes the expected conditions. When these conditions aren't met, and sensor data isn't meaningful, users request changes to the plan.
Files of instrument calibration	• A separate file usually records how data is recorded, so future users of the data may use different calibration techniques based on later information and understanding • Including this file and the data to which it was applied in the archival data records makes sure we don't lose information
Records of archival data	• The main product of this function ensures others can use the information for generations to come. The archival data record includes the basic sensor data, the ancillary data, and the calibration files.

Table 3.26. Products of Payload Data Processing. (Continued) These are typical post-launch products for a scientific mission. Commercial and defense missions will have similar products, except for the press releases.

Product	Comments
Hard-copy products	• Hard-copy products are still used in many missions. Electronic versions are becoming more common, but missions still use high quality, specialized photo processing.
Press releases (for science missions)	• NASA missions use the output from payload processing to inform the public. Although military missions don't usually have press releases for their payloads, they do have similar requirements for briefings. • These products are for a specialized audience so they require a different approach compared to those we use to understand information from payload sensors. We must understand these requirements early rather than after launch.

3.8.4 Key Considerations

We can lower costs for processing payload data by addressing the topic early in the mission-concept study and then emphasizing them while developing the mission operations concept. During the MOS conceptual design (see Chap. 4), we include these concepts for processing payload data:

- Required final products
- Diagrams of end-to-end data flow
- The ancillary data required to process the payload data and its sources
- Robustness of the payload processing to data loss or drop out. How will the proposed data compression and data formatting react to the noisy environment?

The mission operations system can make the analysis more difficult if it loses, or doesn't collect, data. This means we need to pay special attention to the size, quality, and completeness of our ground system and how these characteristics will affect processing of the payload data. For many payloads, it's good enough for the ground system to complete 95% of the processing, but that depends on where the 5% loss occurs. Data lost as a block may be acceptable, but if the 5% loss is from compressed data across all the instruments, it may be unacceptable.

When designing and doing payload data processing, we should

- **Use Existing Tools.** As discussed under payload planning and analysis, using existing tools is an obvious way to save resources. Within the scientific community, many payload users now take advantage of standard processing packages.
- **Understand Requirements for Ancillary Data.** We have to understand and have available the ancillary data needed to

interpret payload data. If we discover after launch that this data isn't readily available, solving this problem will be expensive.
- **Ensure Data Processing Robustness.** We must understand early how well the system can process data without losses. It must be able to transport data smoothly, taking into account such factors as data compression, data formatting, and the characteristics of the link performances. It is now common to simulate these characteristics early in the design phase to ensure that the data-processing system is robust.

3.9 Archiving and Maintaining the Mission Database

The payload archive for processed data is usually separate from the mission database that supports controlling and acquiring mission data. But because these entities are closely related, we treat them as one element of the MOS. The *mission database* receives, stores, and delivers data among the nine MOS functions. It can be centralized or distributed, usually receives data in real time, and can be accessed in either real time or at some later time. The *archive* receives selected data for permanent storage and either immediate or historical review.

NASA scientific missions require the payload data to be archived so scientists other than the ones who designed the observation or built the instrument can use it. This data must be in a form others can use without specialized knowledge of the instrument. NASA archives cover several scientific disciplines.

The National Academy of Science did a study on archiving in 1986 and issued their recommendations in the CODMAC report [National Research Council, 1986]. NASA has followed several important recommendations in their archiving program. The report stressed that an active scientific archive should be located where scientists are using and improving the data. An example is the center for infrared processing and analysis on the California Institute of Technology campus. This archive holds the data from the Infrared Astronomical Satellite (IRAS), whose mission ended in 1984. The archive is still actively used. Another example is the Space Telescope Science Institute on the Johns Hopkins University campus. The Science Institute holds the archive data from the Hubble Space Telescope. The mission is still in progress. Scientists around the world use this archive.

The mission database is the repository for all data collected and delivered to the project by data transport and delivery, as well as for data prepared for sending to that transport function. This database contains the controlled files of data required for processing the spacecraft data, plus the raw and processed spacecraft data. The database receives data in near-real time, but people in planning and analysis can use it at any time. Mission databases in the past have been centralized, but with today's technology, they're often distributed—around the world in some cases involving NASA scientific missions.

Active archives contain the instrument data generated during instrument testing and calibration before launch. They provide a valuable source of data for comparison with flight data. The degree of pre-flight testing varies from mission to mission but is often very extensive. After launch, the active archives contain the data products sent to payload analysis from mission operations and the higher-quality data that payload analysis produces.

Dormant archives store data from past missions that is either not used or very rarely used. Active scientists aren't at the storage location.

3.9.1 Information Required to Archive and Maintain the Mission Database

Tables 3.27, 3.28, and 3.29 detail the information and processes required in archiving mission data and maintaining both the mission database and the products of this database.

Table 3.27. **Information Required to Archive and Maintain the Mission Database.** Archives enable use of the data long after the mission is over.

Information Required	Comments	Where Discussed
Telemetry and tracking data	• The data received by the mission database is generally referred to as level 0. The mission database then processes and stores this data as level 1, so we can use it to analyze the mission. • The level 1 data may also be part of the data archived.	Chap. 13 Sec. 3.8
Data from other functions in the mission operations system (MOS)	• Other MOS elements also generate data and store it in the mission database. Examples are command files and pass reports from mission control. • Some of these products are also packaged into data to be stored in the archive. Analysis often requires engineering data for us to understand the payload data effectively. We may correlate this ancillary data to each payload product.	Sec. 3.1–3.8
Final payload products	• Products the MOS generates for payload analysis—called level 3 data—are stored in the mission database and transferred to the project archive	Chap. 13
Higher-level products from the payload	• Data processed by the payload centers—called level 4 data—are also transmitted to the project archive for use by others	Chap. 13

3.9.2 Process for Archiving and Maintaining the Mission Database

Table 3.28. Process for Archiving and Maintaining the Mission Database. The archive allows users to access data from previous missions and enables comparative analysis as well as more detailed analysis of the data.

Steps	Comments	Where Discussed
1. Manage and Retrieve Data	• We acquire or receive the data from the mission database or payload users • We provide cataloging and retrieval, which references the data sets and allows retrieval by many different types of queries • We must ensure the data received adheres to defined, agreed-to, standard formats	Chap. 13 Sec. 3.8
2. Secure Data	• Data security is required as with the overall mission operations system • For scientific missions, we must give certain people access to some data for a time before it becomes available to other researchers. We call this data proprietary because it belongs to the scientist for a specified period.	Chap. 13
3. Notify Users of Data's Arrival	• After placing data products into the archive, identify the new products and notify users of their availability • Usually automated for missions that generate large amounts of data	Chap. 13

3.9.3 Products of Archiving and Maintaining the Mission Database

Table 3.29. Products of Archiving and Maintaining the Mission Database. The products are data—and more data.

Products	Comments
Archived data	• At the top level, two types of data are in the archive: payload data and the ancillary data that describes its attributes • We back up and store data in physically separate facilities
Operational data to and from functions in the mission operations system	• The mission database is a two-way database that allows the mission operations functions to share data and pass data from one function to another • The database is populated in real time, when we contact a spacecraft, as well as after the contact, when we add information received during the contact

3.9.4 Key Considerations

The technology for archives and mission databases is changing rapidly, so our challenge is to keep up with the technology while efficiently and cost-effectively providing access to data. The medium of choice for many scientific missions is now CD-ROMs, whereas several years ago it was still tape. We also need to reconsider the cost/performance ratio of delivering a data set by electronic means instead of by Federal Express.

Besides considering the technology issues in the mission operations concept (see Chap. 4), we also must ask ourselves how the data gets to the user. Does the archive send or does the user request? We must develop the archive plan, including the required ancillary data, as early as possible, so the ground team and spacecraft designers can more easily build a cost-effective archiving system.

3.10 Systems Engineering, Integration, and Test

These next four areas are most often on-going functions that apply equally well before and after launch. These disciplines appear in many fields, but we discuss only those that apply to mission operations.

The engineering function for the ground system receives support from the nine MOS functions previously discussed. Thus, systems engineers understand the overall system and the interfaces, and each of the nine MOS functions provides details. Remember that what may be a sub-system at one level (e.g., the nine MOS functions) will appear as a system at the next level (e.g., activity planning and development). Thus, systems engineering occurs at all levels. But here we're talking about engineering of the mission operations system.

3.10.1 Information Required for Systems Engineering, Integration, and Test

Tables 3.30, 3.31, and 3.32 present respectively the information, processes, and products involved in systems engineering, integration, and test.

Table 3.30. **Information Required for Systems Engineering, Integration, and Test.** The information listed enables this function to work across the nine main MOS functions from which it draws support.

Information Required	Comments	Where Discussed
Project requirements	• Systems engineering keeps the project "vision"— making sure the data system and requirements satisfy this vision.	Chaps. 1 and 4
Outputs from the operations concept	• Systems engineering uses these key outputs: − Operations scenarios − Derived requirements − Timelines − Data-flow diagrams	Chap. 4

3.10.2 Process for Systems Engineering, Integration, and Test

Table 3.31. Process for Systems Engineering, Integration, and Test. These steps are typical for most missions.

Steps	Comments	Where Discussed

1. Consider Typical Systems Engineering Functions	• Consider these typical functions:
	  – Develop system architecture
	  – Generate, review, and control requirements
	  – Define, document, and control interfaces
	  – Monitor the application of standards that the project requires, such as software walkthrough
		Chaps. 1 and 4

2. Integrate	• Usually (today—always), a different group of people does integration • Integrate the software programs and program sets into a system • Integrating software can be a large task depending on the complexity of the system being developed, or the multi-mission system being adapted for a new project • Validate the interfaces • Certify the system ready for operational testing, training, or use	Chaps. 4 and 6
3. Simulate, Test, and Train People for the Mission	• Preparing simulations and testing can be as much work as carrying out the mission itself • Develop training plans • Develop simulation capabilities • Conduct training • Certify operational readiness	Chaps. 4 and 6
4. Evaluate System	• Based on the mission operations concept, discussed in Chap. 4, develop and validate performance requirements • Test the MOS to verify that system performance meets requirements	Chap. 4
5. Ensure Network Is Secure	• Network security is becoming more important as the mission operations systems are becoming distributed and, in many cases, are using public networks • Write and implement the security plan; monitor the network	Chap. 12

3.10.3 Products from Systems Engineering, Integration, and Test

3.10.4 Key Considerations

While the systems engineering processes of mission operations are not all that different from that of any large software system, the requirements are often fuzzy. They also change during the development cycle as the payload users learn more

Table 3.32. Products from Systems Engineering, Integration, and Test. We must make sure all MOS people review and understand these products.

Products	Comments
System architecture	• In the system architecture, we describe the vision of the sponsor and project management. Developing this vision early and keeping it updated are important because it provides a framework for daily decisions.
Specification of software interfaces	• This specification is one of the most important products of systems engineering—especially if the project requires distributed processing or integration of the payload users' software into mission operations.
Integration and test plan	• The integration and test plan describes what to do when integrating new deliveries into the system. Having recursive tests and automated testing is important. Tools today allow even graphical user interfaces to be tested automatically, without a person sitting in front of a console.
Training plan	• The training plan developed for the mission operations manager allows mission operators to show they're ready for launch
Procedures	• Cover integration and test plus training
Security plans and tests	• Increasingly important, system security needs good plans and regular testing

about what they really need. Especially on scientific missions, these payload users have a lot of power with upper management in the project and with NASA. Thus, we must involve the payload users early and use rapid development and prototyping, rather than following the classic process of requirements, design, development, and delivery. Prototyping allows users to see capabilities.

The mission operations manager (MOM) should also

- Ensure the key concept trades for mission operations occur early
- Maintain the big picture

3.11 Computers and Communications Support

Designing and implementing mission operations hardware may involve staff within the project and from other organizations. For example, if the project uses an existing system, hardware development becomes a joint effort between the project and the organization that owns the basic system. Of course, the project manager must oversee hardware development by an outside organization to make sure it operates properly. Other textbooks treat designing and building hardware for information systems, so we don't discuss it in detail in this book.

3.11.1 Information Needed to Support Computers and Communications

In Table 3.33, the required information for computer and communications support is listed, as well as where the information is discussed in this book.

Table 3.33. Information Needed to Support Computers and Communications. These requirements enable engineers to develop the hardware systems a mission needs.

Information Required	Comments	Where Discussed
Data-flow diagrams	• We develop data-flow diagrams first while developing the operations concept, as discussed in Chap. 4. Allocate processes to hardware and software at the same time. This is the starting point for designing the hardware.	Chaps. 4 and 13
Requirements for computers or workstations	• Most mission operations now use distributed workstations or are moving toward them. • The number and location of the workstations depend on data flow, the mission operations architecture, and organization and staffing levels. Systems engineering typically specifies these requirements. • Many of the scientific missions have staff around the world who need workstations and, therefore, computer networking.	Chaps. 7 and 8
Requirements for networking and data communication	• The specific and derived requirements for flowing data around the control center, country, or world demand much more networking and communications equipment. • Commercial open networks such as internet require attention to ensure that critical command and control have enough security to protect them from unauthorized users.	Chaps. 12 and 13
Requirements for voice communications	• Voice communications and video-conference capabilities are important to understand early in the design phase. Designers often overlook these requirements, which are becoming much more important as we move from centralized to distributed control.	Chaps. 9 and 12

3.11.2 How to Support Computers and Communications

Because these processes, shown in Table 3.34, are typical for any computer hardware, we don't discuss them in detail.

3.11.3 Products from Computer and Communications Support

The products listed in Table 3.35 are typical of hardware in any information system.

Table 3.34. **How to Support Computers and Communications.** These steps are similar to those for any information system.

Steps	Comments	Where Discussed
Design and Build Computers and Communications Systems	• Understand existing capabilities and define what you must add to meet requirements • Get the data-system architecture from the EEIS engineer who helped develop the mission concept and from MOS engineers who helped generate the mission operations concept • Based on the hardware architecture and what you must add to the system(s), design and build the added capabilities • Test the mission operations system	Chap. 4
Maintain Computers and Communications Systems	• Maintain all computers and network equipment throughout the mission	--

Table 3.35. **Products from Computer and Communications Support.** The hardware and communications systems allow us to develop software and, later, carry out the mission.

Products	Comments
Computer and communications systems	• The ground hardware for mission operations
Networks and network access to functions of the mission operations system	• The networks and network access for mission operations • Identify open networks and those restricted for command and control functions
Voice-communication system	• The voice and video conference systems for mission operations

3.11.4 Key Considerations

We must understand and include current technology in the design to get the best performance from hardware. Looking at maintenance and its alternative sources helps keep costs down.

3.12 Developing and Maintaining Software

We must develop and maintain software throughout a project's lifecycle:

- The ground system before launch
- Software to correct errors after launch
- Software to incorporate changes in mission requirements after launch

On many of today's missions—especially interplanetary missions which take 2–10 years to reach the outer planets—we plan to develop more software capabilities after launch in order to make the mission successful. In some cases, mission operations people develop flight software before and after launch.

We must understand in advance how much development and maintenance will be necessary after launch, so we can organize and build procedures that will meet operational requirements. Table 3.36 shows what information is needed for developing and maintaining mission software.

Table 3.36. **Information Needed to Develop and Maintain Software.** This information is typical for any software-development project.

Information Required	Comments
New requirements	• Most mission operations systems build on existing ones, but requirements will often emphasize new aspects of the system and be written as if nothing existed. Usually, we can refer to existing capabilities and documentation, but if we're asking for new capabilities, we must clearly state our minimum requirements.
Error reports	• Error reports start maintenance
Control authority for change	• To control costs, we must make sure our change authorizations are efficient. Deciding to approve a change after detailed design is done wastes resources.
As-built documentation	• An output of software development required for maintenance

3.12.1 Information Needed to Develop and Maintain Software

3.12.2 How to Develop and Maintain Software

The steps listed in Table 3.37 are typical of any software-development effort. Chapter 18 is devoted to software. The MOM is responsible for the development of a large, often complete software system and should be very familiar/knowledgeable with software development processes. For scientific space missions, requirements often aren't well defined, and the end user (often the scientist developing one of the instruments on the spacecraft bus) doesn't always pay attention to mission operations until late in the development cycle. To interact with the scientist early and clarify requirements at the same time, we may try rapid prototyping or try delivering the ground data system in increments. The Mars Pathfinder project demonstrated data flow for downlink during concept development. They built on this capability in concept development, so a basic system existed when the project was approved.

Table 3.37. How to Develop and Maintain Software. The mission operations manager must understand that developing a mission operations system mainly means developing a big information system with a lot of software.

Steps	Comments
Manage Software Development	We should: • Understand the requirements and ensure they are testable and agree with the mission operations concept • Develop the software; use rapid prototyping to ensure the system's users understand how fuzzy requirements will translate into hardware • Deliver the ground data system in increments • Plan for and understand that you'll be developing and maintaining software while the mission is operating • Ensure that schedules are compatible with schedules for developing and testing the spacecraft bus • Test the software—often it is developed in several increments • Deliver software to operations • Train on the new capabilities as required • Maintain the software
Maintain Software	• Changes to correct errors • Changes needed because mission requirements change

3.12.3 Products of Software Development and Maintenance

The outputs listed in Table 3.38 are typical of any software project.

Table 3.38. Products of Software Development and Maintenance. We discuss these topics in detail in Chap. 18.

Products	Comments
Plan for software development and maintenance	• The sponsor usually requires this plan, which documents and communicates to the mission operations staff how the project will proceed
Schedules	• Schedules are a part of any software development. We must understand the project schedules and how the software schedule supports the project's milestones: − Launch date − Date the mission operations system (MOS) must be ready − Start of MOS training − End-to-end tests of the system − System tests of the spacecraft bus
Test plans and reports	• These are required for any software development, but we must test interfaces and defined functions of a spacecraft or ground-system element that other agencies will supply

Table 3.38. **Products of Software Development and Maintenance. (Continued)** We discuss these topics in detail in Chap. 18.

Products	Comments
Working software and as-built documentation	• The result of the plans, schedules, and tests • Maintenance requires documentation • The level of documentation varies depending on whether the developers or a different organization does the maintenance
Reports of performance and errors	• Tracking of errors and authorizing their fixes must be efficient

3.12.4 Key Considerations

This area is also rapidly changing. What was developed as unique on the last mission may very well be available off the shelf as commercial software today. Understanding what is available commercially will save money and time. Looking forward—by trying to design software for reuse on future missions—is important. Finally, developers must communicate existing and planned capabilities to the system's users and operators. We have to find a way to involve them in discussions, reviews, demonstrations, and tests. Their inputs are important. The earlier and more involved operators become with the users, the more operable the developed system will be.

3.13 Managing Mission Operations

Management functions for mission operations differ before and after launch. Before launch, we must develop and train so the entire mission operations system is ready for launch. Sometimes, one manager oversees both; sometimes responsibility transfers formally from one organization to another. Names also vary depending on the organization. Before launch, the mission operations manager may be called the ground system or ground-data-system manager. At this point, the MOM works with other project functions to make sure operations are compatible with mission and spacecraft design. This process begins during the study phase and during development of the mission concept and the mission operations concept, but it continues throughout the development phase. After launch, manager names include mission director, mission operations manager, and chief of mission operations.

This section focuses on management functions unique to mission operations and discusses unconventional management practices.

3.13.1 Information Needed to Manage Mission Operations

Table 3.39 lists the information that management needs to be successful. We use the first three inputs to guide development of the mission operations system.

Before launch they guide changes to the mission activities that arise when status deviates from plans.

Table 3.39. Information Needed to Manage Mission Operations. This information enables the mission operations manager to make decisions during the mission.

Information Required	Comments	Where Discussed
Requirements	Includes those placed on the mission operations system and those derived and generated by the mission operations system. The requirements not only are important for developing the MOS but also aid operational decisions during the mission.	Chap. 1
Operations concept	The operations concept is developed early in the lifecycle (study phase) and kept updated. This concept represents a common understanding between all project people and the sponsor(s). Although the mission operations manager is key in generating the operations concept, it's also important to follow the concept every day.	Chap. 4
Sponsor's goals and vision	We must understand what the sponsor is expecting. How does he or she view schedule versus cost versus product quality? Can this project stand alone, or is it an enabling project for a new program? How does the sponsor view operational risk? The MOM's decisions and actions need to complement and enforce the sponsor's goals and vision.	Chaps. 1 and 4
Status versus plans	After launch, managers must make operational decisions when deviations from plans or expectations occur. Understanding the current status versus the current plan is essential to making timely decisions.	Sec. 3.1

3.13.2 How to Manage Mission Operations

Table 3.40 lists management steps, all of which take place before and after launch. Before launch, we focus on developing the ability to support mission operations. After launch, we maintain the system, develop changes as required by changing mission conditions, and—on such projects as long planetary missions—finish developing capabilities needed to operate over longer periods.

3.13.3 Products of Mission Operations Management

The products of management aren't documents like those from many of the first nine functions. The management products listed in Table 3.41 keep mission operations focused on conducting the mission. We often don't have the luxury of analyzing a problem for weeks or months before reaching a decision. We may have to decide in hours or days how to keep mission operations on track, despite problems or surprises.

3.13 Mission Operations Functions 87

Table 3.40. How to Manage Mission Operations. These steps are typical of those mission operations managers do. Each mission will emphasize certain steps. Shading indicates post-launch steps.

Steps	Comments	Where Discussed
1. Define and Develop Operations Organization	The MOM will: • Generate and update the operations concept • Define an operations organization that minimizes interactions between groups and is process oriented • Define clearly the responsibilities for each position and make sure all members understand the operations organization • Monitor the organization's development as it matures • Make changes to improve the efficiency of operations	Chap. 4
2. Manage Interfaces	The MOM will: • Define each of the interfaces for the development organization • Define due dates for the receivables and deliverables that flow across the interface for each area • Manage the performance of the development areas by updating the receivables and deliverables schedule weekly or monthly, as appropriate	Chap. 4
3. Manage Change Control and Program Control	• Standard management texts describe these processes • We must be sure the ground system evaluates the effects of changes to the spacecraft bus before considering them for approval	--
4. Administer Contract Procurement	• Standard management texts describe these processes	--
5. Manage Contingency Reserve	• Before launch, we include performance reserve for tools and processes, in addition to typical reserves for resource development • After launch, we manage contingencies to ensure that added requirements fit within timelines and capabilities	Sec. 3.1
6. Manage Resources of the Mission Operations System (MOS)	The MOM will: • Manage these resources to meet mission objectives within the budget • Manage the MOS to provide a satisfactory return for a fixed or agreed-on cost. In the late 1990s, we managed the MOS to return the very best payload data.	Chap. 4

Table 3.40. **How to Manage Mission Operations. (Continued)** These steps are typical of those mission operations managers do. Each mission will emphasize certain steps. Shading indicates post-launch steps.

Steps	Comments	Where Discussed
7. Manage Operations Schedule	The MOM will: • Manage daily activities to use minimum staff while maintaining a proper and acceptable set of checks and balances • Ensure review and approval of planned command activities are at the lowest appropriate level and contribute value to the planned activity	Chap. 7
8. Manage Miscellaneous Activities	The MOM will: • Approve ad-hoc analysis and allocate resources to these activities • Manage development and maintenance activities needed during operations • Develop processes for placing decisions into a risk matrix to make sure the risk isn't too high compared to the effect on cost or performance	--

Table 3.41. **Products of Mission Operations Management.** The operations manager's most important products are timely decisions and approvals.

Products	Comments
Project policies and guidelines	• The management team, or the MOM, issues policies and guidelines during the mission • These statements encompass all aspects of the mission, but the focus is on guidelines that respond to changes in planned capabilities or performance
Decisions and approvals	• Decisions and approvals cover standard and non-standard aspects of the mission • Making timely decisions about problems or situations that arise during the mission is extremely important • We must make sure an approval adds value to the product or recommended decision. The lower the level of approval, always considering risks to the project, the more efficient the organization.
Reports and status to customers and sponsors	• The MOM reports upward and outward • Sponsors and customers shouldn't be surprised by hearing about problems or successes from third parties • Reports must be meaningful and useful
Goals and visions of the project	• The mission operations manager's goals and vision are just as important to mission operations as the sponsor's goals and vision are to the mission operations concept

3.13.4 Key Considerations

The mission operations manager must understand all aspects of the mission operations well enough to make effective decisions. For example automation and autonomy can save enormous costs during operation, so managers must understand how to use them.

The operations staff at nearly all facilities and programs are being multi-tasked. Specialized operational positions remain on only the most complex missions. A flat organization allows multi-tasking to take place more naturally while also saving money.

Once we've established the first conceptual organization, we identify the organizational interfaces. What are the inputs and outputs of each organizational element and where do they begin or go to? If the diagram looks like a spider web, we don't have an efficient organization.

References

CSP Associates, Inc. August 19, 1993. *Mission Operations Cost Study Results of Phase I.* Prepared for the Flight Projects Office, Jet Propulsion Laboratory, Pasadena, CA.

National Research Council Committee on Data Management and Computation, Space Sciences Board, Commission on Physical Sciences, Mathematics, and Resources. 1986. *Issues and Recommendations Associated with Distributed Computation and Data Management Systems for Space Sciences.* National Academy Press, Washington, D.C.

Chapter 4
Developing a Mission Operations Concept

Gael F. Squibb, *Jet Propulsion Laboratory, California Institute of Technology (Retired)*

Klaus Briess, *German Aerospace Center, Institute of Space Sensor Technology and Planetary Exploration*

4.1	Process for Developing a Mission Operations Concept
4.2	Steps for Developing the Mission Operations Concept
4.3	Mission Operations Concept, Bi-spectral Infrared Detection (BIRD)

A *mission operations concept* describes—in the operators' and users' terms—the operational attributes of the mission's flight and ground elements. It results from the cooperative work of several disciplines. Its development is similar to that of a space mission concept, as discussed in Chap. 1, but the mission operations concept is more detailed and emphasizes the way we will **operate** the mission and **use** the space element (operational characteristics). It is generated in phases and becomes more detailed as our design of the operations system progresses.

The mission operations concept follows from, and is consistent with, the mission concept. It's the most important deliverable from the mission operations manager (MOM) before launch and is crucial in keeping system costs down. It's also the main way a MOM influences the design and operability of the mission and spacecraft. It often results in changes to the mission concept to reduce lifecycle costs. The MOM must understand what a mission operations concept is, how to develop one, and what it contains.

We generate a mission operations concept in progressively more detailed layers. It doesn't detail a mission's development or operational costs, but is essential in cost estimation. By combining the operations concept with assessments of operational complexity (Chap. 5), we can determine the probable costs of operations early in mission design.

4.1 Process for Developing a Mission Operations Concept

Figures 4.1 and 4.2 show the process for developing a mission operations concept, along with the information it requires and its products. Table 4.2 shows the process steps and where we discuss them further.

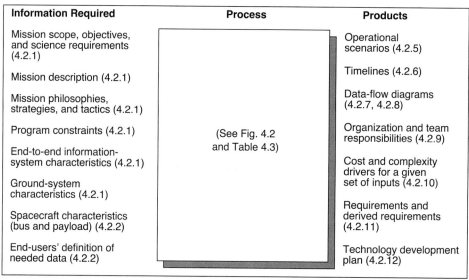

Fig. 4.1.   Developing a Mission Operations Concept. We describe the mission operations concept using a standard set of products.

Developing a mission operations concept requires different disciplines to communicate with each other. These disciplines include

- Designers of the mission, spacecraft, payload, and ground system
- Operators of the ground system
- Users (those who receive data from mission operations)

4.1 Developing a Mission Operations Concept

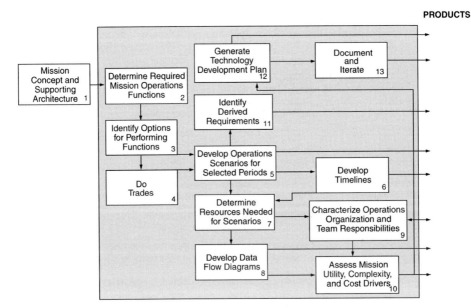

Fig. 4.2. Process for Developing a Mission Operations Concept. The process is iterative and the products contain increasing levels of detail as we move from the study phase to the development phase.

A space vehicle often operates differently from the way its designers had in mind. Early communication between designers, operators, and users shortens the development time because fewer changes are required during development or testing.

The process emphasizes areas in these disciplines for which we should study trades to minimize lifecycle costs and get better information from the mission. The operations concept, when documented, provides derived requirements for developing functions of the mission operations system. Because the mission operations concept responds to top-level mission requirements, we can easily trace these derived requirements to top levels while designing the mission. Each mission concept should have a corresponding mission operations concept. If we develop the two concepts in parallel, we can save time and shorten a project's conceptual and definition phases. That's because the mission operations concept can quickly feed back operational cost drivers into the mission concept development.

To influence development, we need to complete the mission operations concept early, when the interfaces and attributes of the inputs are less defined and trades are possible.

Completing the mission operations concept early often forces us to resolve design incompatibilities, which operations developments or procedures would otherwise have to solve or minimize. Table 4.1 shows how required information solidifies during the project phases. Usually by the time production and deployment begin, only the ground system, people and procedures, or flight software can change.

Table 4.1. Changes to Information Required for the Operations Concept Versus Mission Phase. This table shows when inputs freeze within the project phases.

	Project Phase				
Information Required for Operations Concept	Needs Analysis	Concept Development	Detailed Development	Production and Deployment	Operations and Support
Mission scope, objectives, and science requirements	Changeable	Changeable	Changeable for cause	Frozen	Frozen
Mission description	Changeable	Changeable	Changeable for cause	Frozen	Frozen
Mission philosophies, strategies, and tactics	Changeable	Changeable	Frozen	Frozen	Frozen
Program constraints	Changeable	Changeable for cause	Frozen	Frozen	Frozen
Characteristics of the end-to-end information system	Changeable	Changeable	Changeable	Existing frozen; New changeable for cause	Existing frozen; New changeable for cause
Capabilities and characteristics of the ground system	Existing frozen; New changeable	Existing frozen; New changeable	Existing frozen; New changeable	Existing frozen; New changeable for cause	Existing frozen; New changeable for cause
Spacecraft capabilities and characteristics (bus and payload)	Changeable	Changeable	Changeable for cause	Frozen	Frozen
Flight software	Changeable	Changeable	Changeable	Changeable for cause	Existing frozen; New changeable for cause
End-users' definition of needed data	Changeable	Changeable	Changeable	Frozen	Existing frozen; New changeable for cause

Because the mission is continually changing and maturing, updating the mission operations concept during the project is also important.

Chapter 3 describes the 13 mission operations functions. People representing these functions need to help develop a mission operations concept. Often, in the

4.1 Developing a Mission Operations Concept

early phases (conceptual), one person will represent several functions. One representative will usually lead discussions and be the custodian of the concept. This person often comes from the system engineering or mission planning functions and typically has experience working with information systems.

Because the mission and mission operations concepts are closely linked (see Sec. 4.2.5), some people participate in both activities. Table 4.2 relates the mission operations functions described in Chap. 3 to the main engineering disciplines involved in developing them and to the inputs to and outputs from the mission operations concept. This table follows the order of putting information into the mission operations concept, as shown in Fig. 4.1.

Table 4.2. Providers of Information Required to Develop a Mission Operations Concept. Functions of the Mission Operations System (MOS) are listed in the order they are needed. Products include use and description of new or existing capabilities, timelines, and requirements and derived requirements for each function.

MOS Function	Discipline	Information Provided by MOS Function	Where Discussed
Mission planning	Mission-planning engineer	• Mission scope, objectives, and payload requirements • Mission description • Mission philosophies, strategies, and tactics	Sec. 3.1
Managing mission operations	Project manager	• Program constraints and requirements • Sponsor's goals and vision	Sec. 3.13
Systems engineering, integration, and test	End-to-end information systems (EEIS) engineer Ground systems (GS) engineer	• EEIS characteristics • GS characteristics	Chaps. 6, 11, and 12 Sec. 3.10
Payload planning and analysis	Payload designer Payload-operations engineer	• Payload characteristics • Payload operations	Chaps. 8 and 14 Sec. 3.7
Payload data processing	End-users (recipients of the data products)	• Data-product definition • Processing algorithms	Chap. 8 and 13 Sec. 3.8
Spacecraft planning and analysis	Spacecraft designer Spacecraft operations engineer	• Spacecraft characteristics • Spacecraft operations	Chaps. 8 and 15 Sec. 3.6
Activity planning and development	Sequence designer and operations engineer	• Mission plan • Mission rules • Test and integration requirements	Chaps. 7 and 8 Sec. 3.2

Table 4.2. **Providers of Information Required to Develop a Mission Operations Concept. (Continued)** Functions of the Mission Operations System (MOS) are listed in the order they are needed. Products include use and description of new or existing capabilities, timelines, and requirements and derived requirements for each function.

MOS Function	Discipline	Information Provided by MOS Function	Where Discussed
Mission control	Mission-control designer and operations engineer	• Mission plan • Mission rules	Chap. 8 Sec. 3.3
Data transport and delivery	Data-transport designer and operations engineer	• GS capabilities • User requirements • Data format and volume	Chaps. 11 and 12 Sec. 3.4
Navigation planning and analysis	Navigation designer and navigation operations engineer	• GS capabilities • Requirements for trajectory and attitude data	Chaps. 8 and 10 Sec. 3.5
Archiving and maintaining the mission database	Archive designer and operations engineer	• GS capabilities • User requirements • Data format and volume	Chaps. 8 and 13 Sec. 3.9
Computers and communications support	Computers and communications engineer	• Data-flow diagrams • Computer, network, and voice requirements • GS capabilities	Chap. 11 Sec. 3.11
Developing and maintaining software	Software-development engineer	• Software requirements • Existing software capabilities • Change control authority	Chaps. 6 and 18 Sec. 3.12

4.2 Steps for Developing the Mission Operations Concept

Figure 4.2 depicted how to generate a mission operations concept; Table 4.3 lists the major steps. We describe the process in detail below.

4.2.1 Identify the Mission Concept and Supporting Architecture

We begin developing the mission operations concept by examining the mission concept and supporting architecture. By obtaining the information listed in the rest of this section, or by making assumptions, we can describe the mission in users' and operators' terms. When information is not available—the design is not yet started or not specified in the mission architecture—the operations concept team assumes information and gives this input to the person responsible for it. The operations concept is then valid until the assumed information changes. We can

4.2 Developing a Mission Operations Concept

Table 4.3. Developing a Mission Operations Concept. This table expands Fig. 4.2 by identifying key items and pointing to chapters with additional information.

Step	Key Items	Where Discussed
1. Identify the mission concept and supporting architecture; gather information.	• Characteristics of information system • Characteristics of payload • Characteristics of spacecraft bus • Definition of data product • Ground system	Chap. 1 and 2
2. Determine functions needed for mission operations.	Functions usually vary for each mission concept and related architecture. We can combine or eliminate some functions	Chap. 3
3. Identify ways to carry out functions and whether capability exists or must be developed.	• Where conducted (space or ground) • Degree of automation on ground • Degree of autonomy on spacecraft • Software reuse (space and ground)	Chaps. 1–3
4. Do trades for items identified in Step 3.	Options are often selected before developing operational scenarios. These trades are done within the operations element, which includes the flight software.	Chaps. 1–5 Table 3.1
5. Develop operational scenarios for the functions determined in Step 2 and the options selected in Step 4.	*Operations scenarios* are step-by-step descriptions of how to do integrated activities. We identify major issues and drivers on the operations system.	Chaps. 4, 6–8
6. Develop timelines for each scenario.	Timelines identify events and how fast and when they occur. They drive the performance parameters for each mission operations function.	Chaps. 4, 6–8
7. Determine the type of resources needed to perform each step of each scenario.	The allocation to hardware, software, or people is based on what, how quickly, and for how long steps must be done	Chap. 4
8. Develop data-flow diagrams.	Data-flow diagrams form the basis for the ground- and flight-data systems and the command, control, and communications architecture.	Chaps. 4, 11, and 13
9. Characterize responsibilities of the organization and team.	We need to identify organizations involved and their structure, product responsibility, interfaces and number of people. To improve cost-effectiveness of operations, we should minimize the number of organizations and interfaces.	Chaps. 4 and 5

Table 4.3. Developing a Mission Operations Concept. (Continued) This table expands Fig. 4.2 by identifying key items and pointing to chapters with additional information.

Step	Key Items	Where Discussed
10. Assess mission utility and complexity, as well as the cost of mission operations.	The cost estimates include both development and operations costs. We refine them each time we update the mission operations concept.	Chap. 5; Chaps. 3 and 20 [Larson and Wertz, 1999]
11. Identify derived requirements and cost and complexity drivers; negotiate changes to mission concept if necessary.	We document derived requirements and ensure consistency with top-level requirements.	Chaps. 1, 2, and 4
12. Generate technology development plan.	The technology to support the mission concept may or may not exist.	Chap. 6 Sec. 4.12
13. Iterate and document.	Iteration may occur at each step.	--

determine the cost of this change by modifying the operations concept as needed and then re-evaluating the complexity and cost of the mission operations using the model described in Chap. 5.

Mission Scope, Objectives, and Payload Requirements. We must state mission objectives in terms of data the payload can get through effective operations. These objectives specify what the spacecraft must do to achieve the mission's scientific, commercial, or defense goals. This input defines and describes the users of the payload data and their level of sophistication. Our system will be different for a scientist who can use state of the art computer analysis tools and for a politician who wants to access information from the Earth Observation System. We need to know how people will use the payload's processed data. Potential uses are research, commercial, defense, education, or public information. We also need to know the timeliness requirements for the payload's processed data and the success criteria (percentage of total data actually returned).

Mission Description. This input contains information about the trajectory, such as launch date(s) and window, trajectory profile, maneuver profile needed to meet mission objectives, mission phases, and a description of activities required during each phase. Observation strategies describe how we'll gather the mission data. We either define and finalize the observation strategies and mission description before launch or adapt them to data gathered during the mission.

Mission Philosophies, Strategies, and Tactics. These items are rules not associated with the health and safety of the mission. They may relate to the mission objectives or to the background of the mission designers and those developing the mission concept. It's important to determine if the item is associated with the mission or a person's preferences.

Examples are:
- Maximize real-time contact and commanding or maximize onboard autonomy and data-storage capability
- Maximize the involvement of the educational institutions and teach science students important aspects of issues like operations or space physics
- Make sure a central authority approves all commands
- Limit the image budget to 50,000 images

Program Constraints. The sponsor of the mission and the project manager impose these non-technical constraints. Operators must follow them until the project manager is convinced they are increasing the cost of the mission. Examples of program constraints are:
- Limit mission cost and cost profiles
- Use a specific tracking network
- Use existing flight hardware
- Use existing ground-system capabilities and design the spacecraft to be compatible with them
- Centralize or distribute operational teams
- Use multi-mission versus project-dedicated teams
- Involve students versus dedicated professionals
- Involve educators and the educational community

Capabilities and Characteristics of the End-to-End Information System (EEIS). The EEIS engineer will have helped develop the mission concept, which describes the spacecraft and ground elements in terms of information flow and processes. System-level requirements include:
- Using information standards (layered protocols or CCSDS standards)
- Locating capabilities and processes (includes both space and ground)
- Characterizing the information system's inputs and outputs

The EEIS engineer should specify these requirements in terms operators and users can understand, not in terms used by computer scientists and programmers.

Capabilities and Characteristics of the Ground Element. Most missions are designed around the ground system of a specific agency, such as the AFSCN, NASA, or European Space Agency. Each ground system has standard services which will support the mission at low (or even no) cost if the mission meets specified interfaces and standards. We follow these requirements on the flight system until they keep us from meeting mission objectives. We must ask:
- What standards does the ground system require the spacecraft to meet?

- How do we carry out the standard and what requirements does this place on the spacecraft?

Some standards, such as CCSDS, are broad and allow for interpretation. Thus, we might meet CCSDS standards for telemetry and commanding but choose a way of meeting them that our ground system won't support.

4.2.2 Determine Functions Needed for Mission Operations

The mission concept drives top-level functions, but the spacecraft's capabilities determine the detailed ones. A completely autonomous spacecraft requires few operations, whereas a spacecraft payload that can't compute or store data on board requires continuous ground control or ground automation. To determine what we must do, then, we have to understand the spacecraft's characteristics. The mission concept may state these characteristics if the mission concept team includes operators. Otherwise, we'll develop them as part of the mission operations concept. Through discussions, the operators and developers define the characteristics we describe below.

Capabilities and Characteristics of the Payload. Including payload designers and mission planners in the mission concept process leads to timely definition of the payload characteristics. People working on the concept need to develop information that answers the types of questions we ask below. These questions come from past missions and often start payload designers thinking early about how the payload will operate and the differences between testing it on the ground and operating it in space. (See Chap. 14 for more information.)

To understand how a payload will operate, we must describe what the payload will do during an observation period.

- What are the payload's attributes?
- What is the commanding philosophy? Is the payload commanded by using
  - Buffers?
  - Macro commands?
  - Tables?
- Does the payload use default values?
- Will some commands degrade or damage the payload?
- Do some mechanisms depend on previous commands?
- Does the payload use position commands or incremental commands to control rotating or stepping mechanisms?
- What classes of commands does the payload use?
  - Real-time?
  - Stored program?
- What processing occurs within the instrument?

- How can we describe the payload in terms of
  - CPU/Memory?
  - Closed-loop functions?
  - Predictive commanding versus event-driven commanding?
  - Requirements placed on the spacecraft bus for instrument control?
  - Mechanical power and thermal attributes?
  - Avoidance areas (Sun, Earth, South Atlantic Anomaly, Venus, or Moon)?
  - Requirements for controlling the spacecraft bus?
- Are the payload apertures larger than the spacecraft bus's pointing control?
- What are the user-specified parameters for observation?
  - How are these converted into instrument commands?
- What is the instrument heritage?
- What ground processing and analysis are required to support the instrument's operation?

These questions may seem obvious, but they come from missions in which the spacecraft had been designed and, in some cases, built before anyone asked them.

The example below shows how considering the payload's operation will help us define the mission operations concept.

An instrument was designed with an aperture that has a field of view of ten arc-seconds. At the same time, the spacecraft designers designed a spacecraft bus with a control authority (the ability to point with a certain accuracy) of 20 arc-seconds. The operators couldn't be sure that the celestial object they commanded the spacecraft bus to observe would end up in the instrument's field of view. The solution was to

- Command the spacecraft to the desired position
- Observe with a different instrument that had a wider field of view and see how far the spacecraft was off from the desired position
- Generate attitude commands to move the spacecraft just a bit (tweak commands) until the actual attitude corresponded to the desired attitude
- Verify the attitude errors were gone
- Select the instrument with the narrow aperture and observe as specified

---

This one design issue on the Infrared Space Observatory (ISO) caused routine real-time involvement of operations, decision making, and commanding for this mission. This involvement required more ground software, controllers trained in routinely commanding the attitude-control system, and more people whenever the instrument was used. After launch the accuracy of the flight attitude control system far exceeded the design requirement and the ground software and procedures that were developed pre-launch were not used.

The same design issue was present on the Space Infrared Telescope Facility (SIRTF), launched in April 2003. The solution that SIRTF chose was to develop flight software that would do this task. The choice was not based on the cost of flight software development versus the cost of ground operators performing the function, or the efficiency of the observatory based on one solution versus the other. Rather the flight software solution was selected because the high gain antenna of the observatory is fixed, and when observing there is no communication with the ground, unlike the earlier example of the Infrared Space Observatory that always had ground contact. In the case of SIRTF, a trade should have been made during the mission design phase when the type of antenna, the accuracy of the instrument / science return and the costs of flight software and ground procedures and software could have been analyzed and a system decision made. The goal of generating a mission operations concept is to identify incompatibilities and cost drivers early, before designing and building any hardware.

It's a good idea to ask the designer of a payload instrument how to go from an observer's requirement to a set of commands for the instrument. Sometimes the answer is simple, but in other cases instruments have been designed for a laboratory rather than for space and the process is complex. Let's look at another example.

An astronomer knows the celestial object we need to observe and its estimated brightness within the wavelength we'll look at. A spectrometer is the specified instrument. What must the operations staff do to generate the commands for this observation?
- Make sure the previous observation has an intensity lower than the one we'll observe
- Add a calibration before the observation
- If the observation is longer than five minutes, add short, internal calibrations every five minutes until the observation is complete
- Add a calibration after the observation

This instrument design requires us to sort the observations and do them in order of increasing intensity, even though it may not be the most efficient way to move the telescope from one object to another.

For this design, the astronomer's requested observation time drives how long we must look at the object, which in turn determines the total number of commands our spacecraft requires. Also, until we've established the instrument's characteristics in space, we can't set the calibrations and durations. So we'll probably need to recalculate after launch.

4.2 Developing a Mission Operations Concept

Capabilities and Characteristics of the Spacecraft Bus. As is true for the payload's capabilities and characteristics, timely definition of the spacecraft's characteristics depends on including spacecraft designers and mission planners in the mission concept process. (See Chap. 15 for more information.) Again, people working on the concept have to answer the following types of questions for the overall spacecraft and its subsystems:

- What are the spacecraft's operational attributes?
  - What commands and parameters are sent from the ground?
  - How are the values of these commands determined?
  - How many commandable states are there?
  - Are engineering calibrations required? What are the purpose, frequency, and schedule constraints of the calibrations?
- How many engineering channels need monitoring?
- Do these channels provide subsystem-level information to the operators, or must operators derive information about subsystems?
- For the attitude-control system
  - Are guide stars used? If so, how are they selected?
  - How does the pointing-control accuracy compare to instrument requirements?
- What types of margins exist and what must be monitored and controlled?
- What expendables need monitoring during flight?
- Does the spacecraft subsystem use any onboard, closed-loop functions?
- What are the attributes of the spacecraft's data system?
- What processing must we do on the ground to support spacecraft operations?
- What is the heritage for each of the spacecraft subsystems?

Consider an example of how a design decision affects operations.

The Galileo spacecraft was designed to take heat from the Radioisotopic Thermoelectric Generators (RTGs) and use it to warm the propulsion system. This design saved weight and power and cost less to develop. But the spacecraft's operational characteristics tied together subsystems for propulsion, health and safety, thermal transfer, and power. Operators had to check each command load to see how it changed power states and affected the propulsion subsystem. Engineers from power, thermal, and propulsion had to check each sequence, even if only the payload instrument's states changed. For example, turning an instrument off (or on) caused the heat output of the RTG to change.

End-Users' Definition of Needed Data. End-users inform mission operations by defining how, and how often, they will use data from the payload. By understanding these data products, ground-system engineers can start designing how they'll get them.

We must understand how confident the end-users are about the products. Often, they don't know what they want until they see how the instrument works in flight and what it observes. In these cases, engineers can prototype processes before launch and finish them after launch, once definitions are complete.

We also need to define the products' relationship to the payload data by answering the following questions:

- Is the product based on the payload's raw data or must we remove the payload instrument's signatures?
- Must the data be calibrated? How? Does it involve processing special calibration observations? At what rate do we expect the calibration files to change?
- Does the data need to be converted into geophysical units? How? Where do the algorithms for this conversion come from? Must the project generate them and update the mission database as they become more refined?
- What are the formats and media of the payload's data products? Is there a community standard, such as the Astrophysical community's use of an Flexible Image Transport System (FITS) format on all of NASA's astrophysics missions?
- What ancillary data must we provide so the end-user can interpret the payload data?
  – Spacecraft position?
  – Spacecraft attitude?
  – Ground truth data?
- Who processes the payload data?
  – Project?
  – End-user?
- How does the processed data get into the archives?
  – Through the project?
  – Through the end-user?
- What, if anything, must the project archive after the flight phase is over?

Generate a List of Top-Level Functions Operators Must Do. Referring to Chap. 3, we can address each process within the 13 operational functions to generate this list using the information we've developed in Sec. 4.2.1 and 4.2.2.

When time constraints or requirements are available, we may add them to our early design of a mission operations system for a given mission concept.

4.2.3 Identify Ways to Accomplish Functions and Whether Capability Exists or Must Be Developed

At this level, many steps—and the ways to do them—will be straightforward. For example, to track an interplanetary spacecraft, we use NASA's Deep Space Network. For other steps, we'll have to identify several options and describe them. For example, to determine a spacecraft's orbit, we might use global-positioning satellites and automated procedures on board, or we might track the spacecraft and calculate its orbit on the ground.

To understand options, we may build a table that contains the operations functions that apply to the mission's database and avionics. Then, we identify whether the avionics (automated) or the ground system will do each function. If on the ground, we further determine whether the hardware, software, or operators will complete it. If a check goes in more than one place, we describe which functions are done in each place and whether options exist. Table 4.4 shows how such a table would look.

If the ground hardware and software do something, we ask, "Could the avionics partially or completely do this function and lower the mission's lifecycle costs?" For example, if we were considering orbit determination, we'd ask, "Could we plan and analyze position location within the avionics?" Then, we'd look at the accuracies of the GPS system, check the cost of GPS receivers that are flight qualified, and do a first-order estimate of the change in lifecycle costs. We have to remember that the costs of tracking facilities are important in this type of trade.

4.2.4 Do Trades for Items Identified in Step 3

For the options identified in step 3 that will drive either performance or cost, a small group of engineers needs to do trades and decide how to carry them out. In some cases, engineers may develop an operations scenario for each option to describe it in detail.

4.2.5 Develop Operational Scenarios

Operational scenarios are key to an operations concept. A *scenario* is a list of steps, and we can describe an operations concept with about a dozen top-level scenarios. Typically, planners generate three types of scenarios during study and design, with each increasing in detail:

- User scenario
- System scenario
- Element scenario

Table 4.4. Identifying Where to Carry Out Functions. Using a table similar to this one will help the MOM identify options for carrying out mission operations. We assume functions not included in table are done on the ground. As you evaluate each function, place a check mark in the table to indicate where you carry out the function. The functions listed correspond to the functions in Fig. 3.1 that may be part of the spacecraft avionics. The example is for a hypothetical but typical deep space mission.

MOS Function	Where to Do the Function		
	Spacecraft Avionics	Ground Hardware/ Software	Operators
Activity planning and development	Sequencing engine	Generate sequences	Run software. Validate output prior to sending to spacecraft
Mission control	Correct reception of sequence load	Transmit commands, alarm check engineering telemetry and instrument health and safety telemetry	Notify operators of correct load and notify staff if alarms
Data transport and delivery	Receive uplink data, transmit downlink data, store telemetry, playback telemetry, format and encode	Transport data from tracking stations to control center. Transmit/receive data from control center to other operation areas or science areas.	Monitor and validate system performance
Navigation planning and analysis	Optical navigation closed loop	Processes radiometric data for Doppler and angles data. Generate trajectories, generate maneuver requests.	Monitor and analyze processing system and onboard optical navigation performance
Spacecraft planning and analysis	Capture last minute of data before safe mode or error state. Generate subsystem alarm summary.	Process spacecraft data and generate summary and status information	Review processed data; make requests for calibration or analysis spacecraft sequences
Payload planning and analysis	None	Generate payload commands for science acquisition and instrument calibration sequences	Generate software inputs, analyze program outputs, validate commands requests
Payload data processing	Process optical navigation images and generate corrected trajectory information	Process outputs of spacecraft optical navigation	Analyze outputs of optical navigation

Each of these scenarios has a corresponding timeline and data-flow diagram, as discussed in Sec. 4.2.6 and 4.2.8.

During the early study phases of a mission, users develop a scenario to show how they want to acquire data and receive products from the payload. For a NASA mission, the user would be the principal investigator or science group or, for a facility spacecraft such as the Hubble Space Telescope, an individual observer.

We create a system scenario after we've developed the operations architecture—during the operations-system design. Here, we emphasize the steps within a process needed to conduct the mission. Finally, during element design, we expand these system scenarios to include more detailed elements and subsystems.

The mission concept, mission operations concept, and design of the space and ground elements are closely related. As design proceeds, costs are always in planners' minds. The costs of development and operations are both important, so we must estimate them during study and design. As designs for the space and ground elements mature, we have to do trade studies to get the lowest cost and best design within some cost cap. Whenever we expect to exceed this cap or believe we can reduce costs by changing the mission concept, we study possible mission trades. Figure 4.3 depicts this process.

Typical operational scenarios, along with questions to ask or items to include, appear below.

Develop a scenario for the total mission based on the trajectory and the mission goals. Define the trajectory in terms of activity periods and mission phases. (See Sec. 3.1.)

- Launch
- Spacecraft bus checkout
- Payload checkout
- Key payload periods, data gathering, and inactive or quiet periods
- Maneuvers
- End of mission

For each phase specified, we describe in words the steps (inputs, processes, and outputs) required to go from the user(s), request for an observation to the payload's acquisition of requested data. We also address how to handle review and approval.

The scenarios discussed below are part of the uplink process:

How Do Users Plan the Request? A typical request from an astronomer could be: Use a particular instrument to observe a celestial object with an estimated flux and located at a given position; obtain a 4×4 mosaic image.

- What parameters are used to describe the request?
- What tools are required to help the user make the request?

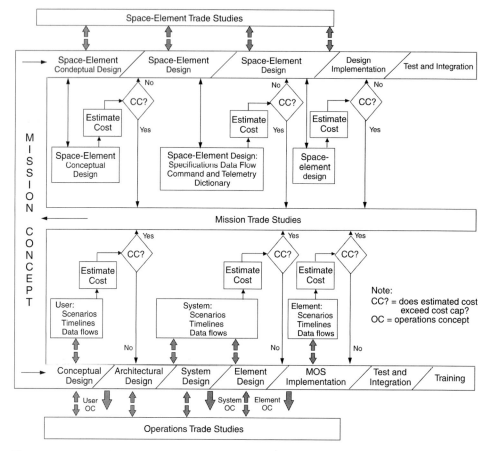

Fig. 4.3. Trade Studies for Mission Operations and Space Element. Designing to cost requires frequent cost estimates and trade studies within mission operations and the space element.

- Where is the user?
  - At the control center?
  - At a remote site?
- What are the form and content of the request?
- Is the request part of a mission plan that has constrained the user?
- How does the mission operations system support the user?
- What type of user(s) are involved in the mission?
  - Single user?
  - Multiple users?

- Where and how is the request is delivered?
  - Spacecraft?
  - Activity planning group?
  - Mission control?

For many missions today, users stay at home but submit requests to the control center. Commands are interactive or non-interactive. The non-interactive commands don't need monitoring or checking. The control center can create and send them directly to the spacecraft during the current or next track.

How Are the Activity Requests Integrated? Integration resolves conflicts among users, tracking facilities, and spacecraft resources. Fewer requirements for integration mean lower costs plus shorter and more adaptable sequences.

- Who is requesting the activity?
- What are the relative priorities for these requesters?
- Where do inputs come from?
  - User(s)?
  - Mission control?
  - Spacecraft analysts?
  - Payload analysts?
- What are the flight rules?

The spacecraft and payload designers should list flight rules they expect activity planners and developers to validate before sending commands to the spacecraft.

How Are the Integrated Requests Converted to Command Mnemonics and Binary Files? We may need to create one command for each activity, or we may need to expand blocks into many commands. We also must describe how to generate and maintain the command dictionary that will control all conversions.

How Are the Binary Files Wrapped with Ground and Spacecraft Protocols and Transmitted to the Spacecraft? Getting messages from the ground to the spacecraft will be more difficult if the ground system uses a standard that the project hasn't imposed on the spacecraft.

- Who is responsible and what steps are required to get the commands into the spacecraft?
- What steps will get the commands from the control center to the tracking station?
- What ground resources are used for this transmission?
- What are the uplink characteristics?
  - Frequencies?
  - Command rates?

- What are the allowable bit and expected-error rates?

We have now transitioned into activities that occur on board the spacecraft:

How Does the Spacecraft Receive and Verify the Command Files and Store or Execute the Commands?

- What are the interfaces and processes between the spacecraft's avionics system and the payload's data system for various commands?
  - Immediate execution?
  - Stored commands (within the spacecraft)?
  - Commands stored by the payload instruments?
- How are the commands verified?
  - Command receipt?
  - Command execution?
- Is the verification real time or delayed?
- Who supplies the information that will enable the commands to be verified?

The following scenarios are for downlink—relaying data from the payload to the recipient:

What Are the Processing Functions and Interfaces From Payload to User?

- What happens to the data as it goes from the payload instrument to spacecraft storage?
  - Is the data compressed?
  - What are the payload output rates?
  - How does the spacecraft collect the payload data? Synchronous or asynchronous?
  - What type of storage is used? (Solid-state memory is replacing tape recorders on most missions.)
  - What are the storage and retrieval characteristics? Is the data retrieved by location or by a request for a file?

Today most data is retrieved by location or first-in, first-out (FIFO) from an area of storage. The next generation of missions will store and retrieve data by files, just as we do today on our personal computers.

- What happens to the data as it goes from the storage area to the spacecraft transmitter?
  - Is the data encoded to improve signal-to-noise characteristics of the link? If so, what type of encoding?
  - Is the data encrypted for security?

- What transmit rates are available on the spacecraft?
- Can stored data be transmitted at the same time as real-time data?
* How do we transmit data to storage in the ground database (level 0 processing)?
  - Must the spacecraft or spacecraft antenna be pointed prior to transmission? How?
  - What transmitter and receiver characteristics determine the link performance?
  - With these characteristics plus the encoding, what are the error rates and allowable data-transmission rates?
  - What are the transmission rates for the various mission phases?
  - What are the processing and storage steps from receiving data at the ground station to storing this data in a project database?

We have now transitioned from the spacecraft to the ground system.

* What steps are necessary from receiving the data in the project database to having the content ready for delivery to the end user? We must specify processing volume and speed to properly size the system—usually an extensive section that's different for each mission.

How Is the Data Formatted, Analyzed, and Delivered to Meet the User's Requirements?

* What data must we analyze in real time for
  - Health and safety?
  - Payload quality?
* What data must we analyze later to
  - Control the mission?
  - Determine quality and content of payload data?

How Is the Data Archived? What processing must we do to meet the requirements for

* Format?
* Quantity, quality, and continuity?
* Frequency of updating the archive?

How Does the Processed Data Change the Mission Plan or Activity Plan? (Closing the Loop)

* How is the processed data compared to the mission plan?
* What are the criteria for changing the mission plan?
* How do we change the plan?

- How do we deliver the modified plan to the operations-system element for implementation?

How Does the Payload-Calibration Plan get Generated, Verified, and Modified?

- What is the payload-calibration plan?
- How are descriptions of calibration activities generated? Is specialized software necessary? Does the payload instrument have special modes only for calibration?
- How are the calibration observations analyzed? By whom? How fast?
- How are new observations generated?

What Are the Scenarios for Top-Level Contingencies? These scenarios address what happens when spacecraft-bus or payload anomalies occur. We must assign categories of anomalies, primary responsibility, (ground or space element), and required response times.

Other scenarios may be helpful. It takes imagination and knowledge of the mission to develop them.

4.2.6 Develop Timelines for Each Scenario

Now we can add times needed to do each set of steps and determine which steps can be run in parallel or must be serial. This information becomes a source of derived requirements for the mission operations system's performance.

Various agencies, universities, and companies own timeline tools. There are no standard tools, but many are modified from commercial, off-the-shelf software. Most missions use the same timeline tools for operational scenarios and activity planning.

4.2.7 Determine the Types of Resources (Hardware, Software, and People) Needed to Perform Each Step of Each Scenario

Once we've developed scenarios we may assign machines or people to do each step. This choice will be obvious for many steps, but others may be done by people or machines, depending on performance requirements and available technology.

Having allocated resources, we turn steps assigned to hardware and software into data-flow diagrams. For steps assigned to people, we develop an operational organization and assign steps and functions to teams.

At this point, we should examine each step to which we've assigned an operator and ask, "Can this process be automated to eliminate the operator? How?" We should allow a person to do something only when lifecycle costs mandate human involvement. We can't accept the idea that we need a person because we've always done it that way. Technology is advancing so rapidly that a machine may very well do now what a person had to do on the last mission.

Through this process, we'll also discover areas on which to focus R&D funding for possible automation in future missions. This concept of "justified operation" was developed at a NASA workshop held in January, 1995. [NASA, 1995]

4.2.8 Diagram Data Flow

System-engineering tools can convert the machine steps into data-flow diagrams showing processes, points for data storage, and interrelationships.These tools also generate a data dictionary which ensures a unique name for each process or storage point in the data flow. These computer-aided systems engineering (CASE) tools then generate information we can use for development.

4.2.9 Characterize Responsibilities of the Organization and Team

Once we've defined processes, we gather the people steps and form an organization around them. We assign teams to the steps and analyze the organization to establish operational interfaces. Generally, the more inputs required from different teams, the more complicated, costly, and slow the operations organizations will be.

4.2.10 Assess Mission Utility, Complexity, and Cost Drivers of Mission Operations

Chapter 5 describes methods and concepts for identifying mission complexity and cost drivers. Development and operations (post-launch) costs for a given set of inputs are key aspects of an operations concept. Chapter 3 describes the top-level cost drivers for the 13 operations-system functions.

4.2.11 Identify Derived Requirements

We relate the scenarios we've described to the top-level requirements. We can then use the steps within a scenario as the source of derived requirements. The operations concept is a good place to keep and document the relationship between the top-level requirements and the derived requirements related to these scenarios.

4.2.12 Generate a Technology Development Plan if Appropriate

The technology to support a mission concept may not exist or may not be focused and prototyped to a level that is appropriate for the mission approval. Identifying the technology needed to support a mission operations concept, along with the schedule and needed funding, is an important output of generating the mission operations concept.

4.2.13 Iterate and Document

We document results of our effort to develop an operational concept so others can benefit from this information. Following is a suggested structure for this document. While developing the concept, we should keep this information in electronic form and available to all members of the project, so people can review and critique it.

- Inputs
- Scenarios
  - Describe operational scenarios—those listed in this chapter or more detailed ones based on them
- Timelines
- People and procedure functions
  - Organization and team responsibilities
- Hardware and software functions
  - Data-flow diagrams
- Requirements and derived requirements

The mission operations concept is the operations manager's most important product before launch. It enables the manager to discuss changes to the mission concept based on quantitative data. The earlier the first mission operations concept appears, the greater the leverage for minimizing lifecycle costs. It's important to keep the mission operations concept current because it's the best top-level description of how the mission will be flown and the tools we need to fly it.

4.3 Mission Operations Concept, Bi-spectral Infrared Detection (BIRD)

The following mission operations concept was developed by Klauss Briess based on the mission that Deutsches Zentium für Luft-und Raumfahrt (DLR) is flying. DLR is flying an experimental payload to determine the sensor capabilities for detecting forest fires. This operations concept was developed after launch based on the process described in Secs. 4.1 and 4.2.

4.3.1 Mission Scope, Objectives, and Science Requirements of BIRD

For hot spot events as forest and vegetation fires, volcanic activity or burning oil wells and coal seams a dedicated space instrumentation does not exist. Sensors now being used for the observation of these events have some drawbacks because they are not designed for the hot spot investigation. For the near future there are missions planned with a new generation of cooled infrared array sensors. The German BIRD (Bi-spectral Infrared Detection) mission will answer a lot of technological and scientific questions related to the operation of a compact bi-spectral infrared push-broom sensor on board a micro satellite and related to the detection and investigation of fires from space. Therefore, the BIRD primary mission objectives are:

- Test of a new generation of infrared array sensors adapted to Earth remote sensing objectives
- Detection and scientific investigation of High Temperature Events such as forest fires, volcanic activities, and coal seam fires
- Test and demonstration of new small satellite technologies, such as new board computers, star sensors, reaction wheels, on board navigation system, and others

The secondary mission objectives are:

- Test and demonstration of on board classification by means of a neural networks circuit
- Demonstration of a low-cost user ground station

BIRD is a science and technology demonstrator mission with no strict timeliness requirements for the payload processed data.

BIRD Mission Description

The functional requirements on the orbit are shown in Table 4.5. They follow from the mission objectives and the performance requirements on the payload system. But the driver for the orbit definition consists in the cost constraints and the resulting piggyback launch strategy. So a range of different orbits should be appropriate for the BIRD mission. For the selection of a suitable piggyback launch opportunity primary and secondary requirements on the orbit are defined. The requirements with the highest priority are:

- Flight over the regions of main interests as listed in Table 4.5
- Altitude between 450 km and 900 km, preferably 470...500 km
- Inclination ≥ 53° (to cover Germany)

A Sun-synchronous orbit is desirable but not compelling. The requirements on the orbit injection precision and the stability of the ascending node have got secondary priority and should not be very strong. The repeat time has no priority for the orbit definition.

The duty time of the payload is 10 minutes in one orbit. The data of one duty cycle can be stored in the 1Gbit mass memory and will be transmitted during the next pass to a German ground station. Simultaneous data take and down-link are possible, too.

About one and a half years before the planned launch date the launch contract was signed and so the orbit parameters are fixed (Table 4.6).

Table 4.5. Top-level Functional Mission Requirements on the Orbit. Flexible requirements increase options and help to keep costs down.

Priority	Parameter	Requirements on Orbit
Primary Orbital Requirements	Mainly spatial coverage of	Germany, boreal forests in Eurasia and Canada, tropical rain forests, savannahs, subtropical vegetation zone, rice fields in South-East Asia, fire endangered forest areas in USA and Europe, volcano chains
	Time coverage	As often as possible
	Orbit altitude	LEO: 450 km...900 km, preferably 450...500 km
	Inclination	$\geq 53°$
	Main duty periods	June till September at the northern hemisphere July till October at the southern hemisphere
Secondary Orbital Requirements	Illumination conditions	Sun-synchronous desired
	Required local time	Between 1 p.m. and 3 p.m. desirably
	Global spatial coverage	Min. 100 km swath, all land areas between and including the northern and southern temperate climatic zones
	Eccentricity	0.000 desirably, but not cogent necessarily
	Accuracy of orbit injection	≤ 40 km
	Stability of ascending node	±30 min solar time

Table 4.6. Orbit Parameters of the BIRD mission. These parameters meet the requirements listed in Table 4.5.

Parameter	Base Line
Circular orbit radius	6949.8 km
Orbit inclination	97.67 degree
Orbit altitude	568 km
Orbit type	Polar, sun-synchronous
Time of equatorial crossing in descending node	10:30 Local Time

Mission Philosophies, Strategies and Tactics

The BIRD small satellite mission shall demonstrate the scientific and technological value and the technical and programmatic feasibility of the combination of ambitious science and new, not yet space-proofed advanced technologies with a small satellite mission concept under low-budget constraints.

Some iteration loops between total mission costs, technical plan, scientific objectives and operational aspects were carried out to define a mission with

- Ambitious science, i.e. new problem solutions and no duplication of data products already available
- Important potential for application of the new problem solutions
- User needs and urgency to get the data
- Implementation of a low-cost and quick end-to-end data flow
- Innovative character of methods and technologies developed for BIRD with a high potential for technology transfer to industry
- Technology experiments and space proof of new technologies for small satellite missions
- An affordable access to space by piggyback launch of the satellite
- Technical and programmatic feasibility of the mission under low-budget constraints
- Clear limits of the payload and mission operations and of the observation capabilities because of a very limited operational time (duty cycle)

Despite the drawback of the very limited mission operations time (duty cycle) the unique combination of the

- New scientific data
- New technologies and methods to gather and process the data
- Singular thematically processed data
- New technological solutions to get the data

with low-cost constraints and a fast development time, the BIRD mission has become a pacemaker within the space application of a new generation of infrared sensors for Earth remote sensing objectives.

Program Constraints

Small satellites have to meet a big challenge: to answer high performance requirements by means of small equipment and especially of small budgets. It becomes feasible by using new commercial off the shelf technologies in spacecraft and payload systems. The higher risk due to the lack of space qualification and heritage are reduced in the BIRD mission by using:

- State-of-the-art technologies
- A mixed strategy in the definition of the quality level of the EEE parts and components
- A dedicated quality assurance plan
- A risk management system

- Extensive redundancy strategies
- Extensive tests, especially on system level
- Large designs margins (over-design)
- Robust design principles
- Robust flight procedures and robust spacecraft safe mode

The constraints for the mission design are described in Table 4.7. The main constraint and the design driver for the mission consists in the launch strategy as auxiliary or piggyback payload. That assures a low-cost approach. The launch strategy has the drawback that the orbit can not be optimized related to the functional requirements and the sensor systems. The orbit follows from the appropriate launch opportunity and the mission has to be designed according to the pre-determined orbit.

Table 4.7. Constraints for the Mission Design. Constraints represent trade-offs among cost and performance requirements.

	Constraints
Lifetime	1 year in orbit
Mission design philosophy	Design to cost
Launch constraints	Launch as piggyback or as auxiliary payload into LEO
Instruments	Wide angle optical subsystem (WAOSS) + infrared sensors
Mission type	Micro-satellite mission with scientific and technological objectives
Co-operation	Co-operation between different DLR institutes and the Technical University of Berlin
Environmental conditions	Space conditions within the Van Allen Belt, Sunspot maximum in the year 2001, Dosis (Si): < 7krad behind 2mm Al (spherical geometry) < 100 krad at a surface of spacecraft O+-fluence: 10^{25} atoms/m^2
Interfaces	Compatibility of the scientific down-link to the DLR ground control station Weilheim and the DLR main data receiving ground station in Neustrelitz
Funding	DLR

End-to-end Information System Characteristics for BIRD

The requirements of the specific scientific objectives on data delivery and complementary data analysis by the end user are summarized in Table 4.8. The data level and time requirements and the planned complementary data characterize the end to end information system. It is documented in the BIRD Utilization Concept.

Table 4.8. Requirements on BIRD Data Delivery and Complementary Data Records. (L0, L1A, L1B refer to levels of data product processing.)

Objective	BIRD Data Delivery Requirements		Complementary Data Record Requirement	
	Required Delivery Level (L0 – L1B) (See Table 4.9 for Definitions of Levels)	Possible Delay Between Record and Delivery	Planned Date Records from Simultaneous Spaceborne Sensors*	Acceptable Time Difference to Complementary Sensor Records
Vegetation fire recognition and quantitative analysis	L 0 – L 1B, depending on application	1 d (optimum), 1-3 month (acceptable)	EOS/ MODIS, NOAA/ AVHRR,	Few minutes!!!
Burned area monitoring	L 1B preferred	< 1 month	NOAA, SPOT, ERS, LANDSAT	< 6-8 h
Volcanic crater and lava flow monitoring	L 0 – L 1B, depending on application	< 1 d (L 0 / L 1A for alert & quick look), 2-3 weeks for L 1B	EOS/ ASTER & MODIS LANDSAT/ ETM	< 24 h
Coal seam burning recognition	L 1B preferred	< 1 d (L0 quick look), 2-3 weeks for L 1B	ERS/ SAR, LANDSAT/ ETM	< 24 h
Urban heat island monitoring	L 1B preferred	< 1 d (L 0 quick look), 2-3 weeks for L 1B	EOS-ASTER, LANDSAT / ETM	< 2 h
Monitoring of inland water thermal pollution	L 1B preferred	< 1 d (L 0 quick look), 2-3 weeks for L 1B	EOS-ASTER, LANDSAT / ETM	< 2 h
Optical monitoring of volcanic and fire plumes (aerosols)	L 0 – L 1B, depending on application	As soon as possible (a.s.a.p.)	EOS/ ASTER & MODIS LANDSAT/ ETM	< few hours

* ASTER = Advanced Spaceborne Thermal Emission and Reflection Radiometer; AVHRR = Advanced Very High Resolution Radiometer; EOS = Earth Observing System; ERS = European Remote Sensing Satellite; ETM = Enhanced Thematic Mapper; LANDSAT = Land Remote Sensing Satellite; MODIS = Moderate Resolution Imaging Spectroradiometer; NOAA = National Oceanic and Atmospheric Administration; SAR = Systems Analysis Recording; SPOT = Systéme pour l'Observation de la Terre.

Ground Systems Characteristics for BIRD

The general mission and communication architecture are depicted in Fig. 4.1. The space segment is controlled completely by the German DLR ground segment. Only for the initial acquisition phase an external ground station support has to be coordinated. The telemetry link has to be compatible with CCSDS standard. The primary ground station for the BIRD data reception is located in Neustrelitz (DLR).

The ground station receives both real time data and recorded data via direct transmission. The primary ground station from mission operations is located in Weilheim and is controlled by the German Space Operation Center of DLR in Oberpfaffenhofen. The BIRD mission is open for implementation of other secondary data reception ground stations anywhere in the world. They can apply for BIRD data reception and then be taken into account in connection with the mission timeline and scheduling. In this case the received BIRD data should be sent to the BIRD main ground station in Neustrelitz for archiving purposes. Today a ground station in Argentina is included within the BIRD ground segment. Besides the DLR main ground stations in Weilheim and Neustrelitz (Germany) a mini ground station is implemented in Berlin-Adlershof for experimental purposes. The science team has the task to conduct field campaigns for validation and for support of interpretation of the remote sensing data using airborne and ground truth measurements to be performed simultaneously with BIRD overpasses.

BIRD Spacecraft Characteristics (Bus and Payload)

According to the need to decrease the spacecraft costs and keep high performance characteristics of the spacecraft bus, the BIRD satellite demonstrates new developed technologies at moderate costs in space. These are in particular:

- Compact micro-satellite structure with a high stability, fitting to different launch adapters
- High peak power supply for the duty cycle (> 200 W) at average power supply > 60 W
- Use of state-of-the-art technologies for passive thermal control (radiators, heat pipes, heaters, sensors) in the low-cost price range
- Use of S-band communication system for high and low bit rate command and data transmission
- State-of-the-art spacecraft bus computer with integrated latch-up protection and error correction systems
- High performance attitude control system using new developed technologies
- Integrated onboard navigation system

BIRD's Payload System

The BIRD payload system is characterized by

- A bi-spectral infrared sensor system (Medium Wave Infrared-MIR, Thermal Infrared-TIR) in combination with a 3-phase-angle vegetation sensor (Red, Near Infrared - NIR)
- Tremendous high dynamic range (20 bits) of the infrared sensor system

4.3 Developing a Mission Operations Concept

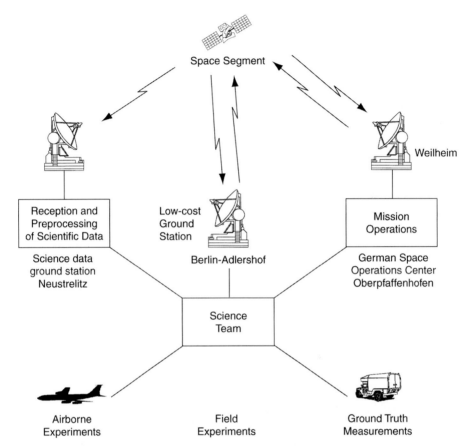

Fig. 4.4. BIRD Ground System Characteristics. The mission and communication architecture ties together the space and ground elements of the mission.

- Sub-pixel resolution by means of a particular system design and processing conception
- Test of thematic onboard data processing by means of a neural network chip

End-user's Definition of Needed Data

The BIRD data will be processed and archived in the ground station Neustrelitz. They generate data of level 0 and level 1 according to Table 4.9.

The primary ground station will send BIRD data of level 1a to scientific co-investigators according to their request. Data processing beyond level 1 and distribution of these products is the responsibility of the users and limited to non-

Table 4.9. Level of BIRD Data Products. Users determine what level of data products they need.

Level 0	Unprocessed instrument/payload data in combination with spacecraft data, communication frames/headers removed
Level 1A	Time-organized single sensor raw data with an appendix of • Instrument housekeeping data • Radiometric and geometric calibration coefficients • Geo-referencing parameters (ephemeris data) • Other ancillary information
Level 1B	Radiometric and geometric processed level 1A data to sensor units (radiometric and geometric calibrated data)
Level 2	Interpreted geophysical parameters (hot spot temperatures, hot spot extension, vegetation indices, cloud parameters)

commercial purposes. All raw data of BIRD (level 0) will be archived in the DLR ground station Neustrelitz. In most of the cases, data products of level 1 will be also archived in Neustrelitz or in other places of level-1 users. BIRD level 0 or level 1 data products will be distributed in a timely manner according to the requirements of the co-investigators.

Operational Scenarios

There are different experiment scenarios corresponding to the different scientific and technological objectives of the BIRD mission. The mission operations have to consider the following main experiment scenarios:

- Fire experiment scenario
- Hot spot sounding scenario
- Volcano sounding scenario (monitoring mode)
- Onboard classification scenario
- Technology scenario

As an example, the classification scenario is described. The onboard classification scenario is characterized by a synchronized data-take of all spectral channels over a defined target region or test site during daylight. The data will be stored within the mass memory. The onboard classification will be carried out in an off-line mode, i.e., after finishing the measurement sequence and storage of the data within the mass memory the data will be read out, processed, classified and stored again for the down-link in Germany. The verification and evaluation of the classification should be supported by means of airborne measurements, ground truth measurements or/and field trips. The exemplary classification should be carried out for the following experiment objectives:

- Classification of hot spots and suppression of false alarms
- Classification of vegetation areas and estimation of burned areas

- Identification of clouds
- Identification of objects that have to be rejected within the onboard classificator
- Onboard thematic data reduction and generation of thematic maps

Timelines

Data acquisition of BIRD will be directed by the mission objectives. Because of the limited mission operation time of 10 minutes in one orbit and 5 orbits per day the data acquisition will be selected and scheduled very carefully by the experiment manager. The co-investigator and the science team members of BIRD will give their requirements to the experiment manager. The experiment manager collects the midterm and short-term requirements for measurement sequences by a form sheet. He develops a first proposal for a mid term and short-term observation plan and discusses it in the science team in Berlin-Adlershof. After this evaluation the schedule will be given to the German Space Operation Center for a detailed preparation of a mission operation plan. This detailed observation plan takes into account different constraints like power budget and mass memory budget and makes a detailed schedule for the mission operations by means of the DLR's own timeline tool. The results are sent back to the science coordination in Berlin-Adlershof for confirmation. After confirmation the German Space Operation Center prepares the mission operations, time-line and scheduling.

Data Flow Diagrams

The data flow diagrams of the control center and systematic process are shown in Figs. 4.5 and 4.6. The data flow diagram of processing, calibration, high level product generation and archiving of the BIRD data is depicted in Fig. 4.5. The systematic processing of the BIRD data (processes) are drawn in rectangular or 6-angular boxes; the data (inputs and outputs) and its storage and archiving are drawn as drums.

Organization and Team Responsibilities

The BIRD phase E activities are organized in a small team. There are five main work packages: Management, Science and Mission, Mission Operations, Spacecraft and Technology Demonstration, Data Reception and Processing.

The Mission Operations are conducted by the Mission Operations Director (MOD) and are carried out in the German Space Operations Center of DLR: But the complete mission activities are led by the Leader of the BIRD mission supported by the Deputy Project Team Leader, the Science Coordinator, the Flight System Engineer, the Ground System Engineer and the Project Manager.

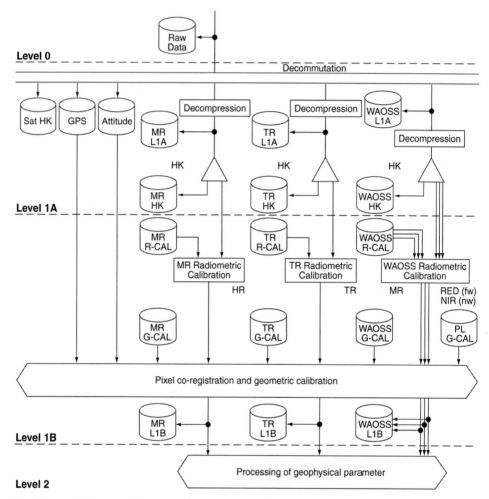

Fig. 4.5. **Data Flow Diagram of the Systematic Processing of the BIRD Science Data.** We illustrate here the processing for data products through Level 1B. (HK = housekeeping data; L1A = defined in Table 4.9; L2A = defined in Table 4.9; GPS = global positioning system; MR = medium wave infrared radiometer; TR = thermal infrared radiometer; R-Cal = radiometric calibration; WAOSS = wide angle optical subsystem.)

Cost and Complexity Drivers for a Given Set of Inputs

The BIRD phase E budget is fixed from the beginning. The cost drivers are: The science team, the science activities like field trips, airborne experiments and ground truth measurements, the data processing and archiving efforts, the mission operation preparation and the mission operation activities. Other temporary cost drivers are the additional ground station support during the initial acquisition phase and the expenses for the sub-system engineers supporting the operations team during the initial acquisition and the commissioning phase of BIRD.

Requirements and Derived Requirements

As an example, the derived requirements for the onboard classification scenario, which are derived from the requirements of the onboard classification experiment, are listed in Table 4.10.

Table 4.10. Requirements of the Onboard Classification Scenario. This table includes both data and operational requirements.

	Requirements
Experiment operators	BIRD payload operators, remote sensing experts, fire-ecological experts
Operational main regions	Region of Berlin, forests, agricultural areas, industrial regions and coastal zones in Europe
Main duty periods	Vegetation period
Instruments and devices	Maps, Geographic Information System (GIS)
Facilities	• Test site or well known reference area (local) • Airplane with a spectrally modified Wide Angle Airborne Camera in combination with the infrared sensor system is desirable but not necessary
Spacecraft/payload operational interfaces	• Ground station for receiving of down-linked scientific data (including instrument housekeeping data) • Down link of raw data and high level data products (classification results) • Link to the mission control center for preparation of payload operations

Technology Development Plan

The technology development plan includes different hardware and software tools for test, training, verification and simulation of mission operations. Solution consists in using the BIRD engineering model, the attitude hardware-in-the loop-test facility or the BIRD flight model for tests, training and verification of the mission operation concept. Figure 4.6 shows the block diagram of the BIRD ground telemetry system as an example for technology development related to mission operations. To start parallel actions in the detailed design and manufacturing phase of the project a simulator for the Data Acquisition System (Direct Ingest System) was developed and used instead of the receiving station and the RF link with the spacecraft. So the telemetry processor could be developed and tested in parallel with the spacecraft integration. In a next step the engineering model and later the flight model of the spacecraft are linked via the Data Acquisition System of the ground station with the telemetry processor. It is already developed and tested in the framework of the technology development plan mentioned above.

All of this information about the BIRD mission is captured in just a few pages. As stated earlier, the ideal time to generate the operations concept is during the conceptual or design phase. It is always a useful exercise no matter where in the lifecycle the project is. It is also a good way for a person just coming on to a mission to either read the operations concept for the mission, or if one has not been generated, to generate an operations concept for the mission by following the process. Each of the sections can be expanded as more information becomes available and trades can be identified for many of the areas that are described above.

4.3 Developing a Mission Operations Concept

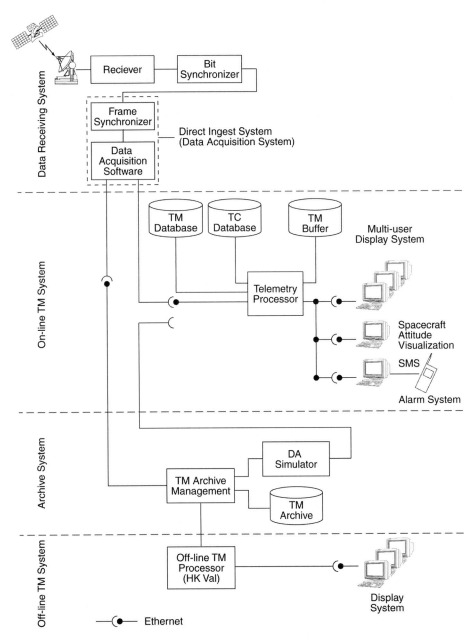

Fig. 4.6.   Simplified Block Scheme of the Ground Telemetry System of the BIRD Mission in Connection With the Experimental Ground Station. Here we depict data flow once the ground system has received the telemetry.

References

Future NASA Miniature Spacecraft Technology Workshop. Feb. 8–10, 1995. Pasadena, CA.

Chapter 5
Operations Cost Estimations

John Carraway, *Jet Propulsion Laboratory, California Institute of Technology (Retired)*

Mark K. Jacobs, *Science Applications International Corporation*

5.1 Assessing Operations Complexity
5.2 Operations Cost Modeling

One of the most difficult jobs of the MOM is to estimate accurately the cost of post launch operations. This is usually done by past experience, intuition or just plain guessing. This chapter contains two different ways to estimate post launch operations costs. Section 5.1 discusses how the complexity of operations drives costs. We list 95 complexity factors shown to drive operations costs. We must understand how to use these complexity factors during the design phase of the mission, spacecraft, and payload. The MOM that can discuss these factors intelligently with the spacecraft manager or project manager, early in the lifecycle, can influence significantly the design of the mission and spacecraft. Otherwise, he or she may end up designing an operations system that supports a mission and spacecraft design that did not consider operations. This section also shows how spacecraft design and orbit can influence costs. Section 5.1 constituted all of Chap. 5 in the first edition of CESMO. Section 5.2 discusses how the concepts in Sec. 5.1 have been turned into a cost model that NASA has developed and is now using. These discussions both give insight into the complexities of estimating the costs of operations early in the project lifecycle. The MOM may use elements from either or both of these sections to estimate post launch costs. Even if the MOM does not use these cost models, he or she must understand the factors discussed and how they influence costs. This will help to ensure that the operations that the MOM designs are cost effective.

5.1 Assessing Operations Complexity

Understanding operations costs based on measurements of operations complexity is of interest to a mission operations manager (MOM) who wants to see how design requirements for the spacecraft, payload, mission, and ground system drive operations costs. If lowering costs is also a priority, we list 95 complexity parameters to negotiate with operations users (sources of operations requirements) to do so. Finally, we present a model for converting operations complexity into predicted operations costs.

If we understand operations complexity we can see why operations costs are so much higher for some missions than for others. During the project-requirements phase, understanding operations complexity helps us define and negotiate requirements that are compatible with low costs. Concurrent design can be effective if project engineers can show how their requirements affect operations complexity and costs. During design, assessing operations complexity gives us rules for trade studies to reduce operations costs with less expensive designs for the spacecraft, instruments, or mission. Even after launch, operations complexity metrics suggest changes in designs for the mission, flight software, and ground system.

Quantifying metrics of operations complexity establishes numbers for comparing costs between missions, helps identify areas for trades to reduce operations costs, and gives us a numerical model that predicts the operations costs of a mission.

We organize the 95 metrics for operational complexity into four categories:

1. Mission design and planning
2. Flight system
3. Risk avoidance
4. Ground system

These four categories correspond to the four main sources of requirements on operations. We want to relate operations costs to requirements placed on operations by the mission plan, the spacecraft and instrument design, the project's policies on operational risk, and the ground system. We've placed the metrics into high, medium, and low ranges so we can do early estimates while the project design is still flexible enough to lower operational costs.

We present a cost model at the end of this section. It uses these complexity metrics to compute four complexity factors corresponding to the four categories above. These complexity factors allow us to predict the workforces for 14 Jet Propulsion Laboratory missions. For each mission, we show the metrics, computed values for the complexity factors, and predictions for the workforces. Model predictions are accurate to within 25% of actual values for 13 of the 14 cases studied. Figure 5.1 is a block diagram of the model.

Sections 5.1.1 through 5.1.4 present tables of metrics for operations complexity, interpretations of the tables, and suggested design rules for low-cost operations

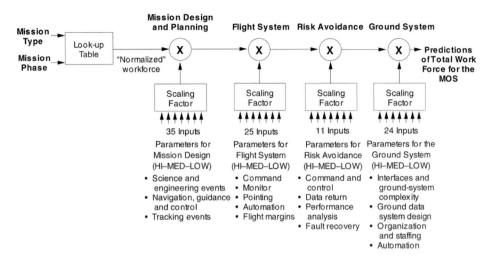

Fig. 5.1.   Block Diagram of a Cost Model for Mission Operations. Metrics predict operations complexity for four categories. By comparing predicted complexity with similar missions, we estimate the workforce for our mission.

based on them. Section 5.1.5 presents metric data from various missions and prediction results from the cost model. We can use the model to estimate the operational cost of a new mission if the new mission corresponds to one of the four mission types listed in Table 5.5. If it doesn't, we can still use the model to compute relative costs and to determine how changing one or more requirements changes our estimated cost.

5.1.1   Complexity Metrics for Mission Design and Planning

How we design and plan a mission has a strong influence on operations costs. Mission design imposes requirements on operations in such areas as levels of mission activity, timeline margins, and planning horizons. We can lower operations costs by choosing orbits or mission events that match an operations work day, or better yet, a single shift. The metrics in this section suggest ideas like designing timelines with margins, mission events with re-try opportunities, and planning strategies for efficient, flexible operations.

Mission activities that are important to operations costs include engineering and science events for the flight system, guidance and control events, and tracking events. The complexity metrics in this section characterize activity levels, criticality, and planning requirements for these events. Table 5.1 presents 35 measures that establish complexity metrics for mission design and planning.

Table 5.1. Complexity of Mission Design and Planning. Generous margins in the timeline, multiple re-try opportunities, and planning flexibility reduce complexity.

	Low	Medium	High
Science and Engineering Events			
Frequency • Number per day?	Fewer than 10	10 to 20	More than 20
Criticality • Science re-try opportunities • Engineering re-try opportunities	More than 2 More than 2	1 or 2 1 or 2	None None
Complexity • Timing accuracy • Instrument pointing events • Spacecraft pointing events	Hours or more None Fixed	Minutes Platform Articulated antenna, panels	Seconds or less Spacecraft maneuver Spacecraft maneuver
Data return • Number of routine downlink data modes • Real time vs. playback	1 or 2 Playback: FIFO*	3 to 6 Playback: Non-FIFO	7 or more Real time
Science and engineering planning • Event repetitiveness • Number of consumables and margin constraints routinely planned • Timeline duty cycle margin • Plan execution time • Plan development time • Response time to late change request	 Highly repetitive None More than 50% Less than 1 day Less than execution time More than 50% of development time	 Some repeat, some unique 1 or 2 5 – 50% 1 day to 1 week 1 to 2 times execution time 10% to 50% of development time	 Many unique 3 or more Less than 5% More than 1 week 3 times or more than execution time Less than 10% of development time
Navigation, Guidance, and Control			
Maneuver frequency • Slews per day • Propulsive maneuvers per quarter	None 1 or fewer	1 to 5 2	More than 5 3 or more
Maneuver criticality • Re-try opportunities	More than 3	1 or 2	None
Maneuver complexity • Timing accuracy • Maneuver accuracy	Days 1 sigma or less	Hours 2 sigma	Minutes or less 3 sigma or more

Table 5.1. Complexity of Mission Design and Planning. (Continued) Generous margins in the timeline, multiple re-try opportunities, and planning flexibility reduce complexity.

	Low	Medium	High
Navigation, Guidance, and Control (Continued)			
Navigation data return • Number of navigation data types • Navigation data per day	1 or 2 Less than 1 hour	3 or 4 1 to 4 hours	5 or more More than 4 hours
Maneuver planning • Maneuver design repetitiveness • Number of consumables and margin constraints routinely planned • Mission delta V margin	Highly repetitive None More than 50%	Some repeat, some unique 1 or 2 5% – 50%	Many unique 3 or more Less than 5%
Ephemerides • Number of objects needing ephemerides	Fewer than 3	3 to 5	More than 5
Tracking Events			
Frequency • Number of passes per day • Hours of link coverage per week	Fewer than 2 Fewer than 8 hours	2 or 3 8 to 56 hours	More than 3 More than 56 hours
Criticality • Re-try opportunities	Playback data with replay option	Routine real-time data	Critical real-time data
Complexity • Station configuration changes per pass • Peak data return rate • Simultaneous, multi-station coordination	None Less than 1 kbps None	1 or 2 1 kbps–100 kbps Fewer than 5% of tracks	More than 2 More than 100 kbps More than 5% of tracks
Tracking facility planning and scheduling • Coverage repetitiveness • Track duration time margin • Schedule iterations • Tolerance of late losses in scheduled coverage	 Repeating weekly pattern More than 50% 1 or 2 More than 20% of tracks	 Some repeat, some unique 5% – 50% 3 or 4 5% to 20% of tracks	 Every day unique Less than 5% More than 4 Fewer than 5% of tracks

* FIFO = first in, first out. Data stored first is played back first.

Let's define some terms and look at examples to help us understand and use Table 5.1.

- **Events.** We define a *science or engineering event* as a complete set of self-contained actions involving the spacecraft, instruments, and (sometimes) ground system. It involves configuring the spacecraft or instruments and collecting data. Science observations or engineering calibration are typical examples. Maneuvers and slews are examples of navigation, guidance, and control events; specified, scheduled DSN or TDRSS passes are examples of tracking events.
- **Frequency.** Activity level or number of events per unit time.
- **Criticality.** Fewer opportunities to retry means higher criticality.
- **Complexity.** Specification accuracy and requirements for pointing, maneuvering, or tracking.
- **Data Return.** How complex is the data-return system? How many down link data modes; real time or playback?
- **Planning.** Influences on event-planning complexity include the event's repetitiveness, constraints on planning, timeline margin, plan duration, plan-development time, and response time for replanning.

The design rules in Table 5.1 suggest that, to operate at low cost, our mission design must

1. Consider the effect of activity levels on operations staffing
2. Allow multiple chances to retry events
3. Allow margins in the accuracy of event timing
4. Allow simple pointing for the spacecraft and engineering instruments
5. Allow simple data capture and return
6. Use repetitive events as much as possible
7. Minimize the number of design rules, constraints, and parameters that we must plan for and manage as resources
8. Provide margins in the planning timeline
9. Plan over short durations
10. Use comparable amounts of time to develop and carry out the plan
11. Allow late requests for replanning, not by changing the current plan at the last minute, but by incorporating the requests into the next plan

5.1.2 Complexity Metrics for the Flight Systems

Designs for spacecraft and instruments significantly affect operations costs. Designs involving tight coupling and high interaction between spacecraft subsystems, or between instruments, are more complex and more expensive to operate. Designs also cost more to operate if they have in-flight consumables with low margins that operators must carefully plan for, protect, and manage. Designs with numerous flight rules and constraints complicate operations; reduce operational options and robustness; and require expensive, close to zero-defect, techniques for simulating and validating commands.

We don't expect operational margins to go forever unused. Rather, we want operators to allocate them during flight to respond to in-flight surprises and faults, to complete bonus science or mission goals, or to allow more efficient or lower-cost operations. Flight margins are to operations teams as design margins are to design teams. Lack of margins results in limited options, inflexibility, and high costs.

Understanding operational complexity for a spacecraft leads to design rules and recommendations for better operations. These rules make spacecraft and instrument designers more sensitive to operability issues. The problem is that most of the spacecraft attributes recommended to achieve operability cost money. Spacecraft designers can estimate with fair precision how operability adds to mass, development dollars, development schedule, and test time. Before we decide to spend these resources, we must predict operations cost savings with the same precision. Until we get credible models of how spacecraft attributes affect operations cost, most missions will tend to reject spending more money now for the promise of poorly understood savings in the long term.

Table 5.2 presents 25 metrics for the flight system.

Table 5.2. Complexity of the Flight-System Design. The complexity metrics in this table characterize our ability to command, monitor, point, automate, and develop flight margins for the spacecraft and instruments.

	Low	Medium	High
Command			
Commandable states	Fewer than 50	50 – 200	More than 200
Flight rules and constraints	Fewer than 10	10 – 20	More than 20
History-dependent commandable states	None	1 – 2	More than 2
Onboard tables routinely updated	Fewer than 10	10 – 20	More than 20
Number of instruments	Fewer than 3	3 – 5	More than 5
Monitor			
Telemetry channels	Fewer than 100	100 – 1000	More than 1000
Ambiguous states	None	1 – 2	More than 2
Pointing			
Attitude control	Gravity gradient	Spinner	3-axis
Pointing accuracy	More than 0.1°	0.1° – 0.01°	Less than 0.01°
Independent fields-of-view	Fewer than 3	3 or 4	More than 4
Articulating devices	None	1 – 3	More than 3
Constraints on hazard pointing of instruments	None	1 – 2	More than 2

Table 5.2. Complexity of the Flight-System Design. (Continued) The complexity metrics in this table characterize our ability to command, monitor, point, automate, and develop flight margins for the spacecraft and instruments.

	Low	Medium	High
Automation			
Unattended safing of flight system	More than 1 week	2 – 7 days	Less than 2 days
Flight-system command states requiring routine ground command for safing	None	1 – 10	More than 10
Flight Margins			
Onboard consumables managed by operations	1 or 2	3 – 5	More than 5
Onboard data storage	More than 3 times the DL* period	2 – 3 times the DL period	Less than 2 times the DL period
Onboard command-file memory	More than 3 times the plan duration	2 – 3 times the plan duration	Less than 2 times the plan duration
Onboard flight-software memory	More than 30% margin at launch	20%–30% margin at launch	Less than 20% margin at launch
Speed of the flight computer	More than 30% margin at launch	20%–30% margin at launch	Less than 20% margin at launch
Time to uplink full plan	A fraction of a pass	Approx 1 full pass	More than 1 pass
Time to downlink planned data storage	A fraction of a pass	Approx 1 full pass	More than 1 pass
Real-time downlink bandwidth (data rate)	More than twice the data-capture rate	1 – 2 times the data-capture rate	Less than the data capture rate
Telecom link margin	More than 3 dB	1.5 – 3 dB	Less than 1.5 dB
Power margin	Power available more than peak load	Power available approx equal to peak load	Power available less than peak load
Thermal margin	No thermal constraints on pointing or power	1 or 2 thermal constraints on pointing or power	More than 2 thermal constraints on pointing or power

* Downlink

Let's define some terms and show examples to help us understand and use Table 5.2.

- **Commandable states.** The number of flight-system states that are commandable. The number of commandable state specifications we'd need to define a unique operational configuration of the spacecraft and instruments. For example, one commandable filter wheel with four positions counts as a single commandable state.
- **Flight rules and constraints.** The number of operational flight rules and constraints we must obey in commanding the spacecraft and instruments. (See Sec. 3.1)
- **History-dependent commandable states.** The number of commandable states whose command response depends on initial conditions. A

two-position switch controlled by a toggle command is one example. A filter wheel commanded by "step forward 3 positions" rather than "go to filter # 1" is another example.

- **Telemetry channels.** The number of onboard measurements routinely downlinked. This number includes status telemetry for the flight system's commandable states, as well as temperature, voltage, current, and other sensor-measurement values throughout the flight system.
- **Ambiguous states.** The number of commandable states for the spacecraft and instruments that telemetry channel values don't specify uniquely.
- **Independent fields-of-view.** The number of flight-system elements having independent pointing requirements or constraints. Examples are a solar panel, a high-gain antenna, instrument sensors, and radiator fields-of-view. Several instruments sharing a common boresight would count for just one field-of-view.
- **Articulating devices.** Things we must move routinely during operations. Solar panels, articulated antennas, scan platforms, filter wheels, and instrument covers are examples.
- **Constraints on hazard pointing of instruments.** Instruments that have flight rules disallowing us from pointing them in certain directions (e.g., at the Sun).
- **Unattended safing.** The amount of time the flight system can safe itself without ground intervention and not suffer irreversible damage.
- **Routine safing commands.** The number of states the ground staff must routinely command to prevent irreversible damage.
- **Onboard consumables.** *Flight consumables* include propellent, number of thruster firings, start-stop cycles of the tape recorder, tape across the head, and battery charge-discharge cycles. *Replenishable consumables* include battery charge, wheel momentum, and onboard memory.
- **Onboard data storage.** Low-cost missions have enough margin in onboard data storage to make up for occasional missed downlinks.
- **Onboard command-file memory.** The margin for command-file storage is based on the capacity required to store a typical full uplink plan.
- **Onboard software memory.** We define software margin in terms of unallocated memory at launch.
- **Speed of the flight computer.** The onboard processing margin for the flight computer(s) calculated in terms of millions of instructions per second (MIPS) margin at launch.

- **Time to uplink full plan.** The uplink bandwidth margin calculated in terms of how much time it takes to uplink a full plan.
- **Time to downlink planned data storage.** The downlink bandwidth margin calculated in terms of how much of a pass it takes to downlink a full quantity of planned data storage.
- **Realtime downlink bandwidth.** The downlink bandwidth (downlink bit rate) margin calculated in terms of the flight system's nominal data-capture rate.
- **Telecom margin.** The downlink margin calculated in terms of the downlink's signal-to-noise ratio. (See Sec. 11.3.)
- **Power margin.** Power margin calculated as a function of peak load and total power available—both generated and stored.
- **Thermal margin.** Thermal margin established in terms of the number of power and pointing constraints.

The design rules in Table 5.2 suggest that, to operate at low cost, our spacecraft and instrument design must:

1. Minimize the flight system's controllables and observables (command states and telemetry points) that operators must routinely manage
2. Minimize the flight system's operational rules and constraints
3. Minimize history-dependent command states
4. Minimize ambiguous states
5. Simplify devices for pointing, controlling attitude, and articulation
6. Provide onboard automation that safes the spacecraft and instruments for long periods without ground interaction
7. Provide operational flight margins for consumables, onboard data storage, onboard sequencing memory, flight-software memory, flight-computer speed, time to uplink full command load, time to downlink full memory, telecom link, power, and thermal

5.1.3 Complexity Metrics for Operational Risk Avoidance

Policies for avoiding operational risks can have a considerable impact on operations costs. If we decide not to tolerate command errors, we may need a zero-defect uplink with elaborate procedures for simulation, constraint checking, validation, and approval. If we can't tolerate data losses, we'll need a zero-defect downlink with elaborate procedures for scheduling and for acquiring, detecting, recalling, validating, and archiving data.

Policies on using onboard automation and fault protection can drive operations costs in several ways. Planners sometimes so mistrust certain event-triggered algorithms on the spacecraft that they instruct operators to fly the spacecraft so these algorithms won't get triggered. As a result, modeling, analysis, simulation, and human oversight intensify because mission policy considers triggering of onboard automation an operational failure.

Risk policy requirements on operations to predict and prevent onboard failures, rather than to simply detect and respond to failures once they have occurred, results in more modeling, trending, in-flight testing, and performance analysis. These tasks drive up operations costs significantly.

Table 5.3 presents 11 complexity metrics for risk avoidance.

Table 5.3. **Project Operational Risk Avoidance.** These metrics quantify our tolerance for risk, error, and failure.

	Low	Medium	High
Command and Control			
Command errors tolerated per week	More than 2	1 or 2	None
Onboard adaptive algorithms— tolerated entries per week	More than 2	1 or 2	None
Activity simulation	None	Simulation by software model	Simulation by flight-system testbed
Command reviews and approvals	Fewer than 2 per command file	2 or 3 per command file	More than 3 per command file
Data Return			
Amount of lost science data tolerated	More than 5%	0.5% – 5%	Less than 0.5%
Amount of lost engineering data tolerated	More than 5%	0.5% – 5%	Less than 0.5%
Performance Analysis			
Routine performance validation	Fewer than 50 parameters	50 – 200 parameters	More than 200 parameters
Routine trend analysis and prediction	None	For up to 5 performance parameters	For more than 5 performance parameters
Routine performance modeling	Simple algorithms	Configuration-controlled software models	Spacecraft-system testbed
Model calibration	Fewer than 2 times per year	2 – 4 times per year	More than 4 times per year
Fault Recovery			
Tolerated timeliness in ground response to an anomaly	More than 1 day	8 – 24 hours	Less than 8 hours

Let's look at some terms and examples to help us understand and use Table 5.3.

- **Command errors.** Includes commands that inadvertently violate a flight rule or constraint, cause unintended consequences, and wouldn't have been approved under more rigorous constraint checking and review.

- **Number of onboard adaptive algorithms tolerated per week.** This is an attempt to quantify operational use of onboard automation. It distinguishes between missions that use automation and missions that have onboard automation but avoid risk by not using it, thus creating an uplink process meant to avoid triggering these algorithms.
- **Activity simulation.** A software-simulation model emulates the flight system's behavior and allows activities to be run on the ground (often in faster than real time) to validate expected results before being uplinked. An alternative approach is to validate activities using a testbed of flight-like hardware.
- **Command reviews and approvals.** The number of authorities or teams that must review and approve a command file before it may be uplinked. For this metric, review by several members on the same team counts as just one review.
- **Amount of lost data tolerated.** The percentage of data we expect to acquire and downlink but lose because we don't have zero-defect commanding and downlinking. The loss may be permanent or temporary (until a recovery activity and downlink is scheduled and carried out).
- **Performance validation.** The number of onboard parameters that operators routinely monitor and report on including onboard sensor measurements and configuration and status measurements.
- **Trend analysis and prediction.** The number of onboard parameters for which we routinely predict and analyze future states.
- **Performance models.** The number of onboard parameters whose performance we routinely analyze by running formal ground models or hardware-simulation models.
- **Model calibration.** The number of special in-flight tests we do to provide a performance model. A maneuver activity to calibrate an antenna is an example.
- **Anomaly response time.** The time in which operations must respond to an onboard anomaly measured from the time of the anomaly to the time of an uplinked response. This time includes two-light-time delays as well as detection delays due to scheduled no-track durations.

The design rules in Table 5.3 suggest that, to operate at low cost, we must

1. Tolerate faults in the uplink
2. Trust onboard automation and allow its use
3. Permit simple simulating and validating of activities and commands
4. Minimize review and approval of commands

5. Tolerate faults in the downlink
6. Reduce the number of onboard parameters that operators must routinely analyze for performance
7. Permit operators to respond to observed failures, rather than requiring them to predict and prevent failures
8. Require minimum modeling and engineering calibration to ensure performance
9. Allow reasonable ground response times to observed anomalies during flight

5.1.4 Complexity Metrics for the Ground System

A complex ground system certainly influences operations costs. Operations tools and displays are part of this complexity, but so are the geographical, program, and institutional environments. Having to schedule and compete for limited ground resources may save development or institutional money, but it increases operations costs. In the same way, inherited ground systems can save development dollars, but they drive up costs for operations, which must adapt to new users' requirements. Adapting to new flight software or supporting new flight-systems can also drive up operations costs.

Table 5.4 presents 24 complexity metrics for ground systems.

Table 5.4. Complexity of Ground Systems (GS). We group metrics by interfaces and shared resources, data-system design, organization, and how much major operational tasks are automated.

	Low	Medium	High
Interfaces and GS Complexity			
Geographical distribution of operations and ground system	1 site	2 or 3 sites	More than 3 sites
Institutions needed to operate the GS	1 – 3	4 – 6	More than 6
Science teams	1 or 2	3 – 6	More than 6
Shared GS components	None	1 or 2	More than 2
Users sharing instruments	1 or 2	3 – 5	More than 5
Shared operations teams	None	1 or 2	More than 2
Shared interval for resource-scheduling	Hours	Days	Weeks or more
Command and control data	High-order command language	Combination of blocks and individual commands	Individual command level

Table 5.4. Complexity of Ground Systems (GS). (Continued) We group metrics by interfaces and shared resources, data-system design, organization, and how much major operational tasks are automated.

	Low	Medium	High
Design of the Ground System			
GS designed to what requirements	Designed for single user project	Designed for many users, tailored for single user	Designed for many user projects
GS built, maintained, upgraded by	Project - user	Combination of project and other	Another institution
Development of flight-system software after launch	None planned	Up to 10%	More than 10%
Organization and Staffing			
Staffing schedule for most operations positions	Prime shift only	2 shifts per day	3 shifts per day
Number of management levels	1 level	2 levels	More than 2 levels
Tasking strategy	Project specialists, multi-tasked	Combination of multi-task and multiproject	Task specialists, multi-project
Automation			
Number of separate steps requiring project operator action(s) in:			
Alarm monitoring from decommutation through notification	1 or 2	3 – 6	More than 6
Level 0 data capture from receipt from tracking facility to project data base	1 or 2	3 – 6	More than 6
Real-time command from entry to transmission	1 or 2	3 – 6	More than 6
Activity generation from entered requests through command file generation	1 or 2	3 – 6	More than 6
Scheduling of tracking coverage	1 or 2	3 – 6	More than 6
Number of transfers of non-electronic data in above processes defined			
Alarm monitoring	1 or 2	3 – 6	More than 6
Data capture	1 or 2	3 – 6	More than 6
Real-time command	1 or 2	3 – 6	More than 6
Activity generation	1 or 2	3 – 6	More than 6
Scheduling of tracking coverage	1 or 2	3 – 6	More than 6

The metrics in Table 5.4 are intended to lower operations costs for a single project; they may or may not do so across a large set of projects. For instance, shared operations may compensate for an individual project's higher complexity costs because the project has to schedule and compete for shared resources, accept unneeded capabilities, and pay its share of overhead for institutional management.

Here are some terms and examples to help us understand and use Table 5.4:

- **Ground system (GS).** The hardware, software, people, and procedures needed on the ground to operate a mission.
- **Institutions needed to operate the ground system.** The number of geographically separate operations sites (5 miles or more apart) that must cooperate to carry out routine operations. This metric doesn't account for travel and per diem costs that may be required to centralize operations.
- **Science teams.** The number of science teams that participate in routine operations.
- **Shared components in the ground system.** The number of major elements in the ground system shared among multiple users and controlled by organizations other than the project.
- **Users sharing instruments.** The number of science teams that don't control dedicated instruments, but rather share instruments controlled by someone else. In our model, this metric is the number of science teams, not scientists.
- **Shared operations teams.** The number of operations teams shared among multiple users and controlled by organizations other than the project.
- **Shared resource scheduling.** The lead time required for scheduling shared resources.
- **Command and control language.** The level of command and control language for the flight system. High-level language contains reusable block or macro command expansions that carry out complex functions. Low-order language requires us to specify each executable command.
- **GS design heritage.** This metric categorizes ground systems by purpose: built for the user project, built for various user projects but tailored to each user, or built for various user projects with no tailoring permitted.
- **GS maintenance.** Who maintains the ground system? Besides monitoring and repairing the ground system, maintenance includes deciding what and when to upgrade, scheduling rebuilds, and delivering new versions or configurations.
- **Postponed development on the flight system.** Has all the planned flight-system software been developed, integrated, and tested before launch? Or must the ground system develop or change a lot of flight software during operations?
- **Staffing schedule.** The number of 8-hour shifts staffed per day.

- **Management levels.** The number of hierarchical levels in the operations organization chart. An example of a three-level hierarchy would be operators who report to team chiefs, who report to office managers, who report to the mission operations manager.
- **Tasking strategy.** Do operators work several tasks for this project only? Or are they task specialists who support multiple projects?
- **Automation—operator actions.** How many actions must an operator routinely take for an operations task? For example, an alarm-monitoring design could require only one action from the operator if it had incoming data automatically compared to standard alarm limits and, when a limit was violated, had an auto-dialer phone call people until it gets an acknowledgment (the single human action). An alternative, less automated, design might require an operator to (1) activate alarm monitoring; (2) load that day's special alarm limits; (3) see when an alarm triggers; (4) look up whom to notify for the particular alarm; (5) look up the phone number; (6) dial the call; (7) report the alarm; and (8) obtain acknowledgment.
- **Automation—non-electronic data transfers.** How many transfers of data aren't electronic? As automation decreases, data will require more reading, keyboard entry, or physical handling and logging (like hard-copy schedules, timelines, or tapes)—all examples of non-electronic interfaces.

The design rules in Table 5.4 suggest that, to operate at low cost, our design for a ground system must

1. Minimize geographical distribution. This rule doesn't consider travel and per diem costs required to co-locate operations.
2. Minimize institutional interfaces
3. Minimize constrained resources such as shared data-system elements and shared operations teams
4. Permit quick scheduling to turn around shared resources
5. Use a command language that exploits high-order, reusable commands (block or macro)
6. Develop as little flight software as possible after launch
7. Schedule operations to keep staff low on off-prime shifts
8. Minimize management levels in the operations organization
9. Use multi-task experts within a project rather than task experts working several projects
10. Automate by decreasing the number of human steps needed in routine operations
11. Automate by lowering the number of non-electronic transfers of data for routine operations

5.1.5 Using Complexity Metrics to Predict Operations Costs

Figure 5.1, at the beginning of this chapter, is a model for predicting operations costs based on complexity metrics. The model predicts operations costs in terms of workforce size. This approach allows each project to compute dollar costs by multiplying the predicted workforce by their unique salary rates, burden rates, inflation, and duration.

We obtain the workforce prediction from the product of four numerical factors for operations complexity, scaled by an empirically determined constant that depends on mission type and phase. We compute each complexity factor by counting the project's high, medium, and low complexity metrics in each category; multiplying by high, medium, and low weighting factors (empirically determined); and then computing the average value. We compute the average value by adding the products of the complexity metrics and weighting factors and dividing by the total number of metrics. The weighting factors are the same for all mission types and are determined empirically by adjusting them to achieve a best fit of the data used to calibrate the model. Table 5.5 lists the mission-type constants and weighting factors.

The model user specifies the mission type and phase and then fills in the 95 metrics on operations complexity arranged under high, medium, and low parameters. Based on these user inputs, the model predicts an operations workforce that matches the project's operational complexity.

The model predicts costs for operating large or small uncrewed missions and either Earth-orbiter or planetary missions. Model calibration data is based on NASA's JPL missions. Table 5.5 shows each mission's complexity metrics and computed values for the complexity factors. The last three columns of Table 5.5 show model predictions to be within 25% of the actual mission workforce except for one mission (Galileo Cruise). In the future, we'll expand the calibration database to include Goddard and DOD missions.

For example, we predict the FTEs for the Voyager mission by first identifying the mission type. Voyager is a cruise mission, so the Mission Type Constant is nine. We next determine the 95 complexity metrics (high, medium or low) for Voyager. We find the numerical complexity factor for the four categories:

Mission Design = $(7 \times 1.4 + 18 \times 1.6 + 7 \times 4) / (7 + 18 + 7) = 2.08$
Flight System = $(6 \times 1.2 + 11 \times 2.5 + 8 \times 3.6) / (6 + 11 + 8) = 2.54$
Risk Avoidance = $(1 \times 0.7 + 5 \times 1.4 + 5 \times 2) / (1 + 5 + 5) = 1.61$
Ground System = $(8 \times 0.7 + 12 \times 1.7 + 3 \times 2) / (8 + 12 + 3) = 1.39$

Next, we calculate the overall complexity factor by multiplying the four individual factors

Factor = $2.08 \times 2.54 \times 1.61 \times 1.39 = 11.83$

Finally, we calculate the predicted FTEs by multiplying the Mission Type Constant by the complexity factor:

FTEs = $9 \times 11.83 = 107$

Table 5.5. Cost Model for Mission Operations. Predicted versus actual operations staffing in full-time equivalents (FTEs). Note: The numbers of project parameters (low, medium, and high) for each metric aren't always the same for different projects. Some project counts are less than 95 because they were unable to answer all the questions. Other project counts are more than 95 because they used an earlier version of the model with more questions.

	Mission Design (MD)			Flight Systems (FS)			Risk Avoid (RA)			Ground System (GS)			Mission Type Constant	Complexity				Factor	FTEs Predict	FTEs Actual	Error of Predict
	Low	Med	High	Low	Med	High	Low	Med	High	Low	Med	High		MD	FS	RA	GS				
Weighting	1.4	1.6	4	1.2	2.5	3.6	0.7	1.4	2	0.7	1.7	2									
Cruise																					
Voyager pre-enc	7	18	7	6	11	8	1	5	5	8	12	3	9	2.08	2.54	1.61	1.39	11.83	107	107	0%
Ulysses	17	12	6	6	17	2	5	6	0	11	12	1	9	1.91	2.28	1.08	1.25	5.91	53	65	-22%
Ulysses-polar pass	9	14	12	5	16	4	5	5	1	9	13	9	9	2.37	2.42	1.14	1.5	9.74	88	70	20%
Voyager intersteller	13	16	6	6	11	8	4	3	4	15	8	1	9	1.94	2.54	1.36	1.09	7.3	66	52	21%
Galileo	12	12	11	2	7	16	3	4	4	7	10	7	9	2.29	3.1	1.43	1.5	15.13	136	330	-142%
Austere Cruise																					
Voyager-Austere	13	16	6	5	12	8	7	2	2	16	7	1	5	1.94	2.59	1.06	1.05	5.59	28	33	-18%
Pluto	18	9	9	15	9	6	16	4	0	15	13	2	5	2.1	2.07	0.84	1.22	4.45	22	18	19%
Orbital*																					
MGLN-radar map	7	9	19	7	6	12	1	5	5	7	15	2	11	2.86	2.66	1.61	1.43	17.59	193	230	-19%
MGLN-gravity	9	14	12	8	6	9	4	3	4	9	15	0	11	2.37	2.48	1.36	1.33	10.62	117	125	-7%
TOPEX	9	12	17	6	8	13	7	4	0	6	9	6	11	2.63	2.74	0.95	1.5	10.31	113	85	25%
IRAS	17	13	5	11	11	3	2	8	1	11	9	3	11	1.85	2.06	1.33	1.26	6.36	70	72	-3%
SIRTF	17	7	7	11	9	4	6	3	2	15	5	3	11	2.03	2.09	1.13	1.09	5.2	57	45	21%
Flyby																					
Galileo-prime	5	8	22	3	5	17	2	3	6	5	9	10	18	3.08	3.09	1.5	1.62	24.63	443	420	5%
Galileo-playback	6	11	18	3	6	16	2	3	6	5	9	10	18	2.8	3.05	1.6	1.62	22.08	397	420	-6%
Your Mission																					

* Note: MGLN = Magellan, TOPEX = Ocean Topography Experiment, IRAS = Infrared Astronomical Satellite, SIRTF = Space Infrared Telescope Facility.

5.1 Operations Cost Estimations 147

The model and the complexity metrics will work throughout the project-development cycle. While analyzing preliminary requirements, we could use them to predict how science and mission requirements would affect operations cost, as well as to support descoping studies for cost-capped missions. They could support establishing an operations concept early in the project design, when we have the best chance to influence designs for the payload, spacecraft, and mission. While designing a project, we could use the model to support trade studies such as evaluating design options that save money on spacecraft development but reduce operability. Finally, during the systems-design phase, when more accurate cost estimates based on detailed design are available, we could use it to support "what-if" trade studies, employing a fraction of the time and effort required by detailed cost estimates based on point design.

The model doesn't predict costs for:
- Pre-launch development or for postponed development after launch. The model applies to operations, not development costs.
- Processing science data beyond level 0 because instruments tend to be unique and require unique data processing. The model **does** include costs for processing science data by producing on the ground a complete, error-corrected, time-ordered replica of the science data produced on the spacecraft.
- Space-link operations, such as DSN and TDRSS, which traditionally have been free to the project. It **does** include project costs of scheduling and interfacing with these services.
- Major in-flight anomalies and failures
- Managing the project, program, or science. It **does** include costs for managing the project's operations organization.
- The project's Public Information Office

The model is causal. Although cost-modeling techniques vary, the two basic approaches are the associative and the causal.
- *Associative models* predict costs based on parameters we correlate or associate with costs. Because associative parameters may have little or no direct causal influence on operations costs, such models aren't very useful when we're trying to decide how to reduce costs. For example, one traditional associative operations model predicts costs based on spacecraft and payload mass. However, we can't use it to claim that operations costs will go down if the spacecraft structure is redesigned to a lower mass. Another more trivial example would be that we could probably predict operations costs as a function of the operations manager's age because more expensive missions tend to use older, more senior personnel. But we couldn't claim for a given project that the way to reduce operations costs would be to replace the operations manager with a three-year-old.

- *Causal models* predict outcomes based on inputs that influence the outcome directly. An example of a causal model from the world of health and medicine is a quiz that attempts to predict your life expectancy based on how much you exercise, how many packs of cigarettes you smoke per day, how many pounds overweight you are, etc. Each of these questions identifies parameters believed to be causally associated with life expectancy. Besides using these "design rules" for predictions, we can also apply them to change our lifestyles and live longer. In the model we present here, the operations-complexity inputs are all believed to be causally related to operations costs. Thus, reducing the complexity value of any parameter should reduce operations costs. So we can use the operations-complexity parameters not only to predict costs but also as an organized set of design rules to achieve low-cost operations.

The model is based on requirements. We intend it to work for various operations designs, team structures, staffing strategies, and ground data systems. It does so by defining the operations-complexity metrics to **model operations costs in terms of the requirements on, rather than the design of, the operations system.** It's designed this way for several reasons. One reason is that it must be able to predict costs in very early design phases, before the operations design has been settled. A second, more important reason is that operations costs are driven the hardest by requirements imposed on operations, and only to a lesser extent by the efficiency of the operations-system design. The difference between a 200-person operations team and a 20-person team arises mainly from a difference in performance requirements, and only secondarily from a difference in the efficiency of an operations design. We can reduce operations costs most by negotiating cost-effective operations requirements early in a project. This model supports such negotiating by covering the four main sources of requirements on the operations system: the mission design, the flight-systems design, the project's policies on operational risk, and inheritance and constraints on the ground. It's no accident that the model's four factors correspond to these four most significant sources of operations requirements.

5.2 Operations Cost Modeling

Cost is always an important consideration for the MOM, so we should understand how to interpret and use results from cost models. The accuracy of most operations models is not much better than ±20%, which may be unacceptable growth to a project far along in development. For the MOM, the best use of these models is as an independent check of estimates from a bottoms-up, or grass-roots, approach. Elements with significant discrepancies should be looked at more closely and the bases for the different estimates should be reviewed and compared to identify which is most credible. Generally, the approaches used for bottoms-up, or grass-roots, estimates are far better than those in a cost model, but models sometimes identify effects that may have been inadequately accounted for. We describe here the approach used for NASA's Space Operations Cost Model (SOCM). The MOM should understand the functions, processes and attributes that this model predicts will drive costs.

5.2.1 Using Complexity Metrics for Cost Modeling

Section 5.1 described operations complexity metrics and gave an example of how to use these to generate an estimate of operations staffing requirements. The model was a good predictor of past planetary missions, but applying this type of model during early mission concept formulation is difficult, given the level of detail needed to characterize all 95 metrics. Furthermore, we don't know how well it would predict staffing requirements for future missions or how the approach accounts for the latest technology advances. Computing and communications technologies are advancing rapidly and have a significant impact on the staffing required to support many mission operations functions. These and other technology advances also underlie new capabilities for operating remote spacecraft and collecting and distributing data on the ground.

This section describes a model developed by NASA to estimate operations costs for robotic space science missions. The new model is named the Space Operations Cost Model (SOCM) and has been used extensively by cost analysts, mission planners, and operations experts since 1997. Although much of the foundation for SOCM came from the complexity metrics described in Sec. 5.1, the model had to be simplified to be used in early concept formulation stages, where little is known about specific implementation details. So NASA developed a multi-level approach, with the first level designed to provide a rough estimate with minimal input requirements, and additional levels requiring more detailed inputs to improve accuracy. The MOM should pay particular attention to the inputs to this cost model and how they drive the costs.

Multi-level Approach

The goal of the multi-level approach was to enable development of rough Mission Operations and Data Analysis (MO&DA) cost estimates with a minimum set of inputs and to provide capability to refine estimates as additional details become available. This is shown schematically in Fig. 5.2.

In this approach, a minimum set of inputs, called "Level 1" inputs, describes the project's architecture (spacecraft, instrument, and ground data system designs), mission, and programmatic requirements in sufficient detail to provide a reasonable basis for estimating costs. Additional details can be incorporated using a Cost Driver Influence Matrix reflecting judgment from operations experts as to how lower level inputs can affect requirements. Input selections for the lower levels can either increase or decrease the Level 1 estimate, depending on how the

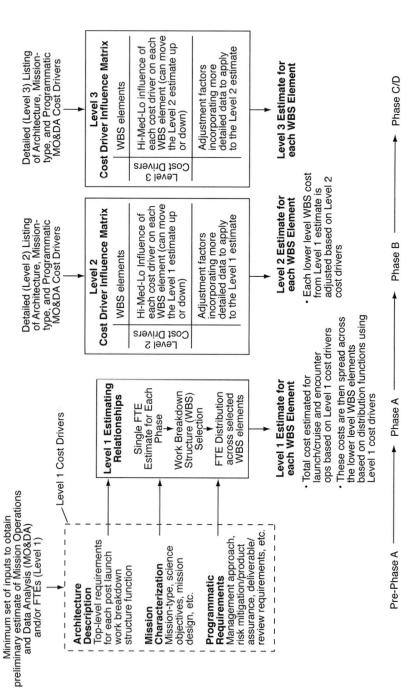

Fig. 5.2. Schematic of Multi Level Modeling Approach. As the project progresses through the various phases, more detailed inputs/data become available to refine cost estimates and narrow the cost uncertainty range.

project's approach compares to standard practices. The Level 1 estimate is a starting point that we can later refine by entering inputs for the lower levels.

Like the model described in Sec. 5.1, SOCM initially estimates staffing requirements. We can then combine these with duration inputs to calculate full-time equivalents (FTEs); one FTE represents labor for one person working full-time for one year. Then we can spread the FTEs across a Work Breakdown Structure (WBS) and determine the costs based on specific labor rates and labor cost to total cost conversion factors. As more information becomes available, we can incorporate it into the estimate to improve accuracy. This is shown in Fig. 5.3.

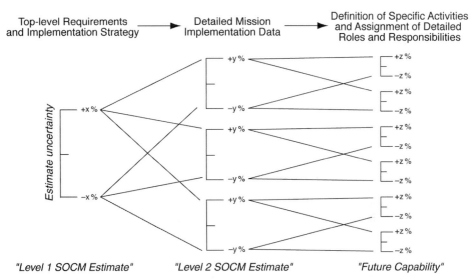

Fig. 5.3. **Estimate Uncertainty Ranges.** Detailed lower level inputs refine the high-level cost estimates. This tends to reduce uncertainty and improve accuracy.

The starting point Level 1 estimate has a rather large uncertainty that we can improve significantly with more detailed input data. An inherent assumption for a Level 1 estimate is that the implementation details that determine the lower level inputs are consistent with current standard practices. As shown in Fig. 5.3, estimate results from the lower levels can be outside the uncertainty range associated with a Level 1 estimate. This can occur if most of the lower level inputs are substantially more or less challenging than standard practices. The SOCM model is currently working with only two operating levels. We hope that with more levels, the uncertainty associated with an estimate at Level 2 or lower could be narrowed further.

Level 1 Description

The Level 1 approach was developed to minimize the number of required user inputs while still yielding a reasonable estimate of operations requirements. Figure 5.4 shows a block diagram identifying key Level 1 elements.

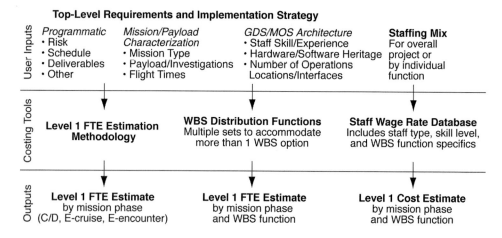

Fig. 5.4. Block Diagram for SOCM Level 1. SOCM uses top-level requirements and implementation strategy, together with project staffing mix, to form the Level 1 cost estimate. (GDS/MOS = Ground Data System/Mission Operations System; WBS = Work Breakdown Structure; C/D = Development/test and launch; E = Post Launch Operations.)

The inputs are high-level descriptions of requirements and implementation strategy and the staffing mix to support each operations function. The requirements and implementation strategy inputs are determined by programmatic, mission design, spacecraft and payload design, and the ground data system (GDS) or mission operations system (MOS). Staffing skill/experience inputs include six labor types for each operations function with options for senior (+10 yrs experience) and junior (<10 yrs experience) staff working for a university, government organization, or industry. Labor costs are different for each staff type. In addition, we use fixed factors for each operations function to convert labor costs to total costs for that function. These labor-to-total cost factors are substantially higher for functions like mission control, which uses and maintains an infrastructure of GDS/MOS equipment and software tools, than for functions like management, which are mostly direct labor.

After assigning all the Level 1 inputs, the model can provide a detailed breakout of staffing and costs for all required operations functions.

Level 2 Description

Level 2 of the model is intended to reflect more detailed inputs describing the various project elements and how the planned implementation compares to current standard practices. Figure 5.5 shows a block diagram of the Level 2 model elements.

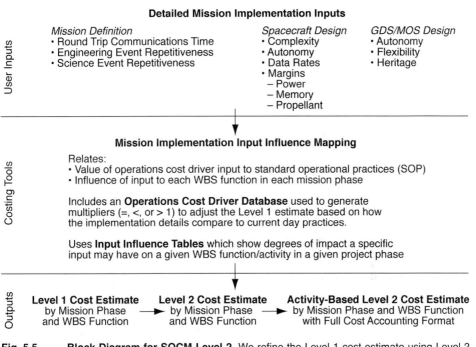

Fig. 5.5. Block Diagram for SOCM Level 2. We refine the Level 1 cost estimate using Level 2 inputs as they become available.

All Level 2 inputs are defined in terms of how they relate to standard practices. We use the Level 2 inputs to calculate factors which go into modifying the Level 1 estimate. After running a Level 1 estimate, all Level 2 inputs are defaulted to standard practices.

Every Level 2 input goes into one of nine categories, each with a different degree of influence on the model estimates. We modify the Level 1 estimate using the model's Cost Driver Database, which includes tables defining the potential influence of each input. If the planned approach uses a significantly simpler (or less costly) approach than standard practices, the Level 1 estimate is reduced, and vice versa. If many of the inputs defined as having a significant influence are simpler (or more complex) than standard practices, the Level 1 estimate may be substantially decreased (or increased).

The model does not rely on a complete definition of all Level 2 inputs to get an improved Level 1 estimate. We assume all unknown inputs to be similar to standard practices unless specified otherwise.

Accommodating Different Mission Types

SOCM uses the same input categories and output elements for all missions, but Level 1 and 2 inputs are tailored to specific mission types. This reflects the differences between missions operating in Earth orbit or in near Earth space and missions operating in deep space and at distant planetary bodies. For Earth orbiting missions, we place a greater emphasis on the GDS/MOS elements, since these missions usually require frequent data downlinks and a highly capable data handling system on the ground to collect and distribute large amounts of science data. For planetary missions, trajectory and launch vehicle requirements typically dictate a tailored spacecraft design to minimize mass, which requires close observation and control of various spacecraft functions and resources. Science data downlink rates and daily volumes for planetary missions are generally much lower than for Earth orbiting missions, so planetary GDS/MOS designs tend to focus more on spacecraft support.

Past, current, and future NASA robotic science missions were studied to gain better understanding of different operational requirements and are shown in Table 5.6. Mission candidates included examples of conventional approaches, recent low-cost projects that represent standard operational practices (SOP), and future missions that incorporate state-of-the-art (SOA) technology advances. This candidate mission set represents five different NASA centers.

Because the focus of the new modeling approach was to estimate operations costs for current and future projects, many past missions could not be used. Although the approaches used for some of these missions were very advanced at the time, technology improvements now allow us to carry out many of the functions much faster, with less support, and at lower cost.

The Reference Mission Set includes many different types of missions, but for SOCM we classify them into two groups – Planetary and Earth orbiting (EO). The Planetary missions are those targeted to our moon or other deep space objects or locations. These missions have low-activity cruise phases followed by more intense encounter phases. Science data collection during cruise is usually minimal. Earth orbiting missions operate in Earth orbit or near-Earth space and typically collect science data continuously. Earth orbiting missions may have multiple operating modes or phases, each of which we must evaluate individually.

5.2.2 Relating Inputs to Current Practices and Technologies

Data from the Reference Mission Set was studied to identify trends and correlate improvements to related technology advances. The study formed the

Table 5.6. Candidate Missions for SOCM Reference Mission Set. Studying these missions yielded insight on how advances in technology influence mission requirements, and hence, costs.

NASA Center	Conventional Approach (Past Projects)	Use of Low Cost Modern Business Practices (Current Projects/SOP)	Future Missions (Future Projects/SOA)
Goddard Space Flight Center (GSFC)	• Gamma Ray Observatory (GRO) • Hubble Space Telescope (HST) • Energetic UV Explorer (EUVE)	• Advanced Composition Explorer (ACE) • Far UV Spectroscopic Explorer (FUSE) • Solar, Anomalous and Magnetospheric Particle Explorer (SAMPEX) • X-Ray Timing Explorer (XTE)	• Midex (MAP, IMAGE) • SMEX (TRACE, FAST, SWAS, WIRE) • ESSP • EOS • NMP EO Missions
Marshall Space Flight Center (MSFC)		• Advanced X-ray Astrophysics Facility (AXAF) • Space Station	• Mars Exploration • Advanced Launch Vehicles
Johnson Space Center (JSC)	• Shuttle Orbiter	• Space Station	• Mars Exploration
Jet Propulsion Laboratory (JPL)	• Galileo • Magellan • Voyager	• Discovery Program (Mars Pathfinder, NEAR/APL) • Mars Global Surveyor (MGS)	• New Millennium DS Missions • Discovery Program (Lunar Prospector, Stardust) • Pluto Flyby
Kennedy Space Center (KSC)	• Shuttle Orbiter		• HRST, RLV

basis for how SOCM uses model inputs to generate operations cost estimates. This effort improved our understanding of the relationship of technology to mission requirements and is described in this section.

Assessing Impacts from Advanced Technology Applications

Analysts frequently apply cost models to assess the effects of potential technology investments to improve performance. We need to understand the planned application of the performance improvement before assessing cost impacts. Performance improvements that add capabilities usually increase costs as well. We can save money by using improvements to meet requirements with simpler or smaller systems if the mass savings are enough to let us use a less costly launch vehicle. Performance improvements can also increase margins, thus reducing operations costs. To understand fully how advanced technologies affect the overall cost, we need to consider system requirements for development and operations.

The relationship between overall cost, or lifecycle cost, and performance is shown schematically in Fig. 5.6. Trends indicate that adding or removing a certain amount of performance to a mission can cause costs to escalate or diminish, respectively. The range over which performance applications have the most benefit is unique to each mission based on its requirements.

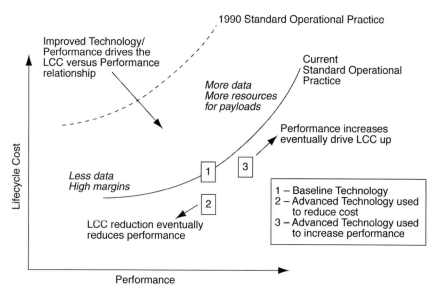

Fig. 5.6.   Relationship of Lifecycle Cost (LCC) and Performance. Depending on how we use them, technology advances can increase or decrease overall mission costs.

"Constructive" Estimating Relationships (CERs)

A lack of available data makes it more difficult to evaluate the complexity associated with incorporating the most recent technology advances. A project may rely on a detailed assessment of a technology to determine what the potential cost impacts might be. For SOCM, we use expert judgment regarding operations technologies and approaches as well as data from missions in the Reference Mission Set to generate estimates. The method incorporates a set of "tuning" inputs to help match results from the expert judgment input to the requirements of the missions in the Reference Mission Set. This is a "constructive" approach, where we use the tuning inputs to construct methods relating the expert judgment to staffing and costs.

Figure 5.7 is a diagram showing how SOCM works and what each element includes. Level 1 and Level 2 have separate tuning controls. Level 1 tuning includes definition of minimum and maximum staffing levels for high-level functions and weighting inputs, the operations cost influence for each input type, for the Level 1 inputs. We use headcount distribution tables to spread the Level 1 estimate across the 13 WBS functions. For Level 2, tuning includes assigning factors to each Level 2 input that determine how much each input can affect the estimates. The Cost Driver Database holds the expert judgment data defining the relative impact for each Level 2 input.

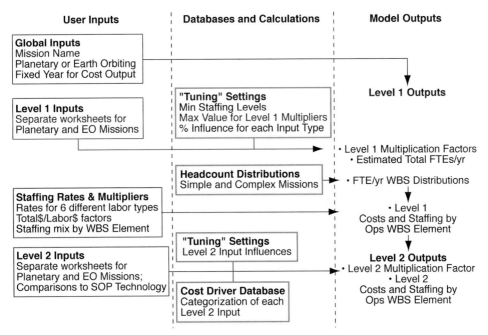

Fig. 5.7. SOCM Methodology Diagram. Tuning helps to incorporate expert judgement into the cost estimates. This is especially important when we lack detailed information on the cost effects of new technology advances.

The "tuning" process is unique to SOCM and can be seen as a brute-force approach to making the model work. The process, shown in Fig. 5.8, involves many iterations, each one testing different tuning settings. The iteration cycle continues until all the missions in the Reference Mission Set are being estimated to within 30% with only the Level 1 inputs and to within 20% after assigning all Level 2 inputs. When difficulties crop up, operations experts are consulted to identify possible changes to the inputs and input weightings.

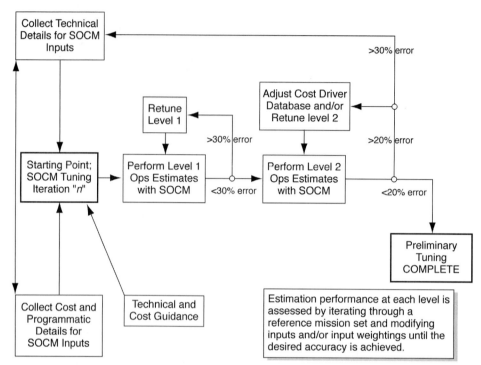

Fig. 5.8. **SOCM "Tuning" Process.** We adjust the tuning inputs until cost estimates for all the reference missions are within the given error margins: ±30% for Level 1 and ±20% for Level 2.

Figure 5.9 provides a sample of the "tuning" worksheet for Planetary missions. The FTE Starting Points and Limits define the range of estimates that can come from the Level 1 inputs. Level 2 inputs are first split into three high-level categories (I, II, III) and then assigned one of three lower level sub-categories, resulting in a total of 9 options. This worksheet has similar categories for Earth orbiting missions, but different values.

The initial Level 1 estimate result is staffing requirements for three teams: the Flight Operations Team, the Science Operations Team, and the Navigation Team. We distribute these initial estimates across the 13 WBS functions using reference distributions for a simple example and a complex example. The resulting staffing distribution functions determine where in the range between a very simple and very complex mission a given concept is. Table 5.7 shows these ranges for planetary missions and Table 5.8 shows the ranges for Earth orbiting missions.

SOCM's Cost Driver Database holds data representing expert judgment and includes descriptions for each input and definitions of standard practices and

FTE Starting Points and Limits

Minimum/ Limit FTE for Development	Minimum/ Limit FTE for Cruise	Minimum/ Limit FTE for Encounter
10/120	8/96	12/144

Percent Influence for Each Level 1 Input

	Mission Type	Target	Program Risk/Schedule	Spacecraft Design	GDS/MOS Characteri- zation	Science Payload
Development	16.7	16.7	16.7	16.7	16.7	16.7
Cruise	25.0	20.8	12.5	16.7	8.3	16.7
Encounter	16.7	20.8	12.5	16.7	8.3	25.0

Impact of Level 2 Cost Driver Inputs (used internally by model with Figs. 5.10 and 5.11)

	% Influence
I	15
II	9.375
III	3.75

Low	65%
Med	85%
High	100%

% Influence	Low	Med	High
I	9.75	12.75	15
II	6.09	7.97	9.375
III	2.44	3.19	3.75

Fig. 5.9. **Sample of Tuning Inputs for Planetary Missions.** Tuning inputs for Earth orbiting missions fall into similar categories, but have different values.

high-cost and low-cost approaches. Examples of entries for planetary missions are shown in Table 5.9 and for Earth orbiting missions in Table 5.10.

All Level 2 inputs are initially set to the medium value, which represents SOP. If a Level 2 input is assigned the high value, representing SOA, the Level 1 estimate is increased by an amount determined from the input ranking. If a Level 2 input is assigned the low value, it reduces the Level 1 result. If many of the Level 2 inputs are assigned a low or high value, the change to the Level 1 estimate may go outside the Level 1 range set up on the "tuning" sheet. For most projects, some of the Level 2 inputs are high and some are low, resulting in an estimate relatively close to the Level 1 result.

SOCM's modeling approach can be updated by redefining the Level 2 inputs and input range definitions using data defining the most recent technologies and approaches, and then "retuning" the model with a more current Reference Mission Set. Because technology advances continually, the model will probably need updates every 4–5 years.

Table 5.7. Staffing Percentage Distributions Planetary Missions. We base these distributions on the Level 1 estimates for staffing requirements.

Operations WBS Elements	Simple Cruise	Complex Cruise	Simple Encounter	Complex Encounter
Management	8.1	20.9	10.9	5.0
Mission Planning and Integration	0.0	2.0	0.0	2.9
Command/Uplink Management	0.0	7.4	0.0	10.5
Mission Control and Operations	13.7	10.2	18.4	15.9
Data Capture	14.5	7.9	9.8	6.4
Position/Location Planning and Analysis	8.1	15.3	10.9	13.4
Spacecraft Planning and Analysis	22.6	17.4	22.8	13.0
Science Planning and Analysis	3.2	6.7	2.2	17.2
Science Data Processing	4.0	0.5	2.7	1.8
Long-term Archives	1.6	0.9	1.1	0.7
Systems Engineering, Integration, and Test	9.7	7.4	10.3	6.5
Computer and Communication Support	2.4	1.8	2.7	1.9
Science Investigations	12.1	1.5	8.2	5.4
	100	100	100	100

5.2.3 Definitions of Inputs and Outputs

Inputs

SOCM uses five input categories, which are the same for Levels 1 and 2 and all robotic science mission types. The elements under each category are important in predicting the operations cost of the mission and the MOM should understand these elements. These input categories include:

1. *Mission*—mission orbits, targets, number of spacecraft, phase durations, others
2. *Programmatics*—risk policies, schedule, management mode, contract type, others
3. *GDS/MOS*—past experience for lead organization, maturity and heritage, number of support organizations, data handling requirements, autonomy, others
4. *Science Payload*—number and type of instruments, pointing requirements, conflicts, operating modes, complexity, science team requirements, others
5. *Spacecraft Design*—heritage, design complexity, autonomy, resource margins, others

Table 5.8. Staffing percentage Distributions for Earth-orbiting Missions. The difference between the distributions shown here and those of Table 5.7 reflect the inherent differences between EO and planetary missions.

Operations WBS Elements	Simple Nominal	Complex Nominal	Simple Extended	Complex Extended
Management	4.0	4.3	4.0	4.3
Mission Planning and Integration	7.0	3.0	7.0	3.0
Command/Uplink Management	16.0	9.0	16.0	9.0
Mission Control and Operations	27.0	20.5	27.0	20.5
Data Capture	5.0	18.5	5.0	18.5
Position/Location Planning and Analysis	1.0	0.3	1.0	0.3
Spacecraft Planning and Analysis	3.0	8.0	3.0	8.0
Science Planning and Analysis	3.0	13.0	3.0	13.0
Science Data Processing	7.0	8.0	7.0	8.0
Long-term Archives	1.0	3.0	1.0	3.0
Systems Engineering, Integration, and Test	20.0	7.0	20.0	7.0
Computer and Communication Support	5.0	1.5	5.0	1.5
Science Investigations	1.0	4.0	1.0	4.0
	100	100	100	100

We can think of the Level 1 inputs as high-level representations of each of these input categories. The Level 2 inputs provide more specific implementation details covering each of these input categories and how the planned approach compares to standard practices. We describe inputs separately for Planetary and Earth orbiting missions.

Output Options

Determining the best output categories for modeling operations turned out to be much more complex than expected. Organizations have different ways of providing support and tracking operations staffing and cost. Adding to the complexity are unique requirements associated with specific mission types.

The selected work breakdown structure (WBS) for SOCM does a reasonable job of mapping to many different robotic science mission types and project organizations. Table 5.11 defines each of the 13 WBS elements and the functions it includes. These WBS elements can be mapped into higher-level categories that cover the Flight Operations Team, the Science Operations Team, and the Navigation Team to support development of the Level 1 estimating methodology.

Table 5.9. Cost Driver Database for Planetary Missions. This database enables us to incorporate expert judgements into the cost estimate.

Cost Driver	Mission Type	Affected System	Units	High Dollar Value	Medium Dollar Value	Low Dollar Value	Impact	Driver Level	Description
Hardware Redundancy	Planetary	Programmatics		Full redundancy with rapid switchover	Selected redundancy	Limited or no redundancy	Low	II	GDS/MOS system redundancy
Software Redundancy	Planetary	Programmatics		Full redundancy	Selected redundancy	Limited or no redundancy	Low	II	Measure of number of ways to perform observations with software
Crosstraining/ Staffing Overlaps	Planetary	Programmatics		Limited crosstraining	Crosstrained within functions	Full crosstrained	Low	II	Number of staff assigned/trained to perform same function
Telemetry Downlink Data Rate	Planetary	Spacecraft	kbps	>2	0.8–2	<0.8	Low	II	Nominal downlink data rate for telemetry
Spacecraft Complexity	Planetary	Spacecraft		Dual-spin/3-axis+	3-axis	Spinner	Low	I	Type of attitude control and number of articulating components
MOS Flexibility	Planetary	GDS/MOS		High	Medium	Low	Med	I	Measure of the ability to change operations plans in flight
Science Event Complexity	Planetary	Mission		Constrained, complex coordination of science observations (multiple science modes)	Few constraints, moderate science observation complexity	Few instruments, simple and few science observation modes	Med	I	Number of unique science instrument command sequences

Table 5.9. Cost Driver Database for Planetary Missions. (Continued) This database enables us to incorporate expert judgements into the cost estimate.

Cost Driver	Mission Type	Affected System	Units	High Dollar Value	Medium Dollar Value	Low Dollar Value	Impact	Driver Level	Description
Average Light Time	Planetary	Mission	min	>30	10–30	<10	Med	II	Time required to send commands to the spacecraft from Earth
Engineering Event Complexity	Planetary	Mission		Risky events and/or sign, real-time contact	Repetitive and non-hazardous events	Only routine, non-hazardous events	Med	II	Number of unique engineering command sequences
Instrument Support Complexity	Planetary	Payload		Constrained, complex instrument interaction (with other instruments and spacecraft)	Routine calibrations, few scheduling constraints	Simple payload with few science observation constraints	Med	I	Relates to number of instruments, conflicts, flight rules for instrument operation
Risk Plan—Spacecraft	Planetary	Programmatics		Class A or B program, high priority and low/med risk	Class C program, medium priority and med/high risk	Class D program, small spacecraft without redundancy	Med	I	Measure of the spacecraft operational risk based on design implementation
Verification Requirements	Planetary	Programmatics		Stringent	Nominal	Relaxed	Med	II	Imposed reliability through system testing
Staff Experience	Planetary	Programmatics		New OPS team	1 to 2 similar previous missions	More than two similar previous missions	Med	III	Experience of operations staff with similar systems

Table 5.9. Cost Driver Database for Planetary Missions. (Continued) This database enables us to incorporate expert judgements into the cost estimate.

Cost Driver	Mission Type	Affected System	Units	High Dollar Value	Medium Dollar Value	Low Dollar Value	Impact	Driver Level	Description
Spacecraft Autonomy	Planetary	Spacecraft		Minimal onboard capabilities, significant contact with ground	Moderate onboard telemetry monitoring capabilities	Significant onboard capabilities, minimal ground contact	Med	I	Ability of the spacecraft to operate without ground control
Data Return Margin	Planetary	Spacecraft		<1	1–2	>2	Med	II	Ratio of max downlink rate to max rate that the spacecraft science instruments collect data
Power Margin	Planetary	Spacecraft		<1	1–1.2	>1.2	Med	II	Ratio of max available power to peak power demand
Memory Margin	Planetary	Spacecraft		<2	2–3	>3	Med	II	Ratio of onboard storage capacity to max quantity of data to be downlinked in a single pass
Propulsion Margin	Planetary	Spacecraft	%	<5	5–10	>10	Med	III	Percent additional propellant available beyond planned burns and maneuvers

Table 5.9. Cost Driver Database for Planetary Missions. (Continued) This database enables us to incorporate expert judgements into the cost estimate.

Cost Driver	Mission Type	Affected System	Units	High Dollar Value	Medium Dollar Value	Low Dollar Value	Impact	Driver Level	Description
MOS Heritage	Planetary	GDS/MOS	%	<60	75	>85	High	III	Degree to which the GDS/MOS uses existing hardware and software
Encounter Criticality	Planetary	Programmatics		Unique opportunities	Nominal	Partial success	High	I	Measure of ability to perform observations if there is a failure at initial encounter
MOS Autonomy	Planetary	GDS/MOS		Several complex safe modes or experimental autonomy	Simple and robust safe mode/onboard telemetry monitor	Reuse of proven, sophisticated autonomy	High	I	Degree to which software tools have been automated for operations
Risk Plan—Instruments/Payload	Planetary	Programmatics		Complex, redundant payload	Few hazardous operations, limited redundancy	Simple payload, no redundancy	High	I	Measure of the instrument/payload operational risk based on design implementation
Risk Plan—GDS/MOS	Planetary	Programmatics		Minimize risk to mission safety, data loss <2%	Minimize risk to mission safety, accept 5% data loss	Accept moderate risk to mission safety, data loss >5%	High	I	Measure of the GDS/MOS operational risk based on design implementation

Table 5.10. Cost Driver Database for Earth Orbiting Missions. The differences between the categories in this database and those in the planetary missions database reflect the inherent differences between the two types of missions.

Cost Driver	Mission Type	Affected System	Units	High Dollar Value	Medium Dollar Value	Low Dollar Value	Impact	Driver Level	Description
Data Processing—Heritage/Reuse	Earth orbiting	GDS/MOS	%	Less than 60%	75%	More than 85%	Low	II	% of ground data processing system based on existing designs
Engineering Event Complexity	Earth orbiting	Mission		Risky events/significant real-time contact	Repetitive/no hazardous events	Routine, non-hazardous events	Low	I	Number of unique engineering command sequences
Crosstraining/Staffing Overlaps	Earth orbiting	Programmatics		Limited crosstraining	Crosstrained within functions	Fully crosstrained	Low	II	Number of staff assigned/trained to perform same function
Operations Type	Earth orbiting	Mission		Targeted and/or constrained	Orbit-driven/activities based on orbital events	Survey	Low	I	High level characterization of operation concept
Command Frequency—Generation Time	Earth orbiting	GDS/MOS		Less than one day before upload	One day before upload	More than one day before upload	Med	II	Time allowed to generate commands to modify/affect mission operations
Command Frequency—Real-time Commands	Earth orbiting	GDS/MOS		Special commands on some passes	Routine commands on most passes	No commands on some passes	Med	II	Frequency of real-time commands for uplink
Data Processing—Maximum Downlink Rate	Earth orbiting	GDS/MOS	Mbps	10s to 100s	1 to 2	less than 1	Med	II	Maximum downlink data rate accommodated

Table 5.10. Cost Driver Database for Earth Orbiting Missions. (Continued) The differences between the categories in this database and those in the planetary missions database reflect the inherent differences between the two types of missions.

Cost Driver	Mission Type	Affected System	Units	High Dollar Value	Medium Dollar Value	Low Dollar Value	Impact	Driver Level	Description
Data Processing—Maximum Bits/Day	Earth orbiting	GDS/MOS	Mb	>1000	500–1000	<500	Med	II	Maximum number of bits downlinked per day
Data Processing—On-line Storage	Earth orbiting	GDS/MOS	GB	<2	2–8	>8	Med	II	Size/capacity of onboard data storage system
Data Processing—Storage/Playback Frequency	Earth orbiting	GDS/MOS		Once per orbit	Several times per day	Once per day or less	Med	II	Number of days that data can be stored without downlink
Science Event Complexity	Earth orbiting	Mission		Constrained/multiple observation modes	Few constraints	Survey	Med	I	Number of unique science instrument command sequences
Payload Flight Heritage	Earth orbiting	Payload		New instruments; payload includes advanced technology	Most instruments have flight heritage	Most instruments have flown together, no advanced technology	Med	I	Measure of individual instruments and total payload package flight experience

Table 5.10. Cost Driver Database for Earth Orbiting Missions. (Continued) The differences between the categories in this database and those in the planetary missions database reflect the inherent differences between the two types of missions.

Cost Driver	Mission Type	Affected System	Units	High Dollar Value	Medium Dollar Value	Low Dollar Value	Impact	Driver Level	Description
Instrument/ Payload Operating Modes	Earth orbiting	Payload		Several instruments with multiple operating modes; 3+ observing modes	Fewer than 3 operating modes per instrument; 2–3 observing modes	2–3 operating modes per instrument; single observing mode for all instruments	Med	I	Identifies number of operating modes for each instrument and observing modes for total payload; modes include calibration
Hardware Redundancy	Earth orbiting	Program-matics		Full redundancy with rapid switchover	Selected redundancy	Limited or no redundancy	Med	II	GDS/MOS system redundancy
Data Return Margin	Earth orbiting	Spacecraft		< 1.1	1.1–2	> 2	Med	III	Ratio of max amount of data that can be downlinked to the average amount required per downlink
Power Margin	Earth orbiting	Spacecraft		< 1	1–1.2	> 1.2	Med	III	Ratio of max available power to peak power demand
Memory Margin	Earth orbiting	Spacecraft		< 1.2	1.5–2	> 2	Med	III	Ratio of onboard storage capacity to maximum quantity of data to be downlinked in a single pass
Data Processing— Autonomy	Earth orbiting	GDS/MOS		Minimal	Nominal	Extensive	High	II	Measure of the degree of autonomy in ground data handling system

Table 5.10. Cost Driver Database for Earth Orbiting Missions. (Continued) The differences between the categories in this database and those in the planetary missions database reflect the inherent differences between the two types of missions.

Cost Driver	Mission Type	Affected System	Units	High Dollar Value	Medium Dollar Value	Low Dollar Value	Impact	Driver Level	Description
Command Frequency— Sequences	Earth orbiting	GDS/MOS		Loaded more than once per day	Daily	Loaded less than once per day	High	II	Frequency of developing sequences for uplink
Data Processing— Data Completeness	Earth orbiting	GDS/MOS	%	>98%	95%	<90%	High	II	Measure of data return requirement vs. minimal acceptable data return
Data Processing— Data Delivery Time	Earth orbiting	GDS/MOS	hrs	24 or less	24 to 48	More than 48 hours	High	II	Time allowed to deliver data products after raw data is downlinked
Instrument Support Complexity	Earth orbiting	Payload		Constrained operation, complex instrument interactions	Routine calibrations, few schedule constraints	Simple instrument with few operations	High	I	Relates to number of instruments, conflicts, flight rules for instrument operation
Staff Experience	Earth orbiting	Programmatics		New OPS team	1 or 2 similar missions	More than 2 similar missions	High	II	Experience of operations staff with similar systems
Risk Plan— Spacecraft	Earth orbiting	Programmatics		Redundant spacecraft, several $100M development	Class C, $100M flight system development	Small spacecraft, no redundancy, tech demo mission	High	II	Measure of the spacecraft operational risk based on design implementation

Table 5.10. Cost Driver Database for Earth Orbiting Missions. (Continued) The differences between the categories in this database and those in the planetary missions database reflect the inherent differences between the two types of missions.

Cost Driver	Mission Type	Affected System	Units	High Dollar Value	Medium Dollar Value	Low Dollar Value	Impact	Driver Level	Description
Risk Plan—Instruments/Payload	Earth orbiting	Programmatics		Complex, redundant spacecraft	Few hazardous OPS, limited redundancy	Simple payload, no redundancy	High	II	Measure of the instrument/payload operational risk based on design implementation
Risk Plan—GDS/MOS	Earth orbiting	Programmatics		Accept minimal risk to efficiency and data loss <1%	Accept mod risk to efficiency and data loss <5%	Accept minimal risk to mission safety, and mod data loss	High	II	Measure of the GDS/MOS operational risk based on design implementation
Spacecraft Autonomy	Earth orbiting	Spacecraft		Several complex safe modes or experience approach	Simple robust safe mode; onboard telemetry monitor	Proven sophisticated autonomy	High	II	Ability of the spacecraft to operate without ground control
Maneuver Frequency	Earth orbiting	Spacecraft		Once a month or more	A few times per year	Once per year or less	High	III	Frequency of spacecraft maneuvers over nominal operations period

5.2 Operations Cost Estimations

Outputs for each WBS element include steady-state headcounts, equivalent FTEs, and costs for each operations phase. A project can also have multiple operations phases to cover different operational strategies. Examples include planetary missions with long low-activity cruise periods followed by intense encounters and other missions that have multiple observation modes with significantly different operational requirements. WBS breakouts can be provided for each phase.

Table 5.11. SOCM WBS Elements and Lower-Level Functions. The work breakdown structure for this model lends itself to a wide variety of mission types. (SRM = Safety, Reliability, and Manufacturability; TDRSS = Tracking Data Relay Satellite System; DSN = Deep Space Network.)

WBS Element	Lower-Level Functions Included
1.0 Management	Project management/administration; mission operations management; interface coordination to non-project support; project scientist; program control; SRM and QA support; contract/procurement administration; contingency
2.0 Mission Planning and Integration	TDRSS/DSN requirements generation, planning and negotiation; integrate engineering and science observation requests; orbit planning/mission design; guidelines and constraints development; project contingency planning; ground system engineering, development and maintenance; team management
3.0 Command/Uplink Management (Sequencing)	Command generation analysis and assessment; command development and product generation; ground system engineering, development and maintenance, team management
4.0 Mission Control and Operations	Coordinate institutional support; monitor/control real-time ground data systems elements; real-time spacecraft and instrument commanding and monitoring; generate integrated sequence of events; produce spaceflight operations schedules; negotiate/schedule short-range institutional facility support; mission support area operations; support to planning and analysis teams; ground system engineering, development and maintenance; team management
5.0 Data Capture and Accountability	Telemetry operations; engineering data record generation; experimenter data record/supplementary experimental data record generation; common data file generation; data accountability, expediting, and product delivery; tape library support; ground system engineering, development and maintenance; team management
6.0 Position/Location Planning and Analysis (Navigation)	Orbit determination; optical navigation; maneuver design and analysis; attitude/pointing determination and planning; trajectory analysis; ground system engineering, development and maintenance; team management
7.0 Spacecraft Planning and Analysis	Mission planning/command generation support; engineering observation and calibration planning; non real-time spacecraft performance assessment, trend analysis and engineering calibration analysis, anomaly investigation support; ground system engineering, development and maintenance; team management

Table 5.11. SOCM WBS Elements and Lower-Level Functions. (Continued) The work breakdown structure for this model lends itself to a wide variety of mission types. (SRM = Safety, Reliability, and Manufacturability; TDRSS = Tracking Data Relay Satellite System; DSN = Deep Space Network.)

WBS Element	Lower-Level Functions Included
8.0 Science Planning and Analysis	Mission planning/command generation support; instrument observation and calibration planning; quicklook analysis; instrument calibration and first-order science analysis; non real-time instrument performance assessment and trend analysis; ground system engineering, development and maintenance; team management
9.0 Science Data Processing	Image processing support; image analysis; science data processing (SDP); image tape library, photo-lab; science library support; SDP preparation for delivery to science archives; ground system engineering, development and maintenance; team management
10.0 Long-Term Archives Support	Science data archives maintenance and support; special request processing and other support services to researchers; ground system engineering, development and maintenance; team management
11.0 Systems Engineering, Integration and Testing	GDS systems engineering; systems/software integration; mission simulation, test and training support; facility engineering and support; management support
12.0 Computer and Communications Support	Computer operations; communications operations support; ground system engineering, development and maintenance; management support
13.0 Guaranteed (Principal Investigator) and Guest Science	Data analysis and science interpretation

An activity-based WBS, shown in Table 5.12, was developed in parallel to the 13 WBS elements. For this WBS option, seven generic activities are identified for each of four major project elements. This type of output may give improved insight into what really drives mission operations costs.

Table 5.12. Activity-based Operations WBS. This WBS option provides an alternate way to view operations cost drivers.

	Major Project Elements (nouns)			
Activities (verbs)	Spacecraft	Science Payload	Ground System	Nav/Orbital Tracking
1.0 Plan				
2.0 Command				
3.0 Monitor	Allocation of activity requirements to major project elements			
4.0 Analyze				
5.0 Develop				
6.0 Provide Data Services				

The model uses the activity-based WBS to provide an alternative view of the outputs and is related to the 13 WBS functions as shown in Table 5.13.

Table 5.13. Mapping of the 13 WBS Elements to the Activity-based Operations WBS. The numbers in the second through fifth columns correspond to the numbered WBS elements from Table 5.11. (PI = Principal Investigator; GI = Guest Investigator.)

	Spacecraft	Science	Ground System	Navigation
Planning	2. Mission planning 7. Spacecraft planning and analysis	2. Mission planning 8. Science planning and analysis	2. Mission planning	2. Mission planning 6. Navigation planning and analysis
Command	3. Command 4. Mission control	3. Command 4. Mission control	3. Command 4. Mission control	3. Command 4. Mission control
Monitor	4. Mission control	4. Mission control	4. Mission control	4. Mission control
Analysis	7. Spacecraft planning and analysis	8. Science planning and analysis 13. PI & GI science	--	6. Navigation planning and analysis
Data Services	5. Data capture 10. Data archives 12. Computer and communications	5. Data capture 9. Science data processing 10. Data archives 12. Computer and communications	5. Data capture 10. Data archives 12. Computer and communications	5. Data capture 10. Data archives 12. Computer and communications
Development	--	--	--	--
Overhead Services	1. Management 11. Sys Eng + Integration and Test	1. Management 11. Sys eng + Integration and Test	1. Management 11. Sys eng + Integration and Test	1. Management 11. Sys eng + Integration and Test

SOCM Examples. We present two hypothetical example missions here—a planetary mission and an Earth orbiting mission. SOCM input templates show input options and selections for each modeled mission. These templates are nearly identical to the model's user interface. In each example, SOCM outputs follow the Levels 1 and 2 inputs.

5.2.4 SOCM Planetary Mission Example

Planetary Level 1 Inputs. Figure 5.10 shows the Level 1 inputs arranged by the five input categories described in the previous section. Because there are so many different planetary mission types, there are many inputs for Mission Characterization. Since the types of science payloads are also diverse, an independent study examined costs for operation and costs for data analysis of many different instrument types. For example, radio science instruments may be relatively inexpensive to operate but require much data analysis support. Instruments like sample collection devices are more difficult to operate, but do not require much data analysis support. The Level 1 instrument ranking is based on a combination of costs for operation and costs for data analysis.

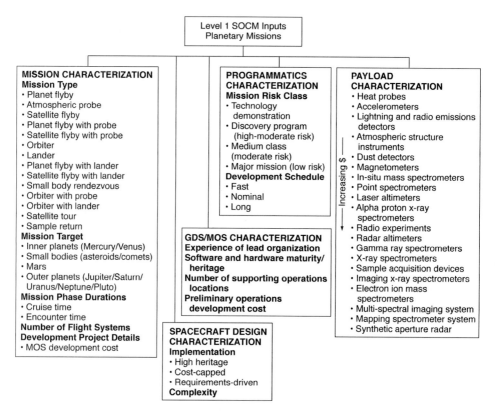

Fig. 5.10. SOCM Level 1 Inputs for Planetary Missions. The items checked off for a given mission go into the Level 1 cost estimate.

Planetary Level 1 Example—Mars Orbiter. For the Mission Characterization inputs in this example, we assume a substantial science payload operating on a Mars orbiter for two years. The primary goals of the mission are to collect different sets of multi-spectral images and perform gravity mapping experiments. We'll call this mission MOE (Mars Orbiter Example). Other assumptions include a 14-month cruise through orbit insertion and six months after the end of flight operations for additional post-flight data analysis (PFDA). Table 5.14 shows how we apply these inputs to this specific mission.

5.2 Operations Cost Estimations

Table 5.14. Mission Characterization (Planetary Level 1). MOE is a Mars orbiter with a single flight system.

Mission Name: ____MOE____

Total Cruise Phase duration (mo.): __14__   Total Encounter Phase duration (mo.): __24__

Total Post-Flight Data Analysis duration (mo.): __6__

MISSION CHARACTERIZATION (Planetary Level 1)

Mission Type	Target	# of Flight Systems
____ Planet Flyby	____ Inner Planets	__1__ enter # of Identical Flight Systems
____ Atmospheric Probe	____ Small Bodies	
____ Satellite Flyby	__X__ Mars	
____ Planet Flyby with Atmos Probe	____ Outer Planets	
____ Satellite Flyby with Atmos Probe		
__X__ Orbiter		
____ Lander		
____ Planet Flyby with Lander		
____ Satellite Flyby with Lander		
____ Small Body Rendezvous		
____ Orbiter with Atmos Probe		
____ Orbiter with Lander		
____ Satellite Tour		
____ Sample Return		

For Level 1, there are only two programmatic inputs. For MOE, we assume that the importance of the science places it in the "low risk" Mission Risk Class. Since the science payload is very advanced and requires an advanced spacecraft design, we assume the development schedule (start through orbit checkout) is more than four years. Table 5.15 shows these inputs for MOE.

For the GDS/MOS inputs we assume MOE is operated by a very experienced organization, but much of the ground system will be new to handling very high data rates and science data volumes. We also assume there will be multiple locations that provide major operations support. These include one serving as mission control to collect downlinked data and to upload commands, another as a science operations center collecting science data and generating payload command sequences for upload, and a third location providing support for spacecraft health monitoring and analysis. For the GDS/MOS inputs to this mission, see Table 5.16.

Table 5.15. Programmatics Characterization (Planetary Level 1). Spacecraft design and science payload on MOE drive these inputs.

PROGRAMMATICS CHARACTERIZATION (Planetary Level 1)

Mission Risk Class	Development Schedule
____ Technology Demonstration (tech > sci)	____ Fast (< 2.5 yrs)
____ Discovery, moderate risk	____ Moderate (2.5-4 yrs)
__X__ Medium, low risk	__X__ Long (> 4 yrs)
____ Major, minimum risk	

Table 5.16. GDS/MOS Characterization (Planetary Level 1). Ground system and software requirements offset the high level of organizational experience.

GDS/MOS CHARACTERIZATION (Planetary Level 1)

Lead Org Level of Exp	MOS Software Maturity/ Heritage	# of Supporting Ops Locations
__X__ Extensive	____ Extensive	_3_ enter #
____ Average	____ Average	Of supporting ops
____ Low	__X__ Low	Locations

The next set of inputs is Payload (Science) Characterization. To do the required science observations, four separate multi-spectral imaging instruments and an Ultra Stable Oscillator (USO) form the science payload. Table 5.17 lists these inputs for MOE.

Table 5.17. Science Characterization (Planetary Level 1). The number and types of instruments determine these inputs.

SCIENCE CHARACTERIZATION (Planetary Level 1)

__ Heat Probes	__ Point Spectrometers	__ Sample Acquisition Devices
__ Accelerometers	__ Laser Altimeters	__ Imaging X-Ray Spectr
__ Lightning and Radio Emissions Detectors	__ Alpha Proton X-Ray Spectrometers	__ Electron Ion Mass Spectrometers
__ Atmospheric Structures Instruments	_1_ Radio Experiments (Ultra Stable Oscillator)	_4_ Multi-Spectral Imaging Systems
__ Dust Detectors	__ Radar Altimeters	__ Mapping Spectr. Systems
__ Magneto-meters	__ Gamma Ray Spectrometers	__ Synthetic Aperture Radar
__ In Situ Mass Spectrometers	__ X-Ray Spectrometers	

The final set of inputs comes under Spacecraft Design. The spacecraft for MOE is designed specifically to support the science payload requirements and will be a 3-axis stabilized design. For the Design Complexity input, we generally consider spin-stabilized spacecraft to be "low," 3-axis to be "medium," and more complex

5.2 Operations Cost Estimations

systems to be "high." An example of a more complex system is Galileo, a Jupiter orbiter, that has a spin platform but is 3-axis stabilized. We must also consider pointing and other requirements when assigning the Design Complexity input. Table 5.18 gives the Spacecraft Design inputs for this mission.

Table 5.18. Spacecraft Design Characterization (Planetary Level 1). The single Design Complexity input actually reflects several design requirements.

SPACECRAFT DESIGN CHARACTERIZATION (Planetary Level 1)

Spacecraft Design Implementation	Design Complexity
_____ High Heritage	_____ Low (minimal # of flight rules)
_____ Cost-Capped	__X__ Medium
__X__ Requirements-Driven	_____ High (several unique engineering requirements)

For staffing assumptions, we use the SOCM model defaults, which include a mix of junior and senior government, university, and industry staff.

Planetary Level 2 Inputs. Level 2 inputs are shown in Fig. 5.11. For each item, there are three selection options, which include standard operating procedures (SOP) (this is the default), simpler or lower cost than SOP, and more complex or state of the art (SOA). More descriptive information and quantitative ranges are provided for each input to help select the proper option. As technology advances, most of these inputs may not change, but the definition of what is SOP, lower cost than SOP, or SOA probably will change.

The Planetary Level 2 inputs include many inputs describing programmatic requirements and spacecraft design and margins. For most planetary missions, the spacecraft are not standard "off-the-shelf" designs, and mass constraints force tightening up of resource margins. Although this approach helps fit the flight system on a reasonable size launch vehicle, it can complicate operations, especially with new designs, due to the added support required to monitor and control spacecraft performance. This is a classic development cost versus operations cost trade. If operations are not a substantial portion of the overall costs, it may be a less expensive approach to minimize mass, enable use of a lower cost launch vehicle, and have more expensive operations. If planned operations are a more substantial portion of the overall costs, it could be less costly to add in performance margins or make design changes to reduce operations costs.

Planetary Level 2 Example—Mars Orbiter. After entering all Level 1 inputs, we can assign any known Level 2 inputs to refine the Level 1 result. For the Mission Characterization inputs to MOE, we assume the high cost value for Science Event Complexity to cover the multiple observing modes that will be required. Since the mission is an orbiter, Engineering Event Complexity seems consistent with the medium cost value. Table 5.19 summarizes these inputs.

For the Programmatic inputs, we assume the medium cost value for most inputs except for the Risk Plans for the spacecraft and GDS/MOS, Verification

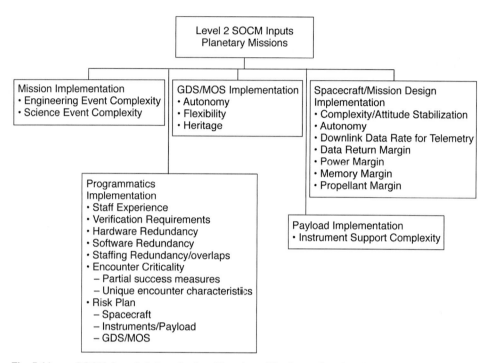

Fig. 5.11. SOCM Level 2 Inputs for Planetary Missions. Level 2 inputs address how we implement the Level 1 inputs.

Table 5.19. Mission Implementation (Planetary Level 2). MOE has highly complex science events, but its engineering is SOP.

MISSION IMPLEMENTATION (Planetary Level 2)			
Cost Driver	Low Dollar Value	Medium Dollar Value	High Dollar Value
Science Event Complexity	Few instruments, simple and few science observation modes	Few constraints, moderate science observation complexity	_X_ Constrained, complex coordination of science observations (multiple science modes)
Engineering Event Complexity	Only routine, non-hazardous events	_X_ Repetitive and non-hazardous events	Risky events and/or significant real-time contact

Requirements, and Hardware Redundancy. We assign the high cost value to all of these, to reflect the importance of returning all data reliably. We also give the operations staff credit for conducting several similar projects and assign the Staff Experience input the low cost value. Table 5.20 lists these inputs.

Table 5.20. Programmatics Implementation (Planetary level 2). Requirements associated with data return drive up costs of several inputs in this category.

PROGRAMMATICS IMPLEMENTATION *(Planetary Level 2)*			
Cost Driver	Low Dollar Value	Medium Dollar Value	High Dollar Value
Encounter Criticality	___ Partial Success	_X_ Nominal	___ Unique Opportunity
Risk Plan— Spacecraft	___ Class D program, small spacecraft without redundancy	___ Class C program, medium priority and med/high risk	_X_ Class A or B program, high priority and low/med risk
Risk Plan— Instruments/ Payload	___ Simple payload, no redundancy	_X_ Few hazardous operations, limited redundancy	___ Complex, redundant payload
Risk Plan—GDS/ MOS	___ Accept moderate risk to mission safety, data loss > 5%	___ Minimize risk to mission safety, accept ~5% data loss	_X_ Minimize risk to mission safety, data loss < 2%
Verification Requirements	___ Relaxed	___ Nominal	_X_ Stringent
Hardware Redundancy	___ Limited or no redundancy	___ Selected redundancy	_X_ Full redundancy with rapid switchover
Software Redundancy	___ Limited or no redundancy	_X_ Selected redundancy	___ Full redundancy
Crosstraining/ Staffing Overlaps	___ Fully crosstrained	_X_ Crosstrained within functions	___ Limited crosstraining
Staff Experience	_X_ More than 2 similar previous missions	___ 1 to 2 similar previous missions	___ New operations team

For the GDS/MOS inputs, we assume the high cost value for MOS Heritage, due to operations center design changes to support high data rates and volumes from the science payload. The other inputs will be assigned the medium cost value. We list these inputs in Table 5.21.

To account for the complexity and number of science instruments, we assign the high cost value to the Instrument Support Complexity input, as shown in Table 5.22.

For the spacecraft design, we assume standard practices for all the Level 2 inputs except the Downlink Data Rate, which is assigned the high cost value because of the need to support a substantial science payload. Table 5.23 gives these inputs for MOE.

Table 5.21. GDS/MOS Implementation (Planetary Level 2). Significant changes to the MOS lead to a high-cost MOS Heritage input.

GDS/MOS IMPLEMENTATION *(Planetary Level 2)*			
Cost Driver	Low Dollar Value	Medium Dollar Value	High Dollar Value
MOS Autonomy	Reuse of proven, sophisticated autonomy	_X_ Simple and robust safe mode/onboard telemetry monitor	Several complex safe modes or experimental autonomy
MOS Flexibility	Low	_X_ Medium	High
MOS Heritage	> 85%	~ 75%	_X_ < 60%

Table 5.22. Science Implementation (Planetary Level 2). This input stresses instrument constraints and interactions.

SCIENCE IMPLEMENTATION *(Planetary Level 2)*			
Cost Driver	Low Dollar Value	Medium Dollar Value	High Dollar Value
Instrument Support Complexity	Simple payload with few science observation constraints	Routine calibrations, few scheduling constraints	_X_ Constrained, complex instrument interaction (w/ other instruments and spacecraft)

Table 5.23. Spacecraft Design Implementation (Planetary Level 2). These inputs include not only flight characteristics but spacecraft interactions with the ground.

SPACECRAFT DESIGN IMPLEMENTATION *(Planetary Level 2)*			
Cost Driver	Low Dollar Value	Medium Dollar Value	High Dollar Value
Spacecraft Complexity	Spinner	_X_ 3-axis	Dual-Spin/3axis+
Spacecraft Autonomy	Significant onboard capabilities, minimal ground contact	_X_ Moderate onboard telemetry monitoring capabilities	Minimal onboard capabilities, significant contact with ground
Telemetry Downlink Data Rate	< 0.8 kbps	0.8 – 2 kbps	_X_ > 2 kbps
Data Return Margin	> 2	_X_ 1 – 2	< 1
Power Margin	> 20%	_X_ 0 – 20%	< 0%
Memory Margin (storage capacity)	> 3 × 1 pass	_X_ 2 – 3 × 1 pass	< 2 × 1 pass
Propulsion Margin	> 10%	_X_ 5 – 10%	< 5%

5.2 Operations Cost Estimations

Outputs for Planetary Example. Table 5.24 shows results from the SOCM Level 2 estimate for MOE. The Level 2 estimate is substantially higher than the Level 1 estimate because so many Level 2 inputs are assigned a high cost value and few are assigned the low cost value. The increase is just over 50% and is outside the 30% accuracy of the Level 1 output. This highlights the importance of working through as many of the Level 2 inputs as possible, even when little is known about the concept being modeled. The Level 2 result is considered substantially more accurate. These results do not include costs for the DSN Tracking Network service.

Table 5.24. SOCM Level 2 Outputs for the Mars Orbiter Example (MOE). The large number of high-cost Level 2 inputs accounts for this estimate being more than 50% higher than the Level 1 estimate.

	Level 2 Mission Operations Estimate—Phase E			
	Costs are FY 2002			
Mars Orbiter Example (MOE)	Cruise	Encounter	Post-Flight Data Analysis	Totals
Annual FTE/$ Estimates				
Flight Operations	63.9	88.4		
Nav/Tracking Operations	10.6	16.0		
Science Operations	12.7	26.4	15.6	
Total FTEs/yr	87.2	130.8	15.6	
Annual FTE Cost	$14.9	$21.5	$2.9	
Annual Operations Serv.				$2.9
Summary				
Phase duration (mo)	14.0	24.0	6.0	44.0
Total Operations Services				
Total FTE $M	$17.4	$43.1	$1.5	$62.0
Total $M				$62.0

	Level 2 Mission Operations Cost Estimate			
	2002 Constant FY $K			
Mars Orbiter Example (MOE)	Phase E Cruise	Phase E Encounter	Post-Flight Data Analysis	Phase E Total
1.0 Mission Planning and Integration	201.7	749.2		950.9
2.0 Command/Uplink Management	746.3	2712.7		3459.0
3.0 Mission Control and Operations	1394.2	5176.7		6571.0
4.0 Data Capture	3707.7	6931.0		10,638.7
5.0 POS/LOC Planning and Analysis	2196.3	5713.9		7910.2
6.0 Spacecraft Planning and Analysis	2341.6	5282.1		7623.7
7.0 SCI Planning and Analysis	754.2	3956.3		4710.5
8.0 Science Data Processing	288.5	790.8	197.7	1277.0
9.0 Long-Term Archives	412.0	770.1	192.5	1374.6
10.0 System Engineering, Integration, and Test	1529.9	3816.6		5346.5
11.0 Computer and Comm Support	718.5	1965.2	491.3	3175.0
12.0 Science Investigations	865.5	2372.4	593.1	3831.0
13.0 Management	2230.1	2862.3		5092.3
Project Direct Total	**17,386.5**	**43,099.4**	**1,474.6**	**61,960.5**

Table 5.24. SOCM Level 2 Outputs for the Mars Orbiter Example (MOE). (Continued) The large number of high-cost Level 2 inputs accounts for this estimate being more than 50% higher than the Level 1 estimate.

Mars Orbiter Example (MOE)	Level 2 Mission Operations FTE/yr Estimate			
	Phase E Cruise	Phase E Encounter	Post-Flight Data Analysis	FTE/yr Average
1.0 Mission Planning and Integration	1.0	2.1		1.7
2.0 Command/Uplink Management	3.6	7.6		6.1
3.0 Mission Control and Operations	10.2	22.2		17.8
4.0 Data Capture	9.5	10.3		10.0
5.0 POS/LOC Planning and Analysis	10.6	16.0		14.0
6.0 Spacecraft Planning and Analysis	17.2	22.6		20.6
7.0 SCI Planning and Analysis	4.5	13.7		10.3
8.0 Science Data Processing	1.8	2.9	2.9	2.9
9.0 Long-Term Archives	1.1	1.1	1.1	1.3
10.0 System Engineering, Integration, and Test	7.4	10.7		9.5
11.0 Computer and Comm Support	1.8	2.9	2.9	3.0
12.0 Science Investigations	5.4	8.6	8.6	8.8
13.0 Management	13.3	9.9		11.2
Project Direct Total	**87.2**	**130.8**	**15.6**	**117.2**

5.2.5 SOCM Earth Orbiting Mission Example

Earth Orbiting Level 1 Inputs. Figure 5.12 shows the Level 1 Earth orbiting inputs. Inputs for Payload Characterization are not as detailed as for planetary missions and only discriminate between imaging and non-imaging payload types. Since these missions generally collect substantially more science data than planetary missions, we include an input for GDS/MOS Characterization to indicate the role of the science team in the overall operations support plan.

Earth Orbiting Level 1 Example—Orbiting Telescope. The Earth orbiting mission example is a search for Earth-like planets in other solar systems using a telescope operating from a libration point. The telescope will be directed at specific targets defined by the science team. We'll call this mission the Planet Finder Example (PFE) and assume it will operate for four years plus an additional year after flight operations for post flight data analysis. Since the mission is looking for specific targets in deep space, the Mission Type is Targeted-Space Science and the Tracking Network is the DSN.

We assume the Mission Risk Class is comparable to past Explorers: a long development schedule, management by NASA, and a mix of in-house and contracted hardware and services.

PFE will be operated from a multi-mission control center and use heritage systems from past government projects. The science team will be responsible for data processing and analyzing instrument health, but the mission control team will develop final sequences.

PFE will have only one instrument, the telescope. Pointing requirements will not be too stringent. We'll assume a medium sized guest investigator program and

5.2 Operations Cost Estimations

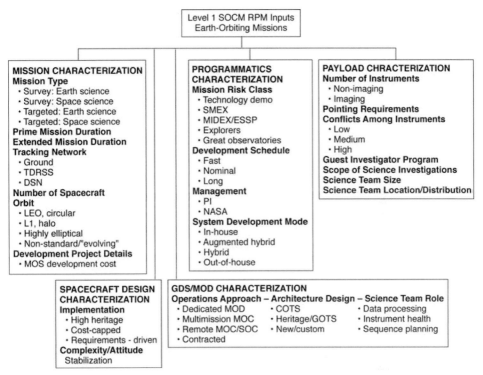

Fig. 5.12.   SOCM Level 1 Inputs for Earth Orbiting Missions. The science team participation has a considerably greater effect on costs for an Earth-orbiting mission than for a planetary one.

5–10 separate science investigations using the data. The science team will include about 25 members, who will obtain data from a central science operations center. For this mission, we assume a high-heritage 3-axis stabilized spacecraft. For staffing assumptions, we use the SOCM model defaults, which include a mix of junior and senior government, university, and industry staff. Tables 5.25 through 5.29 summarize the Level 1 inputs as they apply to PFE.

Earth Orbiting Level 2 Inputs. Figure 5.13 shows Level 2 inputs for Earth orbiting missions. Details describing the GDS/MOS Implementation dominate the Level 2 inputs and identify performance requirements and expectations for the entire ground system.

Earth Orbiting Level 2 Example—Orbiting Telescope. For Mission Implementation, the only input not assigned a medium cost value is Operations Type, which is targeted. See Table 5.30.

We assume the multi-mission operations staff have supported more than two similar projects and are fully cross-trained. See Table 5.31.

Table 5.25. Mission Characterization (Earth Orbiting Level 1). A high-cost mission type can drive up the costs for other Mission Characterization inputs.

Mission Name: ____PFE____				
Total Nominal Mission duration (mo.): _48_				Total Extended Mission duration (mo.): __0__
Total Post-Flight Data Analysis duration (mo.): ___12___				
MISSION CHARACTERIZATION (Earth Orbiting Level 1)				
Mission Type		Tracking Network	Orbit	# of Flight Systems
___ Survey - Earth Science		___ Ground	___ LEO, circular	_1_ enter number of identical Flight systems
___ Survey - Space Science		___ TDRSS	_X_ L1, halo	
___ Targeted - Earth Science		_X_ DSN	___ Highly Elliptical	
_X_ Targeted - Space Science			___ Non-Standard/Evolving	

Table 5.26. Programmatics Characterization (Earth Orbiting Level 1). The PFE mission involves fairly complex, hence costly, programmatics.

PROGRAMMATICS CHARACTERIZATION (Earth Orbiting Level 1)			
Mission Risk Class	Development Schedule	Management Mode	Contract Type
___ Tech. Demo (tech > sci)	_____ Fast (< 2.5 yrs)	_____ PI	_____ In-House
___ SMEX	_____ Moderate (2.5-4 yrs)	__X__ NASA	_____ Augmented Hybrid
___ MIDEX/ESSP	__X__ Long (> 4 yrs)		_X_ Hybrid
_X_ Explorers			_____ Out-of-House
___ Great Observ			

Table 5.27. GDS/MOS Characterization (Earth Orbiting Level 1). PFE takes advantage of heritage from previous missions to lower the GDS/MOS cost estimate.

GDS/MOS CHARACTERIZATION (Earth Orbiting Level 1)		
Operations Approach	Architecture Design	Science Team Role
_____ Dedicated Mission Operations Center (MOC)	_____ Commercial Off-The-Shelf (COTS)	_____ Data Processing
__X__ Multimission MOC	__X__ Heritage/Government Off-The-Shelf (GOTS)	__X__ Instrument Health
_____ Remote MOC/Science Operations Center (SOC)	_____ New/Custom	_____ Sequence Planning
_____ Contracted		

Table 5.28. Science Characterization (Earth Orbiting Level 1). Level 1 inputs include not only instruments but features of the science teams involved in the mission.

SCIENCE CHARACTERIZATION *(Earth Orbiting Level 1)*			
Number of Non-Imaging Instruments	*Number of Imaging Instruments*	*Pointing Requirements*	*Conflicts Among Instruments*
_____ Enter number	__1__ Enter number	_____ Low	__X__ Low
		__X__ Medium	_____ Medium
		_____ High	_____ High

Scope of Guest Investigator Program	*Number of Separate Science Investigations*	*Science Team Size (not all Full Time)*	*Science Team Location/ Distribution*
_____ Small	_____ Fewer than 2	_____ Fewer than 10	_____ Collocated at 1 facility
__X__ Medium	_____ From 2-5	_____ 10-20	_____ Central SOC with 1-2 remote
_____ Large	__X__ 5-10	__X__ more than 20	__X__ Central SOC with 2+ remotes
	_____ > 10	_____ more than 50	_____ Central SOC with wide distribution
			_____ 2 - 3 SOC locations
			_____ Multiple SOCs with wide distribution

Table 5.29. Spacecraft Design Characterization (Earth Orbiting level 1). These input categories are the same as for a planetary mission.

SPACECRAFT DESIGN CHARACTERIZATION *(Earth Orbiting Level 1)*	
Spacecraft Design Implementation	*Design Complexity*
__X__ High Heritage	_____ Low (minimal number of flight rules)
_____ Cost-Capped	__X__ Medium
_____ Requirements-Driven	_____ High (several unique engineering requirements)

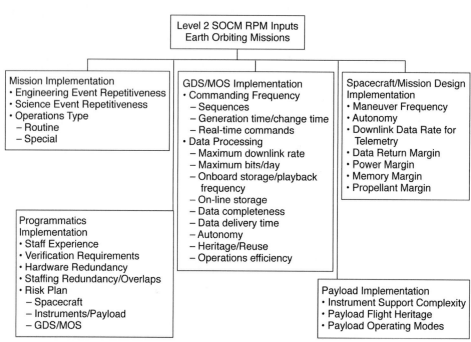

Fig. 5.13.   SOCM Level 2 Inputs for Earth Orbiting Missions. Since the main focus is on data delivery, GDS/MOS inputs receive special emphasis.

Table 5.30.   Mission Implementation (Earth Orbiting level 2). The high-value Operations Type Input for PFE is consistent with the Level 1 Mission Type Input.

MISSION IMPLEMENTATION *(Earth Orbiting Level 2)*			
Cost Driver	Low Dollar Value	Medium Dollar Value	High Dollar Value
Engineering Event Complexity	Routine, non-hazardous events	_X_ Repetitive/no hazardous events	Risky events/significant real-time contact
Operations Type	Survey	Orbit-driven/Activities based on orbital events	_X_ Targeted and/or Constrained
Science Event Complexity	Survey	_X_ Few Constraints	Constrained/multiple observation modes

Table 5.31. Programmatics Implementation (Earth Orbiting Level 2). The emphasis here are on risk tolerance and staff qualifications.

PROGRAMMATICS IMPLEMENTATION *(Earth Orbiting Level 2)*			
Cost Driver	Low Dollar Value	Medium Dollar Value	High Dollar Value
Staff Experience	_X_ More than 2 similar missions	___ 1 or 2 similar missions	___ New operations team
Risk Plan – Spacecraft	___ Small spacecraft, no redundancy, technology demo mission	_X_ Class C, ~$100M flight system development	___ Redundant spacecraft, several $100M development
Risk Plan - Instruments/ Payload	___ Simple payload, no redundancy	_X_ Few hazardous operations, limited redundancy	___ Complex, redundant spacecraft
Risk Plan - GDS/MOS	___ Accept minimum risk to mission safety, and moderate data loss	_X_ Accept moderate risk to efficiency and data loss < 5%	___ Accept minimum risk to efficiency and data loss < 1%
Crosstraining/ Staffing Overlaps	_X_ Fully crosstrained	___ Crosstrained within functions	___ Limited crosstraining
Hardware Redundancy	___ Limited or no redundancy	_X_ Selected redundancy	___ Full redundancy with rapid switchover

For the spacecraft design, we assume more than 20% power margin and onboard memory sufficient to allow missing a downlink without loss of data. See Table 5.32

We assume mostly standard practices for the PFE GDS/MOS. The exceptions include using the high cost value for Maximum Bits per Day offset by the low cost value for On-Line Storage. Science downlinks are assumed to be no more than once per day. See Table 5.33

For the science payload, we'll assume standard practices for all inputs. See Table 5.34

Output for Earth Orbiting Example. Table 5.35 show results from the SOCM Level 2 estimate for PFE. For this example, many of the Level 2 inputs were assigned the low cost value and only a few assigned the high cost value. The Level 2 estimate is substantially lower than the Level 1 result, but within the 30% Level 1 claimed accuracy. These results do not include costs for the DSN Tracking Network service.

Table 5.32. Spacecraft Design Implementation (Earth Orbiting Level 2). Positive margins and relative design simplicity help to keep costs down.

SPACECRAFT DESIGN IMPLEMENTATION *(Earth Orbiting Level 2)*			
Cost Driver	Low Dollar Value	Medium Dollar Value	High Dollar Value
Spacecraft Autonomy	Proven sophisticated autonomy	_X_ Simple robust safe mode; Onboard telemetry monitor	Several complex safe modes or experimental approach
Maneuver Frequency	Once per year or less	_X_ Several times per year	Once a month or more
Data Return Margin (max rate/avg rate)	> 2	_X_ 1.1 – 2	< 1.1
Power Margin	_X_ > 20%	0 – 20%	< 0%
Memory Margin (storage capacity)	_X_ > 2 × 1 pass	1.5 – 2 × 1 pass	< 1.2 × 1 pass

Table 5.33. GDS/MOS Implementation (Earth Orbiting level 2). These inputs represent the primary focus of the mission and thus require careful attention.

GDS/MOS IMPLEMENTATION *(Earth Orbiting Level 2)*			
Cost Driver	Low Dollar Value	Medium Dollar Value	High Dollar Value
Command Frequency – Sequences	Loaded less than once per day	_X_ Daily	Loaded more than once per day
Data Processing – Data Completeness	< 90%	_X_ ~ 95%	> 98%
Data Processing – Data Delivery Time	More than 48 hours	_X_ 24 to 48 hours	24 hours or less
Data Processing – Autonomy	Extensive	_X_ Nominal	Minimal
Data Processing - Heritage/Reuse	More than 85%	_X_ ~ 75%	Less than 60%
Cost Driver	Low$ Value	Med$ Value	High$ Value
Command Frequency - Generation Time	More than one day before upload	_X_ One day before upload	Less than one day before upload
Command Frequency - Real-Time Commands	No commands on some passes	_X_ Routine commands on most passes	Special commands on some passes

5.2 Operations Cost Estimations

Table 5.33. GDS/MOS Implementation (Earth Orbiting level 2). (Continued) These inputs represent the primary focus of the mission and thus require careful attention.

GDS/MOS IMPLEMENTATION *(Earth Orbiting Level 2)*			
Data Processing – Max. Downlink Rate	Less than 1 Mbps	_X_ 1 to 2 Mbps	10s to 100s Mbps
Data Processing - Max. Bits/Day	< 500 Mb	500 – 1000 Mb	_X_ > 1000 Mb
Data Processing - On-Line Storage	_X_ > 8 GB	2 – 8 GB	< 2 GB
Data Processing - Store/Playback Freq.	_X_ Once/day or less	Several times/day	Once per orbit

Table 5.34. Science Implementation (Earth Orbiting Level 2). These inputs concern instrument heritage and use.

SCIENCE IMPLEMENTATION *(Earth Orbiting Level 2)*			
Cost Driver	Low Dollar Value	Medium Dollar Value	High Dollar Value
Instrument Support Complexity	Simple instrument with few operations	_X_ Routine calibrations, few sched constraints	Constrained operation, complex instr interactions
Payload Flight Heritage	Most instruments have flown together; No advanced technology	_X_ Most instruments have flight heritage	New instruments; Payload includes advanced technology
Instrument/Payload Operating Modes	2-3 operating modes per instrument; Single observing mode for all instruments	_X_ Fewer than 3 operating modes per instrument; 2-3 observing modes	Several instruments with multiple operating modes; 3+ observing modes

As we said, it is important to compare the output of the cost model with the grass roots estimate. We need to understand any significant differences. We can also use the model to vary input parameters and see how sensitive the operations costs are to variations in the input parameters.

5.2.6 Goals for Improvement

The SOCM tool has been in use for many years and has generated reasonable estimates for most mission candidates. The approach does a good job with a limited set of inputs (Level 1) and in modifying the estimate based on additional input details (Level 2). NASA plans to improve the model to account for the latest technology advances and most recent mission experience.

Table 5.35. SOCM Level 2 Outputs for the Planet Finder Example (PFE). The many low cost value inputs for Level 2 contributed to an overall estimate that is significantly lower than that for Level 1.

PFE	Level 2 Mission Operations Estimate—Phase E			
	Costs are FY 2002			
	Nominal	Extended	Post-Flight Data Analysis	Total
Annual FTE/$ Estimates				
Flight Operations	27.2	13.6		
Nav/Tracking Operations	0.3	0.1		
Science Operations	53.8	26.9	38.2	
Total FTEs/yr	**81.3**	**40.7**	**38.2**	
Annual FTE Cost	**$13.3**	**$0.0**	**$6.5**	**$12.0**
Annual Operations Serv.				
Summary				
Phase duration (mo)	48.0	0.0	12.0	60.0
Total Operations Services				
Total FTE $M	**$53.4**	**$0.0**	**$6.5**	**$59.9**
Total $M				**$59.9**

PFE	Level 2 Mission Operations Cost Estimate			
	2002 Constant FY $K			
	Phase E Nominal	Phase E Extended	Post-Flight Data Analysis	Phase E Total
1.0 Mission Planning and Integration	1459.9	0.0		1459.9
2.0 Command/Uplink Management	3416.3	0.0		3416.3
3.0 Mission Control and Operations	3898.2	0.0		3898.2
4.0 Data Capture	3108.1	0.0		3108.1
5.0 POS/LOC Planning and Analysis	201.9	0.0		201.9
6.0 Spacecraft Planning and Analysis	571.8	0.0		571.8
7.0 SCI Planning and Analysis	9819.9	0.0		9819.9
8.0 Science Data Processing	14,516.2	0.0	3629.1	18,145.3
9.0 Long-Term Archives	6569.1	0.0	1642.3	8211.4
10.0 System Engineering, Integration, and Test	4112.7	0.0		4112.7
11.0 Computer and Comm Support	1920.3	0.0	480.1	2400.3
12.0 Science Investigations	3008.4	0.0	752.1	3760.5
13.0 Management	749.7	0.0		749.7
Project Direct Total	**53,352.4**	**0.0**	**6503.5**	**59,855.9**

Table 5.35. SOCM Level 2 Outputs for the Planet Finder Example (PFE). (Continued) The many low cost value inputs for Level 2 contributed to an overall estimate that is significantly lower than that for Level 1.

PFE	Level 2 Mission Operations FTE/yr Estimate			
	Phase E Nominal	Phase E Extended	Post-Flight Data Analysis	FTE/yr Average
1.0 Mission Planning and Integration	2.0	1.0		2.0
2.0 Command/Uplink Management	4.8	2.4		4.8
3.0 Mission Control and Operations	8.4	4.2		8.4
4.0 Data Capture	2.3	1.2		2.3
5.0 POS/LOC Planning and Analysis	0.3	0.1		0.3
6.0 Spacecraft Planning and Analysis	1.2	0.6		1.2
7.0 SCI Planning and Analysis	17.0	8.5		17.0
8.0 Science Data Processing	26.4	13.2	26.4	33.1
9.0 Long-Term Archives	4.9	2.4	4.9	6.1
10.0 System Engineering, Integration, and Test	5.8	2.9		5.8
11.0 Computer and Comm Support	1.4	0.7	1.4	1.8
12.0 Science Investigations	5.5	2.7	5.5	6.9
13.0 Management	1.3	0.7		1.3
Project Direct Total	**81.3**	**40.7**	**38.2**	**90.9**

One of the biggest weaknesses in SOCM is the inability to estimate costs for pre-launch development of GDS/MOS elements. These costs are usually estimated with separate development models, most of which do not adequately incorporate the effects of many operations cost drivers.

Better communication of cost model results to the MOM and operations teams is needed. Cost analysts should study model results carefully to identify and focus attention to potential problems. The analyst and the MOM should understand how the models work and how specific approaches affect them. Feedback from the MOM and other operations experts on the cost modeling process substantially enhances the quality and utility of future cost model development efforts.

References

Adams, M. and W. B. Gray. "Design and Implementation of the Mission Operations System Cost Model." Jet Propulsion Laboratory D-3119 March 15, 1986.

Carraway, John. "FY'94 End-of-Year Report for Technical Infrastructure Task #962-93201-0-3170-MOS Cost Model." Jet Propulsion Laboratory IOM #JBC-317-02-16-95, February 16, 1995.

Carraway, John. "Lowering Operations Costs Through Complexity Metrics." November 15, 1994. Space Ops '94. Greenbelt, Maryland.

Carraway, John. "MO&DA Cost Estimates for Three 2012 Mission Set Scenarios." Jet Propulsion Laboratory IOM # JBC-317-12-11-92, December 11, 1992.

Kohlhase, C. E. "Criteria for Identifying and Funding Cassini Operability Improvements." Jet Propulsion Laboratory IOM # CAS-CEK-04-93, January 21, 1993.

Squibb, Gael. "MO&DA Cost Model Technical Inputs." Jet Propulsion Laboratory IOM # TMOD.GFS.94.003:jmm, July 15, 1994.

Chapter 6

Defining and Developing the Mission Operations System

David E. Kaslow, *Lockheed Martin*
Jeffrey K. Shupp, *Lockheed Martin*

6.1 Definition and Development Process
6.2 Monitoring Definition and Development
6.3 Scenarios and Subsystems to Monitor

The mission operations system consists of the space and ground assets needed to accomplish mission objectives. The system divides into components we call elements—in categories such as space, ground, and communication. One element is mission operations, which includes tasking to meet mission objectives, carrying out the mission, and processing and distributing mission data [Larson and Wertz, 1999].

In this chapter we describe how to develop concepts, requirements, and designs for the system and its elements. Mission concepts, plus specified requirements for the system and interfaces, describe how the system and its elements interact. In the same way, each element has mission operations concepts and element specifications. This chapter focuses on defining and developing the ground element for mission operations, with examples of mission execution. Mission execution includes collecting mission data and maintaining the spacecraft and ground equipment.

Early control of requirements and design for mission operations and the entire system is critical if we want to develop and operate at the lowest effective cost. Delays in understanding and stating requirements for mission operations become much harder, and more costly, to deal with as the project moves into development. Yet such delays are typical. In this chapter, we present a process that the mission operations manager (MOM) can use to successfully define and develop a space system.

The MOM leads the Mission Operations System through definition and development into operations to produce:

- Definition and development phases that are within cost and schedule constraints
- Efficient and cost-effective operations
- Operations that meet mission objectives

This chapter presents the information MOMs need to manage and judge the adequacy of definition and development, based on their evaluation and approval:

- The definition and development plans, written at the beginning of the definition and development phases. These plans present in detail the management and technical work that must be done, along with the corresponding schedules.
- The major design reviews, which include lower-level reviews and formal presentations to establish approved baselines for concepts, requirements, and design specifications.

We have organized this chapter into the following three major sections covering the four topics illustrated in boxes in Fig. 6.1:

- Section 6.1: Definition and Development Process

    In this section we look at how the system and mission operations element evolve through definition and development. (See first box in Fig. 6.1.) We outline the process, showing how the mission statement evolves into system requirements and how concepts evolve into element design and development. To clarify, we exemplify the definition, decomposition, and design trades needed to achieve a successful preliminary design review for mission operations.

- Section 6.2: Monitoring Definition and Development

    In this section we discuss how the MOM monitors definition and development of mission operations and the system. This monitoring includes ensuring that the system requirements are properly incorporated into the design of the mission operations element. We also discuss which attributes should be monitored. (See second and fourth boxes in Fig. 6.1.)

- Section 6.3: Scenarios and Subsystems to Monitor

    In this section we show and discuss example scenarios and subsystems that need monitoring. (See third box in Fig. 6.1.)

6.1 Defining and Developing the Mission Operations System

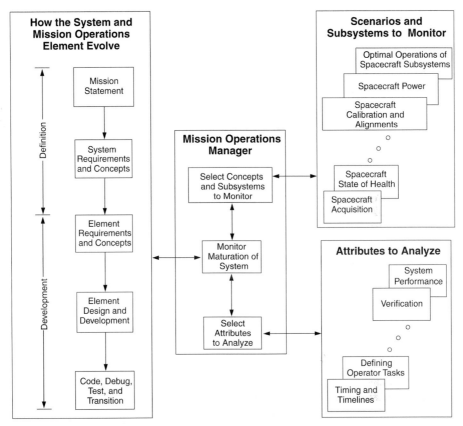

Fig. 6.1. Monitoring Definition and Development of a System. Analyzing attributes of key scenarios and subsystems underpins the mission operations element's evolution throughout definition and development.

6.1 Definition and Development Process

Definition and development for both the system and ground operations element begin with a mission statement and end with the design of the software, hardware, and operations. We'll review the process as outlined in Fig. 6.1 and then present a lower-level overview and more detailed discussion of the components illustrated in Fig. 6.2.

- Define system concepts and requirements
- Define element concepts and specifications
- Define configuration items

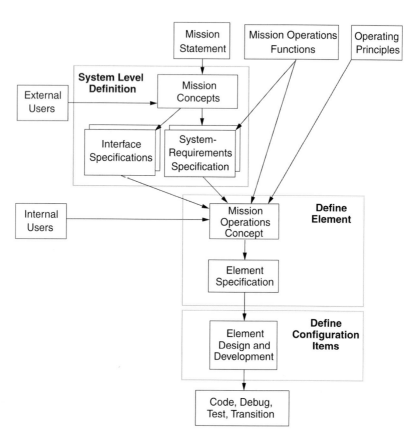

Fig. 6.2.   Evolution of System and Mission Operations Elements. The mission concept and mission operations concept drive how we define the system and mission operations element, respectively. The high-level functions establish the basic processing entities of the element. The operating principles provide the basic approach to mission operations on the ground.

The first three processes can be accomplished in a water-fall schedule or by

- Concurrent definition of system and elements

Taking a system from a mission statement to designing software executables and hardware components spans two very different periods in a system's or element's development. The first period, as illustrated on the left of Fig. 6.1, covers the definition phase for the system and mission element. We start this phase by issuing the mission statement. We end with the system defined well enough that requirements for the system and interfaces are contractually binding and we can accurately assess the cost to develop the mission operations element. In other words, definition proceeds until the system-requirements specification articulates all system objectives defined in the mission concept. A *requirements specification*, as used in this chapter, is a document that organizes requirements and contains information on verification of the requirements, verification method, and any necessary performance results. The requirements in the system specification have to be clear enough that designers of the mission operations element can unambiguously interpret and break them out in a way that maintains the system's intent. This step is necessary to put together a not-to-exceed cost for the element's required capabilities. It ends when we review the element's functions and validate the designer's interpretation and response to system requirements.

Although we don't discuss costing in this chapter, the MOM must have accurate and complete costs. High estimates can discourage projects, while low estimates will result in overruns which demand changes to requirements and designs in order to recover the overrun costs. Estimating the cost to develop a ground element for mission operations is very difficult. First, we develop an element architecture, based on element requirements. The architecture includes hardware, software, and operations components; operations and support-computer architecture; support infrastructure such as maintaining software and managing configurations; and operations and support facilities. Then we estimate the cost to develop the element architecture. The cost includes managers, engineers, and development personnel and facilities.

The second period covers development of the mission operations element. The phases of development are shown in Table 1.2 of Chap. 1. We start with the element specification and end with software, procedures, and hardware installed and transitioned to operations at the ground station. This process evolves based on results of identified design trades, as concepts and designs mature. This phase comprises many distinct sub-phases. At a high level, the sub-phases are preliminary design, which ends in the preliminary design review (PDR); critical design, which ends in the critical design review (CDR); code, debug, and test; and installation, checkout, and transition to operations. This chapter focuses on the PDR and CDR because they encompass the design trades and analysis needed for cost-effective design of the mission operations element.

In Fig. 6.2 we illustrate a classical waterfall approach to definition and development. In this model, understanding the mission capabilities moves from the system level, to the element level, to the configuration item level. The advantage of this approach is that concepts and requirements mature at one level before flowing down to a lower level. But compared to the concurrent approach presented in Sec. 6.1.4, it requires more time for definition and development and it is difficult to feed back changes from lower levels to higher levels. At the top of Fig. 6.2 is the mission statement, which defines the system's purpose. In the system-level definition, the mission concepts describe how the system's elements react to various influences to meet the mission objectives defined in the mission statement. In doing so, they manifest the relationships between elements, as well as the roles of the elements themselves in meeting the mission objective. Recall that the system is made up of several elements; our focus here is on the ground element for mission operations. For each element, mission concepts treat the other elements as external users, with the roles of the element defined in high-level terms. We stipulate the role of each element in the system's requirements specification for that element, whereas we stipulate its relationship to the other elements in the interface specifications. We can construct requirement specifications for each element in a topical or a functional organization, each with its own advantages and disadvantages. We provide examples in this section to illustrate the issues involved in identifying the system's abilities in the system-level specifications.

In defining the elements, system designers use the mission operations concept to turn the system's requirements specification into the element design. This concept guides how designers define the context within which the element meets the system-level concepts. Stated another way, we use the mission operations concepts for element interpretations of upper-tier specifications. As we show in Fig. 6.2, the mission operations concepts derive not only from the upper-tier specifications but also from the needs of the element's internal users (operators and maintainers), facility constraints, and basic processing procedures used as operating principles. Through analysis, the element designers determine how the element design will satisfy the higher-level definition of the system's requirements. This is true regardless of whether a single task within the element or several tasks working together embody the upper-tier requirement. The mission concepts also define the role of each of the element's tasks (software, hardware, and operators) in meeting the upper-tier requirements. Equally important, these concepts define the interfaces between those tasks. The element specification then specifies all roles and relationships.

A *configuration item* (CI) is a logical grouping of tasks that we develop and test as a unit. When element designers and developers define configuration items, they produce a more mature element design, designating tasks as capabilities and finally developing them into software executables and hardware components. They complete development through a long period of coding, debugging, and testing before handing the CIs off to operations.

Definition and development proceeds by phases but is also characterized by a see-saw effect of "what" versus "how" steps. In general, steps that involve requirement specifications describe what the system, element, or executable must do; steps that involve concepts and design describe how the system, element, or executable meets those requirements. To clarify how these steps operate, we offer a concept for operating spacecraft's state of health from the spacecraft and ground perspectives, for the system and its elements.

6.1.1 Define System Concepts and Requirements

As we highlighted in Fig. 6.2, the system-level definition covers the mission concepts and the requirements specifications for systems and element interfaces.

6.1 Defining and Developing the Mission Operations System

In Table 6.1 we illustrate the attributes and characteristics of this definition, which breaks out into five main parts:

- Establish Mission Concept
- Specify System Requirements
- Organize System Requirements—Topically
- Organize System Requirements—Functionally
- Specify Interfaces

In this subsection, we'll discuss each of these parts and then run through an example showing how to develop concepts and specifications.

Table 6.1. System-Level Definition. The mission concepts, system-requirements specifications, and interface specifications define all the elements at the system level.

Document	What the Document Does
Mission Concept • Definition of elements • Inputs and outputs of system	Allocates tasks and timelines to elements
System Requirements Specification	Allocates requirements to each element
Interface Specification • Inputs and outputs among elements • Data format, frequency, and content • Constraint information	Describes data flow, control flow, and constraints among elements

Establish Mission Concepts. The mission concepts establish the system design at the highest level. Here, system designers break the system into elements and determine a high-level concept of the role each element is to play within the space-mission architecture. Designers organize elements logically within data and control boundaries, as well as along space and ground boundaries. The mission concepts describe the inputs and outputs of the system and specify at a high level how the system responds to satisfy the mission statement.

Specify System Requirements. System designers must specify system requirements, based on the mission concepts, for each of the elements in the system. To each element, every other element is an external user—either providing input to, or receiving output from, another element. The mission operations element on which this chapter focuses is a central clearing house of data and control for the mission architecture. In this role, it communicates with space and ground assets to conduct the spacecraft's mission and state of health operations. It receives tasking data which drives control of the payload and it prepares data for correlating the spacecraft-mission data to the tasking data. Virtually all mission operations elements share high-level responsibilities, or functions, as discussed in Chap. 3. The system-requirements specification for the mission operations element identifies all of these functions, but we may organize the requirements levied against them in several different ways.

Organize System Requirements Topically. The topical approach states how the element must respond to each system stimulus. It complements a systems-analysis technique developed by McMenamin and Palmer and known as *Essential Requirements Analysis*. [McMenamin, 1984] This technique involves analyzing the system's inputs and outputs to determine the essential nature (no design constraints imposed) of the processing required to respond to each of them. This approach has the advantage of leaving the internal-element design more open and not implying control and data relationships between the element's functions. It also tends to define the element requirements at a higher level, which allows the element designers more freedom in developing the element without constraints on the technology to be used. Its main disadvantage is that the requirements against an element's function are distributed throughout the document. Also, a requirement assigned to a function within a particular element may be repeated in the system-requirements specification under different topics. For example, two completely different input events to the mission operations element may require us to extract data from the downlinked telemetry. A topical organization would place a telemetry-processing requirement under two topics. But this requirement redundancy may add choices to the element design. Although the need for telemetry-based data may be the same under the two topical areas, one of the events may require a much faster response than the other. This difference in performance requirements provides a design option to the element that might not be apparent if we organize the system requirements another way.

Designers could take the topical approach to the extreme by specifying system requirements for an element based only on inputs and outputs of the functions relative to external interfaces and operator actions. This kind of topical specification views the element as a black box, discussing only the element's responsibility in meeting external interfaces, without explicitly placing requirements on the element functions to generate the data. Rather than specifying the element's internal operations, it leaves all of these issues to element design. For example, the mission operations element might be required to provide its mission-plan information to another ground element at a particular time each day. The system-requirements specification will discuss the timing of that message, the recipient element, and the responsibility to deliver the plan during defined contingencies. But the specification wouldn't get into the expected timelines for producing the plan or the inputs that influence it. These details are left for the element designers to resolve.

Organize System Requirements Functionally. In this type of organization, system designers specify in one place all system-level requirements against a particular function. Organizing the specification this way allows everyone to see easily how much processing each function requires. But the functional approach has two main disadvantages. First, it tends to specify the relationship between functions, a task system designers normally leave to element designers. By specifying the relationship, we ensure that all the inputs and outputs to a function are delineated. Second, the specification tends to lose the reasoning for a particular requirement because the requirement becomes divorced from the topic that drove it. Going back to the telemetry-processing example above, a functional specification may stipulate only the most stringent processing timeline, instead of the multiple timelines for specific driving events, thereby eliminating the possibility of a design option for the element.

Sometimes the system-requirements specification will have topical and functional organization. But designers should avoid this hybrid form if possible, for it carries the disadvantages of both schemes without any added value. Further, because the upper-tier specification looks and feels like a functional approach, element designers may lose requirements against functions embedded in the topical sections.

Specify Interfaces. The interface specifications include all requirements for data and control passing into or out of the element, as shown in Table 6.1. Stated another way, the mission operations element shouldn't need access to, or knowledge of, any other element's requirements to operate properly within the system. These specifications tell exactly what data is shared between the element and each of its external users. They show how often, and at what times, data is passed, and they specify how the data is to look, so the interfacing elements can read and write the information in a format everyone agrees to.

The interface specification also contains constraint information so one element's processing doesn't cause a processing problem for its interfacing element. This is especially important for the interface specification between mission operations and the spacecraft. The specification includes constraints dealing with command timing, rate and acceleration, and antenna motion. If sensors on board the spacecraft detect a violation of these constraints, the spacecraft could be permanently damaged unless we place it into a safe operating condition. Operating constraints in the interface specification will help the spacecraft meet its on-orbit life expectancy by specifying limits on certain items, such as duty cycle or switching frequencies for equipment that will keep it operating safely. Designers achieve this kind of detail by placing high-level concepts for each element in the mission concept and then specifying constraints as designs for each pair of elements mature.

Example Showing How to Develop Concepts and Specifications. Looking at the spacecraft's state of health is a good way to see how the mission concept helps mold relationships among elements throughout the system. Less sophisticated spacecraft, like Tracking and Data Relay Satellite, are based more on real-time command and control, with very little onboard processing of detected errors. In these cases, the mission operations' role is to recognize the error condition for the spacecraft by interpreting the telemetry as it arrives and to take the appropriate actions, such as quickly sending the proper commands to place the spacecraft into a safe condition. Recovery to normal operations is simple as well. More sophisticated spacecraft, such as the Hubble Space Telescope or the Magellan spacecraft, can detect more error conditions on board and place the spacecraft autonomously into a safe condition. Mission operations' role with this kind of spacecraft is to ascertain, after the fact, what the spacecraft detected that caused the error, to operate the spacecraft in the safe condition for an extended time, and to recover the spacecraft to normal operations. Also, because the spacecraft may put itself into a particular kind of safe condition for each type of error, the recovery is more complex. System designers must analyze the abilities and limits of each spacecraft configuration to determine how much sophistication the spacecraft needs to meet mission and maintenance objectives for the system.

If systems analysis of the spacecraft's state of health shows that the more sophisticated spacecraft better meets the system's objectives, designers must expand the system-requirements specifications and interface requirements to support it. The element specification requires the onboard processing to detect error conditions, such as overlapping commands and violations of the spacecraft's acceleration limits or temperature ranges. The onboard processing must autonomously put the spacecraft into one of several potential safe operating conditions and transmit the needed information to the ground in the telemetry. The requirements specification for mission operations includes requirements to detect when the spacecraft has entered a safe operating condition and to operate the spacecraft in that condition indefinitely. Mission operations must recover the spacecraft to its normal operating condition within a specified time of the go-ahead signal. The interface specification contains details of all of the spacecraft's safe operating conditions as well as the kinds of error conditions the spacecraft must detect.

6.1.2 Define Element Concepts and Specifications

As we show in Fig. 6.2, the system-requirements specification, along with interface specifications for external users, drives what the element does. Element designers specify the element's response to these drivers by defining the element while developing the mission operations concept and the element specification. Table 6.2 lists the attributes and characteristics of the element-level definition. This subsection addresses

- Mission Operations Concept
- Internal Users and Mission Operations Concept
- Definition of High-Level Functions and Mission Operations Concept
- Operating Principles and Mission Operations Concept
- Design Trades and Mission Operations Concept
- Element Specification

6.1 Defining and Developing the Mission Operations System

Table 6.2. Defining the Mission Operations Element. The mission operations concept and element-requirements specifications define the mission operations element.

Document	What the Document Does

Mission Operations Concepts	Allocates tasks and timelines to hardware and software configuration items and operations
	• Establishes processing and operations tasks for the element
	• Lists information required and products
	• Responds to internal users
	  – Mission operators
	  – Data-systems operators
	• Implements high-level functions through processing tasks
	• Defines key operating principles
	• Acts as source of design trades

Element-Requirements Specification	Allocates requirements to hardware and software configuration items and operations • Captures requirements from concepts • Assigns tasks to configuration items • Reduces system requirements to individual configuration items • Identifies interface specifications and constraints

Mission Operations Concept. Element designers use the mission operations concept to set forth the highest-level design at the element level. Here, they show generally how each processing and operator task will work within the high-level element functions. The tasks, which we'll discuss later, are grouped logically into the Computer Software Configuration Item (CSCI) and Hardware Configuration Item (HWCI) definitions, with operator tasks and procedures (Ops) added where necessary. In general, designers group tasks into CIs to minimize data and control flow between tasks of different CIs while keeping tasks of the same CI relevant to each other. Mission operations concepts describe the element's inputs to, and outputs from, each CI. They also describe, at a high level, how the element's tasks work together to meet its system-requirements specification. For example, one concept will define how plan update, a task assigned to activity planning, tells command generation, a task assigned to activity scheduling, that the plan has changed. The mission operations concepts also overview the operators' roles and the kind of process interaction the operators will see in the operational element. To provide enough insight into the processing and operator tasks, these concepts delve into characteristics within and between CIs, as necessary, to indicate the element's response to a system requirement.

Internal Users and Mission Operations Concept. As we show in Fig. 6.2, another influence on the element's requirements, and ultimately its design, is how the element responds to its internal users. For a ground element, these users include operators, people that maintain hardware and software, engineers that tune the element's performance in operations, and support services, such as platform hosting and layout, network management, and data-center operations.

The most visible of a ground element's internal users are its mission operators. As with the interface specifications between two elements, designers lay out internal-user interfaces very early and then iterate them as the element evolves. Even though these interfaces are extremely important to cost-effective element design and requirements, designers often don't take them as seriously as the external-user interfaces. Although design drives the ultimate look and feel of an element's operations, they also depend on concepts for internal-user interfaces at this level. Additionally, items such as limits on station staffing for operations and maintenance, training for station staff, and the amount of operator involvement versus software automation, will strongly influence the ground element's design.

Technologies for computing and database management are changing rapidly, and it may take five or more years to define and develop systems. As a result, the computers and data-systems architectures established in the definition phase drive a ground element's design. Thus, element designers must decide whether to use technologies that meet general industry standards or to develop new ones, keeping a careful eye on lifecycle costs.

Definition of High-Level Functions and Mission Operations Concepts. By defining functions at a high level, designers provide the seeds for a good mission operations concept, which must define all of the element's processing tasks. Then, using these definitions of high-level functions to define tasks, element designers can determine the scope, control, and flow of data for each function. That means they can analyze and trade design attributes of the element long before we create software executables and hardware components. Further, by putting together element tasks, the mission operations concept helps to complete the element definition. This definition is complete when it covers the system requirements plus the requirements levied by or derived from the needs of internal and external users.

The processing tasks for mission operations depend directly on the language that describes the element's requirements in the system-requirements specification. As an example, mission operations calculates the latest spacecraft ephemeris and may also have to send the ephemeris to other elements. From this very straightforward requirement, we can quickly identify or derive several tasks across the element functions—for example, orbit determination, ephemeris propagation, and external-message handling. Assuming the updated ephemeris is based on information passed through telemetry, we also need tasks for processing telemetry, handling internal messages, and managing databases. The mission operations concept then ties all of these tasks together to describe how the element generates the ephemeris data, constructs the outgoing message, and sends it out.

Operating Principles and Mission Operations Concept. To develop an effective mission operations concept, element designers must also identify key operating principles that the element will follow as it responds to the system requirements. An example is the relationship of activity planning to scheduling.

The designers can look at planning and scheduling from three different perspectives. They can distinguish the two functions by timeframe—for example planning takes care of all activities up to four hours from the present time, and scheduling operates only the first four hours. Or planning handles all general aspects of operations, and scheduling does all detailing. Or planning handles activities that require a lot of operator negotiation before approval, whereas scheduling deals with activities that don't require this negotiation. Designers must work out these high-level operating principles before the mission operations concept begins to mature, because they structure how we identify and analyze tasks that will become part of this concept. If we don't define and thoroughly describe a principle early on, it will take a lot of money to correct this omission later.

Design Trades and Mission Operations Concept. Besides showing how an element's tasks are related, the mission operations concept defines situations and then presents responses, including a timeline for each response. Thus, developing the mission operations concept provides the definitive source of design trades for the element, which in turn may greatly affect requirements for the element's software, hardware, and operations.

Trades based on the mission operations concept can also spawn extensive analysis of software algorithms and changes to requirements. Working with the software engineers, designers of the mission operations concept describe generally how the element will react to a given situation; continuing analysis of the software methods must validate that concept. Failure to do the necessary analysis results in one of two expensive situations for a program: (1) the element will oversolve the problem, putting in complex functions that will rarely, if ever, be used, or (2) the element won't put enough into the design, unrealistically relying on operator procedures to handle what was believed to be a small-chance situation but turns out to be a regular occurrence. The second circumstance is more expensive than the first because we spend more to fix a finished product than to include a new capability during development. The results of the trades are incorporated as specific design concepts and scenarios into the mission operations concept.

Trades on the mission operations concept fall into five categories:

1. Operator- versus software-directed reactions to situations
2. Mission objectives versus timeline margin
3. Task-unique versus generic processing
4. Database versus message-based control and data flow
5. Ability to operate during and recover from a contingency versus loss of the mission objectives during that contingency

These categories are typically the most sensitive areas of the mission operations element's development and lifecycle costs because the trades involve increasing or reducing the mission software's complexities, processor speed and sizing, and data-management volume and complexities. We discuss each of these trade categories below.

Trade 1: Operator- Versus Software-Directed Reactions to Situations

Allocating tasks to operators or software to detect errors and recover from software-executable faults is a prime example of a design trade. Designers can choose software or operators to detect and recover from certain faults, with software providing data to the operator. The trade is between allowing software to gather data and display tasks and then relying on operator tasks to assimilate and isolate the error to its source versus a more automatic approach, which allows the operator to stay focused on the mission. Designers must analyze the error condition, discussing each option in an operations concept in enough detail to understand fully the issues of each option. For an operator-driven solution, the concept states the kind of data the software should collect and display to the operator—for example, having missed receiving an output. It describes how operators determine the source of the error. It also describes any follow-on actions the operator may take to recover from the problem and reconfigure the receiving process to go with an old version of the input data. It establishes online procedures to help the operator isolate the source of an error. Of course, designers have to consider how much they're loading the operator, what's happening with the main task while an operator isolates the error condition, and whether there's enough time to recover mission processing.

Trade 2: Mission Objectives Versus Timeline Margin

Certain tasks may require solutions of very high quality. In these cases we need high-fidelity modeling or complex algorithms for mathematical searches. Defining a detailed, finite-element thermal model to predict a spacecraft subsystem's temperature is an example of such a complexity. Complexity, plus the need for a quick solution, creates an opportunity for a concept trade. One way to resolve the issue is to add data-processor speed or capacity by using parallel processors instead of serial processors. Another way is to use a two-stage solution: a coarser model does most of the processing, and a detailed model does final processing. A third way is to trade processing time from another task done earlier or later, adding this time to the task requiring the detailed model. Element designers weigh the pros and cons of each of these approaches to arrive at the most effective concept.

Trade 3: Task-Unique Versus Generic Processing

Often designers don't take advantage of synergistic processing when dealing with tasks that have only some unique qualities. For example, the processing tasks that respond to a request for a high-priority activity in activity planning are very similar to those supporting normal-priority requests, although response times and optimizing techniques may differ. Unique tasks offer independence and allow us to tune both processing tasks so we take advantage of these unique attributes. But there's a trade between duplicating tasks to handle variants and creating a common processing task with unique drivers. In general, the more generic the processing tasks, the more cost effective the overall design because duplicating similar tasks often results in duplicating the developed software. Designers must weigh generic processing against the overhead (including timeline and control) for modifying the software to fit the tasks.

Trade 4: Database Versus Message-Based Control and Data Flow

In the software-processing architectures of the 1970s, the data was persistent, and processing tasks existed long enough to read in the data, manipulate it, and put out new results. Data didn't "flow" much because it really didn't move. Since the 1990s, the roles of data and processing tasks have reversed. The processing tasks are now persistent and distributed, whereas the data is temporary. The data goes from one processing task to another over a LAN. The volume of this data can become so immense that it can drive a design because of delays in transmitting the data or the complexity required to keep track of all destinations. Still, a lot of data is extracted from a database instead of being routed. The trade involves determining what data is best sent in a message, and what is best kept in a database. For example, many processing tasks require spacecraft ephemeris. The frequency of the ephemeris time points varies from task to task: some require points at no more than one-minute intervals, while others require time-point intervals of one second or less. The more volume put into a message the longer the delay time from source to destination. Element designers must also consider any limits on storing archives if the element logs and stores its message traffic for troubleshooting.

Trade 5: Ability to Operate During and Recover from a Contingency Versus Loss of the Mission Objectives During that Contingency

Depending on how critical a mission objective is, mission operations may need to remain highly available by providing rapid recovery to backup data processors. The trade is between satisfying a mission objective and adding development and operations effort to provide recovery. For example, designers may need to develop ways to maintain process state, detect contingency situations, and reconfigure software and process state on backup data processors. For operations, software maintains process state while completing normal tasks and provides backup data processors. An element designer must trade off the severity of losing a mission objective with the chance of its occurring. Another factor in the trade is the response time needed to detect the failed primary processors and recover to the backup processors.

Element Specification. Element designers use requirements language in the element specification to encapsulate roles and relationships from the mission operations concept. The specification includes each task, including operator tasks. First, they assign tasks to software, hardware, and operators. Then, they analyze the software and hardware tasks to group them into CIs and the operator tasks to group them into preliminary positions.

Many CIs may go into meeting one system-level requirement. The requirements in the element specification are not simply restated from the system-requirements specification. Element designers analyze requirements based on the mission operations concept, show how each CI meets the system requirements, and show how data and control flow between CIs. If done properly, this step will contribute to a program's success by keeping its costs down. Otherwise, the program will require a lot of effort to redefine the tasks later because further design details will lack the proper insight into what is needed to support the element concepts. This situation doesn't occur until testing begins at the element level. The subsequent requirements and design steps, which we discuss in Sec. 6.1.3, can concentrate on requirements for the individual CIs, rather than constantly backtracking to the element and system-level specifications to understand what a requirement meant.

The element specification, as well as the lower-level specifications we'll discuss next, must explicitly reflect the requirements for interface specifications at the system level, as well as the constraints for which the element is responsible. Furthermore, designers must break down these requirements and constraints whenever necessary to understand fully what the element must do. For example, details about the content of messages exchanged between elements may impart requirements on functions that generate or respond to the external message. If the designers don't stipulate these interface specifications in the element specification and lower documents, they'll have to rework the element's design at the worst time in development—the final stages of element testing.

6.1.3 Define Configuration Items

As highlighted in Fig. 6.2, we further design and develop mission operations by defining an operations concept, requirements specifications, and interface specifications for all CIs. Table 6.3 shows the attributes and characteristics of this definition. This subsection covers

- Operations concepts and capabilities for CIs
- Software and hardware components
- Element-design maturity

Operations Concepts and Capabilities for Configuration Items (CI). Design and requirements for an element see-saw through "how" and "what" steps at two more levels: CI design and physical layout of the software and hardware. The first step below the element specification is the CI design, which includes requirements specifications for accompanying software, hardware, and operators, as well as those for the interfaces between CIs. Just as the mission operations concept describes how the element responds to its requirements, so we must describe how a CI responds to its requirements. The CI operations concept helps define the CI design in that it specifies how the hardware, software, and operations components of the CI interact to satisfy the concepts and requirements specification for the mission operations

Table 6.3. Definition of Configuration Items (CI). We develop operational concepts and requirements specifications for each CI.

Document	What the Document Does
Configuration Item Operations Concept	• Allocates requirements to capabilities • Establishes how configuration items meet element requirements • Assigns capabilities to tasks
Configuration Item Requirements Specification	• Allocates requirements to executables • Form executables from capabilities • Shows design and methods trades

element. This first step builds an understanding of what a CI must do to meet its role as defined in the mission operations concept and element specification.

Configuration item designers then break out CI requirements into capabilities. A capability, in this context, refers to a concept for processing that encapsulates several related requirements. For example, we may follow similar criteria to place different calibration activities onto a planning timeline. Each calibration activity will have a time budget reserved in activity planning, with specific calibration tasks generated in activity scheduling. In the mission operations concept, element designers identify the task as placing the activities into a plan and allowing the operator to interact as necessary. The CI designer may derive a generic capability to do many such activities so they meet the task objectives stated at the element level.

A CI-level operations concept is vital. It describes how the capabilities work together to satisfy requirements. If designers don't develop a CI operations concept, they can't turn requirements into capabilities or form executables. Moreover, the operations concept at the CI level provides a common way to discuss the CI requirements, tasks, and ultimately, capabilities, in terms of how they support the mission objectives. Analyzing capabilities doesn't involve specifying software or hardware design. That will come later. The designers must start developing the CI operations concept early in the definition phase and complete it before the preliminary design review, so they can develop and allocate corresponding requirements at the Preliminary Design Review (PDR).

Usually, the tasks of the CI identified in the mission operations concept reduce to more than one capability. Designers derive more capabilities until the CI can do what is assigned to it. As an illustration, let's go back again to activity placement. While analyzing how this capability works within the CI, designers must determine what triggers it. For example, if the CI must start placing activities based on time, they need a capability to manage this timing, so they should add it to complete the CI operations concept.

Software and Hardware Components. The second level of CI design and requirements specification represents the physical layout of the software and hardware. The software executables and hardware components take their requirements from the CI-requirements specification. At this level, we must define very specific data and control flows between executables. As in all preceding steps, we must analyze requirements to understand fully their impact on the design. And we must work through a set of trades to arrive at the most cost-effective design. CI designers must consider the CI's operations concept within the element and not just "bin" requirements from capabilities to CI executables. If the executable software doesn't behave consistently with its higher-level operations concept, the element design won't close.

Element-Design Maturity. The element design and requirements specifications take the element to the critical design review, which represents the end of a program's design phase. The final output to support the Critical Design Review (CDR) is the detailed design for software, hardware, and operations, as well as components of the data and computer system. It takes the element design down to individual software modules. Defined at this time are prologues, all variables set or used by the software modules, detailed processing logic, actual displays that operators of the element will use, and all static and dynamic data with initial values. The hardware units and modules are at final design and ready for manufacture. The operator tasks and displays are defined, and operator loading is at acceptable levels. A performance analysis demonstrates that timing and timelines close and that we've designed and properly sized the data and computer system. A risk analysis shows there are no unacceptable design or production risks. The element design must be at this level of maturity by the program's CDR, for during the next 1–2 years, the entire element-design team will be absorbed in converting the detailed design into actual code and then testing from the CI level to the element level. When designers of the mission operations element build several hundred thousand to several million lines of code, they can ill afford to go back to the element- or system-level requirement and renegotiate its interpretation.

6.1.4 Concurrent Definition of System and Elements

Figure 6.3 shows a variant, known as the concurrent approach, on the waterfall approach to definition and development we've presented. The concurrent approach has parallel system and element design phases with feedback; the waterfall approach has sequential, non-overlapping system and element design phases. One advantage to the concurrent approach is that the time to produce an element design from a mission statement is reduced because the system-level definition matures in parallel with the element-level design. Another advantage is the ability to work system-level design trades as the element design matures, with least-cost solutions defined before the element reaches the critical design phase. But the concurrent approach does risk having an immature element design because functional and performance requirements may not be complete at the system level. We reduce the risk by using conceptual system requirements. These conceptual requirements aren't contractually binding because they're not in system specifications, but they do provide enough direction for the element designers to do analysis and further assign

6.1 Defining and Developing the Mission Operations System

capabilities to requirements. Then, if design difficulties arise, the system and element designers can work together to solve them.

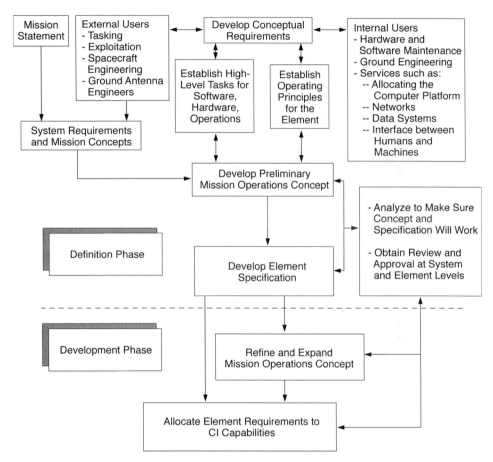

Fig. 6.3. Concurrent Development of Requirements for the Mission Operations Element. The concurrent process is a cost-effective and efficient approach to defining and developing the ground element. Note the parallel design of the system and element, as well as the chances for feedback.

This subsection addresses the following aspects of concurrent definition:

- Conceptual system requirements and external or internal users
- Preliminary operations concepts, high-level tasks for elements, and element operating principles
- Element specification

- Completing definition and starting development
- Benefits and risks of concurrent definition and development

Conceptual System Requirements and External or Internal Users. The source of the element's requirements in concurrent development is the conceptual system requirements established by the external and internal users of the element. The conceptual system requirements express, in high-level terms, what the system should achieve.

Developing conceptual requirements is iterative, based on the element designers' assessment of requirements that drive the conceptual requirements. Each iteration refines the requirements to the point at which we can generate mission concepts, system-requirements specifications, and interface specifications. We commonly trade on roles and responsibilities between elements.

The conceptual requirements associated with external users, as identified in Fig. 6.3, are easier to develop because they're characterized by the inputs, outputs, and timelines required to meet mission objectives. An example of such a requirement at the system level is the throughput of spacecraft mission data, which affects spacecraft processing and transmission, data processing on the ground, space-to-ground communication paths, and activity scheduling.

For each element involved, system designers might write this conceptual requirement as: "Provide a capability to generate, process, or limit processing of up to x bits per second of mission data." They might also describe conditions that inhibit the specified throughput. Rather than specifying the details of how each element responds to this requirement at the system level, the concurrent approach allocates the requirement to all of the affected elements and allows them to determine how they would meet it.

Conceptual system requirements also reflect the needs of the element's internal user, as identified in Fig. 6.3. This is the more difficult set of requirements to develop because they don't deal with specific mission objectives. Instead, they consider the more esoteric desires of element operations, such as operability, maintainability, and testability, and the element's general capabilities, such as data management. This difficulty applies especially to mission operations because this element interacts so much with operators and engineering support. Therefore, element designers must design carefully to make sure conceptual requirements cover all aspects of mission and support operations.

Preliminary Operations Concepts, High-Level Tasks for Elements, and Element Operating Principles. Element designers now define the element's high-level tasks and operating principles, in turn, to support the conceptual requirements. The requirements reflecting the needs of an element's internal and external users significantly influence internal processing. The system's conceptual requirements for speed and complexity determine whether ground processing is highly automated or highly manual. For example, the conceptual requirements from external and internal users will address reacting to a situation that affects the spacecraft's state of health. The speed at which mission operations must react to the situation and establish contingency communications links to the spacecraft will greatly influence its processing tasks and operating principles for spacecraft command and control.

The preliminary mission operations concept responds to conceptual requirements, high-level tasks for elements, and basic operating principles. Preliminary concepts define the general data flow, processing, and timelines that satisfy the requirements. Element designers then define subsets of these preliminary concepts during the definition phase and develop and finalize the rest in the development phase, as the lower-level design fleshes out the concepts. This approach is risky because, if their subset for early development doesn't identify the correct architecture and design for the rest of the element, they'll have to do a lot of redesign when they should be finishing the mission operations concept. Thus, they must place most concepts for design and architecture in the preliminary mission operations concept. Later in this chapter, we'll offer ways to identify key concepts that drive the element's design, effectively taking this risk out of development.

The mission operation concept is also preliminary during definition because system designers are developing the system-requirements specification at the same time. Thus, our mission operations concept becomes a tool for analyzing and negotiating the evolving system requirements. To support this effort, the mission operations concept also feeds into, and is fed by, the element designers' analysis of whether or not the requirements are workable. As a result, we can't finalize the mission operations concept until we completely specify the system requirements and are sure they'll work.

Element Specification. Next, element designers define the element specification based on the conceptual requirements, the system-level requirements, and the preliminary mission operations concept. Designers analyze requirements at a high level and allocate them to the element's CIs, as discussed in Sec. 6.1.2. These requirements define what data and capabilities the element must provide to the external users, to the internal support organizations, and to internal processing. In essence, the element requirements are the contract between the element and the external and internal support organizations.

Completing Definition and Starting Development. Once element designers begin development, conceptual system requirements no longer drive element requirements and processing or operator tasks. Instead, the designers begin focusing on the system-requirements specification and associated interface specifications. The rest of development depends more on analyzing and allocating requirements into functions and tasks, as discussed in Sec. 6.1.3. However, as mentioned previously, the preliminary mission operations concept addresses only the driving subset of the element concepts. Designers must define the rest of the concept, but this step usually overlaps with the early part of steps that establish lower-level requirements and reduce the design to its basic elements. We tolerate this overlap to help us understand all of the tasks driving operations of the element, not just those that influence decisions about architecture and tasks.

Therefore, early in the development phase, designers expand and refine the preliminary mission operations concept to support the element requirements.

Finally, the element design begins. It, too, is an iterative process: defining the lower-level design components (CSCI, HWCI, and operations), analyzing element requirements, deriving additional requirements, and allocating these requirements to the lower-level design components. Element designers complete it based mainly on analyzing the design at these lower levels. Issues related to speed of processing or algorithmic complexity may oblige them to reexamine the operations concept for other options.

Benefits and Risks of Concurrent Definition and Development. The biggest risk with this approach is that system designers may not understand the operations concept of a new system requirement before they allocate it to the elements. System-level trades are also much more difficult because parallel efforts are going on within the system and elements. Ultimately, though, the quality of the analysis used in dealing with system and element trades is much higher because the lower-level design steps are involved early to offer insight into a potential problem that a higher-level step might not otherwise uncover. If designers carry out this approach well, they can attain a cost-effective element design in the least amount of time. But if they don't address the right questions, their incomplete knowledge can greatly drive up cost and schedule overruns.

6.2 Monitoring Definition and Development

The key to the success of any space-ground system lies in refining concepts and requirements for the system and elements. If we analyze and design at the right levels, following the process discussed in Sec. 6.1, the system will have fewer issues to resolve as it approaches operation. As illustrated in Fig. 6.1, mission operations managers (MOMs) participate in system definition and development from the mission statement, through the mission concept and requirements specifications, to the operations concept. They do so by focusing on refining key concepts and configurations or subsystems, some shown in Fig. 6.1.

In this section we'll discuss how MOMs identify early all concepts that drive the element design, analyze them thoroughly, and make them consistent with the conceptual system requirements that drive them. Of special importance are the concepts covering system requirements that are central to satisfying the mission objective, and the concepts that span several elements or several CIs within an element. We also discuss criteria to determine key systems and element concepts. Hence, in this section, we're identifying concepts that must be scrutinized more closely and thoroughly during definition and development.

6.2.1 Design Attributes that Need Monitoring

Figure 6.1 shows that, as the system and element designs begin to mature, we must select several attributes of the designs to monitor during each phase of definition and development. We have to examine each of these attributes in each phase to determine if the element design will comply fully with requirements. Two of these attributes deal with the qualitative characteristics of the element, and three deal with the performance (quantitative) characteristics. One attribute concerns design operability; another concerns testability and maintenance. In the next subsections, we discuss these attributes:

- Processing Concepts
- Definition of Operator Tasks
- Timing and Timelines
- Definition of System Performance
- Optimal Operations of Spacecraft Subsystems
- Definition of Displays, Alarms, and Procedures
- Verification

We discuss each attribute in three parts. The first gives an overview of the attribute. The second covers the criteria that make the attribute key—one we must monitor. The third discusses the MOM's role in monitoring the attribute.

The introduction to this chapter stated that MOMs must lead mission operations through definition and development into operations. We do this by first requiring that plans for technical and management tasks be written and approved at the start of definition and development. We must review those plans in detail to see if element designers and engineers have defined all the tasks, schedules, and people needed to carry out missions operations on the ground. Tables 6.4 through 6.10, representing the attributes respectively, list some of these tasks. Then, through informal and formal reviews of the elements, we make sure the tasks are done and issues resolved according to the attributes discussed below.

Common to Tables 6.4 through 6.10 are the tasks for developing the preliminary and final design. This design specifies in a set of controlled baseline documents how to build the components for hardware, software, and operations. As MOMs, we must approve these documents. We also must require thorough reviews of the design baseline at different phases of the program and approval of the design before proceeding to the next phase.

As part of reviewing the definition and development plans, we should ensure a robust approach to developing the mission operations concepts. These concepts are extremely important—they are the basis for defining, developing, testing, and operating the element. It is very easy to underestimate the amount of work needed to develop a correct and complete set of concepts. We should review the definition and development plans for an approach for developing, reviewing, and approving the concepts; schedules; and manpower estimates. The approach should be detailed enough to cover all aspects of the ground and mission operations element. The schedules should have enough detail so we can see that the concepts and scenarios cover all mission operations on the ground and track incremental progress. The staffing estimates should reflect review and analysis of requirements, discussions with engineers and users, development of alternative concepts, and re-evaluation and refinement of concepts and scenarios as overall knowledge matures.

Processing Concepts. Table 6.4 highlights tasks for developing processing concepts throughout the development cycle up to the critical design review. This attribute is often overlooked at the system and element levels. It shows that we need to understand how components work together at the system and configuration-item levels.

Table 6.4. Data-Processing Concepts. We must establish early and then refine the processing concepts, which define how components of the system and ground-mission element work together.

Phase	Tasks to Cover
Definition	• Establish paradigms for system-level processing • Analyze paradigms to clarify element roles and responsibilities
Development (Preliminary)	• Develop element-level processing concepts consistent with system paradigms
Development (Preliminary Design Review)	• Develop processing concepts for configuration items based on operator task, processing scheme, and concepts for organizing data
Development (Critical Design Review)	• Develop final configuration items and element-level concepts

The element designers must address at least the following areas. They need to establish concepts for starting and stopping processes, as well as for detecting and resolving failures in processes and processors. They must develop criteria for determining if data transfer should be by database, messaging, or memory. Controls based on time or events are necessary for all processes. They need to establish criteria for determining whether we should retain data in files or in relational databases.

Example of System Paradigm: Data Flow and Access among Elements. At the system-definition level, as the element requirements are generated, the processing concepts establish how one element deals with the others. The processing concepts show how the element is viewed as a partner of operations with the other elements. They specify how data passes between elements rather than identifying specific data. For example, several elements usually exist in a ground station. Spacecraft engineers may work alongside mission operations engineers but may have developed their own data-processing software to access the same physical data. A system-level processing concept might require mission operations to send any data needed by others (a *data-push paradigm*), or might allow others to come get any data they need (a *data-pull paradigm*). But the elements' roles in providing or obtaining the necessary data are radically different.

Example of Element-Level Paradigm: Data Flow and Access within an Element. Element designers have to choose a processing paradigm at the element level. For example, they need to determine how other tasks provide data to the software tasks. One paradigm has data embedded within a message. Another sends notification messages to interested processes and has them query back to the source process. Still another is to have time be the trigger, with either access to a common data element or a query to the generating process. The element designers must decide on a processing paradigm to guide later concepts for processing within configuration items.

Example of a Paradigm within a Configuration Item (CI): Control and Data-Flow Mechanisms. At this level, the CI designers determine if they need unique software for processing each activity and constraint or if they can define generic software that is driven by an activity with a unique database and constraints. CI designers also need to decide whether database constructs should define the processing rules, perhaps by using an expert system, or whether these rules should be explicit in the software. Further, element designers need to show how the software or hardware interacts to satisfy a higher-level task, establishing first its capabilities and then its executables. To show interactions within a CI, designers must consider such things as what triggers the processing, how data flows through the CI toward a final product to be passed to another CI, and how various components control the processing. Finally, they must define how the software or hardware reacts to operator control and intervention. For example, operators may require delays in the processing because they know input data isn't yet available. In this case, the software must provide a way for the operator to hold the processing flow.

When to Monitor Processing Concepts. This attribute is always key to system and element design. All processing supports some form of work to satisfy element and system requirements. The processing concepts describe how to do that work. For CI design especially, it's not as important to know that an executable is made up of several components as it is to know how groups of executables work together to meet an objective or performance capability or to react to an operator's direction.

How to Monitor Processing Concepts. As MOMs, we monitor this attribute of design from the beginning of system definition to provide system paradigms, where required, that resolve the individual elements' roles in providing data or service. We further monitor the system paradigms during element design to ensure the element meets them as it assigns tasks for the hardware, software, and operators. We also monitor this attribute as the design moves to configuration items. This ensures that the CI designers are designing all aspects of the executable details, including how database information is constructed and used by the processes, and how the CIs work together to do element-level tasks.

If we don't have enough time or expertise to monitor the processing concepts properly, we should establish an internal program authority and an external expert panel to oversee them.

Definition of Operator Tasks. Table 6.5 shows how we develop operator tasks throughout the development phase up to the critical design review.

Table 6.5. Definition of Operator Tasks. We must start determining the operators' roles and staffing levels in the definition phase.

Process Phase	Tasks to Cover
Definition	• Establish the system's conceptual requirements related to operator involvement, training, and number of positions • Establish initial set of operator requirements based on the system's conceptual requirements
Development (Preliminary)	• Define operator tasks based on operations requirements • Define operator positions • Assign tasks to operator positions—preliminary
Development (Preliminary Design Review)	• Assign tasks to operator positions—final
Development (Critical Design Review)	• Evaluate operator loading • Evaluate operations teams

As stated in Sec. 6.1.3, the system's conceptual requirements reflect internal and external users of the element. The operators are internal users. At the system level, conceptual requirements may direct a certain maximum number of people for operations, or it may otherwise direct the expected operator involvement in mission operations. Other operator requirements will come from analyzing the element's processing tasks, where all or some of those tasks are assigned to operations. Using this set of initial operation requirements, element designers can define the operator tasks.

Operator Tasks. Operators either monitor or control—before, during, or after data processing. For example, if an operator starts processing data when certain conditions occur, these tasks are for pre-processing control. Examining telemetry measurands is monitoring during processing. These categories help identify the operator's relationship to the software executing the mission. The software supporting these operator tasks has different responsibilities based on the categories. For control tasks, the software waits for the operator's explicit direction at the designated control points before proceeding. For monitor tasks, the software provides information and continues to process; it doesn't wait for an operator input. Therefore, element designers must understand the operator's role in mission operations.

Operator Positions. Once element designers define the operator tasks, they combine them into operator positions that cover all aspects of operations, with like tasks assigned to the same operator position. For example, a real-time operator position should have tasks related to the spacecraft's state of health, such as monitoring telemetry, uplinking commands, and generating specific trending reports. Designers shouldn't assign this position to monitor or control the processing of ephemeris generation.

Operator Loading. Element designers also examine the operator positions for operator loading—how much work the operator has to do in a given period. Does the operator constantly monitor, control, and interact, or are breaks scheduled into the routine? Designers may need to change the operator's workload or add people to off-load a single operator during peak periods. For example, they may have required the software to allow an operator to generate commands manually, examine the generated commands, and analyze telemetry to support uplinking a new command load to the spacecraft. Loading analysis may show the operator can't do all of these tasks in the defined timeline. Designers must then choose either to reduce the operator's interactions, relax the timeline constraints, or add a second operator who can do some tasks, such as analyzing the telemetry, while the first operator prepares the new command loads. Some tasks, such as starting and overseeing development of the activity plans, are required only periodically. Designers usually assign these positions only to the day shift as opposed to the continuous shift for real-time operators. Even so, they have to analyze these positions for loading because control and analysis normally dominate activity planning.

When to Monitor Definition of Operator Tasks. The MOM would monitor this attribute of design when the level of operator staffing is an issue addressed in the conceptual requirements for the system. For systems that continue operating on orbit for a long time, the operations and maintenance costs of a program far outweigh the development costs. In these cases, managers must pay very close attention to the staffing required to operate and maintain the ground and space assets. This attribute is also considered key when the element's operations are operator intensive and operability of the system becomes an issue. The relationship between software processing and operators leads to a natural tension. The more tasks an operator must do, the more prone the operations environment is to human error, especially when unusual situations arise. But software maintenance and upgrading become more difficult and require more people whenever we automate a ground element more to off-load the mission operator. As a compromise position, designers can leave the automatic decision processing out of the software but make it easier to control by providing enhanced automated procedures, such as call-up menus that are easy for operators to use.

Monitor Goals for Staffing. We must monitor this attribute for several conditions. The first is to make sure staffing meets our goals. Training and maintaining a skilled operations staff for any operational position costs a lot of money for extra people and the infrastructure to keep people trained. As operations managers, we ensure the element designers are sensitive to this issue as they analyze task and position loading.

Monitor Operability. We also have to make skilled operators available to evaluate thoroughly the tasks and operability of all operator positions. The operator tasks must match the education and training of the actual operators. For example, operator tasks shouldn't imply that, in order to control the processing, an operator must first know the underlying logic path, or algorithm, that the software took to get to the decision point. Most operators aren't trained to know the details of algorithms; they're trained to know what conditions to look for and what to do next when a particular condition arises.

Monitor Operator Loading. Finally, we make sure the operator positions are properly loaded by examining the results of task analysis and determining whether tasks are appropriate and logically grouped into positions. If we find improper loading, we make sure element designers do appropriate design trades to alleviate it. These trades may include converting operator tasks into software requirements, writing software-assisted procedures, adding people, or moving tasks to other operator positions.

Timing and Timelines. In Table 6.6 we show tasks for developing timing and timelines throughout the development cycle up to the critical design review.

Table 6.6. Timing and Timeline Attributes of Design. We must allocate to the lowest level each timing and timeline requirement within the ground element of mission operations.

Process Phase	Tasks to Cover
Definition Phase	• Allocate system requirements, events, and timelines to each element • Establish reasonableness of timing and timeline allocation based on high-level functions for the element's hardware, software, and operations
Development Phase (Preliminary)	• Convert element's timing and timelines to computer software configuration items, hardware configuration items, and operations • Evaluate the element's ability to meet timing and timelines based on its functions
Development Phase (Preliminary Design Review)	• Confirm the ability to meet timing and timelines based on preliminary executables and initial processing traffic
Development Phase (Critical Design Review)	• Confirm the ability to meet timing and timelines based on final executables and processing traffic

Initial Estimates—Data Volume and Steady-State Processing. System designers need to determine which elements are involved in meeting required timelines and then allocate part of the system timeline to each of these elements. The timeline may be completely allocated to a single element if we need only one element to carry out the system task. Element designers then have to allocate each element's timeline to its processing and operator tasks and ensure this allocation is correct through analysis and design trades. Early in a system and element design, there is a lot of uncertainty as to how much processor power and data flow will be needed to achieve a defined performance objective. Thus, designers must not try to analyze the timeline in detail. Instead, they need "back of the envelope" analysis of the volume of input and output data and steady-state processing. They begin by projecting known data flow and measures of processor use from similar tasks onto current tasks. Even with this immaturity of the data supporting the analysis, obvious choke points in processing and data flow surface quickly. With this information, the element designers can support trades on configuration-item responsibilities, server sizing, and relief for timeline requirements.

Refined Estimates—Detailed Analysis and Prototype. As the design matures through preliminary, PDR, and CDR subphases of development, the uncertainties in the timeline analysis decrease, so element designers need to replace static models of processing and data flow with dynamic ones. That is, coarse software and hardware tasks in the static models become executable components, with processing based on more detailed analysis or prototype evaluations. An important transition of static to dynamic models occurs where the element's mission applications interact with the services that support the mission. For example, during the preliminary element design, designers might apply a static overhead to reading a block of data from a database. As the design develops, they replace this static overhead with more realistic data, based on whether the data resides in flat files or a database and on how well software can access the database through a network. With this extra fidelity, the element designers can find hidden choke points in the design or processing architecture not revealed by the simpler static models.

When to Monitor Timing and Timelines. Timelines need monitoring whenever there is a risk of not satisfying timing and timeline requirements, particularly when resources contend in their use of the network or database, or whenever the processing workload forces the design to a top-end machine or to more exotic processor architectures, such as parallel processing machines. If element designers discover these trip points, they must consider assigning another processing paradigm or renegotiating the requirements.

Monitor Modeling of Timeline Requirements. We have to monitor how the element design evolves to ensure proper progress in the timeline analysis and trades. We need to identify all timeline issues by the program's PDR and resolve them during the PDR and CDR. This resolution includes transitioning static models to more dynamic ones as the design matures. That is, timeline analysis must track with the element's evolving design. If element designers don't keep the design current with the element's concepts for passing data, control, and processing characteristics, the entire analysis may be useless.

Monitor Modeling of the Interaction between Mission and Service Layers. We monitor this aspect to determine if we've modeled mission applications enough to draw out conflicts between, and overhead for, application services. Computer architectures use distributed computing and data servers with an application service layer one notch above the operating system. This service layer insulates mission applications so they don't need to know where the data comes from or which processor holds a particular application. But element designers must model very carefully the time penalty associated with this layering. If an application calls a service many times, and that service requires twice the time originally allocated, the designer of a configuration item may have to reconsider whether the mission can afford to use such a service. Our modeling of this dynamic behavior between the mission and service layers must be detailed enough to discover potential problems in our timelines before we code and test the final product.

Monitor Modeling of Mission Operations Scenarios. Finally, timeline analyses of individual requirements and system events must work together in the context of real operations. So element designers have to synchronize their timeline analysis with the element's evolving concepts for processing to yield an accurate "day-in-the-life" scenario for operations. For example, if we have to produce a schedule within a certain time, our timeline model must show data arriving and being processed on the same frequency. If processing is based on ending contact with the spacecraft, the timeline model must script all contacts in a day and set events into its dynamic triggers based on ending the contact. This effort can be quite extensive, but it's our only assurance, before starting operations, that mission operations will work within its specified timelines and within the defined capacities for data flow and processing. Following is an example of monitoring the modeling of one mission scenario.

Monitoring Timeline to Update the Spacecraft's Execution of a Mission Activity. Let's examine a system requirement to update the spacecraft's execution of a mission activity within 30 minutes of a triggering event. Normally, the spacecraft is loaded with mission instructions once per day. We'll assume, for this example, that the triggering event occurs 10 hours after the nominal update.

30-Minute Total Allocation: 20-Minute for Mission Operations + 3-Minute for Establish Link + 7-Minute for Update Spacecraft. First, we analyze the requirement at the element level. Because the command-uplink rate is pre-established, it takes 7 minutes to update and enable an existing onboard memory with new instructions. It also takes 3 minutes to establish a space-ground link. Therefore, of the original 30 minutes, mission operations has 20 minutes in which to prepare for the new contact and uplink with the spacecraft.

20-Minute Mission Operations Allocation: 5-Minute for Mission Planning + 10-Minute for Activity Planning + 5-Minute for Mission Control. Three functions work together to prepare the new sequence: mission planning, activity planning, and mission control. We have to update the mission plan with a new contact time and an activity-level description of what has to be uplinked. We'll begin the analysis by allocating five minutes to mission planning for this effort. Activity planning then is alerted that we've updated the mission plan and now must prepare the actual command-load update. Because we don't know the exact nature of any one update, we allocate ten minutes to activity planning. Finally, the last five minutes of the timeline is allocated to mission control so we can prepare for and establish a space-ground link, and then uplink the new commands.

Actual Mission Operations Need: 15-Minute for Activity Planning and 10-Minute for Mission Control. After analyzing the individual tasks within the element design, we discover that mission control requires at least ten minutes to support a space-ground link. Further, activity planning requires 15 minutes to prepare the new commands. At first, then, it appears we can't meet the system timeline.

Solution: Design for Concurrency between Activity Planning and Mission Control. But after further examination, we discover that, if we send the plan directly to control and activity planning, we can do the real-time processing while generating the new commands. With this small change to the element design, the new requirement is fully satisfied.

Defining System Performance. Table 6.7 shows how we define tasks related to system performance throughout the development phase up to the critical design review.

Table 6.7. **Defining System Performance.** We must define thoroughly the work required to design and demonstrate the system's performance requirements.

Process Phase	Tasks to Cover
Definition	• Establish the system's conceptual requirements for performance (quality, throughput, and response) • Investigate potential constraining points in the system that mission operations must deal with • Analyze the system's conceptual requirements to establish the ground-processing capacities necessary to support the desired system performance
Development (Preliminary)	• Convert element performance requirements and timelines to computer software configuration items, hardware configuration items, and operations • Identify areas of defined system performance that require complex algorithms, state-of-the-art processing, or data management • Establish performance tests that demonstrate the element's ability to meet objectives for system performance
Development (Preliminary Design Review)	• Prototype and analyze processing to ensure we can meet test metrics for performance
Development (Critical Design Review)	• Finalize algorithm and processing design based on results of prototyping and analysis

Factors that Limit Satisfying the Mission Objective. Conceptual requirements for the system, which eventually become specifications of system requirements, state the quality and quantity of mission data that the system must support, as well as responsiveness to input stimuli the system requires to meet the mission objective. The quality of data influences, and is influenced by, characteristics of the spacecraft payload and operational orbit. The quantity and responsiveness of data produced and processed by the system depends on the number of mission spacecraft; the amount of time the spacecraft can communicate with the ground; the data rates (including mission, command, and telemetry) of the space-ground links; and the throughput capacities for ground processing. System and element designers must fully analyze and understand these factors early in the definition phase.

We may create constraining points while upgrading the system to better meet the mission objective. For example, as a system matures, advances in technology create a capability for increased quality, quantity, or speed in one element that other elements can't handle. Also, at times, we may have more spacecraft within communications line-of-sight with the ground than ground elements to receive or process the data. Managing these constraining points requires more complex mission operations design.

Identifying Limits. The system and element designers' initial analyses of the system constellation, spacecraft-bus design, spacecraft payload, and ground processing determine where constraints prevent a system or element from meeting system performance objectives. When they identify a constraint on ground processing, they have to solve it—typically by adding complexity, processing power, or future technology. Proper analysis of the system requirements will identify all potential areas of increased complexity and lead to system approval of an approach the ground element will take when the situation occurs.

Designers have to avoid overdesigning or underdesigning element processing to cope with these situations by balancing the number of occurrences against the severity of any single occurrence. It's sometimes difficult to quantify the severity of a situation for such a trade. But they must try, so they don't choose an exotic solution for situations that don't occur often enough to warrant the cost of developing and maintaining the solution.

Analyzing Trades. System designers need to do system-level analysis and simulations as early as possible to determine the system sizing and spacecraft capabilities required to meet the mission objective. These include parametric studies of sizing for the spacecraft constellation and ground receiver, orbit parameters, command rate, telemetry rate, mission-data rate, mission-sensor capabilities, and if appropriate, tasking throughput. These studies compare benefits (better satisfaction of one or more mission objectives) to costs (increased on-orbit or ground hardware, increased processing speed) for these kinds of variables. Usually, the parametric analysis establishes a clear point for the best ratio of benefit to cost—the break-even point on which designers should base requirements for system performance. They need to keep these parameters in mind because technology advances often move this break-even point.

As an example, assume that a primary mission objective for a system that examines global deforestation is to minimize the time between looks to a given geographic location. There is an altitude limit at which sensors can resolve an area, given the current technology. We may decide to study a technology trade on sensor resolution versus number of spacecraft required. Because advanced technology is always a few years ahead of current technology, and systems have a design life of a decade or more, later spacecraft may use the newer technologies.

Figure 6.4 illustrates the results of this parametric study. Curve 1 shows that, with current technology, increasing the number of spacecraft initially achieves large gains in the mission objective. But curve 2 shows that, after a point, the cost of building and concurrently operating multiple spacecraft becomes prohibitive. Advanced resolution sensors, which allow for increased orbit altitudes, are the more cost-effective solution to increased performance, as shown by curve 3. The advanced sensors allow the same resolution at a higher altitude, and higher altitude increases the spacecraft's coverage. Therefore, fewer spacecraft are required to achieve the same time between looks to a particular area. The system's break-even point could occur at four spacecraft with current technology sensors, but at three spacecraft with advanced sensors. The system designers, who are responsible for the system requirements, must be careful not to overspecify the ground element to operate four spacecraft, all with advanced sensors.

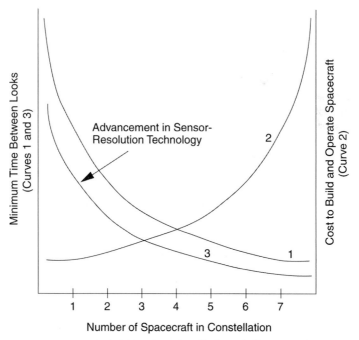

Fig. 6.4. **Illustration of Parametric Analysis for System Performance.** Parametric analysis of system performance results in a design that meets current and future needs. We can increase coverage by increasing the number of spacecraft (curve 1), which increases operating costs (curve 2). But we can also increase coverage by improving sensor resolution and increasing spacecraft altitude (curve 3).

When to Monitor Definition of System Performance. Because of the direct influence this attribute of element design has on the cost of an element, it is always key. Further, system and element designers must revisit the need for these system-performance requirements when analysis shows that meeting the specification adds cost (in required processing power or algorithmic complexity).

Monitor Definition of System Performance. The MOM must review the definition and development plans to make sure they include the tasks in Table 6.7. We also ensure that tasks are carried out and issues resolved according to the preceding discussions.

We monitor this attribute of the element design to ensure that analysis and simulations are correct and complete. But more importantly, we make sure the mission objective isn't over-satisfied, resulting in an over-designed spacecraft or mission sensor, or leading to a spacecraft constellation and ground-processing capacity that's larger than necessary.

Finally, because monitoring this attribute is so important, we should understand performance requirements, simulation, and analysis. Then we can develop our own conclusions, rather than relying solely on those of the system and element designers.

Optimal Operation of Spacecraft Subsystems. Table 6.8 shows how we achieve optimal operations of spacecraft subsystems throughout the development phase up to the critical design review.

Table 6.8. Optimal Operation of Spacecraft Elements. We must analyze thoroughly the space and ground elements and their interactions for limits on optimal operations.

Process Phase	Tasks to Cover
Definition	• Establish conceptual requirements related to system performance (quality, throughput, and response) • Determine which spacecraft subsystems constrain mission operations
Development (Preliminary)	• Convert element requirements and timelines to computer software configuration items, hardware configuration items, and operations • Analyze the design and methods of mission operations to fully use spacecraft subsystems • Use analysis and prototyping to do trades between optimal operations and complexity of ground processing
Development (Preliminary Design Review)	• Select "break point" in trade space
Development (Critical Design Review)	• Develop final methods based on analysis and prototyping

Constraining Subsystems and Operations. To establish conceptual requirements, element designers need to analyze the spacecraft's components to determine which spacecraft subsystems and operations influence mission objectives. The analysis of system performance (refer to previous section, Defining System Performance) may show that ground operations must use all of the spacecraft's capabilities for power, loading, and payload to meet the mission objectives for data quality or throughput. In other words, we have to push the spacecraft to its limits without violating constraints. For best use of the spacecraft and its subsystems, system and element designers must study trades on gains in system performance versus complexity of ground operations. These trade studies include the effect of increased dependence on processor speed and capacity, as well as on algorithmic complexity.

Fidelity of System Modeling. Another area for trades is the fidelity of the spacecraft subsystem modeling required to get the additional performance from the spacecraft, which must take into account how much uncertainty is in the model's input data. This is probably the parameter least understood in optimizing trades. For example, if the power expenditure has to be planned over a day so that the most important mission objectives are met at all times, the element's planning tasks may have to model power expenditure. However, if the equations necessary for accurate modeling of the subsystem require knowledge of the subsystem's dynamic behavior, we have to statistically generate the data or simplify the equations. Having high-fidelity equations with low-fidelity data leads to false expectations, and possibly to continual changes in order to isolate more of the statistically outlying data. Through this analysis, designers determine a break point in the trade space—the point at which adding more complexity is not worth pursuing.

When to Monitor Optimal Operation of Spacecraft Subsystems. This attribute needs monitoring whenever the spacecraft's abilities limit meeting the mission's performance objectives. The analysis and methods designed to solve these problems will consume the single largest engineering effort in developing a ground element. Therefore, we must reduce the complexity of the ground element as much as possible by closely monitoring this attribute.

Monitor Mission-Objective Satisfaction and Design Complexity. We examine the analysis and trade studies to make sure the system meets mission objectives well enough to satisfy users. This may entail revisiting and renegotiating the system requirements. We also ensure that the mission objective isn't over-satisfied as a result of designers adding unnecessary complexity to operations, methods, and design.

To avoid complex design, we must not commit an element's functions to overcome several of the spacecraft's limitations at the same time. The spacecraft's subsystems tend to drive different, almost mutually exclusive, solutions from the ground to compensate for these limitations. Trying to sort out all of these features at the same time puts tremendous pressure on the organization for element design and often causes development costs and schedule slips.

Definition of Displays, Alarms, and Procedures. Table 6.9 shows how we define displays, alarms, and procedures throughout development up to the critical design review. They're the bridge between the operator and the mission software and hardware in the ground element. Operators need to assess, monitor, and control not only the mission but also its underlying software and hardware. How the operator receives data to support mission execution or processing of abnormalities is critical to the overall element design.

Table 6.9. **Definition of Displays, Alarms, and Procedures.** We must start meeting operators' needs early in the definition phase.

Process Phase	Tasks to Cover
Definition	• Establish conceptual requirements related to system performance (quality, throughput, and response) • Investigate driving display characteristics to support conceptual requirements for the system
Development (Preliminary)	• Convert element requirements to computer software configuration items, hardware configuration items, and operations • Develop preliminary displays • Prototype and evaluate displays for operability • Develop concepts for alarms and procedures
Development (Preliminary Design Review)	• Define operational displays, alarms, and procedures based on defined operator tasks
Development (Critical Design Review)	• Refine displays, alarms, and procedures to complement needs of operator tasks and software interfaces for configuration items

Initial Concepts for Display. Even in the earliest phases of a system's development, the look and feel of the environment for operational processing begin to take shape. Human Machine Interface (HMI) concepts can help determine if a display can reasonably achieve a system's conceptual requirement. For example, one conceptual requirement may be to visually inspect the entire mission plan from a single display. Depending on the size of the constellation and the kind of data to be displayed, this may not be feasible. An early concept of a plan display will show whether it can reasonably present all of the spacecraft activities placed on a timeline in the constellation. This work may change the perspective of how data are presented for review, and hence, influence the interface between the operators and software.

Display Definition and Operability. Element designers define operator displays. However, because this attribute is the most visible sign of the element design, the end users—the flight operators—are invited to help review and critique the displays from the beginning of design. The designers prototype displays as needed to demonstrate features and to get approval from system and operations people.

Displays must be designed for operability because the way operators interact with the mission software and hardware truly affects the system's efficiency. Also, end users measure the quality of a ground element's design based on operator interaction. If they see the interaction as inefficient, all software and hardware processing, including complicated algorithms employed in the processing, is suspect. Designers must create displays that allow operators to navigate through them efficiently. They must also establish standards that provide a common look and feel for operator interaction, at least across the tasks associated with a given position, if not across the entire suite of positions.

Alarm Presentation to Operator. In general, alarms follow an evolution similar to that of displays. But because alarms bring urgent information to the operator's attention, element designers give alarm processing special consideration. Alarms must get an operator's attention without obscuring current work. For example, if telemetry indicates a spacecraft's state of health condition is critical, operators need to know which of the spacecraft's subsystems is having the problem. However, they are still monitoring the rest of the spacecraft's operations through other displays. If an operator isn't examining the faulted subsystem's components at the time of the fault, he or she either reduces that display to an icon or places it behind the other displays. When the alarm for that subsystem occurs, it shouldn't immediately obscure the operator's view of other displayed measurands (for example, by automatically bringing the display for the faulted subsystem to the foreground). Changing the color of either the display border or the icon to red, with an adjoining message on a continuously viewable message-center display, gets the operator's attention without jeopardizing other work.

Alarm Levels and Displays. Alarm concepts, including the various levels of alarms and the corresponding audio/visual displays, are established early in development. For example, the most visually perceptive way to illustrate a link fault of some kind is to change the link color on a continuously displayed graphic. The graphic may normally show the signal strength numerically next to a line connecting the spacecraft to the ground receiver. Whenever that signal strength goes below some level, the line and value change color. This immediately draws the operator's attention to the source of the out-of-limit condition. Alarm levels significantly influence the design of software and hardware because element designers must understand the severity of an alarmable event and build an appropriate reaction.

Alarm Overrides. Coupled with alarm levels is a feature allowing operators to override or reset alarms. Element designers must carefully examine conditions for overriding an alarm because doing so improperly may jeopardize a spacecraft's state of health operations. The software and hardware underlying these alarms will vary depending on their ability to continue processing when the operator overrules the alarm. Designers can best select these concepts through iterative prototyping and evaluation, in which element designers sit with operators to work out the most instructive ways to show and respond to alarms.

Procedures. Procedures are slightly different from displays and alarms. Element designers develop procedures to guide operators through mission activities or problem resolution. The procedure may be as simple as an online handbook that operators refer to when they see a condition they're not familiar with, or it can be as elaborate as a software-based script that can make certain data-based decisions for the operator until it reaches a step that requires human action. For example, an operator may use a dynamically executed procedure to isolate a fault in the telemetry-processing path. All the operator knows at first is that telemetry isn't updating on the measurand displays. The operator carries out the procedure for isolating a telemetry fault. The procedure script may have queries to the last few seconds of telemetry to see if signal strength fell off, indicating a possible spacecraft problem. Or it may have messages sent to the ground software and hardware components, with healthy components responding, thus isolating the faulted components.

Procedures and Mission Software Design. All procedures follow a basic flow that is displayed to the operator. The procedure will either find the source of the failure or reach points in the procedure requiring an operator's help. Obviously, the more we tie the procedure to the mission software and hardware executables for information, the more those executables must be aware of what the procedure needs. But even procedures in online handbooks will influence the software design. The handbook may ask the operator to check if a measurand is set one way or another, thus requiring software to retrieve and display this information quickly. Data-retrieval speed, organization, and display characteristics of results may depend on what is in the procedures, especially those that help operators maintain the spacecraft's state of health.

When to Monitor Displays, Alarms, and Procedures. Displays, alarms, and procedures are always key to element design because they directly influence the definition of operator tasks. As stated earlier, these attributes bridge what the operator does to control and monitor the element with the underlying processing by software and hardware. If the bridge is missing or improperly thought out, there's little chance the underlying processing will meet the operators' needs. More importantly, displays, alarms, and procedures establish the element's operability in the eyes of the system and the end users. Buyoff for the entire element design lies largely in the operators' ability to believe the element is operable.

Monitor User Needs. We monitor this attribute to assure operators, mission engineers, and spacecraft engineers that our prototypes and evaluations satisfy the technical and operability requirements for the human interface. Further, we make sure the mission and service-layer design needed to generate displays is in place at the program's PDR.

Verification. Table 6.10 shows how we verify results throughout the development phase up to the critical design review. Even though we don't verify element and system requirements until late in the development cycle, element designers need to engineer the element properly to meet verification needs beginning early in definition and development. Verification and testing of an element are often overlooked in the rush to get the mission operations understood and designed.

Table 6.10. Verification as an Attribute of Design. We must design the element and define testing methods to support verification within the ground element.

Process Phase	Tasks to Cover
Definition	• Establish and assign verification categories for each system and element requirement • Analyze requirements for verification ambiguity and appropriate level of verification
Development (Preliminary)	• Convert element requirements and timelines to computer software configuration items, hardware configuration items, and operations • Develop concepts for conducting and transitioning between tests
Development (Preliminary Design Review)	• Analyze design's ability to be tested and to have its requirements verified at configuration item and element levels • Develop preliminary test plans for verifying configuration items and element
Development (Critical Design Review)	• Develop final test plans for verifying configuration items and element

Verification Methods. We build verification into the design starting with the actual language of requirements in the system definition. For each requirement, designers of the element and its configuration items must determine the effort needed to show the requirement is satisfied. Using verbs such as process, use, and perform disallows verification. For instance, requirements such as "the element shall use ephemeris in its planning functions" is so open-ended as to permit almost endless possibilities for verification. A better requirement would be "the planning function shall generate ephemeris for the purpose of pointing the ground tracking antennas to the spacecraft with an accuracy of 0.1 degrees."

After the requirements are stated in a verifiable language, designers assign verification methods and a time frame. Typical verification methods are test, demonstration, analysis, and inspection. Verification can occur during testing of individual CIs, testing of the integrated element, and system-level demonstrations. Testing operates the component under specific conditions and verifies a quantitative requirement. Demonstration operates a component and qualitatively verifies its functions. Analysis is verification using models or other mathematical methods to show requirements are satisfied. Inspection consists of reviewing designs or documents to ensure they meet requirements.

Most of the mission operations element's requirements pertain to control and processing decisions and are verified by a formal test. Test inputs are generated and read into the software or hardware, and the output is examined and compared to expected results. A certain class of requirements involves a spectrum of conditions, such as specified orbit characteristics, that can't be tested to meet all possible combinations of specified inclinations or altitudes. In these cases, verification by analysis is appropriate. That is, designers should show analytically that the equations won't produce unpredictable results.

Another way to verify wide-ranging requirements is to define a specific test agreed to by designers of the system and element. This test should have a well-defined set of input data as well as expected processing results. Usually, complex algorithms meet these requirements, which correspond to limits on system or element performance.

An area often overlooked is validating algorithms after they're developed but before they're implemented in software. Element designers should analyze them and do prototyping if necessary to prove that the methods satisfy requirements. Examples of methods that require validation are orbit determination and spacecraft power models.

Designing for Testability. As the element design begins to mature, designers and engineers must address issues that appear only in a test environment, such as the ability to set a test date and time into the future from the current clock date and time, the ability to stop a test in the evening and resume the next morning, and the ability to transition an element from a test to operations. As the requirements are analyzed within the CIs and eventually specified as executable components, designers also have to examine how to verify them. This is especially true for requirements satisfied by complex algorithms. In some instances, multiple algorithms may process data sequentially or with control and feedback. Verifying that we've met the CI requirements is most difficult in these cases because we must show theoretically that the lead-off algorithm needs verifying first, before any subsequent algorithms. Unfortunately, this approach consumes a lot of time. Therefore, CI designers must configure their executables with test-only ports for input and output data, so they can verify multiple algorithms one at a time. If we don't understand this effort as the design matures, verification will cause cost and schedule slips.

Element designers create test plans for the element and its CIs as the CI development matures toward PDR. They should base these test plans in part on the mission operations concepts and articulate the test scenarios needed to verify each of the selected requirements. At this point in program development, element and CI designers must understand how many different kinds of tests and what efforts to set up, obtain, and analyze data we need to verify certain requirements.

When to Monitor Verification. Because verifying an element's requirements is a costly attribute of its development, verification is always key in all of the subphases of definition and development. It is also key because verification occurs not only after the design is complete but also after all of the code and hardware are built. To find that we can't verify the design at this late stage puts a great strain on the development schedule for delivery of the final product. MOMs must pay close attention to the requirements that may be difficult to satisfy. One example could be timing or timeline requirements for spacecraft acquisitions. Another example is spacecraft health or safety requirements such as avoiding depleting the batteries in power-constraining situations.

Monitor Verification. We monitor this attribute to ensure the element designers understand verification for all system and element requirements. We also ensure they understand how the element will verify the requirements, first at a high level and, ultimately, down to individual algorithms. We must check the requirements specification, test plans, and test procedures for clear, specific details of verification—precise statements about how designers generate test input data and then obtain and process test output data. For example, they should either provide or explicitly reference all mathematical methods required to support verification of requirements. Finally, we make sure test-unique requirements are placed, as needed, on the CI design to allow testing.

6.2.2 Ensuring Completeness and Maturation

Definition and development is a holistic process, constructed in layers. The layers represent two views at the same time. First, they represent the level of uncertainty in design and requirements—uncertainty that must be peeled away as the system and element designs evolve. Secondly, they represent the subphases of definition and development. The most cost effective way to develop a space-ground system is to peel away each successive layer completely, addressing all attributes of the design while handling uncertainties in the design and requirements at each layer. The MOM's role is to monitor definition and development of the system and element so that (1) requirements and concepts cover all the ground element's operations, and (2) key concepts and subsystems are mature in the early phases. At the end of each of the steps discussed in Sec. 6.1.3, the program should be able to document and present a consistent level of completeness and maturity in the evolving system and element designs.

Maturation of Requirements, Methods, and Timeline. Too often, a program's maturity and completeness are measured by a schedule of calendar events instead of an understanding of concepts, requirements, and design. In this mode, a program becomes focused more on ensuring that a set of requirements is documented, than on ensuring that the program exhibits proper understanding behind the requirements and design. To avoid this situation, we must be aware of the symptoms. For instance, suppose an audit of the requirements on an element reveals that an upper-level requirement is repeated verbatim in the next-lower layer of requirements with neither analysis nor an operations concept to describe how the element meets the upper-tier requirement. In this case, the element design is not maturing. In fact, the element designers are merely "binning" requirements, not analyzing and designing the element. Another symptom manifests itself in methods and timeline analysis. The element design isn't maturing if the early stages of a methods development can't describe all of the inputs to the method or how a method operates on this input data. Also, if timelines allocated to an element or its tasks can't be proved or disproved by analysis or prototyping, the element design is not maturing. If any of these symptoms are present in the element design and the program still remains on schedule, the program is headed for a major failure later in the development phase.

Design Completeness. Another error that element and CI designers tend to make in developing an element is working locally or examining in detail a particular design attribute, while completely ignoring other attributes. For example, the CI designers may become too focused on how the software executables of a computer software configuration item are grouped and constructed, while completely ignoring the operations' concept of these processing tasks. We must ensure that this fine-focused approach is not taking place at any of the definition and development steps, and especially not as the program approaches the preliminary and critical design reviews. The symptoms of this situation are difficult to detect until the actual design reviews—when all attributes of an element undergo extensive review. Early warning signals of an impending problem surface in the areas of timing and timeline analysis, verification, and processing concepts. Suppose we can't answer basic questions on starting and triggering processes, CPU requirements for executing a task, and ways to verify requirements. Yet, detailed methods are already under way. If so, the element's design attributes aren't progressing equally.

6.2.3 System Reviews—Concepts, Requirements, and Design

Review Mechanisms. As we discussed in Sec. 6.1.3, concurrent definition and development isn't free of risk. We must ensure proper system-level review and approval of the preliminary mission operations concepts and element specification during definition, as well as of the refined mission operations concepts during preliminary development. This is to prevent a program from missing crucial concepts and definition in the most formative stages when the parallel activity offers the highest risk. We must review carefully the element's operations concepts and requirements specifications at the early stages because they set the tone for all operations and the design architecture. People who do this review must be expert in conducting operations, system definition and interface specifications, software and hardware maintenance, and computer and data processing technologies.

If possible a program should avoid formal reviews lasting two to four days because so much data is presented in such a short time that the opportunity to critique and examine issues is virtually eliminated. Instead, we should organize reviews as requirements and design are developed. The element's concepts to support the system's conceptual requirements are the central theme of the review. We also discuss each attribute of the design. This part of the review is essential to an element's maturation because it shows that the element designers understand all of the issues of a given design before development proceeds to the next subphase.

Review of Requirements Maturation. Also discussed at the reviews are the requirements allocated to the element or the function within the element, depending on the review phase. This should not be a simple restatement of the requirements from the upper tiers; it is to be a discussion of what the requirements mean at the design level being reviewed and of how they will be satisfied. We keep reviewing requirements until the program's preliminary design review, when requirements have become executables. For requirements we consider key, the review must discuss how we've analyzed the concept or design to prove that it satisfies the requirements. When software algorithms are to satisfy requirements, the review should discuss them in enough detail to assure reviewers they'll do so.

Finally, the review must show that the design activities are addressing the uncertainties we're working or have newly identified. The review of these to-be-resolved (TBR) and to-be-determined (TBD) issues shouldn't simply state that we have to work an issue by a given date; it should include what specific steps will close the TBD/TBR, such as analytical analysis, prototyping, or interviews with operators.

Role of the Mission Operations Manager (MOM). For such a review to be successful, MOMs need to play an integral part in definition and development. As operations managers, we need to review the design internally, as the subphase evolves, so that the designers can answer the kind of questions that surface in the external review. Further, we must make sure the element-design organization treats the review as constructive, not something that has to be done because it is on the schedule or something they can prepare for in a couple of weeks. We need to ensure that the element designers make available enough materials and that enough time and resources are committed for the review. The element designers can't provide the documentation to the reviewers just ahead of the review dates. Because the review should be constructive, we must be able to secure a commitment from the external review groups to participate for months, not for a few weeks here and there. Finally, MOMs and participants need to see these reviews as a learning opportunity and a way to correct a wayward concept or design.

6.3 Scenarios and Subsystems to Monitor

The key to efficient and operable ground mission operations is understanding (1) all required operations and maintenance activities, and (2) all necessary effort in definition and development. MOMs must make sure the operations and maintenance activities are analyzed broadly and deeply so appropriate definition and development tasks are carried out. Failure to identify how much effort definition and development will take has two results. The first is that the work isn't done and thus mission operations aren't fully defined. The second is that the need for the work is identified after the contracts have been negotiated, so we'll have to absorb the work within the cost of the current contract or change the contract.

In this section we use the principles discussed in Sec. 6.1 and 6.2 to show how element designers should analyze mission operations scenarios with regard to definition and development tasks as well as operations. Whenever possible, they should do this analysis before starting work on definition and development. If limitations on knowledge and time make an early start impossible, they should carry out the analysis during the early parts of definition and development while refining the operations concept. This timing will ensure a properly focused effort. We provide only a few examples here. MOMs should require review of all mission and maintenance activities and analysis of critical ones as illustrated in the following subsections.

This section illustrates the conceptual and analysis issues for scenarios that system and element designers must address to be confident that the element design is complete at the end of definition and development. These scenarios show how broad and deep details must be to gauge the design's maturity. This section also illustrates the level of engineering effort required to understand fully the mission operations element's responsibilities in refining scenarios involving other elements—in this case, the spacecraft element.

We'll discuss these spacecraft scenarios:

- Acquisition
- Spacecraft bus and payload calibrations
- Spacecraft safemode operations
- State of health operations
- Command and control
- Power

For each example, the discussion is divided into three parts. The first part discusses the attributes that make the scenario a key concern and require monitoring by the MOM. The second focuses on what we need to support the definition phase, indicating the kind of analysis element designers must do, as well as the level of detail required at the end of the definition phase. The third part discusses the issues and analysis for the development phase, following the same format as the definition phase.

6.3.1 Spacecraft Acquisition

It's important to monitor the spacecraft acquisition scenario because of major concerns from three design attributes. First is the critical timing and interaction between the ground and space subsystems. Second is the need to establish and obtain approval for operator tasks. Third is the need to establish displays, alarms, and procedures for normal and contingency situations.

In the definition phase, system designers establish the spacecraft acquisition events and timeline and allocate them to both the spacecraft and mission operations elements. The acquisition events start with the pre-pass checks and end with establishing the communications link from the ground or relay spacecraft to the mission spacecraft. System and element designers carry out analysis to demonstrate that the parts of the timeline allocated to each element are reasonable. Determining reasonable timeline allocations requires that each element establish the high-level processing and operator tasks needed to support spacecraft acquisition. The analysis specifies the different types of links (widebeam and narrowbeam), antenna scan strategies, and equipment configurations needed for acquisition. The definition phase can't be considered complete until details in four areas clearly establish element responsibilities. The first area is understanding the spacecraft's performance characteristics, such as antenna slew rates and the time needed to configure the equipment supporting acquisition. The second area involves major constraints, such as line-of-sight limitations between the ground and spacecraft, or antenna travel limitations relative to the spacecraft body. The third area is understanding the spacecraft's autonomous processing requirements for acquisition, such as the criteria for going to autotrack, or loss thresholds for losing a lock signal. The final area pertains to messages, data, or other information, such as oral or written directives passed between the spacecraft and people in the mission operations element.

In the development phase, element designers analyze the spacecraft acquisition requirements and timelines and allocate them to the software and hardware processing tasks and operator tasks through the mission operations concept. They analyze to show that the timeline allocated to each task is reasonable. Preliminary alarms and procedures for operator displays are defined for normal and possible contingency acquisition. In the latter part of the development phase, which supports the preliminary and critical design reviews, further analysis confirms that the element tasks, as broken down into capabilities and executables, can meet their timelines. Element designers assign the operator tasks defined for acquisition to operator positions and then analyze the operator positions to ensure operators aren't overly tasked. They also prototype operator displays and alarms to demonstrate operability.

6.3.2 Spacecraft Bus and Payload Calibrations

The MOM monitors the scenario for calibrating the spacecraft bus and payload because of major concerns from two design attributes. First, calibration methods affect algorithm complexity. Second, flight operators must interact and coordinate with mission and spacecraft engineers and the processing tasks that support this interaction.

In the definition phase, system designers define the requirements for calibrating the spacecraft subsystems. The requirements cover spacecraft attitudes, maneuvers, equipment configurations, geometric constraints, and frequency of occurrence or other trigger mechanisms. System and element designers determine whether the calibrations need to be part of the mission plan or can wait for the activity plan. Trades are done between the spacecraft processing the calibration data on board and the mission operations element processing the data with subsequent uplink of updated parameters. Also, designers should think through how to distribute the results of the processing to spacecraft engineering. One more consideration in the trade is whether the ground processing needs to be online as part of mission operations, or whether spacecraft engineering can do it off-line. The definition phase is complete when these trades have been resolved and each element fully understands its role in calibration and data processing.

In the development phase, element designers develop ways to incorporate calibration activities into the activity plans. If the activity is to be planned, they must cover calculating the placement and duration of calibration activities, as well as criteria for including them in the activity plan. If the activity is to be scheduled, the methods need to cover calculating the specific attitudes, maneuvers, and equipment commands. The attitudes can be relative to the Sun, stars, or other objects. Maneuvers can be specified by required rates or a required number of star crossings. Also, in this phase, designers define the interaction between flight operators, mission engineers, and spacecraft engineers in processing, reviewing, approving, and updating onboard parameters and ground databases.

6.3.3 Spacecraft Safe-mode Operations

This scenario covers entry into, operations within, and recovery from a safemode. It focuses on spacecraft operations when the spacecraft isn't in any imminent danger but is unable to support mission-payload operations. We monitor the scenario to ensure proper requirements analysis and maturity in support of the spacecraft's safe operations.

In the definition phase, system designers specify the events that cause entry into a safemode, such as a constraint being violated or the onboard mission load area being empty because a new load wasn't uplinked. They define the spacecraft activities that occur in the safemode, such as sunbathing to maintain power levels and planned acquisitions. These activities allow us to contact the spacecraft with the least effect on the mission plan and provide chances for loading. The various levels of safing are established from the upper level, which uplinks a set of activities to the lower level for autonomous command execution. Finally, designers establish the safing period—the number of days the spacecraft can operate in this state before requiring a new load from the ground. Because these loads can span many days of operations, scenarios must be developed for maintaining (generation and uplink) the "out-day" safing information. The definition phase is complete when the interface specifications between the spacecraft and mission operations elements are defined well enough so the ground element knows how to operate the spacecraft in an extended safemode state. The interface specifications must also articulate the procedures and uplink commands necessary to recover the spacecraft from its various safe mode levels to normal mission operations.

In the development phase, element designers define ways to support planning, scheduling, and loading of safing activities. They also refine the recovery procedures from the various levels of operator roles and establish any displays needed to fully support the recovery. Finally, they carry out analysis to verify that the safing scenarios and activities result in the spacecraft's safe and efficient operation.

6.3.4 Spacecraft's State of Health Operations

The operations scenario to maintain a spacecraft's state of health ranges from immediate to near-term to long-term. The scenario covers areas in which mission operators' actions are critical to avoid long-term loss of spacecraft capabilities. Immediate state of health covers areas such as constraint violations, telemetry out of range, and equipment malfunction. Near-term state of health includes battery voltage approaching minimum limits and equipment temperatures nearing upper or lower limits. Long-term state of health includes propellant levels and cumulative equipment cycling limits.

We monitor this scenario as we do the safing scenario to ensure proper analysis and maturity of requirements, especially in the area of safe operations of the spacecraft during contingency or anomalous conditions. We also monitor the scenario to ensure that element designers address operability of displays, alarms, and procedures for immediate spacecraft situations. Finally, we assess how data is made available to the spacecraft engineers without affecting ongoing mission operations.

In the definition phase, system designers analyze the spacecraft design to determine how it will handle state of health over the immediate, near, and long terms. They analyze each part of the design related to state of health, so they can determine the level of monitoring and protection assigned to the spacecraft's software and hardware and to the ground's software and operators. They also identify responses to each design factor for state of health. Responses range from onboard autonomous commanding, to operator real-time commanding, to generating and uplinking replacement loads. Finally, designers determine data needed by spacecraft engineers to assess the spacecraft's state of health over the long term. The definition phase is complete when the interface specifications reflect all responses and their required timelines.

In the development phase, element designers design the displays that an operator will need to monitor the spacecraft's state of health. Because displays are critical to spacecraft safe operations, they're designed for operability, including the ability to navigate efficiently through the displays and to access data quickly and assess the spacecraft's state of health. Designers must establish alarm scenarios, including the various levels of alarms and corresponding audio and visual displays. They determine how much an operator can override or reset alarms and develop operator procedures, including online executables that respond to state of health problems. They analyze displays, alarms, and procedures for operability. This analysis includes both prototype and formal evaluations involving operators and spacecraft engineers. They develop methods to support developing and uplinking replacement loads that respond to spacecraft's state of health problems (for example, defining and uplinking a sunbathing activity if power levels drop too low). They develop ways to assess long-term state of health and to access data, so spacecraft engineering's need for frequent or large amounts of data won't interfere with normal operations.

6.3.5 Spacecraft Command and Control

This scenario involves two fundamentally different types of spacecraft characteristics—spacecraft execution of commands previously uplinked from the ground and spacecraft execution of real-time commands from the ground.

Spacecraft Execution of Previously Uplinked Commands. This includes discussions on loading parts of the spacecraft's memory for execution by the spacecraft's onboard software, strategies the ground element will use for timely updates to the load, and the ground processing tasks necessary to prepare future loads. The loads generated must take into account the most recent knowledge of spacecraft state.

Spacecraft loading is one of the most difficult and complicated interfaces to work out between spacecraft and ground engineering. The MOM monitors this scenario because we want spacecraft elements to operate as well as possible.

In the definition phase—through interface exchange meetings with spacecraft engineering—element designers determine exactly how the spacecraft software executes out of its own memory. To do so, they must know exactly how the onboard computer interprets the instructions and the specific timing of command execution. They analyze strategies to keep the spacecraft loaded with data for the mission and state of health operations. Considerations include the spacecraft's ability to accept updates to loads, the risk to mission success should an update not occur, the available onboard memory, and the ability to include the latest tasking information in the updates. Element designers need to get system approval for the loading strategy they choose because many factors influence this strategy, including sending and receiving messages involving other elements.

The element designers analyze the ground processing tasks necessary to produce the load and allocate them to a processing timeline. They go through a series of scenario trades that cover several processing paradigms. We may load the spacecraft ad hoc or at discrete fixed points either in time or in the spacecraft orbit. The processing tasks may execute in parallel or in series, depending on the interdependence or independence of the tasks and the ground processing capacities. This scenario depends on the currency of tasking requests for onboard execution of the load. Designers must develop scenarios for operator interaction, review, and approval of the load preparation and uplink.

Element designers must also develop scenarios for simulating the spacecraft execution of command loads to test whether the ground loads are correct and to verify that the ground response to certain telemetry signals is appropriate.

In the development phase, element designers understand and analyze the software complexities associated with the ground processing timeline. They finish the detailed interfaces and the spacecraft software. The CI designers finish their algorithms for load generation, including command sequencing strategies and mechanisms to detect and resolve violations of command constraints.

Spacecraft Execution of Real-Time Commands. This scenario differs from the scenario for executing previously loaded commands in that the real-time commands are carried out as soon as the spacecraft receives them from the ground. We use this kind of commanding for less sophisticated spacecraft that have neither onboard memory nor extensive capacity for onboard processing. We also use it for sophisticated spacecraft when the spacecraft's state of health operations are at risk or when a ground operator needs to act faster than generating a load update allows. This scenario yields a simpler interface than loaded commands because the spacecraft executes the real-time commands directly (for example, to configure equipment or point antennas). These real-time commands aren't interlaced with commands generated by the spacecraft. For spacecraft with the less sophisticated onboard processing of real-time commands, there is no need to update parameters in the spacecraft's software, which would require a well coordinated effort between spacecraft and mission operations engineering.

In the definition phase, the element designers develop the scenarios for uplinking the real-time commands. They trade generating the commands in real-time versus having the commands pregenerated and stored, but with the ability to select and uplink them in real time. The former approach requires fast processing to generate the appropriate commands in the spacecraft's format in real time, whereas the latter approach requires a very efficient ability to search and retrieve information from a database. Like the other commanding capability, real-time commanding requires simulation, as well as scenarios for operator interaction, review, and approval. But these areas are more critical in real-time commanding because of the very nature of the relationship between command uplink and execution.

In the development phase, element designers define the preliminary and final designs for spacecraft fault detection and isolation, command lookup, and subsequent uplink. As part of the development effort, the element and CI designers conduct studies that examine whether we can use "smart" software, such as rule-based-inference engines, to generate commands. They also explore the responsiveness and database complexities associated with using pregenerated commands that are retrieved based on real-time events.

6.3.6 Spacecraft Power

As mission operations managers, we monitor this scenario when the spacecraft's power capacity limits satisfying the mission objective. We monitor the scenario for two design attributes: (1) the algorithm complexity that accompanies developing methods to support best operation of the spacecraft, and (2) the processing scenarios required to describe the roles of and relationships among the mission operations tasks for planning, scheduling, and determining spacecraft state.

Because modeling the subsystem involves a large range of complexity, system and element designers use the definition phase to explore and decide how simple or sophisticated the models need to be to satisfy the mission objectives. System designers determine the solar-array performance, onboard subsystems and payload power levels, and battery capacities. They also determine how much power-gathering activities, such as sunbathing, are needed to maintain adequate battery state-of-charge and voltage for mission activities. Finally, they establish the battery's operating requirements, including minimum allowable voltage, necessity for full-charge periods, and periodic battery reconditioning. With this information, the system and element designers determine, by analysis, if power limits the mission objective. Power may become a significant issue in meeting the mission objective only when efficiencies degrade in the battery or solar array. This information is critical in determining the ultimate complexity of the ground element design needed to counter this limitation.

In the development phase, if power is a resource we must allocate to spacecraft activities, element designers develop and validate the supporting methods for planning and scheduling. They determine a strategy for allocating power to activities that maximizes the spacecraft's power margin while flexibly responding to the dynamics of planning and scheduling. Designers also develop ways to handle power as a spacecraft constraint. They include the processing required to recognize and compensate for a low state-of-charge in a battery before the spacecraft acts on its own. They do trades to determine how much mission planning, activity planning, and real-time command and control contribute to satisfying the scenario for spacecraft power.

References

Larson, Wiley J. and James R. Wertz. 1999. *Space Mission Analysis and Design.* Third Edition. Netherlands: Kluwer Publishing.

McMinneman and Palmer. 1984. *Essential Requirements Analysis.* Englewood Cliffs, NJ: Prentice-Hall Inc.

Chapter 7

Activity Planning

David E. Kaslow, *Lockheed Martin*
Jeffrey K. Shupp, *Lockheed Martin*

7.1	Evolution of Activity Planning
7.2	Understanding the Dimensions of the Planning Components
7.3	Analyzing Components of the Activity Plan

One of the most visible pieces of a mission operations element is its activity plan. An *activity* represents a time reservation within which we satisfy a maintenance or mission objective. We must understand the many aspects of activity planning to develop a plan that meets the needs of the mission objective most effectively. This chapter presents the process for developing activity planning within mission operations from definition through preliminary development. The chapter focuses mainly on developing concepts and requirements for the system and the mission operations element in simple and complex planning environments. We also address those aspects of a concept or requirement that require resolution at the system level rather than within mission operations. We include alternate operations concepts, along with factors that influence the selection of an operations concept through requirements and design trades. We also discuss how to develop the methods for software tasks and procedures for operator tasks that support the activity plan.

Figure 7.1 breaks out the three sections in this chapter and includes several common terms:

Mission objectives. The mission objectives determine the data's content, quantity, and quality. The system tasking organizations provide these objectives to mission operations.

Maintenance objectives. The maintenance objectives determine what we need to do to make sure the space and ground equipment remains calibrated, aligned, and

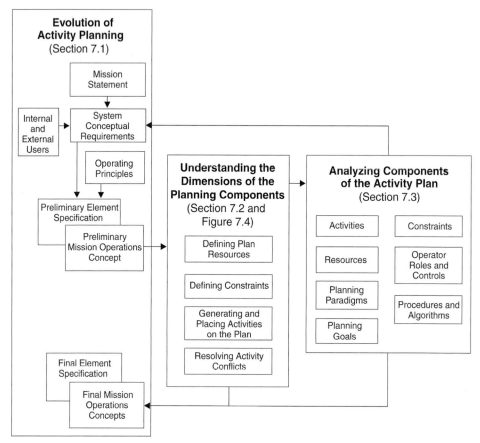

Fig. 7.1. Perspectives on Activity Planning. Applying the methods presented in Chap. 6 results in cost-effective definition and development and efficient operations for activity planning.

in good working order throughout its operational life. Spacecraft and ground system engineering provide the maintenance objectives to mission operations.

Constraints. Constraints define the operational boundaries that mission operations must observe in meeting the mission and maintenance objectives. One class of constraints defines the safe operating conditions for space and ground elements. These constraints are determined by spacecraft and ground system engineering. The second class of constraints, defined by the mission tasking organization, determines the geometric and temporal conditions that the mission objectives must satisfy.

Resources. Resources are mission components that contribute directly to generating, transmitting, and processing mission data. Examples are ground receivers, spacecraft sensors, or communications-link bandwidths. The activity plan deals with a resource when it is in limited supply and therefore needs to be prudently or optimally allocated to satisfy the mission objective. Spacecraft and ground-system engineering determine which components are planning resources.

Activity. Constraints apply within or between activities. We assign resources to activities. We satisfy the mission and maintenance objectives when corresponding activities are planned and carried out in a timely and effective manner.

Plan. The effectiveness of the activity plan depends on its ability to satisfy the mission and maintenance objectives, while ensuring that the spacecraft and ground elements operate safely, efficiently, and effectively. All activity plans have the same three planning components: activities, resources, and constraints. The ultimate effectiveness of activity planning depends strongly on how designers of the system, mission operations element, and configuration items define and develop ways to handle the interactions of these components.

7.1 Evolution of Activity Planning

Activity planning evolves within the mission operations element, as we see in Fig. 7.1 and Fig. 7.2. Our discussion here follows on the definition and development process described in Chap. 6, beginning with a mission statement and continuing through concepts for the mission and for operations of configuration items (CI).

This section is intended as a guide for defining and developing activity planning. It shows how activity planning progresses from the mission statement to the high-level operations concept. Our presentation is at a high level, but it covers the breadth of activity planning. Actual definition and development should go into much greater detail. We must be sure the plan components, conceptual requirements, operating principles, and high-level concept for operations incorporate all the items in Fig. 7.2 and that we address the matters in the following paragraphs.

Mission Statement. As shown in Figs. 6.1, 6.2, and 6.3, the system mission statement is the basis for defining and developing the system. Similarly, as Figs. 7.1 and 7.2 show, we use a mission statement to start developing the activity plan. The mission statement is "The activity plan consists of the activities necessary to support normal operations and planned maintenance of the mission operations element. Activities are arranged on a timeline to satisfy the mission and maintenance objectives, while ensuring that the spacecraft and ground system adhere to constraints and operate effectively."

Conceptual Requirements. As shown in Fig. 6.3, we use conceptual requirements to start developing preliminary mission operations concepts and requirements. We derive these conceptual requirements from the mission statement, and they mature as the system concepts and requirements mature. We

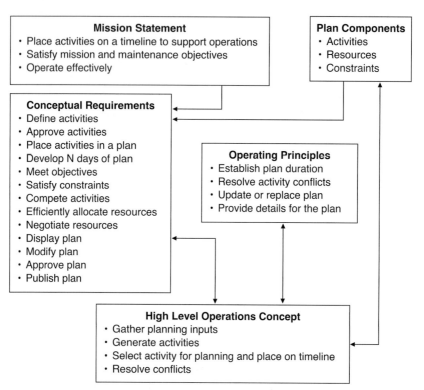

Fig. 7.2. **Defining and Developing the Activity Plan.** We must define the plan components, operating principles, and conceptual requirements before we can develop a preliminary operations concept. This process is iterative. As we add more detail to our operations concept, we modify components, requirements, and principles to increase mission effectiveness.

list the conceptual requirements for our mission statement in Fig. 7.2 and explain them in the following paragraphs.

The mission statement is reflected, and augmented, in several of the mission's conceptual requirements. Engineers and designers include requirements to place activities in a plan so they satisfy mission and activity constraints while meeting the mission and maintenance objectives. They include other requirements to make sure the plan allocates efficiently the resources needed to satisfy the mission and maintenance objectives. Requirements are added to allow negotiation on resources for which there is competition.

As we discussed in Chap. 6, there are both internal and external users of the mission operations element's products (in this case, the activity plan). The mission's conceptual requirements identify the needs that are critical to mission operations and the system. For the internal users of the activity plan, we include requirements that give operators an overview of space and ground operations.

Often the plan is displayed so operators can see the spacecraft's upcoming activities at a glance. This display can be graphical or in text, be arranged by spacecraft revolution or time, and be a workstation display or a wall projection. All of these characteristics would be included at a high level in the display's conceptual requirements. Other requirements supporting the internal users are to provide a way to support the organization's authorization of the planned activities and for an operator to be able to modify and approve a plan. One of the requirements supporting the external users is publishing the plan periodically to people that use the plan to establish their data-processing timelines, such as exploiting data from the mission downlink.

Operating Principles. We must agree on several operating principles before we can define a mission operations concept that supports the conceptual requirements. We list the operating principles in Fig. 7.2 and present them in detail in the following paragraphs.

The duration of the activity plan, which typically covers three to fourteen days into the future, will drive many other mission operations characteristics, such as the accuracy of the ephemeris and the desired amount of operator insight. One important decision is whether the plan can carry conflicts. A plan carries a conflict if we assign time and resources to an activity in such a way that it violates a constraint with another activity or if the resources are supporting another activity. If the only way to resolve the constraint violation degrades one or more activities sharing the conflict, it may be better to inform an operator and resolve the conflict later. The plan may then contain unresolved conflicts, but the part of the plan—usually one to two days—that supports near-term scheduling and commanding must always be valid—with no unresolved conflicts.

We must also determine how often to revisit the plan based on new information and how one plan transitions into another. While we're preparing a new candidate, the operational plan and its updates are in effect. Finally, we need to consider the plan's level of detail. This depends on how operations uses the plan. In general, the plan requires more detail when we use it to operate the spacecraft in an automated fashion. Conversely, it requires less detail if it's used as a basis for a detailed schedule or simply to guide the operators. Stated another way, we must specify many details about the activities within the plan before translating them into spacecraft commands. These details include activity start and stop times, equipment configurations, and spacecraft orientations. If we use the plan to operate the spacecraft in an automated fashion, the planning function must develop the details; otherwise, an operator or the scheduling function can specify them.

Preliminary Concept for Mission Operations. With the mission's conceptual requirements and the operating principles in mind, CI designers can begin to develop the preliminary concept for activity planning. Figure 7.2, in the "High level Operations Concept" box, lists the four high-level tasks needed to produce an activity plan. In Fig. 7.3 we show how these tasks interact.

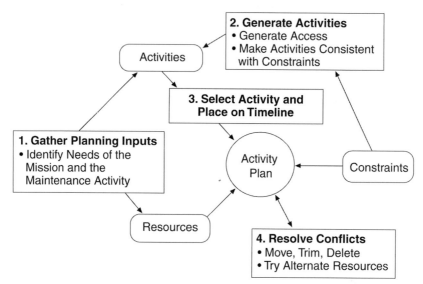

Fig. 7.3. **Tasks Required for Activity Planning.** Activity planning consists of placing spacecraft and ground activities on a timeline and assigning planning resources to those activities while satisfying resource constraints.

The first task involves gathering all inputs that define the mission and maintenance objectives that the plan must satisfy. Next we generate candidate activities based on activity objectives and constraints. Constraints can be geometric, time, subsystem limits, or inter-activity exclusions. Third we select a candidate activity and place it in the plan, while checking constraints and proper use of resources. Then, if we can't place the activity as defined, the fourth task is exploring resolution strategies. We use such strategies as modifying (trimming, moving, or deleting) the activity or competing activities already in the plan. Efficient strategies won't need a lot of runtime for exhaustive enumeration, but they will increase the complexity of the planning algorithms. We repeat the last two tasks until the activity meets mission and maintenance objectives, until we can't explore other placement options because of constraints, or until we've used up the time allotted for the search. Finally, we include opportunities for operators to monitor, influence, and approve the activity plan as it develops.

With this preliminary high-level concept for activity planning, configuration-item (CI) designers use analysis to develop and prototype ways to carry out activity planning. Trades at the CI level to support a method or algorithm may influence the mission operations element's, or even the system's, definition of the requirements.

7.2 Understanding the Dimensions of the Planning Components

As we illustrate in Fig. 7.4, there are four aspects of generating an activity plan. Each aspect provides several ways of viewing the planning components, and each way influences how we generate and use the plan. In other words, there are dimensions to defining components of the activity plan. We highlight these dimensions for each of the four aspects in Fig. 7.4 and then detail each aspect in the following paragraphs.

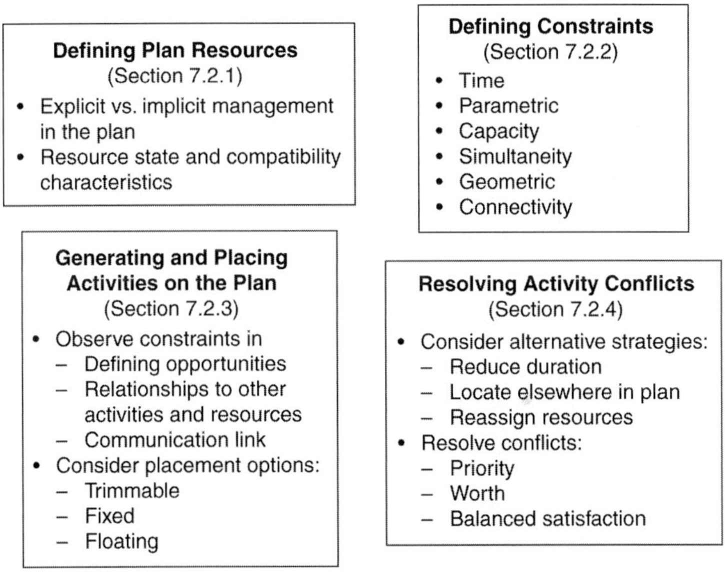

Defining Plan Resources
(Section 7.2.1)
- Explicit vs. implicit management in the plan
- Resource state and compatibility characteristics

Defining Constraints
(Section 7.2.2)
- Time
- Parametric
- Capacity
- Simultaneity
- Geometric
- Connectivity

Generating and Placing Activities on the Plan
(Section 7.2.3)
- Observe constraints in
  - Defining opportunities
  - Relationships to other activities and resources
  - Communication link
- Consider placement options:
  - Trimmable
  - Fixed
  - Floating

Resolving Activity Conflicts
(Section 7.2.4)
- Consider alternative strategies:
  - Reduce duration
  - Locate elsewhere in plan
  - Reassign resources
- Resolve conflicts:
  - Priority
  - Worth
  - Balanced satisfaction

Fig. 7.4.   Dimensions of the Activity-Planning Components. We must analyze different approaches to defining resources and constraints, generating and placing activities on a timeline, and resolving conflicts.

7.2.1 Defining Plan Resources

As stated in the introduction, a plan resource is a system component that is in limited supply and, therefore, needs to be allocated to satisfy the mission objective. We may define planning resources as physical assets, such as ground receivers, mission and relay spacecraft, and ground telemetry, tracking, and command (TT&C) hardware. They may also be the expendables of an asset, such as the power on a spacecraft. They may be paths or the bandwidth of a communication stream.

Space- and ground-system engineers must identify which components of the space and ground elements to treat as resources and also identify, develop, and validate the concepts and methods for allocation.

Explicit and Implicit Resource Management. We use one of two ways to manage resources in a plan. The first is to assign the resource explicitly to particular activities at specific times. We use this method if assigning activities depends on knowing which resource will be needed. For example, we assign activities to a specific ground receiver if that receiver is the only one able to support the required downlink. The second way to manage resources is to treat them implicitly as constraints that mission and maintenance activities must observe and then placing them in the plan. In this context, the number of resources available—not any specific resource—is important.

Resource State and Compatibility Characteristics. Resources have characteristics we must take into account when assigning activities to them. Properties of the state of the mission spacecraft and relay and ground receivers are vital to the protocols for establishing communication links that support the activities. The amount of time a spacecraft takes to acquire a relay or ground receiver may be different depending on the state of the spacecraft or the ground receiver at the time of acquisition. For example, if the spacecraft is making its first acquisition after a scheduled orbit adjustment, its position at the time of acquisition is not very well known. Therefore, we may need to use a unique procedure for this acquisition that searches a wider than usual spatial area. We must reflect this state properly in allocating the acquisition time to the plan. Similarly, we must allocate each component of the communication link from the spacecraft to the ultimate source or destination of the communications data.

Another matter that arises when assigning activities to resources is compatibility. There may be multiple generations of spacecraft and ground assets in the system, with some assets being at least partly incompatible. In this case, we may not meet, or may not meet fully, certain mission objectives if we assign the activity to particular resources.

7.2.2 Defining Constraints

In the following paragraphs, we discuss the various constraint categories, as listed in Fig. 7.4, and their application to activity planning. As mission operations managers, we should make sure the space and ground elements are analyzed thoroughly for constraints, with the results documented in the element specifications and interface-control documents.

Time Constraints. Constraints take on many forms in a plan. One form is the relationships of activities to other activities, to resources, or to defined points in time or orbit. For example, we may prohibit a certain spacecraft activity until the spacecraft has exited the orbital umbra by at least five minutes and separate it from subsequent activities by at least fifteen minutes. Time constraints can be either prohibitive (the activity must not be closer than a defined period) or inclusive (the activity must not be farther away than a defined period). A special case of the inclusive constraint is known as the co-existence constraint, which requires that a certain activity be completely within another activity, with defined margins.

Spacecraft activities normally require a coexistent contact between the spacecraft and ground receiver, plus additional time for acquiring and dropping the signal before and after the activity.

Parametric Constraints. These constraints require a parameter to stay above or below a particular value throughout the plan. For example, a spacecraft constraint may require a subsystem's temperature to stay below a specified value at all times. For this kind of constraint, we have to use a mathematical model to translate planned activities from a time placement and duration into the parameter being constrained. Sometimes the parameter being constrained is time, which requires no parametric translation. Limits on the spacecraft subsystem's duty cycle fall into this special case of parametric constraint. Other examples of parametrically defined constraints are power, defined by a minimum state of charge, and momentum limits, defined by a not-to-exceed torque during an orbit revolution.

Capacity Constraints. These constraints specify a number of occurrences of an activity that are required or are not to be exceeded over a defined time interval. For example, we may constrain the spacecraft by limiting its acquisitions during any single orbit revolution. Other examples include the rate at which tasking requests can be accepted and the rate at which mission data can be processed.

Simultaneity Constraints. A special case of the capacity constraint applies when the time interval involved is set to zero. This is known as a simultaneity constraint. We usually apply it to resources expressed as a number of resources per unit of time, as opposed to those explicitly assigned to activities on a timeline. Examples include the number of ground hardware units available in TT&C to support concurrent acquisition and the number of wideband downlink streams that can be processed at one time.

Geometric Constraints. These constraints require specific spacecraft or payload orientations to carry out the activity. For example, some activities might require the spacecraft's aperture to traverse certain star patterns to determine the spacecraft's attitude and position. Other examples include restrictions on the movement of spacecraft and ground-receiver antennas, line-of-sight restrictions imposed by the horizon for ground receivers, and specific attitudes required of the spacecraft to collect the most solar energy or to transfer the most heat through thermal radiation.

Connectivity Constraints. Connectivity constraints are meant to satisfy several activities or resource conditions at the same time. An example is a constraint on the communications link. In a communications link, the spacecraft must have an established contact with a ground receiver. Sometimes, it communicates directly with the ground. At other times it must use a relay spacecraft, such as Tracking Data Relay Satellite System (TDRSS). In the latter cases, the relay spacecraft must first contact the ground receiver and then contact the mission spacecraft. So we must establish a link between the ground receivers and the TT&C hardware in the mission operations element. While ensuring all communications links are in place, we check the plan to make sure the link resources aren't being used by another spacecraft, in maintenance, reserved, or otherwise restricted for use at the planned contact times.

When to Place Constraints in the Plan. We apply all constraints whenever we're trying to place an activity in the plan, but we start with those that deal with the given activity. For example, geometric constraints usually apply to one particular activity and therefore fall into this category. The second set of constraints includes those that affect neighboring activities or resources in a local area of the plan. Connectivity, simultaneity, and time constraints fall into this category. Finally, we check constraints that deal with a large timeframe in the plan. Parametric and general-capacity constraints fall into this category. We use this strategy so we need less computing capacity to produce a plan free of constraint violations. In general, parametric and capacity constraints require more sophisticated methods than the other kinds.

7.2.3 Generating and Placing Activities in the Plan

Engineers must analyze thoroughly and identify all activities in the space and ground elements and document them in the element specifications as they decide which ones to include in the activity plan.

Activities that Make up a Plan. Normal mission operations consist of many activities. These activities include establishing communication with the spacecraft—acquiring the spacecraft, configuring the spacecraft to send or receive data, uplinking commands, downlinking mission and maintenance data, and droplinking. Other activities satisfy the mission and maintenance objectives—commanding the mission payload, thermal and momentum dumping, adjusting the orbit, collecting power, and calibrating and aligning the spacecraft bus and payload. Still others process the collected mission and maintenance data on the ground—analyzing and distributing mission data and analyzing calibration, alignment, and maintenance data. Finally, normal operations include preventive-maintenance activities, such as running diagnostics and replacing worn parts on a scheduled basis for special-purpose hardware in the TT&C subsystem or the system ground receivers.

Activity Placement. Each of the activities that goes into a normal operations mission plan has certain characteristics that determine where it can be satisfied on the timeline. Typically, the activity is described as a start and end time or, equivalently, a start time and duration. Activities usually have several types of constraints associated with them, some for defining opportunities, others for defining the relationships of the activities to each other and to the resources assigned to support them. Some activities require a command and telemetry link to the mission operations element's TT&C hardware. Other activities require an additional link with a high data rate to a data-processing element, while still others don't need a communications link. Finally, engineers assign activities a set of importance factors that they use when activities compete for a particular resource.

Trimmable Activities. Depending on an activity's nature, we may have some options for when it occurs in the plan. Activities that support the mission objectives directly must occur when the spacecraft's position in the orbit allows an access that satisfies these objectives. Access in this context means the spacecraft meets geometric constraints imposed by the mission. If several accesses will satisfy the mission objective, we may place the activity on the plan at the best access point, or we can move it to another one so we can satisfy other objectives. Satisfying the mission objective means meeting its specific characteristics. These include time of day, month, or year; the quality or quantity of the data collected; and any unique spacecraft-objective geometries. This means that some mission activities aren't movable. In other words, any change to the start or end time of the mission activity will cause it to fail in some or all of its mission objective. But most mission activities can be shortened without failing the entire mission objective. We refer to these activities as *trimmable*.

Fixed Activities. Maintenance activities may or may not be similarly restricted. Certain calibrations and alignments for the spacecraft bus or payload require definite Sun- or star-to-spacecraft geometries or, as in the case of adjusting a spacecraft orbit, require the activity to occur in a very definite part of the orbit, such as perigee. We consider these activities immovable. Furthermore, because of the nature of data collection for maintenance activities, we must place their entire duration in the plan for them to be satisfied. For example, certain calibrations for a spacecraft subsystem may require a series of physical component settings with data collection at each setting. If we don't collect data at all of the settings, the calibration is useless. Because these activities are immovable and untrimmable, we call them *fixed activities*.

Floating Activities. Other maintenance activities may simply require the spacecraft to meet special geometries, such as pointing radiators to deep space to dump thermal energy, or pointing solar panels towards the Sun for power collection. These less restrictive constraints allow activities to be satisfied over some continuous interval, so we consider the activities movable and call them *floating activities*. Like the fixed activities, floating activities still require a certain amount of time to be satisfied.

7.2.4 Resolving Activity Conflicts

In this subsection, we look at how to place activities in a plan when they conflict with other activities or violate constraints. As operations managers, we must make sure conflict-resolution strategies are in the specification for mission operations on the ground. System managers have to approve matters of priority, worth, and balanced satisfaction. We should also ensure that mission engineers identify, develop, and verify methods for resolving constraint violations.

Alternative Strategies for Placement. Once we've determined opportunities for placing activities, we begin putting them into the plan against specific start and end times and resources, while watching for constraint violations and conflicts.

Constraint violations differ from conflicts. A *constraint violation* means that two or more activities' placements or resource assignments exceed a constraint limit. In this case, we should try to change where we place the offending activities. A *conflict* indicates that the constraint limit is exceeded and we can't change placements to clear the violation. Our alternatives for placement depend on whether the activities are fixed, floating, or trimmable. We may also have some latitude in resource assignments to clear a constraint violation, especially for communication links. The spacecraft may be able to use several different paths to satisfy the activity. Different arrangements of the link assignments may free up a resource for a particular spacecraft that has a limited number of paths, while still fully satisfying another spacecraft's needs.

Resolving Conflicts. Our plan has conflicts whenever we've unsuccessfully exhausted every alternative to place activities so they're free of constraint violations. We may allow the plan to carry the conflict until an operator manually deletes or modifies the activity allocation in the plan to clear the constraint violations. Or we may use software to eliminate the conflicts automatically. To remove conflicts when no more placement options are available, we eliminate or trim activities of least importance to the system until the plan meets all constraints.

Resolution by Priority and Worth. Engineers can identify the least important activities in many ways. One way is to define a priority scheme. If two activities are in conflict, the activity with the higher priority gets the resource and time interval. All of the lower-priority activities in conflict with it are trimmed, moved to another time interval, or deleted. A second way is to define a worth to each activity which may vary with time over the course of the activity. Then, an activity of higher worth wins out when the combined worth of the competing activities is lower but loses out when that combined worth is higher.

Resolution by Balanced Satisfaction. If engineers can't assign worth to all system activities, they have to use other schemes to resolve conflicts and another type of metric to resolve ties for competing resources. One concept, called balanced satisfaction, involves finding a level of satisfaction for the group of competing activities that best satisfies the overall activity plan. We can use the duration of the competing activities as a metric for balanced satisfaction. Other considerations can determine tie breakers in a conflict, along with or instead of duration. They include whether the activity is overdue for execution and how many future opportunities exist to carry it out.

7.3 Analyzing Components of the Activity Plan

The hardest goal to achieve in activity planning is defining inputs, methods, and outputs that offer the most effective design. An effective design is one that satisfies the requirements and fulfills the concepts without unnecessary or extra design that adds cost and schedule risk in development and requires more effort in maintenance and operations. We can achieve this goal by analyzing concepts and requirements for the system and for mission operations. Figure 7.1 lists aspects of activity planning that we must analyze to define requirements, operations concepts, operator procedures, and software methods.

We must make sure that the definition and development plans identify analysis tasks for activity planning and that the analysis is carried out in a timely, correct, and thorough manner. We must also ensure the activity-planning concepts and specifications reflect the results of the analysis. Finally, we work with all of the external and internal organizations to ensure the activity planning meets their needs.

External Organizations. The mission operations element is the hub of the space-ground system. Here we transform mission requests into spacecraft activities. These activities become commands that are uplinked to and executed by the spacecraft, with the resulting mission data routed to the data-processing center on the ground. Because the activity plan coordinates these activities, many organizations external to mission operations influence the definition and development of activity planning. These organizations participate in tasking, product exploitation, spacecraft engineering, ground-system engineering, and, of course, operations.

Internal Organizations. The designers of configuration items within activity planning work closely with external organizations. The CI designers develop the requirements and operations concepts for activity planning; they also develop and validate the software methods used to generate the activity plan. The operations engineers also work closely with external organizations. They define and analyze operator tasks, define and analyze loading on operator positions, and prototype displays and operator interactions.

We address the following aspects of activity planning in the next sections:

- Activities
- Constraints
- Resources
- Operator role and control
- Planning paradigms
- Planning procedures and algorithms
- Planning goals

7.3.1 Activities

Table 7.1 shows the issues we must consider and deal with to define and place activities in the activity plan.

Mission Activities. At the system level, the tasking organization and planning engineers for the system and mission operations work together to develop conceptual requirements that govern the planning and placement of mission activities. They address such processing issues as defining access, satisfying mission objectives, defining quality of access, and establishing geometric constraints for the activity. They also determine how to manage the input-tasking information, including timelines, frequency, and throughput volume.

Table 7.1. Activities. We must define and analyze activities to develop placement strategies.

Process Step	Issues to Address
Identify System Conceptual Requirements	• Mission-access definition, quality, and constraints • Management of input tasking information • Maintenance activity triggers and relationships to mission
Develop Element Concepts and Methods	• Role of planning versus scheduling for access accuracy • Options for placing maintenance activities • Automated versus manual planning of activities

Determining the Mission Activity's Access. In response to the conceptual requirements for mission access, the CI designers develop ways to calculate the access interval, quality, and duration needed to satisfy the mission objective at the required level of performance. The *access interval* is the time in orbit that the spacecraft is in view of the mission subject. The quality is a measure of the mission data that can be collected—for example, the minimum ground distance or stellar angular resolution for collecting optical data. The required performance is the accuracy of the access calculations and the maximum allowed computer execution time, if timelines are important. Conceptual trades may help resolve issues concerning required performance for calculating mission access. For example, the accuracy of equations should never exceed the accuracy of the input data. Designers determine the accuracies of the calculations by the context in which they apply. If the access information is also used by scheduling, we may need increased accuracy, which in turn requires more exact and more complex equations.

Other trades define how planning and scheduling support the mission activity. The main purpose of planning is to allocate the system resources to support the mission and maintenance objectives, with data at a definable level of accuracy. But activity planning may be more or less involved in supporting the scheduling. At one extreme, it may act as a mission-data filter for scheduling. This concept requires more exact and complex equations to ensure that acceptable accesses are not incorrectly filtered out during planning. At the other extreme, planning employs much simpler, maybe even empirically constructed, equations. In this case, scheduling must recalculate all of the mission accesses based on more accurate data. This approach requires significant validation to ensure that the simplified equations are within the error of the remaining input data for planning and to ensure that the more accurate methods meet the timelines in scheduling.

Maintenance Activities. Engineers in system or mission operations planning will join spacecraft and ground-systems engineers to develop requirements for planning and placing maintenance activities. They address processing questions, such as what maintenance activities need to be planned and what activities can wait until scheduling. They also discuss and decide on how frequently particular maintenance activities must occur, what spacecraft and ground circumstances trigger the need for these activities, and the relationships of these activities to the mission activities.

Planning Maintenance Activities and Resolving Conflicts. The CI designers for planning develop concepts and methods to identify and resolve conflicts when maintenance activities are added to the activity plan. Design trades for maintenance activities may take care of some of the matters. For example, as discussed in Sec. 7.2.3, some maintenance activities can occur at various times and still meet the maintenance objective. We can consider this freedom of placement a simplification or an added complexity. As operators, we'd find it much simpler to do a maintenance activity over some interval in the defined planning period. Complexity arises when the placement choices, computed from the possible maintenance accesses, all compete with other activities on the plan. Also, some maintenance activities may occur infrequently in operations; yet finding placement options for them can be difficult. These activities are good candidates for operator-controlled planning, instead of planning for software to carry them out automatically. In this case, we have to trade the added operator workload against the added software complexity needed to minimize that workload.

7.3.2 Constraints

Table 7.2 shows what we must consider and deal with to define and use constraints in the activity plan.

Table 7.2. Constraints. We must analyze constraints to determine the best way to model and incorporate them in the activity plan.

Process Step	Issues to Address
Identify System Conceptual Requirements	• Handling of hazardous versus non-hazardous constraints • Planning of activities using simpler constraints
Develop Element Concepts and Methods	• Parametric modeling of constraints versus time budgets • Highly accurate methods

Identifying Constraints. Spacecraft- and ground-system engineers work with designers of the mission operations element to develop constraints on the spacecraft and ground equipment. They discuss the types of constraints and whether these constraints affect the spacecraft's or ground's immediate or long-term state of health. They also determine which commands are hazardous. For example, any commands that activate spacecraft thrusters are considered hazardous—so much so that they can't be uplinked to the spacecraft until the spacecraft has cleared its launch vehicle.

Approach to Satisfying Constraints. Engineers and designers can do many design trades when dealing with constraints. Generally, the simpler the criteria for planning an activity to meet its constraints, the better the design. For example, the spacecraft may have certain equipment-warmup cycles that depend on the time since the equipment last operated. There are two ways to meet this constraint. The first way is to incorporate the dependencies in the activity-planning constraints and check to see if too much time has elapsed before each use of the equipment. A much simpler way is to plan to operate the equipment regularly for a short period, whether an activity needs it or not, in order to keep the warmup time to a minimum. In this way, planning doesn't need to know the details of the equipment states or how much time has elapsed since the last time the equipment was used. Planning also doesn't need to react to situations in which planned use of the equipment has been canceled, thus changing the time dependency.

Approach to Modeling Parametric Constraints. Engineers and designers must be extremely careful in specifying parametric constraints. By their nature, parametric constraints tend to apply to large portions (e.g., 24 hours) of the planned activities and require equations to translate the activities' start and end times into units of the parameter being measured. A battery's state of charge is an example of a parametric constraint. Again, as for the warmup constraint discussed above, we must trade complexities in the design to arrive at the most effective way to meet the constraint. For this example, a more exact approach would model all of the equipment characteristics in terms of their effect on the battery's state of charge, and would then model how the equipment components support each activity in the plan. Many times, however, the equations that model the constraint require more accurate data than we can get from planning. On the other hand, if they model each activity based on historical performance to have a constant rate of change of battery charge, this rate times the activity duration gives the change in battery state of charge. This approach dramatically reduces the complexity of the design needed to meet the constraints, while still keeping operations or future spacecraft needs very flexible.

Accuracy of Constraint Methods. Both of these examples of constraint trades need approval by the system designers and element designers of the resources imposing the constraints. To support that approval, CI designers develop and validate methods to meet these constraints. They must analyze carefully to offer a design with the best balance between algorithmic complexity and conservatism. A highly accurate method produces the most conservative solution but with an associated increase in algorithmic complexity. Some things can easily destroy the perceived accuracy of the planning data and make the method useless. These include the probability of not executing what was planned, variation in the parameters within the constraint equations between the predicted and post-activity values, and the willingness of operators to operate the spacecraft very near the constraint limits.

7.3.3 Resources

Together, the system and element designers, as well as the spacecraft- and ground-system engineers, determine which system components are activity planning resources. They must determine whether the component is available enough to support fully the mission objective. Available means able to carry out its intended tasks. Downtime for maintenance or recovery to operations-support levels are part of this availability.

Sometimes we can analyze best by prototyping the parts of the system to measure the level of use. A trade that can reduce the complexity of resource planning examines whether we must explicitly model the resource in the plan and specify activity assignments, or whether we can implicitly handle it through a set of capacity constraints. With this analysis and trade decisions in hand, CI designers develop and validate ways to allocate the resource.

7.3.4 Operator Role and Control

Operator Role in the Planning Process. Suppose activities are few, objectives are simple, and competition among activities for the system resources is light. In this case, an operator can construct a plan manually by using a timeline tool that checks objectives, constraints, and resources. The operator changes an activity manually to resolve conflicts. But if activities are numerous, objectives are complex, or competition is heavy, we need a more automatic planning algorithm. In this case, element designers develop a suite of algorithms to construct a plan, with those algorithms carrying out the same reasoning and actions the operator would use in constructing the plan manually. The operator's role in planning also depends on his or her skill and knowledge of how to satisfy the mission and maintenance objectives.

In either instance, the operator needs to have certain tools, controls, and procedures to construct a plan. Controls allow the operator to influence the plan in areas such as activity placement, conflict resolution and alternative solutions, stability and responsiveness, analysis and resulting changes, and approval. For example, the operator constructs a plan by selecting an activity and then determining the start time and duration that satisfy the activity's objective and constraints. We usually use automated tasks to help the operator select times within the access interval of the activity. If the activity also requires a link resource from the system, we must place it on the plan's timeline so it satisfies system constraints. Several placement options may be available to satisfy the activity, but some may conflict with existing activities or result in contention for resources. The operator has to modify either the new activity or one or more existing activities to clear the constraint violation or resource contention.

Automated algorithms will always have an absolute metric with which to gauge the quality of one set of options over another. The problem is that activity planning, as part of mission operations, deals with metrics we often can't express absolutely. So the mission operator must be able to manipulate the plan directly or indirectly to get operational approval. The control mechanisms available to an operator depend somewhat on the processing architecture. For example, in a batch architecture, operator controls are part of process execution. But in a transactional architecture, event triggers govern operator controls.

Plan-Visualizing Tools. These tools help an operator place an activity in the plan. They typically display activities on the timeline from multiple perspectives. One view is of the activities relative to the resource links, including mission spacecraft, relay spacecraft, ground receivers, and ground hardware. Another focuses on specific resources. For example, the display may show activities from all mission spacecraft that use a certain ground receiver. Other tools help the operator determine how well an activity satisfies its objective. Displays will show messages when an activity's placement violates a constraint relative to other activities or when resources contend. The display should also show the extent of the violation and which other activities or resources are involved. The visualizing tool is usually a series of specialized displays using commercial, off-the-shelf (COTS) products and mission-specific software. The COTS products provide constellation views, including lines of sight from spacecraft to spacecraft and from spacecraft to ground receiver, as well as activity placements. The mission-specific code is usually reserved for mission-specific calculations and displays of the activity objective, activity constraints, and system-constraint parameters.

7.3.5 Planning Paradigms

Table 7.3 illustrates the range of possibilities for three planning issues.

Table 7.3. Planning Paradigms. We must establish early on and then continue to develop certain paradigms that determine the operation's approach to planning.

Process Step	Issues to Address

Identify System Conceptual Requirements	• Creating the plan
	  – Build a new plan versus update the old plan
	• Operator involvement
	  – Create the plan manually versus automatically

Develop Element Concepts and Methods	• Processing approach – By batch versus by transaction

The operational look and feel of activity planning depend on decisions about processing paradigms in these three areas. They influence the operator's role in developing the activity plan, the software algorithms that assist or automate the operator's planning choices, and the approaches for ad hoc changes to the plan.

Develop a New Plan or Change an Old One. A plan made by updating a current plan offers a very stable environment for operators. Changes occur only when they add measurable value to the mission and maintenance objective. But if inputs to the plan are constantly changing, or if we're optimizing over long periods, building a plan from scratch yields better solutions. To choose the correct paradigm, we must analyze to get the proper mixture of best performance and stability.

Determine the Operator's Role. Operator-directed placement with automated constraint checking gives operations the most control over the mission. However, if operators can rely on sophisticated automated processing to generate the activity plan, they're better able to oversee and analyze it. Also, if activity-planning inputs are too complex, trying several solutions may be cumbersome and time-consuming.

Choose Batch or Transactional Processing. *Batch planning* uses a series of processes to place items into a plan. These processes have a defined task duration, with persistent data being passed from one task to the next. External updates to and feedback within them are limited, and their algorithms tend to be complex and monolithic. Because batch-processing functions have a strict process flow, they are inherently controllable. On the other hand, in *transactional planning*, processes converse with one another to arrive at a plan. Transactional processing uses continuous processes, easily incorporates updates, and normally includes feedback. For transactional processing, algorithms tend to be simpler individually but more difficult to coordinate as a group. It also requires controls, so responses to updates don't cause unwarranted changes to the already planned activities.

7.3.6 Procedures and Algorithms

Table 7.4 shows the issues we must consider and resolve for the planning procedures and algorithms that generate and update the activity plan.

Table 7.4. Planning Procedures and Algorithms. Planning procedures and algorithms require thorough analysis, development, and validation.

Process Step	Issues to Address
Identify System Conceptual Requirements	• Define plan transition times • Coordinate plan update and extension • Trade plan stability versus responsiveness to change

Develop Element Concepts and Methods	• Consider time to generate the plan
	  – Changes in shifts for plan personnel
	  – Input/output and feedback timelines
	  – Time to generate and carry out plan
	  – Currency of plan relative to inputs
	• Respond to dynamics
	  – Unexpected processing results
	  – New and updated inputs
	  – Updated information on the spacecraft's state
	• Consider complexities introduced by
	  – Spacecraft and mission objectives
	  – Resource limitations and tight constraints
	  – Accuracy and responsiveness

Define Plan-Transition Time. We have to update the activity plan regularly as planning needs change or time passes. The plan must always be operational, so we apply updates to a candidate version of the plan. Transition of the candidate to the operational plan must be seamless. That is, the parts of both plans up to and just beyond the transition time must be the same. How far into the plan the transition time is from the current time depends on the amount of the plan we've already committed to activity schedules and the loads already uplinked to the spacecraft. If we generate a day's worth of activity schedule at a time, the normal transition time is 24 hours into the future. But if we're commanding the spacecraft in real time and generating the schedule an hour at a time, our transition time can be just two hours into the future. The closer the plan transition time is to the present, the more responsive the activity plan can be to near-term changes.

Coordinate Plan Update and Extension. Updates to the plan occur in two parts of the candidate plan. The first part includes updates from the transition time to the end of the current plan, usually three to fourteen days ahead. We must coordinate updates to an existing plan with all the ground components that use the plan data. Additionally, updates to an existing plan shouldn't greatly perturb the current plan. The second part to update is the new section, usually a day long. The extension is a new part of the plan and must integrate smoothly into the end of the current plan, so we have to assign resources in the current plan to allow this.

Operator Shift and Inputs/Outputs. The development cycle for our activity plan should run end to end during a single shift, if possible. If it does, a single operator can monitor the creation and transition of the candidate plan into operation, rather than having to coordinate that task with other operators. CI designers must also consider the timeline for inputs from users because updates to the activity plan can't be approved until we know all of these inputs—from inside or outside the mission operations element. Updates also depend on the timing of feedback that occurs because we've processed data from completed mission or maintenance activities. Finally, they also have to allow time to get output to users.

Generating and Carrying Out the Plan. As stated earlier, the plan transition time depends on the time needed to produce an activity schedule based on the activity plan, coupled with the time for generating and uplinking spacecraft commands to carry out mission or maintenance activities. Also, the processing for updates to the plan must occur within the time from the latest inputs being available and the earliest results produced. CI designers must analyze carefully to ensure there's enough time to process the inputs and produce a new candidate plan ready for transition to operations.

Plan Currency. How current must our plan be? Mission inputs often have continuous or regular updates. Trying to reflect all changes as they occur demands too much of our computing resources because most changes don't influence the overall activity plan and the ones that do, occur at unpredictable moments in the planning day. Engineers and designers need to analyze the system and mission operations to balance stability against responsiveness to change. Then we must set the plan's currency properly.

Respond to Dynamics. Planning is dynamic because we must be able to respond to unexpected results from current mission and maintenance activities, to incorporate new or updated inputs from users, and to allow for increased accuracy of the predicted orbital geometries and our knowledge of the states of the spacecraft subsystems.

Complexities of Plan Development. The complexity of procedures and algorithms we use to develop our plan reflects the complexity of the mission and maintenance environment—spacecraft and mission objectives, resource allocations and tight constraints, and accuracy and responsiveness.

Complexities of Spacecraft and Mission Objectives. One complex situation might be a spacecraft with multiple mission payloads that share common components and spacecraft resources. In this case, we plan activities to satisfy mission objectives while efficiently allocating the spacecraft's resources, such as power, to the payloads. We must also observe intra- and inter-payload constraints, such as deployment restrictions and ability to do concurrent processing. Complexity may also arise if the spacecraft can process data faster than the downlink and ground station can. In this case, we must plan the mission and maintenance activities to satisfy these capacity limits. Other spacecraft limitations, such as memory, power, or uplink and downlink rates, may introduce complexities into the planning procedures and algorithms. Finally, the mission objective itself may impose complexities. If an activity has few opportunities to satisfy mission objectives, we'll need more complex placement options and conflict-resolution strategies.

Complexities of Resource Limitations and Tight Constraints. Conflict resolution to satisfy a mission objective can become even more complex when resources are limited. In this case, activity planning has to make the best use of the limited resource to satisfy the best subset of activities. Algorithms involving complex mathematical techniques, known collectively as operations-research techniques, may be required to satisfy the activity best in a reasonable computation time. A highly constrained environment can result in the same algorithm complexities as a resource-poor one because the constraints allow so few options. Also, testing the plan against these constraints for each combination of activity assignments requires a lot of processing.

Complexities of Plan Accuracy and Responsiveness. If planning must employ detailed models to stay within given parameters, or if an accurate solution requires a complex equation, planning becomes more complex. As stated earlier in our discussion of constraint considerations, engineers and designers must analyze carefully to ensure that highly accurate specifications are warranted. Complexities also arise when a deadline is so short that only the most efficient strategies to search for placement options can produce a new plan in time. This complexity is opposite from the complexity of trying to find the most accurate solution.

Reducing Complexity. Some of the design trades that the CI designers in activity planning might use to deal with these complexities include giving some of the most complex decisions to the operator. Of course, they must take care not to overload operators with tasks that keep them from monitoring the rest of activity planning. We can also reduce or eliminate complexities if the system and mission operations element can agree on a conservative way to allocate activities to resources, satisfy constraints, and optimize mission objectives. Finally, designers can reduce the algorithm complexities if they take a converging, interactive, iterative approach rather than trying to examine everything at the same time.

7.3.7 Planning Goals

Satisfy Objectives Best. Activity planning—and procedures supporting it—must reflect the desired "goodness" of the plan. The best plan satisfies all mission and maintenance objectives, observes all system and activity constraints, and uses resources efficiently. It also produces a very good answer that satisfies all mission and maintenance objectives in minutes to hours rather than an absolutely best answer in hours to days. If the system limits our plan, designers must do trades to find the best possible way to meet mission objectives—based on the priority or worth of each activity. To determine an activity's priority or worth, we look at its importance among other competing activities, how often it must access the mission objective, and whether it must continue over a shorter or longer term.

Provide a Flexible Plan. A plan that leaves room for changes is an important goal. Developing a plan that uses all of the resources and drives constraints to their limits doesn't allow for additional tasking without large changes. For example, if we plan to use all available power for a given day, and a new activity is needed, we have to delete some of the plan to meet the new activity's power requirements. It's more operationally sound if we plan to do less than the spacecraft's absolute limits, so this kind of situation can draw on planned reserves.

7.3.8 Goal Based Sequencing

Most science mission spacecraft execute a predetermined command load that specifies every operation from spacecraft orientation and payload configuration to mission data gathering and downlink. Predetermined command loads are used in Earth, Moon and Sun orbiting missions and planetary orbiting and flyby missions. Ground-based software and operators construct the command loads based on the mission objectives and targets of opportunity as well as the spacecraft capabilities and constraints.

These missions execute on well defined cycles: mission activity planning and scheduling, command load construction and uplink, and mission data gathering and downlink. Command loads also direct support activities such as payload calibrations and alignments, power gathering and battery maintenance, communication acquisition and droplink, and safe mode operations. This paradigm works well since the targets of opportunity are well-defined and the spacecraft state, e.g., power, thermal, position, and orientation, is known in real-time or very near real-time.

However, other missions such as Mars landers and rovers, as well as asteroid and comet orbiters, landers and flybys, may execute command loads that are developed on board the spacecraft. This is goal-based sequencing or autonomous commanding. We use goal-based sequencing when knowledge of the mission state, i.e. spacecraft state, environment, and target of opportunity, is uncertain or even unknown to Earth-based operators and software. This happens when 1) the spacecraft is so distant from Earth that the communication time lag is very large, or 2) there is insufficient communication bandwidth to support the quantity of data needed for a complete specification of the mission state, or 3) there is limited communication time, such as when using the Deep Space Network.

The onboard software can be responsible for many aspects of the mission, including the journey to the target of opportunity (e.g. a specified object on the Martian surface or a rendezvous with an asteroid), acquisition of data (e.g. imaging with a camera, spectrometer or magnetometer; collection of a soil or rock sample), and a sample return to Earth or an in-situ analysis.

Autonomous navigation to asteroids and comets requires that a spacecraft determine its position and orientation, calculate the thrust needed to maintain the required trajectory, and command firings of the thrusters. The spacecraft also needs to locate and determine the range to the target object during final rendezvous.

Autonomous landing on an asteroid or on a planet requires that a lander

- Locate the area of investigation
- Assess the ground terrain with respect to slopes and obstructions
- Maneuver to an acceptable landing site
- Determine distance to the ground and rate of descent
- Land using some combination of rockets, parachutes and landing bags

Navigation across a planetary or asteroid surface requires

- Assessment of the terrain with respect to the rover's capability to negotiate the irregularities, slopes and obstructions
- Selection of targets of interest
- Determination of waypoints
- Navigation from one waypoint to the next

This is an extremely complex process, one that must be orchestrated—but not controlled—by an operator. The rover should have dual cameras to provide three-dimensional imaging of the terrain and some level of autonomy to negotiate obstacles when traveling between waypoints.

We select the area of investigation based on scientific interest but it must 1) support communication between the lander and rover and other mission components (e.g. orbiting relay or Earth ground station) and 2) have sufficient sunlight to power the rover and recharge the batteries if the lander and rover are operating off of solar power. Scientists should to be able to identify the landing site by ground imaging of distinctive features or some other autonomous method of determining position.

The design of the software and payloads for goal-based sequencing must be able to handle all contingency situations. Redundant and alternate components, error detection and failover capabilities, state-of-health determination and maintenance, and safe mode operations must all be able to respond to any anomalous situation. They do so by recovering the spacecraft/lander/rover to normal operations or by leaving it in a state with sufficient capability to await and react to communications from Earth-based mission operations. The commanding, telemetry, and mission data sent back to mission operations must be inclusive enough for the operators to assess thoroughly the mission state and to reproduce the past actions in order to investigate the anomalous situation.

These autonomous capabilities must be developed from a set of requirements that incorporate all the anticipated operating conditions and constraints. This requires considerable research and analysis since these types of missions are exploring mostly unknown territory. NASA and others who are starting to use goal-based sequencing are struggling with how to test these autonomous software systems before launch in a way that exercises fully all conditions and constraints. Considerable work is needed in this area, but software experts and research scientists are addressing this issue. However, there are currently no agreed upon methods or approaches in this field.

Chapter 8

Conducting Space Mission Operations

Eileen Dukes, *Lockheed Martin (retired)*

8.1 A Day in the Life of Operators
8.2 Uplink Process
8.3 Downlink Process
8.4 Planning and Analysis Functions

In the previous chapters we discussed the functions required for mission operations. In this chapter we put the functions together to show how the operations flow. We discuss details of activity planning and development and mission control—concentrating on the uplink and downlink tasks, but also showing how the other functions, such as planning and analysis, feed into the flow. This chapter gives the person who has never been directly involved in operations a feel for the day-to-day activities as well as a sense of how we implement an operations timeline. We also provide recommendations for making operations more efficient and hence more cost-effective.

One of the first challenges facing a mission operations manager (MOM) is matching the operations style to the mission requirements. There are two extreme perspectives on spacecraft operations. One extreme maintains that people skilled in operations should do all operations, with little or no regard for the particular spacecraft they're operating. The other extreme maintains that design engineers who built the spacecraft should be the ones to operate it. Both extremes, of course, have their pluses and minuses, and most successful operations mix operations experts and spacecraft experts. Figure 8.1 shows some of the factors we use to determine the correct mixture. Consider, for example, system maturity. When a system and the spacecraft are new, the balance tends to tilt more towards the design experts. As the system matures, the spacecraft or family of spacecraft becomes well understood and the balance tends to shift toward operators.

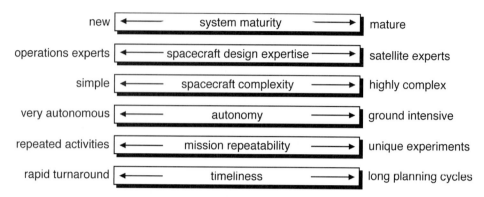

Fig. 8.1. **Factors That Affect the Balance Between Operations and Spacecraft Expertise.** We use these factors to select the appropriate operations style for the mission.

Systems engineers or subsystems experts can provide the expertise. Systems engineers have a broader understanding of the spacecraft and how the subsystems interact. Subsystems experts have a narrow but deep view, usually restricted to their subsystem. Using systems engineers is usually more efficient because we can cover a broad range of spacecraft knowledge with fewer people. But if the mission is unique or very complex we need some subsystem experts. A good compromise is to supplement systems engineers with subsystem experts only for the operationally complex subsystems, such as command and data handling, attitude control, and the payload.

Finding the correct balance is not always easy. Not having enough spacecraft expertise readily available can result in more down time and sometimes in loss of mission. Conversely, a staff with insufficient operations experience will tend to be slow and can sometimes be more interested in using the spacecraft as an engineering testbed than in meeting the mission objectives. Also, design engineers tend to get bored with, and may be too expensive for, routine operations. We must balance these abilities to get the best ratio for the given mission.

Routine operations occur when the processes are well established, which happens when procedures are fairly static, people are fully trained, timelines are usually met, and products are being regularly delivered. It doesn't mean that nothing unusual ever happens; if that were the case, we wouldn't need skilled operators. Normally, we consider routine operations less risky because we understand and practice them. But as repetitiveness becomes monotony, details can get overlooked. As the old saying goes, "Familiarity breeds contempt." One of our challenges is to keep the operations job interesting and operators alert as the mission matures.

On the other hand, we consider *special operations* more risky because they're unique and often critical. They require special procedures, special training, and possibly, rehearsals or dry runs. These operations include one-time events, such as launch or a peculiar calibration, and may also include operations during or following an anomaly.

One way we can find the correct balance is through evolving operations. Although procedures can't change continually, operations need to adapt smoothly and quickly to changes. When a system and spacecraft are new, more engineers familiar with the design are needed to characterize the new spacecraft, calibrate it, and work out the procedures. These engineers then apply their knowledge to updated processes and procedures. Eventually, we train more operators to replace the engineers. The operators understand how to do the mission and begin to streamline and improve the operations.

We need to encourage operators to improve their productivity by using computers to automate their tasks as they become standardized and repetitive. Increased automation of repetitive tasks can lead to a gradual reduction in staff. Besides, the current explosion in computer technology can make a system outdated in five years and obsolete in ten years. Gradually incorporating the new technology allows for upgrades without disruptive change.

Let's consider an extreme example that illustrates this point:

Consider the operator whose job is to sort through line-printer outputs for the past 24 hours, extract the values for eight of the spacecraft's engineering measurements, and add these values to a cumulative plot. Suppose it takes an hour per measurement so this task comprises the entire eight-hour shift. This job is a prime candidate for automation. As a first step, this person could research spreadsheet or graph programs that would eliminate the manual plotting. These programs usually also include basic statistical functions that could enhance the task. At the same time, by adding to his or her personal tools, the operator has a new challenge and a chance for personal growth.

As the next step, we can investigate an automated link between the telemetry processing and the plotting tool, rather than printouts. Again, the operator can improve by learning more about telemetry processing in order to define the requirements and participate in building an automated trending interface.

Once the task is automated, what used to take eight hours may now require only one hour. The operator now has seven hours to analyze trends more deeply, help others automate similar tasks, or reduce overall staffing. Furthermore, the operator has had a chance to exercise some engineering skills by defining the requirements, doing trade studies, perhaps designing the implementation, testing, and incorporation of new methods into procedures.

People can be reluctant to work themselves out of a job if that is how they perceive it. But most people will take the chance to grow by exercising skills or acquiring new skills. We can reduce staff through attrition and provide growth opportunities by promoting from within as higher-up or more skilled positions open up. Promoting from within rewards initiative and continual improvement and, very importantly, keeps program expertise high.

Operations before launch are often designed rigidly, with level staffing through the years. But this approach leads to high mission costs and doesn't allow us to take full advantage of the staff's ability to learn. Providing a structure for operations that allows it to evolve can reduce long-term mission costs.

An example of this evolution is the payload control for the Extreme Ultraviolet Explorer (EUVE). After two years of flight, the staff was able to transition to an automated monitoring system for payload control. The EUVE science-operations center built Artificial Intelligence (AI) software to "mimic the monitoring responsibilities of the human science payload controllers. The AI software perpetually monitors the science payload. When a rule is violated by a data point in the telemetry stream, calls are made to external processes that either rectify the problem or sound an alarm. If a problem occurs during an unstaffed shift, the software autonomously pages an anomaly response coordinator." [AI Magazine, 1994] This automation has allowed the mission to reduce costs by reducing shift-coverage requirements.

8.1 A Day in the Life of Operators

Mission planning consists of collecting requests to use the system and spacecraft, balancing these requests against available capabilities and resources, and creating the activity plan. Requests come from two major groups: the product users or science teams and the analysts who operate and maintain the spacecraft and payload. Capabilities include those of the spacecraft, the payload, the ground system, and mission operations. Resources include all of the people, hardware, software, and telecommunications links both inside and outside the project.

During the development phase, we produce a mission operations concept, which contains the general sequence of events, taking into account planned capabilities and resources. The concept includes the purpose of the mission, the desired data return, and goals. Refer to Chap. 4 for a more detailed discussion on developing a mission operations concept.

8.1.1 Mission Plan

During operations, the mission operations concept documents the project's policies and generally structures mission operations. We must then translate this high-level document into a mission plan and establish an appropriate duration for the plan, taking into account the response time required by users, development cycles for activities, planning cycles for all the groups involved, and the fidelity of planning inputs such as accuracy of ephemeris prediction. For a dynamic mission, the plan covers two weeks to two months. For an interplanetary mission, it covers several months to a year. The planning horizon needs to be long enough to allow all of the mission operations functions to do their own scheduling, but not so long that it becomes obsolete.

The mission plan collects information from

- Spacecraft-bus analysts
- Payload analysts
- Navigation
- Ground systems
- Telecommunications-resource scheduling
- User community

Figure 8.2 illustrates these inputs and their corresponding outputs.

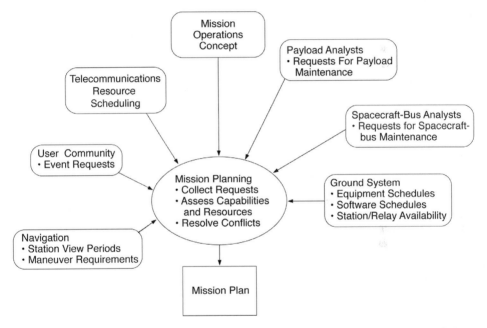

Fig. 8.2. Information Required for Mission Planning. Mission planners must balance resources, users' request, spacecraft analysts' request, and mission objectives when creating a mission plan.

Ideally, mission planning is independent of both the users or requesters and the spacecraft analysts. Users, given a free rein, will tend to drive up mission operations costs by trying to do everything without regard to resources. The spacecraft team may tend to use the spacecraft as an engineering test bed. Mission planners need to be independent and balance the spacecraft needs against operations capabilities, resources, and user requests to meet the mission objectives and stay within the mission operations budget.

A simplified mission plan for a mission with two-week activity plans might look like Table 8.1. The primary events for the activity plans are identified, along with any special requirements or constraints. For this mission, mapping is the primary activity and is the main activity in each command load except for Activity Plan 5. Activity Plan 1 and 2 also contain calibrations of the gyroscopes related to attitude control and system-level pointing. The special consideration for Activity Plan 2 shows that it spans the Christmas holiday. It will act as a guide when we translate the mission plan into an activity plan, so staffing can be minimal on Christmas Day. Because the pointing calibration is a special activity, it should definitely not be scheduled on Christmas. The special consideration in Activity Plan 3 requires more tracking coverage than usual to provide an accurate solution quickly after the orbit-trim maneuver. It directs the scheduler of ground-system resources to schedule more station time during that plan. Spacecraft configuration or celestial geometry may influence events. In the example shown, battery reconditioning can be done only in Activity Plan 5 because it requires full Sun (i.e., no occultation). Finally, mapping can't occur during battery reconditioning because the power requirements are incompatible.

Table 8.1. Example of a Mission Plan. This table is typical of the information and detail contained in a mission plan.

Activity Plan	Primary Events	Geometry	Special Consideration
1	Mapping, gyro calibration	>20 min. occulted	Station 4 maintenance
2	Mapping, pointing calibration	>30 min. occulted	Christmas holiday
3	Mapping, orbit-trim maneuver	>20 min. occulted	Extra tracking required
4	Mapping, special test	<20 min. occulted	Timing update for test
5	Battery reconditioning	No occultation	No mapping

8.1.2 Activity Plan

At some point, we must translate the mission plan into an activity plan. We can do so when the inputs, such as targets, maintenance schedules, and orbital geometry, are accurate and not likely to change. For a dynamic mission, this point would usually be a day to a week before plan activation. For a science mission, such as an interplanetary mission, it's usually six to eight weeks before plan activation.

The inputs for the activity plan are the same as for the mission plan but more detailed. Of course, changes may create new inputs. Such changes might include a broken antenna at one of the stations which will require two months to fix, out-of-tolerance pointing that requires calibration, or a delay in changing telemetry flight software because the telemetry-system upgrade is behind schedule.

The *activity plan* provides the next level of detail, down to the minute; it passes directly to command development for implementation. It's typically a time-ordered listing of events with approximate durations allocated against resources. This plan can be on paper or in an electronic file. Software often helps produce the activity plan, with many of the inputs also produced and delivered electronically.

Planning software uses a database of activity definitions, including required resources and duration estimates. Other inputs include an *orbit-propagation-geometry file*, which contains spacecraft position and timing information, and a *station-allocation file*, which defines the station and times of coverage allocated to the spacecraft. (Note: we use "station" broadly to include ground stations and relay spacecraft.) The software could also include algorithms for resolving conflicts and may provide the operator with selectable options.

A simplified activity plan might look like Fig. 8.3, which shows a ten-hour excerpt that includes the gyro calibration. The calibration is scheduled for two hours on day 1 of this activity plan. A special play-back of the calibration shows up at this level, along with the special station coverage it requires. For a detailed discussion of activity plans, see Chap. 7.

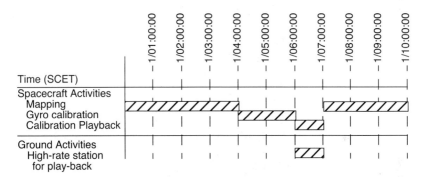

Fig. 8.3.   **Example of an Activity Plan.** Figure shows times in spacecraft-event times (SCET), the planned activities, and their estimated duration.

8.1.3 Shift Operations

Not all positions are shift positions, and different members of the team may have different shift schedules. Positions for people who send commands to and receive telemetry from the spacecraft are usually staffed 24 hours a day. But confining operations to a single shift can greatly reduce operations costs. A high degree of spacecraft autonomy, onboard data storage, and favorable orbital geometry can combine to enable single-shift operations. A few support positions, such as computer system administrators, may also be staffed 24 hours a day depending on the mission's timeline requirements. Alternatively, support people may be on-call through telephones or pagers if required response times aren't too short. Other positions, such as management, long-range planning, analysis, and ground software are typically on a normal work schedule of 8:00 a.m. to 5:00 p.m., five days per week.

Mission requirements strongly influence coverage and shift requirements. A spacecraft in low-Earth orbit which requires short contacts every 90 minutes tends to drive 24-hour operations, whereas a geosynchronous communications spacecraft with long, easily scheduled contacts may fit into a regular work day.

A typical day in operations usually begins between 6:00 a.m. and 8:00 a.m. local time with a briefing from the overnight crew to the incoming day crew. This *hand-over briefing* includes spacecraft status, any commanding performed, anomalies that occurred, ground status such as station outages, and any other significant events. The briefing is usually one-on-one: each position gives a customized, detailed briefing to the counterpart on the next shift.

Actual shift schedules vary widely. Some programs operate on a traditional three-shift rotation, in which each shift spans 8.5 to 9 hours and people rotate between the shifts weekly. Others operate on a two-shift rotation, with 12-hour shifts for four days followed by four days off. People who are expected to work this way need to know in advance that it will be a shift position and what the shift rotation will look like.

Many people prefer the 12-hour days because they get four-day weekends every week. But this schedule's drawbacks include the long day and the decrease in alertness after eight hours, especially when those hours are in the early morning. Adjustments can be difficult for the night shift because they tend to revert to mostly daytime schedules on their days off and never really adjust to a nighttime schedule. This contributes to further fatigue and loss of alertness by the fourth day. Four days off can also be a drawback in a dynamic environment, leading to discontinuity and increased reliance on the quality of the shift-handover briefing when the person returns. In these arrangements, people usually work days or nights for two weeks to a month and then swap after their days off.

On the other hand, some people prefer the rotating, three-shift approach with rotations either weekly or in multiples of weeks. This schedule has the advantage of shorter work days and a more normal work week. However, shift changes are usually separated by only two days, which can be a short turnaround.

When designing the shift schedule, keep in mind natural circadian principles, which dictate that the rotation always be clockwise—rotating from days, to evenings, to nights. Also consider employee preferences and company policy. For instance, the night shift may be much more palatable if the company pays more for it. For a more detailed discussion of circadian principles and shift-work schedules, check "Wide Awake at 3 A.M.," by Richard M. Coleman. A poor shift schedule can lead to low employee morale, decreased productivity, and mistakes that affect the mission. Conversely, a well-constructed shift policy can result in higher productivity, better health, and less time lost.

8.1.4 The Status Meeting

The *status meeting* is the main forum for communications on the operations team. It's almost always a meeting because of the improved communication face-to-face. Other groups not co-located, such as the user community, may be connected by telephone—or by video link, as it becomes more available. There's no substitute for the periodic interaction of all involved parties; still, for efficiency, meetings should be kept to a minimum. This status meeting is often held daily, with a maximum interval of weekly. Twice a week, say on Monday and Thursday, is a good compromise for a mature operation. The frequency of the status meeting may change with mission phase or during special operations. Even a system that has weekly meetings will often have at least abbreviated meetings daily during launch or other critical operations. For example, the Air Force typically conducts the status meeting as a 7:30 a.m. "stand-up" meeting, in the belief that, if everyone has to stand up, the meeting will run efficiently. Ideally, the status meeting lasts a half hour but no more than one hour.

The status meeting includes representatives from all of the operations areas including the control team, ground hardware, ground software, communications, operations, management, payload operations, mission planning, spacecraft engineering, navigation, and any other interested parties. A typical agenda includes reviews by all major elements of their current status, their activity plan, and the mission plan, as appropriate. The master activity plan is presented by mission planning and approved. For daily meetings, we may designate one day per week to review the mission plans, which shouldn't require daily review. But we must be careful not to input requirements at the meeting, except maybe as a heads-up. The team should come into the status meeting with a proposed plan—working it real-time will bog the meeting down.

We review the plan for the spacecraft and the ground plan, including software upgrades, hardware maintenance, and planned outages. Although schedules are coordinated at the lower levels, many conflicts may not be evident until we have the complete picture.

8.1.5 Developing Command Loads

The *command load* flows directly from the activity plan and is often referred to as the "standard" sequence or standard commanding. Sometimes, we'll need commanding that wasn't included in the activity plan—called *non-standard commanding*. These commands can also be stored commands, although they're usually stored in a different area of onboard memory than the standard command loads. Non-standard commands can also be *real-time* or immediate-execution commands: those that execute as soon as the computer or hardware receives them. In general practice, the operations concept always states there will be few or no unplanned or non-standard commands, but they're inevitably required. We need to define a disciplined system to handle non-standard commands and avoid errors, while also responding quickly when anomalies require the commands.

After the activity plan has been reviewed and approved, we begin developing the command load to carry it out. If the timeline is very tight, initial development may start in parallel with the status meeting. The activity planning group carries out the plan. This group or function is variously called planning and scheduling, planning and analysis, sequence design, command-load development, or uplink design. We follow the timeline for activity planning discussed in Chap. 7.

The predominant method of commanding is through *stored command loads*, which are sequences of commands with time tags. These commands reside in an onboard computer to be executed when their time occurs. They're designed to repeat functions while changing only parameters and command details. The structure and function of these loads are validated extensively before launch. The dynamic parameters, such as time and turn angle, are input to execute the specific activity plan. Depending on the commanding complexity and the computer's memory space, stored command loads may cover several hours up to several weeks. Other factors, such as the accuracy of the orbit propagation and prediction, also influence the duration.

The inputs for developing the command load are very similar to those for mission planning (see Fig. 8.4), although they're more detailed and should be electronic. Highly automated development leads to speed and accuracy because electronic interfaces, once validated, are repeatable and reliable.

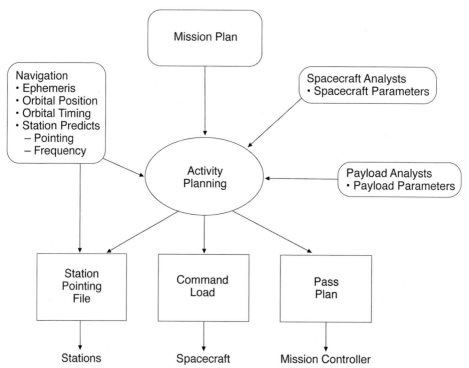

Fig. 8.4.   Information Required for and Products from Activity Planning. Inputs are similar to those for the mission plan, but they require more detail.

Timelines for missions with a daily activity plan are the most strenuous. Extended planning cycles have similar development processes with an extended duration. Extended timelines also allow us to develop command loads through several cycles. Daily development more likely will have a single cycle flow. Multiple cycles allow us to refine the product repeatedly. Users often tend to provide late inputs. In some cases the final plan bears little or no resemblance to the initial cycle. We should try to limit this tendency; it wastes effort reviewing products that change drastically, and the final product has fewer reviews than it should have. In this case, either there are too many cycles or the process starts too early.

Sometimes, we can't avoid late changes because a spacecraft change or a new discovery demands them. But late changes shouldn't be the norm. Ideally, command loads change only when dynamic parameters, such as orbit determination and timing, become more accurate.

Figure 8.5 shows a typical flow for command development. We first expand the plan into its component events using specific timing and parameters. The output of this expansion is an *activity-event file* that contains precise timing information, commands, command parameters, and station information. It may also include spacecraft-status information that shows how commands affect the spacecraft's state and allows state propagation and checking. The activity-event file should be in both electronic and humanly readable formats, though these aren't necessarily separate.

Next, we review and check the activity-event file and validate the commands. We check for resource conflicts that weren't visible at the plan level, violations of mission rules, limits, command timing, and configurations of the spacecraft and ground equipment. Groups that handle the payload, spacecraft, ground element, and planning should participate in the product review.

We can do this review manually or through automated software. The trend is towards more automated checking with software that can encompass rules defining timing, configuration, and limits. More sophisticated software can propagate states and model dynamics to verify spacecraft maneuvering and pointing, thermal characteristics to check temperature limits, power levels to verify power margins, and payload characteristics.

In an ideal world, we'd find no errors during the review, but the world isn't ideal. If we discover errors, we must identify them to the rest of the project and propose corrections. Sometimes, corrections are simple, such as a mistyped parameter. Other times, the error can be very complicated and require a change to the plan.

An example of a major change is a desired calibration we can't do because it violates a power limit. We could delay the calibration until solar occultation is lower, add commands to power off equipment not in use during the calibration, or abbreviate the calibration. For the first option, mission planning must replan the event or possibly add an event to a later activity plan when the resources are available to move the calibration. The second option involves all of the spacecraft's

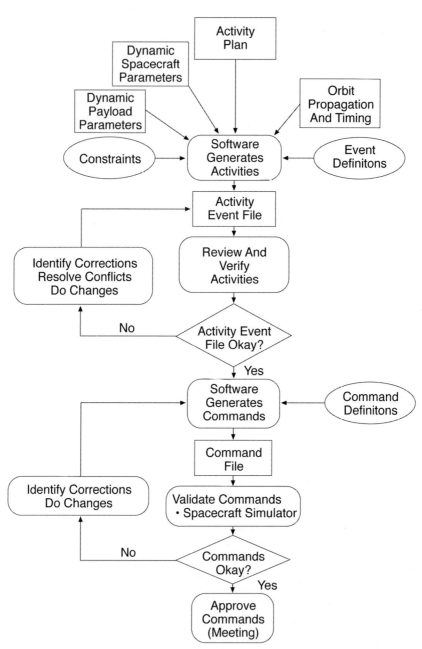

Fig. 8.5. Developing Command Loads. This process generates a command load ready for uplinking to the spacecraft.

subsystems, including the payload, but wouldn't require replanning. It would require adding commands to reconfigure the equipment. The third option affects mainly the requester, who must determine if an abbreviated calibration would suffice; it may require changes to input parameters, or adding or deleting commands.

After the activity-event file has successfully passed all reviews, it's translated into low-level, binary computer instructions. This step requires the command definitions, usually contained in a command database, to translate the command name to the proper computer instructions. It also usually includes translating the time-tags from a standard time format (Universal Time, Coordinated [UTC] or Julian) to the spacecraft clock's time. Time-tags tell the computer when to carry out a command. The translation software also assigns memory locations to the commands in the file. Depending on the complexity of the sequencer or onboard computer, this allocation may be simple or very complex. It may require the command-translation software to maintain a map of the computer memory and propagate the state from event to event. A less reliable method would require the operator to input the desired memory addresses and manually maintain a memory map.

The command-translation software also adds appropriate headers and trailers, which contain such information as identification codes, message type, destination, message size, and the error-detection codes required by the command decoder or the flight software.

Ancillary products may appear at this point in the flow if we didn't already produce them with the activity event file. Flight controllers or ground stations use these products, which are sequences of events or predictions. Examples are the contact-support plan (CSP), integrated sequence of events (ISOE), keyword files, and station control files.

The output of command translation is a *command file* that is ready for uplink to the spacecraft. But before uplinking it, we must validate it. Because command validation is extremely important, we discuss it separately below. Once the commands are validated, we seek command approval. Through *command approval*, we make sure all of the development steps have occurred properly, commands have no remaining errors, all of the necessary coordination has taken place, and all affected groups approve the command upload. We can get command approval in two main ways: a command-approval meeting or authorized persons electronically entering an approval code. Whether through meeting and signing a piece of paper or by electronic means, command approval releases the command file to the mission controllers for uplink to the spacecraft. During operations development, the mission operations managers need to decide what level of authority is required for command approval and who can approve violation overrides and waivers. We must document any violations or waivers, along with the reason for approval.

8.1.6 Command Validation

Command validation consists of the reviews and simulations that ensure the command load is free of errors. *Command* errors are any commands sent to the spacecraft that result in an action other than what was intended. Some command errors are harmless, but others are catastrophic to the mission. When trying to reduce mission operations costs, we should emphasize catching errors that affect, limit, or are catastrophic to the mission. We shouldn't sweat the small stuff. The cost to catch 100% of the errors is very high, so we must decide which errors to catch and which ones to let slide. This way, we reduce the amount of review and validation for the commands we designate "no impact."

A Magellan experience illustrates the problem with judging the severity of command errors. We determined that memory readout commands were harmless and that an erroneous command would result only in our not getting the area of memory we expected. We'd need to re-send a correct command, but the impact would be a delay and a little more work. However, in one case, because of a ground-software limitation, we decided to store a memory buffer for the attitude and articulation control subsystem (AACS) in the command and data subsystem (CDS) memory. We would read it out periodically by transferring the memory contents back to the AACS memory and doing the standard memory readout. Unfortunately, nobody recognized that the dynamic parameters for the gyro-bias estimate were also in this memory buffer. When the old contents stored in the CDS computer were transferred back to the AACS memory, old values for gyro bias were written back into the AACS and used for attitude propagation. Using the old values resulted in the spacecraft drifting from Earth point, which called for several days of recovery operations. We learned two lessons from this incident. First, avoid becoming too complacent about harmless commands because we may be using the command in a different way. Second, never read data back into an active area of memory.

During design and development, we must determine what type of command validation we need and how much is affordable for this mission. The mission operations concept should address when we'll validate, who will do it, whether it will be online or off-line, and whether it will be automated or done manually. This design then becomes part of the mission plan, which defines in greater detail how and when we'll validate commands.

Often, most of command validation occurs before flight using a simulator and the spacecraft. We validate all of the individual commands and many of the stored command files on the spacecraft during ground test. We test the files with a representative set of values or parameters; we can't test all values because of time and budget constraints.

The most common way of validating commands after launch is by simulation. We can do this simulation on a real-time simulator of the type used during development. It usually consists of a flight computer, at least a breadboard version of the spacecraft, a version of the flight software, some or all of the interface hardware, and software models of some or all of the remaining hardware. We also model spacecraft dynamics. This type of simulator provides the most extensive validation, but it's also expensive. It may be time consuming as well, because it executes at the same rate as the actual spacecraft, which makes it impractical for running all commands in large loads.

As spacecraft become smaller and less expensive, highly accurate simulators using flight hardware are less likely for development and are rarely justifiable for operations alone. We can use software simulators that run the flight software, perhaps on a different platform, with software models of all the hardware. The disadvantage of this type of simulator is less accurate timing and interfaces. But one of the great advantages is that these simulations can run on a faster computer than the flight computer, which lets them run faster than real-time.

We may also combine methods. An example would be a software simulation of the flight system for commanding that runs very fast and checks command structure, ranges, and memory management. Large command loads could run as a matter of course. Additionally, the real-time simulator could run all or part of the critical or unique command loads.

In addition to simulation, or along with it, spacecraft configuration and state propagation are important to command validation. Commands or activities may have a prerequisite hardware configuration or precursor commands. Relationships between activities or command loads are very important, but they're also one of the hardest things to keep track of. For instance, activity A requires that box A be powered—its normal state. But an anomaly occurs before activity A executes. Then activity B is uplinked and executed, which results in box A being powered off. Activity A won't execute properly, which shows up as a command error. We must track the spacecraft's state very closely and propagate it forward to make sure we meet expectations and requirements.

8.2 Uplink Process

Uplinking or *commanding* is the process of transmitting commands from a ground computer through a series of ground equipment and transmitters to the spacecraft. The US Air Force uses command centers at the Onizuka Air Force Station and Schriever Air Force Base, with transmission through its network of Remote Tracking Stations (RTS) at eight locations worldwide. NASA's Earth-orbiting spacecraft have various command centers, including the Goddard Space Flight Center, with transmission over the Satellite Tracking and Data Network (STDN)—a world-wide network of ground antennas. Interplanetary spacecraft are controlled mainly from a command center at the Jet Propulsion Laboratory (JPL) with transmission through the Deep Space Network (DSN), which has large (up to 70 m) antennas at three sites: Goldstone, California; Madrid, Spain; and Canberra,

Australia. Various programs also use spacecraft relays. The most widely known of these relay spacecraft is NASA's Tracking and Data Relay Satellite System (TDRSS), which is a constellation of three geosynchronous spacecraft. These are controlled by a command center at White Sands, New Mexico. A relay system such as TDRSS provides the advantage of nearly continuous uplink and downlink for any spacecraft in low-Earth orbit (LEO). Science missions that produce a lot of data, such as the Hubble Space Telescope, use TDRSS extensively. Chapter 12 describes various tracking networks.

The plan that describes the command uplink carries various names, such as contact-support plan or an integrated sequence of events. This plan lays out all of the particulars needed for a given pass or uplink, such as the configuration of the station and its associated ground equipment, the command rate, telemetry and data rates, time for acquiring the signal, and time for losing the signal. The plan can be a computer-executed series of transactions or it may be a paper listing of the required events, which operators do manually. For simplicity, we'll refer to this plan as the *pass plan*.

While the planners and schedulers are building the command load, the control team is working on the pass plan, which includes the contact-support plan. The controllers need to know how long this particular load is going to take to uplink and whether to schedule it over a single pass or multiple passes. A *pass* (sometimes called a contact) is the period between acquiring and losing the spacecraft's signal. For a low-Earth orbiter using remote-tracking stations, a pass may vary from 4 to 12 minutes. A pass can last several hours for a geosynchronous spacecraft. TDRSS or a relay spacecraft also allows a pass of several hours. Elements of the *pass plan* are shown in Table 8.2.

An *uplink* is a steam of contiguous data transmitted from the ground to the spacecraft while the ground transmitter is active and maintains the spacecraft's receiver in-lock. It includes all of the header information required by the spacecraft receiver and decoder, as well as commands or sequences of commands. If the transmitter is active, then off, then active within a single pass, it has done two uplinks. An *uplink window* is the time during which we may send the uplink. We can specify it as no-earlier-than or no-later-than, or we can bound its transmission with earliest and latest times. A backup uplink window over a different station pass may also be specified. We may also schedule several uplinks over a single pass, each with its associated uplink window.

How we determine the uplink window for a given command load or command can be fairly complicated. Command loads or stored commands are usually time-tagged with a spacecraft clock time at which time they should start executing. Based on the available memory in the onboard computer, we can uplink the command load hours or days ahead of its execution time. If memory space is limited, the timing constraints may be quite tight, so we may have to uplink it after the current command load completes but before the new execution time. Immediate-execution commands start when the spacecraft receives them, so the

Table 8.2. Elements of a Pass Plan. We tailor these elements as required for the type of mission.

Pass Plan Element	Description
Station Rise and Set Times	The time that the station is visible from the spacecraft in UTC.
Station Acquisition and Loss-of-Signal Times	The time in UTC that the station is far enough above the horizon to maintain adequate signal lock. Typically, 5–20 degree mask is used. A different mask may be used for uplink and downlink and both times would be included. This time could also vary if the entire time of station visibility isn't being used, which is common for DSN passes.
Uplink Window	The time window in UTC that the planned command(s) can be uplinked to the spacecraft.
Uplink Window Constraints	Information the controller may need to interpret the uplink window correctly.
Command Uplink Duration	The amount of time in seconds that is physically required to transmit the command(s).
Uplink Rate	The uplink rate in bits per second that the ground hardware is set to for commanding.
Downlink Rate	The downlink rate in bits per second at which the spacecraft will be transmitting.
Other ground configuration parameters	Other items that are required to configure the ground system correctly. These items vary depending on the ground system used.

time of uplink controls the execution timing completely. Between stored commands and immediate commands, many other command types exist, which vary with each spacecraft's command architecture. A command handbook should document fully the types of command loads and the requirements and constraints for their use. We can also place the command requirements in software that can automatically determine the type of command load and the applicable rules. Then, by accessing an electronic version of the pass schedule, the software can provide possible command windows to the operator.

Preparing for the Pass

The pass plan actually begins with prepass operations, well before the spacecraft appears. During this period, we configure the ground system and brief everyone involved in the pass. We calibrate the station and do loop-back tests. A *loop-back* test consists of sending a command from the console through the ground-commanding system to the antenna. Instead of actually sending the signal, we loop it back to the controller. Also at this time, the station predictions are loaded or activated at the station. As discussed in Sec. 3.5, these predictions include antenna-pointing angles, frequencies, and timing information.

The *mission controller* is the senior person on the real-time control team and is responsible for carrying out the pass plan. The primary interface to the stations is the *range* or *resources controller*, who coordinates the communications configuration and monitors and reports the station readiness. For LEO spacecraft using the remote-tracking station, a typical prepass operation begins eight minutes before acquiring the signal, as depicted in Fig. 8.6. A typical prepass timeline for a DSN pass is 1.5 hours. The ground-commanding configuration includes such items as the transmitter rate and power, crypto and decoder units, spoof-checking enable, and ranging-signal enable, as described in Chap. 11.

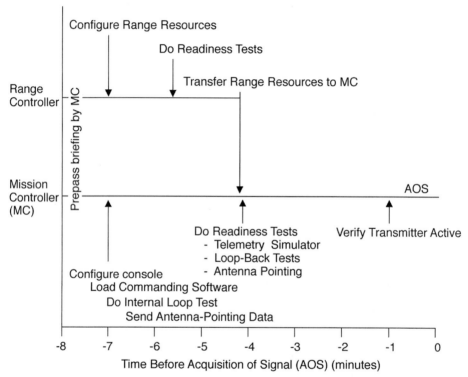

Fig. 8.6. **A Typical Prepass Timeline for a LEO.** This figures illustrates the types and timing of activities to prepare for a pass.

8.2.1 Executing the Pass

A station's *view period* is the time (as defined by the geometry) that the spacecraft is visible from the desired antenna. Chapter 10 shows how to calculate the geometry. A pass may encompass all or part of the view period and is the time that the antenna and associated ground resources are dedicated to that spacecraft.

Acquisition of signal occurs whenever the ground antenna's receiver locks onto the spacecraft's electromagnetic transmission (or carrier signal). Ground equipment then processes the signal into a usable data stream, as described later. The telemetry is *in-lock* whenever spacecraft telemetry is being processed and displayed. We usually reserve the initial period of telemetry for spacecraft status and health checking, as shown in Fig. 8.7. We examine this telemetry to determine the state of the spacecraft and verify that it's in the expected state based on the last pass and any known commands which may have executed while it was out of contact.

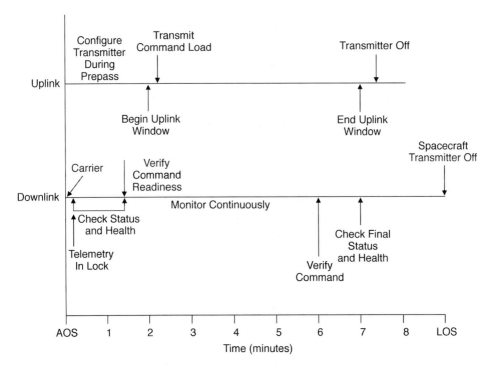

Fig. 8.7.   A Timeline of Typical Activities for a Pass by a Low-Earth Orbiter. Acquisition of signal (AOS) and loss of signal (LOS) define the pass duration.

In a typical LEO pass, the status-and-health period may last 30–90 seconds, depending on the spacecraft's complexity and the level of automation. In a highly automated system, software does all checking based on a set of rules expressed as conditions on the data values. In practice, given the processing ability and technology of modern computers, we can reduce the status-and-health check to a "status OK" message, as discussed for payload monitoring in the EUVE. However, defining the rules for measurement processing may be difficult, and autonomous state prediction is imperfect. Also, a good operator develops a sense about what nominal values should be and may often catch a problem in its early stages, before any rule would have indicated it.

Mission controllers use a console that consists of video-display terminals, a telephone handset or headset tied into the various voice networks, notebooks containing procedures, spacecraft data, a logbook, and anything else they might need close at hand. Each operations position typically has its own console, similar to that shown in Figure 8.8, with the appropriate procedures and data.

Fig. 8.8.   A Mission Controller Console at a NASA Center. Consoles vary, but this configuration is fairly typical.

During the pass, mission controllers check the status-and-health, continuously monitor telemetry, analyze that telemetry in real time, transmit any command loads or commands scheduled for this pass or uplink window, and verify the receipt and execution of transmitted commands. They also coordinate with all other positions and lead anomaly investigations. If they detect any problems during the pass, such as violation of a red-alarm limit during the status check, they abort the command plan (at least temporarily) and start the appropriate response defined by procedures. Chapter 16 discusses fully how to handle anomalies.

In addition to the mission controllers and range controllers, several other people may be involved in the pass. A *deputy mission controller* works much like the mission controller and shares the workload during busy passes or critical events. He or she would also be the second approval for command enable, if required. Another control-team position is the *spacecraft analyst*, who monitors the spacecraft telemetry more carefully and verifies functional commands. Of all the real-time positions, this person knows the most about how the spacecraft works. Depending on the level of spacecraft expertise required for the mission, this person may be either a systems engineer who helped design the spacecraft or an operator who has had extra training on it. Besides the spacecraft analyst, subsystem analysts may monitor telemetry and verify functional commands, especially during critical events.

Once the status-and-health check is successful, commanding may begin. Commands to be sent during this pass are promoted to the command system or included in the command data-base designated for this pass. Requiring a positive action to make the command available is part of a system of checks and balances that reduces the chance of selecting and transmitting an erroneous command. Many operations also require the buddy system, in which at least two people must verify that the selected command is correct before sending it. Operational procedures or the commanding software ensure we use this system.

The command database may be set up so all commands or selected commands are available for each pass. Some commands are restricted so we can't select them in real-time. *Restricted commands* are considered dangerous—inadvertent execution could be catastrophic to the mission. For example, a flag or bit might indicate that the command is restricted and that we must remove or reset this flag during command development and approve it during the command-approval meeting. Examples of restricted commands are igniting solid rocket motors, deploying appendages, and enabling and firing pyrotechnics.

Even though the transmitter was configured and checked before the pass, it remains in a standby state. As part of sending the command, we must apply high power to the transmitter. Applying high power is often referred to as going active on the command. The transmitter is active during the uplink. In spacecraft that use a coherent signal or an active ranging signal, a continuously uplinked carrier signal keeps the spacecraft receiver in-lock and provides a frequency reference for the downlink signal. The command transmission is then modulated onto the carrier signal. We check commands at various points along the transmission path to ensure the correct signal goes to the spacecraft. These checks include parity checks, sequence ID, and packet timing. If a check fails, we abort the transmission and retransmit the uplink, noting the time and duration of the transmission in a paper or electronic log.

The pass plan can have one or more uplinks that may require us to verify telemetry between them. A command that changes the state of the spacecraft may require verification before we send a subsequent command requiring that state. The commanding procedure usually specifies a minimum set of verifications for all commands, as well as additional or specific requirements for certain commands. For non-routine events or commanding, the spacecraft subsystem or payload analysts may also monitor the telemetry in real time and tell the mission controller which commands to transmit.

An example of this real-time interaction is battery reconditioning. A battery has been disconnected from the main bus, discharged, and then connected to a charge circuit. An expert on the spacecraft battery monitors the telemetry on the battery's temperature, voltage, and charge current to decide whether or not the battery should be reconnected to the main bus during this pass. After examining the battery telemetry and deciding the battery is recharged and ready to be reconnected, the battery analyst would use proper protocol to notify the mission controller over a voice net. It might go something like, "Power to mission controller. Battery 1 has completed recharge and is ready to be connected. Power is 'go' for transmit of B1CNT, battery 1 connect." And the mission controller might reply "Copy. Power is 'go' for B1CNT. It is in the queue and will go active in 30 seconds."

An example of multiple uplinks with verifications is a memory load to write-protected memory that requires the write-protects to be disabled so the onboard computer won't reject the memory load and register a fault or alarm. In this case, the commands to disable the write-protects should be sent and positively verified by telemetry before uplinking the memory load.

Command verification takes two basic forms: command acceptance or authentication and functional verification. *Command acceptance* consists of telemetry from the command decoder or onboard computer, which registers receipt of the command through counters, an acknowledge bit, or simply the lack of command-reject errors. The spacecraft design includes command-receipt logic and error detection.

Functional command verification consists of verifying that the intended action of the uplinked command was successful. In the write-protect example, a telemetry measurement containing the status of the write-protects (ENABLE/DISABLE) would be a functional check of the state. There may not be any direct measurements, such as a state, to do the functional verification, so we may need several measurements or indirect measurements. For example, a command goes out to turn on a heater, but the heater has no status bit. An increase in the battery-discharge current of a predicted amount would indirectly verify that the heater did turn on. An increase in the temperature of the item being heated would be a correlating measurement.

Somewhat the inverse of command verification is *fault detection*. During the real-time pass, this detection mainly takes the form of checking alarm limits. Rules for the expected state or value of the telemetry measurement are defined and the current value of the measurement is compared to these rules. If a violation occurs, an alarm notifies the controller. We typically use yellow and red alarms. Yellow alarm limits are set to warn us that the spacecraft analysts should look at something more closely. Red alarm limits are set for action—usually an immediate action by the controller.

The display and formatting of the telemetry at the console is very important, including such human factors as colors, the size of the font, and the amount of information. Using color on the measurement values to show status—green for nominal, yellow for exceeding the yellow alarm limit, and red for exceeding the red alarm limit—is a common technique that helps the controller process and understand the information. If color is not available, we can use boldface letters, reverse video, and audio alarms. We need to avoid overwhelming the controller with information while providing enough information to do what's required. Different controller positions require different displays. Real-time plotting of analog measurements may show trends. Simple schematics may help the controller understand how the measurement works. New, state-of-the-art graphics can include full three-dimensional depictions of the spacecraft with point and click interfaces to bring up subsystem details. But we must resist the urge to make the display interface too flashy if it harms ease of operation and clarity.

After all commands specified in the pass plan have been sent and verified, we do a final health-and-status check. We also note the spacecraft's state. We can't always verify each command functionally, especially in the case of a stored command uplink which may execute over days or weeks, but we should verify command receipt.

Loss of signal occurs whenever the spacecraft is no longer visible to the ground station. If a spacecraft is geosynchronous or uses a relay spacecraft, loss of signal may occur at a specified time.

One of the cost-saving procedures that faster computers and communications technology have made possible is the unattended or automated pass. For these passes, only the computers monitor telemetry and perform health and status checks. If telemetry exceeds alarm limits or error checking detects an error, it triggers a process that pages or calls a responsible person(s) with information regarding the violation or error. Telemetry may also be pushed to a website that the person contacted could access without having to drive to the facility. This approach saves money by not having to staff passes which are during off-nominal working hours while still providing instant response for spacecraft anomalies. It can also improve employee morale since many times the alarm is a false alarm and much time was previously wasted driving to the operations facility in the middle of the night to confirm a false alarm. Automated passes are typically used for downlink only passes but increased automation capabilities are rapidly making automated command passes possible.

The Mars Mission Operations team uses unattended passes very effectively to reduce night and weekend staffing during nominal operations. The team maintains an on-call file with identified points-of-contact for systems and each subsystem with their corresponding pager number. When a red alarm limit violation is detected, a script indexes into the on-call file based on the subsystem designation of the telemetry value in alarm. A text page is sent containing the time of the alarm, the measurement ID, and the current value. In parallel, all of the subsystem telemetry pages are pushed to a website with secure access. After receiving the page, the person can either view the telemetry from home (or other remote location) to determine if the alarm is real and requires immediate action or he or she can proceed directly to the operations facility. If the alarm is real and requires action, a recall procedure is invoked to page more team members and begin anomaly actions.

8.2.2 Post-Pass

After loss-of-signal, we deconfigure ground resources and release shared resources from support. It's important to complete the required paperwork immediately while the memory of the pass is still fresh. We fill out and submit discrepancy reports or failure reports on any anomalies for the spacecraft or the ground system. At this point, downlink post-processing begins. The ranging data is processed and made available to the navigation people. The spacecraft telemetry is stored and made available to the spacecraft analysts for trending and analysis. Payload or science data is routed to the appropriate processing location and made available to the payload analysts.

8.3 Downlink Process

Downlink refers to all data originated at the spacecraft and transmitted to Earth. We describe below the various types of data and their functions.

The spacecraft's main mission is to provide science, mission, or payload data, including everything from the spacecraft images on the evening weather report to measurements of the solar wind. For communications spacecraft, the data is the continuous stream of telephone or picture data.

Operators often don't see mission data because it goes directly to users. For some Earth-orbiting spacecraft, such as communications spacecraft, the mission data travels on a distinct communications link that is dedicated to the mission data and independent of the spacecraft bus. In other cases, we interleave the payload and engineering data before transmitting it all to the ground and downlink it over the same communications network. Ground processing then strips it out.

Engineering or *housekeeping* data comes from the subsystems of the spacecraft bus. It includes such data as temperature, voltage, and attitude. During spacecraft design, the design engineers determine what measurements are important to determine the health, status, and performance of their piece of hardware. (See Chap. 15 for a discussion of spacecraft telemetry.) All of these measurements get collected and traded off against the available space in the data stream to make up the engineering data. Data from the computer shows how the software is working or provides information, such as use of memory, that the software computes.

As onboard computers get more sophisticated, the engineering-telemetry stream is becoming more programmable. There may be more sensors than there is room in the telemetry stream, which may lead to selecting multiple telemetry formats based on mission phase, activity, and hardware configuration. More onboard processing such as filtering or averaging can decrease the amount of data sent to the ground. Increased flexibility on board requires more flexibility on the ground, so we can determine the format in use and process it correctly.

Tracking or *ranging data* provides the position and velocity for the spacecraft's orbit or trajectory. We can collect this data in various ways. The most basic way is by determining where the ground antenna points when it contacts the spacecraft and the delta from where we predicted it would be. *Active ranging* provides more data about velocity. It consists of adding a signal to the uplink, called ranging tones, at a known frequency and measuring the difference in the returned signal. *Doppler ranging* also measures the frequency shift of the downlink signal. Doppler ranging can be totally passive, in which frequency shift is measured relative to a predicted, spacecraft-produced frequency from an onboard crystal oscillator. A more accurate method involves a coherent signal—where the spacecraft's downlink frequency is a fixed multiple of the uplink frequency. By measuring the phase difference between the uplink and downlink signals, we can calculate a distance based on the Doppler shift. For interplanetary spacecraft, we can get even more accuracy by using more than one ground station simultaneously to do very-long-baseline interferometry (VLBI). Earth-orbiting spacecraft often do their own position estimates based on position information from the global positioning system (GPS).

Ancillary data consists of any other information the user needs to process or interpret the mission data. It may include other spacecraft measurements such as the payload's operating temperature or an attitude estimate, or other ground-processed information such as the orbital position. If we process the mission and engineering data separately, we must develop a way to deliver the ancillary data to the user. In the case of interleaved data, we can strip out the appropriate spacecraft measurements and payload data during the original processing.

8.3.1 Processing Telemetry

The first step in processing the downlink data is *synchronizing frames*. A repeating pattern in the serial stream marks the boundary between major frames. Once the processing software finds this frame boundary, it can locate the data slots in the stream as defined by a decommutation map.

Decommutation converts the analog stream to a digital stream of data words. It's the exact reverse of commutation, which takes place on board the spacecraft. At this point, interleaved payload data may be separated from the engineering data if it requires different processing.

Once the data stream is decommutated, it may be stored or displayed in its raw format, but it usually undergoes *decalibration*. This process applies calibration parameters to the data words. These calibration parameters can include binary to hex conversions; polynomials that define a mapping from a binary to analog conversion; or a definition of a state value, such as on or off, depending on how we're going to display or further process the data. To execute the pass, the decalibrated engineering telemetry appears on the operator's console. For a detailed description of the hardware and software needed for telemetry processing, refer to Chap. 13.

After the telemetry is processed, we can store it in a database, display it, or further process it. The flexibility of displays has increased greatly over the years with the advent of window systems and affordable color displays. Displays such as functional schematics with the appropriate data interspersed are now possible. Real-time plots have virtually replaced the old strip-chart recorders, providing a more flexible software equivalent of strip charts (and the pens don't clog).

Improved computers also give operators and analysts more freedom to customize their displays. Some people like to look at the measurement displayed in its engineering units; others prefer hexadecimal. We must control the decommutation and decalibration data to ensure its integrity but should leave the display to the discretion of the person who has to look at it.

8.3.2 Archiving and Retrieval

One requirement we should define during development is how much engineering data we'll store—and for how long. The data can be stored online, where it's immediately accessible, or in long-term media, such as tape or optical disk. The requirement for data storage may drive the type of hardware selected. In general, we must strike a balance between the need for long-term trending and the cost of storing so much information. We usually keep engineering data online for 30–45 days and then archive it until the end of the mission.

Large database systems that can store and process a lot of data have helped lower mission operations costs. This capability has had to keep pace with the more complicated spacecraft and their higher downlink rates. In the past, we stored very little data online, thus requiring us to maintain tape libraries and mount tapes when we needed to access the data.

A quick guideline for determining storage requirements is to expect at least 30 days of data. One day of data is the engineering data rate in bits per second times the average contact length times the number of contacts per day. If we are going to store the data as processed data rather than raw data, some multiplier should account for such things as floating point values requiring more space than bits. However, storing only completely processed data limits flexibility for post-processing later on.

8.4 Planning and Analysis Functions

8.4.1 Spacecraft Planning and Analysis

Spacecraft planning and analysis consists of analyzing and assessing the spacecraft bus. It usually occurs off-line in mission operations and in parallel with real-time commanding and downlinking.

The major tasks are:

- Predicting spacecraft performance
- Assessing the health of the spacecraft by trending
- Assessing actual spacecraft performance against design goals and requirements
- Maintaining spacecraft parameters that may need changing or adjusting
- Making inputs to mission planning regarding future spacecraft activities such as calibrations
- Sometimes, validating stored commands
- Often, maintaining flight software

Chapter 15 discusses these tasks in more detail.

The people who typically do these tasks are the spacecraft experts and are often organized by subsystem. For a new or complicated spacecraft design, the spacecraft-analysis group consists of two or three systems engineers who monitor the spacecraft's overall performance and are usually the most involved in mission planning and commanding. The subsystems represented include command and data, attitude control, power, thermal, propulsion, and telecommunications. We often combine related subsystems, such as power and thermal or attitude control and propulsion, to reduce staffing. Depending on the spacecraft, the flight-software expertise may reside in the command and data subsystem experts, be split between command and data and attitude control, or be a separate group.

Telemetry analysis is the main way to assess and trend the spacecraft's health and performance. This analysis falls under either short-term or long-term trending. In this context, short-term means orbit or day; long-term trending usually occurs over weeks or months. How we determine this time scale helps drive the archive capability discussed earlier. By trending over the two different time scales, we can distinguish between event-driven consequences, such as a new attitude that changes the heating profile, and hardware aging or degradation, such as a decrease in the solar panel's current output caused when silicon solar cells degrade.

The tools we use for this trending analysis include displays similar to the real-time displays, database query and retrieval tools, plotting routines, and programs for statistical evaluation. The tools vary by subsystem and with the particular measurement within the subsystem. Sometimes the telemetry data goes to ground-processing software that calculates derived parameters. For example, attitude error is the downlinked parameter from which the ground software reconstructs the spacecraft's actual attitude. At the point where trending analysis begins looking at derived quantities, it starts to become performance analysis. *Performance analysis* consists of measuring the spacecraft's performance against the design. It often requires us to combine measurements and do more post-processing. Chapter 15 discusses this topic in more depth.

Automation of trending tasks can save a lot of money in mission operations. For example, having ground software that executes a trending script every morning and produces plots or summaries by the time analysts arrive can save the analysts many hours. Database management is one of the areas that lends itself to automation and therefore reduced costs. An analyst with a printout of numbers, a pencil, and a piece of graph paper may spend hours producing a plot that a computer can generate in minutes.

Outputs of trending go into the mission-planning cycle. Examples are requests for calibrations that may be indicated by degrading pointing performance, for more Sun-pointing time because of lower power output, or simply for a parameter update to improve performance. We need a way to place these types of requests into the mission-planning cycle. But we can't always get our inputs into the mission plan. If the input is too late for the mission plan, we want to get it into the activity plan and eventually to the spacecraft through standard command files. However, whenever events need to occur sooner than the standard process will allow, we need a non-standard command process. We can use this same process to do corrections for anomalies, as discussed below.

Predicting the spacecraft's future performance is closely tied to trending analysis. Prediction consists of extrapolating the future performance based on current and past performance. The long-term mission plan is an input to the predictions. These predictions may also be the basis for transferring values of spacecraft parameters, such as command parameters or flight-software variables, that we need to generate activity event files.

Trending and prediction of consumable items are very important. *Consumables* include propellant, battery charge/discharge cycles, mechanical cycles on gimbals or other motors, and anything else that gets used up. For example, predicting when the spacecraft will run out of fuel contributes significantly to mission-life analyses.

Another way to reduce mission operations costs is to develop dual-use software during design. When developing software to analyze the design, considering long-term use will cost very little more. So we should develop software that is compatible with the operations-software environment. For instance, while designing the attitude-control subsystem, we would develop routine software for validating control pointing and performance that also covers analysis and planning during mission operations. That way, we won't have to recreate and validate new software for operations.

Assessing flight software is another part of analyzing the spacecraft's performance. This analysis is typically a by-product of other spacecraft-performance verification or anomaly resolution. In other words, we assess the results of the software execution and analyze the outputs. We may also have some direct measurements of software performance such as measures of CPU use or other onboard diagnostics.

Maintaining flight software consists of changing the flight code based on our assessments of the software's performance, results of anomaly investigations, or new requirements. New requirements can occur if the mission lasts for a long time because people who become familiar with the system want to enhance or expand the spacecraft's capabilities.

Maintaining flight software follows largely the same rules under which it was developed except that we can sometimes speed it up to resolve anomalies. In this case, a requester writes an engineering change notice (ECN), or the equivalent. Anyone who's interested then receives the ECN. We must bear in mind that a change in the flight software can often affect procedures, telemetry displays, and other operations tools. For example, the change may create a new software parameter that we must add to the ground database. We may also have to change procedures to trend and update or maintain this parameter.

After the ECN is approved, we code and test the change. A patch is the most common way to update flight software. Because the onboard software is normally executing while we're updating it, we rarely recompile and reload the flight software totally unless onboard computers are redundant enough for us to reload one computer while executing out of the other. Also, uplink rates are typically slow (<1000 bits per second), so it's very time-consuming to reload the flight software completely. Patches to the code are usually small (<100 words) and can be loaded in spare memory, verified through memory readout, and branched to or moved.

The saying, "No flight-software change is a small change" tells us it's good policy to resist changing flight software. We can't usually test a flight-software patch as thoroughly as we tested the flight load before launch. No matter how good the simulator or testbed is, the patch will run on the spacecraft for the first time when it's operating in space. If problems occur, it's not easy just to reboot.

Often, anomalies bring out problems with flight software. An *anomaly* is any unexpected occurrence on the spacecraft. The consequences of an anomaly range from the annoying, such as doing a memory readout of the wrong address, to catastrophic for the mission. Anomaly resolution has three parts: detection, investigation, and correction.

We can detect an anomaly from a telemetry alarm or from performance trending and analysis. For severe anomalies or ones that require quick response, a contingency procedure usually defines the detection criteria, including correlating evidence, and outlines the steps to be taken. An anomaly reporting and tracking system should be in place and all members of the flight team should use it. Sometimes several small discrepancies could be pointing to a larger anomaly which goes undetected if there is no system to distribute the anomaly information.

The most infamous example of the breakdown of an anomaly reporting system was the loss of the Mars Climate Orbiter (MCO). Although the root cause of the failure was a units problem in ground software that converted spacecraft thruster firings to force, the Navigation team had evidence in-flight that something was wrong. Trajectory solutions following Trajectory Correction Maneuvers did not converge as expected. The navigators were informally working with the spacecraft attitude control group to resolve the "funny". Because the navigators didn't enter the anomaly into the approved reporting and tracking system, the problem didn't receive the required visibility across the program and the issue was not driven to closure before the Mars Orbit Insertion (MOI). Due to the units error in the ground software, the thruster forces were incorrectly modelled in the trajectory prediction and MCO entered the Mars atmosphere and was lost. The MCO Mishap Investigation Board cited the following as a contributing cause of the failure: "An early comparison of these spacecraft-generated data with the tracking data might have uncovered the units problem that ultimately led to the loss of the spacecraft. When conflicts in the data were uncovered, the team relied on e-mail to solve problems, instead of formal problem resolution processes such as the Incident, Surprise, Anomaly (ISA) reporting procedure. Failing to adequately employ the problem tracking system contributed to this problem 'slipping through the cracks.'"

Once we detect a problem that doesn't meet the criteria for pre-established contingency procedures, an anomaly-investigation team looks at the problem, tries to determine the cause, and recommends corrective actions. This team contains members of the flight team and others, as appropriate.

After the anomaly-investigation team has determined the most probable cause of the anomaly and recommended a resolution, we have to correct it. Causes usually fall into such categories as human error, hardware failures, procedures, ground software, flight software, or others. Corrective action can include spacecraft reconfiguration, increased training, procedural updates, or changes to ground or flight software. Chapter 16 covers anomalies in more detail.

8.4.2 Payload Planning and Analysis

Payload planning is closely associated with mission planning. It involves determining what data the payload should take and when, often referred to as *payload tasking*. A dedicated payload group usually does this tasking. For remote-sensing missions, especially military ones, payload tasking may be in near real time. In these cases, the mission plan may have time periods blocked out just for payload activities. The activity plan will then narrow down these opportunities by time or particular sensor. Finally, at activity planning, the precise tasking occurs within the constraints specified by the activity plan. The payload group then resolves tasking conflicts during activity planning.

For science missions, such as the Hubble Space Telescope (HST), a science working group does the tasking and conflict resolution. These working groups meet well in advance, and the tasking plan is integral to the mission plan. When Comet Shoemaker-Levy 93 was discovered, HST went through major replanning of the mission to take data related to the comet's hitting Jupiter. Planners had to bump the science that had been scheduled during that timeframe. Normally we want to stick with the mission plan but we have to be flexible enough to seize such rare opportunities.

Analyzing the payload is very similar to analyzing the spacecraft's performance. The payload, consisting of one or more instruments, is trended and analyzed as other subsystems are.

Payload analysts also use collected data to assess the payload performance. Analysts for the HST wide-field camera were able to look at the processed pictures from the collected data to determine and analyze a system problem. Analysts routinely examine product quality, including the data processing on the ground, to measure the system's performance. This analysis may also be a secondary indication of a spacecraft problem that hasn't shown up in subsystem trending.

8.4.3 Navigation Planning and Analysis

We must also determine the spacecraft's exact location and motion as a function of time, both for where the spacecraft has been and for predicting where it will be. The parameters of the spacecraft's location and motion are known collectively as the *spacecraft ephemeris*.

Knowing where the spacecraft has been, the *reconstructed ephemeris*, is usually required for processing payload data so we can accurately determine the payload's location when it captured the data. Reconstructed ephemeris tends to be very accurate because it's based on direct measurements. Reconstruction also provides a way to predict the future by updating such variables as spacecraft drag or solar torques in the knowledge database.

Knowing where the spacecraft is going to be, the *predicted ephemeris*, is important for long-range planning all the way through to generating pointing predictions for the ground antenna. Predicted ephemeris may also be uplinked to the spacecraft to maintain its knowledge of the position of celestial bodies, such as the Earth and Sun. Normally, prediction intervals for the ephemeris will be tied to the time spans of the products they are required to support: the mission plan, the activity plan, command load duration, and pass duration. The orbit-determination software is fairly complex and can be computationally intensive, so our long-term prediction may be limited.

Routine operations include collecting and processing the tracking data collected during the downlink pass. This processing can occur after each pass, or we can collect and process some number of passes at the same time. Because the orbit-determination software is complex, we usually batch-process the tracking data. Using data from many passes also makes our solution more accurate. We should get ephemeris at least as often as we do activity planning because we need the latest and best position information for accurate payload tasking and subsequent data collection. Chapter 10 includes more detail on navigation.

Another function of navigation is orbit maintenance. As part of ephemeris prediction, we identify the need for future orbital maneuvers, also known as *orbit-trim maneuvers* (OTM). The system design will set requirements on orbit altitude, inclination, or ground track. We must monitor and predict these parameters so we can insert the appropriate OTM into the plan as early as possible.

Once we've identified the need for an OTM, we must have a way to translate that requirement into the commands the spacecraft needs to execute it properly. OTM planning is iterative, involving navigation, attitude control, and propulsion to design the maneuver. The telecommunications, thermal, and power subsystems may also be involved in determining an attitude for the maneuver. If we want or must have real-time telemetry, communications may affect the timing and attitude for the burn. The thermal and power subsystems may constrain the attitude, but even more importantly, we may need thermal modeling to predict temperatures so we can accurately model and predict propulsive performance. Tracking requirements may also increase right after the OTM to provide a solution for the new orbit.

The complexity and risk associated with OTMs vary greatly from mission to mission and depend largely on their frequency and size. Configuring and reconfiguring the spacecraft to do the OTM adds to this complexity. For many communications spacecraft, the thrusters mount on the spacecraft faces and point into or opposite the velocity vector. Thus, we don't need to reorient the attitude, and normal payload activity isn't interrupted. For these types of programs, orbital maintenance is routine. Automating these routine orbital maintenance tasks on board the spacecraft can reduce mission operations costs. By combining GPS capabilities with more capable onboard computers, it is possible to move the orbit maintenance function to the spacecraft. This capability was successfully demonstrated on the UoSat-12 mission using the Orbit Control Kit developed by Microcosm, Inc. In this demonstration, 52 burns were performed over 29 days to maintain tight orbit position requirements. Generally, tight orbit position requirements must be traded against the workforce required to track, predict, plan, and implement many maneuvers, and this usually results in a compromise. Autonomous orbit control can result in increased orbit accuracy and reduced workforce. Other missions such as TOPEX have also experimented with autonomous orbit control.

At the other extreme, once a spacecraft first reaches its operating position, OTMs are risky because they require a lot of spacecraft reconfiguration. The mounting of the propulsion system may require reorienting the attitude, which interrupts the payload's operation. Some spacecraft with long, flexible appendages may need us to lock the appendages in position or even restow them to a launch configuration.

Spacecraft design varies with the required frequency and difficulty of OTMs for the spacecraft. Sometimes unpredicted events, such as unusually high solar activity, may cause us to execute OTMs more often than we designed or planned.

References

AAS 00-071, Autonomous Orbit Control: Initial Flight Results from UoSAT-12, Guidance and Control 2000. Vol. 104, Advances in the Astronautical Sciences, AAS Publications, 2000.

Agrawal, Brij N. 1986. *Design of Geosynchronous Spacecraft.* NJ: Prentice-Hall, Inc.

AI Magazine, Vol. 15, No. 4, 1994. AI Approach Demonstrates Low-Cost Satellite Strategy.

Coleman, Richard M. 1986. *Wide Awake at 3:00 A.M.* New York, NY: W.H. Freeman and Co.

Data System Modernization (DSM) Position Level Training Guide. 1985. Technical Training Division, Air Force Satellite Control Facility, USAF.

Mars Climate Orbiter Mishap Investigation Board Phase I Report, November 10, 1999.

Wall, Stephen D. and Kenneth W. Ledbetter. 1991. *Design of Mission Operations for Scientific Remote Sensing.* London: Taylor and Francis.

Chapter 9

Launch and Early-Orbit (L&EO) Operations

Michael Fatig, *EMS Technologies Defense and Space*

Jeffrey F. Volosin, *Honeywell International*

> 9.1 Handling the Demands of the L&EO Environment
> 9.2 Developing L&EO Operations
> 9.3 Conducting L&EO Operations
> 9.4 Reducing the Cost of L&EO Operations
> 9.5 An Update for Today's Environment

In this chapter we discuss the environment for launch and early-orbit (L&EO) operations, major issues regarding them, a process for developing them, and methods for reducing costs. We also present ways to deal effectively with the demands of L&EO operations and encourage developing plans for them early in the space mission lifecycle. This early planning allows us to coordinate these plans with the designs of the space and ground components for a more effective end-to-end architecture.

9.1 Handling the Demands of the L&EO Environment

9.1.1 The Environment for L&EO Operations

Understanding the demands, rigors, and special characteristics of the environment for L&EO operations is important to planning and operational success. Depending on the program's approach and the space mission's architecture, L&EO activities can range from complex and lengthy to simple and automated. Understanding the L&EO environment and characterizing operations within it allow mission operations managers (MOMs) to make informed decisions about important issues, which in turn help control the cost, performance, and risks of this critical mission phase.

Tough decisions face the MOM and the program team in developing operations for L&EO. It tends to be the shortest phase but is also the most stressful, critical, and demanding on the systems and people. Far too often the mission is designed for normal operations, with the special activities of L&EO operations neither understood nor assessed. The results can be disastrous. The mission design may overlook the demands and risks of L&EO operations until costly late requirements cause risk in developing systems, plans, and procedures. How much should we spend on systems and facilities for the safety and comfort of L&EO operations, when these systems, facilities, and people often won't do anything for most of the mission's lifetime? This is a major issue of cost versus risk.

The environment for L&EO operations is short, intense, critical, and dynamic, so the operations team must have skill and quick reactions to handle

- Increased volume
- Shortened cycle times
- Increased risk

Increased Volume. L&EO operations commonly require resources two to three times that of normal operations. They involve large increases in

- Number of real-time events
- Volume of commanding
- Amount of data processing needed for engineering telemetry
- Number of people involved in operations
- Number of activities that must be scheduled, checked, and conducted

For example, in normal operations we may require one contact with the spacecraft every other orbit—driven by rates of data collection, capability for onboard storage, and requirements for routine commands. However, L&EO operations typically involve a series of critical functions done rapidly after spacecraft separation (such as deploying appendages, stabilizing attitude, powering up critical components, and stabilizing and recharging power). During this period, we may need to contact the spacecraft many times per orbit to respond quickly to these hazardous activities. The number of real-time events increases dramatically, as do the command volume and processing of engineering data.

Because L&EO operations are critical, key design engineers for the spacecraft and ground system are commonly present. It's often necessary and advisable to have these design engineers near command, control, and communications to allow rapid analysis and response to anomalies. The two to four people for normal operations can become a staff of 20 to 30 for L&EO. The design of the operations room and system should match this temporary expansion (additional displays, voice circuits, and work area).

Because L&EO demands many more activities, the volume of commanding and data processing increases greatly. Many non-routine activities take place during L&EO, such as clock adjustments, system configurations and mode changes, system biasing, powering up components, and frequent calibrations. These activities and associated command procedures must be scheduled, compiled into pass plans and command loads, and verified. With more onboard processing and autonomy in today's satellites, the number of ground-controlled activities is decreasing. However, the complexity of a task increases as the relationship moves from a "ground as master, satellite as salve" relationship to "satellite and ground as peers" or even "satellite as master."

Mission planning volume is also much higher in the L&EO operations phase. More activities occur and these activities tend to be unique (rather than repetitive). Further, rapid replanning due to launch slips and anomalies may be necessary, and this drives the overall volume of activity.

The trend towards constellations of virtual satellites and satellites in some form of cross-relationship (such as cross-communications or rendezvous operations) will entail additional demands in the volume of activity.

Shortened Cycle Times. Shortened cycle times result from increased volume and operational dynamics. The increased volume means doing more in the same amount of time, so we have less time to move from activity to activity. During L&EO we may have only minutes to prepare for the next real-time event, so we may need to reconfigure rapidly. Not considering this need could severely restrict L&EO operations or place them at risk. Increased volume also entails shorter cycle times by demanding more resources than normal operations. For example, if we're commanding at two to four times the volume of normal operations, the onboard processor of stored commands may not be adequate for this increase. We may have to generate and uplink stored command loads every eight hours instead of every 24 hours as planned for normal operations. The operations team may therefore have to include staff for mission planning and activity around the clock, instead of day staff only for normal operations.

The dynamics of operations also shorten cycle times. During L&EO operations, critical activities often require validating one step before moving to the next. So we must plan the mission and do real-time operations not in large, contiguous blocks, but in short segments. Two or more options usually follow each short segment and depend on completion and performance of the previous step, so planning and decision making have to be much faster. This approach significantly affects the design of the operations team and the ground system. If the team or system performance doesn't allow these shortened cycle times, we may miss critical events or handle anomalies inadequately.

Increased Risk. Perhaps the most significant attribute of the L&EO environment is risk: the tremendous forces on the rocket during its ascent and during operation of the on-orbit kick motor; the "firsts" associated with each activity and system; the rapid, sequential activities; single-point failures in the system design; and demands on the operations team's reactions. Major problems we may encounter in this phase include

- Off-nominal launch-vehicle performance (improper orbit)
- Damage from the forces of launch (acoustics, g-forces, vibrations, changes in atmospheric pressure, temperature changes, or outgassing)
- Improperly deploying the spacecraft bus's main appendages
- Inability to acquire stable attitude or a power positive mode
- Inability to acquire good rf communications with the ground element, or difficulties in doing so
- Inability to detect and respond to time-critical anomalies
- Improper decisions in stressful, time-critical situations
- Unexpected performance levels in on-orbit systems
- Improper procedures and operator errors
- Unplanned contingencies with quickly recurring hazards
- Problems with software for the ground or flight element

For example, an improper orbit by the launch vehicle drastically changes the timeline for contacting the spacecraft through a ground- or space-based communications network. We must at least rapidly replan and reschedule all operations, then distribute changes to many organizations, facilities, and systems.

Another example is variance in the spacecraft systems' actual versus expected performance on orbit. For example, the power subsystem may not generate as much power as expected. If undetected, under voltage may occur. At least we must redefine the operations plan for the new levels of available power. Another example is the unexpected vibration of the solar array panels as the spacecraft enters the sunlight part of the orbit, as experienced by the Hubble Space Telescope. This "ping" induced vibrations into the attitude-control system, causing target-pointing problems. Other examples include: 1) the unexpected charging of the batteries in the Earth Radiation Budget Satellite due to albedo as the spacecraft was in a 180° pitch configuration, and 2) uplink signal margins that were better than expected on the Solar Anomalous and Magnetosphere Explorer mission, which caused receiver lock on a sidelobe.

Because L&EO operations are risky, we must plan for possible conditions so we can establish a team, system, and procedures to handle these conditions as they occur. This requires a systems approach (space, ground, operations plan, and operations team) across various elements and organizations (Fig. 9.1). Understanding and characterizing the environment for L&EO operations early in mission planning will result in a system, plan, and team that can handle its demands.

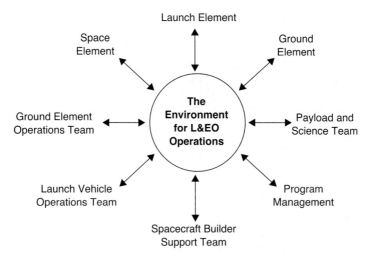

Fig. 9.1. The Environment for L&EO Operations. L&EO operations demand close and rapid coordination of many system elements and organizations.

9.1.2 Phases of L&EO Operations

In this text, we define the *L&EO operations phase* as the period from the beginning of live spacecraft activities during launch countdown to the transition into normal operations in orbit. The *transition to normal operations* is loosely defined as the time when we've largely completed special activation and checkout, made sure the space element works properly and performs well on orbit, and started activities to meet mission objectives. For experimental spacecraft, this would be when we start collecting primary science data. For communications spacecraft, it would be the beginning of communication services to the main user. Figure 9.2 and Table 9.1 define and illustrate the typical phases and subphases for L&EO operations. We define subphases and steps during early planning for L&EO operations, as discussed in Sec. 9.2.1.

We must define L&EO operations in phases and subphases because their activities differ significantly and because this approach allows us to break this complex phase into smaller, more manageable parts. It also provides convenient breakpoints in the operations plan, allowing for replanning periods, decision points, or transitioning responsibility to a new organization or team. It's a vital step in handling the demands of L&EO operations.

9.1.3 Handling the Demands of L&EO Operations

From experience, we define the demands of L&EO early in the design for mission operations and develop ways to handle them. Thus, we can decide where best to handle these demands within the space-mission architecture and

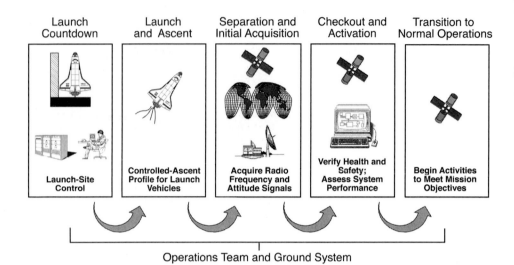

Fig. 9.2. The Phases of L&EO Operations. Defines major groups of activities for further decomposition.

operations team. Without this foresight, the operations team must meet them under stress, often by overusing command, control, and communications. In this section, we'll look at demands on the

- Operations timeline
- Operations team
- Ground element
- Space element

Demands on the Operations Timeline. The timeline for L&EO operations requires much higher levels of activity, many more critical operations, shorter planning, and more data processing than for normal operations. Many missions are too eager to start, often resulting in an aggressive timeline with little room for problems. Experience tells us that, in almost all missions, we'll have trouble maintaining the schedule. Experience also teaches us that a more cautious, slower approach is more likely to start the mission on time—with lower risks and stress. Avoiding these traps produces a more efficient, error-free, and timely L&EO.

Table 9.2 describes some of the demands on the operations timeline and ways to meet them.

Demands on the Operations Team. The team is larger. It's physically dispersed among the control site, the launch site, and possibly other significant areas. The activity level, and therefore the stress level, is higher, and the team must communicate, coordinate, and make decisions rapidly. Table 9.3 identifies the

9.1 Launch and Early-Orbit (L&EO) Operations

Table 9.1. Description of L&EO Operations Phase. L&EO operations has distinct subphases with separate objectives and activities.

Subphase	Steps	Description
Launch Countdown	• Power-up spacecraft • Test spacecraft's functions • Test ground system's readiness • Determine go/no-go • Configure for final launch • Dump and verify on-board memory • Monitor system	Involves the final launch readiness tests of the end-to-end system (short aliveness tests) to support a final go/no-go determination by project management. Once the launch is a "go," places the systems in their final configuration for lift-off. Validate that the correct flight software package has been loaded and that programmable memory areas are loaded as expected
Launch and Ascent	• Lift-off • Ascend through the atmosphere	Entails the pre-programmed, powered-flight profile to achieve orbit. Spacecraft operations during this phase are typically passive, with some automated operations controlled by timers.
Separation and Initial Acquisition	• Separate spacecraft • Start activities immediately after separation • Acquire initial attitude • Acquire initial signal with ground • Transfer to operational orbit (if required)	Involves the spacecraft's separation from the launch vehicle and the critical operations associated with acquiring an initial attitude, settling into a power positive mode, acquiring routine communications with the ground system, and deploying appendages. The separation activities are often automatically controlled by timers, separation switches, or stored commands.
Activation and Checkout	• Activate subsystems • Activate instruments • Check out system performance	Activate the spacecraft's subsystems and instruments; configure for orbital operations; adjust biasing, settling, and drift rate; do checkout. Checkout entails verifying modes and configurations and doing initial system calibrations. Verify overall system performance against predictions, adjust ground models, and modify operating plans.
Transition to Normal Operations	• Configure for normal operations • Begin normal activities	Configure the spacecraft bus and instruments for routine operations to meet mission objectives.

major demands on the operations team and provides guidelines for handling these increased demands. We should use graphic timelines and integrated scripts with easy-to-follow steps to keep all the participants synchronized with one another and with the dynamic activities. We must train and simulate to prepare teams, test communications, build a team approach, and familiarize support engineers with the operations environment. Design and test people should cover all technical disciplines.

Demands on the Ground Element. The many operations activities and the large operations team place extra demands on the mission operations system.

Table 9.2. The L&EO Environment's Demands on the Operations Timeline. Operations are critical, dynamic, and complex.

Demands On The Operations Timeline	How to Handle Demands
Complex, time-critical operations	• Plan operations at a 50% activity level; schedule backup events for critical and demanding activities; develop and test the detailed plan against the spacecraft and simulator • Develop user-friendly graphics and flow charts • Minimize the number of operations that depend on previous operations activities
Many operations depend on previous sequences	• Understand the interdependencies between operations activities and schedule slack in the timeline • Document the interdependencies for use in rapid replanning • Consider using a smart planning tool
Off-nominal space-element performance, higher anomaly rate	• Have readily available expected performance results (from test or models), resource budgets, and operational and survival limits to identify off-nominal performance and determine effects on the resource budget • Have an onsite engineering team for rapid identification and response • Preplan for major anomalies and establish real-time monitors to recognize anomalies. Have flow charts and other aids that help quickly determine a response. • Collect as much engineering data as possible during orbital operations and frequently downlink this data for analysis as needed
More offline data processing, often in a rapid mode	• Estimate the amount and frequency with which to collect engineering data, then size ground systems and operations to handle this data • Establish a separate system and team for continuous data processing
Offline data processing driving the next planned real-time operations	• Make sure the data-processing system and team can process the data in the amount of time needed • Add slack to the schedule and simulate the process
Shorter mission-planning windows	• Establish daily planning cycles, including an 8- or 12-hour shift of "no operations" in which daily replanning can occur without stress
Rapid replanning	• Ensure the system, people, and procedures are in place for rapid replanning at any point in the timeline • Have predefined operations breakpoints and re-entry points—simulate replanning • Design the operations plan in small, self-contained modules to allow operations to be easily altered and repeated • Know the hazards, constraints, and interdependencies of operations activities—have these in readily accessible, concise checklists to use in rapid reviews of the hazards and constraints involved with changes to the operations plans and procedures
High activity level (3 to 5 times greater than normal operations)	• Estimate the volume and phasing of the operations, and ensure the mission operations system and staff are the right size • Try to avoid major fluctuations in work levels and don't build an aggressive timeline (50% activity loading, schedule backup events into timeline)

Table 9.3. The L&EO Environment's Demands on the Operations Team. Requires unique skills for a demanding job.

Demands on the Operations Team	How to Handle Demands
Large, diverse, and often physically distributed operations team	• Organize well and clearly define roles • Identify the team early so we can budget, allocate resources, and retain key engineers whose skills will be critical during acquisition, start-up, and checkout • Ensure the team covers all technical disciplines and includes managers and test personnel that have hands-on experience with the spacecraft • Have clear, concise, and easily interpreted documentation on console for all operators • Conduct all-hands briefings and simulations to make sure everyone knows all the players and to practice team communications
Operations 24 hours a day with long shifts	• Do team-building exercises with the operations team, developing cooperation and relationships • In cases where launch personnel will be rotated in and out of operations on a shift schedule, establish shift 'teams' early on, allowing individuals that will be working together at launch to develop a strong working relationship during pre-launch simulations and dress rehearsals • Meet the physical needs of the operations team (provide food, coffee, and rooms for rest, relaxation, and even sleep)
High levels of activity and high reasoning requirements	• Ensure ready access to technical information • Establish clear, concise responsibilities for monitoring each activity • Establish clear, unambiguous values for identifying off-nominal performance, synchronized to the spacecraft's changing configuration and orbital variations • Automate critical monitoring requirements as much as possible
Rapid, real-time judgments	• Ensure ready access to technical information • Establish a clear process for operations management and decisions—make it fast and as unlayered as possible.
Many communication paths across many groups (voice circuits and face-to-face communications)	• Ensure effective communications paths, good voice protocol, and minimum talk • Establish adequate facility capabilities (identified early and provided as design requirements). Assign communications paths to facilities to support coordination and face-to-face communications over voice circuits—these assignments change from day to day as the focus of the operations for that day changes from subsystem to subsystem.
High visibility from management, the press, and the public	• Establish a room and an information flow so people not directly in the loop will know the mission status • Have a message center for directing incoming calls and outgoing information (avoids interruptions of the console team; ensures the controlled release of information to the media, management, and other interested persons)

Activity levels can be three to five times normal. Not recognizing these demands will stress mission operations systems so they can't handle L&EO requirements. The result is workarounds, which add more complexity and further stress the operations team and process. To produce a capable mission operations system, we must identify to designers early on all L&EO operations, special requirements, and estimated volume, sizing, and phasing. Table 9.4 shows how to deal with some of these demands.

Table 9.4. The L&EO Environment's Demands on the Ground Element. This environment stresses systems and calls for special configurations.

Demands On The Ground Element	How to Handle Demands
More systems to support the larger user community and more frequent ground-to-space communications	• Establish voice circuits that allow the diverse and distributed team to communicate effectively. We often use three to six voice loops and allocate circuits to operations control, engineering discussions, management discussions, and mission-planning coordination. Voice circuits typically don't cost much; yet, inadequate voice circuits can cause serious problems during contingencies.
Unique systems and configurations	• Define, document, and test the configurations needed for L&EO operations • Include volume and reconfiguration requirements in the definition and testing of these unique configurations
Backup systems are often required and must be rapidly configured when failures occur	• Define the critical operations and the possible failures that may jeopardize the mission • Establish backups for these single-point failures • Ensure the system design allows for rapid switching to backup systems during operations • Simulate failures and switching to backup systems
Increased command and data volumes	• Ensure the systems have been designed and tested to the levels of performance and volume expected during L&EO operations
Longer periods of use	• Ensure systems are designed for longer periods of operation • Include provisions in the operations timeline for reboot and reinitializing systems (refresh)
Quick access to data and reference information to respond to the high anomaly rates and requirements for rapid replanning	• Provide for rapid replanning, including the rapid replanning of operations for off-nominal launch times or orbit insertions • Ask the operations team what references they may need and make this information available • Consider online or web based systems for rapid access to technical information • Establish a strong configuration management (cm) process for ensuring adequate tracking of all ground system and spacecraft configuration modifications. An electronic CM system that allows multiple users to have search/sort/view privileges while retaining strict control of changes is preferred. This system should also provide the capability to access pre-launch testing configuration change information for use in anomaly analysis. • Provide a tool for visualizing the spacecraft orbit and attitude at any time with respect to the Earth, Sun, and Moon. The tool should provide information on factors such as: contact zones (with appropriate masking) for Earth or space based communications antennas; orbit day/night transition zones; South Atlantic Anomaly entry/exit points, spacecraft component shading. This tool should be available to all personnel involved in launch operations and can provide critical inputs to anomaly analysis.

9.1 Launch and Early-Orbit (L&EO) Operations 311

Demands on the Space Element. Finally, the conditions surrounding the space element during L&EO operations are also complex, unique, and dynamic. In many ways, these conditions (see Table 9.5) drive demands on the operations timeline, team, and ground system.

Table 9.5. The L&EO Environment's Demands on the Space Element. The spacecraft undergoes stress, operates in the space environment for the first time, and may face many problems.

Demands On The Space Element	How to Handle Demands
Stressful and rapidly changing environmental conditions	• Continuously collect engineering data; downlink and process frequently; monitor critical parameters against expected, operating, and survival limits
Changing spacecraft configurations	• Develop a plan to monitor configurations based on planned operations • Monitor the configuration frequently • Collect engineering data for performance assessments or for each configuration
Higher anomaly rates	• Have engineering expertise on hand, collect engineering data, downlink and process it frequently, and ensure technical references are on hand (drawings, component specifications, simulation and test results)
Deviations from expected values in system performance	• Have expected performance data on hand, establish ways to compare actual to expected performance, compare performance to operating and survival limits, and modify the operating plan for off-nominal performance
Stress is close to design margins for many spacecraft components (batteries, propulsion systems, and thermal subsystem)	• Assess L&EO operations early in design and influence design • Ensure adequate design margins • Identify stressed components and lessen the effects when defining the L&EO plan • Closely monitor operations
Outgassing	• Know outgassing requirements—protect optics and other components sensitive to contamination
Thermal settling and its effect on spacecraft components	• Have expected performance data on hand • Know test results for cold and hot cases; understand the effect of thermal variations on component performance (oscillator frequency drifts, thruster performance, and signal noise); include these effects in the plan for L&EO operations • Have information on-hand describing any autonomous command sequences loaded on the spacecraft that may act if specific conditions are satisfied. This information could be critical in a case where an unplanned configuration change is identified during flight.

We need to understand these conditions, characterize them early in the mission lifecycle, and determine how to handle them (take the risk, add tools and capabilities to the ground system, automate, or add and train a large team). Once we've taken these steps, we can confidently enter the environment for L&EO operations and increase our chances of success.

9.2 Developing L&EO Operations

In this section, we present a structure for planning, testing, and simulating L&EO operations. The degree to which a mission uses the process depends on time, resources, and anticipated mission complexity. Still, it outlines the activities most missions require and serves as a checklist of considerations for all missions.

The process for developing L&EO operations, shown in Fig. 9.3, consists of seven major steps designed to collect, characterize, and determine the detailed plans and procedures for L&EO operations; test the plans and procedures; and prepare the team for actual operations.

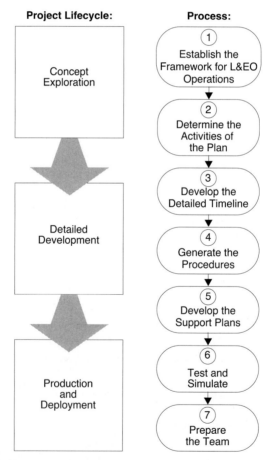

Fig. 9.3. **The L&EO Planning Process.** Begins with concept exploration and evolves towards production and deployment. Early characterization ensures we consider L&EO demands in designing the space mission architecture.

We do steps 1 through 3 (establish a framework, determine activities, develop timeline) as early in the mission lifecycle as possible. We estimate these details during concept development, so we can provide design inputs for systems development. We validate and refine these estimates as the system design matures to ensure a synchronized system and plan.

We do steps 4 and 5 (develop procedures and support plan) after detailed system design and before integration and test. In this way, we can use actual procedures and plans for integration and testing, which will then validate procedures and match testing of systems to their intended use.

We do steps 6 and 7 (test, simulate, prepare team) during integration, test, and dress rehearsal up to the launch day.

9.2.1 The Process Steps

Step 1: Establish the Framework for L&EO Operations

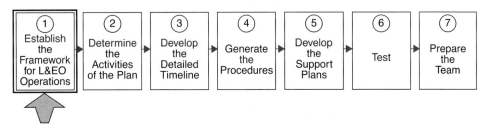

The *operations framework* is the foundation, or set of givens, that guides and bounds planning for L&EO operations. Established by the program's management and systems-engineering team, it's intended to lessen rework by properly guiding the effort with the program view. Experience has found that approaches to L&EO operations differ. Building a consensus reduces the need for redevelopment and reapproval of the L&EO plans. The framework includes

- **Priorities.** Decide the priority of various mission components and objectives (e.g. check out prime instrument first).

- **Duration and Schedule.** Define the broad schedule or duration for L&EO activities. For example, check out the spacecraft bus within two weeks and the instrument after 30 days. There are often technical reasons for schedules or duration, such as a target of opportunity we may miss, orbit decay, or outgassing.

- **Redundancy Philosophy.** Decide whether to check out redundant capabilities as part of the checkout plan, or to forgo it until a failure occurs. Redundancy checkout adds time, complexity, and risk, but it may be necessary if the spacecraft is designed for automatic recovery from failure.

- **Mode-Checkout Philosophy.** Check out all modes before normal operations or as we come to them in the mission plan.
- **Activity Level.** Establish guidelines regarding the pace and level of activity. For example, a project may elect not to pay for the extra staff and resources needed for an around-the-clock operation.
- **Resources.** Issue guidelines regarding use of resources (such as tracking sites, radars, additional facilities, or engineering specialists) to contain costs.
- **Organizational Elements.** Define the organizations involved and a point of contact.
- **Approval Process.** Briefly describe the approval process, if different from other project practices. Because L&EO often involves more than one organizational element, its approval process (typically a joint board or committee) is often different from normal practice.

By specifying these program-level conditions before doing detailed planning, we can avoid problems and rework.

Step 2: Determine the Activities of the Plan

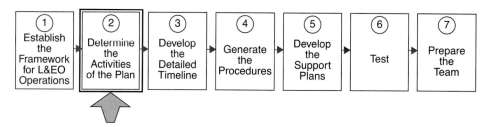

After we establish a framework, the next step is to collect the activities for L&EO operations, determine their attributes (actual or estimates), review and approve the list, and group them logically. These actions provide the foundation for all further planning.

At this point in the program, it's useful to define a common structure for the information we want to collect and generate. As mission complexity increases, so does the amount of information we need for planning and execution. Defining a structure provides a common language for all participants and, more importantly, provides a framework for managing information (moving away from documentation, providing online access to technical information). Figure 9.4 depicts a five-level structure used in many missions. Figure 9.5 shows how to use this structure.

9.2 Launch and Early-Orbit (L&EO) Operations

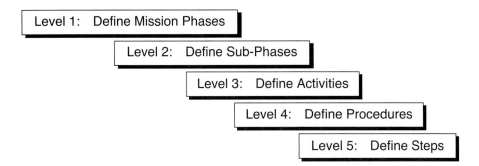

Fig. 9.4. **The Five Levels for Defining a Structured Operations Activity.** With operations complexity increasing, we need good, structured management of operations information.

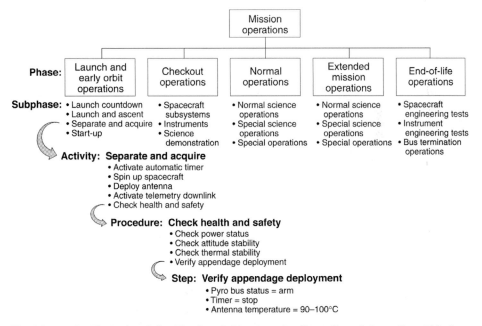

Fig. 9.5. **An Example of the Five-Level Structure for Operations Information.** This figure shows how the structure can be used to break down the L&EO phase into manageable blocks of information.

Assess each spacecraft subsystem and instrument and define all activities for initializing, checking out, and activating the subsystem or instrument. Include the

- Activity name
- Purpose or objective
- Main success criteria
- Major constraints or interdependencies

We compile this information into an integrated list and review it to make sure we've

- Completely defined all items
- Followed the framework
- Identified interdependencies and constraints
- Combined activities for efficiency
- Considered system-level impact of the integrated subsystem plans (loading of systems, overall duration, and data volume)
- Determined that operations won't need more resources than planned

We group checkout activities into logical units to define subphases. We define operations phases and subphases that logically bound how we collect, integrate, plan, and conduct mission operations activities. In this way, we have manageable tasks that contain logical breakpoints (or milestones) for tracking the status of development, implementation, and replanning as needed. With a complete, approved list of activities in hand, we can start grouping them into mission-specific subphases.

To define the subphases, we

- Locate logical transition periods for operations—places where there is a major shift in the activity level, objective, spacecraft configuration, or ground-support posture
- Minimize the dependencies across groups—plan each group separately and reduce the likelihood of changes rippling throughout the plan
- Identify points where we need major acceptance or qualification criteria—breaking up operations at these points allows entry into contingency paths should the qualifications not be met
- Identify major contractual or organizational shifts in responsibilities (if applicable)
- Avoid segments that are extremely large or continuous in time to avoid long, complex plans and procedures; delegate groups to different teams; and provide convenient points for re-entry and schedule changes if you need to replan during operations

9.2 Launch and Early-Orbit (L&EO) Operations

We compile the results of Step 2 into the Integrated Checkout Plan (Fig. 9.6), which is distributed, reviewed, and approved.

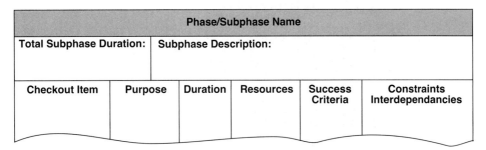

Fig. 9.6. Format of the Integrated Checkout Plan. The integrated checkout plan forces early thinking about the major L&EO activities and their attributes. It helps us build consensus and identify early any factors that affect system design.

Step 3: Develop the Detailed Timeline

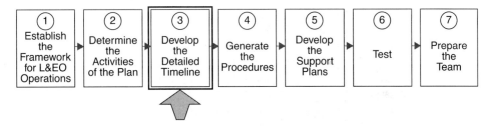

We must collect all detailed information we need to schedule the activities into an integrated timeline. We use the approved integrated checkout plan to schedule and assess each checkout item within a Timeline Assessment Table, shown in Fig. 9.7. Use this table to develop and modify the timeline through successive revisions, including rapid changes during the mission. The table becomes the database for scheduling and carrying out each L&EO activity.

The assessment table addresses such items as:

- **Execution Options.** For most missions, we can control the activity with real-time commands, commands stored on board without real-time visibility, or commands stored on board with real-time visibility. (Future missions may have more options for automation.) But it's still important to define how we'll carry out the activity. We must schedule critical activities that require decisions that can't be automatically checked or predicted for real-time execution. We must also schedule activities that don't require real-time decisions for

		Timeline Assessment Table							
		Execution Option:							
Ref. No.	Checkout Item	Real time	Stored Command	Stored Command with Real-time Visibility	No. Cmds	Freq.	Commands per Stored Command Load	Duration	Time Format	Interdependancies and Constraints

Fig. 9.7. Timeline Assessment Table. This table is the database for scheduling and executing each L&EO activity.

execution by a stored command, do them outside the real-time event (in the blind), and verify them at the next scheduled event. Some critical activities require real-time, go/no-go options but need stored commands because we don't want to risk losing the real-time rf link with the spacecraft. We should schedule these activities for execution by onboard stored commands during a real-time event.

- **Command Volume.** We have to estimate the command volume, including number of commands for a single occurrence, the number of occurrences, and the total number of commands. We must also assess total command volumes for all checkout items to determine if the L&EO plan exceeds the systems' capacities for stored or real-time commands.
- **Telemetry Format.** Various activities often require different telemetry formats. Identify these activities so you can plan them properly, plan telemetry formats, and avoid many format changes.
- **Interdependencies and Constraints.** Identify prerequisites, post-requisites, and constraints for an activity to allow proper planning of activities relative to each other and relative to the support plans (Step 5). Also, assess methods on how to handle the constraint. Three options are software, procedural, or manual checks.

If we're using an expert planning system, we have to take information from the Assessment Table to develop the rule base. But we should maintain the table even after we build the rule base so we can use it in unplanned situations or as a tool to understand and validate the rule base.

Building the Integrated Timeline for L&EO Operations. We use the information in the Timeline Assessment Table to develop an integrated timeline. The timeline illustrates the activities graphically over universal time and mission elapsed time. We can tailor the format of the timeline to each mission, but should consider including

9.2 Launch and Early-Orbit (L&EO) Operations

- All available network coverage based on the planned orbit (ground or space-based telemetry and command systems). As activities are scheduled, we shade the scheduled network resources.
- Telemetry format and changes
- Stored-command-load periods and volume
- Spacecraft activities with some graphical way to distinguish the method of execution (real-time, automatic, or stored command)
- Defined 'back out' plans for relevant activities (e.g., turn off procedures for a component that is being powered on for the first time; rf switch reconfiguration commands for a switching sequence that will connect a transmitter to an antenna for the first time)
- Payload activities
- Record cycles and dumps for onboard data storage
- Relationships between specific command sequences and on-orbit verification test plans
- Relevant orbital events (such as eclipse, South Atlantic Anomaly entry or exit)
- Major ground-based events, such as planning meetings, generating commands, data processing, or any other activity that is important and directly related to integrating and carrying out the L&EO plan

While developing the timeline, we

- Must ensure you have enough time to acquire the signal and check health and safety for real-time activities
- Schedule backup events for critical operations and events that have small time margins
- Consider shift handovers and the operations team's abilities. If we have only one expert resource, we can't schedule around-the-clock activities that depend on this person.
- Consider a 50% loading rate (50% busy and 50% free for contingencies). In most cases we'll use the contingency time or need a break!
- Identify convenient breakpoints in the timeline for replanning and re-entry
- Allow parallel activities, if needed, to expedite the timeline. Scheduling several activities to occur simultaneously through stored commands or with real-time is possible, but we must consider shared resources such as power, telemetry format, attitude position, and people.

Once we've completed the Assessment Table and timeline, we get them reviewed and approved. It may also be useful at this point to develop time-phased loading plots. For example a time-phased plot of the power consumption and battery depth-of-discharge (DOD) may be helpful for power-limited missions. We could plot command volume per stored load to show the size and uplink time of each stored command load. These time-phased plots allow the operations team and engineers to understand system use, stress, and capacity for changes.

When designing the operations timeline, we should consider the ability to change it rapidly during the mission. There are few off-the-shelf tools that can be used directly, but there are numerous relational database tools that can be customized to store all of this information so that it can be easily retrieved and viewed. With a tool such as this, we can quickly change the timeline to keep it accurate.

A well-thought-out timeline graphic becomes one of the most useful tools in presenting an often complex plan for L&EO operations to the large and diverse operations team. On many teams, where personnel are not physically co-located, it may be useful to have a tool that can provide HTML formatted products that can be hosted on a secure web-site where team members can review updated products and submit changes through an on-line form. Strict configuration management should be established so that only a select number of individuals have authority to edit the data. Alternatively, the products could be in a format that can be easily embedded in an e-mail for rapid distribution and viewing by users that have varied computer resources.

This tool can be used to track all planned and unplanned spacecraft and ground system configuration changes. In addition, by populating the tool with pre-launch information collected during spacecraft integration and testing and by continuing to use the tool throughout the on-orbit life of the mission, it can become a complete historical record of activities and configurations changes. This tool will be an invaluable resource during anomaly analyses throughout the life of the mission. Figure 9.8 shows part of the timeline from NASA's Small Explorer SAMPEX mission.

Step 4: Generate the Procedures

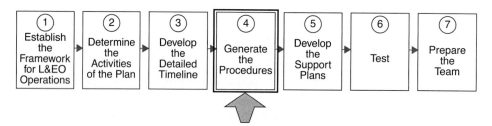

We review the Assessment Table and timeline and identify all procedures required to support the operations. We must define procedures in logical units that conduct discrete events, so we'll have flexibility during the mission. Include

- A launch-configuration table
- Configuration-state tables

9.2 Launch and Early-Orbit (L&EO) Operations

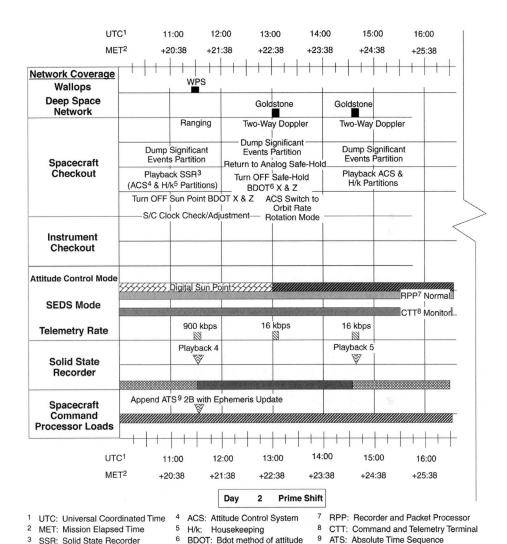

Fig. 9.8. L&EO Operations Timeline. An example of the L&EO timeline for NASA's SAMPEX mission. [NASA, 1992]

- Stored command loads
- Flight-software tables
- Real-time command and control
- Contingencies

Launch-Configuration Table. This defines the spacecraft's configuration for launch. It meets the launch constraints (what can and can't be turned on) and the needs for separation and acquisition on orbit. Set the configuration for minimum commanding once the spacecraft is in orbit and minimum power-discharge levels during ascent. The launch-configuration table includes:

- Component, switch, or software item
- Telemetry indicator
- Desired launch site
- Command to produce the desired state
- Any necessary remarks

Usually, the integration and testing team configures the spacecraft finally at the launch site. The launch-configuration table becomes an agreement between this team and the operations team on the spacecraft's state for launch.

Configuration-State Tables. Throughout L&EO the spacecraft, its subsystems, and its payloads cycle through many modes, so the configuration changes dynamically. By defining configuration states for the various phases, we can use these state definitions for power management, command management, and health and safety checks. With a good set of state definitions, synchronized to the L&EO timeline, the operations team can monitor changing configurations closely.

Stored Command Loads. We recommend defining and generating stored command loads early in mission design. Several factors go into building these loads. First, we segment them into logical units that fit within the system capacity. Second, we segment and schedule them for uplink based on uplink opportunities, including backup opportunities. Third, we make sure they contain any safing or reconfiguration commands necessary to safe the spacecraft if the next load isn't uplinked in time. Finally, we generate, error check, and test them, remembering they often use the stored command processor in ways other than expected for normal operations (larger loads, unusual timing, and sequencing).

Flight-Software Tables. With the increasing use of flight software to control spacecraft operations, we may need to generate, load, and modify from a few to several hundred tables during L&EO operations. Managing, testing, uplinking, and verifying these tables may require responsive systems. Early on we need to construct and test the tables needed to support the L&EO timeline.

Real-Time Command and Control. For L&EO operations, we generate real-time command procedures for all activities designated to execute under real-time command and control. As with stored command loads, we must segment the activities into procedures. Procedures should carry out a given activity, yet not be too lengthy or inflexible. They should have logical break points that allow for return and re-entry, and they should include the interactive steps of commanding and telemetry verification into a single logic flow. Finally, for more complex procedures, we recommend flowcharts. They help develop the procedural logic and are good training tools.

Contingencies. Contingency procedures are perhaps the most difficult and time consuming to develop. First, we define what contingencies may occur that require a preplanned response to avoid serious problems on orbit. To do so, we can assess spacecraft Failure Modes and Effects Analysis (FMEA), spacecraft integration and test anomalies, and experience with similar components or designs on previous missions (Fig. 9.9). These may guide us to where hazards exist and what failures are most common, but the best method is a good brainstorming session with key spacecraft and ground-system engineers. They can help work through the L&EO timeline and assess what could go wrong, how serious it is, whether a quick response is essential (or can we take time to better assess it), and how to respond to it. The result will be a set of contingency paths off the nominal timeline for L&EO operations and the procedures to support them. We also identify points of re-entry into the timeline.

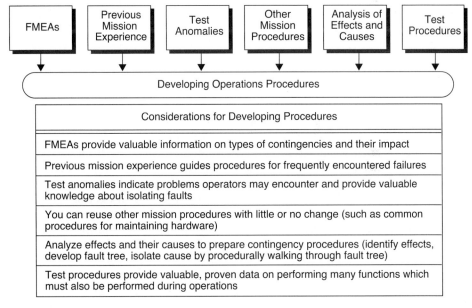

Fig. 9.9. Considerations for Developing Procedures. Considerations beyond the activities timeline develop better procedures, especially contingency procedures. (FMEA = failure mode and effects analysis.)

Since a list of 'possible' contingencies can grow to be fairly large, it is important to prioritize the list to focus the team's attention properly on flushing out the 'most' probable anomaly scenarios. This is critical since limited resources will usually not allow every foreseeable anomaly to be addressed. In addition, the team should address whether individual contingency responses would be best addressed by uplinking commands in real-time or by having pre-loaded event-triggered command sequences placed on the spacecraft.

Most modern spacecraft have event triggered command sequences that can automatically activate when specific conditions are met (e.g., a temperature from a specific thermistor reaches a set value, a voltage reading crosses a set threshold, a switch state changes). In many cases, these onboard sequences can provide a more rapid response to a contingency than ground up-linking commands. In addition, they can reduce the number of contingencies that ground personnel need to be trained to respond to. If sufficient onboard memory is available, it may be possible to load these sequences before launch. If memory limitations prohibit the loading of all planned contingency responses, it may be possible to load/clear some sequences throughout the on-orbit activation period with some sequences only loaded and activated during certain periods (e.g., during initial transition to a new attitude control mode, during initial turn-on of a high-voltage power supply).

Operations personnel should have access to information on these triggers and their effects to facilitate investigations into any unplanned spacecraft configuration changes. Procedure development may vary with the design of the spacecraft, ground system, or timeline for L&EO operations. Still, we need to develop them early, use them to integrate and test the spacecraft and ground system, make sure they're well controlled and documented, and map them to the timeline for L&EO operations.

We may also need to vary the procedures for ground testing. Ground-related constraints may keep us from completing procedures in test (such as thruster and pyro firings). We may also need to change some procedures for flight, especially to avoid problems in orbit. In the end, all command procedures that may be used during flight should be validated in their flight configuration. In cases where they can not be validated properly against the spacecraft, a flight simulator may be used. In cases where the fidelity of available flight simulators is not adequate to validate a specific procedure, we can use a walk through of the procedures with the operations and development engineers as a method of validation.

Step 5: Develop the Support Plans

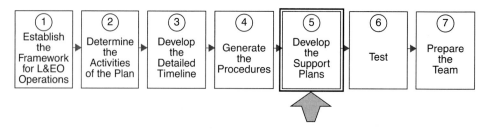

Support plans may include
- Data processing

- Display
- Facility and voice
- Mission planning
- Operations management
- Status and announcement

Data-Processing Plan. Some activities in the L&EO timeline may be interdependent, so we must analyze results of the first step before the second step. It then becomes important to retrieve the data, process it, analyze it, and decide whether to proceed or not. When this is critical, or the timing is tight, we may want a data-processing plan that shows how to handle certain data sets. We should also query each of the many support specialists for data-processing needs and collect results in the plan. In most instances, data processing in support of L&EO operations is more significant than during normal operations. It's best to know in advance rather than to discover needs during operations and be unable to meet them because of constraints on systems or staffing.

Display Plan. We may need a display plan if the number of available displays doesn't allow everyone control over a display device. It defines the available displays and maps them to the timeline. Given the large support staff, a dynamic timeline, and many unique needs during L&EO, we may need special and rapidly changing displays. In addition, we should conduct pre-launch testing with all displays up and running as they would be under the highest load scenario to stress the system and identify any ground system loading issues.

Facility and Voice Plan. L&EO operations often requires support from design engineers for the spacecraft and mission operations system, program managers, and people who do integration and testing. Further, special L&EO facilities are often involved in the operation, so we need to determine who is critically needed, as well as when and where they're needed. The results go into a facility plan. Figure 9.10 lists the facilities used in launching NASA's Cosmic Background Explorer. Note that it used special launch facilities physically distributed across the United States. It was important to decide who would be at each facility, as well as what and how much each could do.

We need to define the operations team, support team, and facility requirements (e.g., floor space, capabilities) early to ensure we've defined interfaces and designed to meet needs. In addition, we must ensure that the individuals located in the primary command and control center are limited to critical personnel only. Overcrowding of this facility can lead to problems with crowd control and noise that may impair real-time operations. Many times, almost everyone associated with a mission will want to be located at the command and control facility since this is where all the 'action' is. Special badging and a VIP viewing area will ensure minimal traffic in operational areas while providing an area where flight status information is available to other observers or engineers involved in off-line assessment in near-real-time.

Goddard Space Flight Center Management Facilities, Building 13

Mission Management Area
• Project Management • Center Management • Directorate Management

Goddard Space Flight Center Mission Operations Facilities, Building 3 and 14

Launch-Control Room	Launch-Support Room	Mission-Operations Room
• Project Management • Operations Approval and Control • Spacecraft Group • Instrument Group • Mission Planner • Payload Operations Control Center Support	• Operations Support • Overflow Area • Analysis Area • Mission Planning Area	• Operations Support • Flight Dynamics Facility Representative • Fault Analysis Support • Attitude Control System Support

Goddard Space Flight Center Science Facilities, Building 7

COBE Science-Data Room	Telemetry Room	Operations Center
• Project Scientist • Science Operations and Analysis • Science Planning Team	• Acquire Science Data	• Science Analysis

Launch-Site Facilities, Vandenburg, CA

Bldg 836	Mission Directors
• Project Management • Integration and Test Team • Spacecraft Group • Instrument Support	• Delta Project • Vandenburg Launch Team • Project Management • NASA Headquarters

Fig. 9.10. L&EO Facilities for Cosmic Background Explorer (COBE) at Goddard Space Flight Center (GSFC). Many diverse facilities are involved in L&EO operations, so we must plan for people and communications. [COBE, 1989]

After assigning facilities, we must assess the communications paths among these people. We define the main groups of people and how they'll communicate; then, generate a set of requirements for the voice system. The goal is to provide enough voice circuits for quick, clear communications between logical groups of people in support of the L&EO activities. Too many people on a single channel may be restrictive, especially when timing is critical. On the other hand, having too many channels makes operations difficult to monitor. One way to design a voice system is to allocate loops for

- Management

- Spacecraft engineering
- Ground-element engineering
- Real-time operations
- Navigation

An additional "all-hands" loop may provide status, direction, and control of the overall timeline. This loop keeps everyone synchronized across the physically distributed operations. Actual voice-loop configurations depend on the number of people and the locations.

Mission-Planning Plan. This plan defines daily planning and decision making. Mission planning during L&EO operations is typically more frequent—with shorter planning periods. Due to the high rate of anomalies and "firsts," plans often require daily refinement. We may have to replan daily: assess results of the current day and review the next day's plans for refinements. During L&EO operations, we may also have a different operations-management team.

Operations-Management Plan. This plan defines the roles, responsibilities, and decision process for the launch countdown and orbital-operations phase.

Launch-countdown decisions usually involve the launch-vehicle team and scheduled go/no-go decisions. Orbital-operations management provides go/no-go decisions for major activities and decisions to solve anomalies. Figure 9.11 shows a typical operations-management process. Due to the large team, diverse locations, and critical nature of operations, we need well-thought-out and defined operations reporting and management to reduce confusion during actual operation.

Status and Announcement Plan. This plan discusses how to release announcements regarding mission status. Typically, many people are interested in this status. We've found that, unless we satisfy these interests, we'll receive many phone calls and visitors to the control complex. Having a defined status and announcement plan will remove this job from the control-center team, thereby reducing the complexity and confusion. Status and announcement plans may consist of press releases, press briefings, phone-message systems, or postings.

All of our support plans must be consistent with the timeline. For example, the status and announcement plan may schedule a press briefing just after a major event. If so, we can schedule daily planning meetings at a time of day when there is a gap in real-time coverage, thereby averting conflicts with real-time operations. The complexity and need for each of these plans vary from mission to mission, so we must develop only what we need.

Step 6: Test

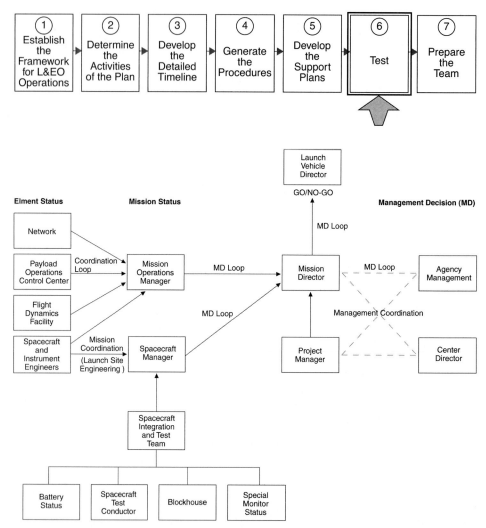

Fig. 9.11. Operations Management for L&EO Operations. Element information usually flows up to mission status, with management decisions based on the status.

The purpose of testing is to validate the systems, procedures, timelines, and people as launch ready. We recommend that the MOM

- Schedule a 'walk through' review with development and operations engineers first

- Test with simulators
- Validate the timeline and procedures with the actual systems and spacecraft
- Provide realistic test environments (loading, phasing)—L&EO operations tend to stress the system and people, so we'll need to load and phase the activities properly to replicate the worst-case scenarios
- Have operators conduct the tests, with developers in support
- Combine (as much as possible) validation and readiness testing for L&EO with integration and test of the ground and space elements
- Plan for retest of modifications and improvements
- Include tests late in the preparation for launch, including launch-site or launch-pad testing
- Test anomalies by using simulators to inject faults
- Define criteria for success

We should develop test plans early in the program so people can build unique tools and interfaces and include the tests in integration and test schedules for the space and ground elements. Test plans should address: maturity of the ground system and spacecraft at any given point; planned spacecraft configuration changes over time due to integration and test activities, and constraints on modifying this configuration at any given time; availability of other elements requiring testing (e.g., ground station, WAN, LAN).

Step 7: Prepare the Team

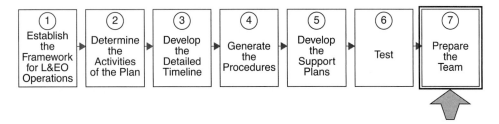

All staff should be aware of their roles, of the integrated team and timeline, and of how to follow along (including tracking the changes that may occur daily). Because the operations team for L&EO often consists of developers, some people won't know console operations and systems. Special training and more simulation may be necessary for them. To ensure people are prepared and aware, we

- **Develop an Integrated Console Script.** This script integrates all activities and keeps everyone synchronized. It orchestrates and controls the activities of the ground, space, launch-vehicle, and operations teams according to the timeline. Optimally, this script should be one of the products from the relational database tool being used to generate the flight timeline as discussed in Step 3.

- **Train People Who Aren't Operators.** They should train in console procedures, voice procedures, and system operations. Console operations in a time-critical environment can be difficult and confusing. The team should be comfortable enough to focus their attention fully on operations rather than how to work the console and systems.

- **Train Operations Personnel.** The spacecraft development team should provide detailed training for all operations personnel supporting launch and on-orbit checkout operations to review spacecraft design features and the results of integration and test activities.

- **Brief All Hands.** We gather all hands (or as many people as we can) and brief them. We summarize all aspects of the L&EO plan including the timeline and support plans, and hand out all launch-ready materials.

- **Simulate All or Parts of the L&EO Operations.** We integrate the actual procedures, people, and systems as much as possible. We also inject faults into the system with a simulator or by voice. This entails simulating actual loading and timing. As launch draws closer, we include as many elements of the end-to-end system as possible for a number of "dress rehearsals." We also have a team monitor the simulation and debrief to go over the results.

9.2.2 Development Team, Products, and Tools for L&EO Operations

Development Team for L&EO Operations. Developing timelines and procedures for L&EO operations requires the support and involvement of many groups across the mission system (Fig. 9.12). The operations team usually takes the lead in developing plans for L&EO operations. A mission operations working group or similar forum develops and reviews the timelines, plans, and procedures outlined in the previous section. The team should meet early in mission development to define L&EO operations in a checkout plan and draft timeline. These early drafts help define the capabilities needed in simulators, interfaces, systems, consoles, and voice circuits. In this way the needs of operations are built into the system, instead of having an L&EO plan that fits operations to a fixed design. Sharing this information through a secure website can provide an easy way to disseminate up-to-date information to all team members.

We'll also need to coordinate with the launch-vehicle team to get the typical inputs shown in Fig. 9.12.

9.2 Launch and Early-Orbit (L&EO) Operations

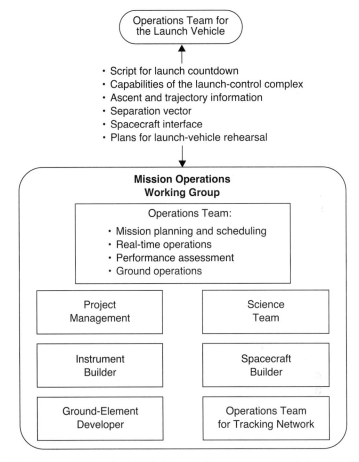

Fig. 9.12. Development Team for L&EO. Integrated team represents all elements. It's established early in mission development and it coordinates all processes and products of L&EO planning.

Products for L&EO Operations. We should publish the results of L&EO development in an L&EO Operations Handbook, which is a quick-reference guide to L&EO operations. We distribute it to all members of the operations and support teams for L&EO, thereby ensuring everyone operates from the same script, timeline, and procedures. The handbook contains all important products and is often indexed and tabbed for quick reference. Figure 9.13 shows typical contents of an L&EO Operations Handbook.

Configuration Control. As with any critical information, we need to control timelines, plans, and procedures for L&EO operations. Depending on the project, the level of configuration management may vary. But it's important to control

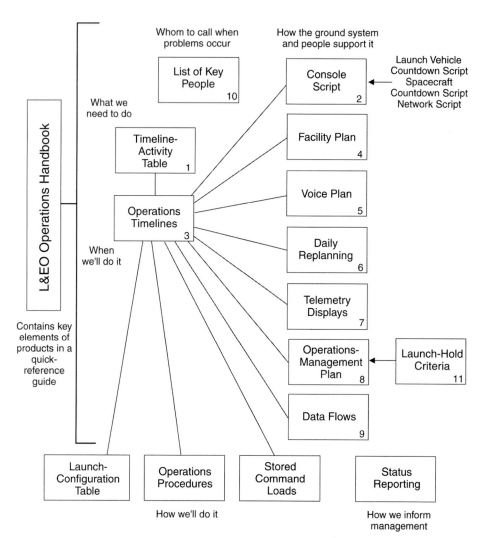

Fig. 9.13.   L&EO Operations Handbook. Provides key information in an easy reference guide to the large, diverse operations team for L&EO.

changes once testing and validation is complete, as well as to label the version clearly (draft, review, in test, approved, or validated), showing revision numbers and dates.

Tools for L&EO Operations. The commercially available tools to develop and carry out L&EO operations are limited. Depending on the complexity of the development and operations task, you may find a timeline-development tool beneficial. For example, planning for the mission to service the Hubble Space Telescope (HST) was a complex task. It required many activities executed in close coordination with Shuttle-crew activities and within short operations periods. Further complicating the planning was the real possibility of a launch delay or slip to later days in the launch window. Consequently, an expert planning tool was developed that contained a rule base for scheduling activities against the timeline. It also allowed rapid replanning to recompute the timeline given a launch delay. This tool proved invaluable during operations.

In contrast, NASA's Small Explorer mission used a simple graphics-drawing tool to develop the timeline, as its operations were simple, with few external variables. We need to assess the complexity and number of external variables before deciding whether to develop a tool. In some cases, a tool will save many hours of manual replanning. In other cases, the planning may be too simple to justify development costs. See Sec. 9.5 for a discussion of new tools now in use.

9.2.3 Special Planning for Multi-Mission Operations and Human Spaceflight

Multi-Mission Operations. With the trend towards smaller missions and multiple spacecraft launches, we may need to plan for operations over multiple missions. Multi-mission planning involves six steps designed to overlay the operations of each mission on the common mission operations elements, assess the effect of multi-mission loading and phasing, and develop strategies for handling the resultant multi-mission plans. Figure 9.14 illustrates this process. Multi-mission planning is useful when launching a single mission using an existing mission system or team, or when launching multiple missions at the same time.

First, we define *operations cases*—distinct mission phases where the operations may vary significantly from the previous case. We characterize and assess data volume, operator tasks, and system use for loading. We should include the total volume and the phasing of activities over time according to the launch and orbit characteristics. Areas of concern include: where loading is beyond planned systems, team capacities, or cost ceilings. We must do pre-launch testing to look at maximum multi-mission loading scenarios to ensure that weaknesses are identified and addressed before launch. Also, since most multi-mission environments generate cost savings by using common ground system equipment to support multiple missions, it is important to use a system for differentiating missions when accessing ground system equipment (e.g., unique color coding of input screen backgrounds for each spacecraft, physical signs). With this knowledge early in mission development, we define various strategies to handle each area, such as automating, migrating to on board, taking risks, adding systems or staff, and so forth. We can then use the most cost-effective strategy.

Implications of Human Flight. L&EO operations for flight crews require complex coordination of spacecraft and crew activities, as well as a new, larger set of responses to anomalies. While human flight adds options for launching the mission, it also adds complexity. Finally, crew safety requires more constraints.

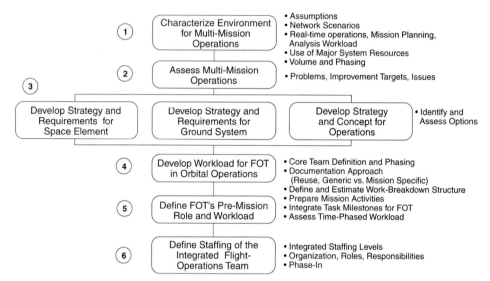

Fig. 9.14. Assessing Multi-Mission Operations. Handling the demands of operations for multiple missions requires special planning. (FOT = flight-operations team) [NASA, 1993]

9.3 Conducting L&EO Operations

Conducting L&EO operations is the team's most stressful yet most exciting activity. It's the culmination of all the work of many people over several years. It's also a technical and emotional activity that taxes systems and people while determining the success or failure of the mission. We've outlined many characteristics of the environment for L&EO operations (Sec. 9.2.1). We've defined the unique demands on the spacecraft, ground system, and operations team. But during actual L&EO operations, these demands are simultaneous, and our planning, testing, and preparation begin to pay off. Here's what to expect:

1. **Operations Team.** People will be anxious, stressed, and possibly tired (depending on the pace of activities leading up to launch). Conflict is a real possibility, but we can lessen it with discipline, a high state of readiness, good operations products that reduce confusion, minimal voice traffic, frequent status messages to keep everyone informed, and calm reactions. Selecting the operations team and shift assignments, skill matching the team to each other and to the timeline, and developing a timeline with a moderate activity level reduce stress and

conflict. Many NASA missions have successfully used a two-shift arrangement. An "A" team, staffed with the most skilled people and experts, operates one 12-hour shift, and conducts all of the critical activities and major checkouts. A "B" team operates the second 12-hour shift. They monitor health and safety, collect and process data, and replan. The B shift is therefore a period to catch up, replan, and re-synchronize while the A team rests. Although the B shift may not have the most experienced personnel from each subsystem/payload, there should be a few overall spacecraft and ground system experts on this team. Establishing the 'A' and 'B' shift teams early on is critical to allowing the teams to work together during pre-launch simulations.

2. **Ground Element.** Nearly all parts of the ground element are online. The ground system during this mission phase is the largest and most complex of the mission, having many people, systems, and interfaces. Data volumes, system use, and system-performance needs are at a peak. We can expect problems with an interface or a system and should be prepared to switch to backup systems or to alter the operations plans. We reduce problems by testing and simulating before launch; having a well defined way to report, discuss, and resolve anomalies; checking a system's readiness often before using it; configuring systems well in advance; and freezing the configuration. Also, the use of 'hot back-up' equipment may be necessary in areas where the time required to start up and configure back-up equipment may be inadequate to meet the timing requirements of operations in the event of failure of a primary unit during a critical period.

3. **Space Element.** The space element undergoes extreme environmental conditions, rapid changes, unusual and new configurations, and the first operation of all systems in space. As with the ground system, we can expect failures and problems. If the team can stay rational and controlled, avoiding quick reactions and hasty conclusions, L&EO operations will be smooth. We can decrease risk and ensure a smooth activation despite variances if we have good reference material handy; know the expected orbital signatures and values of the systems; have good, quick processing systems on the ground for assessing health and safety; and have defined a process for discussing and resolving anomalies.

Although L&EO is the most challenging phase, it's also the most rewarding. We should enter it with optimism, control, confidence, and a readiness to deviate from the plans. After completing L&EO operations, we write a lessons-learned report and give the results to the next mission for process improvement. Optimally, as part of the relational database tool discussed in Sec. 9.5, operations team members should be provided with a form to enter lessons learned throughout the pre-launch, launch and on-orbit checkout phases of the mission. These inputs would then provide the raw material for generating the lessons learned report.

9.4 Reducing the Cost of L&EO Operations

We've discussed ways of succeeding in L&EO operations, but our recommendations and methods have costs, and not all are necessary for each mission. The budget for a given mission may restrict development or operations, so we may need to drop the best approach for a less costly approach, and then further trade development costs against operations costs. As mentioned before, deciding to spend money for a mission phase which can be as short as several days is difficult. Tools and capabilities unique to L&EO will largely sit idle after this short phase. Still, the L&EO phase is often the most critical, most stressful, and most likely to have mission-critical problems.

To reduce the cost of planning for and conducting L&EO operations, we need to determine which of the following options apply to the mission:

1. **Plan Early and Influence Designs for the Spacecraft and Mission Operations.** Characterize the loading and performance requirements of L&EO and determine how to phase activities. Make educated guesses if we have to. Our guesses will at least establish a consistent baseline across the system so the ground elements, spacecraft system, and operations teams are developing to the same activity levels. Go through at least Step 3 of the planning process in the concept phase, influencing design with the results. Good early planning will avert late and changing requirements, which are not only costly but stressful.

2. **Use Common Systems for Mission Operations and for Integration and Testing.** Using the same system for integration and testing and operations can save development money. We often build separate systems and transfer the information, files, databases, and knowledge to operations. Having the same system avoids translations and transfers, lessens risks, reduces validation requirements, and lowers costs.

3. **Share Overflow Facilities and Systems.** As we've mentioned throughout this chapter, L&EO operations demand extraordinary capabilities. Designing mission-unique systems to handle the demands of L&EO is too expensive because much of the system's capability will be under-used after L&EO operations. Identifying resources from an existing mission or operation can alleviate the need to develop unique facilities or to suffer with a team and system impaired because it's sized for normal operations.

4. **Base Integration and Testing on Operations Plans and Procedures.** Combine the integration, testing, and operations test plans; use operations procedures in the test environment; and determine the systems' signatures, values, and behaviors during integration and testing. These actions will help reduce costs (by avoiding duplicate tests and procedures) and risk (by ensuring the systems are tested as they'll be used).

5. **Design Spacecraft with Ample Margins.** Include margins for the demands of launch and early-orbit operations. Ample margins reduce the complexities and risks of planning and operating a limited resource.

6. **Review Plans Frequently and at Critical Points in the Process.** Reviewing plans and procedures frequently and with everyone involved will ensure the planning doesn't go down a path unacceptable to others or inconsistent with systems design.

7. **Automate Spacecraft and Monitor Critical Hazards on Board.** Avoid as much as possible ground control of time-critical, sequence-dependent tasks during the initial orbits. This will reduce complexity and the expense of adding redundant capabilities to the ground system.

8. **Use People from Integration and Testing.** With the increased demands of L&EO operations, we need more hands from the integration and testing teams. In fact, these people don't require much training because they must know the systems well and develop trouble shooting skills before launch. Using them on the operation team provides valuable practical skills.

9. **Use Ground Support Equipment Developed to Support Integration and Testing.** L&EO operations may often require quick processing of instrument data to assess instrument performance. If so, the ground support equipment developed to integrate and test the instrument may provide a quick, proven capability. This equipment can be at the mission-control center, thereby reducing complexity and improving responsiveness.

10. **Reduce Interdependencies.** Design the systems and the timeline to avoid dependencies between activities. If interdependencies are unavoidable, schedule ample time to verify prerequisites.

11. **Automate.** Automate as much as possible the time-critical events, especially the scenarios for separation and early-orbit activation. It's often difficult and expensive to get ground-station coverage at the point when the launch vehicle separates. Further, acquiring a signal reliably just after separation is unlikely.

12. **Place the Team Close Together.** This reduces interface costs, decreases complexity, and promotes the teamwork needed to respond to the large number of possible anomalies.

13. **Reduce the Number of People.** Doing so decreases system and facility requirements, complexity and confusion, and training costs. This reduction can be done by blurring the traditional functional lines between spacecraft and ground system development teams as well as flight operations. Using operations personnel to develop and run spacecraft integration and test scripts, write operations user's manuals, and play a role in the development, testing and population of ground system software elements can greatly reduce the overall team size. In addition, many operations teams are recruiting multi-talented personnel that can monitor and assess spacecraft component health as well as maintain expert system models for the ground system and continue to develop and implement further mission automation. These individuals must have experience in spacecraft engineering and real-time operations, as well as a strong computer programming background.

9.5 An Update for Today's Environment

Changes since the mid 1990's have provided the L&EO mission planner with more tools, yet the fundamental process remains the same. The task remains one of coalescing mission objectives in a risk-managed way within the constraints of the end-to-end mission system and team. The integrated plan is then given to the often broad and dispersed team through various "products." These products communicate and control the orchestrated flow of activities across the end-to-end system in a time-ordered sequence. Unplanned events that require quick reaction and replanning occasionally disrupt this sequence.

The significant changes in the end-to-end mission environment that affect the L&EO mission planner include:

1. **New Tools.** New relational, user interface, and information distribution tools are now available, a result of the continued investment in tools that help people manage and distribute information and knowledge such as:

    a. The use of an integrated relational database tool that can export information in HTML format for web hosting, and the advent of web-based user interface and distribution. This tool serves as a configuration management tool and information for the following types of information:

    – Activity timelines

9.5 Launch and Early-Orbit (L&EO) Operations

- Flight procedures (nominal and contingency)
- Ground system command procedures (with the ability to validate embedded commands against the ground system command database)
- Ground system telemetry and command database
- Console scripts
- Spacecraft configuration management data
- Ground system configuration management data
- Spacecraft/ground system configuration change requests (with automated verification of requested commands against the command database and automated updating of spacecraft configuration status information based on system feedback on successful execution of sequences). Also, e-signature capability for on-line approval.
- Anomaly reports
- On-orbit verification test plans
- Shift reports
- Lessons learned archive

We should begin development of this tool well before launch and begin populating it during spacecraft integration and testing (I&T). By starting early, the tool encompasses the full 'history' of the spacecraft from initial integration through launch and on-orbit checkout. In many cases, we continue to use the tool throughout the life of the spacecraft. By comprising the entire spacecraft configuration change history in a relational database along with the shift reports and anomaly reports, the flight operations team has all of this information at hand, in an easy to search/sort format, for in-depth analysis of spacecraft aging and for assistance in anomaly trouble-shooting. In addition, with the capability to export information in HTML format, the team can use the internet to share information with a physically dispersed community of spacecraft developers, operators, and payload teams.

b. We use graphical tools to provide the launch team with a graphical representation of the spacecraft orbital position and attitude at all times. These tools provide information on such factors as: ground/TDRSS contact periods (including masking), orbit day/night transition points, South Atlantic Anomaly entry/exit points, and spacecraft component shading. In many control centers, the graphic output is projected on a screen in the launch room.

These tools can have a direct and positive impact on the mission planner. However, the mission planner must still deal with the cost-benefit scenario unique to the L&EO environment. Because of the short duration of L&EO operations, investing in the development and use of such tools is a tough decision. While they may mitigate risk and increase overall team proficiency, they may also be used only for a short time.

2. **More Onboard Logic Capabilities.** Increased onboard memory and processing power is leading to more flexibility in reducing ground command requirements. The use of time and event triggered command and macro sequences increases spacecraft autonomy. L&EO sequences may require less direct and less frequent ground interaction, particularly in the risk management area.

3. **The Shift from Specialists to Generalists.** There has been a collapse of traditional launch team roles into a small team of multi-talented operations personnel. The intense drive to lower costs has resulted in more highly skilled but fewer "generalists". Members of the operations team are now building, validating, and populating their own automated expert ground systems, as well as serving as spacecraft engineers beginning during spacecraft I&T. They are operating end-to-end ground systems seamlessly with the operations of the space asset. This overall integration of functions and the resulting decrease in the size of the teams has an impact on the L&EO mission planner. For example, having smaller teams means a quicker path to burn-out and less ability to operate on a 24 hour, 7-day a week schedule.

4. **The Advent of Commercial Satellite Ground Station Services.** The commercial space boom of the mid-90's resulted in the creation of several commercial ground station service firms. These firms have developed world-wide ground station capabilities. They sell satellite uplink and downlink time for a fee. This new capability offers the launch and early orbit planner new cost-effective means by which to obtain the additional real-time coverage needed for the higher demands and critical operations of the launch and early orbit sequence. Planners can procure extra ground station services only as needed for the short period of the L&EO phase. The tough trade-offs of building extra capacity for the demands of the short L&EO phase or letting the operation occur "in the blind" are now somewhat mitigated.

In addition to these changes, new demands on operations will continue to change the L&EO operations environment. Today the driving forces are coming from the military needs (as opposed to the commercial needs of the mid-90's when constellations of low earth orbiting satellites were changing the operations environment). Within the military, space systems are becoming crucial to the battlefield. Space is evolving from a communications support function to the battle commander to a real-time sensing, tasking, and battlefield assessment function directly to the war fighter. With the increase in the utility of space and its direct effect on the battle, space missions of the future must be operationally responsive and must operate in a "mission-assured" mode. The implications for L&EO are just beginning to emerge. It appears that military space mission planners are seeking missions that can be readied for launch in "hours" and initialized in orbit in short order as well. The technologies and operational methods to achieve these are being conceived. Use of the more robust onboard processing capabilities of today's satellite will undoubtedly be a major enabler. Regardless, the task remains fundamentally the same. The mission systems must be analyzed and the initialization process must be defined across space systems, ground systems, and people in order to execute the sequence. It may happen faster and it may happen in a more automated scheme, but the planning must be performed and built into the systems, people, and processes. L&EO will remain a critical phase because of the dynamics of launch, insertion into orbit, and initialization of systems for the first time in the space environment. Good planning will mitigate the risk of failure.

References

AlliedSignal Technical Services Corporation. April, 1997. "An Integrated Product Team Approach for Flight Operations;" Conference on Reducing the Cost of Satellite Operations; Oxford, England.

National Aeronautics and Space Administration. May, 1989. "Cosmix Background Explorer Launch and Early Orbit Operations Development Plan." NASA Goddard Space Flight Center. Greenbelt, MD.

National Aeronautics and Space Administration. July, 1992. "SAMPEX Flight Operations Plan." NASA Goddard Space Flight Center, Greenbelt, MD.

National Aeronautics and Space Administration. September, 1993. "Small Explorer Multi-Mission Assessment." NASA Goddard Space Flight Center. Greenbelt, MD.

Chapter 10

Space Navigation and Maneuvering*

Daryl G. Boden, *United States Naval Academy*
Paul Graziani, *Analytical Graphics, Inc.*

> 10.1 Keplerian Orbits
> 10.2 Orbit Perturbations
> 10.3 Orbit Maneuvering
> 10.4 Launch Windows
> 10.5 Orbit Maintenance
> 10.6 Interplanetary Trajectories
> 10.7 Mission Geometry

In this chapter we provide the tools to calculate some basic parameters for orbit trajectory and mission geometry. We use these parameters to plan for and analyze the navigation function, as described in Sec. 3.5. We also present basic orbital theory, define important terms, and provide basic equations that allow us to calculate many orbital parameters by hand. We begin by explaining terms used to describe orbits and presenting equations for calculating orbital parameters. The next four sections describe aspects of Earth-orbiting spacecraft: orbit perturbations, orbit maneuvering, launch-window calculations, and orbit maintenance. We next describe interplanetary trajectories and a way to estimate velocity budgets for such trajectories. Finally, we describe parameters of orbit geometry for spacecraft in Earth orbit. Several textbooks on satellite orbits and celestial mechanics are available. Some of the most popular are Bate, Mueller, and

---

* Sections 10.1 through 10.5 have been adapted with permission from Larson and Wertz [1999].

White [1971], Battin [1987], Danby [1962], Escobal [1965], Kaplan [1976], Roy [1978], and Vallado [2001].

10.1 Keplerian Orbits

Explaining the motion of celestial bodies, especially the planets, has challenged observers for many centuries. The early Greeks attempted to describe the motion of celestial bodies about the Earth in terms of circular motion. In 1543, Nicolaus Copernicus proposed a heliocentric (Sun-centered) system with the planets following circular orbits. Later, with the help of Tycho Brahe's observational data, Johannes Kepler described elliptical planetary orbits about the Sun. Finally, Isaac Newton solved this system mathematically, basing it on an inverse-square gravitational force.

Kepler spent several years reconciling the differences between Tycho Brahe's careful observations of the planets and their predicted motion based on previous theories. Having found that the data matched a geometric solution of elliptical orbits, he published his first two laws of planetary motion in 1609 and his third law in 1619. Kepler's three laws of planetary motion (which also apply to spacecraft orbiting the Earth) are:

First Law: The orbit of each planet is an ellipse, with the Sun at one focus.

Second Law: The line joining the planet to the Sun sweeps out equal areas in equal times.

Third Law: The square of the period of a planet (its year) is proportional to the cube of its mean distance from the Sun.

10.1.1 Equations of Motion for Satellites

Figure 10.1 depicts the key parameters of an elliptical orbit. The *eccentricity*, e, of the ellipse (not shown in the figure) is equal to c/a; it measures the deviation of the ellipse from a circle.

Isaac Newton explained mathematically why the planets (and satellites) follow elliptical orbits. Newton's Second Law of Motion, applied to a system with constant mass, combined with his Law of Universal Gravitation, provides the mathematical basis for analyzing satellite orbits. Newton's law of gravitation states that any two bodies attract each other with a force proportional to the product of their masses and inversely proportional to the square of the distance between them. The equation for the magnitude of the force due to gravity, F, is

$$F = -GMm/r^2$$
$$\equiv -\mu m/r^2 \qquad (10.1)$$

where G is the universal constant of gravitation, M is the mass of the Earth, m is the mass of the satellite, r is the distance from the center of the Earth to the satellite,

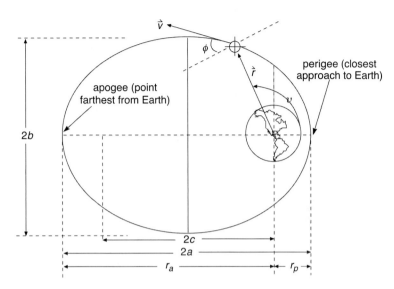

Fig. 10.1. **Geometry of an Ellipse and Orbital Parameters.** ϕ = flight-path angle, the angle between the velocity vector and a line perpendicular to the position vector; $\vec{r}$ = position vector of satellite relative to the center of the Earth; $\vec{v}$ = velocity vector of satellite relative to the center of the Earth; a = semi-major axis of the ellipse; b = semi-minor axis of the ellipse; c = the distance from the center of the orbit to one of the focii; v = the polar angle of the ellipse, measured in the direction of motion from perigee to the position vector. (r_a = radius of apogee; and r_p = radius of perigee.)

and $\mu \equiv GM$. The Earth's gravitational constant, μ, is 398,600.5 km^{3}s^{-2}. Combining Newton's second law with his law of gravitation, we get an equation for the satellite's acceleration vector:

$$\frac{d^2\vec{r}}{dt^2} + (\mu r^{-3})\vec{r} = \vec{0} \tag{10.2}$$

This equation, called the *2-body equation of motion*, is the relative equation of motion of a satellite's position vector as the satellite orbits the Earth. In deriving it, we assumed gravity is the only force, the Earth is spherically symmetrical, the Earth's mass is much greater than the satellite's mass, and the Earth and satellite are the only two bodies in the system.

A solution to the two-body equation of motion for a satellite orbiting the Earth is the *polar equation of a conic section*. It gives the magnitude of the position vector in terms of the satellite's location in the orbit,

$$r = a(1 - e^2) / (1 + e \cos v) \tag{10.3}$$

where a is the semi-major axis, e is the eccentricity, and v is the polar angle or true anomaly, as shown in Fig. 10.1.

A *conic section* is a curve formed by passing a plane through a right circular cone. As Fig. 10.2 shows, the angular orientation of the plane relative to the cone determines whether the conic section is a *circle, ellipse, parabola,* or *hyperbola*. We can define all conic sections in terms of the eccentricity, e, in Eq. (10.3) above. The type of conic section is also related to the semi-major axis, a, and the energy, ε (defined below). Table 10.1 shows how energy, eccentricity, and semi-major axis relate to the type of conic section.

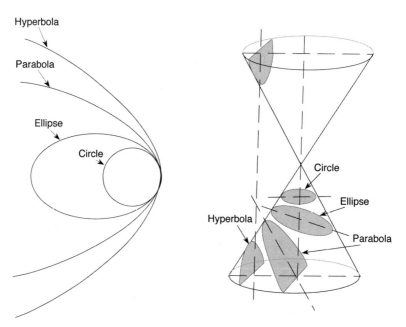

Fig. 10.2. **Satellite Orbits Can Be Any of Four Conic Sections: a Circle, an Ellipse, a Parabola, or a Hyperbola.** We'll use these terms in subsequent sections.

Table 10.1. **Conic Sections.** These conic sections can be described in terms of energy, size, and shape.

Conic	Energy, ε	Semi-Major Axis, a	Eccentricity, e
circle	< 0	= radius	0
ellipse	< 0	> 0	0 < e < 1
parabola	0	∞	1
hyperbola	> 0	< 0	> 1

10.1.2 Constants of Motion

Using the two-body equation of motion, we can derive several constants of motion for a satellite orbit. The first is

$$\varepsilon = V^2/2 - \mu/r = -\mu/(2a) \tag{10.4}$$

where V is the satellite velocity, and ε is the total *specific mechanical energy*, or mechanical energy per unit mass, for the system. It's also the sum of the kinetic energy per unit mass and potential energy per unit mass. Equation (10.4) is referred to as the *energy equation*. Because the forces in the system are conservative, the energy is a constant. The term for potential energy, $-\mu/r$, defines the potential energy to be zero at infinity and negative at any finite radius. Using this definition, we find the specific mechanical energy of elliptical orbits will always be negative. As the energy increases (approaches zero), the ellipse gets larger, and the elliptical trajectory approaches a parabolic trajectory. From Eq. (10.4), we find that the satellite moves fastest at the orbit's perigee and slowest at apogee.

We also know that for a circle the semi-major axis equals the radius, which is constant. Using the energy equation, we discover the velocity of a satellite in a circular orbit is

$$V_{cs} = (\mu/r)^{1/2} \tag{10.5}$$

or, for orbit around the Earth,

$$\cong 7.905\,366\,(R_E/r)^{1/2}$$

$$\cong 631.3481\,r^{-1/2}$$

where V_{cs} is the circular velocity in km/s, R_E is the radius of the Earth in km, and r is the orbit radius in km.

From Table 10.1, the energy of a parabolic trajectory is zero. A parabolic trajectory is one with the minimum energy needed to escape the Earth's gravitational attraction. Thus, we can calculate the velocity required to escape from the Earth at any distance, r, by setting energy equal to zero in Eq. (10.4) and solving for velocity:

$$V_{esc} = (2\mu/r)^{1/2} \tag{10.6}$$

$$\cong 11.179\,88\,(R_E/r)^{1/2}$$

$$\cong 892.8611\,r^{-1/2}$$

where V_{esc} is the escape velocity in km/s, and r is in km.

Another quantity associated with a satellite orbit is the *specific angular momentum*, h, which is the satellite's total angular momentum divided by its mass. We can find it from the cross product of the position and velocity vectors.

$$\vec{h} = \vec{r} \times \vec{v} \tag{10.7}$$

From Kepler's second law, we see that the angular momentum is constant in magnitude and direction for the two-body problem. Therefore, the plane of the orbit defined by the position and velocity vectors must remain fixed in inertial space.

10.1.3   Classical Orbital Elements

When solving the two-body equations of motion, we need six constants of integration (initial conditions). Theoretically, we could find the three components each of position and velocity at any time in terms of the position and velocity at any other time. Alternatively, we can completely describe the orbit with five constants and one quantity which varies with time. These quantities, called *classical orbital elements*, are defined below and are shown in Fig. 10.3. The coordinate frame shown in the figure is the geocentric inertial frame, or GCI. Its origin is at the center of the Earth, with the X axis in the equatorial plane and pointing to the vernal equinox. The Z axis is parallel to the Earth's spin axis (the North Pole), and the Y axis completes the right-hand set in the equatorial plane.

The classical orbital elements are:

- a: *semi-major axis*: describes the size of the ellipse (see Fig. 10.1).
- e: *eccentricity*: describes the shape of the ellipse.
- i: *inclination*: the angle between the angular momentum vector and the unit vector in the Z-direction.
- Ω: *right ascension of ascending node*: the angle from the vernal equinox to the ascending node. The *ascending node* is the point where the satellite passes through the equatorial plane moving from south to north. Right ascension is measured as a right-handed rotation about the pole, Z.
- ω: *argument of perigee*: the angle from the ascending node to the eccentricity vector measured in the direction of the satellite's motion. The *eccentricity vector* points from the center of the Earth to perigee with a magnitude equal to the eccentricity of the orbit.

---

* A *sufficiently inertial coordinate frame* is a coordinate frame that can be considered to be non-accelerating for the particular application. The GCI frame is sufficiently inertial for Earth-orbiting satellites but is inadequate for interplanetary travel because of its rotational acceleration around the Sun.

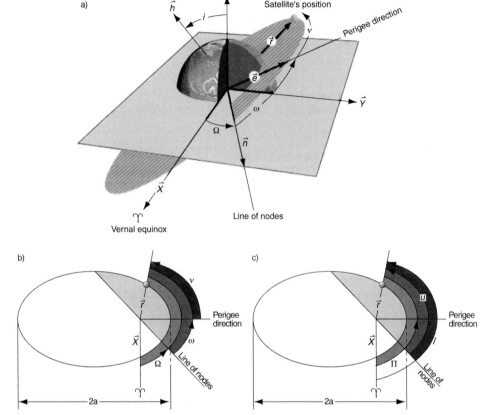

Fig. 10.3. Definition of the Keplerian Orbital Elements of a Satellite in an Elliptic Orbit. Elements are defined relative to the geocentric inertial coordinate frame. ($\vec{e}$ = shape of the orbit; i = orientation of the orbit with respect to the Earth's equator; ω = the low point, perigee, of the orbit is with respect to the Earth's surface; Ω = location of the ascending and descending orbit locations with respect to the Earth's equatorial plane; v = satellite is within the orbit with respect to perigee; $\Pi = \Omega + \omega$ for equatorial orbits; $u = \omega + v$ for circular orbits; l = defines the position of the satellite with respect to the x-axis for circular, equatorial orbits)

v: *true anomaly*: the angle from the eccentricity vector to the satellite position vector, measured in the direction of satellite motion. Alternately, we could use *time since perigee passage, T*.

Equatorial ($i = 0$) and circular ($e = 0$) orbits demand alternate orbital elements (Fig. 10.3) to solve the equations in Table 10.2. For equatorial orbits, a single angle, Π, can replace the right ascension of ascending node and argument of perigee. Called the *longitude of perigee*, this angle is the algebraic sum of Ω and ω. As i

approaches 0, Π approaches the angle from the X-axis to perigee. For circular orbits ($e = 0$), a single angle, $u \equiv \omega + \nu$, can replace the argument of perigee and true anomaly. This angle is the *argument of latitude*; when $e = 0$, it equals the angle from the nodal vector to the satellite's position vector. Finally, if the orbit is both circular and equatorial, a single angle, l, or *true longitude*, specifies the angle between the X-axis and the satellite's position vector.

Given these definitions, we can solve for the elements if we know the satellite's position and velocity vectors. Equations (10.4) and (10.7) allow us to solve for the energy and the angular momentum vector. An equation for the *nodal vector*, n, in the direction of the ascending node is

$$\vec{n} = \vec{Z} \times \vec{h} \tag{10.8}$$

We can calculate the eccentricity vector from the following equation:

$$\vec{e} = (1/\mu)\{(v^2 - \mu/r)\vec{r} - (\vec{r} \cdot \vec{v})\vec{v}\} \tag{10.9}$$

Table 10.2 lists equations to derive the classical orbital elements and related parameters for an elliptical orbit.

Table 10.2. Classical Orbital Elements. For the right ascension of ascending node, argument of perigee, and true anomaly, if the quantities in parentheses are positive, use the angle calculated. If the quantities are negative, use 360° minus the angle calculated.

Symbol	Name	Equation	Check
a	semi-major axis	$a = -\mu/(2\varepsilon) = (r_a + r_p)/2$	
e	eccentricity	$e = |\vec{e}| = 1 - r_p/a = r_a/a - 1$	
i	inclination	$i = \cos^{-1}(h_z/h)^*$	
Ω	right ascension of ascending node	$\Omega = \cos^{-1}(n_x/n)^*$	$(n_Y > 0)$
ω	argument of perigee	$\omega = \cos^{-1}[(\vec{n} \cdot \vec{e})/(n \cdot e)]$	$(e_z > 0)$
ν	true anomaly	$\nu = \cos^{-1}[(\vec{e} \cdot \vec{r})/(e \cdot r)]$	$(\vec{r} \cdot \vec{v} > 0)$
r_p	radius of perigee	$r_p = a(1 - e)$	
r_a	radius of apogee	$r_a = a(1 + e)$	
P	period	$P = 2\pi(a^3/\mu)^{1/2}$ $\approx 84.489\,(a/R_E)^{3/2}$ min $\approx 0.000\,165\,87 a^{3/2}$ min, a in km	
ω_o	orbit frequency	$\omega_o = (\mu/a^3)^{1/2}$ $\approx 631.348\,16\,a^{-3/2}$ rad/s, a in km	

* h_z = component of $\vec{h}$ in $\vec{z}$ direction; n_x = component of $\vec{n}$ in $\vec{x}$ direction.

10.1.4 Satellite Ground Tracks

A satellite's ground track is the trace of the points formed by the intersection of the satellite's position vector with the Earth's surface. In this section we will evaluate ground tracks using a flat map of the Earth.

Although ground tracks are generated from the satellite's orbital elements, we can gain insight by determining the orbital elements from a given ground track. Figure 10.4 shows ground tracks for satellites with different orbital altitudes and, therefore, different orbital periods. The time it takes for the Earth to rotate through the difference in longitude between two successive ascending nodes equals the orbital period. For *direct* orbits, in which the satellite moves eastward, we measure the change positive to the East. For *retrograde* orbits, in which the satellite moves westward, positive is measured to the West.* With these definitions in mind, the period, P, in minutes is

$$P = 4 (360° - \Delta L) \quad \text{direct orbit} \qquad (10.10)$$

$$P = 4 (\Delta L - 360°) \quad \text{retrograde orbit}$$

where ΔL is the longitudinal change in degrees that the satellite goes through between successive ascending nodes. The difference in longitude between two successive ascending nodes for a direct orbit will always be less than 360° and, in fact, will be negative for orbits at altitudes higher than geosynchronous altitude. For retrograde orbits, the difference in longitude between two successive ascending nodes (positive change is measured to the West) is always greater than 360°.

Once we know the period, we can determine the semi-major axis in km by using the equation for the period of an elliptical orbit:

$$a = [(P/2\pi)^2 \mu]^{1/3} \qquad (10.11)$$

$$\cong 331.249\ 15\ P^{2/3}\ \text{km}$$

where the period is in minutes.

Figure 10.4 shows one revolution each for the ground tracks of several orbits with an increasing semi-major axis. The period of a *geosynchronous* orbit (track E in the figure) is 1436 minutes, matching the Earth's rotational motion.

We can determine the orbit's inclination by the ground track's maximum latitude. For direct orbits, the inclination is equal to the ground track's maximum latitude, and for retrograde orbits, the inclination is equal to 180° minus the ground track's maximum latitude.

---

* These convenient empirical definitions don't apply for nearly polar orbits. More formally, a prograde or direct orbit has $i < 90°$. A retrograde orbit has $i > 90°$. A polar orbit has $i = 90°$.

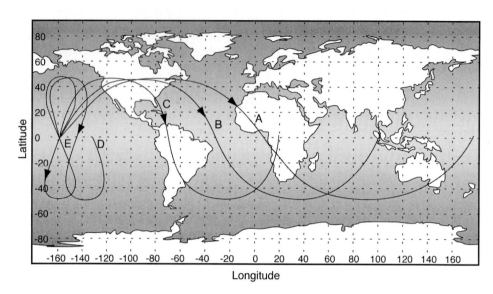

Fig. 10.4. Orbital Ground Tracks of Circular Orbits with Different Periods. (A) $\Delta L = 335°$, $P = 100$ min; (B) $\Delta L = 260°$, $P = 400$ min; (C) $\Delta L = 180°$, $P = 720$ min; (D) $\Delta L = 28°$, $P = 1328$ min; and (E) $\Delta L = 0°$, $P = 1436$ min. [Sellers, 1994]

The orbit is circular if a ground track is symmetrical about both the equator and a line of longitude extending down from the ground track's maximum latitude. The orbits in Fig. 10.4 are circular orbits.

Figure 10.5 shows examples of ground tracks for the following orbits:

A: Shuttle parking orbit, $a = 6700$ km, $e = 0$, $i = 28.4°$;

B: Low-altitude retrograde, $a = 6700$ km, $e = 0$, $i = 98.0°$;

C: GPS orbit, $a = 26{,}600$ km, $e = 0$, $i = 60.0°$; and

D: Molniya orbits, $a = 26{,}600$ km, $e = 0.75$, $i = 63.4°$, $\omega = 270°$.

10.1.5 Time-of-Flight in an Elliptical Orbit

In analyzing Brahe's observational data, Kepler was able to solve the problem of relating position in the orbit to the elapsed time, $t - t_0$, or conversely, how long it takes to go from one point in an orbit to another. To do so, Kepler introduced the quantity M, called the *mean anomaly* and expressed as an angle, which is the fraction of an orbit period that has elapsed since perigee. The mean anomaly equals the true anomaly for a circular orbit. By definition,

$$M - M_0 \equiv n(t - t_0) \tag{10.12}$$

10.1 Space Navigation and Maneuvering

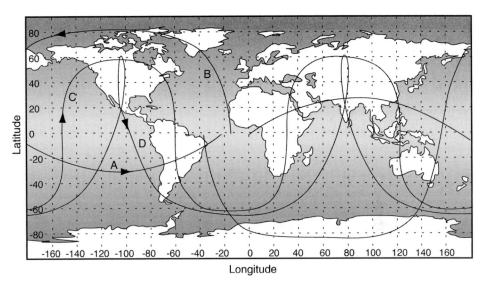

Fig. 10.5. **Typical Ground Tracks.** (A) Shuttle Parking, (B) Low-Altitude Retrograde, (C) GPS, and (D) Molniya Orbits. See text for orbital elements. [Sellers, 1994]

where M_0 is the mean anomaly at time t_0, and n is the *mean motion*, or average angular velocity, determined from the semi-major axis of the orbit:

$$n \equiv (\mu / a^3)^{1/2} \tag{10.13}$$

$$\cong 36{,}173.585\, a^{-3/2} \text{ deg/s}$$

$$\cong 8{,}681{,}660.4\, a^{-3/2} \text{ rev/day}$$

$$\cong 3.125\,297\,7 \times 10^9\, a^{-3/2} \text{ deg/day}$$

where a is in km.

This solution will give the average position and velocity, but satellite orbits are often elliptical, with a radius constantly varying in orbit. Because the satellite's velocity depends on this varying radius, it changes as well. To resolve this problem, we can define an intermediate variable called *eccentric anomaly*, E, for elliptical orbits. Table 10.3 lists the equations relating time-of-flight to orbital position.

As an example, let's find the time it takes a satellite to go from perigee to an angle 90° from perigee, for an orbit with a semi-major axis of 7000 km and an eccentricity of 0.1. For this example,

$v_0 = E_0 = M_0 = 0.0$ rad $\quad\quad t_0 = 0.0$ s
$v = 1.5708$ rad $\quad\quad\quad\quad\quad E = 1.4706$ rad
$M = 1.3711$ rad $\quad\quad\quad\quad\quad n = 0.001\,08$ rad/s
$t = 1271.88$ s

Table 10.3. Time-of-Flight in an Elliptical Orbit. All angular quantities are in radians. The equation for the true anomaly involves the first terms of the series expansion.

Variable	Name	Equation	
n	mean motion	$n = (\mu / a^3)^{1/2}$	
		$\approx 631.348\,16\ a^{-3/2}$ rad/s	(a in km)
E	eccentric anomaly	$\cos E = (e + \cos v) / (1 + e \cos v)$	
M	mean anomaly	$M = E - e \sin(E)$	(M in rad)
		$M = M_o + n(t - t_o)$	(M in rad)
$t - t_o$	time-of-flight	$t - t_o = (M - M_o)/n$	($t - t_o$ in s)
v	true anomaly	$v \approx M + 2e \sin M + 1.25 e^2 \sin(2M)$	

Finding the position in an orbit after a specified period is more complex. For this problem, we calculate the mean anomaly, M, using time-of-flight and the mean motion, using Eq. (10.12). Next, we determine the true anomaly, v, using the series expansion shown in Table 10.3, a good approximation for small eccentricity (the error is of the order e^3). If we need greater accuracy, we must solve the equation in Table 10.3, relating mean anomaly to eccentric anomaly. Because this is a transcendental function, we must iterate to find the eccentric anomaly, after which we can calculate the true anomaly directly.

10.1.6 Orbit Determination

Up to this point, we've assumed we know the satellite's position and velocity in inertial space—the classical orbital elements. But we often cannot observe the satellite's inertial position and velocity directly. Instead, we commonly receive data from radar, telemetry, optics, or the Global Positioning System (GPS). Radar and telemetry data consist of range, azimuth, elevation, and possibly the rates of change of one or more of these quantities, relative to a site attached to the rotating Earth. GPS provides the range and range-rate relative to a set of satellites. Optical data consist of right ascension and declination relative to the celestial sphere. In any case, we must combine and convert this data to inertial position and velocity before determining the orbital elements. Bate, Mueller, and White [1971] and Escobal [1965] cover methods for combining data, so we don't cover them here.

The type of data we use for orbit determination depends on the orbit selected, accuracy requirements, and weight restrictions on the payload. Because radar and optical systems collect data passively, they require no additional payload weight, but they are also the least accurate methods of orbit determination. Conversely, GPS data is more accurate but it requires additional payload weight. We can also use it for semi-autonomous orbit determination because it requires no ground support.

10.2 Orbit Perturbations

The Keplerian orbits discussed above provide an excellent reference, but other forces act on the satellite to perturb it from the nominal orbit. We can classify these perturbations, or variations in the orbital elements, based on how they affect the Keplerian elements.

Figure 10.6 illustrates a typical variation in one of the orbital elements arising from a perturbing force. *Secular variations* represent a linear variation in the element. *Short-period variations* are periodic in the element with a period less than or equal to the orbital period. *Long-period variations* are those with a period greater than the orbital period. Because secular variations have long-term effects on orbit prediction (the orbital elements affected continue to increase or decrease), we'll discuss them in detail. If the spacecraft mission demands that we determine the orbit precisely, we must include the periodic variations as well. Battin [1987], Danby [1962], and Escobal [1965] describe methods of determining and predicting orbits for non-Keplerian motion.

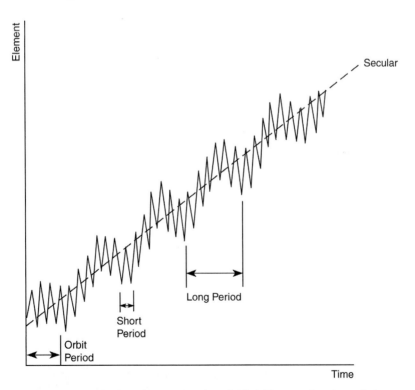

Fig. 10.6. **Secular and Periodic Variations of an Orbital Element.** Secular variations represent linear variations in the element, short-period variations have a period less than the orbital period, and long-period variations have a period longer than the orbital period.

Perturbing forces cause the classical orbital elements to vary with time. To predict the orbit we must determine this time variation using techniques of either special or general perturbations. *Special perturbations* employ direct numerical integration of the equations of motion. Most common is Cowell's method, in which the accelerations are integrated directly to obtain velocity and again to obtain position.

General perturbations solve analytically some aspects of the motion of a satellite subjected to perturbing forces. For example, the polar equation of a conic applies to the two-body equations of motion. Unfortunately, most perturbing forces yield not to a direct analytical solution but to series expansions and approximations. Because the orbital elements are nearly constant, general perturbation techniques usually solve directly for the orbital elements rather than for the inertial position and velocity. They are more difficult and approximate, but they allow us to understand better how the perturbations affect a large class of orbits. We can also get solutions much faster than with special perturbations.

The primary forces which perturb a satellite orbit arise from third bodies such as the Sun and the Moon, the nonspherical mass distribution of the Earth, atmospheric drag, and solar-radiation pressure. We describe each of these.

10.2.1 Third-Body Perturbations

The gravitational forces of the Sun and the Moon cause periodic variations in all of the orbital elements, but only the right ascension of ascending node, argument of perigee, and mean anomaly experience secular variations. These secular variations arise from a gyroscopic precession of the orbit about the ecliptic pole. The secular variation in mean anomaly is much smaller than the mean motion and has little effect on the orbit; however, the secular variations in right ascension of the ascending node and argument of perigee are important, especially for high-altitude orbits.

For nearly circular orbits, e^2 is near zero, and the resulting error is of the order e^2. In this case, the equations for the secular rates of change resulting from the Sun and Moon are

right ascension of ascending node:

$$\dot{\Omega}_{MOON} = -0.003\,38\,(\cos i)/n \tag{10.14}$$

$$\dot{\Omega}_{SUN} = -0.001\,54\,(\cos i)/n \tag{10.15}$$

argument of perigee:

$$\dot{\omega}_{MOON} = 0.001\,69\,(4 - 5\sin^2 i)/n \tag{10.16}$$

$$\dot{\omega}_{SUN} = 0.000\,77\,(4 - 5\sin^2 i)/n \tag{10.17}$$

where i is the orbit inclination, n is the number of orbit revolutions per day, and $\dot{\Omega}$ and $\dot{\omega}$ are in deg/day. These equations are only approximate; they omit the variation caused by the changing orientation of the orbital plane with respect to both the Moon's orbital plane and the ecliptic plane.

10.2.2 Perturbations Because of a Nonspherical Earth

When developing the two-body equations of motion, we assumed the Earth was a spherically symmetrical, homogeneous mass. In fact, the Earth is neither homogeneous nor spherical. The most dominant features are a bulge at the equator, a slight pear shape, and flattening at the poles. For a potential function of the Earth, Φ, we can find a satellite's acceleration by taking the gradient of the potential function. One widely used form of the geopotential function is

$$\Phi = (\mu/r)[1 - \sum J_n (R_E/r)^n P_n(\sin L)] \tag{10.18}$$

where $\mu \equiv GM$ is the Earth's gravitational constant, R_E is the Earth's equatorial radius, P_n are Legendre polynomials, L is the geocentric latitude, and J_n are the dimensionless geopotential coefficients, of which the first several are

$J_2 = 0.00108263$
$J_3 = -0.00000254$
$J_4 = -0.00000161$

This form of the geopotential function depends on latitude, and the geopotential coefficients, J_n, are called the *zonal coefficients*. Other, more general expressions for the geopotential include sectoral and tesseral terms in the expansion. The *sectoral terms* divide the Earth into slices and depend only on longitude. The *tesseral terms* in the expansion depend on both longitude and latitude. They divide the Earth into a checkerboard pattern of regions that alternately add to and subtract from the two-body potential.

The potential generated by the non-spherical Earth causes periodic variations in all of the orbital elements. But the dominant effects are secular variations in right ascension of ascending node and argument of perigee because of the Earth's oblateness, represented by the J_2 term in the geopotential expansion. The rates of change of Ω and ω due to J_2 are

$$\dot{\Omega}_{J_2} = -1.5 \, n \, J_2 \, (R_E/a)^2 (\cos i)(1 - e^2)^{-2} \tag{10.19}$$

$$\cong -2.06474 \times 10^{14} a^{-7/2} (\cos i)(1 - e^2)^{-2}$$

$$\dot{\omega}_{J_2} = 0.75 n \, J_2 \, (R_E/a)^2 (4 - 5 \sin^2 i)(1 - e^2)^{-2} \tag{10.20}$$

$$\cong 1.03237 \times 10^{14} a^{-7/2} (4 - 5 \sin^2 i)(1 - e^2)^{-2}$$

where n is the mean motion in deg/day, R_E is the Earth's equatorial radius in km, a is the semi-major axis in km, e is the eccentricity, i is the inclination, and $\dot{\Omega}$ and $\dot{\omega}$ are in deg/day. Table 10.4 compares the rates of change of right ascension of ascending node and argument of perigee resulting from the Earth's oblateness, the Sun, and the Moon.

Table 10.4. **Secular Variations in Right Ascension of the Ascending Node and Argument of Perigee.** For spacecraft in geostationary orbit (GEO) and below, the J_2 perturbations dominate; for spacecraft above GEO, the Sun and Moon perturbations dominate.

Orbit	Effect of J_2 (Eqs. 10.19, 10.20) (deg/day)	Effect of Moon (Eqs. 10.14, 10.16) (deg/day)	Effect of Sun (Eqs. 10.15, 10.17) (deg/day)
Shuttle	a = 6700 km, e = 0.0, i = 28°		
$\Delta\Omega$	−7.35	−0.00019	−0.00008
$\Delta\omega$	12.05	0.00242	0.00110
GPS	a = 26,600 km, e = 0.0, i = 60.0°		
$\Delta\Omega$	−0.033	−0.00085	−0.00038
$\Delta\omega$	0.008	0.00021	0.00010
Molniya	a = 26,600 km, e = 0.75, i = 63.4°		
$\Delta\Omega$	−0.30	−0.00076	−0.00034
$\Delta\omega$	0.00	0.00000	0.00000
Geostationary	a = 42,160 km, e = 0, i = 0°		
$\Delta\Omega$	−0.013	−0.00338	−0.00154
$\Delta\omega$	0.025	0.00676	0.00307

Molniya orbits are highly eccentric ($e \cong 0.75$) with approximately 12-hour periods (2 revolutions/day). The orbital inclination is chosen so the rate of change of the argument of perigee, Eq. (10.20), is zero. This condition occurs at inclinations of 63.4° and 116.6°. For these orbits, the argument of perigee is typically placed in the southern hemisphere, so the spacecraft remains above the northern hemisphere for approximately 11 hours/orbit. The perigee altitude is chosen to meet the mission constraints. Typical perigee altitudes vary from 200 to 1000 km. We can calculate the eccentricity and apogee altitude using the semi-major axis and perigee.

In a *Sun-synchronous orbit*, the spacecraft's orbital plane remains approximately fixed with respect to the Sun because we match the secular variation in the right ascension of ascending node [Eq. (10.19)] to the Earth's rate of rotation around the Sun. A nodal precession rate of 0.9856 deg/day will match the Earth's rate of average rotation about the Sun. Because this rotation is positive, Sun-synchronous orbits must be retrograde. For a given semi-major axis, a, and eccentricity, we can use Eq. (10.19) to find the inclination that will keep the orbit Sun-synchronous.

10.2.3 Perturbations from Atmospheric Drag

The principal nongravitational force acting on spacecraft in low-Earth orbit is atmospheric drag. Drag acts in a direction opposite to that of the velocity vector and removes energy from the orbit. This reduction of energy causes the orbit to get smaller, leading to further increases in drag. Eventually, the orbit's altitude becomes so small that the spacecraft reenters the atmosphere.

The equation for acceleration due to drag on a spacecraft is

$$a_D = -(1/2)\rho\,(C_D A/m)V^2 \qquad (10.21)$$

where ρ is the atmospheric density, C_D is the coefficient of drag ≈ 2.2, A is the spacecraft's cross-sectional area, m is the spacecraft's mass, and V is the spacecraft's velocity with respect to the atmosphere.

We can approximate the changes in semi-major axis, a, and eccentricity, e, per revolution, and the lifetime of a spacecraft in a circular orbit, using the following equations:

$$\Delta a_{rev} = -2\pi\,(C_D A/m)a^2\,\rho_p\exp(-c)\,[I_0 + 2eI_1] \qquad (10.22)$$

$$\Delta e_{rev} = -2\pi\,(C_D A/m)a\,\rho_p\exp(-c)\,[I_1 + e/2\,(I_0 + I_2)] \qquad (10.23)$$

where ρ_p is the atmospheric density at perigee, $c \equiv ae/H$, H is the density scale height, and I_i are Modified Bessel Functions* of order i and argument c. The term $m/(C_D A)$, or *ballistic coefficient*, is modelled as a constant for most spacecraft.

For circular orbits, we can use the above equations to derive the much simpler expressions:

$$\Delta a_{rev} = -2\pi\,(C_D A/m)\rho a^2 \qquad (10.24)$$

$$\Delta P_{rev} = -6\pi^2(C_D A/m)\rho a^2/V \qquad (10.25)$$

$$\Delta V_{rev} = \pi\,(C_D A/m)\rho a\,V \qquad (10.26)$$

$$\Delta e_{rev} = 0 \qquad (10.27)$$

where P is the orbit period, and V is the spacecraft's velocity.

---

* Tables of values for I_i can be found in many standard mathematical tables.

A rough estimate of the satellite's lifetime L, in revolutions, as affected by drag can be computed from

$$L \approx -H / \Delta a_{rev} \tag{10.28}$$

where, as above, H is the atmospheric density scale height.

10.2.4 Perturbations from Solar Radiation

Solar-radiation pressure causes periodic variations in all of the orbital elements. Its effect is strongest for spacecraft with low ballistic coefficients—light vehicles with large frontal areas such as Echo. The magnitude of the acceleration in m/s² arising from solar-radiation pressure at the Earth is

$$a_R \approx -4.5 \times 10^{-6} (A/m) \tag{10.29}$$

where A is the effective cross-sectional area of the spacecraft exposed to the Sun in m², and m is the spacecraft's mass in kg. For satellites below 800 km altitude, acceleration from atmospheric drag is greater than that from solar-radiation pressure; above 800 km, acceleration from solar-radiation pressure is greater.

10.3 Orbit Maneuvering

At some point during the lifetime of most spacecraft, we must change one or more of the orbital elements. For example, we may need to transfer from an initial parking orbit to the final mission orbit, rendezvous with or intercept another spacecraft, or correct the orbital elements to adjust for the perturbations discussed in the previous section. Most frequently, we must change the orbit's altitude, plane, or both. To change a spacecraft's orbit, we have to change the velocity vector in magnitude or direction. Most propulsion systems operate for only a short time compared to the orbital period, so we can treat the maneuver as an impulsive change in the velocity while the position remains fixed. For this reason, any maneuver changing a spacecraft's orbit must occur at a point where the old orbit intersects the new orbit. If the two orbits don't intersect, we have to use an intermediate orbit that intersects both. In this case, the total maneuver will require at least two propulsive burns.

In general, the change in the velocity vector to go from one orbit to another is given by

$$\Delta \bar{V} = \bar{V}_{NEED} - \bar{V}_{CURRENT} \tag{10.30}$$

We can find the current and needed velocity vectors from the orbital elements, keeping in mind that the position vector doesn't change much during impulsive burns.

10.3.1 Coplanar Orbit Transfers

The most common type of in-plane maneuver changes the orbit's size and energy, usually from a low-altitude parking orbit to a higher-altitude mission orbit such as a geosynchronous orbit. Because the initial and final orbit don't intersect (see Fig. 10.7), the maneuver requires a transfer orbit. Figure 10.7 represents a Hohmann[*] Transfer Orbit. In this case, the transfer orbit's ellipse is tangent to the initial circular orbit at the transfer orbit's perigee and to the final circular orbit at the transfer orbit's apogee. The orbits are tangential, so the velocity vectors are collinear, and the Hohmann transfer represents the most fuel-efficient transfer between two circular, coplanar orbits. When transferring from a smaller orbit to a larger orbit, the change in velocity is applied in the direction of motion; when transferring from a larger orbit to a smaller, the change of velocity is opposite to the direction of motion.

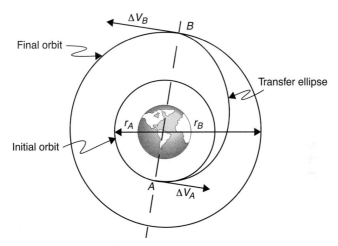

Fig. 10.7. **Hohmann-Transfer Ellipse Showing Orbit Transfer Between Two Circular, Coplanar Orbits.** The Hohmann Transfer is the minimum energy transfer between two circular, coplanar orbits.

The total change in velocity required for the transfer is the sum of the velocity changes at perigee and apogee of the transfer ellipse. Because the velocity vectors are collinear, the velocity changes are just the differences in magnitudes of the velocities in each orbit. If we know the size of each orbit we can find these differences from the energy equation:

---

[*] Walter Hohmann, a German engineer and architect, wrote *The Attainability of Celestial Bodies* [1925], consisting of a mathematical discussion of the conditions for leaving and returning to Earth.

$$\varepsilon = V^2/2 - \mu/r = -\mu/(2a) \qquad (10.4)$$

If we know the initial and final orbits (r_A and r_B), we can calculate the semi-major axis of the transfer ellipse, a_{tx}, and the total velocity change (the sum of the velocity changes required at points A and B) using the steps in Table 10.5. VtxA and VtxB in the table are respectively the velocities for perigee and apogee of the transfer orbit, points A and B in Fig. 10.7. The table also gives an example to illustrate this technique: transferring from an initial circular orbit of 6563 km to a final circular orbit of 42,159 km. As Fig. 10.7 shows, the transfer time is half the period of the transfer orbit.

Or we can write the total ΔV required for a two-burn transfer between circular orbits at altitude r_A and r_B as

$$\Delta v_{total} = \Delta v_A + \Delta v_B \qquad (10.31)$$

$$= \sqrt{\mu}\left[\left|\left(\frac{2}{r_A}-\frac{1}{a_{tx}}\right)^{\frac{1}{2}} - \left(\frac{1}{r_A}\right)^{\frac{1}{2}}\right| + \left|\left(\frac{2}{r_B}-\frac{1}{a_{tx}}\right)^{\frac{1}{2}} - \left(\frac{1}{r_B}\right)^{\frac{1}{2}}\right|\right] \qquad (10.32)$$

where $\sqrt{\mu} = 631.3481$, ΔV is in km/s and all of the semi-major axes are in km. As in step 1, $a_{tx} = (r_A + r_B)/2$.

The above expression applies to any coplanar Hohmann transfer. In the case of small transfers (r_A close to r_B), we can conveniently approximate it in two forms:

$$\Delta V \approx V_{iA} - V_{fB} \qquad (10.33)$$

$$\Delta V \approx 0.5(\Delta r/r)V_{A/B} \qquad (10.34)$$

where

$$\Delta r \equiv r_B - r_A \qquad (10.35)$$

and

$$r \approx r_A \approx r_B \qquad V_{A/B} \approx V_{iA} \approx V_{fB} \qquad (10.36)$$

The two small burns are of nearly equal magnitude.

The result in Eq. (10.33) is more unusual than it might at first seem. Assume that a spacecraft is in a circular orbit with velocity V_{iA}. In two burns we increase the velocity by an amount ΔV. The result is that the spacecraft is higher and traveling more slowly than before by the amount ΔV. An example will clarify this result. Consider a spacecraft in a circular orbit at 400 km such that $r_A = 6778$ km and $V_{iA} = 7700$ m/s. We'll apply a total ΔV of 20 m/s (= 0.26% of V_{iA}) in two burns of 10 m/s each. From Eq. (10.34) the total Δr will be 0.52% of 6778 km or 35 km. Thus, the final orbit will be circular at an altitude of 6813 km. Immediately following the first burn of 10 m/s, the spacecraft will be at perigee of the transfer orbit with a velocity of 7710 m/s. When the spacecraft reaches apogee at 6813 km, it will have slowed according to Kepler's second law by 0.52% to 7670 m/s. We then apply the second burn of 10 m/s to circularize the orbit at 7680 m/s which is 20 m/s slower than its original velocity. We've added energy to the spacecraft, which has raised the orbit and resulted in a lower kinetic energy but enough extra potential energy to make up for the reduced speed and the added ΔV.

Sometimes we need to transfer a spacecraft between orbits faster than the Hohmann transfer will allow. Figure 10.8 shows a faster transfer called the *One-Tangent-Burn*. In this case the transfer orbit is tangential to the initial orbit. It intersects the final orbit at an angle equal to the flight-path angle of the transfer orbit at the point of intersection. An infinite number of transfer orbits are tangential to the initial orbit and intersect the final orbit at some angle. Thus, we may choose the transfer orbit by specifying the size of the transfer orbit, the angular change of the transfer, or the time required to complete the transfer. We can then define the transfer orbit and calculate the required velocities.

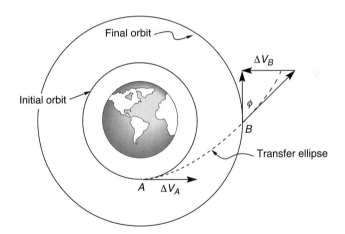

Fig. 10.8. **Transfer Orbit Using One-Tangent Burn between Two Circular, Coplanar Orbits.** The change in velocity, ΔV_B, for the second burn is greater than the ΔV_B for the Hohmann Transfer.

For example, we may specify the size of the transfer orbit, choosing any semi-major axis that is greater than the semi-major axis of the Hohmann-transfer ellipse. Once we know the semi-major axis of the ellipse (a_{tx}), we can use the equations in Table 10.5 to calculate the eccentricity, the angular distance traveled in the transfer, the velocity change required for the transfer, and the time required to complete the transfer.

Table 10.5. Computations for an Orbit Transfer Using a One-Tangent Burn. For example, see Battin [1987].

Quantity	Equation
Eccentricity	$e = 1 - r_A / a_{tx}$
True anomaly at second burn	$v = \cos^{-1}[(a(1 - e^2) / r_B - 1) / e]$
Flight-path angle at second burn	$\phi = \tan^{-1}[e \sin v / (1 + e \cos v)]$
Initial velocity	$V_{iA} = 631.3481 \, r_A^{-1/2}$
Velocity on transfer orbit at initial orbit	$V_{txA} = 631.3481 \, [2/r_A - 1/a_{tx}]^{1/2}$
Initial velocity change	$\Delta V_A = |V_{txA} - V_{iA}|$
Final velocity	$V_{fB} = 631.3481 \, r_B^{-1/2}$
Velocity on transfer orbit at final orbit	$V_{txB} = 631.3481[2/r_B - 1/a_{tx}]^{1/2}$
Final velocity change	$\Delta V_B = [V_{fB}^2 + V_{txB}^2 - 2V_{fB}V_{txB}\cos\phi]^{1/2}$
Total velocity change	$\Delta V_T = \Delta V_A + \Delta V_B$
Eccentric anomaly	$E = \tan^{-1}[(1 - e^2)^{1/2} \sin v / (e + \cos v)]$
Time-of-flight	$TOF = 0.001583913 \, a^{3/2} (E - e \sin E)$, E in rads

Table 10.6 compares the total required velocity change and time-of-flight for a Hohmann transfer and a one-tangent burn transfer from a low-altitude parking orbit to geosynchronous orbit. The semi-major axis a_{tx} of the transfer orbit for the one-tangent burn is 28,633 km.

Table 10.6. Comparison of Coplanar Orbit Transfers from Low-Earth Orbit to Geosynchronous Orbit. Spacecraft in repeating ground tracks will fly over the same locations on Earth within an integer number of days.

Variable	Hohmann Transfer	One-Tangent-Burn Transfer
r_A	6570 km	6570 km
r_B	42,200 km	42,200 km
a_{tx}	24,385 km	28,633 km
ΔV_{Total}	3.935 km/s	4.699 km/s
TOF	5.256 hr	3.457 hr

10.3 Space Navigation and Maneuvering

Another option for changing the size of the orbit is to use a constant-low-thrust burn, which results in a *spiral transfer*. We can approximate the velocity change for this type of orbit transfer by

$$\Delta V = |V_2 - V_1| \qquad (10.37)$$

where the velocities are the circular velocities of the two orbits. Following the example in Table 10.7, the total velocity change required to go from low-Earth orbit to geosynchronous is 4.72 km/s using a spiral transfer. We get this value by subtracting the results of step 3 from the results of step 2 in the table.

Table 10.7. Hohmann Transfer Calculations. A satellite is transferring from low-Earth orbit to geosynchronous orbit.

Step	Equations		Example
1	a_{tx}	$= (r_A + r_B)/2$	= 24,364 km
2	V_{iA}	$= (\mu/r_A)^{1/2} = 631.3481(r_A)^{-1/2}$	= 7.79 km/s
3	V_{fB}	$= (\mu/r_B)^{1/2}$	
		$= 631.3481(r_B)^{-1/2}$	= 3.08 km/s
4	V_{txA}	$= [\mu(2/r_A - 1/a_{tx})]^{1/2}$	
		$= 631.3481 [(2/r_A - 1/a_{tx})]^{1/2}$	= 10.25 km/s
5	V_{txB}	$= [\mu(2/r_B - 1/a_{tx})]^{1/2}$	
		$= 631.3481 [(2/r_B - 1/a_{tx})]^{1/2}$	= 1.59 km/s
6	ΔV_A	$= |V_{txA} - V_{iA}|$	= 2.46 km/s
7	ΔV_B	$= |V_{fB} - V_{txB}|$	= 1.49 km/s
8	ΔV_{TOTAL}	$= \Delta V_A + \Delta V_B$	= 3.95 km/s
9	Time of transfer = P/2		= 5 hrs 15 mins

10.3.2 Changes in the Orbital Plane

To change the orientation of the satellite's orbital plane, typically the inclination, we must change the direction of the velocity vector. This maneuver requires a component of ΔV to be perpendicular to the orbital plane and, therefore, perpendicular to the initial velocity vector. If the size of the orbit remains constant, the maneuver is called a *simple plane change* (Fig. 10.9a). We can find the required change in velocity by using the law of cosines.

$$\Delta V = (V_i^2 + V_f^2 - 2V_i V_f \cos\theta)^{1/2} \qquad (10.38)$$

For the case in which V_f is equal to V_i this expression reduces to

$$\Delta V = (2V_i^2 - 2V_i^2 \cos\theta)^{1/2}$$

$$= 2V_i[(1 - \cos\theta)/2]^{1/2}$$

$$\Delta V = 2V_i \sin(\theta/2) \tag{10.39}$$

where V_i is the velocity before and after the burn, and θ is the required angle change.

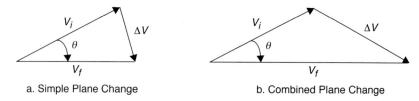

a. Simple Plane Change   b. Combined Plane Change

Fig. 10.9. **Vector Representation of Simple and Combined Changes in Orbital Plane.** For the simple plane change, the initial and final velocities are equal in magnitude.

For example, we calculate the change in velocity required to transfer from a low-altitude ($h = 185$ km), inclined ($i = 28°$) orbit to an equatorial orbit ($i = 0$) at the same altitude as follows:

$$r = 6563 \text{ km}$$

$$V_i = 631.3481 \, (6563)^{-1/2} = 7.79 \text{ km/s}$$

$$\Delta V = 2(7.79) \sin(28°/2) = 3.77 \text{ km/s}$$

From Eq. (10.39) we see that if the angular change is equal to 60°, the required change in velocity is equal to current velocity. Plane changes are very expensive in terms of the required change in velocity and corresponding fuel consumption. To hold down cost, we should change the plane at a point where the spacecraft's velocity is lowest: at apogee for an elliptical orbit. In some cases, it may even be cheaper to boost the spacecraft into a higher orbit, change the orbit plane at apogee, and return the spacecraft to its original altitude.

Orbital transfers typically require changes in both the orbit's size and plane, such as transferring from an inclined parking orbit at low altitude to a zero-inclination orbit at geosynchronous altitude. We can do this transfer in two steps: a Hohmann transfer to change the size of the orbit and a simple plane change to make the orbit equatorial.

This involves adding to the ΔV_{Total} of the Hohmann transfer a ΔV to change the inclination. For example, let's find the total change in velocity to transfer from a Shuttle parking orbit ($r_i = 6563$ km; $i = 28°$) to a geosynchronous, equatorial orbit ($r_i = 42,159$ km; $i = 0°$). Table 10.5 gives the ΔV_{Total} of the Hohmann transfer (before the plane change) as 3.94 km/s. From Eq. (10.39), the plane change ΔV is:

$$\Delta V = 2V_i \sin(\theta/2)$$
$$= 2V_{fB} \sin(\theta/2)$$
$$= 2\,(3.07) \sin(28°/2)$$
$$= 1.49 \text{ km/s}$$

The total velocity change is therefore 3.94 km/s + 1.49 km/s = 5.43 km/s.

A more efficient method (less total change in velocity) would be to combine the plane change with the tangential burn at apogee of the transfer orbit. Continuing with our example, we boost the spacecraft into the transfer orbit with a $\Delta V_A = 2.46$ km/s (Table 10.5, step 6). The velocity at apogee, V_{txB}, is 1.59 km/s. At this point, we must change the velocity vector's magnitude and direction, finding the required change in velocity using the law of cosines (Fig. 10.9B):

$$\Delta V = (V_i^2 + V_f^2 - 2V_i V_f \cos\theta)^{1/2} \tag{10.40}$$

where V_i is the initial velocity of 1.59 km/s, V_f is the final velocity of 3.07 km/s, and θ is the required angle change of 28°. Putting these numbers into Eq. (10.40), we get

$$\Delta V = 1.83 \text{ km/s}$$

As we can see from this example, we can combine a small plane change ($\theta \approx 0$) with an energy change for almost no additional cost in ΔV or propellant. In practice, geosynchronous transfer uses a small plane change at perigee and most of the plane change at apogee.

Another option is to complete the maneuver using three burns. The first burn is a coplanar maneuver placing the spacecraft into a transfer orbit with an apogee much higher than the final orbit. A combined plane change follows when the spacecraft reaches apogee of the transfer orbit. This maneuver places the spacecraft in a second transfer orbit which is coplanar with the final orbit and has a perigee altitude equal to the altitude of the final orbit. Finally, when the spacecraft reaches perigee of the second transfer orbit, another coplanar maneuver places it into the final orbit. This three-burn maneuver may save fuel, but the fuel savings comes at the expense of the total time required to complete it.

10.3.3 Orbit Rendezvous

Orbital transfer becomes more complicated when the objective is to rendezvous with or intercept another object in space: both the interceptor and target must arrive at the rendezvous point at the same time. This precision demands a phasing orbit to do the maneuver. A *phasing orbit* is any orbit which results in the interceptor achieving the desired geometry relative to the target to start a Hohmann transfer. If the initial and final orbits are circular, coplanar, and of different sizes, the phasing orbit is simply the interceptor's initial orbit (Fig. 10.10). The interceptor remains in the initial orbit until the relative motion between the interceptor and target results in the desired geometry. At that point, we inject the interceptor into a Hohmann transfer orbit. The equation for the wait time in the initial orbit is

$$\text{Wait Time} = (\phi_i - \phi_f + 2k\pi)/(\omega_{int} - \omega_{tgt}) \qquad (10.41)$$

where ϕ_f is the phase angle (angular separation of target and interceptor) needed for rendezvous; ϕ_i is the initial phase angle; k is the number of rendezvous opportunities (for the first opportunity, $k = 0$); ω_{int} is the interceptor's angular velocity; and ω_{tgt} is the target's angular velocity. We calculate the lead angle, α_L, by multiplying ω_{tgt} by the time of flight for the Hohmann transfer; ϕ_f is 180° minus α_L.

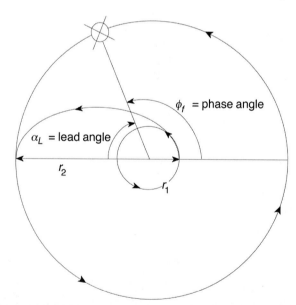

Fig. 10.10. **Geometry Depicting Rendezvous between Two Circular, Coplanar Orbits.** The phase angle is the angular separation between the target and interceptor at the start of the rendezvous, and the lead angle is the distance the target travels from the start until rendezvous occurs.

10.3 Space Navigation and Maneuvering

The total time to rendezvous is equal to the wait time from Eq. (10.41) plus the time-of-flight of the Hohmann transfer orbit.

The denominator in Eq. (10.41) represents the relative motion between the interceptor and target. As the size of the interceptor's orbit approaches the size of the target's orbit, the relative motion approaches zero, and the wait time approaches infinity. If the two orbits are exactly the same size, the interceptor must enter a new phasing orbit to rendezvous with the target (Fig. 10.11). For this situation, the rendezvous occurs at the point where the interceptor enters the phasing orbit. The period of the phasing orbit is equal to the time it takes the target to get to the rendezvous point. Once we know the period, we can calculate the semi-major axis. The two orbits are tangential at their point of intersection, so the change in velocity is the difference in magnitudes of the two velocities at the point where the two orbits intersect. Because we know the size of the two orbits, and therefore, the energies, we can use the energy equation, Eq. (10.4), to solve for the current and needed velocity.

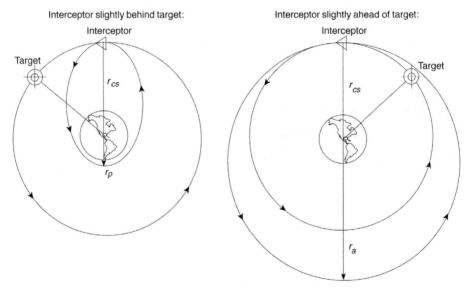

Fig. 10.11. Rendezvous from Same Orbit Showing the Target Leading and Trailing the Interceptor. The period of the phasing orbit is equal to the time it takes the target to travel to the rendezvous point.

We must often adjust the relative phasing of spacecraft in circular orbits. We do this by making the spacecraft drift relative to its initial position. The *drift rate* in deg/orbit for spacecraft in Earth orbit is

$$\text{drift rate} = 1080 \, \Delta V / V \qquad (10.42)$$

where V is the orbit's nominal velocity, and ΔV is the velocity change required to start or stop the drift.

10.4 Launch Windows

Similar to the rendezvous problem is the launch-window problem, or determining the appropriate time to launch from the Earth's surface into the desired orbital plane. Because the orbital plane is fixed in inertial space, the launch window is the time when the launch site on the Earth's surface rotates through the orbital plane. As Fig. 10.12 shows, the time of the launch depends on the launch site's latitude and longitude and the spacecraft orbit's inclination and right ascension of ascending node.

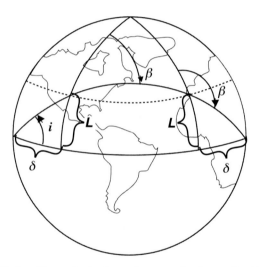

Fig. 10.12. **Launch Window Geometry for Launches near the Ascending Node and Descending Node.** The angles shown are the orbit's inclination (i), launch-site latitude (L), and launch azimuth (β). δ is the angle measured from the nearest node to the launch-site longitude.

For a launch window to exist, the launch site must pass through the orbital plane, placing restrictions on the orbital inclinations, i, that are possible from a given launch latitude, L:

- No launch windows exist if $L > i$ for direct orbit or $L > 180° - i$ for retrograde orbits
- One launch window exists if $L = i$ or $L = 180° - i$
- Two launch windows exist if $L < i$ or $L < 180° - i$

10.4 Space Navigation and Maneuvering

The *launch azimuth*, β, is the angle measured clockwise from north to the velocity vector. If a launch window exists, the launch azimuth required to achieve an inclination, i, from a given launch latitude, L, is

$$\beta = \beta_I \pm \gamma \approx \beta_I \tag{10.43a}$$

where

$$\sin \beta_I = \cos i / \cos L \tag{10.43b}$$

and

$$\tan \gamma = \frac{V_L \cos \beta_I}{V_0 - V_{eq} \cos i} \approx \left(\frac{V_L}{V_0}\right) \cos \beta_I \tag{10.43c}$$

where V_L is the launch site's inertial velocity as given by Eq. (10.47) below, V_{eq} = 464.5 m/s is the velocity of the Earth's rotation at the equator, and $V_0 \approx 7.8$ km/s is the velocity of the satellite immediately after launch. β_I is the inertial launch azimuth, and γ is a small correction to account for the velocity contributed by the Earth's rotation. For launches to low-Earth orbit, γ ranges from 0 for a due-east launch to 3.0° for launch to a polar orbit. The approximation for γ in Eq. (10.43c) is good to better than 0.1° for low-Earth orbits. For launches near the ascending node, β is in the first or fourth quadrant, and the plus sign applies in Eq. (10.43a). For launches near the descending node, β is the second or third quadrant, and the minus sign applies in Eq. (10.43a).

Let δ, shown in Fig. 10.12, be the angle in the equatorial plane from the nearest node to the longitude of the launch site. We can determine δ from

$$\cos \delta = \cos \beta / \sin i \tag{10.44}$$

where δ is positive for direct orbits and negative for retrograde orbits. Finally, the *local sidereal time*, LST, of launch is the angle at the time of launch from the vernal equinox to the longitude of the launch site:

$$\begin{aligned} LST &= \Omega + \delta & \text{(launch at ascending node)} \\ &= \Omega + 180° - \delta & \text{(launch at descending node)} \end{aligned} \tag{10.45}$$

where Ω is the resulting orbit's right ascension of the ascending node.

Having calculated the launch azimuth required to achieve the desired orbit, we can now calculate the velocity needed to accelerate the payload from rest at the launch site to the required burnout velocity. To do so, we use topocentric-horizon coordinates with velocity components V_S, V_E, V_Z:

$$V_S = -V_{bo} \cos \phi \cos \beta_b$$
$$V_E = V_{bo} \cos \phi \sin \beta_b - V_L$$

$$V_Z = V_{bo} \sin \phi \qquad (10.46)$$

where V_{bo} is the velocity at burnout (usually equal to the circular-orbit velocity at the prescribed altitude), ϕ is the flight-path angle at burnout, β_b is the launch azimuth at burnout, and V_L is the velocity of the launch site on the Earth at a given latitude, L, as given by

$$V_L = (464.5 \text{ m/s}) \cos L \qquad (10.47)$$

Equation (10.46) doesn't include losses in the launch vehicle's velocity due to atmospheric drag and gravity—approximately 1500 m/s for a typical launch vehicle. Also, in Eq. (10.46) we assume the azimuths at launch and at burnout are the same. Changes in the launch vehicle's latitude and longitude during powered flight will introduce small errors into calculating the burnout conditions. We can calculate the velocity required at burnout from the energy equation if we know the orbit's semi-major axis and radius of burnout.

10.5 Orbit Maintenance

Once in their mission orbits, many spacecraft need no orbital adjustments. But mission requirements may demand that we maneuver the spacecraft to correct the orbital elements when perturbing forces have changed them. Two particular cases of note are spacecraft with repeating ground tracks and geosynchronous-equatorial spacecraft.

Using two-body equations of motion, we can show that a spacecraft will have a repeating ground track if it has exactly an integral number of revolutions per integral number of days. Its period must therefore be

$$P = (m \text{ sidereal days})/(k \text{ revolutions}) \qquad (10.48)$$

where m and k are integers, and 1 sidereal day = 1436.068 min. For example, a spacecraft orbiting the Earth exactly 16 times per day will have a period of 89.75 min and a semi-major axis of 6640 km.

Next we modify the spacecraft's period to account for the drift in the orbital plane caused by the Earth's oblateness (J_2). We can calculate the rate of change of the right ascension of ascending node, $\Delta \Omega$ due to J_2 from the two-body orbital elements. In this case the new period is

$$P_{New} = P_{Two\text{-}body} + \Delta\Omega / \omega_{Earth} \qquad (10.49)$$

where ω_{Earth} is the angular velocity of the Earth.

Because the nodal drift is based on the two-body orbital elements, we must iterate to find the new orbital period and semi-major axis. Continuing with the previous example, assume a perigee altitude of 120 km and an inclination of 45°.

In this case, we find the compensated period is 88.20 min and the new semi-major axis is 6563 km.

Table 10.8 shows examples of spacecraft placed in orbits with repeating ground tracks.

Table 10.8. Examples of Spacecraft in Orbits with Repeating Ground Tracks.

Satellite	Semi-Major Axis (km)	Revs	Days
SEASAT	7168.3	43	3
LANDSAT 4/5	7077.8	233	16
GEOSAT	7173.6	244	17

The Earth's oblateness also causes the direction of perigee to rotate around the orbit. If the orbit isn't circular, and the mission limits the altitude over specific targets, we must control the location of perigee. One possibility is to select the orbital inclination to be at the critical inclination (63.4° for a direct orbit and 116.4° for a retrograde orbit), so the location of perigee is fixed. If other constraints make this selection impossible, we must maintain the orbit through orbital maneuvers. We can change the location of perigee by changing the flight-path angle by an angle θ. Only the direction of the velocity vector is changing, so we can find the change in velocity from the equation for a simple plane change:

$$\Delta V = 2V \sin \theta/2 \tag{10.50}$$

A final consideration for a low-altitude orbit with repeating ground tracks is the change in the semi-major axis and eccentricity due to atmospheric drag. Drag causes the orbit to become smaller. As the orbit becomes smaller, the period also shortens, causing the ground track to appear to shift eastward. If some tolerance is specified, such as a maximum distance between the actual and desired ground track, the satellite must periodically maneuver to maintain the desired orbit.

We can use Eq. (10.11) to calculate the change in semi-major axis per revolution of the orbit. Given the change in the orbit's size, we can also determine the change in the period:

$$\Delta P = 3\pi \Delta a / (na) \tag{10.51}$$

If there are constraints for the orbit's period or semi-major axis, we can use Eqs. (10.22) and (10.50) to keep track of them until the orbit needs to be corrected. Applying a tangential velocity change at perigee will adjust the semi-major axis when required. We can find the current and needed velocities from Eq.(10.4), because we know the sizes, and therefore the energies, of the two orbits.

Geosynchronous-equatorial orbits also require orbital maintenance. Spacecraft in these orbits drift when perturbations occur from the oblate Earth and from third-body interactions with the Sun and Moon. Matching the period of a geostationary orbit with the Earth's rotational velocity results in a resonance with the J_{22} term in the geopotential. This resonance term results in a transverse acceleration—an acceleration in the orbital plane—which causes the spacecraft to drift in longitude (*East-West drift*). The Sun and the Moon cause out-of-plane accelerations, which make the spacecraft drift in latitude (*North-South drift*).

North-South stationkeeping is necessary when mission requirements limit the drift in latitude or inclination. If not corrected, the orbit's inclination varies between 0° and 15° with a period of approximately 55 years. The approximate equations for the worst-case change in velocity in one year are

$$\Delta V_{Moon} = 102.67 \cos \alpha \sin \alpha \quad (m/s) \qquad (10.52)$$

$$\approx 36.93 \text{ m/s, for } i = 0$$

$$\Delta V_{Sun} = 40.17 \cos \gamma \sin \gamma \quad (m/s)) \qquad (10.53)$$

$$\approx 14.45 \text{ m/s, for } i = 0$$

where α is the angle between the orbital plane and the Moon's orbit, and γ is the angle between the orbital plane and the ecliptic.

The transverse acceleration caused by resonance with the J_{22} term results in periodic motion about either of two stable longitudes: 75° and 255° East. If a spacecraft is placed at any other longitude, it will tend to orbit the closer of these two longitudes, resulting in East-West drift of up to 180° with periods of up to 900 days. Suppose a mission for a geostationary spacecraft specifies a required longitude, l_D. The yearly change in velocity required to compensate for the drift and maintain the spacecraft near the specified longitude is

$$\Delta V = 1.715 \ |\sin(2(l_D - l_S))| \qquad (10.54)$$

where l_D is the desired longitude, l_S is the closest stable longitude, and ΔV is in m/s.

For example, we can find the velocity change required for one year if the desired longitude is 60° west:

$$l_D = -60° \qquad l_S = 255°$$
$$\Delta V = 1.715 \text{ m/s}$$

After the spacecraft's mission is complete, we have several options, depending on the orbit. We may allow low-altitude orbits to decay and the spacecraft to reenter the atmosphere or use a velocity change to speed up the process. We may also boost spacecraft at any altitudes into benign orbits to reduce the probability of collision with active payloads, especially at synchronous altitudes. Because coplanar velocity changes are more efficient than plane changes, we would normally apply a tangential change in velocity. Its magnitude would depend on the difference in energy of the two orbits. For example, the velocity change required to deorbit (drop perigee altitude to 0 km) a satellite in a circular orbit at radius, r, and velocity, V, is

$$\Delta V_{deorbit} \approx V\left(1 - \sqrt{\frac{2R_E}{R_E + r}}\right) \quad (10.55)$$

We don't have to reduce perigee altitude to 0 km. If we choose a less conservative deorbit altitude, $H_{deorbit}$, we can determine the deorbit ΔV from Eq. (10.56) by replacing R_E with $(R_E + H_{deorbit})$. Thus, choosing a 50-km deorbit altitude would reduce the FireSat $\Delta V_{deorbit}$ from 198 m/s to 183 m/s. Note that only perigee is reduced in the deorbit burn. Reducing perigee to 100 to 150 km could result in several orbits over which apogee is reduced before the spacecraft reenters, which might not allow adequate control of the deorbit conditions.

10.6 Interplanetary Trajectories

Many missions of interest require the spacecraft to escape the Earth's influence and travel to other objects (planets, moons, comets, and asteroids) in the Solar System. The spacecraft may pass by the object, enter into an orbit around it, or land on its surface. By adding bodies to the problem, we've greatly increased its complexity. In fact, we can't solve the problem analytically. However, by splitting the trajectory up into parts, we can find an approximate solution that estimates the velocity budget and transfer time for a spacecraft traveling between two bodies in the solar system. We call this method of solution a patched-conic approximation. It divides the interplanetary trajectory into three segments: the hyperbolic escape from the first body, the elliptical transfer orbit about the Sun, and the hyperbolic arrival at the second body. We solve each segment of the trajectory with the previously defined two-body equations of motion and match the conditions from one segment of the trajectory to the next.

For Earth-orbiting spacecraft, we used the Earth-centered inertial (ECI) coordinate frame to describe and solve the equations. The ECI frame is inertial enough to solve Earth-orbiting trajectories, but we must define a new frame for interplanetary problems: the heliocentric, ecliptic coordinate frame shown in Fig. 10.13. By definition, the center of the frame is at the center of the Sun, and the fundamental plane is the plane of the ecliptic—the Earth's orbital plane. We choose the principal direction, $\hat{I}$, as the vector pointing from the Sun to the vernal equinox; $\hat{K}$, as the direction perpendicular to the ecliptic plane; and $\hat{J}$ in the ecliptic plane to complete our right-handed system.

10.6.1 Patched-Conic Approximation

The only forces we consider in this problem are the gravitational forces of the Earth, the Sun, and the target planet acting on the spacecraft. The spacecraft's equations of motion are defined by combining Newton's second law and his law of gravitation to get

$$\sum \vec{F} = m\ddot{\vec{r}} = \vec{F}_{\text{gravity Sun}} + \vec{F}_{\text{gravity Earth}} + \vec{F}_{\text{gravity target}} \tag{10.56}$$

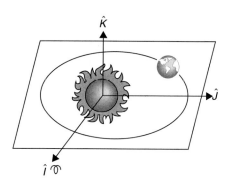

Fig. 10.13. **Heliocentric-Ecliptic Coordinate System for Interplanetary Transfer.** Origin—center of the Sun; fundamental plane—ecliptic plane (plane of the Earth's orbit around the Sun); principal direction—vernal-equinox direction. [Sellers, 1994]

We can't solve this problem directly, but we can solve it if we divide it into the three separate trajectories shown in Fig. 10.14. We define the *sphere of influence* (SOI) to be the point where the spacecraft transitions from the Earth's or the target planet's influence to the Sun's influence, or vice versa. Inside the SOI we consider the two-body equations of motion relative to the planet; outside the SOI, we consider the two-body equations of motion about the Sun.

We first solve the Sun-centered, elliptical transfer orbit from the Earth to the target planet. We must make several assumptions to solve this problem. We assume the Earth and the target planet are in circular, coplanar orbits and the transfer orbit is tangential to the orbits of the Earth and the target planet. In this case, the transfer orbit is a Hohmann transfer, and we use the methods described in Sec. 10.3.1 to solve for the velocities referenced to the Sun. We define these velocities below and provide equations or references for them in Table 10.9:

V_{Earth} is the Earth's circular velocity about the Sun

$V_{transfer\ at\ Earth}$ is the velocity on the transfer orbit near the Earth

$V_{SOI\text{-}Earth}$ is the velocity of the spacecraft relative to the Earth at the SOI

V_{target} is the target planet's circular velocity about the Sun

$V_{transfer\ at\ target}$ is the velocity on the transfer orbit near the target planet

$V_{SOI\text{-}target}$ is the velocity of the spacecraft relative to the target planet at the SOI

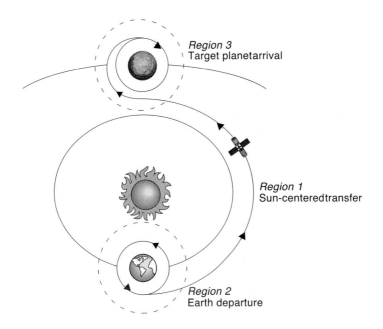

Fig. 10.14. Three Regions of the Patched-Conic Approximation. We can break the trajectory for interplanetary transfer into three distinct regions in which the gravitational pull of only one body dominates the spacecraft. [Sellers, 1994]

Table 10.9. Interplanetary Hohmann Transfer. We assume circular, coplanar orbits for the planets and solve the problem using the patched-conic method.

Step	What to Use
1. Determine velocity of Earth and target planet	Table 10.10
2. Determine velocities on transfer orbit at Earth and target planet	Eq. (10.4), use μ of the Sun
3. Find V_{SOI} at Earth and target planet	$V_{SOI\text{-}Earth} = (V_{transfer\ at\ Earth} - V_{Earth})$ $V_{SOI\text{-}target} = (V_{transfer\ at\ target} - V_{target})$
4. Determine hyperbolic escape/arrival trajectory	Eq. (10.56)
5. Calculate $\Delta V_{hyperbolic\ at\ Earth}$	Eq. (10.58) and (10.59)
6. Calculate $\Delta V_{hyperbolic\ at\ target}$	Eq. (10.61) and (10.62)
7. Calculate time-of-flight	Eq. (10.63)

Table 10.10 lists the orbit radius, circular velocity, and gravitational parameter for the planets in the solar system and the gravitational parameter for the Sun.

Table 10.10. Parameters of the Solar System. We use these parameters to find the various velocities on the interplanetary trajectory.

Planet/Sun	Mean Radius (km)	Circular Velocity (km/s)	Gravitational Parameter (km^3/s^2)
Sun	--	--	1.327×10^{11}
Mercury	57.9×10^6	47.87	2.232×10^4
Venus	10.81×10^7	35.04	3.257×10^5
Earth	14.95×10^7	29.79	3.986×10^5
Mars	22.78×10^7	24.14	4.305×10^4
Jupiter	77.8×10^7	13.06	1.268×10^8
Saturn	14.26×10^8	9.65	3.795×10^7
Uranus	28.68×10^8	6.80	5.820×10^6
Neptune	44.94×10^8	5.49	6.896×10^6
Pluto	58.96×10^8	4.74	8.37×10^2

The velocities of the spacecraft relative to the Earth and the target planet are called the *hyperbolic excess velocities*. They're equal to the difference between the planet's velocity and the spacecraft's velocity on the transfer orbit. We match these velocities with the spacecraft's velocity relative to the planets at the SOI to define the hyperbolic escape and arrival trajectories relative to the planets. If we assume the potential energy of the orbit relative to the planet is zero at the SOI, the specific mechanical energies of the hyperbolic trajectories are defined by the kinetic energies:

$$\varepsilon_H = V_{SOI}^2/2 \tag{10.57}$$

Because we assume two-body motion, the energy at the SOI defines the size of the orbit and the semi-major axis (Eq. (10.4)). Once we know the energy of the orbit, we can solve for the velocity at any other point in the orbit, relative to the planet, from the energy equation:

$$V = \sqrt{2\left(\varepsilon_H + \frac{\mu_{planet}}{r}\right)} \tag{10.58}$$

10.6 Space Navigation and Maneuvering

This value defines the burn-out velocity for departure. If the spacecraft is already in a parking orbit, we can calculate the velocity change needed to start the transfer (see Fig. 10.15).

$$V_{hyperbolic\ at\ Earth} = \sqrt{2\left(\varepsilon_H + \frac{\mu_{Earth}}{r_{park\ at\ Earth}}\right)} \quad (10.59)$$

$$\Delta V_{boost} = |V_{hyperbolic\ at\ Earth} - V_{park\ at\ Earth}| \quad (10.60)$$

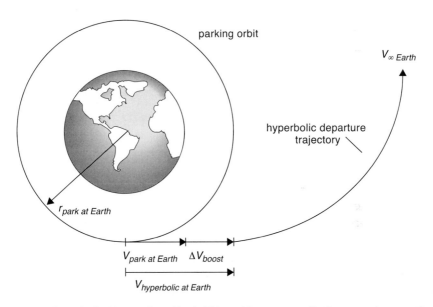

Fig. 10.15.   Hyperbolic Escape from Earth. This problem uses an Earth-centered perspective and requires the spacecraft to increase its velocity by an amount ΔV_{boost}. [Sellers, 1994]

Similarly, we can calculate the velocity at any point along the hyperbolic arrival trajectory relative to the target planet (see Fig. 10.16):

$$\varepsilon_H = V_{SOI}^2/2 \quad (10.61)$$

$$V_{hyperbolic\ at\ target} = \sqrt{2\left(\varepsilon_H + \frac{\mu_{planet}}{r_{retro}}\right)} \quad (10.62)$$

$$\Delta V_{retro} = |V_{park\ at\ target} - V_{hyperbolic\ at\ target}| \qquad (10.63)$$

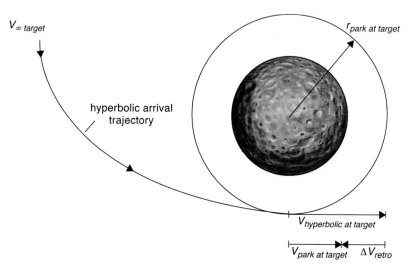

Fig. 10.16. Hyperbolic Arrival at a Target Planet. This problem focuses on the target planet and requires the spacecraft to slow down by an amount ΔV_{retro} to drop into orbit. [Sellers, 1994]

We apply the velocity change at the target planet only if we want the target planet to capture our spacecraft. If we don't decrease the velocity, the spacecraft continues on the hyperbolic trajectory and exits the target planet's SOI with the same hyperbolic velocity it had when it arrived at the planet's SOI. The two velocity changes approximate the velocity budget necessary to complete an interplanetary Hohmann transfer.

The time of flight (TOF) to complete the transfer is about equal to the TOF of the heliocentric, elliptical transfer orbit. Because we assume a Hohmann transfer, the TOF is one half the period of the elliptical transfer orbit.

$$TOF = \pi \sqrt{\frac{(r_{Earth} + r_{target})^3}{8\mu_{Sun}}} \qquad (10.64)$$

Another critical step in solving the interplanetary transfer is ensuring proper phasing so the spacecraft meets the target planet when it arrives at the planet's orbit. We solve this problem as we did the rendezvous problem

10.6.2 Swingby Trajectories

In the previous section we saw how to get from one planet to another using an interplanetary Hohmann-transfer orbit. But suppose we choose not to retro-fire at the target planet to enter orbit or land on the planet. In this case, the spacecraft will continue along the hyperbolic trajectory relative to the target planet and exit the SOI with a relative velocity the same as when it entered the SOI, but the direction of the velocity vector will change. (Figure 10.17.)

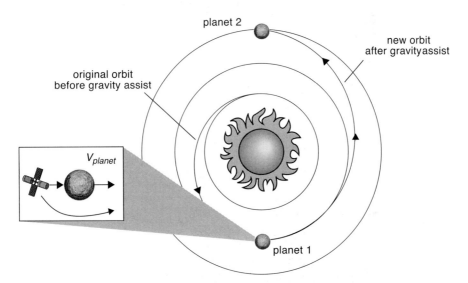

Fig. 10.17. Gravity Assist. During a gravity assist, a planet pulls the spacecraft, changing its velocity with respect to the Sun and thus altering its orbit around the Sun. The planet's orbit also changes, but very little. [Sellers, 1994]

We see in the figure that we can increase or decrease the spacecraft's velocity relative to the Sun by passing behind or in front of the planet relative to the Sun. This maneuver is called a gravity-assist trajectory because it uses the planet's gravitational acceleration to change the spacecraft's velocity. We can consider this velocity change "free" because it doesn't require any onboard propellant to complete the maneuver. The amount of velocity change available is a function of the arrival velocity and the turning angle, d, which is in turn a function of the closest approach distance to the planet. The magnitude of the arrival velocity is determined by the transfer orbit and can't be changed, so the only design parameter is the closest approach distance. We calculate the turning angle from the following equations.

$$e = 1 + \frac{r_\rho V_h^2}{\mu_{planet}}$$ (10.65)

where r_ρ equals distance of closest approach to planet.

$$\delta = 2\sin^{-1}\left(\frac{1}{e}\right)$$ (10.66)

If the spacecraft passes in front of the planet, its velocity relative to the Sun decreases and it drops into a smaller orbit. But if the spacecraft passes behind the planet, its velocity increases relative to the Sun. (Fig. 10.18)

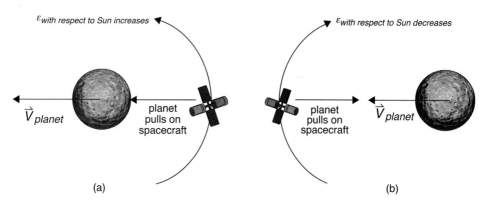

Fig. 10.18. Gravity Assist with the Spacecraft Passing Behind and in Front of a Planet. (a) During a gravity-assist maneuver, a spacecraft's energy will increase with respect to the Sun if it passes behind the planet. (b) During a gravity-assist maneuver, a spacecraft's energy will decrease with respect to the Sun if it passes in front of the planet. [Sellers, 1994]

Gravity-assist trajectories reduce the propellant requirements for interplanetary travel by changing the spacecraft's velocity without using propulsive burns. The Galileo spacecraft used three gravity assists (one from Venus and two from Earth) to travel from Earth to Jupiter.

10.6.3   Lagrange Points

Interesting classes of orbits arise when we consider the restricted three-body problem. This problem considers the gravitational attraction of two distinct bodies, called the primary and secondary, acting on a spacecraft. The restricted three-body problem assumes the primary and secondary are in circular orbits about the center of mass of the system and neglects the mass of the spacecraft and all other orbit perturbations. If we attach a coordinate frame to this rotational system, there are five stationary points, called Lagrange points, shown in Fig. 10.19 for the Earth-Sun System.

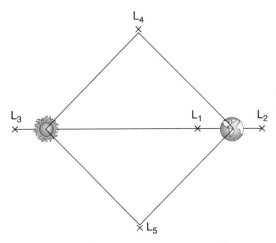

Fig. 10.19. Lagrange Points in the Earth-Sun System. Points L_1, L_2, and L_3 lie on the line connecting the Earth and the Sun and are in unstable equilibrium. Points L_4 and L_5 are equidistant from the Earth and the Sun and are in stable equilibrium for the restricted three-body problem.

Lagrange points exist for all pairs of bodies (e.g. Earth-Sun, Earth-Moon, and Sun-Jupiter). Points L_4 and L_5 are equidistant from the two bodies in the system and are stable equilibrium points. If we place a satellite near one of these two points, but not exactly equidistant, the spacecraft will remain in the vicinity of the Lagrange point and orbit around it. The other three Lagrange points are on the line connecting the two bodies and are unstable equilibrium points. A spacecraft placed near one of the co-linear points, but not exactly at it, will eventually drift away from the co-linear point. Therefore, if we place a spacecraft at one of the co-linear points, we must include propellant for station keeping to maintain the orbit.

Following are approximate equations for the distance of the co-linear points from the center of mass of the system. These equations apply when $M \gg m$.

$$r_1 = R_B\left[1 - \left(\frac{m}{3M}\right)^{\frac{1}{3}}\right]$$

$$r_2 = R_B\left[1 + \left(\frac{m}{3M}\right)^{\frac{1}{3}}\right]$$

$$r_3 = R_B\left[1 - \left(\frac{7m}{12M}\right)^{\frac{1}{3}}\right]$$

where m is the mass of the secondary body, M is the mass of the primary body, and R_M is the distance from the primary body to the secondary body.

Minimum energy transfer orbits from the primary body to the Lagrange Point use Hohmann transfer orbits while transfers from the secondary body use patched-conic approximations.

10.7 Mission Geometry

10.7.1 Coordinate Systems

To discuss mission geometry, we must define several coordinate systems. Different coordinate systems are best for different geometries. We specify a *coordinate system* by defining an origin and the orientation of a set of linearly independent axes. Coordinates are quantities, such as distances and angles, which allow us to specify a location within a coordinate system. We may use different types of coordinates within a single coordinate system.

Table 10.11 shows several coordinate frames used to calculate mission geometries.

10.7.2 Parameters of Mission Geometry

Many software packages are available to calculate parameters of mission geometry. But often we may want a quick method to approximate mission-geometry parameters such as elevation angle, coverage area, and maximum time in view. Figure 10.20 shows the relationship between selected parameters, the spacecraft's orbit, and the Earth.

First, we find the Earth's angular radius, ρ, from the orbit altitude, H.

$$\sin\rho = \frac{R_E}{R_E + H} \qquad (10.67)$$

Next, we must specify either the elevation angle, ε, the nadir angle, η, or the Earth-central angle, λ. If we specify the elevation angle based on some constraint

10.7 Space Navigation and Maneuvering

Table 10.11. Coordinate Frames Used to Calculate Mission Geometries. We specify coordinate frames by listing origin, fundamental plane (plane containing X and Y unit vectors), and the directions of the X and Z axes.

Coordinate Frame	Origin	Fundamental Plane	X-axis	Z-axis	Comment
Earth-Centered Inertial (ECI)	Center of Earth	Equatorial plane	Vernal Equinox	Earth spin axis	X and Z axes are specified on a particular date. The most common are Jan. 1, 1950 or Jan. 1, 2000.
Earth-Centered Fixed (ECF)	Center of Earth	Equatorial plane	Greenwich Meridian	North Pole	Coordinate frame rotates with the Earth
Topocentric Horizon (SEZ)	Point on or near surface of Earth	Local horizon	Local South	Local Vertical	Used to express vectors relative to station coordinates
Vehicle Velocity, Local Horizon (VVLH)	Center of spacecraft	Spacecraft local horizon	Direction of velocity vector projected into plane of local horizon	Nadir	Used for Earth observations

on minimum elevation, we solve for the nadir angle, the Earth-central angle, and the distance, D, from the spacecraft to the target using

$$\sin \eta = \sin \rho \cos \varepsilon \tag{10.68}$$

$$\lambda = 90° - \eta - \varepsilon \tag{10.69}$$

$$D = R_E (\sin \lambda / \sin \eta) \tag{10.70}$$

where R_E is the radius of the Earth, 6378.14 km

Or we can specify the nadir angle and then solve for the elevation angle (solving Eq. (10.68) for ε), central angle, and distance. We may also want to find the elevation angle and nadir angle for a given subsatellite point and target on the Earth's surface. For this problem, we start with the coordinates (latitude and longitude) of the subsatellite point and the target, as well as the altitude of the orbit. We again solve for the Earth's angular radius using Eq. (10.64). If the latitude and longitude of the subsatellite and target are (δ_T, L_T) and (δ_S, L_S) respectively, we solve for the Earth-central angle, nadir angle, elevation angle, and distance from the spacecraft to the target:

$$\cos \lambda = \sin \delta_S \sin \delta_T + \cos \delta_S \cos \delta_T \cos |L_S - L_T| \tag{10.71}$$

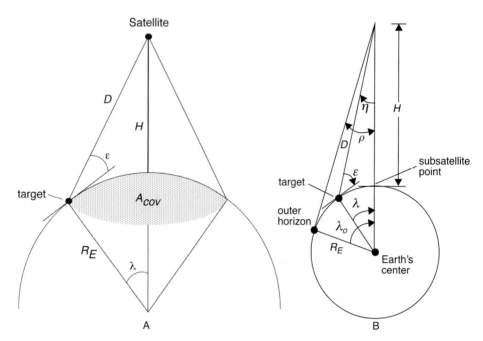

Fig. 10.20. **Mission Geometry.** Part A shows the coverage area for a given geometry, and B depicts the angular relationship between the spacecraft, the target, and the center of the Earth.

$$\tan\eta = \sin\rho \sin\lambda / (1 - \sin\rho \cos\lambda) \quad (10.72)$$

$$\cos\varepsilon = \sin\eta / \sin\rho \quad (10.73)$$

$$D = R_E (\sin\lambda / \sin\eta) \quad (10.74)$$

Note that λ, η, and ε must always sum to 90°.

Next, we calculate the coverage area, A_{cov}, and the maximum viewing time, t_{view}, using the Earth-central angle and the orbit period. The coverage area is the part of the Earth's surface in view of the spacecraft sensor(s), with limits specified by either the nadir angle or the elevation angle. Maximum viewing time is the maximum length of time that a spacecraft is visible from a point on the surface of the Earth at a specified minimum elevation angle, assuming the spacecraft passes directly over the point on the Earth's surface.

$$A_{cov} = (2\pi R_e^2)(1 - \cos\lambda)$$

$$t_{view} = P\lambda_{max} / 180°$$

where P is the orbit period and λ_{max} is the Earth-central angle, which corresponds to the minimum elevation angle.

Please read Larson and Wertz, Chap. 5, for a more complete discussion of mission geometry.

References

Bate, Roger R., Donald D. Mueller, and Jerry E. White. 1971. *Fundamentals of Astrodynamics*. New York, NY: Dover Publications.

Battin, Richard H. 1987. *An Introduction to the Mathematics and Methods of Astrodynamics*. New York, NY: AIAA Education Series.

Danby, J. M. A. 1962. *Fundamentals of Celestial Mechanics*. New York, NY: Macmillan.

Escobal, Pedro R. 1965. *Methods of Orbit Determination*. Malabar, FL: Robert E. Krieger Publishing Co.

Kaplan, Marshall H. 1976. *Modern Spacecraft Dynamics and Control*. New York, NY: Wiley and Sons.

King-Hele, D. 1964. *Theory of Satellite Orbits in an Atmosphere*. London: Butterworths Mathematical Texts.

Larson, Wiley J. and James R. Wertz. 1999. *Space Mission Analysis and Design*. Third Edition. Netherlands: Kluwer Academic Publishers.

Pocha, J. J. 1987. *An Introduction to Mission Design for Geostationary Satellites*. Boston, MA: D. Reidel Publishing Company.

Roy, A. E. 1978. *Orbital Motion*. Bristol and Philadelphia: Adam Hilger.

Sellers, Jerry Jon. 1994. *Understanding Space: An Introduction to Astronautics*. New York, NY: McGraw Hill.

Vallado, David A. 2001. *Fundamentals of Astrodynamics and Applications*. 2nd Edition, El Segundo, CA, Microcosm Press.

Wertz, J.R. 2002. *Mission Geometry; Orbit and Constellation Design Management*. El Segundo, CA, Microcosm Press

Wiesel, William E. 1989. *Spaceflight Dynamics*. New York, NY: McGraw-Hill Book Company.

Chapter 11
Communications Architecture

Gary M. Comparetto, *The MITRE Corporation*
Richard S. Davies, *Stanford Telecommunications, Inc.*

> 11.1 Defining and Evaluating Requirements
> 11.2 Evaluating Design Options
> 11.3 Forming the Architecture

The communications system comprises the set of components that transport and deliver information from a source to one or more destinations. The information may include any combination of data, voice, and video and the sources/destinations may be on the ground, in the air or in space.

Designing the communications system is complicated. It's often done ad hoc by various people that are technically competent in a wide range of disciplines. The whole process can intimidate a mission operations manager (MOM) and, if unstructured, can result in a poorly documented, weakly justified end product that becomes more difficult to change as it ages. The inertia builds up, so to speak, and critical questions often appear heretical. Hence, the design is outdated by the scheduled operational date!

One school of thought contends that trying to design a large, complicated communications system is a lot like trying to invent new products, so it requires unstructured trial and error. Structure, they say, inhibits original thought and creativity and is too cumbersome to incorporate dynamic design iterations and changes. Although some of these arguments may be valid, past experience in designing large communications systems calls for a process by which we document system options, justify them technically, and review them periodically to ensure that they remain cost-effective while meeting the stated system requirements.

In this chapter, therefore, we adopt a top-down, structured approach to designing communications systems. As Fig. 11.1 shows, it has three main phases. In Phase 1, defining and evaluating requirements, the technical design team identifies driving requirements for the communications system.

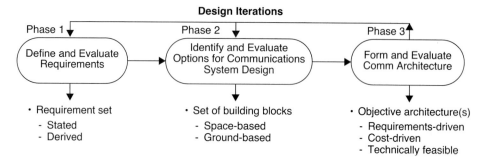

Fig. 11.1. Designing Communications Systems. Phase 1 of communications system design results in requirements that guide us in developing building blocks from which we can form alternative communications systems.

In Phase 2, which considers design options, we evaluate communications system component designs in terms of complexity, power, size, and other factors. The result is a set of building blocks, consistent with the requirements derived in Phase 1, that we can use to construct the communications system.

In Phase 3, forming the communications system, we construct alternatives based on the design characteristics and component design options from Phase 2 and the requirements from Phase 1. Phase 3 is iterative, so we must revisit Phases 1 and 2 often to determine if each alternative can meet program requirements while remaining technically feasible.

11.1 Defining and Evaluating Requirements

Figure 11.2 shows the principle characteristics of requirements definition, during which we identify performance requirements, operational requirements, and system constraints. We then analyze them within program constraints such as existing assets, the technology baseline, program schedule, and allowable system cost. The following paragraphs address the more common performance requirements, operational requirements, system constraints, and program constraints for communications system designs.

11.1.1 Performance Requirements

The main performance requirements for satellite-communications (SATCOM) systems are the data rate, bit-error-rate (BER), end-to-end (E/E) delay, and link availability.

Military Satellite Communications (SATCOM) systems also require anti-jam (A/J) capability and may have to operate in a multi-path fading environment. We discuss these in Sec. 11.2.3.

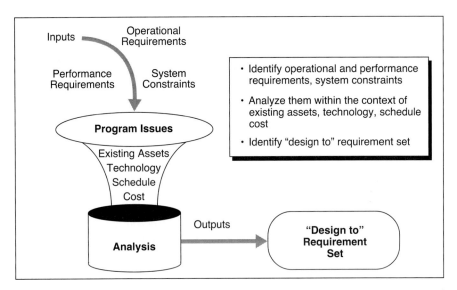

Fig. 11.2. Defining Requirements. In this phase, we use performance and operational requirements plus system constraints to develop "design to" requirements.

Data Rate

Increasing the data rate is almost always desirable. *Data rate* will largely dictate the design (and cost) of two of the space/ground link's main features—the high-power amplifier (HPA) in the ground terminal and the terminal antennas in the spacecraft and the ground terminal. The data rate is proportional to the quantity of information per unit time transferred between the spacecraft and ground station. Everything else being equal, the higher the data-rate requirement, the larger the transmitter power (HPA power) or antenna size required to close the communications link. In general, the HPA's power requirement is a linear function of the data-rate requirement, whereas the antenna-size requirement will vary as the square root of the data-rate requirement. For example, if the data-rate requirement doubles for a given space/ground link and the antenna sizes remain constant, the HPA power needed to support the new data-rate requirement would be double the original design. On the other hand, if the HPA power remains constant, the required antenna diameter (transmit or receive) would need to increase by a factor of 1.414 (the square root of 2).

We often don't have the data-rate requirement for a particular system, so we must derive it using other system requirements. For example, a system may have to support message transfer at a rate of x messages per second. Assuming the messages were each composed of y bytes of information, with each byte represented by eight bits of data, the required data rate would be

$$R_d = x \text{ (messages/s)} \cdot y \text{ (bytes/message)} \cdot 8 \text{ (bits/byte)} \quad (11.1)$$

$$= 8xy \text{ bits/s or bps}$$

So for a required message rate of 25 messages/second and a message size of 80 bytes/message, the resulting data-rate requirement would be 16,000 bps or 16 kbps.

Finally, the concept of link duty factor affects the data rate. The link duty factor is the fraction of time that the communications link is actively supporting the data-rate requirement. For example, a duty factor of 20% for a ground terminal means the terminal is transmitting data only 20% of the time. A duty factor less than 100% could occur because the link must support multiple users or because something obscures the satellite's line-of-sight for part of its orbit. Whatever the reason, a link duty factor of less than 100% means we must transmit at rates higher than the data-rate requirement to maintain throughput. For example, if the required data rate is 20 kbps and the link has a duty cycle of 10%, the effective data-rate requirement would be 200 kbps to compensate for the link's not transmitting 90% of the time—200 kbps for 10% of the time and 0 kbps for 90% of the time results in a throughput of 20 kbps.

Bit-error-rate

Bit-error-rate (BER) is another key requirement that drives communications system design. The BER is somewhat of a misnomer because the term really doesn't represent a rate but a probability of bit error. The BER is defined as the probability that a bit will be demodulated or decoded (if applicable) incorrectly. So, a BER of 10^{-5} would mean that, on average, one bit in 100,000 will be demodulated or decoded in error.

The BER requirement depends on the system application. For example, voice communication can tolerate a BER of 10^{-3}, while a computer-to-computer link may require a BER of 10^{-10}. Picture transmission BERs fall in between and depend upon the amount of data compression.

To define this requirement properly we must define the user-communication system interface. For example, if the user equipment employs error detection circuitry, a higher BER allocation to the communication system would be acceptable for the same system performance.

End-to-end Delay

End-to-end (E/E) delay is the time between message transmission and reception. The E/E delay requirement is critical in duplex links where users are communicating with each other in real time. Voice communications become severely degraded when the E/E delay exceeds 0.5 second.

In a broadcast application, where one user is transmitting a message to a number of simultaneous receivers, the E/E delay is seldom critical.

Digital data is often transmitted in blocks or packets. These blocks contain error detection codes. If no error is detected, the receiver transmits an acknowledgment back to the transmitting terminal. If the received signal contains an error, the user requests the block be retransmitted. Network protocols have been developed to control the sequence of blocks and identify lost and duplicate blocks. An excessive E/E delay (deep space missions may have hours of delay) may cause these protocols to fail. The problem becomes worse at higher data rates where more blocks have to be stored during the E/E delay period. Special protocols have been developed for transmitting high data rates over links with large E/E delay. (See http://www.ccsds.org for space related protocols that handle large E/E delay.)

The E/E delay in a communications system having multiple nodes can be thought of as a composite delay, with components accounting for transmission, propagation, queuing, and processing. The transmission delay is simply the time it takes to transmit a block or packet of data. We compute it by dividing the data block size by the transmission data rate. For example, the transmission delay for an 80-byte block of data transmitted at a data rate of 9.6 kbps would be

$$\tau_{trans} = (80 \text{ bytes})(8 \text{ bits/byte}) / 9.6 \text{ kbps} \quad (11.2)$$

$$= 6.67 \times 10 \text{ s} = 66.7 \text{ ms}$$

Propagation delay, also referred to as the speed-of-light delay, represents the amount of time the transmitted signal takes to traverse the path between the source emitter and the destination receiver. In calculating the propagation delay, we assume the transmitted signal (an electromagnetic wave) travels at the speed of light, c.

As an example, assume a spacecraft is in a geostationary orbit. This orbit is common for communications satellites (COMSATs) because the spacecraft appears to be stationary from Earth. The altitude for a geostationary spacecraft is ~35,785 km. When the ground terminal is directly under the spacecraft, the source-to-destination distance is equal to the spacecraft's altitude. Consequently, we calculate the uplink propagation delay as follows:

$$\tau_{prop} = \frac{(\text{Source-to-Destination Distance})}{c} \quad (11.3)$$

$$\frac{35,785 \times 10^3 \text{ m}}{3 \times 10^8 \text{ m/s}} = 0.12 \text{ s}$$

The one-way, source-to-destination, propagation delay consists of uplink and downlink components and is 0.24 (2 × 0.12) seconds. That's why we commonly refer to the propagation delay of a geostationary spacecraft as a quarter-second delay "per hop," where a hop represents the composite uplink and downlink paths from source to destination.

The *queuing delay* applies to a communications architecture that delivers information from node to node via packets. Queuing occurs when a packet arrives at a processing node (satellite-based or ground-based) when that node is busy processing other packets. The queuing delay is the amount of time that transpires before the packet is processed and depends on several parameters, including the packet's mean arrival rate, the service time for each packet, and the use time—the fraction of time that the server is busy. Typically, the queuing delay is relatively flat and small (i.e., several milliseconds) for widely ranging use factors. Stallings [1985] discusses queuing delays fully along with today's common data-switching methods, including packet switching, message switching, and circuit switching.

The *processing delay* is a catch-all term that encompasses the delays associated with communications equipment. For example, in a typical spacecraft, the received signal undergoes frequency translation, signal amplification, and antenna routing before it's retransmitted. These delays are usually quite small and can be ignored. However, processing delays incurred by error correction coders, interleavers, vocoders, and encryption devices can be significant. We need to consider these delays in designing the communications system.

Link Availability

Link availability is the probability that when a user wants to communicate, the link is available for his use. Commercial telephone networks strive for an availability of 0.9999. Other applications, such as cellular phone systems, settle for a lower availability. Two factors that dominate link availability are rain attenuation (at frequencies above 10 GHz, and radio interference or jamming. (See Sec. 11.1.2.)

Closely related to link availability is link outage time, the average time the link is unable to operate due to its physical environment. Unlike end-to-end link delay, outage time is usually measured in minutes or hours. An example is the duration of a thunderstorm that causes the signal level to drop below threshold. A mobile link outage may occur when the vehicle drives under a bridge, through a tunnel or under heavy foliage. A link outage may also occur in a nonsynchronous satellite system due to gaps in satellite coverage. The maximum outage time that can be tolerated is determined by the operational use of the system. If continuous (or near-continuous) coverage is not required, we may use a store-and-forward satellite configuration.

Another factor that affects overall link availability is the reliability of the communication system network and equipment. A low-orbit multisatellite system such as IRIDIUM can provide an alternate path should a satellite fail. The satellite may have redundant transponders, and the ground terminal may have redundant equipment.

11.1.2 Operational Requirements

The main operational requirements areas for satellite communications systems include security, standardization, backward compatibility, spacecraft orbit, spacecraft mobility, user-terminal characteristics, and anti-jamming capabilities. The user or sponsor usually specifies these requirements, which must be reflected in the design of the communications system.

Security

Communications systems have two main security components: communications security (COMSEC) and transmission security (TRANSEC). *COMSEC* means disguising the actual transmitted data content and typically involves data encryption. *TRANSEC* means disguising the transmission method and normally involves generating security keys and variables that support spread-spectrum techniques. We can separate the communications system design from TRANSEC if we focus our attention on the spread-spectrum technique rather than how we generate the security keys. We take this approach here. The following paragraphs give a brief overview of common techniques for COMSEC and TRANSEC.

We typically encrypt blocks or data streams. With *block encryption*, the plaintext (data to be encrypted) is segmented into blocks of fixed size and each block is then separately encrypted. With *data-stream* (or simply stream) *encryption*, blocks don't have a fixed size and each plain-text bit is separately encrypted. Block encryption is more common; the two most widely used techniques are the data-encryption standard and public-key encryption.

The data-encryption standard uses a 56-bit key and operates on 64-bit blocks. The same key is used to encrypt and decrypt the data. The algorithm employs a technique described as *product ciphering*, in which we repeatedly perform simple encryption in tandem, resulting in an overall cipher system that is more powerful than any individual part. In effect, the complexity of the product of the encryption components is greater than their sum.

Public-key encryption techniques differ in that we use two different keys, one for encryption and one for decryption. In public-key systems, the encryption key can be revealed publicly without compromising the system's security. The encryption and decryption keys are developed in pairs and only the decryption key is kept secret. Deriving such sets of keys depends on complicated mathematical functions but involves the relatively simple premise that certain mathematical functions are easier to compute in one direction than in the reverse direction. For example, it's easier for most of us to calculate the cube of six in our heads than to calculate the cube root of 216! Sklar [1988] is one of the few texts that discuss both the data-encryption standard and public-key systems, as well as other encryption techniques.

Within the context of this chapter, having to provide COMSEC in a communications system will alter the communications-terminal design by calling for additional equipment to perform the encryption. Additional equipment means increased terminal weight and cost. Processing delays may also increase, which may prevent the system from meeting its end-to-end delay requirement. And the options for data routing may be limited if we must decrypt the full data packet completely at each intermediary communications node.

As stated, we'll focus our discussion of TRANSEC on the spread-spectrum techniques used to achieve transmission security rather than on generating security keys. The two main spread-spectrum techniques are frequency hopping (FH) and direct sequence (DS) or pseudorandom noise (PN). These are discussed in Sec. 11.2.3.

Standardization

Standardization supposedly

- Maximizes a system's operational effectiveness and minimizes its lifecycle costs
- Enhances the system's performance, interoperability, maintainability, portability, reliability, and availability

An additional "ility" sometimes given in number two above is scalability, which is a system's ability to get bigger easily without having to scrap the current system.

Standardizing means we must

- Limit excursions from current design techniques and do detailed technical justifications if we vary them
- Emphasize existing standards (government and commercial) and processes
- Emphasize commercial, off-the-shelf (COTS) equipment and non-developmental items (NDI)

We should certainly consider standardizing when we develop a communications system, but we must be sure it doesn't become the primary driver. If it does, we may not be able to meet other performance and operational requirements.

Standardizing can cost us more downstream if we have to extend a communications system into the future. In most cases, to reduce lifecycle costs, we have to accrue certain costs up front by incorporating new equipment, processes, and techniques that will save money later. The up-front development and acquisition costs can be substantial, and planners will often rule against an apparently expensive system that will have lower lifecycle costs than the alternatives.

Backward Compatibility

In most cases, communications systems aren't built from scratch. Instead, we extend an existing infrastructure of some sort into the future. If this is the case, and depending on the size of the infrastructure, all types of ground and space assets may not have reached their useful lifetime. Requiring the users to procure new equipment so it will conform to the new communications system would cost too much. Thus, we have to develop a plan that addresses using existing equipment, capabilities, and facilities throughout their lifetimes. In other words, we have to look at backward compatibility.

Backward compatibility can significantly burden the system designer and often precludes the best solution. For example, it's almost always possible to develop a new communications system that meets all stated requirements (except, perhaps, for cost!) using a clean sheet of paper, but we may not be able to meet all performance requirements if we must integrate old equipment into the new system. One way to address backward compatibility is to develop a solid plan that phases in the newly developed system to match major program milestones for assets in the existing infrastructure.

Spacecraft Orbits

Spacecraft orbits include three key parameters that will set requirements on communications-system design [Gagliardi, 1984]:

1. *Coverage area, A_{cov}*—The part of the Earth's surface that can receive the spacecraft's transmission, assuming the ground terminals are operating at an elevation angle larger than a prescribed minimum. (Note: The elevation angle of a ground terminal is the angle defined between a line tangent to the Earth at the ground terminal's location and the pointing direction to the spacecraft).

2. *Slant range, S*—The line-of-sight distance from a fixed point on the Earth to the spacecraft

3. *Viewing time, τ_{view}*—The length of time that a spacecraft is visible from a ground terminal, assuming a prescribed elevation angle

The relationships among the coverage area, elevation angle, slant range, and viewing time are discussed in Sec. 10.7.

Characteristics of the spacecraft's orbit will directly affect the design of the communications system. (See Sec. 11.2.) We may choose higher spacecraft altitudes to increase the coverage area per spacecraft and viewing times for a given minimum elevation angle. However, the cost to place a spacecraft into orbit increases dramatically as the altitude increases and may outweigh the potential benefits of increased coverage area and viewing time. Instead, we may need to place more spacecraft into lower orbits, with each one having a smaller coverage area and viewing time than the geostationary spacecraft. But together, as a composite constellation, they'll be able to meet the requirements at lower cost for the overall system.

User Terminals

User terminals are either fixed, mobile, or transportable. Fixed users typically employ large, stationary ground terminals. They aren't easily relocated and represent a large fraction of the communications system's overall cost. However, they can transmit a large amount of data because of the high achievable gains of their antenna designs.

Mobile users usually use small terminals that can communicate while in motion. Examples of this type of terminal include cellular phones, airplanes, and car radios. Note that larger mobile terminals do exist to support maritime communications between ships (mobile, but relatively slow) and shore stations.

Transportable users represent the middle ground. That is, they employ medium-size terminals that are too large to support communication links while moving but are small enough to be packaged straightforwardly and relocated. Transportable terminals are common in the military, where theaters of operation move quickly during a conflict, and the communications system must keep up with the advancing columns.

Each user category has associated characteristics that affect the communications system design. For example, fixed terminals normally employ large antennas that support requirements for high communications throughput (a plus) but are limited in their ability to track spacecraft in low-Earth orbit (a minus). As another example, mobile terminals, because of their small size and mobility, normally employ omni-directional antennas that are simple to design and manufacture (a plus) but which can't generate any appreciable signal gain (a minus). For omni-directional antennas at ground terminals, the spacecraft must compensate for this lack of gain by increasing the size of its transmit antenna, increasing the output transmit power level, or decreasing the slant range between the spacecraft and the mobile user (low-Earth orbits versus geostationary orbits).

Recently, however, there has been significant interest in programs such as the Future Combat Systems Communications (FCS-C) program led by the Defense Advanced Research Development Agency (DARPA) in developing communications systems that support mobility using directional antennas. The focus of this program is to provide the technology and system integration and demonstration to develop a communications system capable of achieving high data rate/low latency communications, having robust anti-jam and low probability of detection (LPD) characteristics. DARPA's approach to meeting these opposing constraints is through a multi-tiered mobile ad hoc network using directional antennas. The use of directional antennas in such an environment is quite challenging due to the pointing requirements imposed on the communications system.

In summary, we must consider the characteristics of the user's terminal when forming options for the communications system.

Jamming

Jamming is the presence of unwanted signal components at the receiver of a friendly terminal—intentional (as in electronic warfare) or not (as in radio-frequency interference, or RFI). Intentional jamming is usually much more challenging to design against in a communications system. A requirement to operate in the presence of a jamming threat can have a major impact on the communication system design.

Anti-jam (A/J) capability refers to a system's ability to overcome the harmful effects of a jammer. A requirement for A/J will mainly affect our design of the communications waveform (the choice of modulation format, coding, interleaving, and spread-spectrum technique) and the terminals (the antenna's design and the amplifier's power rating).

Jammer types include:

- Wideband (or broadband) noise jammer—The jamming signal's power is transmitted evenly over the communications signal's total bandwidth
- Partial-band noise jammer—Similar to wideband noise jammers except that only part of the signal bandwidth is jammed
- Pulse jammer—A high-energy signal repeatedly transmits over a very narrow bandwidth
- Tone (multi-tone) jammer—A high-energy signal continually transmits over single (multiple) frequency tone(s)
- Repeat-back (or frequency-follower) jammer—The jammer intercepts the transmitted signal, determines the transmitted frequency, then re-transmits interference at this particular frequency. It must finish this process before the friendly signal changes to a new frequency.

The main goal for each of these jammers is to degrade the received BER at the friendly terminal. Techniques to combat jamming are described in Sec. 11.2.3.

11.1.3 System Constraints

In the early phases of design, we must identify all system constraints so that we can bound the communications system's trade space. By doing so, we won't consider non-viable alternatives and will shorten the process for designing a communications system.

Frequency Spectrum Allocation

The carrier operating frequency affects the ground-terminal antenna's achievable size and beamwidth. In turn, these factors affect spacecraft size, mass, and complexity indirectly. The carrier operating frequency also determines how much attenuation we can anticipate due to rain and other benign- and stressed-channel characteristics. Doppler shift, which increases with carrier frequency and relative velocity between transmitter and receiver, is a significant factor in systems employing mobile terminals and low orbit satellites.

Regulations constrain our choice of operational frequency, the usable data bandwidth, and the allowable power flux density that impinges on the Earth's surface. Table 11.1 lists the frequency bands allocated to space communications. These assignments originated with the International Telecommunications Union (ITU) and the World Administrative Radio Conference (WARC). In the United States they are administered by the Federal Communications Commission (FCC) for commercial users and by the Interdepartmental Radio Advisory Committee (IRAC) for military users.

The system designer must apply for and receive from the appropriate agency permission to operate at a specified frequency, orbit, and ground-terminal location. This procedure is often time-consuming. For an excellent summary of this complex subject, see Morgan and Gordon [1989]. Also shown in Table 11.1 are the maximum allowed power flux densities that may be radiated by the spacecraft onto the Earth. These limits, set by the ITU, are necessary to avoid interference with existing terrestrial services, such as microwave relay links.

Table 11.1. The International Telecommunications Union's Limits on Frequency Bands and Flux Densities. Power-density limits are for elevation angle > 25°. They are about 10 dB less for lower angles. [Table 13-12, Larson and Wertz, 1999]

Frequency Band	Frequency Range (GHz)		Service	Limit on Power Flux Density for Downlinks (dBW/m^2)
	Uplink	Downlink		
UHF	0.2–0.45	0.2–0.45	Military	--
L	1.635–1.66	1.535–1.56	Maritime/Nav	−144/4 kHz
S	2.65–2.69	2.5–2.54	Broadcast	−137/4 kHz
C	5.9–6.4	3.7–4.2	Domestic Comsat	−142/4 kHz
X	7.9–8.4	7.25–7.75	Military Comsat	−142/4 kHz
Ku	14.0–14.5	12.5–12.75	Domestic Comsat	−138/4 kHz
Ka	27.5–31.0	17.7–19.7	Domestic Comsat	−105/1 MHz
SHF/EHF	43.5–45.5	19.7–20.7	Military Comsat	--
V	~60		Satellite Crosslinks	--

One criterion for allocating frequency bands is possible interference between links. Extensive analysis is required when applying for a frequency band and orbit to avoid interference with, or by, existing services, such as terrestrial microwave links and ground-based radar operations. Especially significant are the ground antenna's off-axis sidelobe power level limits and their impact on the communications system design. Table 11.2 includes the ITU maximum off-angle equivalent isotropic radiated power (EIRP) density limits for ground-to-space communications links operating at C-, Ku- and Ka-band frequencies. As shown, the allowable EIRP density at Ka-band is much more stringent than at Ku-band (~20dB). This dramatically impacts the use of small aperture antennas in Ka-band communications system designs since it results in wider beamwidths and, correspondingly, higher off-axis power levels that may exceed the ITU limits.

Table 11.2. ITU Maximum Off-Angle EIRP Density Limits for Ground-to-Space Communications Links Taken From ITU-R Recommendation S.534-7. Off-axis EIRP density limits may require the use of signal spreading techniques to support small-antenna operations.

Transmit Frequency (GHz)	Off-Axis Angle	Maximum EIRP Density
5.9–6.4	$2.5° \leq \varphi \leq 7°$	$(32 - 25 \log \varphi)$ (dBW/4 kHz)
	$7° < \varphi \leq 9.2°$	11 (dBW/4 kHz)
	$9.2° < \varphi \leq 48°$	$(35 - 25 \log \varphi)$ (dBW/4 kHz)
	$48° < \varphi \leq 180°$	-7 (dBW/4 kHz)
14.0–14.5	$2.5° \leq \varphi \leq 7°$	$(39 - 25 \log \varphi)$ (dBW/40 kHz)
	$7° < \varphi \leq 9.2°$	18 dB (dBW/40 kHz)
	$9.2° < \varphi \leq 48°$	$(42 - 25 \log \varphi)$ (dBW/40 kHz)
	$48° < \varphi \leq 180°$	0 (dBW/40 kHz)
29.5-30	$2° \leq \varphi \leq 7°$	$(19 - 25 \log \varphi)$ (dBW/40 kHz)
	$7° < \varphi \leq 9.2°$	-2 (dBW/40 kHz)
	$9.2° < \varphi \leq 48°$	$(22 - 25 \log \varphi)$ (dBW/40 kHz)
	$48° < \varphi \leq 180°$	-10 (dBW/40 kHz)

Two geostationary spacecraft in approximately the same orbit location servicing the same ground area may share the same frequency band by (1) separating adjacent spacecraft by some minimum angle (typically 2°) that is larger than the ground station's beamwidth; or (2) polarizing the transmitting and receiving carriers orthogonally, which allows two carriers to be received at the same frequency without significant mutual interference. Right-hand and left-hand circular polarization are orthogonal, as are horizontal and vertical linear polarization. Commercial systems use these frequency-sharing techniques extensively [Morgan and Gordon, 1989].

An alternative to operating in the radio frequency spectrum is to use optical frequencies (see Sec. 11.2.8). At present there are no restrictions in using this band (except possibly power limitations due to personal safety considerations).

Physical Constraints

Many aspects of the communications-architecture system design depend on system physical constraints, such as

- Maximum antenna size at the ground terminal or spacecraft (based on mobility and maneuverability requirements, costs, limits on real-estate, and requirements for launch-vehicle integration)
- Maximum transmit power (based on requirements for cooling the high-power amplifier, available technology, and environmental limitations)
- Maximum spacecraft weight (based on launch costs and the launch vehicle's lifting capacity)
- Restrictions on basing ground stations at overseas locations (based on survivability concerns, operational cost, frequency licensing, and length of use)
- Limitations concerning transmission media, based on existing communication infrastructure to which the satellite communications system may be connected

Constraints imposed by the spacecraft and ground terminal designs are discussed elsewhere in this book.

Channel Constraints

Of major importance are the constraints imposed by the ground-satellite transmission channel. We discuss the channel characteristics below.

Channel characteristics affect the communications system directly in almost every aspect—from the modulation format, to the antenna design, to the number and location of ground terminals and spacecraft. The *channel* is the medium through which we communicate. For conversations, the channel is air. For local area networks, it's twisted cable. For satellite communications, it's a combination of atmosphere and space. Channel characteristics can be natural or man-made. We usually refer to a channel consisting of natural phenomena as a *benign channel* (although communications in a benign channel can be quite challenging depending on the natural phenomena!). In contrast, a channel consisting of man-made phenomena (e.g., nuclear detonations and electronic jamming) is referred to as a *stressed channel*.

Atmospheric Absorption

The one-way attenuation due to atmospheric absorption depends on operating frequency, the Earth's surface temperature, humidity, and the Earth-station antenna elevation angle. The International Telecommunications Union has calculated theoretical one-way attenuation for a United States standard atmosphere. This calculation is for July at 45° North latitude and for frequencies between 7 and 50 GHz. From this data, Schwab [1980] has approximated the vertical (zenith) path attenuation (in dB) with the following polynomial expression:

$$L_{atm}(f, zenith) = e^{(k_1 f + k_2 f^2 + k_3 f^3 + k_4 f^4)} - 1 \tag{11.4}$$

where f is the operating frequency in GHz, zenith refers to a vertical (90° elevation) path, and the coefficients (k_1, k_2, k_3, and k_4) are given in Table 11.3. Values of atmospheric attenuation for elevation angles (ε) other than the antenna's zenith pointing angle are given by

$$L_{atm}(f, \varepsilon) = \frac{L_{atm}(f, zenith)}{\sin(\varepsilon)} \tag{11.5}$$

Table 11.3. Coefficient Values for the Schwab Polynomial Approximation. We can calculate the atmospheric absorption through Eq. (11.4) for the given frequency ranges.

Coefficients	7 GHz ≤ f ≤ 22 GHz	22 GHz < f ≤ 50 GHz
k_1	−0.00617564	0.13030320
k_2	0.00432368	−0.00816987
k_3	−0.00044445	0.00015648
k_4	0.00001358	−0.00000074

Using Eqs. (11.4) and (11.5), we generated the atmospheric absorption values in Table 11.4 for several elevation angles and for frequencies ranging from 20 to 50 GHz.

Table 11.4. Atmospheric Absorption (in dB) as a Function of Frequency and Antenna Elevation Angle (ε). The atmospheric absorption generally increases with increasing frequency and decreasing elevation angle.

	Elevation Angle, ε (deg)			
f (GHz)	90°	30°	20°	10°
20	0.25	0.50	0.73	1.44
30	0.20	0.40	0.58	1.15
40	0.30	0.60	0.88	1.73
44	0.60	1.20	1.75	3.46
50	1.79	3.58	5.23	10.31

Ionospheric Attenuation

Jamming isn't the only source of disturbance to the communication signals. We've known since the late 1950's that spacecraft signals suffer propagation disturbances in the natural ionosphere [Yeh, 1959]. Many reports state that these signals undergo even greater disturbances following high-altitude nuclear detonations (NUDETs) [King, 1980]. In either a natural or an artificially disturbed (resulting from a NUDET) ionosphere, random fluctuations of electron density cause radio waves to be scattered as they propagate through such regions, producing random variations in the received signal's amplitude and phase. Such signal variations are called scintillations or fading. Middlestead [1987] discusses the extensive research on nuclear detonations in the ionosphere and their effects on signal propagation.

These studies show that requiring a communications system to operate through a nuclear-disturbed region (a nuclear-scintillation requirement) will strongly affect choices on almost every characteristic of its design. These include modulation, coding, interleaving, and spread-spectrum choices in a signal-waveform design. We may have to employ satellite-to-satellite crosslinks to avoid transmitting through a nuclear-disturbed region. Or we may choose the spacecraft's orbits so they're less susceptible to the physical and communications-channel effects of NUDETs detonated at particular altitudes. The link margin may have to increase drastically to overcome the fading effects of scintillation, so the signal will have enough power at the receive terminal (received SNR) to meet the required BER. A nuclear-scintillation requirement will always have a considerable effect on our choices in developing the communications system; thus, we should validate it before moving the design forward.

Rain Attenuation

Rain attenuation has long been recognized as a principal cause of unwanted signal loss in satellite communications (SATCOM) systems operating from 3 to 300 GHz. Signal attenuation of tens of dB is possible depending on the specific characteristics of the rain event. So we have to account for signal attenuation from rain in designing the communications system to ensure that the communications link works properly.

Rain attenuation can have a significant effect on link availability, especially at frequencies above 10 GHz. In developing the communications system design we should specify a link margin equal to the worst rain attenuation we anticipate on a given link to compensate for the rain-induced power fades.

The link margin for rain can be significant and depends directly on the requirements for link availability. Figure 11.3 shows the expected rain attenuation as a function of frequency and elevation angles for link availabilities of 98.0% and 99.5%. We generated the data in this figure using the Crane model—a set of tables and equations based on observed climatic data used to estimate rain attenuation [Crane, 1980]—and assuming the climate of the United States' northeastern section. The link availability for a specific rain attenuation will be higher in the Midwest and lower in the Southeast, compared to those shown in the figure. Note that the Crane rain model attempts to model average weather conditions empirically over a period of years; it doesn't account for higher or lower rainfall in a particular year.

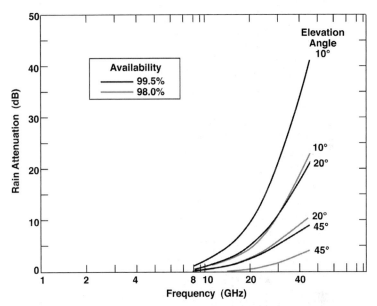

Fig. 11.3. **Rain Attenuation Predicted by Crane Model for Climate Typical of the Northeastern United States.** Ground-station altitude = 0 km, latitude = 40°. For other cases, see Ippolito [1986]. [Larson and Wertz, 1999]

As shown in Fig. 11.3, rain attenuation increases substantially for space-to-ground frequencies over 10 GHz. Figure 11.3 also shows that rain attenuation increases rapidly as the antenna's elevation angle decreases below 20°. When operating at frequencies above 10 GHz, a good rule of thumb is to limit the spacecraft's minimum elevation angle to 20°, especially in high rainfall areas. But increasing the minimum elevation angle reduces the coverage area significantly, which may mean the communications system won't meet other system requirements. We must balance all major performance and operational requirements.

Foliage

Another important contributor to unwanted signal loss in a communications system is foliage. Of all of the signal propagation effects typically evaluated for communications links operating in the millimeter wave regime, signal attenuation due to foliage is probably the most difficult to quantify accurately. The wide diversity in the types and density of foliage makes the estimate of attenuation highly variable. Most studies performed to date to quantify the impact of signal attenuation due to propagation through foliage have been concerned with horizontal propagation paths (i.e., parallel to the ground) and have only considered frequencies below 1250 MHz (see, for example, Tamir [1977], Horwitz [1979], and Tewari [1990]). In this region, the leaves are small compared to the wavelength and the forest is treated as a homogeneous layer above the Earth. But we can't assume homogeneity in the higher frequency regimes (3 – 100 GHz), where the leaves and the spaces between the leaves are large compared to the signal wavelength.

It's no wonder that the inconsistency in characterizing foliage from one study to another has made empirical models inconsistent. Comparetto [1993] gives a broad overview of various empirical models and data on foliage attenuation generated over a wide frequency range. That study determined foliage attenuation was difficult to quantify accurately. But it also showed that foliage attenuation could severely degrade, and potentially preclude, satellite communications regardless of the available link margin. The signal attenuation was 3 to 4 dB per meter of foliage in the 44 to 20 GHz operating regime but much less severe at frequencies below 1 GHz.

The Lincoln Laboratory at the Massachusetts Institute of Technology (MIT/LL) reported the results of a series of experiments carried out to characterize the attenuation due to foliage of a SATCOM signal operating at a frequency of 20 GHz. They used two experimental setups: (1) a transportable SATCOM terminal that received a frequency-hopped downlink signal from a satellite at geosynchronous altitude, and (2) a terrestrial setup composed of a transmitter operating at 20 GHz in the continuous wave (CW) mode with a compatible receiver. The conclusions were as follows:

1. Foliated trees are essentially opaque to 20 GHz signals from a communications standpoint (i.e., it is not practical to design a system with sufficient link margin to overcome the attenuation).
2. The "rf edge" of a tree is roughly equivalent to its "visible edge." Thus, one can expect to receive signals in the gaps between trees if the geometry is correct.
3. Unfoliated trees can also attenuate 20 GHz signals significantly; however, they are not opaque to communications. Diffraction, scattering, and multipath all play a part in signal loss, giving rise to attenuation levels which are highly variable and dependent upon the exact geometry of the received antenna and the tree(s).

Based on the severity of foliage attenuation in the millimeter wave regime, communications through any length of foliage should be discouraged. It is predominantly through operational adjustments and procedures (e.g., operating through holes in the foliage), and not link margin design, that communications through foliage can be assured with any degree of confidence. Thus, we might consider alternate frequencies or not using satellite communications if we have to operate through foliage.

Dust Attenuation

Most studies to date concerning the attenuation of SATCOM signals from dust have assumed the dust results from a nuclear detonation. However, experience from Desert Storm operations has shown we must consider signal attenuation from local dust storms in designing communications systems. SATCOM signals incident on dust particles in the atmosphere undergo some absorption and scattering, depending on the particles' size, shape, and complex dielectric constant as well as the signal's wavelength (or frequency). A thorough discussion of dust attenuation is beyond the level of this text. Refer to Comparetto [1993] for an overview of this topic. Comparetto shows the signal attenuation from dust is about 0.5 dB/km at 45 GHz and 0.2 dB/km at 20 GHz. Clearly, a requirement to operate through dust in the benign channel may reduce the communication link's performance significantly, depending on the operating frequency, and may require significant excess link margin in the communications system design to overcome this effect.

Program Constraints

Program constraints, such as cost, schedule, and the state of current and projected technology also affect our design. For example, the best ways to meet performance and operational requirements might end up being untenable when we consider cost. One example of a system for which cost became a driving factor was the Milstar COMSAT system. It was originally designed to operate through severe nuclear and electronic jamming environments. Planners emphasized meeting these and other operational and performance requirements without rigidly constraining costs. The result was a military-communications system design that met its operational and performance requirements very well—but at an unacceptably high cost. As a result, the design of this system underwent intense scrutiny because of cuts in defense spending. A major effort was made to re-validate the operational and performance requirements which resulted in major modifications to the communications system design. In a sense, the ground rules were changed in mid-stream by shifting emphasis from meeting requirements to minimizing cost. As a result, the Milstar program probably suffered unfairly.

Schedules also drive our design. For example, stipulating short- or near-term schedules for reaching initial operational capability (IOC) on a particular system may drive the system designers to use existing and proven technology in order to reduce risk. As a result, the system may not be able to meet certain performance requirements without future upgrades to the system. This approach can lead to short-term solutions for long-term problems, thereby fueling the "stove pipe" method commonly found in communications-systems. We must be sure short-term solutions don't preclude including enhanced technology as the system matures. This practice is commonly called Pre-Planned Product Improvement or PPPI.

As mentioned earlier, an important constraint on communications systems is the requirement to incorporate an existing infrastructure. Of course, this constraint depends on the size of the infrastructure, but it always drives a number of design decisions during development. An example is the Air Force Satellite Control Network (AFSCN). The vision since the mid 1990's has been to develop a satellite-control system for the DoD that responds to Air Force, Army, and Navy requirements. This, in itself, is a challenging task given the services' diversity. Worse yet, each service performs some degree of spacecraft control and therefore has some infrastructure to carry out their missions. In some cases, this existing infrastructure is substantial. For example, the AFSCN has been operating since about 1960 and represents a yearly commitment of more than $500M. The common-user segment of the AFSCN includes a number of remote tracking stations and space/ground antennas, two major operational control nodes, and various dedicated and support facilities and assets. Obviously, all future spacecraft-control systems must address the current AFSCN infrastructure if they're to be considered fiscally sound and technically doable. Although using this infrastructure may save operations costs, it may also keep the communications system from meeting operational, cost and/or performance requirements due to constraints imposed by the infrastructure.

Finally, the proposed lifetime of the communications system drives both cost and design decisions in its development. For example, if the proposed lifetime of a particular spacecraft exceeds the projected lifetime of the communications system's components, we may need to use redundant components to meet requirements for end-of-life (EOL) reliability. In many spacecraft systems today, we use redundancy for traveling-wave tube amplifiers, onboard power supplies, modems, multiplexers, and frequency up/down converters to meet the lifetime requirements. Costs therefore go up because of the number of units and the cost of launching their greater weight and volume. Also, the specified lifetime will affect the design choices made in terms of projected technology. For example, if the system lifetime includes several generations of the communications system's main components (e.g., computers), the system must be upgradeable without requiring us to redesign it.

11.2 Evaluating Design Options

Figure 11.4 shows the key characteristics of this phase, in which we evaluate the design characteristics of the ground terminal and spacecraft payload, together with the performance and operational requirements identified in the requirements-definition phase. To do so, we must consider the system constraints (including system complexity, power, size, and other factors) while identifying options for the communications system. We keep the most favorable options as the building blocks we'll use to develop the communications system. The following paragraphs discuss the more common options related to the spacecraft and ground terminal.

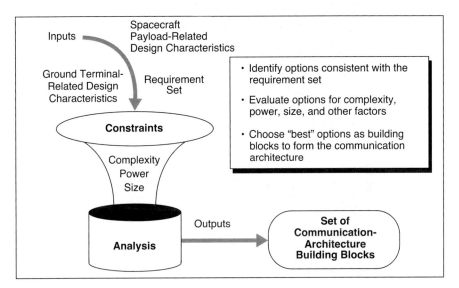

Fig. 11.4. Evaluating Communication System Design Options. In the design option phase, satellite payload and ground terminal based design characteristics are combined with the Phase 1 requirement set to develop a set of building blocks for the system.

11.2.1 System Configuration (Architecture)

A satellite communications system consists of multiple ground stations interconnected by satellite links. Most communication satellites today use repeater transponders (more commonly referred to as *bent-pipe transponders*). In a *bent-pipe configuration*, the spacecraft receives the uplink signal, translates the frequency of the received signal to the appropriate downlink frequency, amplifies the signal, and then retransmits the signal on the downlink. The bent-pipe transponder incorporates proven technology and is highly reliable and straightforward to use. However, bent-pipe transponders do have several significant disadvantages when the spacecraft receives multiple signals at the same time.

First, bent-pipe transponders are inherently non-linear. That is, the amount of amplification depends on the input-signal level. Normally, we want to operate the transponder in the linear region, where a step change in input-signal level will result in a corresponding step change in output-signal level. We do so to avoid generating intermodulation (IM) products and to reduce signal-suppression effects. The IM products interfere with the uplink signals and, essentially, rob part of the transponder's available transmit power.

A spacecraft's transponder, when operating in the non-linear regime, will tend to amplify stronger signals more than weaker signals, in a sense suppressing the transponder's output of the weaker signals. We call this effect *signal suppression*. To make sure the transponder operates linearly, we back off its output power about 3 to 6 dB from the maximum potential output power, thus reducing the transmitter's power output to roughly one-half to one-fourth of the maximum (saturated) power. Comparetto [1989 and 1990] further discusses IM products and signal-suppression effects in non-linear transponders.

An alternative to the bent-pipe transponder involves onboard processing using a *regenerative transponder*. In this case, the signal is demodulated on board the spacecraft and then routed to the appropriate downlink modulator/transmitter or antenna beam.

When ground terminals at the source and destination are within line-of-sight (LOS) of the same spacecraft, communications is straightforward, as shown in Fig. 11.5A. But if they're not, we need a data relay to support *source-to-destination connectivity*. The two main ways to relay data in satellite communications are ground-terminal relay (Fig. 11.5B) and satellite crosslinks (Fig. 11.5C).

Ground-terminal relay involves a double hop and an increased delay which can affect certain types of communications (e.g., the round-trip propagation delay is about one second, which is quite noticeable in voice communications). Ground terminals may also drive up costs, depending on their design and location. In many cases, we need an overseas ground terminal, which requires the host nation's approval before it can operate. Despite these drawbacks, ground-terminal relays are a proven and reliable way to support connectivity beyond line-of-sight in satellite-communications systems.

To date, very few satellite crosslinks have been incorporated into communications. The reasons vary but typically include

- Lack of technological maturity
- Limited availability of the required space-qualified hardware
- Difficulty of space-based antenna tracking, especially for low-Earth orbits
- Increased weight to the spacecraft payload from the satellite crosslink system. (This weight increases launch cost and may exceed the launch vehicle's ability to insert the spacecraft into orbit.)
- Favorable economics for the ground terminal relay to date

The last item above was clearly demonstrated with the IRIDIUM system. IRIDIUM is a space-based, mobile, satellite-communications system that employs 66 spacecraft in low-Earth orbit. The constellation consists of six orbital planes, with 11 spacecraft per plane. The satellites are in a near-polar orbit at an altitude of 485 miles (780 km) and they circle the earth once every 100 minutes. Each satellite is cross-linked to four other satellites; two satellites in the same orbital plane and two in an adjacent plane. Originally developed by Motorola, IRIDIUM was successful from a

11.2 Communications Architecture

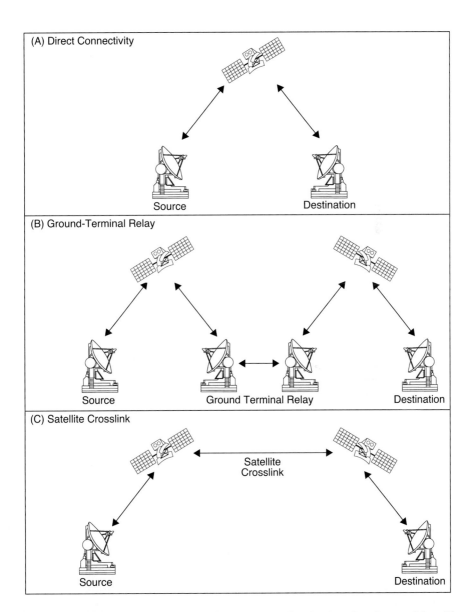

Fig. 11.5. Various Configurations of Sources and Destination for Connectivity. Direct transmission, ground-terminal relay, or satellite crosslinks can satisfy source-to-destination connectivity.

technological standpoint, providing source-to-destination connectivity using a sophisticated satellite crosslink design. IRIDIUM began offering service in 1998; however, the system could not overcome financial problems due to unexpectedly high operating costs and a lower-than-expected market share. The result was bankruptcy - IRIDIUM was then acquired by Iridium Satellite LLC in December 2000 and is currently providing global, wireless personal communications services on a much smaller scale than originally anticipated. Comparetto [1993 and 1994] discusses the IRIDIUM system, as well as two competing designs: Odyssey (by TRW, Inc.) and Globalstar (by Loral/QUALCOMM Inc.).

Requiring a ground terminal to support continuous spacecraft-to-ground connectivity (as opposed to a store-and-forward system) drives the choice of orbits in a communications system. The only way truly to support continuous connectivity is through geostationary orbits. In a geostationary orbit, the inclination is nominally zero degrees (positioned at the equator), and the period is one sidereal day (23 hours, 56 minutes, 4 seconds). Systems have been designed and deployed to provide high rate internet access using satellites in geostationary orbits. Hughes Network Systems and DirecTV, Inc., offer a service called DirecWay which is a satellite-based broadband service offering high-speed Internet service bundled with over 225 digital entertainment channels of DirecTV programming.

The newest application of satellites in geostationary orbits to provide communications to a large market involves the provision of satellite-based radio services. Sirius Satellite Radio, based in New York, and competitor XM Satellite Radio, based in Washington, D.C., have launched satellites to provide over 100 radio channels directly to your car. The XM Satellite Radio network comprises two satellites, appropriately called Rock and Roll, and the Sirius network comprises three satellites. All satellites are in geostationary orbit and are currently providing services.

Although a satellite in geostationary orbit provides connectivity to a large geographical area on the Earth's surface (and potential market), it is expensive to insert a spacecraft into such an orbit. One other way to support "quasi" continuous connectivity between the ground and a spacecraft is through a network of ground stations, in which at least one ground terminal is in view of the spacecraft at any given time. We use the term "quasi" because the individual spacecraft-to-ground contacts are periodically broken. We'd need a geographically dispersed global network for truly continuous coverage.

If continuous connectivity isn't necessary, we may use a store-and-forward approach. The satellite, in a sub-stationary orbit, receives and stores messages when in view of the transmitting terminal. The satellite transmits the messages later when it passes over a receive terminal. For example, we may employ a single ground station on the continental United States (CONUS) and support spacecraft-to-ground connectivity whenever the spacecraft is in view of the ground station as it completes it orbital journey. This approach would save a lot of money over the continuous-coverage case because

- We can dispense with overseas ground stations to relay messages between satellites, thereby reducing operating cost and eliminating host-nation approval
- Launch cost per spacecraft decreases (no need to use a geostationary orbit)
- Flexibility increase for choosing the launch vehicle because more launch vehicles can insert a spacecraft into a sub-geostationary orbit

In summary, we must carefully validate a requirement for continuous spacecraft-to-ground connectivity because it will increase the cost and complexity of the SATCOM system.

11.2.2 Modulation and Coding

There are numerous modulation and coding techniques to choose from. The cost and complexity of the communications system are highly dependant on this choice.

Modulation is the process by which an input signal varies the characteristics of a radio frequency carrier (usually a sine wave). These characteristics include amplitude, phase, and frequency. *Coding* is the process by which we add redundancy to the transmitted data stream to improve power efficiency for a given modulation technique, though at the expense of bandwidth. That is, the required received signal-to-noise ratio (SNR) to achieve a given bit-error-rate (BER) is lower by an amount referred to as the coding gain for a system that employs Forward-Error-Correction-Coding (FECC) techniques.

Demodulation measures the received carrier's amplitude, frequency and/or phase to estimate the transmitted data stream. *Decoding* removes the data redundancy from the received signal to detect and correct transmission errors. The most common form of FECC is convolutional encoding with Viterbi decoding. The theory behind FECC techniques is beyond the scope of this chapter, so we'll focus on common modulation techniques used in satellite-communications architectures. See Lin and Costello [1983], Proakis [1983], and Sklar [1988] for detailed discussions of FECC techniques along with the supporting theory.

Digital modulation, versus analog modulation, is predominant in most communications systems today. There are three primary categories of digital modulation to consider: amplitude shift keying (ASK), frequency shift keying (FSK), and phase shift keying (PSK). In each case, the amplitude, frequency or phase, respectively, is varied as a function of the binary sequence input to the modulator. ASK seldom appears in spacecraft systems because spacecraft normally use non-linear transponders or amplifiers. That is, these transponders do not amplify uniformly over the full range of input signal levels. This introduces uncertainty in the mapping of the received amplitude level to the input binary sequence. Additionally, ASK results in the generation of unwanted intermodulation (IM) products which "rob" some of the available transponder power (see Comparetto [1989 and 1990]). Both FSK and PSK techniques are quite common in space applications because they result in a constant signal envelope (i.e., constant transmit power level), allowing the satellite transponder to operate at or near saturation without generating IM products. This results in maximum transponder power efficiency.

Figure 11.6 depicts the most common modulation techniques used in spacecraft systems today. *Binary phase-shift keying* (BPSK) is most common. It consists of setting the carrier phase to 0° to represent a binary 0, and to 180° to represent a binary 1. In *quadrature phase-shift keying* (QPSK), we examine two bits and identify one of four symbols (00, 01, 10, and 11). We then map each symbol to one of four carrier phases (0°, 90°, 180°, and 270°). Note that for QPSK, the symbol rate is equal to one half of the bit rate, which decreases the required transmission bandwidth by one half.*

Frequency-shift keying (FSK) is another common modulation scheme. In FSK we assign two separate carrier frequencies, F1 and F2, to a binary 0 and a binary 1, respectively. The separation between F1 and F2 must at least be equal to the data rate in order to avoid performance loss that would otherwise result from mutual interference. Thus, the required transmission bandwidth is at least twice the width of the spectrum generated by BPSK (a minus), but the demodulation techniques available for FSK are less complicated than those for BPSK or QPSK (a plus). In M-ary frequency shift keying (MFSK), we map M separate frequencies to the M combinations of the transmitted bit stream, taken $\log_2 M$ bits at a time. For example, in a typical 8-ary FSK system, the input bit stream is sampled 3 bits at a time ($\log_2 8$), resulting in eight possible combinations (000, 001, 010, 011, 100, 101, 110, 111). Depending on the 3-bit combination observed, we assign one of the 8 possible frequencies to the carrier (F1, F2,..., F8). For 8-ary FSK, the symbol rate is one-third the bit rate (only one symbol is transmitted for every three input bits), and the required transmission bandwidth is approximately 8/3 the bit rate. The factor of eight results from using a frequency spacing between each of the eight possible frequencies that is equal to the transmitted data rate. Less common modulation schemes include minimum shift keying (MSK), offset QPSK (OQPSK), and 8-ary PSK.

One modulation that is gaining popularity in high data rate SATCOM applications is the M-ary Quadrature Amplitude Modulation (QAM) family. The M-ary QAM modulation schemes result in improved bandwidth efficiency (good for high data rate applications), albeit with a corresponding price of higher required received SNR to achieve a given BER. Sklar [1988] describes these modulation schemes.

---

* One of the most common FECC techniques used today is the rate 1/2 convolutional code. The rate 1/2 means that for every data bit that enters the encoder, two coded symbols exit. As a result, the system enjoys a coding gain (a plus) at the expense of doubling the required transmission bandwidth (a minus). Rate 1/2 convolutional encoding is often used with QPSK to offset this drawback.

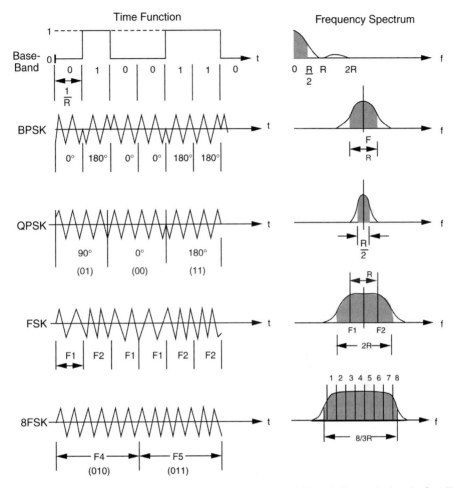

Fig. 11.6. Modulation Types Commonly Used for Digital Signal Transmission in Satellite Communications. R is the data rate. The shaded region is the required bandwidth. [Larson and Wertz, 1999]

To demodulate a received digital data bit reliably, the amount of received energy per bit, E_b, must exceed the noise power spectral density, N_o, by a specified amount. In other words, we need a minimum E_b/N_o to achieve a given BER. Communications theorists derive analytical expressions that relate the received E_b/N_o to the BER as a function of the modulation/demodulation type and FECC/decoding techniques applied. This area can become quite complex for M-ary modulation schemes combined with FECC techniques. We'll simply acknowledge the relationship between modulation type and bandwidth, as well as FECC's effect on overall system performance. These concepts are significant in developing a communications-link design, as we discuss in the following section.

The BER is a function of the received signal-to-noise ratio (SNR). Depending on how we modulate and code, as well as the characteristics of the transmission channel, the relationship between the BER and the received SNR can be complex. The expression below provides the BER, or probability of bit error (P_b), as a function of the received SNR for binary-phase-shift key (BPSK) modulation if we use coherent demodulation (including phase information) but don't use forward-error-correction coding (FECC):

$$P_b = (1/2) erfc\left(\sqrt{\frac{SNR}{2}}\right) \tag{11.6}$$

where $erfc(x)$ is the complementary error function, defined as

$$erfc(x) = \frac{2}{\sqrt{\pi}} \int_x^\infty e^{-t^2} dt \tag{11.7}$$

Several excellent references address this topic, including Sklar [1988], Ha [1986], Gagliardi [1984], Proakis [1983], Simon, et. al. [1985], and Spilker [1977]. The actual derivation of the BER equations is complex and involves probability, random variables, and stochastic processes. Papoulis [1965] has served as a benchmark reference in these areas for years.

With increases in carrier frequency and effective isotropic radiated power (EIRP) in both ground and satellite terminals, very high data rates are possible with moderate size antenna dishes. In these situations, data rates become high enough that power and SNR considerations are no longer the limiting factors. Instead the transponder bandwidth may become the main issue, assuming that traditional QPSK signals are in use. Under these conditions, data rates can be further increased by changing to higher order, bandwidth efficient modulations, such as 8-PSK, 16-QAM (quadrature amplitude modulation) or 64-QAM. Some of these higher order modulations, such as the QAM family, are not constant envelope, and therefore are not normally considered for operation with satellites with nonlinear "bent pipe" transponders. However, if the transponder is used in a frequency division multiple access (FDMA) mode of operation (discussed later in this chapter), with more than 10 carriers, and no one signal dominating, constant envelope modulations are not required, since amplitude variations are preserved through the transponder. On the other hand, non-constant envelope modulations should not be used in a Time Division Multiple Access (TDMA) mode of operation - constant envelope modulation such as 8-PSK must be used.

The use of higher order modulations provides a great advantage for bandwidth efficiency; however, these modulations generally have worse power performance. Hence, changing to a higher order modulation may increase the available data rate, but may cause the system to become limited by power and SNR considerations again. Consequently, we must make a power/bandwidth trade-off and for a given transponder EIRP and earth terminal capability, there will be an optimum modulation format that maximizes achievable data rate.

One way to overcome this effect is to use forward error correction coding (FECC) techniques to decrease the SNR required for a particular BER. Reed-Solomon encoding and convolutional encoding are in use on many of today's satellites. Turbo codes are a powerful new error correction technique. Turbo codes use the concept of "product codes" to minimize the complexity of the decoder. A simple example of a product code is shown in Fig. 11.7.

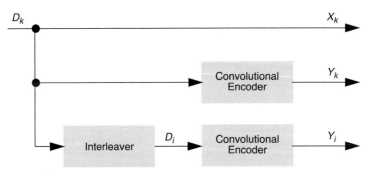

Fig. 11.7. **Example of a Generic Rate 1/3 Turbo Encoder.** The primary innovation of turbo codes is the interleaver, which permutes the original information bits before input to the second encoder.

As shown, the turbo encoder comprises two rate 1/2 convolutional encoders separated by an N-bit interleaver. The result is a rate 1/3 turbo encoder. The received signal is then sequentially decoded in an iterative fashion.

The primary benefits of turbo coding techniques include improved BER performance at lower SNR (referred to as the coding gain) which could provide additional rain margin and/or a decrease in the required terminal EIRP or receive figure-of-merit (i.e., G/T) to support a given communications system. Studies have shown that it is possible to achieve near-Shannon limit performance using turbo code techniques. The data demonstrate that turbo codes offer significant coding gain over conventional coding schemes - up to 4 dB over soft decision convolutional codes and up to 2 dB over a Reed-Solomon (RS) code concatenated with a convolutional code. Additionally, the application of turbo coding techniques effectively "spreads" the transmitting signal, which could help meet potentially limiting future International Telecommunications Union (ITU) Earth terminal transmission criteria.

The performance of turbo code implementations is a function of the interleaver size and design, the constituent codes used to encode the data and the number of decoder iterations. Turbo code implementation has been shown to achieve a BER of 10^{-5} at an E_b/N_o of 0.7 dB for an additive white gaussian noise (AWGN) channel with 18 decoder iterations.

Along with the benefits identified above, there are several important issues associated with the use of turbo codes in a communications system. First, the design of turbo codes is complex and computationally expensive. A number of code parameters must be optimized, including the decoder structure (parallel versus serial concatenation), encoder polynomial, code rate (i.e., number of component encoders and whether or not to apply puncturing techniques), interleaver structure, the number of iterations, and the decoder algorithm. Second, the decoding delay can be significant with turbo codes. The decoding delay is a function of the interleaver size and the data rate. Finally, signal acquisition and tracking can be difficult with the low SNR values that can be supported with turbo codes. The "theoretical" received SNR required to provide acceptable BER levels with turbo code implementations can be in the 1 dB range or lower. This is especially important in communications system applications in which signal acquisition and tracking will be challenging, such as mobile communications nodes or deep space communications.

Finally, Fig. 11.8 shows the relationship between the probability of a bit error, or BER, and the received E_b/N_o, or SNR, assuming BPSK modulation with and without using several FECC techniques. As shown, the BER can vary significantly with only small variations in the received E_b/N_o.

11.2.3 Anti-jamming Techniques

Various design techniques have arisen over the years to combat jamming:

- Spread-spectrum processing, which decreases the jammer signal's effect on the received friendly signal (e.g., frequency hopping [FH] and direct-sequence or pseudo-random noise [PRN])
- Antenna nulling, in which we detect a jammer's presence and sharply decrease the antenna's gain in its direction
- Forward-error-correction coding (FECC), which compensates for the additional error jamming causes

In an FH system, the frequency of the transmitted signal hops across a wide bandwidth (relative to the data bandwidth) at a rate fast enough to keep a frequency-follower jammer from intercepting the signal, determining the frequency, and retransmitting an interfering signal before the next hop. FH systems can use an effective spread bandwidth in the GHz range, with hop rates in the khps (kilo hops/s) range.

In a DS or PRN system, the transmitted data are modulated with a very high-rate PRN code. The result is that the transmitted signal occupies a much wider instantaneous bandwidth than is actually needed to transmit the given data but at a lower average energy across the transmitted frequency spectrum. In this way, the

11.2 Communications Architecture

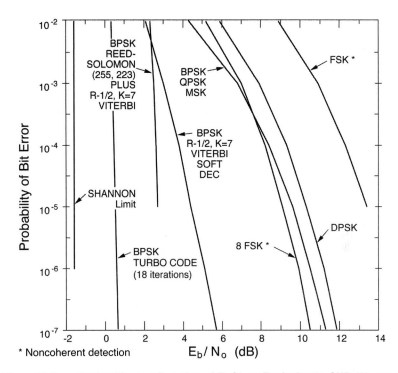

Fig. 11.8. Bit Error Probability as a Function of E_b/N_o or Equivalently, SNR. We can approach the theoretical performance (Shannon) limit by using error-correction coding. [Larson and Wertz, 1999].

transmitted signal appears as background noise to the unintended receiver. The spread-bandwidth limit for DS systems is about 100–300 MHz. Dixon [1976] and volume one of Simon [1985] are comprehensive references on the topics of spread-spectrum systems using FH and DS.

Onboard satellite processing is preferable to the bent-pipe transponder to despread the received uplink signal before retransmission on the downlink. Otherwise the uplink jamming signal will take over most of the satellite transmitter's power, leaving little for the signal.

Torrieri [1992] and Simon [1985] treat jamming and jamming countermeasures in more detail.

11.2.4 Multiple Access

Some missions require multiple uplinks or downlinks, especially for communications systems that integrate a number of spacecraft and ground stations into a single network. In such systems, it costs less to share the limited amount of satellite link capacity among users. Figure 11.9 shows three basic techniques for sharing link capacity: frequency-division multiple access (FDMA), time-division multiple access (TDMA), and code-division multiple access (CDMA).

In *FDMA* systems, a set of Earth stations transmits modulated uplink carriers for the spacecraft to relay simultaneously to various downlink Earth stations. Each uplink carrier is assigned a frequency band within the available rf bandwidth of the spacecraft. In a bent-pipe transponder, the entire rf frequency spectrum appearing at the spacecraft input is frequency-translated to form the downlink. The destination ground station receives the source ground station's signal by tuning to the proper band in the downlink spectrum. FDMA represents the simplest way to achieve multiple access. The required system technology and hardware are readily available in today's communication market [Gagliardi, 1984].

In *TDMA* systems, a single time slot in each time frame is assigned to a single input channel. A digitized input signal is sampled and stored in buffer memory. These samples then transmit as short bursts within the assigned time slots. The bit rate during the burst is high, therefore requiring a high peak transmitter power. At the receiver, the samples are sorted, stored, and then read out at the original rate. These samples are then converted to an analog signal, if required, and smoothed to obtain a replica of the original input signal. If the spacecraft uses a multiple-beam antenna, we may use a switching matrix on the spacecraft to route each time-slot burst to the desired downlink antenna's beam. NASA's Advanced Communications and Technology Satellite (ACTS) system uses this technique, known as Satellite Switched (SS) -TDMA [Naderi, 1988].

If onboard processing is used on the spacecraft, different multiple access techniques can be used on the uplink and downlink. For example, the FLTSATCOM EHF package (FEP) is a system that employs FDMA on the uplink and TDMA on the downlink [McElroy, 1988].

In *CDMA* systems, each user is assigned a unique pseudorandom noise (PN) code, with all users transmitting simultaneously and at the same frequency. The data is first phase-modulated by a carrier (normally using BPSK or QPSK techniques), with the resulting carrier then biphase-modulated with a PRN code that is at a much higher rate than the offered data traffic. This results in the generation of a wide bandwidth, low spectral density spread spectrum signal. Each user undergoes the same process and, in effect, behaves as a low level interferer to every other user in the system. The maximum number of CDMA users is limited by their aggregate background noise level. The received signal is then "de-spread" by applying the same PRN code to the received data stream. If the codes are synchronized the user signal is recovered and the other CDMA user signals simply contribute to the background noise level. This is because of the high auto-correlation and low cross-correlation characteristics of the PRN codes used in CDMA designs.

The number of simultaneous users that a CDMA system can handle is limited by the low-energy noise the undesired users' signals generate. Assuming all CDMA users' code rates and data rates are the same and that their carrier powers at the receiver input are equal, the question becomes: How many users can the

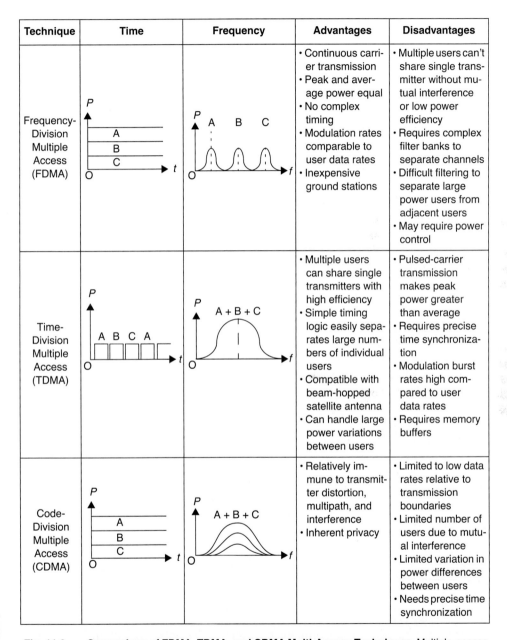

Fig. 11.9. **Comparison of FDMA, TDMA, and CDMA Multi-Access Techniques.** Multiple-access techniques allow different users to share the same transmission channel. The plots represent power (P) as a function of time (t) or frequency (f) for multiple users. [Larson and Wertz, 1999]

system support? Equation (11.8) estimates the maximum number of users, N_{users}, the system can handle as a function of the energy-per-bit to noise power density ratio, E_b/N_o (Note: E_b/N_o equals the received SNR):

$$N_{users} = 1 + R_{ss}\left(\frac{1}{(C/N_o)_{Req'd}} - \frac{1}{(C/N_o)_{Act}}\right) \quad (11.8)$$

$$(C/N_o)_{Req'd} = R_d(E_b/N_o)_{Req'd} \quad (11.9)$$

$$(C/N_o)_{Act} = R_d(E_b/N_o)_{Act} \quad (11.10)$$

where R_{ss} is the spread bandwidth (normally taken as the PRN-code chip rate, R_c), R_d is the data rate (R_c is much greater than R_d), $(C/N_o)_{Req'd}$ and $(C/N_o)_{Act}$ represent the required and actual carrier-to-noise power density ratios, respectively, and $(E_b/N_o)_{Req'd}$ and $(E_b/N_o)_{Act}$ represent the required and actual signal-to-noise ratios, respectively. Note that C/N_o and E_b/N_o are expressed as power ratios, not in dB, in Eqs. (11.8) through (11.10). Typical values for the chip rate (R_c) range from 1 to 100 megachips per second (Mcps).

The Global Positioning System (GPS) is an example of a system employing CDMA. Each spacecraft transmits a PRN code with a different time phase. The PRN chip rate, R_c, is 1.023 Mcps; the data rate, R_d, is 50 bps; $(E_b/N_o)_{Req'd} = 10$ dB = 10 (when not expressed in dB); and $(C/N_o)_{Act} = 38.6$ dB-Hz = 7,244 Hz (when not expressed in dB). Substituting into Eqs. (11.8) and (11.9),

$$N_{users} = 1 + 1.023 \times 10^6 \, (1/(50 \times 10) - 1/7244) = 1905.8$$

we find that the system can support 1,906 users. Another example of an operational CDMA system is NASA's Tracking and Data Relay Satellite System (TDRSS). TDRSS offers a CDMA-based multiple access service than can support the simultaneous reception of 20 return signals from a low Earth orbit platform through TDRSS to a centralized ground terminal.

CDMA systems suffer from the complexity associated with synchronizing the transmit and receive PRN codes and, as shown above, from a limitation in the maximum number of users supportable within a given bandwidth assignment. CDMA is usually less bandwidth-efficient than FDMA or TDMA, but it is less susceptible to interference, including multipath caused by reflections from buildings or other objects. This characteristic makes CDMA attractive for satellite-communications systems using mobile terminals. For more information on multiple-access techniques, see Gagliardi [1984], Sklar [1988], or Ha [1986].

If accurate time and frequency synchronization is maintained between transmitter and receiver, orthogonal PRN codes can reduce interchannel interference. Extensive work in multiple access techniques is underway to increase the capacity of wireless systems; see Hikmet [2000] and Hara [1997].

11.2.5 Communication Networks

Two broad types of communications networks exist within the context of a communications system: switched communications networks and broadcast communications networks. In a *switched communications network*, data is transferred from source to destination through a series of intermediate nodes. These nodes aren't concerned with the content of the data; rather, they switch data from node to node until it reaches its destination. *Broadcast communications networks* have no intermediate switching nodes. At each node a transmitter/receiver communicates over a medium shared by other nodes. A transmission from any one node is broadcast to and received by all other nodes within the transmission footprint. Most SATCOM networks employ broadcast-communication techniques in that the transmitted data are receivable by any ground terminal within the spacecraft's footprint; however, future SATCOM systems employing satellite-to-satellite crosslinks will lend themselves to switched-network techniques that could enhance the system's network performance.

Two common types of switched communications networks include circuit-switched networks and packet-switched networks. In a *circuit-switched network*, a dedicated communications path is established between the source and destination node through a number of intermediate nodes in the network. Essentially, the path is a connected sequence of physical links between nodes. On each link, a logical channel is dedicated to the connection. Data generated by the source node is transmitted along the dedicated path continuously. No routing or switching delay accrues along the path (other than the initial circuit-setup delay) because the path is fixed. The most common example of a circuit-switched network is the telephone network.

In a *packet-switched network*, it's not necessary to dedicate transmission capacity along a dedicated path within the network. Rather, data goes out in a sequence of small bundles, called packets. Each packet is independently routed through the network from node to node, traveling from source to destination. At each node, the entire packet is received, briefly stored, and then transmitted to the next node. Message switching is a form of packet-switching in which the transmitted packet consists of the entire message to be transmitted. One benefit of message switching is that we don't need to reconstruct the received data packets because each packet received comprises the entire message. The primary drawbacks to message switching are that the transmission delay may be quite large, depending on the message size, and may vary if the message sizes aren't uniform.

Satellite-to-satellite crosslinks are needed to exploit the advantages of switched satellite communications networks. Currently, IRIDIUM is the only operational satellite system that employs crosslinks in support of a packet switched communications system design for the space segment. For more detail on network communications, see Stallings [1985].

11.2.6 Spacecraft Antennas

Selection of the spacecraft antennas is perhaps the most crucial task in the communications systems design. The antenna size, weight and pointing requirements have a significant effect on the spacecraft design.

A number of antenna design options, listed in Table 11.5, are available. The parabolic reflector antenna is best suited for applications where the peak gain is above 20 dB and beamwidth is less than 15°. For lower gain wider beam applications, we may use other types of antennas that are lighter in mass and simpler to implement, especially at frequencies below 1 GHz. Such an application is the earth coverage satellite antenna, where the beamwidth is made just big enough to illuminate the earth. At geosynchronous altitude, this beamwidth is 18°. INTELSAT uses a simple horn antenna at 4 GHz for this application. The helix is often lighter weight and easier to mount on a satellite structure when the frequency is below about 2 GHz. A single helix is used for the FLTSATCOM UHF antenna, while INTELSAT uses a quad-helix L-Band antenna.

The basic antenna design options are reflectors, lens, and phased array. The reflector is desirable for satellite use because of its light weight, low complexity and cost, and design maturity. Offset feeds are often used to simplify the satellite structure. The feed is mounted within the satellite structure and aimed at the reflector, which is deployed after launch. Another advantage of the offset feed is it minimizes aperture blockage, which reduces the sidelobes levels.

To support high data rates with low satellite power, the antenna beamwidth should be narrow to achieve high antenna gain. (After all, why waste satellite power transmitting to an area where no receivers are located?) However, the narrow beam may provide insufficient coverage to establish links between widely separated ground stations. We solve this problem by using antennas that generate simultaneous multiple beams or a single beam that is scanned (hopped) over the earth's surface using time multiplexing between channels. We thus provide high antenna gain and broad area coverage at the same time.

Table 11.5 lists several approaches for generating multiple or scanning beams. One way is to switch the transmitter to a feed located off the antenna axis. However, a reflector antenna's gain decreases when scanned off axis. This loss can be offset by a shaped secondary reflector, but scan angles greater than about 10° are difficult to achieve. The lens antenna can be designed for good scanning performance, but its mass is generally larger than the reflector (plus feed) when the diameter exceeds about 0.5 m. Scanning is controlled electronically by switching between feed elements or by varying the amplitude or phase of each element. For example, the MILSTAR uses a lens with a switched feed array, and the ACTS uses a reflector with an offset feed array.

A phased array antenna can generate one or more beams simultaneously. The beams are formed and steered by controlling the phase and/or amplitude of each radiating element of the array. Currently under development are phased array

Table 11.5. Antenna Design Options Used in Spacecraft, Usually at Frequencies Above 2 GHz.
Note the characteristics of each type of antenna.

Antenna Type	Characteristics
Parabolic Reflector Center-Feed	• Aperture blockage raises sidelobe level. • Simple, lightweight structure. • Feed-mounted equipment exposed to environment. • Long transmission line from feed reduces efficiency.
Parabolic Reflector Cassegrain	• Aperture blockage raises sidelobe level. • Lightweight structure. • Short, low-loss transmission line. • Feed-mounted equipment accessible behind reflector. • Shaped subreflector increases efficiency (Increases gain by ~1.5 dB).
Parabolic Reflector Off-set Feed	• Same as Center-fed Parabolic Reflector except low aperture blockage reduces sidelobe level and increases efficiency. • Convenient for satellite mounting with feed embedded inside satellite.
Parabolic Reflector Off-set Shaped Subreflector with Feed Array for Scanning	• Light weight. • Low aperture blockage reduces sidelobe. • Limited scan angle ≈ 10°.
Lens with Switched-Feed Array	• Good aperture efficiency (no blockage). • Mismatch at lens surface causes losses. • Heavy, especially low-frequency applications (used at frequencies above 15 GHz with diameters below 0.5m).
Phased Array	• High aperture efficiency. • Multiple independently steerable beams. • High reliability (distributed active components). • High cost, weight. • Higher losses in feed distribution system. • High EIRP obtained from many small transmitters (space combining).

antennas in which the solid state power amplifiers are integrated with the radiating elements.

11.2.7 Power Amplifiers

A crucial design option is the selection of the transmitter power amplifiers for the ground terminal and satellite. The ground terminal power amplifier impacts the size, mobility and cost. The satellite power amplifier drives the spacecraft weight and cost.

Four categories of power amplifiers are currently in use: the traveling-wave tube amplifier (TWTA), the solid-state amplifier, the klystron, and the microwave power module (MPM). Of the four, TWTAs are the most extensively developed. Their theory is also well understood, and they've been used successfully in all types of space missions. For these reasons, TWTAs have emerged as the universal choice in power amplifiers for Earth stations and satellites operating at 30 GHz and beyond.

The *klystron* is used mainly in ground-terminal designs, which require high transmission powers (kilowatts) at high frequencies (>30 GHz). They are large and typically require cooling to reduce the system noise temperature that results from random electron motion. For these reasons, they're seldom used on board satellites.

In recent years, research and development on *solid-state power amplifiers* (SSPA) has accelerated. The two most common types of SSPAs include Gallium Arsenide, field-effect transistor (GASFET) amplifiers and impact-avalanche transit time (IMPATT) diode amplifiers. Solid-state amplifiers aren't as efficient as TWTAs or klystrons, and power-combining techniques are typically required to attain output power of 10–20 watts for operating frequencies up to 44 GHz.

But solid-state amplifiers offer significant advantages over TWTAs and klystrons, including decreased size and weight, ruggedness, compactness, and longer life. Of course, all these advantages translate into cost savings for the system.

We expect solid-state amplifiers to become the amplifier of choice for satellite applications as their efficiency, output power, reliability, and component availability increase over time. Under development are techniques for integrating multiple solid state amplifiers with elements of a phased array antenna or antenna feed. These will perform both power combining and antenna beam steering functions.

Finally, microwave power modules (MPMs) may be an attractive alternative to SSPAs and TWTAs for multi-band ground terminals. MPMs contain a mini-TWTA, solid state driver and power supply integrated into a smaller and lighter package than that of a traditional TWTA, and with higher power and efficiency than a SSPA. The solid state driver can be operated well below saturation without compromising overall efficiency, resulting in good linearity. A high-voltage power supply is required, but the voltage is typically half (4 kV) that required by a TWTA. Power-combining MPMs should allow us to realize power levels of 340W at Ka-band frequencies and 225 W at EHF. Advances in millimeter wave power combining and packaging techniques could be reasonably expected to result in weight estimates of well below 25 lbs by 2015. Power-combining MPMs would also result in graceful degradation in terminal capability in the event of a single tube failure. Although MPMs are an attractive solution, technology investments will be required to improve output power at higher frequencies and to develop efficient and lightweight millimeter-wave packaging and power combining techniques.

11.2.8 Optical Communication Links

In recent years, lasers generating narrow-band energy at optical frequencies have provided an attractive alternative to microwave-frequency transmission. Unfortunately, clouds and rain seriously attenuate optical links, so these links have limited application in satellite-Earth communications. However, an optical link is well suited for crosslink communications between satellites. Intersatellite links have been proposed using optical links with capacities above 300 Mbps.

Optical crosslinks are superior to microwave crosslinks for high data rates because they can support these rates using relatively small antenna diameters and system weights, as shown in Figs. 11.10 and 11.11. On the other hand, the narrow optical beams typical of laser communications (beamwidth is inversely proportional to operating frequency) are difficult to acquire and point accurately, requiring complex and sometimes heavy pointing mechanisms. Figure 11.11 compares rf and laser crosslinks, showing that rf links are usually better for data rates less than about ~100 Mbps because of their lower mass and power. But development of more efficient lasers with lighter and steerable optics may some day make lower-rate optical links attractive.

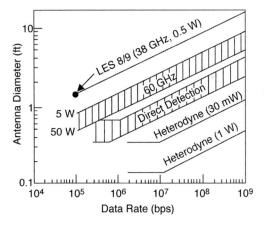

Fig. 11.10. Antenna Diameter and Data Rate for Several rf and LASERCOM Systems. Optical systems (direct-detection and heterodyne) require smaller antenna diameters compared to rf crosslinks. [Larson and Wertz, 1999].

One of the first successful demonstrations of the potential for laser communications occurred in 1980 with the Airborne Flight Test System (AFTS) sponsored by the U.S. Air Force. A 1 Gbps data rate link was established between a KC-135 aircraft and a ground station located at White Sands, New Mexico employing a cavity pumped, frequency doubled Nd:YAG laser. This program demonstrated that a reliable link could be established between a dynamic platform (aircraft) and a receiving ground station. The equipment was not packaged for

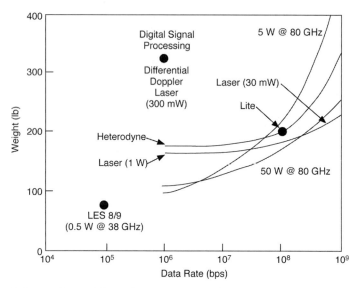

Fig. 11.11. A Comparison of Crosslink Package Weights. At data rates greater than about 100 Mbps, an optical system provides the lightest package [Larson and Wertz, 1999].

compactness, with rack-mounted electronics and a "flying optical bench" making up the hardware, although the program was successful enough that the Air Force decided to proceed with a production lasercom effort for satellite-to-satellite communications.

In 1994 Japan launched the Laser Communication Experiment (LCE). A small laser terminal was to provide 1 Mbps up and down link capability to the ETS-VI satellite. Unfortunately, the satellite was placed in a highly elliptical orbit, making communications difficult. Limited data was collected by Japanese researchers at their optical ground site outside Tokyo. A group at the Jet Propulsion Laboratory (JPL) was able to secure special funding to conduct a series of tests with LCE from their facility at Table Mountain, California. These were the first communications from a laser ground station to a satellite. Although the LCE terminal was not built with state-of-the-art hardware, it provided valuable knowledge of the challenges of integrating the diverse technologies needed to produce a space-qualified laser terminal.

Another application of optical links between spacecraft and Earth is the blue-green laser link being developed by ARPA and the US Navy for submarine communications [Weiner, 1980]. The laser frequency of 6×10^{14} Hz (equivalent to a wavelength of 0.5 μm) was chosen for its ability to penetrate sea water. Even so, the water loss can range from 5 to 50 dB or more, depending on the actual depth of the submarine. In addition, loss due to cloud scattering is 4 to 14 dB. We overcome these losses by using low data rates of 10 to 100 bps, advanced coding techniques, and high-gain, narrow-beam optics.

One driver for using laser communications is the desire to operate at higher frequencies relative to rf communications. The carrier operating frequency affects the ground-terminal antenna's achievable size and beamwidth. In turn, these factors indirectly affect spacecraft size, mass, and complexity. The carrier operating frequency also determines how much attenuation we can anticipate due to rain and other benign- and stressed-channel characteristics (see Sec. 11.1.2).

11.3 Forming the Architecture

Figure 11.12 shows the main characteristics of this phase, in which everything comes together. We consider the existing infrastructure, defined requirements, and the components identified in the design option phase to put together potential communications systems. Much to the chagrin of communications-system engineers, past experiences (both good and bad) and politics often affect the outcome of this phase. We must evaluate each communications system alternative for operational, technical, schedule, and cost feasibility.

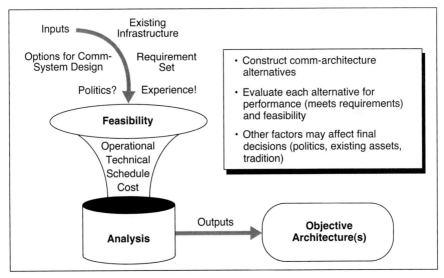

Fig. 11.12.   Forming the Communications System. We now integrate the results of the previous phases to generate communications system alternatives.

In this section, we describe the more common types of communications architectures (i.e., the way in which the communications systems are connected) and the process for designing the space-to-ground communications link. An example of a design iteration needed to meet a change in the stated performance requirements concludes this section.

11.3.1 Select System Architecture and Orbit Configuration

The first step in forming the communications system is to select the communications-system architecture. We must do this to define the spacecraft orbits so that the ground-to-space links can be designed.

One way of categorizing communications-system architectures is by describing the spacecraft orbits or the main way of distributing the data throughout the network. Figure 11.13 shows several examples.

The *store and forward* architecture (Fig. 11.13A) for relaying communications data through spacecraft appeared in 1960 when the US Army launched the Courier satellite [Mottley, 1960] and is still an important part of modern-day satellite systems. ORBCOMM, a satellite system that provides global data services (similar to 2-way paging or email) via low-Earth orbit satellites, is based upon a store and forward communications architecture.

In store and forward satellite architectures, the spacecraft is generally positioned in a low-altitude orbit (under 1000 km), receives data, and stores it in memory until it moves within view of a receive ground station, at which time it transmits the stored data. This architecture results in a relatively low launch cost because we're using low-altitude orbits. The spacecraft cost also stays relatively low because of a simpler communications payload. One example uses omni-directional (or, at least, wide-beamwidth) antennas to close the communications link. The larger beamwidth results in lower antenna gain, but the nearness of the low-altitude spacecraft decreases free-space loss to compensate for this effect. Additionally, the wide-beamwidth antenna greatly decreases the complexity required for pointing and stabilizing the antennas, and spacecraft stationkeeping is normally not required. The main disadvantage of this architecture is its long access time and transmission delay, perhaps hours, the result of waiting for the spacecraft to pass within view of the user's ground station before delivering its data.

Virtually all communication spacecraft, as well as many others, use a geostationary orbit (Fig. 11.13B). The spacecraft enters a near-zero-degree inclination orbit at an altitude of approximately 35,785 km. The period of the orbit is exactly equal to the period of the Earth's rotation (one sidereal day or 23 hours, 56 minutes, 4 seconds), making the spacecraft appear stationary when viewed from the ground. The cost of ground stations is usually less for this architecture because it requires little or no antenna-pointing control. A geostationary network is far easier to set up, monitor, and control compared to a dynamic network containing non-geostationary spacecraft. There is no need to switch from one spacecraft to another, for the spacecraft is always in view of the ground station. The main disadvantages of this architecture include the lack of coverage available above 70° latitude and the high cost to launch a spacecraft into a geostationary orbit. Furthermore, the one-way, source-to-destination propagation delay for a geostationary spacecraft was shown earlier to be approximately 0.25 second, which can sometimes cause problems in communications-satellite systems. Problems

11.3 Communications Architecture 431

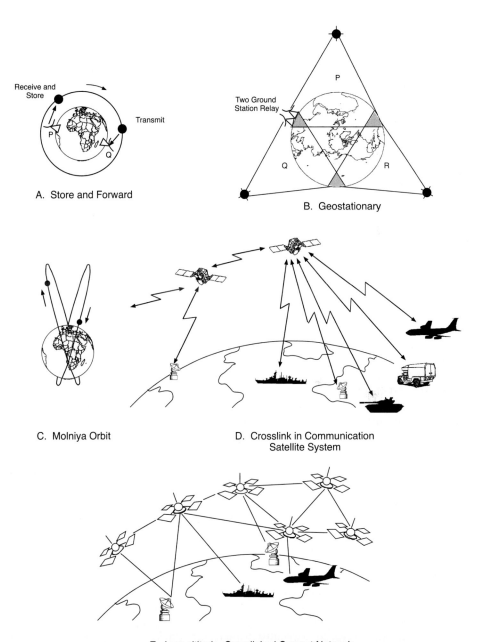

Fig. 11.13. Typical Communications Architectures Used to Satisfy Different Mission Requirements. Orbit configurations or space elements categorize the communications architectures. [Larson and Wertz, 1999]

include voice echoes in voice-communications systems and acknowledgment delays in stop-and-wait message protocols.

Examples of spacecraft systems employing geostationary orbits are INTELSAT, INMARSAT, the Galaxy satellite system, the Defense Satellite Communications Systems (DSCS), the Tracking and Data Relay Satellite System (TDRSS), and the Fleet Satellite Communications System (FLTSAT). Refer to Comparetto [1993] for a more detailed description of the INTELSAT and INMARSAT systems.

The Russian space program uses a *Molniya-orbit architecture* (Fig. 11.13C) to cover the northern polar regions. The spacecraft are in highly elliptical orbits with an apogee of 40,000 km, a perigee of 500 km, and an inclination angle of 63.4°. The apogee is over the North Pole to cover northern latitudes. The period of the orbit is 12 hours, but because it is highly elliptical, the spacecraft spends about eleven hours of each period over the northern hemisphere and approximately eight hours in site of Russian ground stations. Two or more spacecraft orbit in different planes, phased so that at least one is always in view from all northern latitudes. Unfortunately, the Molniya orbit requires continuous changing of antenna-pointing angles at the ground station and switching links between spacecraft as they move into and out of view of the ground station. This makes ground terminals more complex. Furthermore, although this architecture covers the northern latitudes well, it covers the southern latitudes rather poorly.

Satellite-to-satellite crosslinks can support source-destination connectivity for those cases in which the respective ground terminals are not within view of the same spacecraft. Satellite crosslinks were addressed in Sec. 11.2.1. Figure 11.13D shows a configuration in which satellite crosslinks are used between geostationary spacecraft (as opposed to the low-orbiting spacecraft in Fig. 11.13E). In crosslink configurations, an intermediate spacecraft relays the data to the target spacecraft, which contrasts with doing a double hop using two adjacent ground stations, as shown in Fig. 11.13B. The crosslink configuration performs better than the double-hop configuration because it has a smaller propagation delay (the double hop's delay is about 0.5 s). Crosslinks also don't require landing rights from a foreign country for ground-terminal operations (for communications architectures needing global connectivity). Foreign-based ground terminals can be costly and decrease system survivability and security. The obvious disadvantage of a satellite-crosslink architecture, however, is having to develop and implement the crosslink package, which can greatly increase the system's overall complexity, risk, and cost.

The *low-altitude, crosslinked architecture* (Fig. 11.13E) is basically an excursion from geostationary crosslinks. Multiple spacecraft are placed in low-altitude (500 to 3000 km) orbits, and satellite-to-satellite crosslinks support a richly connected network. The data would typically transfer in packets of a few hundred or thousand bits (packet switching), with each packet time-stamped and labeled with its destination. The data packets may arrive at the destination node by different paths exhibiting different propagation delays, depending on the spacecraft/

ground station geometry at the time of transmission. The receiving station must be able to sort and reassemble the packets in the correct order to obtain the original message.

The low-altitude, cross-linked architecture is highly survivable because it has many paths between a given source and destination. The number of spacecraft, together with their low-altitude orbits, improves resistance to jamming from the ground. The reason is that each spacecraft can only be seen by a relatively small segment of the Earth's surface so that many jammer terminals (which means a prohibitively high cost) would be necessary to disrupt network communications effectively. Finally, the uplink transmitter power needed at the ground terminal is lower because the ground terminal and spacecraft are closer together. This closeness also decreases the probability of unauthorized reception. On the other hand, this architecture requires complex network synchronization and spacecraft control. One example of this type of architecture that has received notoriety is the IRIDIUM system discussed previously in Sec. 11.2.1.

We can categorize satellite-communications architectures by two main functions: collecting and distributing data and communicating. Weather satellites are a good example of the first function. Using sophisticated sensing equipment, they measure atmospheric conditions, photograph the atmosphere and cloud activity, and then send the data to ground stations. On the ground, the data goes through further processing and, perhaps, further distribution to interested users such as the military, local television stations, or civil-emergency agencies. We can distribute data in various ways, including direct broadcast, relay transmissions, multiple hops, and interfaces with ground-communications networks. The actual method isn't important, as long as the architecture collects or generates data and delivers it to the user.

All spacecraft communicate, but some live to communicate. An example of this type of architecture is the INTELSAT system [Comparetto, 1993]. An international organization was formed in August, 1964, to produce, own, manage, and use a global communications-satellite system—the INTELSAT system. It consists of over 140 member nations and supports direct communications links among 200 countries, territories, and dependencies using more than 1,300 antennas located at over 800 Earth stations. As of June, 2004, the INTELSAT space segment included 28 active spacecraft, able to support up to 24,000 voice circuits each! The last launch of INTELSAT 10–02 in June 17, 2004, is the most sophisticated and is expanding the INTELSAT system to 24 orbital locations around the world. As stated in the INTELSAT Agreements (the governing documentation for this body), the prime objective of the INTELSAT system is:

> To provide, on a commercial basis, high quality, reliable international public telecommunications services.

INTELSAT's only purpose in life is to support telecommunications services, which they define as telephony, telegraphy, telex, facsimile, data transmission, and radio/television. Thus, they chose a spacecraft, antennas, ground-terminal locations, and orbital configurations to produce these services.

Telemetry represents the spacecraft's health and status data. Tracking involves transferring and updating the orbital-element set and other data to maintain accurate data on the spacecraft's position and motion. Commanding is the process of transmitting specific command sequences to the spacecraft to carry out certain actions (e.g., fire a thruster, move an antenna). There are basically two philosophies for doing telemetry, tracking, and commanding (TT&C) in a communications-system architecture: keep the complexity on the ground or put it in the spacecraft.

In keeping the complexity on the ground, operators at ground stations control the mission in (near) real time by transmitting commands directly to the spacecraft. One advantage of this approach is flexibility to changing requirements because changes are easier in the ground element than in space. Other advantages are greater reliability and a less complex, lower-cost spacecraft. The disadvantages include greater vulnerability to human error and the costs associated with the ground-control element.

If the complexity is on board the spacecraft, the spacecraft itself does TT&C using onboard data sensing and programmed decision making. This arrangement replaces ground control, is highly survivable, has fast response time (communication link delays are eliminated), excludes errors introduced by human operators, and reduces ground equipment and operations cost. However, this method is less responsive to changing or unanticipated requirements, and the spacecraft itself is more complex, more costly, and potentially less reliable. Even with an autonomous control architecture, a ground station is usually needed to collect data from the spacecraft and to back up the onboard control system. In the future, however, we expect the spacecraft to handle more functions, such as stationkeeping, to reduce dependence on control from the ground station. Chapter 12 discusses the TT&C functions in more detail.

11.3.2 Design the Ground/Space Communications Link

We design the ground/space communications link by specifying and changing various parameters in the communications system to achieve the performance specified in the user requirements. This equation allows us to calculate the communications link performance for a given set of parameters. A link-budget equation quantifies the available received signal-to-noise ratio (SNR) on a given link. The received signal-to-noise ratio is also commonly referred to as the energy per bit to noise power density ratio, or E_b/N_o. Once we determine the received SNR, we compare it to the SNR required to achieve a required bit-error-rate (BER) and refer to any excess SNR as *link margin*. By definition, a negative link margin tells us the communications-link design won't support the given data rate at the required BER. A link margin of 0 dB tells us the received SNR is just enough

11.3 Communications Architecture

to close the link, and any further, or unforeseen, link degradations will reduce link performance.

The link-budget equation in its most general form is

$$SNR_{avail} = (E_b/N_o)_{avail} = EIRP + G/T - L_{fs} - L_{other} - k - R_d \quad (11.11)$$

where
- $EIRP$ = effective isotropic radiated power of the transmit terminal
- G = receive antenna gain
- T = system noise temperature of the receive terminal
- L_{fs} = free space signal loss
- L_{other} = term that accounts for the other link loss terms (antenna pointing loss, rain, atmospheric absorption and implementation)
- k = Boltzmann's constant (−228.6 dBW/K-Hz)
- R_d = data rate

Note that each of the terms identified in Eq. (11.11) is expressed in units of dB.*

We then compute the link margin, M, as follows:

$$M = (E_b/N_o)_{avail} - (E_b/N_o)_{req'd} \quad (11.12)$$

where $(E_b/N_o)_{avail}$ is the SNR available at the receiver and $(E_b/N_o)_{req'd}$ is the SNR required to achieve a given BER. $(E_b/N_o)_{req'd}$ is a function of the modulation format, the presence of forward-error-correction coding (FECC), and whether the operational environment is benign or stressed.

The required link margin is a function of the required link availability. The communications link is typically designed so that the received SNR results in the required BER for a benign environment. A margin is then added to accommodate rain, jamming or other ill effects in the transmission channel. Additional link margin may be required to compensate for other expected operating conditions or equipment degradation.

For a given communications system design, we can usually juggle the $EIRP$, G/T, and R_d terms to meet all of the performance and operational requirements for the communications link. The L_{fs} term is a calculable parameter, whereas we typically use the L_{other} term to account for most of the link uncertainties. We discuss each of these parameters further below.

We can compute the EIRP simply by adding the transmitted power to the gain of the transmitting antenna, with both values expressed in dB. Suppose we have a ground terminal employing a 20 W high-power amplifier (HPA) with an antenna providing a gain of 15 dB. The EIRP would be

---

* For more detail on how the link-budget equation is derived, see either Gagliardi [1984] or Sklar [1988]. Both are excellent.

$$EIRP = 10 \log(20) + 15 = 28 \text{ dBW} \tag{11.13}$$

Note that the antenna gain was already in units of dB in Eq. (11.13).

We usually know the receive G/T of off-the-shelf ground terminals. But we may have to develop the G/T requirement for future ground terminal designs while meeting other system requirements. The receive G/T is simply the receive antenna's gain (not in dB) divided by the receive system's temperature (also not in dB). In calculating G/T for a given ground-terminal design, we commonly treat the terms individually and generate their values in units of dB. The G/T value, or figure of merit, is then the difference between the receive antenna's gain and receive system's temperature, with both values expressed in dB:

$$\text{Figure of Merit} = G/T = G(\text{in dB}) - T(\text{in dB}) \tag{11.14}$$

The maximum antenna gain, G, for a directional antenna is given by:

$$G = \eta\left(\frac{4\pi A}{\lambda^2}\right) \tag{11.15}$$

where η is the antenna efficiency (typically 0.55 for parabolic-dish antennas), λ is the carrier wavelength, and A is the antenna aperture area, which is $\pi D^2/4$ for a parabolic-dish antenna having a circular aperture of diameter D. The carrier wavelength is related to the carrier frequency by $\lambda f = c$, where c is the speed of light (3 × 10^8 m/s). Let's express the antenna gain in terms of the carrier frequency, substitute $\pi D^2/4$ for the antenna-aperture area, and express the antenna gain in units of dB. Eq. (11.15) then becomes

$$G = \text{Constant} + 10 \log(\eta) + 20 \log(f) + 20 \log(D) \tag{11.16}$$

where the constant is a function of the units used for the carrier frequency and antenna diameter, as shown in Table 11.6.

Table 11.6. Constants for the Antenna-Gain Equation as a Function of the Units Used for Frequency and Antenna Diameter. The antenna gain is a function of the square of the frequency and the square of the antenna diameter.

Constant (dB)	Units of Frequency, f	Units of Antenna Diameter, D
−159.6	Hz	meters, m
20.4	GHz	meters, m
−169.9	Hz	feet, ft
10.1	GHz	feet, ft

11.3 Communications Architecture

As we see from Eqs. (11.15) and (11.16), the antenna gain is proportional to frequency squared (doubling the carrier frequency quadruples the resulting gain). Also, antenna gain is proportional to the antenna's diameter squared (doubling the antenna's diameter quadruples the resulting antenna gain). These relationships are significant drivers of a communications system architecture.

As an example of using Eq. (11.16), assume an antenna diameter of 2 ft. and a carrier frequency of 19.5 GHz—these are reasonable downlink operating characteristics of a small receive terminal operating in the Ka-band frequency range. The resulting antenna gain using Eq. (11.16) would be:

$$G = 10.1 + 10 \log(0.55) + 20 \log(19.5) + 20 \log(2) = 39.32 \text{ dB} \quad (11.17)$$

Another antenna characteristic that will affect our design is the antenna beamwidth. The *antenna beamwidth* describes the angular spread of the transmitted electromagnetic wave as it traverses the path from the source (or transmitter) antenna to the destination (or receiver) antenna. The beamwidth is a first-order indication of the accuracy with which we must point the antenna toward its target. A common antenna beamwidth prescribed for antennas is θ_{3dB}—the angle across which the antenna gain is within 3 dB (or 50%) of the maximum gain. Recall that Eqs. (11.16) and (11.17) provide the maximum antenna gain for a parabolic-dish antenna. Gagliardi [1984] defines the 3 dB beamwidth as

$$\theta_{3dB}(\text{in degrees}) = \frac{(1.02)\lambda}{D}\left(\frac{180}{\pi}\right) = \frac{(1.02)c}{fD}\left(\frac{180}{\pi}\right) \quad (11.18)$$

where c is the speed of light (3×10^8 m/s), f is the carrier frequency in Hz, D is the antenna diameter in meters, and the term ($180/\pi$) is used to convert from radians to degrees. So, for the example given above, we calculate the 3 dB beamwidth using Eq. (11.18) to be ~1.47°.

The second term in the figure-of-merit expression, G/T, is the *system noise-temperature term*, T. In practice, the system noise temperature consists of three primary components: antenna noise temperature, line-loss noise temperature, and receiver noise temperature. Table 11.7 shows typical values for these three components, together with the system noise temperature (which is the sum of the three components) as a function of frequency for satellite systems using uncooled receivers. To calculate the receive G/T or figure-of-merit for a given ground receiver, we simply subtract the value of the system noise temperature given in Table 11.7 from the calculated antenna gain.

For example, let's use the antenna gain of 39.32 dB calculated previously for a 2-ft Ka-band ground terminal and a value of 28.6 dB-K for the system noise temperature from Table 11.7. (Note the downlink frequency is 19.5 GHz.) In this case, the receive G/T is 39.32 − 28.6 or 10.72 dB/K. If this value of G/T isn't large enough to support the link-budget calculations, one alternative is to cool the receive terminal, thereby decreasing the system noise temperature. The result

Table 11.7. Typical System Noise Temperatures in Satellite-Communication Links in Clear Weather. The temperatures are referred to the antenna terminal. Cooling techniques can reduce the antenna noise temperature, which affects the ground terminal's figure of merit, G/T. [Larson and Wertz, 1999]

	Frequency (GHz)					
	Downlink			Crosslink	Uplink	
Noise Temperature	0.2	2–12	20	60	0.2–20	40
Antenna Noise (K)	150	25	100	20	290	290
Line Loss Noise (K)	35	35	35	35	35	35
Receiver Noise (K)	190	492	592	1728	970	1505
System Noise Temp (K)	375	552	727	1783	1295	1830
System Noise Temp (dB-K)	25.7	27.4	28.6	32.5	31.1	32.6

would be a larger value for G/T. As we see in Eq. (11.11), the larger G/T would increase the available or received SNR and consequently increase the link margin from Eq. (11.12).

The *free-space loss*, L_{fs}, is defined by

$$L_{fs} = \left(\frac{4\pi S}{\lambda}\right)^2 \quad (11.19)$$

where S is the source-to-destination range (slant range), and λ is the carrier wavelength.

Note that the product of the carrier wavelength, λ, and the carrier frequency, f, is equal to the speed of light, c. By applying the appropriate constants to account for the desired units, we can express the free-space loss in units of dB as

$$L_{fs} = \text{Constant} + 20 \log(f) + 20 \log(S) \quad (11.20)$$

where the constant depends on which units we use for the carrier frequency and slant range, as shown in Table 11.8.

As seen from Eqs. (11.19) and (11.20), the loss from free-space attenuation is proportional to frequency squared (doubling the carrier frequency quadruples the resulting loss). Also, loss from free-space attenuation is proportional to distance squared (doubling the slant range quadruples the resulting loss). These relationships significantly drive communications system designs.

Finally, as an example of using Eq. (11.20), assume a slant range of 35,785 km (vertical distance to a geostationary satellite) and a carrier frequency of 6 GHz (typical uplink frequency for commercial satellites). Using Eq. (11.20), we find the free-space loss to be

11.3 Communications Architecture

Table 11.8. Constants for the Free-Space Loss Equation as a Function of the Units Used for Frequency and Slant Range. The free-space loss varies as a function of the square of both frequency and slant range.

Constant (dB)	Units of Frequency, f	Units of Slant Range, S
96.58	GHz	Statute Miles
92.45	GHz	km
97.79	GHz	Nautical Miles, nm
32.45	MHz	km

$$L_{fs} = 92.45 + 20 \log(6) + 20 \log(35,875) = 199.11 \text{ dB} \quad (11.21)$$

The *other-loss term*, L_{other}, is a catch-all term that represents all of the signal loss mechanisms in the benign channel, including rain, atmospheric absorption, dust, foliage, antenna pointing, and system implementation. The losses due to rain, atmospheric absorption, dust, and foliage can be quite severe depending on the carrier frequency. We discussed them in Sec. 11.1.3 under channel characteristics. Here we'll briefly discuss losses from antenna pointing and implementing the system.

The *antenna-pointing loss* accounts for inaccuracies in aligning the transmitter and receiver antennas. The loss is a function of the 3 dB beamwidth, θ_{3dB}, and the antenna pointing error, θ_e. For a parabolic dish antenna (see Gagliardi [1984]), and assuming a Gaussian antenna gain pattern, this loss may be estimated as

$$L_{pointing} = 10 \log\left(e^{-2.76(\theta_e/\theta_{3dB})^2}\right) \quad (11.22)$$

The antenna-pointing error, θ_e, depends on the antenna design and fabrication process and is typically specified for an antenna. We can calculate the 3 dB beamwidth, given the antenna diameter and operating frequency, by using Eq. (11.18). Let's continue with the example of the 2-ft Ka-band ground terminal operating at a downlink frequency of 19.5 GHz. We previously calculated the 3 dB beamwidth to be 1.47°. Assuming an antenna-pointing error of 0.25°, the pointing loss, using Eq. (11.22), would be:

$$L_{pointing} = 10 \log\left(e^{-2.76(0.25/1.47)^2}\right) = -0.35 \text{ dB} \quad (11.23)$$

The *implementation loss* is a term that accounts for most of the non-ideal qualities of communications-link equipment. These typically include uplink-waveform distortion, timing errors, frequency errors, data-detection-matched filter losses, non-ideal payload hardware (phase noise and filtering), and interference between adjacent channels. The implementation-loss term, as a whole, doesn't lend itself to analytical calculations. We can use analysis to calculate some components, such as adjacent-channel interference, but most others require testing. Thus, we usually estimate the implementation loss and then, after testing the communications-link equipment operationally, use test data to verify it.

11.3.3 Example Communications System Design

We can't show here a detailed example of how to design the communications system for a complicated scenario. In practice, teams do this work over months to years. But we can use a simplified example to develop a basic understanding of how we might develop more complicated communications systems. We'll design a space-to-ground communications link while addressing each phase of the communications-architecture design.

Our task is to design a satellite-to-ground downlink. (We assume the effect of the uplink on the system performance is negligible.) Since a continuous non-interruptible link is required, we select a geostationary spacecraft orbit. We also assume the communications system requirements are already defined, as shown in Table 11.9. Listed are performance requirements (P1 through P6), operational requirements (O1 through O9), system constraints (SC1 through SC3), and program constraints (PC1 and PC2). We'll discuss below each requirement and constraint in Table 11.9, showing how it would affect the design.

We derived the data-rate requirement of 5 Mbps using Eq. (11.1) and the link's being required to support a message rate of 3,912 messages/s, with a message size of 160 bytes.

As shown in Table 11.9, we used *message error rate* (MER) instead of the BER in this example. The MER is simply the probability that a received message (or packet) will contain an error after it has been demodulated and decoded. It's a function of the message size (in bits) and the BER:

$$\text{MER} = \text{BER} * \text{Message Size (in bits)} \tag{11.24}$$

In this example, the MER requirement is 0.002 with a message size of 160 bytes or 1280 bits. We can derive the required BER by rearranging the terms in Eq. (11.24):

$$\text{BER} = \frac{\text{MER}}{\text{Message Size}} = \frac{0.002}{1280} = 1.56 \times 10^{-6} \tag{11.25}$$

To be conservative, we set the BER requirement at 1×10^{-6}.

Continuing through Table 11.9, we must minimize E/E delay, which may affect our design but remains to be determined.

The link availability is set at ≥ 98% for a minimum antenna-elevation angle of 10°. This requirement will affect the rain-loss part of the L_{other} term in the link-budget equation (see Eq. (11.11)). No A/J, nuclear scintillation, security, or

Table 11.9. Summary of Requirements for an Example Communications-System Architecture. This table compiles the performance requirements (P1–P6), operational requirements (O1–O9), system constraints (SC1–SC3), and program constraints (PC1 and PC2) for a simple communications-system architecture.

ID	Required Label	Description	Impact
P1	Data Rate	3,912 Messages/s; (160 bytes message)	5 Mbps
P2	BER	MER < 0.002	10^{-6}
P3	E/E Delay	Minimize	TBD
P4	Link Availability	≥ 98.0% at minimum elevation angle of 10°	Will affect rain margin; function of frequency and geographic region
P5	A/J	N/A	None
P6	Nuclear Scintillation	N/A	None
O1	Security	N/A	None
O2	Standardization	N/A	None
O3	Backward Compatibility	N/A	None
O4	Access	Single Access	No FDMA, TDMA, or CDMA
O5	Satellite Orbital Characteristics	≥ 33% of Earth's surface, elevation angle ≥10°	Geostationary satellite
O6	Satellite Mobility	No requirement	None
O7	User Terminal Characteristics	Fixed	No need for satellite tracking antenna
O8	Data Source	Digital	No sampling required
O9	Channel Characteristics	Rain, Atmospheric Absorption, No Dust or Foliage	Must be accounted for in link-budget calculations
SC1	Parabolic dish for satellite antenna; 0.3 m ≤ diameter ≤ 4 m		
SC2	Maximum satellite transmit power of 50 w		
SC3	Commercial broadband ISP applications - Use Ka-band frequency		
PC1	Use currently available technology - No sophisticated engineering		
PC2	Use non-development items to maximum extent practical		

standardization requirements are identified. There are no backward compatibility requirements. We'll use single access, so we don't need to consider FDMA, TDMA, or CDMA techniques.

The orbital-characteristics requirement states that the spacecraft must be able to view at least 33% of the Earth's surface, assuming a minimum elevation angle of 10°. So we choose a geostationary orbit which allow the spacecraft to view ~34% of the Earth's surface. Table 11.9 lists no spacecraft-mobility requirements. The user's ground terminal is fixed.

The data source is digital, which means we don't need to address analog data-sampling techniques. The channel characteristics include rain and atmospheric absorption but not dust or foliage. Once we choose the downlink frequency, we can quantify the amount of rain and atmospheric absorption.

Table 11.9 identifies three system constraints. The first two limit the size of the spacecraft antenna and the amount of spacecraft transmit power, respectively. Both will come into play during the link-budget calculations. The last one, SC3, indicates that the communications link will be used by commercial SATCOM applications at Ka-band. Because of this requirement, and the allocation of frequency bands previously identified in Table 11.1, we decide to use the Ka-band frequency band allocated for space-to-ground communications. We select a frequency of 19.5 GHz.

Finally, the two program constraints are fairly general and won't drive our design choices much. But note that, if SC1 weren't identified, the program constraints in Table 11.9 would probably have driven us to choose a parabolic dish antenna for the spacecraft anyway.

We've already identified most of the data we need to design the space-to-ground communications link in this simple example. But let's go on briefly to the design-option phase to generate any other design characteristics we need. We've addressed the design-option areas below in the order originally discussed in Sec. 11.2.

As stated in Sec. 11.1.1, the E/E delay consists of transmission, propagation, queuing, and processing delays. The transmission delay is a function of the message size and the data rate, both of which are fixed in this example, so it too is fixed. The propagation delay is a function of the source-to-destination distance and the speed of light, both of which are also fixed. Thus, the propagation delay is fixed. A queuing delay occurs only in multiple user/server networks, so it doesn't apply. Therefore, the delay for onboard processing is the only one we can influence in this example.

Performance requirement P3 from Table 11.9 calls for minimum E/E delay. Consequently, we decide to use a bent-pipe transponder instead of onboard processing to eliminate the signal-processing delay on the spacecraft. Of course, there may be other factors in a more complicated communications system that could cause us to use onboard processing even though it might violate requirement P3.

Because of requirement O5, we choose a geostationary orbit to support continuous space-to-ground connectivity.

For the reasons stated above, the transponder design will be a non-regenerative or bent-pipe design. System constraint SC2 requires maximum output power to stay below 50W. Fortunately, TWTAs at 19.5 GHz can easily produce 50W of output power, so a TWTA design would be acceptable. Klystrons aren't an option because they're typically limited to ground applications and SSPAs may not be able to generate the full 50W if required by the link analysis.

We assume a single dedicated user in this example.

We choose rf over lasercom because of the program constraints identified in Table 11.9. One constraint requires available technology, and lasercom is still considered a maturing technology. Another requires us to use non-development items (NDI) as much as possible, and only a few space-qualified lasercom components are readily available.

We're now able to combine the stated and derived requirements, along with the identified design characteristics, into a target communications system design. In this example, the architecture consists simply of the space-to-ground communications link. Figure 11.14 summarizes the communications-link design to this point. As shown in the left hand part of Fig. 11.14, the required BER is 10^{-6}, the required data rate is 5 Mbps, and the frequency is 19.5 GHz. The spacecraft is in a geostationary orbit, and the slant range, S, between the ground terminal and the spacecraft is calculated to be 40,585 km assuming a minimum elevation angle of 10°. We calculate the free-space loss, L_{fs}, to be 210.4 dB, using Eq. (11.21) with the data in Table 11.9, and assuming an operating frequency of 19.5 GHz and the slant range calculated above.

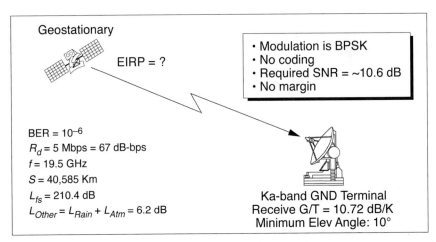

Fig. 11.14. Communications-Link Design Summary. We use the performance and operational requirements, along with system and program constraints, to develop a candidate architecture communications system design for the communications link. The only task left in this example is to solve for the required satellite EIRP.

In the paragraphs above we used the performance and operational requirements, along with system and program constraints, to develop a candidate communications system design for the communications link. The only task left in this example is to solve for the required satellite EIRP.

The loss term in Eq. (11.11), L_{other}, consists of rain and atmospheric absorption. We estimate the rain loss to be ~5 dB, using Fig. 11.3 and assuming a required link availability of 98.0% at a minimum elevation angle of 10° and an operating frequency of 19.5 GHz. We calculate the atmospheric absorption to be ~1.22 dB, using Eqs. (11.6) and (11.7) with the data in Table 11.1.

The top right part of Fig. 11.14 identifies several assumptions. The modulation we chose for this design is BPSK—consistent with the program constraints in Table 11.9, which calls for available technology and NDI where practical. BPSK modems are widely used in SATCOM and are supported by a mature technology base. We won't use FECC techniques so we can keep the design as simple as possible. We can introduce FECC if we need greater power efficiency even though it will reduce bandwidth efficiency. The received SNR required for a BER of 10^{-6} using BPSK modulation is ~10.6 dB. We can calculate this value directly using Eq. (11.6) or estimate it using Fig. 11.8.

There was no constraint placed on the ground terminal, other than that it be fixed. Let us assume a ground antenna dish diameter of 15 inches. This makes the antenna transportable and, with its 3 degree beamwidth, easy to point. With a receive noise temperature of 340°K (worst case during rain), a G/T of 10.7 dB/K is achieved.

We put no margin into this link, so it will be a bare-minimum design.

As shown in Fig. 11.14, the only parameter left to complete the design is the spacecraft EIRP required to close the link. We calculate the EIRP by rearranging Eq. (11.11) as follows:

$$\text{EIRP} = (E_b/N_o)_{avail} - (G/T) + L_{fs} + L_{other} + k + R_d \qquad (11.26)$$

$$= 10.6 - 10.7 + 210.4 + 6.2 - 228.6 + 67.0$$

$$= 54.9 \text{ dBW}$$

Note that the EIRP is simply the sum of the transmit power and the transmit-antenna gain (assuming both terms are expressed in dB). We can now size both parameters based on the required spacecraft EIRP of 54.9 dBW. If we assume the spacecraft transponder's output power is 25W (equivalent to ~14 dBW), the resulting transmit-antenna gain must be 54.9 – 14 or 40.9 dB. We can then use Eq. (11.16) and the data in Table 11.4 to determine the required size of the spacecraft's dish antenna. Using units of GHz for the transmit frequency and meters for the antenna diameter, we can rewrite Eq. (11.16) to solve for the required antenna diameter:

$$D = 10^{\left(\frac{G - \text{Constant} - 10 \log(\eta) - 20 \log(f)}{20}\right)} \qquad (11.27)$$

11.3   Communications Architecture

$$= 10^{\left(\frac{40.9 - 20.4 - 10\,\log(0.55) - 20\,\log(19.5)}{20}\right)}$$

$$= 0.73 \text{ meters}$$

Note that our assuming 25 W for the spacecraft's transmit power and the resulting diameter of the spacecraft's antenna are consistent with the first two system constraints (SC1 and SC2) in Table 11.9. Thus, the resulting design is acceptable: it meets the stated and derived requirements, is consistent with the system and program constraints, and includes the design options and characteristics identified in the design-option phase.

The beamwidth of the spacecraft's antenna is only 1.5°. Therefore we can't meet the ground coverage requirement of 33% of the earth's surface unless we use a steerable (or scanning) antenna. Such antennas, using electronic scanning techniques, have been developed, and are quite suitable as long as the spacecraft knows the location of the receive terminals. Suppose, however, that the requirement is to provide continuous coverage of the entire 33% of the earth's surface? The spacecraft's antenna beamwidth would then have to be increased to 17.1°, reducing its peak gain to 19.5 dB. The antenna, no longer steerable, points to the center of the earth, and its edge-of-coverage gain is 3 dB less, leading to a reduction of EIRP of 24.4 dB! How do we compensate for this loss gain? The most likely approach is to increase the gain of the ground terminal by 24.4 dB, a factor of 275. We do this by increasing the diameter by a factor of $(275)^{1/2} = 16.6$. This makes the ground antenna diameter 6.32 m (20.7 ft), no longer easily transportable, and more difficult to point, but technically feasible.

What happens if the requirement set changes after we've designed the communications architecture? Depending on the amount of change, we may have to repeat many steps in the process and revisit each phase to develop a new communications architecture that responds fully to the changed requirements. We'll discuss one minor change to the requirements so we can see the effect on our design example. Let's assume the message-rate requirement (P1 in Table 11.9) doubles from 3,912 messages/s to 7,824 messages/s. The resulting data rate required to support this new message rate is 10 Mbps. How can we change our design, in the least painful way, to satisfy the increased data rate?

First, let's examine the relationship between data rate and the other parameters in the link-budget Eq. (11.11), which we'll rewrite to solve for the data rate*:

$$R_d = \text{EIRP} + G/T - L_{fs} - L_{other} - k - (E_b/N_o)_{req'd} \qquad (11.28)$$

---

* We assumed a link margin of 0 in developing the above equation, so we could set $(E_b/N_o)_{avail}$ equal to $(E_b/N_o)_{req'd}$.

Now the design change requires us to double (increase by 3 dB) the achievable data rate, which is what Eq. (11.28) solves for. In examining Eq. (11.28), with an eye to doubling the achievable data rate, we see that we have several options:

1. The EIRP can be increased by 3 dB. This would entail doubling the transmit power to 50 W or doubling the spacecraft antenna's gain by increasing its diameter by a factor of 1.414 (the square root of 2), or a combination of the two. While this option impacts the spacecraft weight, it does not violate the system constraints (SC1 and SC2 of Table 11.7).

2. An alternative is to increase the receive figure-of-merit, G/T, by 3 dB. We can do this by either increasing the antenna diameter by a factor of 1.414 or reducing the receiver noise figure (perhaps by cooling) to decrease the receive-system temperature by 3 dB. Or we can do a combination of both that results in an overall G/T increase of 3 dB. (However, decreasing the receive system temperature may not be practical due to external noise, especially during rain and at low antenna elevations.)

3. The free-space loss is essentially fixed because the slant range and operating frequency aren't negotiable.

4. L_{other} is also essentially fixed because the operating frequency drives it.

5. Boltzmann's constant is fixed.

6. Another approach is to reduce the required E_b/N_o by 3 dB by introducing FECC into the design. However doing this usually increases the required rf spectrum requirement, and makes signal acquisition and tracking more difficult.

We carry out the uplink design in a similar manner, except that we calculate the carrier-to-noise (C/N) power density ratios using Eq. (11.10) with R_d replaced by the transponder bandwidth, B. Since we are using a bent-pipe transponder, the C/N in the uplink should be made large (~10 dB) so as not to degrade the downlink performance. In our example this should not be a problem because the EIRP of a fixed ground terminal can be made sufficiently large.

In summary, the option that would least affect the overall design would be to increase the spacecraft's transmit power from 25 W to 50 W assuming that the available satellite transponder could be operated up to 50W as identified in SC2 of Table 11.7.

Obviously, designing communications systems is complicated. It consists of various technical areas, each requiring significant expertise. That's why teams of qualified people normally design all but the simplest systems.

At the beginning of this chapter, we argued that designing communications systems requires a structured approach to make sure the resulting options are well documented, technically justified, and cost-effective. Consistent with this philosophy, we presented a process that included phases for defining requirements, evaluating options, and forming the candidate architecture. We described each phase and discussed key components and technical areas in terms of how they could affect the cost and performance of space-to-ground communications. For a complete view of communications architectures, please see Chap. 12 (ground systems) and Chap. 13 (data processing).

We couldn't cover every detail of a communications system here, so the reader should consult the references mentioned in each section for more detail and theory.

Of course, no matter how structured we try to make it, designing communications systems isn't a clear-cut, step-by-step process that results in a single correct answer. Often, it's iterative, and the result is a family of options. The main thing to keep in mind is that we must tie the resulting architecture options to specified and derived requirements and constraints, and must document the process carefully with detailed technical justification. If we don't, a mission operations manager can easily discard and discount the staff-years of effort on one design in favor of another product, often basing this decision more on opinion than on fact.

References

Altshuler, E. E.1983. "The Effects of a Low-Altitude Nuclear Burst on Millimeter Wave Propagation." *Rome Air Development Center In-House Report.* RADC-TR-83-286, December.

Chan, V.W.S. 1988. "Intersatellite Optical Heterodyne Communications Systems." *The Lincoln Laboratory Journal.* 1(2):169-183.

Comparetto, G. M. 1994. "Global Mobile Satellite Communications: A Review of Three Contenders." Presented at the *1994 AIAA 15th International Communications Satellite Systems Conference.* San Diego, CA., 27 Feb – 3 March.

Comparetto, G. 1993. "A Technical Description of Several Global Mobile Satellite Communications Systems." *J. of Space Comm.*, Vol. 11, no. 2, October: 97–104.

Comparetto, G. and Maj. M. Kaura. 1993. "Using Global Mobile Satellite Communications Systems to Support Army Requirements." Presented at the *AIAA Space Programs and Technologies Conference and Exhibit.* Huntsville, AL., 21–23 September.

Comparetto, G. 1993. "On the Use of INTELSAT and INMARSAT to Support DoD Communications Requirements." Presented at *MILCOM '93*, Paper number 1.2, Boston, MA., 11–14 October 1993.

Comparetto, G. 1993. "The Impact of Dust and Foliage on Signal Attenuation in the Millimeter Wave Regime." *J. of Space Comm.*, Vol. 11, no. 1, July: 13–20.

Comparetto, G., and W. Foose. 1990. "An Evaluation of the Gaussian Approximation Technique Applied to the Multiple Input Signal Case of an Ideal Hard-Limiter." Presented at *MILCOM '90*, Monterey, CA., 30 September – 3 October, Paper # 26.4.

Comparetto, G. 1989. "Signal Suppression Effects in an Ideal Hard-limiter for the Many-Carrier Case." *Int. J. of Sat. Comm.*, Vol. 7, No. 5, December.

Comparetto, G. and D. Ayers. 1989. "An Analytic Expression for the Magnitudes of the Signal and IM Outputs of an Ideal Hard Limiter Assuming "n" Input Signals Plus Gaussian Noise." *Int. J. of Sat. Comm.*, Vol. 7, No. 1, January.

Crane, R. K. 1980. "Prediction of Attenuation by Rain." *IEEE Trans. on Commun.*, Com-28(9): 1717–1733.

Davies, R. 1992. *Space Mission Analysis and Design*. Chapter 13 "Communications Architecture." Netherlands: Kluwer Publishing.

Dixon, R. C. 1976. *Spread Spectrum Systems*. John Wiley, New York, NY.

Gagliardi, R. 1984. *Satellite Communications*. Van Nostrand Reinhold Co., New York, NY.

Ha, T. 1986. *Digital Satellite Communications*. Macmillan Publishing Co., New York, NY.

Hara and Prasad. *Overview of Multicarrier CDMA*. IEEE Communications Magazine, December, 1997, 126–133.

Horwitz, G. M. 1979. "Optimization of Radio Tracking Frequencies." *Trans. App. Phys.*, Vol. 27, May: 393–398.

Ippolito, L. J. 1986. *Radiowave Propagation in Satellite Communications*. New York: Van Nostrand Reinhold.

Jasik, Henry, ed. 1961. *Antenna Engineering Handbook*. New York: McGraw-Hill.

King, M. A. and P.B. Fleming. "An Overview of the Effects of Nuclear Weapons on Communications Capabilities." *Signal*, Jan 1980: 59–66.

Larson, Wiley J., and James R. Wertz. 1999. *Space Mission Analysis and Design*. Third Edition. Netherlands: Kluwer Publishing.

Lin, S. and J. Costello. 1983. *Error Control Coding: Fundamentals and Applications*. Prentice-Hall, Inc., Englewood Cliffs, NJ.

McElroy, D. 1988. "The FEP Communications System." *AIAA 12th International Communication Satellite Systems Conference Proceedings*, 395–402.

Middlestead, R. W., et. al. "Satellite Crosslink Communications Vulnerability in a Nuclear Environment." *IEEE Journal on Selected Areas in Communications*, col. SAC-5, no. 2, Feb 1987.

Mie, G. 1908. "A Contribution to the Optics of Turbid Media, Especially Colloidal Metallic Suspensions." *Ann. Phys.*, Vol. 25: 377–445.

Morgan, Walter L. and Gary D. Gordon. 1989. *Communications Satellite Handbook*. New York: John Wiley & Sons.

Mottley, T.P., D. H. Marx, and W. P. Teetsel. 1960. "A Delayed-Repeater Satellite Communications System of Advanced Design." *IRE Trans. on Military Electronics*, April-July: 195–207.

Naderi, M. and P. Kelly. 1988. "NASA's Advanced Communications Technology Satellite (ACTS)." *AIAA 12th International Communication Satellite Systems Conference Proceedings*, 204–224.

Papoulis, A. 1965. *Probability, Random Variables, and Stochastic Processes*. McGraw-Hill, Inc., New York, NY.

Proakis, J. 1983. *Digital Communications*. McGraw-Hill, Inc., New York, NY.

Rafuse, R. P. 1981. "Effects of Sandstorms and Explosion-Generated Atmospheric Dust on Radio Propagation." *Technical Report Number DCA-16*, Massachusetts Institute of Technology Lincoln Laboratory, 10 Nov.

Sari, H., F. Vanhaverbeke, and M. Moeneclaey, *Extending the Capacity of Multiple Access Channels*. IEEE Communications Magazine, Vol. 38, No. 1, January 2000, 74–82

Schwab, L. M. 1980. "A Predictor Model for SHF and EHF MILSATCOM System Availabilities in the Presence of Rain." *Lincoln Laboratory Technical Note 1980-15*, Lincoln Laboratory, Massachusetts Institute of Tech., Lexington, Massachusetts, February.

Simon, M., et. al. 1985. *Spread Spectrum Communications Volumes I, II, and III*. Computer Science Press, Inc., Rockville, MD.

Sklar, B. 1988. *Digital Communications Fundamentals and Applications*. Prentice-Hall, Inc., Englewood Cliffs, NJ.

Spilker, J. 1977. *Digital Communications By Satellite*. Prentice-Hall, Inc., Englewood Cliffs, NJ.

Stallings, W. 1985. *Data And Computer Communications*. Macmillan Publishing Co., New York, NY.

Tamir, T. 1977. "Radio Wave Propagation Along Mixed Paths in Forest Environments." *Trans. App. Phys.*, Vol. 25, July: 471–477.

Tewari, R. K., S. Swarup, and M. N. Roy. 1990. "Radio Wave Propagation Through Rain Forests of India." *IEEE Trans. on Ant. and Prop.*, Vol. 38, No. 4, April: 433–449.

Torrieri, D. J. 1992. *Principles of Secure Communications Systems*. Artech House, Norwood, MA.

Weiner, Thomas F. and S. Karp. 1980. "The Role of Blue/Green Laser Systems in Strategic Submarine Communications." *IEEE Transactions on Communications*. Com-28(9): 1602–1607.

Yeh, K. C. and G. W. Swenson. "The Scintillation of Radio Signals from Satellite." *J. of Geophysics Research*, Vol. 64, No. 12, Dec 1959: 2281–2286.

Chapter 12

Ground Systems

Matthew J. Lord, *Loral Space and Range Systems*
Edward B. Luers and Richard B. Miller,
Jet Propulsion Laboratory

12.1 Defining the Ground System
12.2 Hardware, Software, and Staffing Requirements
12.3 New Systems and Existing Networks
12.4 Operational Concerns

Bound by the Earth, humans operate spacecraft from the ground. The success of any mission hangs on the ability to extract information from space and get it to the user, who may be anywhere on the globe. Equipment, facilities, and communication links—collectively called the *ground system*—enable us to control the space element.

The ground system provides all the necessary tools to conduct space mission operations. It presents spacecraft operators with the information required to maintain spacecraft health and allows them to control the vehicle's attitude and orbit. It provides knowledge of the payload to mission planners and allows them to alter mission parameters. And it collects, processes, and routes data to users.

This chapter will map the path from space mission requirements to a ground architecture able to meet them. We'll investigate and trade these requirements along the way, discuss the merits of using an existing ground network, and point out when it would be more cost-effective to build a new system. But first, we need a better understanding of the ground system.

12.1 Defining the Ground System

12.1.1 Ground-System Functions and the Single-Station Model

The ground system has three main functions: telemetry, tracking, and commanding, or *TT&C*. *Telemetry* literally means "measurement from a distance." On the ground, we acquire and process engineering or payload telemetry from a spacecraft. *Engineering telemetry* gives operators details of the condition of the spacecraft and its subsystems. Spacecraft operators monitor the spacecraft's health and status using such data as the temperature and voltage of spacecraft batteries. *Payload telemetry* contains data pertaining to the spacecraft's mission. Payload telemetry may be processed by the ground system or sent to (sometimes recorded for) the user for processing. Engineering and payload telemetry may combine in a single data stream, requiring only one set of receiving and data-handling equipment on the ground. Older spacecraft designs used two or more simultaneous telemetry streams, which require the same number of equipment strings in the ground system. Using multiple subcarriers or even multiple carriers is usually a very inefficient use of the space-ground link, particularly in systems that must operate near channel capacity (i.e., with high efficiency). Telemetry systems use packetized instead of Time Division Multiplex telemetry. "Packetized" systems involve moving "messages," or blocks of data, instead of bit streams, and have many advantages, including easy "channelization." Each channel of data can share the same telemetry stream and associated equipment yet be easily separated on the ground. This is particularly true if we use international standards set by the Consultative Committee for Space Data Systems (CCSDS) in designing the packet and telemetry frame formats.

Tracking is acquiring (detecting) a spacecraft and following its motion through space. We track a spacecraft by gathering data on its location and using the data to control ground antennas. We can use different coordinate systems and frames of reference to describe an orbit, but to swing a ground antenna toward a spacecraft, we must know the elevation (measured in degrees from the horizon), the azimuth (measured in degrees from true north), and the time when the spacecraft will be at that location. In addition, we must correct the frequency of the spacecraft receiver and transmitter for current spacecraft mode and temperatures, and for the Doppler effect of its motion.

Commanding, or controlling a spacecraft from the ground, occupies separate tasks and equipment within the ground system. We command a spacecraft by uplinking data to it. The data may contain operations the spacecraft must do, changes to onboard software, or information that spacecraft instruments will reference. The spacecraft may act on each command immediately or store it for execution at a future time, as directed by the command. Some spacecraft operate using onboard sequencing that can control the spacecraft for hours, days, even months (depending on the mission needs). In those cases, "commanding" can consist of sending up new sequence loads that can be thought of as software programs.

The single-station model describes the minimum set of elements we need to do ground-system tasks. It consists of spacecraft operators, hardware and software systems, and the facilities to support them. In the model shown in Fig. 12.1, we can do all critical tasks at one site. The station supports telemetry, tracking, and commanding for one spacecraft at a time. It acquires telemetry with the antenna, receives and demodulates the rf signal, processes the raw data, and displays or stores it for the user. It generates commands, modulates and amplifies them for uplink, and then transmits them to the mission spacecraft. It collects tracking data with the antenna and processes it for orbit planning.

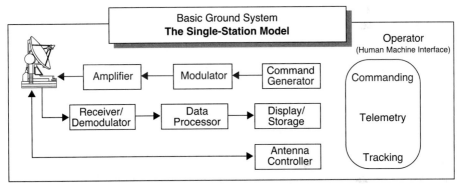

Fig. 12.1. The Single-Station Model. A basic ground system does telemetry, tracking, and commanding. These functions correspond to data transport and delivery, navigation planning and analysis, and mission control, as described in Chap. 3.

This model has limitations that are unacceptable to most missions. A single ground station lacks redundancy, making the spacecraft it supports vulnerable to any number of environmental or human threats that could disrupt ground operations. Tracking data collected from a single station generates less accurate ephemeris. A single station also has a limited field-of-view, providing poor coverage for spacecraft in orbits other than geosynchronous. Even geosynchronous spacecraft will need more ground-observation points during the critical launch and early-orbit-operations phases until they reach geostationary orbit. In addition, a single station is ill-equipped to handle spacecraft anomalies, when the vehicle needs more contacts throughout its orbit for troubleshooting.

Some missions can accept these limitations. Those that can't may still be able to use a single station if the spacecraft is designed for compatibility with other networks, so operators can tap these external resources during special activities. Most missions require more complex ground systems.

12.1.2 Ground-System Elements and a Typical Ground-System Model

We can't fully describe a ground system's abilities with the single-station model. Figure 12.2 shows a more complete network. In this model we can support several spacecraft simultaneously by sharing network resources. Multiple ground stations provide the coverage needed for various orbits and mission architectures. To the single-station model we've added subsystems for recording and timing and have defined the hardware more explicitly. This model of the typical ground system doesn't describe the best configuration for every mission, but it's convenient for discussing the ground system's functions and abilities.

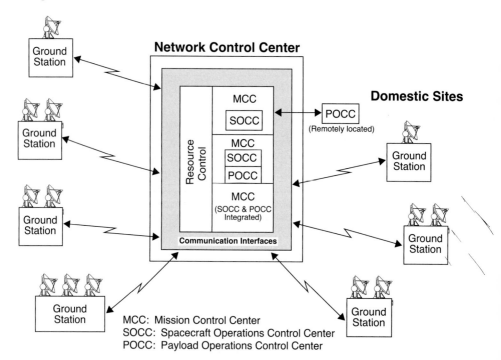

Fig. 12.2. **The Typical Ground System.** This system supports many spacecraft simultaneously with multiple control complexes and remotely located ground stations. We can organize the control complexes in different ways, as shown, but the real network won't have this variety.

We can organize the typical ground system functionally into three areas: ground stations, control centers, and communication links. A *ground station* is an installation on the Earth comprising all the equipment needed to communicate with a spacecraft [Williamson, 1990]. Antennas and TT&C hardware at the ground station transmit and receive the rf signals that operate the spacecraft. The ground station receives commands from the control center, then modulates and formats them for uplink to the spacecraft. It receives telemetry from the spacecraft and relays it to the control center (or directly to the data user). The ground station also tracks the spacecraft, providing data to the control center for orbit determination and prediction. We can double the support capacity of a ground station without doubling the cost by adding a second antenna and suite of TT&C hardware. A ground network may require several ground stations dispersed around the Earth to achieve complete coverage. For deep space missions three equally spaced sites can provide continuous coverage.

The *Network Control Center* is the central node for controlling the ground stations. It maintains the communication systems that pass data to and from the spacecraft-control centers. It also provides host services for the spacecraft-control centers, including building and ground maintenance, utilities, and security. Central control of mission operations at one site minimizes costs for facilities, equipment, and operations.

The *Mission Control Center* (MCC) operates the mission. Engineers and operators within the MCC create a mission plan, schedule spacecraft resources, and select ground resources to meet mission objectives. They determine the spacecraft's orbit and attitude and they send predicted values to the ground station for tracking acquisition. For simple spacecraft, the MCC carries out the mission plan. For more complex spacecraft, a separate facility, the *Spacecraft Operations Control Center* (SOCC) controls the spacecraft's subsystems and processes its data. The SOCC generates commands and passes them to the ground station. It monitors spacecraft-maintenance telemetry received from the ground station.

The SOCC analyzes data only from subsystems that affect the spacecraft's health or attitude. The *Payload Operations Control Center* (POCC) analyzes the mission data. Because it's responsible for the payload, the POCC helps form the mission plan. The POCC will request command sequences to control payload instruments, but to protect the common spacecraft bus, sometimes only the SOCC or the MCC is allowed to transmit these commands.

The *Resource Control Center* (RCC) schedules and monitors the use of all network resources. It accepts requests from the MCCs, and assigns blocks of time for using ground stations, communication links, and equipment strings. The RCC assigns resources to spacecraft based on priority, as described in Sec. 12.4. In smaller networks the MCC controls resources.

The MCC, SOCC, POCC, and RCC can be co-located at the Network Control Center, each occupying a room or several rooms. Some missions choose to place the POCC outside the network at the payload user's site. The SOCC can be integrated with the POCC or MCC. It is often efficient for the contractor that built the spacecraft to provide the SOCC role, either integrated at the MCC site or at a site used by the spacecraft contractor to support several missions. Smaller missions combine some or all of the centers into a single room.

12.1.3 Communication Links

Rarely are all ground system elements co-located, so ground communication links are needed to transport data among them. Figure 12.3 shows the kind of information to be transmitted, depending on the location of facilities. The best type of link for a particular data signal will depend on the type of information being transmitted, how it will be used, and the data bandwidth and time criticality. There are several options for moving information, so we must choose between private and commercial services, satellite and terrestrial links, and switched and dedicated lines. The best solution depends on whether requirements call for voice or data signals; whether distribution is point-to-point or multi-drop; and whether speed, distance, security, or survivability are design constraints. A satellite-control network will likely employ more than one type of link to connect ground-system elements. Powers [1990] provides a good introduction to communications.

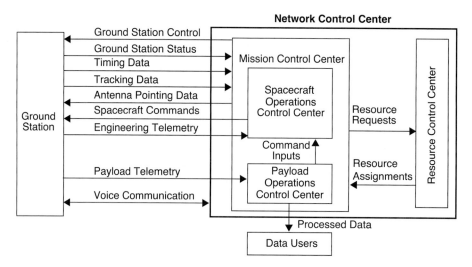

Fig. 12.3.   Ground Communication Links. We must transmit various data among ground-system elements.

Private Data Links. Between buildings on the same campus, or over other short distances, we usually use private communication links. These links can be installed by the operating agency or by an outside contractor. Optical fiber has replaced copper cable as the most popular option for requirements under a few kilometers.

Although inexpensive and simple to install, copper cable is limited by frequency and distance. Optical cable offers increased performance and higher bandwidths at prices that are becoming competitive with copper. A fallout benefit of fiber is the security derived from optical isolation. Fiber cables suppress any unwanted signal riding on the transmitted data, and having no magnetic field, they are much more difficult to tap than their copper cousins. Because most data systems are electrical, signals must be converted to optical for transmission over fiber and back to electrical at the receiving end. As the advantages of fiber continue to persuade the communications industry, more vendors will offer fiber interfaces for their equipment.

The largest cost in a new cable installation is often for constructing the cable's passage (conduits, trays, or poles). Given an unobstructed, line-of-sight path between transmission points, terrestrial microwave radio is a cost-effective alternative to cable. Microwave is effective for high-bandwidth transmission, but frequency coordination is critical to avoid signal interference. The Federal Communications Commission regulates all microwave radio installations.

Table 12.1 compares microwave, fiber, and optical systems. See McClimans [1992] for more information.

Table 12.1. Sample Comparison of Private Communications Media. The values represent the capabilities of moderately priced systems. We assume a small installation. As the size of the project increases, there are crossover points at which fiber or microwave is more cost-effective.

Specification	Copper Cable	Fiber-Optic Cable*	Microwave Radio
Data rate–Mbps	10 to 16	150	90
Distance between repeaters	1 km	10 to 20 km	40 to 50 km
Relative hardware cost	Low	Moderate	High

* For multi-mode fiber. Single-mode fiber systems can transmit several gigabits per second, and unrepeated distances of nearly 100 km are achievable.

Terrestrial Commercial Communication. When transmission distance makes a private link impractical, we use commercial services. For data rates less than 20 kbps, a simple dial-up telephone line connects two modems; if we need dedicated service, we can lease the line. For higher data rates, a popular service to link two points in North America is the T-carrier. A T1 circuit passes 1.544 Mbps, and a T3 circuit passes 44.736 Mbps. These circuits can be used for one signal or be shared by many signals with a multiplexer. Public carriers are working to increase the throughput of communication links. Emerging services such as the synchronous optical network (SONET), using advanced protocols like asynchronous transfer mode (ATM), promise data rates above two Gbps. More modest rates from tens to hundreds of Mbps over ATM are becoming available today. The emergence of packet-based services like ATM represents a trend in ground communications—a departure from dedicated links like the T1 in favor of non-dedicated, broadband services.

A potential disadvantage of commercial communications to a user like the military is the risk of not having control of all its assets. The user depends on the commercial carrier. But when dollars count, the reliability and cost-effectiveness of commercial services are convincing. It now seems to be a risk the military is willing to take.

Satellite Communication. Spacecraft can be the medium as well as the mission. To relay high-bandwidth data across long distances, such as between ground stations and the network control center, the spacecraft has traditionally offered the most cost-effective solution.

Today, the spacecraft competes with fiber and other media whose bandwidth exceeds that of spacecraft, without some of the problems of satellite communication. Geostationary satellites, in particular, are encumbered with a long transmission path, giving rise to a one-way delay of 0.25 seconds, which is annoying for voice communication and disruptive for computer networks. But spacecraft have unique advantages over other services. Satellite communication is mobile and flexible. Cost is largely independent of distance and location of ground terminals. The link doesn't depend on the public-telecommunications infrastructure. And spacecraft can broadcast to an unlimited number of users.

We distinguish communication satellites by frequency, as well as the number and bandwidth of transponders. A typical transponder bandwidth for C-band is 36 MHz, which can be leased in whole or part. Ku- and Ka-band transponders offer greater capacity. The cost of a ground terminal can be moderate to high, depending on the bandwidth, which drives the size and complexity of the antenna and its subsystems. Chapter 11 describes satellite communications in detail.

12.1.4 Describing the Subsystems

We can better understand the typical ground system by examining its subsystems: the *antenna, TT&C, timing,* and *command and data processing.* Figure 12.4 describes the relationships among these groups.

Antenna Subsystem. As we see in Eq. (11.11), antenna gain is one of the few ground factors that can affect link margin. This gain to noise ratio (G/T) is pivotal to deciding whether a ground station can support a particular spacecraft in a particular orbit. Transmit gain is represented in a term called the *effective isotropic radiated power* (EIRP), which is simply antenna gain multiplied by transmit power. Receive gain is represented in the term G/T (receive antenna gain divided by system noise temperature). Together, EIRP and G/T describe an antenna's performance. For a given frequency, gain depends on antenna size, type and efficiency. With different frequencies for uplink and downlink, the same antenna can be used for transmission and reception. These frequencies are offset by a fixed non-integer ratio.

The uplink rf signal passes through the feed and radiates from the reflector. Downlink signals are collected by the reflector and passed through the feed to the low-noise amplifier. The antenna is controlled by servos, drive motors on each

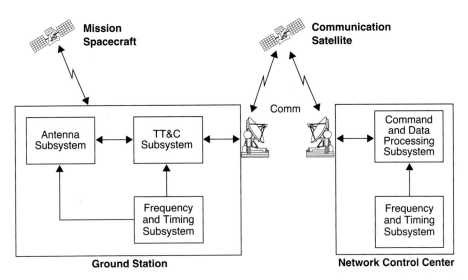

Fig. 12.4. Relationship of the Four Subsystem Groups. Here a local area network replaces the communication link in a co-located network. The next four figures show the four subsystems in more detail.

axis, a position measurement system, and a control unit. Figure 12.5 shows antenna components in the *prime-focus* configuration. An alternate configuration, called *Cassegrain*, uses a subreflector to locate the feed on or behind the main reflector (with the rest of the electronics), eliminating some cable loss. A third configuration has additional optics behind the antenna and focuses the rf beam to a room. This is called a Beam Wave Guide antenna and provides a larger, easier maintenance environment for the feed and low noise amplifier, systems, and room for multiple frequencies.

We control the antenna through the TT&C subsystem in one of three modes: auto, slave, or manual. *Auto-tracking*, which is preferred whenever possible, uses the signal strength of the downlink at the receiver to lock onto and track the spacecraft. We employ *slave tracking* to acquire the spacecraft first before we receive the downlink or as a fall-back method when the signal strength is too low. For this method, a computer programmed with predicted ephemeris drives the antenna by sending it time-dependent pointing angles. A typical acquisition technique is to slave track the antenna into position ahead of the spacecraft track and then switch to autotrack as the spacecraft flies through the antenna beam. For geostationary spacecraft tracking isn't required so the antenna is pointed manually. The spacecraft orbit and its required accuracy determine the sophistication of tracking and ephemeris systems. Higher frequency space-ground links also require more accurate antenna pointing.

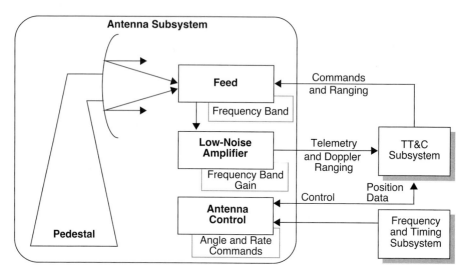

Fig. 12.5.   Antenna Subsystem. Options and considerations are boxed below each hardware component. For example, low-noise amplifiers differ by the frequency bands they pass and the amount of amplification versus noise temperature they provide.

TT&C Subsystem. The TT&C subsystem defines the rf interface between the space and ground elements. The frequency and modulation technique supported by ground hardware must be compatible with the TT&C subsystem on board the spacecraft. (Section 12.2 describes compatibility.) Located within the ground station, the TT&C subsystem includes a high-power amplifier to transmit the uplink signal; modulators, demodulators, synthesizers, receivers, and bit synchronizers to convert signals between rf and digital; and recorders to record and play back the spacecraft's data. Figure 12.6 shows only one of each device but most ground systems use several strings of equipment, cross-strapped through switches, for multiple, simultaneous contacts and redundancy. The equipment is configured before the spacecraft contact, after which it runs with little or no operator control. The computers used to control these systems can be either local at the ground station, or remote at the network control center. The same computers can automate testing of the equipment and the circuits that connect them.

Modern telemetry systems often use coding techniques to increase the efficiency of the space ground link. Coding can range from a simple parity scheme that enables labeling data as suspect, to various *error detection-correction* (EDC) codes that enable virtually error free telemetry. Such EDC codes allow higher data transmission rates at lower signal to noise ratios, as well as increasing the potential for automating the subsequent data processing.

The TT&C subsystem includes hardware to gather orbit data. What we require depends on whether we need tracking data simply to locate the vehicle or to

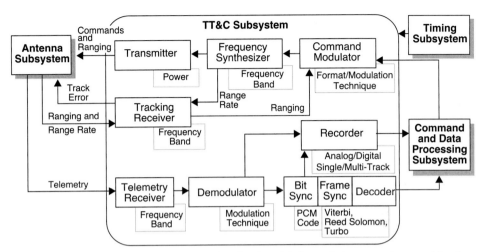

Fig. 12.6. **TT&C Subsystem.** Frequency band and modulation technique are prime considerations for selecting TT&C hardware. Options and considerations are boxed below each component.

calibrate spacecraft sensors. In some cases, the mission needs nothing more than the coarse pointing angles the antenna provides. Most missions need more precise orbit data. We can determine the spacecraft's position from the ground by using optical, radar, and radio techniques. Optical and radar work only for low-Earth orbit. Optical tracking works only if the spacecraft is sunlit and visibility conditions are good, but radar works in any weather. Optical and older radar systems treat the spacecraft as a passive reflector; with modern radar systems, the spacecraft receives and retransmits the signal. Radar is typically used by sites independent of the ground system. Most ground stations use the spacecraft's rf link to measure the distance, or range, and radial velocity, or range rate, of the vehicle relative to the ground station. *Range*, r, is calculated by measuring the round trip delay, t_r, of a signal transmitted on the uplink and returned on the downlink:

$$r = ((t_r/2) - t_p)c \qquad (12.1)$$

where c is the speed of light, and time t_p, accounts for ground and vehicle processing time and other delays.

Range rate is calculated by measuring the Doppler shift of the downlink frequency. The Doppler shift, Δf, is directly proportional to the component of the spacecraft's velocity, $\dot{r}$, in the direction of the ground station

$$\dot{r} = \Delta f \cdot \lambda \qquad (12.2)$$

where λ is the wavelength of the downlink frequency at the spacecraft transmitter. If our spacecraft's downlink frequency is f_d, and we measure a Doppler-shifted frequency on the ground of f_m, we can define the Doppler shift as

$$\Delta f \equiv f_d - f_m \qquad (12.3)$$

and we can solve again for range rate (recall that $c = \lambda f$):

$$\dot{r} = (1 - (f_m/f_d))c \qquad (12.4)$$

Note that the distance and accuracies needed for deep space navigation require using the relativistic form of the equation.

In practice, the ground system does more to calculate range rate than Eq. 12.4 implies, mostly because we can't rely solely on the downlink frequency. When range-rate measurements aren't being taken, the spacecraft can operate in *non-coherent mode*, in which an onboard oscillator will generate the downlink carrier frequency. This onboard oscillator isn't stable enough to provide accurate Doppler measurements on the ground. When we're gathering range-rate data, the spacecraft must be configured for *coherent mode*, in which the spacecraft transmitter is phase-locked to the received uplink carrier. In coherent mode, the spacecraft will shift the received uplink frequency by a fixed ratio and modulate the downlink signal onto this carrier. Table 12.2 lists the downlink-to-uplink frequency ratios and the types of signals used for range measurement at three existing networks.

Table 12.2. Range and Range-Rate Data for Three Existing Networks. Networks operate in the coherent mode to provide accurate range and Doppler measurements.

Network	Range Measurement Signal	Downlink-to-Uplink Carrier Frequency Ratio, r^*
Air Force Satellite Control Network (AFSCN)	1 Mbps pseudorandom noise (PN) code	256/205
NASA Tracking and Data Relay Satellite System (TDRSS)	3 Mbps PN code	240/221 (S-Band) 1600/1469 (Ku-Band)
NASA Deep Space Network (DSN)	Sequential codes consisting of 7 major clocks (1 Mhz to 16 kHz) with up to 20 ambiguity resolving frequencies	240/221 (S-Band) 880/221 (X/S-Band) 800/749 (X/X-Band) 3344/221 (Ka/S-Band) 3344/749 (Ka/X-Band)

* That is, f_d (downlink) = $r \cdot f_u$ (uplink)

Timing Subsystem. Many ground-system functions are time critical. Tracking data is of little use unless we know when the data were collected, and to track a spacecraft by computer we must tell the antenna when to move. We must time-tag telemetry at the recorder and telemetry processor (unless the spacecraft has embedded a time code in the data itself). A mission operator needs accurate time displays to coordinate events during the contact. The timing subsystem does these tasks.

Subsystems on the ground have been synchronized with calibrated atomic clocks and with land-based transmission systems such as LORAN C. The Global Positioning System (GPS) has reduced the cost and increased the accuracy and simplicity of the timing function by an order of magnitude. Racks of timing gear have become just a GPS receiver and time-code generator, connected as shown in Fig. 12.7. (An optional frequency standard allows the system to maintain accurate time if no GPS signal is present.) GPS satellites provide an accuracy to within a few hundred nanoseconds of a world time standard, such as Coordinated Universal Time (UTC). When ground stations are remote from the network control center, each will have a separate timing subsystem to account for the transmission delay between them. We also need a precision source of frequency to support precision Doppler and ranging measurements as well as to provide a precise local time source.

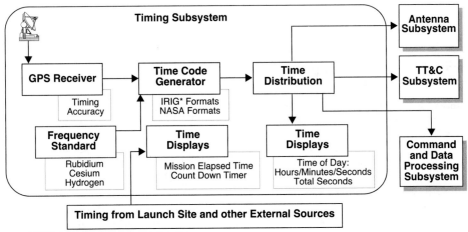

* IRIG: Inter-Range Instrumentation Group

Fig. 12.7. Timing Subsystem. Various time codes and display options allow distribution to the many systems that require accurate time. Today, we can buy the GPS receiver, frequency standard, and time-code generator as a single unit. Options are shown in boxes below each component. Deep space applications require a precision beyond a GPS based subsystem, as do radio astronomy sites.

Command and Data Processing Subsystem. This subsystem defines the data-format interface between the space and ground elements. Commutation, coding, and encryption schemes employed on the ground must match those on the spacecraft. Located within the MCC, computers and supporting systems generate commands, compute orbit dynamics, and process baseband telemetry from the spacecraft and payload into user-required formats. The type and complexity of systems vary greatly but often include the tasks illustrated in Fig. 12.8 and discussed below.

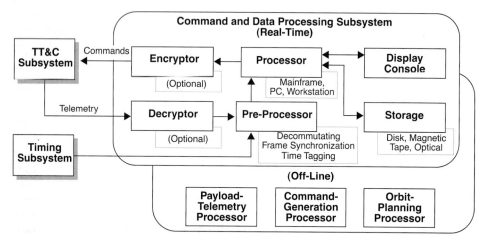

Fig. 12.8.   Command and Data Processing Subsystem. This subsystem consists of real-time and off-line equipment. We may process payload telemetry in real time or after the spacecraft contact. Most modern systems connect the workstations and their peripheral devices through a local-area network. Options are shown in boxes below each component.

The broadcast nature of satellite communication is especially vulnerable to unwanted data monitoring, or worse, illegal commanding. Whenever we need security, we encrypt signals before transmission (usually in the MCC), and decrypt after reception (in the MCC, POCC, or at the user's location). The spacecraft decrypts commands and encrypts telemetry. The military routinely uses encryption; NASA and ESA do not.

Before we can process telemetry, we must format and time-tag it. Formatting includes frame synchronizing, decoding, packet extraction, virtual channel extraction and decommutating, which converts serial bit streams into discrete logical measurands.

The volume and complexity of the telemetry will drive the ground processors' cost and complexity. Large mainframe computers had long helped us do the core tasks of mission operations. Today, the flexibility, computing power, and low cost of desktop workstations commonly drive us away from a centralized (mainframe) architecture. A distributed architecture on a local-area network (LAN) allows us to install and maintain ground processors in modules. Standard interfaces allow us to integrate commercial off-the-shelf hardware and software products from various vendors, and when necessary, a local custom system.

12.2 Hardware, Software, and Staffing Requirements

With an understanding of the ground system's tasks and components, and with a mission in mind, we may begin to bring mission requirements to the ground. This highly iterative process must balance requirements for the spacecraft and mission objectives with cost and risk. It must weigh factors such as desired orbit, data rates, and system complexity against one other. The far-thinking mission operations manager will trade requirements between ground-system components and also between the space and ground elements. Table 12.3 lists the steps that lead to a ground system that can support mission objectives. The overlap in trading system requirements suggests we'll run through the table several times before reaching the most cost-effective solution. Table 12.4 summarizes the key influences on the assessment of the ground system.

Table 12.3. Specifying a Ground System Based on Mission Requirements. This process defines a ground architecture (existing or new) to support the mission objectives.

Step	Description	Reference
1. Characterize the mission	Class and type of mission, type of supports	Chap. 2 Sec. 12.2
2. Define major requirements	Type and frequency of supports, geography, availability, timeliness, bandwidth, compatibility	Chap. 11 Sec. 12.2
3. Identify what the network and associated ground elements must do	Number of ground stations, type of comm links, user connectivity	Chaps. 3 and 11 Sec. 12.2
4. Do key trades	See Tables 12.4 and 12.5	Chap. 2 Sec. 12.2
5. Choose existing or new network	AFSCN, TDRSS, NAVSOC, DSN, or other	Sec. 12.3
6. Evaluate complexity and cost	Calculate the initial cost and model the operational cost	Chap. 5
7. Document and iterate		

If we choose an existing system, we establish compatibility between the space and ground systems and determine how much we must modify the existing system to meet the needs of the mission. If we decide to build a new system, we begin the detailed design. We focus here on developing the architecture—designing a ground system isn't within the scope of this chapter.

12.2.1 Defining the Mission

Ultimately, the mission drives all ground-system requirements and its size, cost, and complexity. Missions differ by sponsor, type, orbit, complexity, and other factors.

Table 12.4. Mission Requirements. Thoroughly characterizing the mission and analyzing key system requirements lay the foundation for developing the ground-system architecture. In some cases a trade for one requirement becomes a system requirement.

	Parameter	Options	Key Trades
Mission Definition	Mission class	• Military • Commercial • Scientific	• System complexity • Security/survivability
	Mission type	• Data collection • Communication	• User connectivity
	Communication and orbit architecture	• Constellation: single or multi-spacecraft • Orbit: LEO, GEO, Molniya	• System complexity • Frequency of supports
	Mission complexity*	• Simple • Complex • Autonomous	• Frequency of supports • Ground system complexity
System Requirements Analysis	Type of supports	• Launch and early orbit operations • Routine on-orbit operations • Anomaly operations	• Frequency of supports • Staffing and resources
	Frequency of supports	• Once per orbit • Once per day • Variable	• Spacecraft design • Geography • Orbit • Staffing and resources
	Geography	• Domestic/foreign • Ground stations • User locations	• Orbit • Frequency of supports
	Availability of resources	• Shared network • Dedicated network	• Network augmentation
	Timeliness of data processing	• Real-time • Store and forward • Record at ground station	• System complexity • Communication bandwidth • Geography
	Communication bandwidth	• Variable data rates and modulation techniques	• Power and antenna size • Equipment complexity • Communication links

* See Chap. 5 for a list of complexity metrics for the ground system.

Mission Sponsor. Missions are either military, commercial, or scientific. Each type has its own philosophy of operations. Military missions have historically favored mission objectives over cost. As defense budgets decrease, this philosophy will shift, but military missions will continue to require ground systems that are impervious to natural disasters, terrorist attack, power outages, and equipment failure. The ground system reflects this security and redundancy, having multiple ground stations, geographically separated back-up control centers, and redundant communication links and equipment strings. On the other hand, profit drives commercial missions just as it does other commercial ventures. Commercial missions operate their spacecraft with far fewer people than do military missions, using a philosophy that is good advice for any mission: don't contact the spacecraft any more than you have to. Scientific missions aim to maximize information return but often operate under strict budgets. Indeed, the limited funding for some scientific missions severely restricts ground-system operations. These missions may emphasize end systems that can process and archive large amounts of data.

Mission Type. The type of mission affects how ground systems look, including user connectivity. Chapter 2 defines five mission types: communications, navigation, remote sensing, scientific, and interplanetary exploration. In a data-collection mission, the user may be connected directly to the network to share data conditioning and processing hardware. The users of a communications or navigation mission, on the other hand, need real-time connectivity, so they will operate transmit-and-receive or receive-only stations that are independent of the ground system that controls the spacecraft.

Mission Orbit. This includes both the number of spacecraft in the constellation and the orbits. Examples of mission architectures for communications satellites include store-and-forward, three-satellite geostationary, and Molniya-orbit, (see Fig. 11.13). The number of spacecraft needed to do the space mission strongly affects the quantity and complexity of ground resources. Equally significant is the orbit. The orbit determines how much time the spacecraft is in view of a ground station and how long it will be before the next pass.

Mission Complexity. The complexity of operations and of spacecraft subsystems influences the ground system's complexity and the number of operators. For example, the level of support goes up when the mission requires frequent orbit maneuvers but goes down when the spacecraft uses autonomous navigation. Mission designers trade complexity in the ground system for complexity on the spacecraft. Spacecraft subsystems that process data on board greatly reduce or eliminate requirements on the ground system.

12.2.2 Analyzing System Requirements

By analyzing system requirements, we can describe the ground-system architecture at a high level. Below are the main system requirements that drive a ground system.

Type of Supports. The ground system must support the type of routine and special activities required during the life of the mission. Special activities demand resources beyond those for routine operations. Some assets needed for launch, for example, may not be necessary during the rest of the mission. Controllers and systems must be prepared to support planned special activities, as well as anomalous conditions.

Frequency of Supports. The ground system must support the number of spacecraft contacts required to operate without significantly restricting mission objectives. The quantity and perishability of data to be downlinked, spacecraft design, and orbit contribute to the frequency of routine operations. Special activities such as orbit and attitude maneuvers or resolving anomalies increase the frequency of supports, whereas spacecraft designed for autonomy depend less on ground commands to do mission tasks. How often we must contact the vehicle drives the number of ground stations and staffing levels.

When downlinking of mission data drives contact frequency, we may trade the spacecraft's data rates and storage capacity with orbit and the number of ground stations to reach the desired level of support. As altitude decreases, station visibility decreases, leaving less time per contact. With a large amount of data to downlink, a spacecraft in low-Earth orbit requires more contacts than do those at higher altitudes. When increasing the altitude is too expensive, or when we can't design for a different orbit without significantly altering mission objectives, we may trade contact frequency with the data rate. We can define the number of contacts, C, required to downlink a given amount of data, Q, in terms of the average duration of a spacecraft's visibility, V, and its data rate, R, as shown below:

$$C = Q/(RV) \qquad (12.5)$$

For example, let's consider a spacecraft in LEO with an average visibility to a ground station of ten minutes. Our spacecraft can store one gigabyte (or eight gigabits) and fills our storage media every three orbits. The data rate is five Mbps. The number of contacts required before spacecraft memory overflows is therefore 2.67, or nearly one contact every orbit. If the ground system doesn't have the availability for this contact frequency but can support higher data rates, we should design the spacecraft for a higher telemetry rate. Note that the available time to downlink data is normally less than the total duration of visibility. Acquiring the spacecraft, sending commands, and checking its health can reduce the available time by several minutes. At higher altitudes, this reduction is less significant.

Geography. Ground stations must provide enough coverage to satisfy mission requirements. As discussed above, the frequency of ground supports dictates the number of ground stations, whereas the spacecraft's altitude and inclination largely determine their locations, within geographical and political constraints. A single station can fully cover only spacecraft in geosynchronous orbit. All other orbits will likely require several dispersed stations for full coverage. We like to construct ground stations in remote locations to avoid frequency interference with terrestrial microwave transmissions. But we want control centers in populated areas to take advantage of cheaper public telecommunications and support services. We discuss coverage more fully in Chap. 10.

Availability of Resources. The ground system must have enough resources to make the required contacts at or near the time desired. The ground system's availability depends on the quantity of resources, other spacecraft supported by the network, the ability to conduct multiple contacts and process multiple data streams, and the entire system's reliability. Reliability depends on the likelihood of components or systems failing and the time required to repair them. When a network formally defines its availability as, say, 0.9999, it is referring strictly to its reliability. An availability of 0.9999 means the network experiences 53 minutes of cumulative downtime per year, but not that each user has access to all network resources the rest of the time.

We can calculate availability from figures on mean time between failure (MTBF) and mean time to repair (MTTR) for components and the system. Most ground systems increase availability with designed redundancy in all elements, including the power subsystem. An uninterruptable power system (UPS) uses batteries and often a diesel generator to provide from 30 minutes to continuous backup power to critical equipment during an outage. We must carefully weigh availability requirements because they impact system costs strongly.

A decision that greatly affects resource availability is whether to use a shared or dedicated network. Spacecraft supported by shared resources must be scheduled based on equipment requirements, station availability, and mission priority, as described in Sec. 12.4. When requirements exceed the ground stations' abilities, we can install more equipment, communication links, or even ground stations.

Timeliness of Data Processing. The ground system must meet the users' requirements for timely data routing and processing. Missions with perishable data must have support from system elements able to communicate, process, and display data in sufficient time, and at a sufficient rate, so ground buffers and processors don't overflow and data isn't delayed. Weather data, for example, is almost useless if more than an hour old. This requirement can tax ground-system coverage severely, or more likely, result in an architecture in which users receive data directly from the spacecraft at stations independent of TT&C on the ground.

Data that isn't time critical allows more flexible, and thus less expensive, ground control. We can record this type of data on board the spacecraft and downlink it less often (a mission architecture called *store and forward*). Or we can record it at the ground station and play it back after the contact or mail tapes to the users. If a communication link doesn't exist, or the bandwidth isn't available, real-time transmission can be expensive. Whenever possible, mission planners should investigate alternatives to real-time data communication. A user requirement for relaying telemetry in real time may actually mean a need for one piece of data that an existing voice network can relay in near real time.*

---

* *Real-time* data is less than a few seconds old, or the time it takes for transmission, processing, and display. *Near real-time* data has been further delayed, as in the case of telemetry data processed at the control center and then reformatted and transmitted to a remote user.

Regardless of the mission, TT&C operations are normally in real time. To maintain the spacecraft's health, the control center must be able to monitor spacecraft subsystems and change parameters while in contact with the vehicle.

Communication Bandwidth. The ground system must have enough bandwidth to transport TT&C signals to and from the spacecraft, as well as to support the user's data routing and processing needs. A spacecraft's data rates for TT&C can't be outside the ground system's maximum and minimum rates for generating and transmitting commands or for receiving and processing telemetry. They must also match the communication bandwidth between segments for data relay. As data rates increase, so do antenna size, transmitter power, and costs for equipment and communication links. We can lower data rates with more frequent contacts, but this approach places more stress on system resources and increases staffing levels. Techniques for data compression or onboard processing lower data rates but add complexity and cost to both the space and ground elements.

The required bandwidth of the space-to-ground link depends mainly on the telemetry data rates. The number of spacecraft instruments to monitor and the amount of payload data to collect affects the downlink bandwidth, which is normally much greater than the uplink bandwidth.

The required bandwidth of the ground-to-space link, essentially the command rate, depends on the spacecraft's complexity. Most spacecraft need only a very low command rate—a few hundred to a few thousand bits per second.

The required bandwidth of the communication link between the ground station and the control center (or user) is the sum of telemetry streams, tracking data, voice, and ground-system status. When telemetry rates exceed this bandwidth, some alternatives may be cheaper than installing or upgrading a communication link. A research spacecraft to be supported by the AFSCN, for example, has a five Mbps telemetry stream that can be received by the ground station but not relayed in real time over the 1 Mbps satellite link to the control center in California. The solution is to record the telemetry at the ground station and mail the tapes to the control center. Another option is to play the data back to the control center at a slower rate (in this example, one-fifth the rate at which the data was recorded). This option keeps processing speeds down but ties up resources longer. Modern packetized systems allow splitting a lower rate channel out at the ground station for real time transmission to the control center while storing and forwarding the higher volume payload data channel(s).

12.2.3 Functional Requirements and Compatibility

The high-level requirements discussed thus far help determine the ground system's basic structure, such as number and location of ground stations or control centers and the size of the communication links between them. The next step in developing the ground architecture is to derive and trade the low-level requirements. By analyzing what the ground system must do, we can determine what we need on the ground. Table 12.5 lists some of the more important functions and related design issues. Of course, we must always consider and trade cost. Refer to Chap. 3 for a detailed description of each function.

Table 12.5. Ground-System Functions. The functions required for a mission drive the ground-system design.

Function	Considerations and Constraints	Key Trades
Data Transport and Delivery	• Quantity and rates of data • Location of ground system elements • Compatibility between space and ground elements	• Process telemetry at ground station vs. control center • Choose type of communication links • Design spacecraft for compatibility vs. modify ground system
Mission Control	• Complexity of mission • Operations and maintenance philosophies	• Shared vs. dedicated resources • Redundancy vs. allowable system downtime
Spacecraft Planning and Analysis	• Complexity of spacecraft bus • Orbit	• Level of ground automation • Sophistication of software
Payload Planning and Analysis	• Type of payload • Orbit	• Level of onboard autonomy • Level of ground automation
Data Processing	• Location of users (co-located or external) • Quantity of payload data	• Process data in Mission Control Center vs. dedicated Payload Operations Control Center • Process data in real time vs. post-pass
Navigation Planning and Analysis	• Orbit • Required knowledge of orbit	• Internal vs. external orbit determination (e.g., NORAD) • Ground vs. onboard processing (e.g., GPS) • Antenna angle data only vs. ranging and Doppler systems
Archiving	• Quantity of data • Compatibility with existing recorders (e.g., at the user's facility) • Duration of storage	• Store raw vs. processed data • Type of storage media • Type of distribution and location of storage (transportability)

We begin the design process by analyzing functions. For example, if we must process payload data at a remotely located POCC in real time, we can begin looking at suitable data-processing systems, as well as communication systems to connect the POCC to the network. And if the mission requires us to archive this data in the POCC, we can look at options for recorders. By analyzing all critical functions in this way, we begin to shape the ground system. From here, we can eliminate existing networks that don't meet our needs, begin investigating ones that do, and compare their cost with the cost of building our own.

Compatibility. Before we know whether the ground architecture can support the mission spacecraft, we must establish compatibility between the two. It may help to understand compatibility if we look at what happens to commands and telemetry as they pass between the spacecraft and the ground system. The goal of any transmission is to receive what was sent and be able to understand it. To achieve this goal, satellite transmission requires us to encapsulate information in successive layers. We first code raw information and format the data, possibly encrypt and encode it, and then modulate the data onto a carrier frequency. Table 12.6 describes this process in seven layers. Note that the spacecraft and ground system must be compatible at every layer. A single incompatibility means we can't contact the spacecraft.

Table 12.6. Seven-Layer Model for Satellite Transmission. Each layer becomes an element of compatibility. We define these layers in the text.

Layer	Element	Options	Where Discussed
7	Data rate	Different rates for engineering and payload telemetry, commanding	11.3 and 12.2
6	Data code	PCM codes: NRZ (L, S, M), Biphase (L, S, M), Miller	11.3 and 12.2
5	Data format	Binary, framed, packetized	12.2 and 13.3
4	Encryption	No encryption or various military and commercial algorithms	11.1 and 12.2
3	Encoding	No encoding or convolutional (with Viterbi decoding), Reed-Solomon	11.2 and 12.2
2	Modulation technique	Analog: AM, FM, PM Digital: FSK, PSK, BFSK, BPSK, QPSK, GMSK	11.3 and 12.2
1	Frequency band	UHF, L-Band, S-Band, X-Band, Ku-Band, Ka-Band, SHF/EHF, V-Band	11.2, 12.2, and 12.3

Layer 7: Data Rate. The range of data rates supported by a ground system is fixed by existing equipment. If the command and telemetry rates for a mission are far outside those supported by an existing ground system, changing the ground system probably won't be practical. The maximum and minimum data rates supported by a ground system are an integral part of too many devices to allow for an easy upgrade.

Layer 6: Data Code. Telemetry data collected from subsystems on the spacecraft bus, or from the payload, are digitally coded into one of several formats. Pulse-code modulation (PCM) offers a standard set of codes, the simplest of which is "non return to zero-level" (NRZ-L). The spacecraft may use codes other than NRZ-L to increase bit density (a measure of the number of transitions in a binary stream) and thus improve the link integrity. Because ground processors expect to see data in the NRZ-L format, all other codes must be converted to NRZ-L at the bit synchronizer or elsewhere in the ground system. Code conversion is a relatively simple operation, so most ground systems will support many PCM codes.

Layer 5: Data Format. After individual telemetry points have been PCM coded, they combine in one of several formats, each with its own protocol. Two generic telemetry formats are framed and packetized. *Framed telemetry* allocates specific locations within a repeating frame to data collected on the spacecraft. Each frame will look like the last, with changes only in the value of the data points. *Packetized telemetry* can use the downlink more efficiently by sending only those data points usable on the ground. A packet may vary in length or be fixed. Packetized telemetry may follow such protocols as synchronous data-link control (SDLC) or be unique to the spacecraft. The well established packet-based telemetry format of the Consultative Committee for Space Data Systems (CCSDS) has found application in well over 100 space missions and is now supported by the majority of commercially available telemetry processing hardware and software. CCSDS Telemetry standards are supported by all the emerging commercial tracking networks.

Layer 4: Encryption. Both the uplink and downlink are sometimes encrypted for security. Many devices and algorithms are available, supporting various data rates and operational configurations. We must ensure compatibility not only between devices but also between the keys* loaded into those devices. Every spacecraft link has its own unique key, so we can test the transmit and receive keys for a particular link only against each other.

Layer 3: Encoding. A communication link, and the ground system as a whole, will normally specify a bit-error-rate below 10^{-5} (1 of every 100,000 bits in error) for voice and between 10^{-6} and 10^{-8} for data. To improve link performance (at the expense of bandwidth) signals are sometimes encoded for transmission using convolutional, Reed-Solomon, or other error-correction methods. Turbo codes are new high efficiency codes. For deep space missions that operate with very weak signals, coding actually increases the data return when a low error rate is needed. Selecting codes in the CCSDS standards will widen the number of networks available with compatible decoding systems. If a spacecraft encodes its downlink or expects to decode its uplink, we have to install compatible hardware on the ground.

Layer 2: Modulation Technique. To transmit a signal over the high-frequency spacecraft link, we must superimpose information on the carrier frequency using modulation. We use amplitude modulation (AM), frequency modulation (FM), and phase modulation (PM) for analog signals, but PM is most common and most efficient with weak signals. Modulating digital signals requires frequency-shift keying (FSK) or phase-shift keying (PSK). To improve efficiency, we may use the more complex binary phase-shift keying (BPSK) or quadrature phase-shift keying (QPSK). We trade modulation complexity with its use of bandwidth and its bit-error-rate performance. New modulation standards have been developed that enable reducing the rf bandwidth for higher data rates with only a small penalty in efficiency (E_b/N_o required). This is important because of spectrum utilization

---

* Keys are codes necessary to encrypt and decrypt data successfully on a particular link. [See Chap. 11]

regulations that space missions must meet. We also have to consider the presence of subcarriers and the modulation index. The modulation index, measured in degrees, is the amount of phase shift used in modulating data onto the subcarrier. Most ground systems allow various modulation techniques. Some receivers can handle several modulation types by employing plug-in cards.

Layer 1: Frequency Band. The most important layer of compatibility is in the frequency band used to carry signals between the spacecraft and the ground station. Antenna systems and TT&C hardware have a narrow tuning range, so ground systems support only a select number of frequencies. Spacecraft that will be fully supported by the AFSCN, for example, must communicate within SGLS frequencies[*]. ITU regulations specify which frequency bands can legally be used based on a mission's purpose (military, science, commercial, communications) and the orbital altitude. In most cases, it's not practical to change an existing ground system to support new frequencies.

Compatibility Testing. It's difficult or impossible to recover from an incompatibility after launch. And with so much invested, it's a tremendous disappointment when the only thing collected from the mission is the insurance. We must discover any problems with the space-to-ground interface before launch, and preferably, before the spacecraft has been delivered to the launch site. No interface-design document or computer simulation will reveal as much about a ground system's ability to support a mission as an actual test. The compatibility test proves that commands, telemetry, and ranging signals can be communicated through rf channels and correctly interpreted by the end systems.

We commonly do compatibility testing twice: first at the spacecraft factory, then again at the launch site. Some networks have mobile compatibility systems that can go to the spacecraft integration facility. Sometimes a test version (also called a "suit case") of the spacecraft rf and TT&C systems can travel to a network test facility. Finding incompatibilities is much cheaper at the factory, where the components, tools, and designers are in place to make changes. Launch-site testing ensures systems are still working and allows last minute adjustments to the hardware.

A simple test is to transfer telemetry and commands by recording the unmodulated data and playing it back directly into the baseband equipment on the spacecraft and at the control center. Most missions would want more elaborate tests that involve the rf equipment end to end. A good test will validate compatibility for

- Command and telemetry formats, modes, and coding
- Control and display equipment
- Cryptographic equipment

---

[*] SGLS, the Space-Ground Link Subsystem, specifies 20 uplink frequency channels between 1.75 and 1.85 GHz, and 20 downlink frequency channels between 2.2 and 2.3 GHz.

- Radio frequency interface
- Ranging
- Recording

12.2.4 Staffing

In evaluating staffing requirements for a ground system, we must consider both the level of support and staff skills. Large, existing networks come fully staffed with operations and maintenance (O&M) people 24 hours per day. Small or dedicated networks may require less than 24-hour support. A smaller network may save money by using contractors for tasks such as maintenance, rather than supporting a full-time staff on site.

The frequency of spacecraft supports and the complexity of space and ground systems determine the required size and skill of the operations staff. Complex spacecraft that lack autonomy require more people, although more complex ground systems can reduce this number. People are expensive. If we can reduce the number of operators by automating acquisition and tracking, data recording, routing, and processing, we can save a lot of operating money.

We should also seek ways to reduce the skill levels required of operators. One way to do this is by using computer-based expert systems to make many of the decisions. Expert systems help spacecraft operators and those who control and monitor the ground network. Systems programmed with detailed knowledge of the spacecraft or ground system can now interpret data instead of operators having to do so. Simple text-based displays are giving way to graphical displays that correlate data in strip charts, bar charts, or cross plots. Expert systems require more up-front software development, to trade against lower costs in staffing and fewer operational errors.

Besides requiring O&M people, a ground system needs engineers, technicians, and support staff to modify and test systems, upgrade facilities, and train operators.

12.3 New Systems and Existing Networks

Early in a space mission the mission operations manager must decide whether to use existing ground resources or build a new system. This decision affects all facets of the mission, from spacecraft design to the operations philosophy. In the early stages of the mission it's easy, but unwise, to ignore the ground system for what appear to be more pressing concerns of the spacecraft and its launch vehicle. Constructing a new ground system can be as long and complicated as constructing the spacecraft. And with an existing system, it's better to know the interface requirements before we design the spacecraft. Without the ground system in mind, we can make an arbitrary decision about the spacecraft design that later forces a costly change to the ground system.

Some missions are born to a ground system. At the first mission-concept meeting for NASA's Cassini probe, for example, no one had to ask if the Deep Space Network would support the vehicle. But other missions should begin investigating support alternatives by asking

1. Is an existing network able and eligible to support the mission?
2. What are the costs of operating in that network?
3. What constraints does that network place on the mission?

For most missions, the constraints of compatibility with an existing network don't harm the mission, provided we define the interface early enough. Other missions can't match any existing network without severely changing the mission's intent and function. These missions must build a dedicated ground system, designed specifically around the space element, for more specialized support.

12.3.1 New Systems

A good argument for constructing a new ground system is that the mission requires little operational support. A spacecraft that needs only infrequent monitoring from a single ground station may be able to afford independence. If we accept the risks and limitations of a single ground station (discussed in Section 12.1.1), building such a system can be more cost-effective than using an existing network. But this is true only if it takes advantage of modern technology.

Missions that can build their own ground systems avoid the burden of history: the legacy of old systems. Once, we had no alternative but to custom design the pieces of a ground system. The hardware was unique, and the software wasn't portable. Operations were built around a central mainframe or mini-computer, and everything was proprietary, including data storage and printer interfaces. It was expensive and inflexible. Microprocessor-based systems have undergone an explosive increase in computing power, with equally dramatic reductions in size and cost. At the same time, open architectures with standard interfaces have been promoted. We're no longer forced into ground-up design with special-purpose components, and we're no longer locked to a single vendor. Today, we can build an entire ground system out of commercial, off-the-shelf (COTS) products.

COTS products keep development and operational costs in check. One costly driver of ground-system design is software development. The arduous task of designing, debugging, and documenting thousands of lines of code doesn't mate well to a success-oriented schedule. We can minimize the amount of software to develop if we buy commercially available software for command and control, mission planning and analysis, and orbit prediction, graphical window managers, and databases.

Another way to reduce development costs is to adapt existing software from previous missions. A better idea still is to restrict vehicle-specific information to the databases and keep the code itself generic. This way the same software can support different spacecraft at the same time.

Turn-key systems take reliance on COTS products a step further. They're complete, ready-to-use systems for spacecraft command and control. They're generic and reusable, and they leave little for the user to develop. Many vendors will even offer to install their systems or help the user integrate them with existing ground hardware. Although most turn-key systems don't include front-end equipment, such as receivers and bit synchronizers, we can buy an entire transportable ground station. Such a system will include a foldable antenna and TT&C systems that can be easily shipped and deployed. A good turn-key system will

- Adapt from pre-launch testing to launch and on-orbit operations
- Tailor to different spacecraft and existing ground systems
- Run on a distributed network to allow easy expanding and sharing of peripherals
- Support open-architecture standards to allow third-party peripherals and software
- Provide security features such as access control, with special restrictions for operations like commanding

Users of turn-key systems depend more on their vendors than users who design and build their own systems. For this reason, a good turn-key vendor provides comprehensive testing and training at installation, as well as strong technical support throughout the mission.

12.3.2 Existing Networks

In most existing networks resources are shared among many users. This is both a blessing and a curse. When we share resources we also share the cost of operations. What we lose in such a network is full control of scheduling and changes without arbitration. Without a powerful reason for independent control, the cost-effectiveness of shared resources persuades most missions to use an existing network.

Existing networks have other advantages. There is less risk to the spacecraft's development schedule, and ultimately the launch date, by using a ground network with known capabilities and interface requirements. In addition, training time and costs are lower. Choosing an existing network to support a space mission means contracting for a service that provides nearly everything. With the facilities and ground hardware comes the staff that will operate the spacecraft and help plan the mission.

The cost for the service from existing networks can be little to nothing. The AFSCN charges are identified after a Programs Project Requirements Document (PRD) is reviewed and support agreements are reached. Other networks have a defined cost. TDRSS charges non-NASA users on a per-minute basis, depending on the type of service. Cost information is available in NASA's Service Catalogs.

Determining which existing network to use depends somewhat on network abilities and loading but mostly on eligibility. Exclusions have always kept commercial spacecraft from contracting with a military network for operational support, but this view is changing. When there were fewer networks the division was clear: NASA and related scientific missions were supported by the Spaceflight Tracking and Data Network (since replaced by TDRSS), or for planetary missions, the Deep Space Network. Department of Defense missions flew mostly out of the AFSCN. While this tradition still holds, missions have crossed these lines. Many of these networks are connected to each other, allowing for hybrid use of their resources. In addition to the "government" ground networks there are commercial entities that are entering the scene. Companies such as Universal Space Network and Honeywell's Datalynx in the United States, and the Vega Group in Europe are offering ground network services both to government agencies and to commercial users.

Hybrid ground architectures combine different existing networks and sometimes dedicated assets to exploit the advantages of each. A spacecraft launched in 1995 is a good example of the kind of mission that benefits from a hybrid ground system. Its payload has a data rate of 25 Mbps, which can be downlinked once every orbit. But maintaining the spacecraft's health requires contacting it throughout its orbit. The AFSCN does the TT&C to maintain the spacecraft, while a dedicated facility in Southern California receives and processes the payload data.

12.3.3 Changing Existing Systems

After deciding to use an existing ground system, the mission operations manager has another decision: use only the resources this system provides or augment it with other resources. There are benefits to designing a spacecraft for 100% compatibility with all existing ground-system elements. Eliminated are the costs, risks, and scheduling that come from system modifications. Added, though, are potential compromises to mission objectives. When these compromises become too great, we must change the system.

An existing system may not meet the needs of a particular mission for many reasons, such as incompatible telemetry formats, data rates, encryption, or complexity of command or telemetry systems. Because the core ground system may have been installed long before we design a spacecraft, the spacecraft commonly employs newer technology than the ground system. Large ground systems undergo major upgrades every 10 to 20 years, in a slow evolution that provides predictable, stable services to their users. Upgrading the entire network with every change in technology is too expensive.

Regardless of the modification, we must allow enough time for planning. Because shared ground systems support many missions at the same time, we must evaluate changes to the network for their effects on other users. Networks strive to reduce costs by combining requirements from different users into one comprehensive solution.

12.3.4 The Air Force's Satellite Control Network (AFSCN)

The AFSCN is a global network of remote ground stations and control centers to command and control US spacecraft. Under the direction of US Air Force Space Command, the AFSCN supports crewed and uncrewed, DoD and non-DoD, space missions.

Operations are directed from two network control centers located at Onizuka Air Station in Sunnyvale, California, and Schriever Air Force Base in Colorado Springs, Colorado. The Secretary of Defense recommended, during the 2005 Base Realignment and Closure (BRAC) process, that Onizuka Air Force Station be closed and the node moved to Vandenberg AFB. His recommendation will go to the BRAC Commission for review. Currently, most missions use only one network control center but some missions use one center as primary and the other as backup. An integrated mission-control center supports TT&C for the bus and the payload. Less often, payload data goes from the control center to an external payload-operations control center for processing following defined interface specifications.

The control centers connect through Domestic Satellite (DOMSAT) links and land-lines to fifteen ground stations at eight geographic locations. In addition, the Eastern Vehicle Checkout Facility (EVCF) is located at Cape Canaveral, Florida, and is used for pre-launch vehicle checkout. It has an Automated Remote Tracking Station (ARTS) core, an antenna that can be manually oriented, Unified S-Band (USB) capability and various interface lines to the AFSCN nodes and other Cape locations. Table 12.7 shows the antenna characteristics for these ground stations, and Table 12.8 lists some of the network's characteristics. In the early 1990s the Air Force upgraded most of the ground stations to Automated Remote Tracking Stations (ARTS), thereby increasing their capacity and availability and reducing costs for operation and maintenance. In 2003, the AFSCN began a 10-year project of migrating toward full automation. During this upgrade, antennas and other specifications will be upgraded; one upgrade is expected to be USB support at all ground stations.

The AFSCN supports SGLS and non-SGLS spacecraft that have one uplink and up to three simultaneous downlinks; three receivers, one can handle only SGLS (subcarrier) downlinks and two additional receivers that can handle SGLS or non-SGLS (FM or direct modulation) downlinks. Ground stations can record telemetry rates up to 5 Mbps but can relay telemetry to the control center in real time at no more than 1.024 Mbps. Higher data rates can be supported with the user buying the extra communications bandwidth.

12.3.5 The Naval Satellite Control Network (NSCN)

The NSCN operates spacecraft for the Naval Space Command. Control of the network is centered at the Naval Satellite Operations Center Headquarters (NAVSOC HQ) at Point Mugu, California. The NAVSOC operates three ground stations at Prospect Harbor, Maine; Laguna Peak, California; and Finegayan,

Table 12.7. Characteristics of AFSCN Antennas. Operations are directed from Schriever Air Force Base, Colorado.

Site	Location and Altitude	Size (m)	G/T (dB/°K)	EIRP (dBW)
New Hampshire (Manchester, NH)	42° 56.9' N, 71° 37.6' W 201 m 42° 56.7' N, 71° 37.8' W 191 m	10 14	21.6 24.6	74.1 77.3
Vandenberg (Lompoc, CA)	34° 49.4' N, 120° 30.1' W 269 m 34° 49.6' N, 120° 30.3' W 266 m	18 14	27.1 25.2	83.7 77.2
Hawaii (Kaena Point, Oahu)	21° 33.8' N, 158° 14.5' W 428 m 21° 34.1' N, 158° 15.7' W 318 m	18 14	26.1 24.6	79.1 77.6
Guam	13° 36.9' N, 144° 52.0' E 218 m 13° 36.9' N, 144° 51.3' E 209 m	18 14	26.6 23.1	83.8 76.5
Thule (Greenland)	76° 31.0' N, 68° 36.0' W 132 m 76° 30.9' N, 68° 36.0' W 132 m 76° 30.9' N, 68° 35.0' W 132 m	7 14 10	20.4 24.4 21.6	73.5 75.4 76.6
Oakhanger (England)	51° 06.8' N, 00° 52.7' W 144 m 51° 06.3' N, 00° 52.9' W 91 m	10 10	21.6 21.5	74.8 75.2
Colorado (Colorado Springs, CO)	38° 50.1' N, 104° 49.2' W 1959 m	10	21.6	74.8
Diego Garcia	07° 16.0' S, 72° 22.4' E 5 m	10	21.6	74.1

Table 12.8. Characteristics of the AFSCN Network. The network supports both space-ground link subsystem (SGLS) and non-SGLS spacecraft.

Parameter	Uplink	Downlink
Frequency	1750–1850 MHz 2025–2120 MHz (USB) (Diego Garcia and EVCF only)	2200–2300 MHz
Modulation	SGLS: AM/FSK/PM Non-SGLS: BPSK or BPSK/PM USB; NRZ-I, M, or S on a 16 kHz subcarrier	SGLS: BPSK, BPSK/PM Non-SGLS: BPSK, QPSK, FM, FM/FM, PM
Data rate	SGLS: 1,2, or 10 kbps Non-SGLS: 100 bps to 100 kbps (BPSK) 100 bps to 256 kbps (BPSK/PM) USB; 250 bps to 2 kbps	SGLS: 1.024 Mbps and 1.7 Mbps Non-SGLS: 5 Mbps (to ground station only)

Guam. The NSCN can do three simultaneous, independent contacts—one at each ground station. The network provides hardware and software for standard data formats. Users with unique data formats must provide their own ground equipment that will either integrate into the NSCN or have its own communication link if located elsewhere. Tables 12.9 and 12.10 show the NSCN's characteristics.

Table 12.9. Characteristics of NSCN Antennas. Operations are directed from the Naval Operations Center Headquarters at Port Mugu, California.

Site	Location and Altitude	Size (m)	G/T (dB/°K)	EIRP (dBW)
Laguna Peak, CA	34°06' N, 119°04' W 450 m	18	25	66
Detachment A (Prospect Harbor, ME)	44°24' N, 68°01' W 7 m	5	15	62
Detachment C (Finegayan, Guam)	13°34' N, 144°50'E 151 m	5	12	N/A

Table 12.10. Characteristics of the NSCN Network. The NSCN is highly compatible with the AFSCN.

Parameter	Uplink*	Downlink
Frequency	1750–1850 MHz	2200–2300 MHz
Modulation	PM	BPSK, PM
Data rate	1 kpbs or 2 kbps	2 Mbps maximum

* Uplink capability currently not available at Detachment C.

Like the AFSCN, the NSCN uses SGLS to communicate with spacecraft. But unlike the AFSCN, the NSCN does most data processing and commanding at the ground station, not the control center. The Integrated Satellite Control System, a network based on microcomputers, evaluates the telemetry. It strips from the state-of-health telemetry the redundancy that results from slowly-changing spacecraft measurements such as battery temperatures. Only changed telemetry points go to the NAVSOC HQ for analysis and archival. In addition, NAVSOC ground stations are fully automated, thus eliminating the need for voice communication between the ground station and NAVSOC HQ. As a result, the NSCN can reduce its communication-link requirements between the ground stations and the control center to a full-duplex, leased line operating at 56 kbps.

12.3.6 NASA's Tracking and Data Relay Satellite System (TDRSS)

A shortfall of any moderately sized ground system is that it covers only a small percentage of the orbit for low-altitude spacecraft. NASA overcame this deficiency with the TDRSS Network. Although the TDRSS Network has both a space and ground segment, it satisfies our definition of "ground system." TDRSS monitors and controls spacecraft, and provides mission data to customers by employing multiple geostationary spacecraft rather than multiple ground stations.

The TDRSS spacecraft fleet is clustered in three main geostationary longitudinal locations. The TDRSS East cluster operates from 41°, 47°, and 49° West longitude. The TDRSS West cluster operates from 171° and 174° West longitude. Operating a Tracking and Data Relay Satellite (TDRS) spacecraft over the Indian Ocean at 275° West longitude has closed the TDRSS Zone of Exclusion (ZOE). The combined TDRSS fleet provides 100% coverage for subsynchronous spacecraft. A TDRS satellite processes no customer data and as such, it's simply a "bent-pipe" frequency translator or repeater.

Three ground terminals, two located near Las Cruces, New Mexico (the White Sands Ground Terminal (WSGT) and the Second TDRSS Ground Terminal (STGT)) and one on Guam (Guam Remote Ground Terminal (GRGT)) support all TDRSS satellites through the use of six Space-to-Ground Link Terminals (SGLT). All the TDRSS satellite flight operations are conducted from the White Sands Complex (WSC) TDRSS Operations Control Center (TOCC). The SGLTs typically communicate customer-formatted data to and from the TDRS relay using pseudorandom noise (PN) spreading below 300 kbps and non-spread above 300 kbps. A SGLT forward link transmits commands and PN ranging code to the TDRS and down to user spacecraft.

On the return link, telemetry and the PN turn-around code are received by the TDRS from customer spacecraft and relayed to the ground terminals at White Sands and Guam. The GRGT customer data is rate limited and is communicated via the WSGT to provide a common embarkation and debarkation point at WSC. The WSC and Guam SGLTs also provide customer spacecraft support in the Goddard Space Flight Center (GSFC) Ground Network (GN) mode of operations (non-PN spread spectrum), generally limited to lower data rates.

NASA's Goddard Space Flight Center in Greenbelt, Maryland manages and operates the TDRSS Space Network. The TDRSS customer services at WSC and Guam are controlled via the Data Services Management Center (DSMC) at White Sands. The DSMC functions include scheduling and configuring the Space Network resources, monitoring various telemetry streams in real time, and providing orbital tracking data to the Flight Dynamics Facility (FDF) at GSFC. TDRSS services at WSC are connected to the customer Mission Operations Centers (MOC's) by using the NASA Integrated Space Network (NISN) Internet Protocol (IP) Operational Network (IONet). Customers can locate their MOC at the GSFC and use the Closed IONet, or, with appropriate communication links, use the Open IONet at a customer's remote site location.

The TDRSS Network offers three basic services, as summarized in Table 12.11. The Single-Access (SA) Service relays forward and return data at a high rate to and from a single spacecraft. For SA service each TDRS has two steerable, 4.9-meter, parabolic antennas that operate in S-Band (tunable) and Ku-Band (fixed) or Ka-Band (tunable). The Multiple-Access (MA) Service relays forward and return data at a low rate to and from many spacecraft operating on a fixed S-Band frequency, using a body mounted phased-array antenna on the TDRS spacecraft. The Demand Access Service (DAS) relays MA return data only at a low rate from many spacecraft at once (generally on a continuous 24x7 basis), using the TDRS body mounted phased-array antenna which operates on the fixed MAR S-Band frequency.

Table 12.11. TDRSS Service Summary/System Capability. Capabilities are typically defined in type of access, frequency band, data rate, and number of channels. (WSC = White Sands Complex; TDRS = Tracking and Data Relay Satellites.)

	Frequency	Service	Max. Data Rate	Services per TDRS[*]	WSC[†] Capability (Channels)	Guam Capability (Channels)
Single Access S Band	2020.4 MHz - 2123.3 MHz	Forward	300 kbps	2	10	2
	2200 MHz - 2300 MHz	Return	6 Mbps	2	10	2
Single Access Ku Band	13.747 GHz - 13.802 GHz	Forward	25 Mbps	2	10[‡]	2
	14.887 GHz - 15.119 GHz	Return	300 Mbps	2	10[‡]	2
Single Access Ka Band	22.55 GHz - 23.55 GHz	Forward	25 Mbps	2	6[‡]	0
	25.25 GHz - 27.50 GHz	Return	300 Mbps / 800 Mbps	2 / 1	6[†] / 0[**]	0
Multiple Access S Band (TDRS 3-7)	2103.1 MHz - 2109.7 MHz	Forward	300 kbps	1	4	1
	2284.5 MHz - 2290.5 MHz	Return	300 kbps	5	20	2
(TDRS H,I,J)	2284.5 MHz - 2290.5 MHz	Return	3 Mbps	5	15	0
Demand Access (MAR only) S-Band (TDRS 3-7)	2284.5 MHz - 2290.5 MHz	Return	150 kbps	50[††]	8[††]	8[††]

[*] Fully operational spacecraft
[†] 4 SGLTs at White Sands Complex (WSC) are capable of supporting the TDRS H,I,J spacecraft. WSC Ka-Band and SMA capability reflect the 6/15 respectively because there are only three TDRS H,I,J spacecraft.
[‡] Ku and Ka FWD/RTN service is shared on the TDRS/WSC SGL. Simultaneous Ku and Ka Service are not possible from a single SA antenna.
[**] Ka-Band (TDRS H,I,J) 800 Mbps is not currently supported on the ground (WSC).
[††] Initial DAS capability is for eight channels expandable to 50 channels.

12.3.7 NASA's Ground Network

The Goddard Space Flight Center's Ground Network comprises two main components, orbital and sub-orbital operations. Supporting organizations include Engineering, Logistics, and Administration. The diverse customer list includes

NASA projects and other U.S. Government agencies, academia, commercial companies and foreign governments and organizations.

Orbital tracking ground stations include the Merritt Island Launch Annex (MILA) in Florida, Wallops Ground Station (WGS) in Virginia, Alaska Ground Station (AGS) and the Alaska SAR Facility (ASF) in Alaska, and the McMurdo Ground Station (MGS) in Antarctica. The existing government owned ground tracking stations are being replaced with ground tracking services provided by commercial facilities and services. The current expanding list of commercial service providers includes DataLynx (PF1) in Alaska, Universal Space Network in Hawaii and Pennsylvania, the University of Santiago (AGO) in Chile, and Space Data Services (SDS) in Longyearbyen, Norway. The NASA Ground Network station characteristics are summarized in Table 12.12.

Table 12.12. Characteristics of NASA's Ground Network. Services are provided to NASA Projects, other U.S. Government agencies, academia, commercial companies, and foreign governments and organizations.

Station	Antenna Diameter	Transmit Frequency (MHz)	EIRP (dBWi)	Receive Frequency (MHz)	G/T (dB/K)	Location	User Tracking
SGS (Norway)	11.3 m	2025-2120	66	2200-2400 8000-9000	23 35.4	78°N 15°E	1- & 2-Way Doppler, Angle
WGS (WFF)	11.3 m	2025-2120	66	2200-2400 8000-9000	23 35	38°N 75°W	1- & 2-Way Doppler, Range, Angle
LEO-T (WFF)	5 m	2025-2120	59	2200-2300	17	38°N 75°W	—
TOTS (WFF)	8 m	2025-2120	62	2200-2400	21	38°N 75°W	1- & 2-Way Doppler, Angle
MGTAS (WFF)	7.3 m (2)	—	—	1435-1535 1670-1720 2200-2400	11 12.5 15.5	38°N 75°W	1-Way Doppler, Range, Angle
SATAN (WFF)	Arrays (3)	147-155	59	136-138	−7.03	38°N 75°W	Range
SCAMP (WFF)	Array	147-155	57	—	—	38°N 75°W	Range (with SATAN)
METEOSAT (WFF)	7.3 m	—	—	1685-1710	12	38°N 75°W	—
MGS (Antarctica)	10 m	2025-2120	63	2200-2400 8025-8400	21.1 32.5	78°S 193°W	1- & 2-Way Doppler
ASF (Alaska)	10 m	—	—	2200-2400 8025-8400	21.1 32.5	65°N 148°W	—
	11.3 m	—	—	2200-2400 8000-9000	23 35	65°N 148°W	—

Table 12.12. Characteristics of NASA's Ground Network. (Continued) Services are provided to NASA Projects, other U.S. Government agencies, academia, commercial companies, and foreign governments and organizations.

Station	Antenna Diameter	Transmit Frequency (MHz)	EIRP (dBWi)	Receive Frequency (MHz)	G/T (dB/K)	Location	User Tracking
LEO-T (Alaska)	5 m	2025-2120	59.2	2200-2300	17	65°N 147°W	—
TOTS (Alaska)	8 m	2025-2120	62	2200-2400	21	65°N 147°W	1- & 2-Way Doppler, Angle
AGS (Alaska)	11.3 m	2025-2120	66	2200-2400 8000-9000	23 36	65°N 147°W	1- & 2-Way Doppler, Angle
MILA (Florida)	9 m (2)	2025-2120	63	2200-2300	24	29°N 81°W	1- & 2-Way Doppler, Range, Angle
PDL (Florida)	4.3 m	2025-2120	58	2200-2300	11	29°N 81°W	—

Suborbital tracking, data acquisitions, and communications include services from the list of facilities provided below. Suborbital services maintain a semi-permanent facility at the University of Alaska's Poker Flat Research Range in Chatanika, Alaska. Other mobile support capability identified below represents a completely self-contained capability to establish mobile launch range capability worldwide.

- Wallops Telecommunications Instrumentation
    - Data acquisition activities from a variety of apertures
    - Data routing between support assets
    - Timing and command destruct
    - Mobile support capability
- Wallops Radar Instrumentation
    - Fixed precision radars
    - Surveillance radar
    - Mobile support capability
- Wallops Control Center and Data Reduction Facilities
    - Launch control displays
    - Command destruct capability
    - Mobile support capability
- Wallops Optical, Photographic, and Video Facilities
    - Fixed camera sites
    - Photographic lab
    - Mobile support capability

- Wallops Meteorological Services Facility
  - Surface observations
  - Weather forecasting office
  - Mobile support capability

12.3.8 NASA's Deep Space Network (DSN)

Spacecraft navigating the solar system require more powerful ground antennas than most satellite control networks can provide. NASA's DSN operates such antennas. The DSN supports tracking and communication for all of NASA's interplanetary spacecraft.

The DSN consists of three multi-station complexes. Each of these sites—Goldstone (California), Canberra (Australia), and Madrid (Spain)—is equipped with one 70 m, several 34 m, and one 26 m antenna, described in Table 12.13. The 70 m and 34 m antennas mainly help us communicate with spacecraft at distances greater than two million km from Earth. The 26 m and other smaller antennas typically support Earth-orbiting satellites. At each site, the 70 m antenna can array with the 34 m antennas for improved telemetry performance. Table 12.14 lists some of the DSN's characteristics.

Table 12.13. **Characteristics of DSN Antennas.** Operations are directed from the Jet Propulsion Laboratory, California.

Site	Antenna No.	Location and Altitude	Size (m)	G/T (dB/°K) S-band X-band, Ka-band	EIRP (dBW)* S-band X-band, Ka-band
Goldstone (California)	14	35° 25.6' N, 116° 53.4' W 1002 m	70	51.0, 62.9, N/A	105.8 116.1, N/A
	15	35° 25.3' N, 116° 53.2' W 974 m	34	40,.2, 54.0, N/A	N/A, 110.1, N/A
	16	35° 20.5' N, 116° 52.4' W 945 m	26	31.7, N/A, N/A	91.4, N/A, N/A
	24	35° 20.4' N, 116° 52.5' W 952 m	34	41.3, 52.8, 64.4[†, ‡]	99.1, 109.9, N/A
	25	35° 20.3' N, 116° 52.5' W 961 m	34	N/A, 53.7, 64.1	N/A, 110.0, 108.5
	26	35° 20.1' N, 116° 52.4' W 970 m	34	N/A, 55.5^3, 65.7[‡]	N/A, 109.9, N/A
	27	35° 14.3' N, 116° 46.6' W 1053 m	34	34.7, N/A, N/A	77.1, N/A, N/A
Canberra (Australia)	34	35° 23.9' S, 148° 58.9' E 693 m	34	40.8, 53.7, 64.4[‡]	99.1, 109.9, N/A
	43	35° 24.1' S, 148° 58.9' E 690 m	70	50.9, 62.8, N/A	105.8, 116.2, N/A
	45	35° 23.9' S, 148° 58.7' E 675 m	34	40.2, 54.0, N/A	N/A, 110.1, N/A
	46	35° 24.3' S, 148° 58.9' E 678 m	26	31.9, N/A, N/A	91.4, N/A, N/A
Madrid (Spain)	54	40° 25.5' N, 4° 15.2' W 838 m	34	41.0, 53.7, 64.4[†, ‡]	99.1, 109.9, N/A
	55	40° 25.5' N, 4° 15.2' W 808 m	34	N/A, 55.3, 64.4[‡]	N/A, 110.0, N/A
	63	40° 25.9' N, 4° 14.9' W 866 m	70	50.9, 63.1, N/A	105.8, 116.3, N/A
	65	40° 25.6' N, 4° 15.0' W 835 m	34	39.6, 54.0, N/A	N/A, 110.1, N/A
	66	40° 25.8' N, 4° 15.5' W 851 m	26	31.8, N/A, N/A	91.4, N/A, N/A

* Calculated for a 20 kW power amplifier at the 70 m and 34 m antennas, and a 10 kW power amplifier at the 26 m antenna. A 400 kW power amplifier is also available for the 70 m antenna.
† Future implementation planned.
‡ Estimated value.

Table 12.14. Characteristics of the DSN System. Available data rates are low for the DSN because of the large distance from the spacecraft to the ground station.

Parameter	Uplink	Downlink
Frequency	2025 to 2120 MHz 7145 to 7190 MHz 7190 to 7235 MHz* 34,200 to 34,700 MHz	2200 to 2300 MHz 8400 to 8500 MHz 31,800 to 32,300 MHz
Modulation	Carrier: PCM/PSK, PCM direct	PM
Data rate	1.0 bps to 4000 bps on a subcarrier (100 to 16,000 Hz), 8000 to 64,000 bps direct on carrier	8.0 bps to 2.6 Mbps

* Only available on the 34 m Beam Wave Guide antennas.

Command and control of many deep space missions are centered at the Jet Propulsion Laboratory in Pasadena, California. JPL normally has people present 24 hours per day, every day, to monitor spacecraft and manage ground systems.

Saying that DSN antennas are heavily scheduled is an understatement. Years before launch, mission designers will investigate the predicted loading of DSN assets for the period of their mission, hoping to minimize contention for antenna time. The best launch period for an interplanetary mission depends mostly on solar-system geometry, but planners will adjust the launch date to align the spacecraft in a part of the sky that has less competition with other spacecraft.

12.3.9 The European Space Agency's Network

The European Space Agency (ESA) is an international body of 13 member nations with a common goal: to give Europe independent access to space. ESA's ground system provides TT&C, as well as Doppler and ranging services, for Earth-orbiting satellites.

The network consists of eight ground stations and three control centers. Two of these control centers, at Redu (Belgium) and Villafranca (Spain), are co-located with ground stations. The primary control center is at the European Space Operations Center (ESOC) in Darmstadt, Germany. Table 12.15 lists the characteristics of ESA's ground system.

12.3.10 International Tracking Stations

In addition to the ground networks described in the previous section there are numerous tracking stations throughout the world as listed in Table 12.16. Information on the characteristics of tracking stations operated by many of the space agencies around the world can be obtained at the CCSDS Web site http://www.ccsds.org. The document of interest is CCSDS 411.0-G-3:Radio Frequency and Modulation – Part 1:Earth Stations.

Table 12.15. Characteristics of the ESA's Antennas. Control centers are at Redu, Belgium; Villafranca, Spain; and Darmstadt, Germany.

Site	Location and Altitude	Size (m)	G/T (dB/°K)	EIRP (dBW)
Kiruna (Sweden)	67° 51.4' N, 20° 57.9' E	15	S-band: 28.9 X-band: 34.5	S-band: 71.0
Kourou (French Guiana)	5° 15.1' N, 52° 48.3' W	15	S-band: 30.1 X-band: 38.2	S-band: 73.0
Perth (Australia)	31° 48.2' S, 115° 53.1' E	15	S-band: 28.2 X-band: 38.2	S-band: 79.0
Redu (Belgium)	50° 1.0' N, 5° 8.4' E	15	S-band: 30.1	S-band: 73.0
Villafranca (Spain)	40° 26.5' N, 3° 57.2' W 40° 26.8' N, 3° 57.1' W	15-1 15-2	S-band: 30.1 S-band: 29.0	S-band: 73.0 S-band: 79.0
New Norcia* (Australia)	31° 2.9' S, 116° 11.4' E	35	S-band: 39.8 X-band: 51.1	S-band: 97.6 X-band: 109.7

* Values are preliminary.

Table 12.16. International Tracking Stations for Earth-Orbiting Satellites. This table lists additional stations for Earth-orbiting spacecraft. All sites provide telemetry and commanding.

Site/Agency	Antenna Diameter (m)	Locations		Frequency Band
		Latitude	E. Longitude	
RSA (Russia)				
Evpatoria*, Ukraine	25 (2)	45° 11' N	33° 11'	$P^\dagger$/C
Ussuriisk, Russia	25 (2)	44° 00' N	131° 45'	P^2/C
Tshelokovo, Russia	25 (2), 12	56° 01' N	37° 52'	P^2/C, P^2
St.Petersburg, Russia	12	68° 02' N	33° 09'	P^2
Jusaly, Kazakhstan	12	45° 19' N	64° 03'	P^2
Kolpashevo, Russia	12	58° 12' N	82° 35'	P^2
Ulan-Ude, Russia	25, 12	51° 33' N	107° 24'	P^2
Petropavlovsk, Russia	25, 12	53° 18' N	158° 26'	P^2
CNES (France)				
Kerguelen Islands, France	10	49° 21' S	70° 15'	S
Aussaguel, France	11	43° 5' N6	01° 30'	S
Kourou, French Guyana	11	05° 06' N	307° 21'	S

Table 12.16. International Tracking Stations for Earth-Orbiting Satellites. (Continued) This table lists additional stations for Earth-orbiting spacecraft. All sites provide telemetry and

Site/Agency	Antenna Diameter (m)	Locations		Frequency Band
		Latitude	E. Longitude	
Hartebeestoek, S. Africa	12	−25° 53'	27° 42'	S
NASDA (Japan)				
Masuda, Japan	18, 13, 10	30° 33'	131° 01'	S
Katsuura, Japan	18, 13, 10	35° 12'	140° 18'	S
Okinawa, Japan	18 (2), 10	47° 53'	11° 06'	S

* Operated jointly by Russian Space Agency and Ukrainian Space Agency.
† 157/184 and 745/930 MHz (uplink/downlink).

12.4 Operational Concerns

After launch, a new set of issues becomes important. We'll look at three ground-system concerns that become critical during the operational phase of the mission: ground anomalies, maintenance, and resource scheduling.

12.4.1 Ground Anomalies

Anomalies are the bane of the mission operations team. An anomaly, by definition, is a problem or an event with an unknown cause. The detection of a problem jerks the operations team from routine operations and propels it into a search for the cause. If the anomaly is severe enough, the team won't be able to proceed with the mission until the problem is resolved.

When an anomaly occurs, we sometimes don't know right away whether the problem is in space, on the ground, or somewhere in between. When the satellite downlink drops out, do we suspect the satellite transmitter? Do we blame the ground receivers and demodulators? Or do we look at the rf compatibility between the two? The answer, of course, is that we look everywhere at once. Solving the mystery of an anomaly is much like solving any mystery: begin with the most probable causes and successively eliminate the innocent. Chapter 16 treats spacecraft anomalies in considerable detail. We'll look briefly at ground anomalies here.

Ground anomalies are fundamentally different from spacecraft anomalies—given enough time and money, we can always resolve a ground anomaly. There is more time, more information, and less panic in the investigation of a ground anomaly. The ground system has no safe-hold mode. Usually, it's easier to find the root cause of a ground anomaly because maintainers can access all ground equipment for testing. They can visit even a remote, unstaffed system during the investigation. Common types of ground anomalies are

- Hardware failures

- Configuration errors (human errors)
- Compatibility problems between space and ground
- Flaws in the software design or database errors

A hardware failure can happen at any time. So can configuration errors, but these, like compatibility and software problems, are more likely to occur during launch and early orbit, when procedures are new and systems undergo operations for the first time.

We attempt to cover future anomalous situations with contingency plans. These tell the operator what steps to take when data isn't at a terminal, or whom to call if a communication link seems to have failed. It's wise to be prepared, but these plans do little to calm the mission manager during the support. Even the simplest of problems can cause the loss of a spacecraft support when the pass is only ten minutes long. Instead, we should concentrate on prevention. Table 12.17 summarizes how to prevent and resolve common ground anomalies.

Table 12.17. Ground Anomalies. Prevention is the goal.

Ground Anomaly	Prevention Technique	Resolution Action	Relative Cost (Typical)
Hardware failure	• Redundancy	• Replace or repair	Medium
Configuration error	• Training • Automation	• More training	Low
Compatibility problems between space and ground	• Testing • Use proven systems	• Replace ground equipment • Change operation procedures	High
Flaws in software design or database error	• Testing • Use commercial products	• Code or data base modification	Low – medium

Redundancy in the design of all ground elements helps to prevent anomalous conditions. With the ability to patch around a suspect piece of equipment, or route data through a backup communication link, we can quickly bring a system back on line. The operations team can then resume their tasks while the maintenance staff investigates the problem.

It's impossible to prevent a human operator from ever making a mistake. We can reduce errors by designing systems with simple, logical human interfaces and training people well for less simple systems. If we can automate these tasks, we avoid the problem altogether.

Perhaps the most difficult anomalies to resolve are compatibility problems. On orbit, they are difficult to find and more difficult to fix. Through pre-launch testing we can discover and prevent incompatibilities, as described in Sec. 12.2. We can also minimize their occurrence by favoring proven technology in the TT&C subsystems of the space and ground segments.

Software won't fail spontaneously, but its design may be faulty. More likely, the database will contain incorrect parameters, or parameters in the wrong locations. Thorough pre-launch testing should unearth such errors before they cause an anomaly. We can reduce the probability of flaws in the software by moving away from custom coding and relying more on commercial, off-the-shelf (COTS) products (if they are well proven) and by multimission re-use of custom generated systems. See Chap. 18 for a more in-depth discussion of mission software.

The ground system itself can help a lot, or not at all, in detecting and resolving anomalies. Older and simpler ground systems don't help: humans must discover all problems. Discovery typically comes through a secondary characteristic, so people have to trace the root cause of the problem. Modern systems will detect malfunctions and report them to their human attendants. A control-and-monitor system that monitors equipment remotely through independent signal lines is one example. More sophisticated systems will detect and resolve equipment failures and then notify humans of the action. This kind of automated redundancy is common at the equipment level. For example, many devices will have redundant power supplies within a chassis because they have a high failure rate. Failure of one power supply requires no human intervention to keep the system on line. Automated redundancy is less common at the system level, but as the cost of such systems falls, replacing human maintainers will be cheaper.

12.4.2 Maintaining and Sparing

Often overlooked in planning and designing a ground system is long-term maintenance. The cost of maintenance can be high, but the cost to the mission of not maintaining a system can be even higher. The level of acceptable risk determines the level of required maintenance. We define this risk by the amount of time a system can be down before the impact to the mission becomes unacceptable, and we weigh this downtime against cost. Different users will have different requirements for system availability. Some military users can't afford more than 0.01 percent downtime, whereas some university-sponsored missions can't afford to provide the 100% sparing such a low downtime percentage would require. For missions in which real-time data processing isn't critical, we can relax maintenance requirements for data relay and processing, provided the ground station records the data.

Steps in developing a maintenance plan include determining acceptable downtime, recognizing existing redundancy that we can exploit, and calculating the cost of each option. The costs include people, training, equipment, and documentation. We should develop the maintenance plan before installing the system because it often affects the design.

We can define three levels of maintenance, as shown in Table 12.18 and illustrated in Fig. 12.9. In Level 1, downtime is unacceptable. An equipment failure must not block data on its way to the user. We achieve this level of maintenance with redundant strings of equipment and hot spares. We install hot spares in the same way as the equipment they are meant to back up, but we don't use them unless the primary device fails. They can be switched into the circuit with minimal delay, sometimes automatically.

Table 12.18. The Three Levels of Maintenance. We must trade initial cost with acceptable risk, using mission requirements.

Level	Allowable Downtime	Characteristics	Initial Cost	Risk
1	None	Redundancy, hot spares	High	Low
2	Short (0 – 24 hours)	100% on-the-shelf spares	Medium	Medium
3	Long (days to weeks)	Repair as required	Low	High

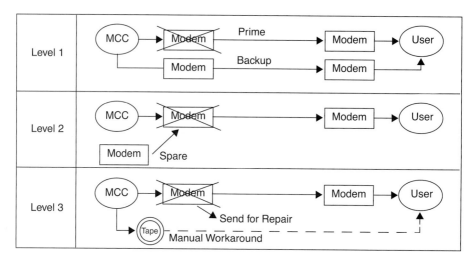

Fig. 12.9. The Three Levels of Maintenance Illustrated with a Modem Link. In this example, users must determine how critical the data from the Mission Control Center is before they can decide which level they prefer.

Level 2 quickly turns around an equipment failure, usually within hours, by using spare hardware on the shelf and an on-site maintenance staff. For many missions, this compromise between cost and risk is the best maintenance solution.

Level 3 provides no sparing but repairs or replaces components as required. We can use this plan for non-critical elements having operational workarounds. In Fig. 12.9, for example, users can wait while a tape with recorded data is shipped to their site.

For all levels of maintenance a well documented system is the key to troubleshooting and resolving component failure successfully. Unfortunately, documentation can also make for a blown budget. It shocks taxpayers to read about the military's $500 hammer, but those familiar with strict documentation requirements have accepted this news. They know it's not unreasonable to pay more to document a component than to buy the component itself. Still, we must temper the tendency to document too much by paring down redundant documents and eliminating unnecessary ones. Today, even the government understands the need to reduce documentation. To keep in step with a shrinking budget, the Department of Defense now requires that government agencies and their contractors use commercial specifications and documentation whenever possible, rather than creating their own [Perry, 1994].

12.4.3 Scheduling Resources

In a shared network, spacecraft must compete for resources. We invariably lack antennas, communication links, and data-handling hardware to support every desired spacecraft contact without some compromise. Therefore, we must schedule all ground resources, usually days in advance (in the case of the DSN, months), to use them best. Resource scheduling ensures that, although users may not get every contact they want, they will get every contact they need.

We assign ground resources to spacecraft based on availability and mission priority. The user submits a request for resources within some window, and in most cases, we can simply adjust the schedule to meet it. When two or more missions vie for the same resources in the same period, mission priority determines which one gets them—in a process called deconflicting. No absolute priority scheme guides planners to the most efficient schedule. A rough mission hierarchy gets planners started: (1) crewed space flight; (2) launch and early orbit operations; (3) normal operations, in the order of low orbit, medium orbit, then high orbit; (4) pre-flight testing or training. Physics gives spacecraft in lower orbits a derived priority because they have fewer chances to contact a ground station.

In addition to these categories, some missions are regarded as having more importance to the network than others; for example, a spacecraft supporting national defense will influence the schedule more than a spacecraft researching the atmosphere. But none of these guidelines will cover every case. In the end, only the event's priority—the relative importance of that particular spacecraft contact—matters. In the civilized arguments of deconflicting, users (with an arbitrator's help) determine whose event is really more important.

Two ground-system specifications influence the availability of a network. *Turnaround time* is the time required between successive contacts to reconfigure computer databases and equipment settings and to change recorder tapes. We must consider this figure, typically several minutes, when scheduling back-to-back contacts. *Maintenance downtime* is the scheduled removal of a system from operational status for calibration, cleaning, or repair. The time required varies greatly from system to system but can reduce the availability of a network by several hours per month.

We use hardware and software tools to decrease the demands on the operators tasked to schedule spacecraft contacts. Commercially available software that graphically displays orbits and ground-station visibilities decreases the required skill level of the operator, and more sophisticated systems can eliminate the need for an operator. Like so many other ground system functions, scheduling is a candidate for automation. In an automated system, users submit requests for resources and let the computer schedule their contacts. The need to resolve conflicts between two missions however, may force some networks to keep a person in the loop.

References

DataLynx http://www.honeywell-tsi.com/DataLynx/index.htm

Hartebeesthoek http://www.sac.co.za/gse/

Klements, H. D. 1992. *Air Force Satellite Control Facility Space/Ground Interface.* TOR-0059(6110-01)-3. El Segundo, CA: The Aerospace Corporation.

McClimans, Fred J. 1992. *Communications Wiring and Interconnection.* New York, NY: McGraw-Hill.

National Aeronautics and Space Administration. 1994. *Mission Requirements and Data Systems Support Forecast.* 501-803. Goddard Space Flight Center, Greenbelt, MD: National Aeronautics and Space Administration.

National Aeronautics and Space Administration. 1988. TDRSS Users' Guide. STDN No. 101.2. Goddard Space Flight Center, Greenbelt, MD: National Aeronautics and Space Administration.

National Aeronautics and Space Administration. 2001. Deep Space Network/Flight Project Interface Design Handbook. JPL-DSN 810-005, Revision E. Jet Propulsion Laboratory, Pasadena, CA: National Aeronautics and Space Administration.

Naval Satellite Operations Center. 1993. *Naval Satellite Control Network to Satellite System Interface Document.* Point Mugu, CA: Naval Satellite Operations Center.

Perry, William J. 1994. "Specifications & Standards--A New Way of Doing Business." Memorandum issued June 29, 1994. Washington, DC: The Secretary of Defense.

Powers, John T., and Stair, Henry H. 1990. *Megabit Data Communications.* Englewood Cliffs, NJ: Prentice Hall.

Pratt, Timothy and Charles W. Bostian. 1986. *Satellite Communications.* New York, NY: John Wiley & Sons.

Space Data Services, Norway http://www.spacetec.no/index.php

Universal Space Network http://www.uspacenetwork.com/

University of Santiago, Chile http://www.cee.uchile.cl/public/home_en.html

Williamson, Mark. 1990. *Dictionary of Space Technology.* New York, NY: Adam Hilger, IOP Publishing.

Chapter 13

Processing Data and Generating Science Data Products

William B. Green, *California Institute of Technology (Retired)*

> 13.1 Characteristics of Science Data Products
> 13.2 Instrument Development and Impact on Data System Design
> 13.3 Relationship Between Uplink System and Science Data Processing System
> 13.4 Downlink Data Processing Flow
> 13.5 Design Considerations for Science Data Processing Systems
> 13.6 Science Data Processing System Development and Test
> 13.7 Science Data Processing System Architecture
> 13.8 Science Data Processing System Implementation
> 13.9 Science Data Archive Considerations
> 13.10 Examples

This chapter describes the production of science data products from space missions. We design payload instruments to achieve various scientific objectives for each mission, and the science requirements dictate requirements on the ground data processing systems that process the data and generate a set of data products for each mission. We must structure the development process to produce a data system that is responsive to the science requirements of the mission, within the constraints imposed by cost and schedule considerations. The development process involves a partnership between the ground system design team, development and test team, the instrument developers, and the scientists that will use the data products to support their research and analysis. We must hold adequate reviews to ensure that the requirements on the ground data system are correct and complete, that the data system design will meet mission requirements, and that the integration and test program can demonstrate that the ground system is ready to support flight operations. This chapter addresses these issues, starting with a description of a standardized method for defining the various types of science data products generated by space missions.

13.1 Characteristics of Science Data Products

Data products generated from instruments flown in space generally represent different levels of processing applied to the instrument data. Several different types of data products can be generated from data produced by a single instrument flown as part of the scientific payload on a space mission. In 1982, the Committee on Data Management and Computation (CODMAC) of the National Research Council established definitions for various levels of science data products produced by space missions [CODMAC, 1982]. These definitions have been widely adopted by the scientific community, and are shown in Table 13.1.

Table 13.1. Science Data Product Levels Defined by National Research Council Committee on Data Management and Computation.

CODMAC Data Product Level	CODMAC Definition
0	Raw Data. Data set corrected for telemetry errors and decommutated. Data are tagged with time and location of acquisition.
1A	Edited data. Unresampled data that are still in units produced by the instrument, but have been corrected so that values are expressed in, or are proportional, to some physical unit.
1B	Resampled data. Data that have been resampled or reprocessed in such a way that the original edited data cannot be reconstructed.
2	Derived data products containing geophysical variables at the same resolution and location as the Level 1 source data.
3	Variables mapped on uniform space-time grid scales, usually with some completeness and consistency.
4 and above	Model output or results from analyses of lower level data (e.g., variables derived from multiple measurements).

13.1 Processing Data and Generating Science Data Products

Level 0 data is produced by the ground data system from telemetry data received at a ground receiving station. The spacecraft data system generally packages data for transmission to the ground. Production of the Level 0 data requires that the telemetry data from the spacecraft be decoded (decommutated) as necessary, and aggregated into a set of logical data records, each of which corresponds to an instrument observation. Level 0 data records must contain sufficient additional information (called the "ancillary data") to enable unique identification of the observation.

Design teams on each mission will analyze the science requirements and define data products to be produced by the ground data system beyond Level 0. Most projects will process data at least to level 1A, where the data have been converted into meaningful physical units and ancillary data (including instrument and spacecraft engineering data relevant to the observation) have been included. Projects often produce data products beyond level 1A, depending on the science objectives of the mission and the nature of the user community that will be using the data in their research and analysis. Identification of the various types of data products at the different levels is an important part of the requirements definition phase when developing a science ground data processing system.

Data products produced and distributed from the Multi-angle Imaging Spectro Radiometer (MISR) instrument, flying on NASA's earth orbiting Terra satellite, provide an example of how data products are defined for one specific space mission. The MISR instrument consists of nine separate cameras that acquire imagery as the spacecraft orbits the earth. Figure 13.1 shows the top level definition of the MISR data products available to the science community, taken from the technical information provided by the MISR website at the Jet Propulsion Laboratory.

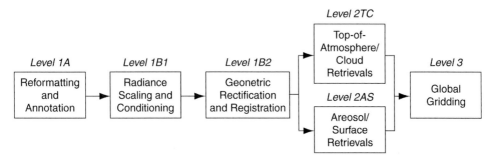

Fig. 13.1.   **Data Products from Multi-angle Imaging Spectro Radiometer (MISR).** The type of instrument determines the form of each level of data product.

13.2 Instrument Development and Impact on Data System Design

Once we define a mission and establish science objectives, we select one or more payload instruments based on them. Government agencies fund the development of flight instruments in several ways, and they generally fall into one of two categories:

"PI (Principal Investigator) Instruments." We develop instruments in this category by contracting with a Principal Investigator to design, test and deliver an operating instrument, and then participate in final testing of the instrument with the spacecraft before launch. The PIs are generally scientists who have competitively bid to provide an instrument for a given mission. Selection of the PI and an instrument depends on many factors, including the capability of the instrument, the degree to which the proposed instrument supports the mission objectives, and the realism of the development plan as outlined in the proposal. PIs may propose performing the work at their home institutions, or may elect to contract out all or part of the development and test effort to other organizations, including commercial companies. The PI has the right to decide how the instrument will be developed and tested prior to spacecraft integration, and the PI's proposal will include a description of the proposed development and test plan, along with cost and schedule information.

"Facility Instruments." The responsibility for developing and testing instruments in this category resides with a government agency. The agency may elect to develop the instrument at government facilities, universities, or commercial companies, or some combination of these entities. The agency may select a team of scientists (referred to as a "science team" or a "project science working group") as advisors to help guide instrument design and development activities to ensure that we meet mission and science objectives.

The fact that a particular instrument is a PI or facility instrument affects the design of the data system that processes the flight data from that instrument. For PI instruments, the government agency can decide that the PI is responsible for some or all of the data processing for that instrument, and for producing a defined set of data products and a schedule for delivering them to the sponsoring agency. Alternatively, the government agency can elect to provide the facilities for processing the instrument data, and give the PI preferential access to the instrument data relative to the general scientific community. An example of benefits given to PIs that provide instruments but do not process the data during operations includes early access to the science data before public release. Another example, implemented on some NASA astrophysics space observatory missions, is the commitment of a certain amount of "guaranteed observing time" during the first few years of flight operation of the observatory to PIs and scientists involved in instrument development.

All of these decisions profoundly affect the design of the science ground data system. If the instrument is a facility instrument, the sponsoring agency must develop a data system to provide rapid access to data by the user community. If the instrument is a PI instrument, where the PI is responsible for the science data processing, the sponsoring agency must design a data system that delivers Level 0 data rapidly to the PI facility for further processing. It must approve the PI's plans for data processing, and then monitor processing during operations to ensure that data product delivery schedules are being met. Often the government agency will be responsible for the long term archiving of the data products, even when the PI produces the data products. If the instrument is a PI instrument where the data processing responsibility resides with the sponsoring agency, the science data processing system must provide rapid access to the mission data for the PI and their associated teams. It must also incorporate systems to control data flow consistent with any agreements made with the PI regarding public data release and dissemination.

13.3 Relationship Between Uplink System and Science Data Processing System

Two areas of uplink operations can affect the processing of science data, and they are (i) mission planning and scheduling and (ii) sequence generation. It is important that science data processing system designers become involved early in the design of the software and procedures that support these two uplink processes.

In mission planning and scheduling we develop the overall plan for spacecraft operations. This process generates the list of activities for the spacecraft and the science instruments to carry out. We often refer to that list as a "mission plan." We must also define a basic logical unit for the science observations acquired by each instrument. For an imaging system, the logical unit of data acquisition is an image. For a spectrometer, the logical unit may be a spectrum or a series of spectra. The mission planning and scheduling system will generate a list of observations to be made. It should also assign a unique identifier to each observation from each instrument on the spacecraft, and should determine the order in which the observations are taken. The identifier travels with all information sent to and received from the spacecraft that relates to that observation.

The sequence generation process converts the mission plan into individual commands for the uplink system to send to the spacecraft. Science data system designers often get involved in designing the set of individual commands sent to each instrument. In many applications, the sequence design engineers establish sets of commands that are executed frequently. These sets of commands, often called "blocks", may be stored on board the spacecraft, so that a single command sent to the spacecraft results in execution of the stored command string. Block design can incorporate parameters, in which case the command to execute a block might include a set of parameters that are then used to control various functions of the instruments (exposure time for an observation is one example). Downlink data system designers should become involved in designing onboard blocks in order to develop a detailed understanding of the instrument functions that will be used in flight.

We establish the unique identifier for each instrument observation during the mission planning and scheduling process. The sequence generation system then develops a command string that incorporates the unique identifier, which will travel through the onboard data system and be embedded in the downlink data stream associated with each instrument observation. The science data processing system will use the same logical identifier as one of the index keys to manage the science instrument data records produced from a given observation or set of observations.

Each project must allocate the responsibility for mission planning and scheduling early in the development of the mission operations concept. For some missions, the mission operations team may receive requests from the Principal Investigators or the science teams associated with each of the payload instrument providers for a set of observations, and have the responsibility of developing a mission plan that best meets the (often conflicting) requirements delivered by one or more instruments. On other missions, such as observatory class missions that solicit periodic observation proposals from the science community (e.g., the Hubble Space Telescope and the Spitzer Space Telescope), the science data system may incorporate the mission planning and scheduling support functions [Green, 2002] [Adler, et. al., 2002]. In this case, a science operations team delivers a completed mission plan to the mission operations teams, who perform the sequence generation and uplink functions.

It is important to note that the science data system is generally the only system that can verify correct implementation of the uplink systems supporting mission planning and scheduling systems and the sequence generation systems prior to launch. As a part of an end-to-end test program, the science data system can verify that the correct observations were commanded on board the spacecraft, and that the correctly formatted data has been provided in the downlink data stream. In addition, science data system developers are very familiar with the instrument characteristics, and can recommend test sequences that correspond to the most frequently used instrument modes for use in pre-flight test and verification activities.

Support for Instrument Engineering Activities. The science data system may have to support some uplink functions associated with instrument operations. Engineers and scientists involved with support of in-flight instrument operations often work at the science facilities that produce the science data products, rather than at the mission operations center. Instrument scientists and engineers can provide feedback to the mission operations teams regarding changes in operational procedures, commands, or blocks, necessitated by actual in-flight instrument performance. This feedback is generally based on analysis of the science data products generated from the instrument observations. The science data system may incorporate a variety of analysis tools to support these functions, and may have a direct interface to the mission operations or uplink data systems to transfer information to the mission operations teams [Spitzbart and Wolk, 2002]. We should generate requirements for support of these functions during the early stages of mission design, with the involvement of science data system design engineers.

13.3.1 Key Considerations

Here are typical issues regarding the relationship between the uplink elements of the data system and the science data processing system that we should address during development of the mission operations concept:

- Establish the basic logical unit for science observations for each instrument, and define the format for the unique identifier associated with each science observation.
- Decide whether the mission planning and scheduling function will be the responsibility of the mission operations or science operations teams, based on the characteristics of the mission, and allocate requirements to specific elements of the ground data system based on that decision.
- Analyze planned mission sequences and determine if blocks will be stored on board the spacecraft. If so, assign responsibility for development and verification of onboard blocks.
- Allocate responsibility for instrument related engineering activities to a specific operations team, and allocate requirements for supporting those activities to a specific element of the ground data system.

13.4 Downlink Data Processing Flow

There are generally three major elements of the End-to-End Information System involved in producing Level 0 science data products: the spacecraft data system, the ground receiving station, and the element of the project's ground data processing system responsible for producing Level 0 data products.

13.4.1 Spacecraft Data System

The downlink data flow path starts on the spacecraft, when the spacecraft and flight instruments begin to execute commands associated with a science observation. Most spacecraft transmit the spacecraft bus or housekeeping data along with the payload data. Production of science data products often requires information from the sensor's and the spacecraft's data streams. Onboard systems consist of science instruments, the computers that collect data, recorders, downlink or telemetry systems, and the spacecraft bus. We need data from many sources to process science instrument data. Spacecraft engineering data may include data from spacecraft systems for power, thermal, electronics, communications, and attitude determination and control. We use the attitude-determination data to process and analyze the instrument data, and provide the spatial reference for each scientific observation.

To map an image or other science observation to the target properly, we need to know the relationship between the sensor optics or antenna pointing and the spacecraft orientation. For some missions, we may be able to determine this relationship prior to flight, and then refine it based on in-flight performance. For other missions, limitations in ground test equipment may make it necessary to model this relationship and then refine it with actual flight data after launch. For imaging systems on earth orbiting missions, ground reference points within a target image can help determine the relationship between spacecraft pointing and instrument orientation, so that we can map the instrument field of view correctly into earth coordinates. For missions that view the sky instead of the earth, we can use known star fields to determine the mapping between spacecraft and instrument pointing, and eventually map the image data into appropriate celestial coordinates. The downlink data system on the spacecraft must return appropriate spacecraft orientation and pointing information with the science observation data.

In addition to spacecraft pointing and orientation information, the spacecraft data system inserts other housekeeping information into the downlink data stream. The spacecraft data system must periodically collect instrument engineering data, such as temperatures and voltages at various locations within the instrument subsystem, and return it to earth along with the sensor data.

Missions generally use several different time references, one of which is spacecraft time, to mark the time that a data set is obtained on the spacecraft. The downlink data stream must include this information, since it may be the only way to distinguish between multiple acquisitions of the same data type. The spacecraft data system also extracts the unique identifier (discussed in the last section) for each observation from the uplink data, and inserts it into the downlink data stream for that observation.

The spacecraft data system is responsible for packaging the data before transmitting it to the ground receiving station. Generally, the spacecraft data system will assemble mission data into "packets". A packet is a logical unit of data, such as a subset of an image readout from a science instrument, or an assemblage of engineering data from a set of sensors on board the spacecraft. The project must determine the appropriate format and content of the packets. In the case of science instruments, data from a single science observation may be transmitted to the ground in more than a single packet. Once the data stream has been "packetized", the spacecraft data system may encode the data using various standards before telemetry transmission to the ground receiving station.

13.4 Processing Data and Generating Science Data Products

Onboard Processing Functions for Sensor Data. The spacecraft data system may process sensor data before transmitting the science data to the ground receiving system. This processing may be done either by the main spacecraft computer, or by other processing elements of the spacecraft data system incorporated in the science instrument package. Typical onboard processing of science instrument data may include:

- Averaging, smoothing, or other functions applied to multiple sensor readouts to produce a single output that is sent to the ground.
- Digital data compression that may be either lossy or lossless. Lossy compression produces a result that cannot be reconstructed to achieve perfect replication of the original data. Data that have been compressed using lossless compression techniques can be decompressed to yield an exact copy of the original observation data. Methods similar to the JPEG standard are often employed for imaging data, and other mathematical compression techniques may be more appropriate for non-imaging instruments.
- Data calibration. In some cases, the initial conversion from instrument units to physical units may take place on board the spacecraft, especially if onboard calibration can minimize the downlink data volume. In these cases, ground commands should enable updating of calibration data files stored on board the spacecraft.

We must also consider onboard processing requirements beyond collecting, storing, and transmitting routine and housekeeping data. For example, it may be possible to calibrate science-sensor data more efficiently on board, especially because the calibration data may come from the sensor itself or from an external source. The advantage of doing this is that users can access the calibrated data as soon at it is received on the ground. Unfortunately, users may not be able to change or improve that calibration as new information becomes available. Onboard calibration requires the ability to change the calibration algorithm as information about instrument performance improves during flight operations. In other words, we have to uplink the calibration routine from the ground to incorporate new knowledge into the calibration procedure. We may also need to communicate new information from data sources external to the spacecraft if they're necessary to improve or modify the calibration.

Sometimes the volume of data generated by the sensors exceeds the downlink capacity, and only onboard data compression can handle it. The compressed data can then be downlinked and decompressed on the ground. In this case, we need to allocate onboard processing capability and processor time to the compression computation. Because compression techniques vary with data and application, we may need to alter the compression algorithm, so we must be able to control this change from the ground. We also have to know the compression algorithm on the ground to decompress the data; it's best to identify the compression technique in the data stream itself.

13.4.2 Ground Receiving Station

Antenna Systems. The ground system has two types of antennas: (1) uplink and control antennas and (2) receive-only, direct-readout antennas. These communicate with the spacecraft data systems through the communications system. We concentrate here on the receive-only ground reception and data-processing system because Chaps. 8, 11, and 12 deal explicitly with the communications, command, and control aspects. Refer to these chapters for questions regarding the data system's radio frequency, command, and control.

We design each space mission with certain objectives in mind. Although the space component (the spacecraft) is likely the costliest and most complex element of the mission, we can't neglect the importance of the ground system and the need to process the sensor data. The receive-only ground system begins with the antennas themselves and extends into data collection and processing. The nature of the antenna depends on the type of orbit the spacecraft is in. For spacecraft in geostationary earth orbit, the antennas can be fixed in position and thus are far less complex than antennas that must move with the spacecraft. Spacecraft in highly inclined earth orbit require antennas that move with the spacecraft as it crosses from horizon to horizon. Deep space missions often use antennas shared among several missions, with highly precise pointing capability.

Functions Performed by Ground Receiving Stations. The basic function of the ground receiving station is to receive the data transmitted by the spacecraft and to deliver properly formatted digital data records to the data system supporting the mission. The receiving station will receive the incoming data stream from the spacecraft, decommutate it, decode it if the data have been encoded on the spacecraft using a standard telemetry format, reconstruct the data packets as transmitted by the spacecraft data system, and format the data into data records in a format negotiated between the receiving station and the mission.

13.4.3 Level 0 Data Processing

The ground data system for a given mission will include an element that produces Level 0 science data records from reconstructed data delivered by the ground receiving station. This system will generally receive packetized data from the receiving station and reconstruct data records that represent individual science observations. The Level 0 data records usually include the following elements:

- The sensor data as received from the spacecraft data system, reflecting the results of whatever onboard processing has been applied to the data (if any)
- A set of ancillary data that must include
  - The unique identifier for the observation

- The spacecraft time at which the observation was initiated
- The ground received time at which the data arrived at the ground receiving station (required to differentiate between multiple transmissions or multiple playbacks of science observation data)
- A set of ancillary data that may include
  - Instrument engineering data (e.g., temperatures, voltages, settings, etc.)
  - Spacecraft engineering (e.g., spacecraft position and orientation at the time of the observation)

Data Accountability. Each mission will establish a requirement for science data completeness, generally expressed as something like "the mission will produce science data records representing X percent of the data acquired by the spacecraft instruments". All elements of the ground data system perform data accountability, starting with the receiving station. The receiving station may generate retransmission requests to the spacecraft to attempt to recover any detected data outages. The Level 0 processing operations team will also evaluate data completeness, and may generate requests to (i) retransmit data from the spacecraft to the ground, (ii) retransmit data from the receiving station to the mission data system, or (iii) repeat an observation on the spacecraft in the case of significant loss of data.

The level 0 data records are the input records sent to the science data processing system that produces level 1 and above data records from the mission. The next section describes the science data processing system. The overall downlink data flow is shown in Figure 13.2.

13.4.4 Key Considerations

Here are the issues we must address relative to the downlink data system during system design and development of a mission operations concept:

- What processing of science instrument data will be done on the spacecraft before transmission of the observation to the ground receiving station?
- Is data compression required? If so, what compression methods meet the bandwidth constraints while preserving the highest quality science data return?
- How is the data to be packetized on board? How many packet formats do we need to accommodate both science observation data and engineering data?
- Definition of the format and content of the data to be transferred from the ground receiving station to the project flight operations system.
- What are the format and content of the Level 0 data records? What ancillary data must be included in the Level 0 data records?

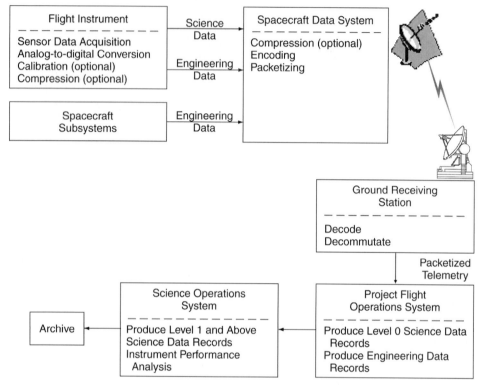

Fig. 13.2. **Downlink Data Flow.** This chart depicts the flow of data from a flight instrument to the ground archive along with the functions/processing that takes place at key locations.

- Which element of the ground data system will perform the Level 0 processing, and which operations team will have the responsibility to produce the Level 0 data records?
- What is the data completeness requirement on the science data and how will we ensure that the mission design meets it?

13.5 Design Considerations for Science Data Processing Systems

This section describes the process of designing a science data processing system that supports the science objectives of a flight mission.

13.5.1 Establish Processing Requirements

The two major categories of science data processing requirements are (i) requirements specifying content and format of the data products generated by the system and (ii) requirements specifying the data throughput and response time requirements that the system must meet. Secondary requirements that can influence system design include requirements for data security, the number of users the system must support, support of multiple data access levels for different users, remote data access and retrieval, and public release of the data.

Establishing the content and format of the data products involves an iterative set of interactions with the science team, and results in an agreed-upon set of data product requirements placed on the science data processing system. The requirements definition process also involves analysis of cost and schedule tradeoffs, ideally involving science team members. The system designers generally interact with the science teams during the requirements definition phase to establish the data processing needed to achieve the mission scientific objectives. The data product requirements are generally specified at the various levels defined by the CODMAC standard definitions. Discussion of the various categories of data products eventually leads to a discussion of the timeline for producing those products.

System engineers should ask the following types of questions of the science teams when defining the requirements on the science data system:

- What data products will you require in order to publish the research results you expect to obtain from this mission?
- What is the timescale on which those products must be delivered? What data do you need to see (i) in real time, (ii) within 24 hours of receipt of data?
- What archival data products do you expect to generate that will form the lasting scientific legacy of this mission?
- How long will it take to approve each type of data product for public dissemination and release to the long term archive? Will we need to reprocess the initially received data to complete the final set of publicly released data products? If so, on what time scale?
- Do you anticipate significant changes in the algorithms used to process the science instrument data during flight operations?
- Do you expect to update instrument calibration files on board and/or on the ground during flight?
- Do we need special engineering data products to support operations either during in-flight instrument certification or during nominal science operations?

- Where will the scientists analyzing the data be located? What data products must be available in what locations and on what time scale for the science teams to meet their analysis needs?
- What ancillary data (instrument and/or spacecraft engineering data) will the science data processing algorithms need to produce the final science data products?
- Does the science data processing system have to support other mission activities, such as planning and scheduling, support of engineering data analysis, education and public outreach, public affairs, etc.?
- What degree of data compression can we accommodate without impacting science analysis?

Once the project managers have held the appropriate budget and scheduling discussions, a final set of science data processing system requirements should be documented in a requirements document placed under project level change control.

13.5.2 Design Lifecycle, Documentation, and Reviews

The science data processing system documentation is usually similar to the project level documentation, except at a more detailed level. The requirements are specified in a System Requirements Document (SRD), which is generally reviewed at a System Requirements Review. We first develop a top level system design, including definition of the various subsystems constituting the science data processing system and their functions and interfaces. We then develop a Functional Requirements Document, which levies requirements across the constituent subsystems at a more detailed level than the SRD.

The design of the science data processing system is reviewed at Preliminary and Critical Design Reviews (PDR and CDR). Preliminary and detailed design documentation should be available for these reviews, along with interface agreements with external elements of the project, signed by both parties. These can be at the same time as the project level PDR and CDR, or may take place on a somewhat different schedule. Often the project level PDR and CDR occur before the ground system PDR and CDR, based on the need for procuring long lead hardware items or other project considerations. It is best, however, that the science data processing system reviews take place on the same schedule as the project level reviews, to ensure that the system design is compatible with other subsystems in the overall flight and ground systems.

Additional documentation relating to the science data processing system might include the following:

- Algorithm Theoretical Basis Documents. An ATBD is peer reviewed by the scientific community and then provided to the system

developers to guide implementation of the algorithms used for processing science data in the production system.
- Users' Guides. These guides, which are generally available on-line and updated periodically during development and operation of the system, provide users with sufficient information on the use of the data system and the content and format of the data products.
- As-built System Descriptions. These documents describe the function and operation of subsystems based on the as-built operational system.
- Security Plan. This plan will include a description of the physical security system that controls access to the operational system. It will also include a description of the levels of access control for the data managed by the science data processing system.
- Project Data Management Plan. This plan may be required by a sponsoring agency (e.g., NASA or ESA). It typically contains a top level description of the science data products that the mission will produce, the policies governing science data access and release, and a description of the responsibilities of various organizations in designing, implementing and operating the science data processing system.

As launch approaches, the science data processing system staff becomes involved in various project level reviews. These can include the Operations Readiness Review, to determine that the total project flight operations system has been tested and verified as ready to support flight operations, and the Launch Readiness Review, to determine that the spacecraft, payload, launch vehicle, and ground support systems are ready for launch and flight operations.

13.6 Science Data Processing System Development and Test

13.6.1 Development Approach

For science data processing system development, it is often useful to establish a phased implementation schedule by defining a set of evolutionary deliveries with specific functionality defined for each delivery. Early deliveries can focus on completion and verification of external interfaces, and early prototyping of science processing algorithms. Later deliveries can reflect increasing maturity based on evolving understanding of instrument characteristics obtained during instrument hardware development and testing. They can also incorporate increasing sophistication in the data processing algorithms. It is often useful to define a critical list of functional capabilities that must be adequately tested before launch, with a secondary list of capabilities that reflect science team or project desires but might be implemented (if resources allow) either before or after the start of flight operations.

Early deliveries are tested internally, within the organization that develops the system, and may not involve other elements of the project. It is advisable to include prototypes of user interfaces as early as possible in the development cycle, so that end users can serve as "beta testers" of the user interfaces with the data system. This allows for early user feedback and responsive evolution of the system's user interfaces.

If early system deliveries incorporate external interfaces that are verified during system level testing, the later deliveries can focus on finalizing the elements of the system that process the science instrument data and produce science data products. These elements are called the *pipeline*.

The hardware components of the science data processing system also evolve during the development cycle. Developers must have adequate levels of hardware support for the early development process. As the system evolves and matures, there may be more than one physical system supporting development and testing. There are three separate versions of the system at any given time:

- The *development* system, which includes processors representative of the final operational environment, networks supporting distributed development, software in various stages of development, an adequate data management system that is probably a reduced version of the final science operations data base envisioned for use during operations, and a configuration management system supporting integration of software into builds for each delivery that are then available for testing.

- The *test* system. This system contains software builds that the developers have completed and certified as being ready for testing at the science data processing system level. This system resides on hardware representative of the final operations environment. A network supporting testing requirements is part of the test system, as is a version of the final data management system envisioned for flight operations, and a configuration management system to track problem and bug reports.

- The *operations* system. This system contains software and hardware that have been tested and verified to be functional at a level corresponding to a major system build. As launch approaches, this system evolves into a fully configured data system capable of supporting launch and flight operations, and fully integrated with other elements of the project's ground data system.

For small missions on tight budgets, all three systems may have to reside within the same hardware configuration. If we adopt this approach, management

13.6 Processing Data and Generating Science Data Products 511

must institute strict configuration management procedures to ensure that all three functions can be shared within the same system without conflict.

13.6.2 Special Test Considerations for Science Data Processing Systems

Testing of the science data processing system during the development phase should follow normal software engineering practice. Once developers have turned over individual modules for incorporation into the next build, an independent test team should assemble the system level build and verify proper operation of each module against requirements. System level verification also takes place at this point, ensuring that all interfaces internal to the science data processing system function correctly. The test team should then work with other project elements to ensure that the interfaces between the science data system and the rest of the project's ground system are functioning correctly. Once that is verified, end to end project level system testing should follow, involving the ground system, spacecraft and instrument simulators, and eventually the flight hardware.

We should impose special test guidelines to verify early the instrument interfaces that are unique to the science data processing system. These include the following:

- **Test Instrument Interfaces as Early as Possible.** Flight instruments can come from many sources—commercial companies, government agencies, and universities being the most common. Instrument level checkout and testing is conducted using Ground Support Equipment (GSE) designed and built by the instrument development team. The GSE operates the instrument in a flight-like manner, and incorporates an extensive set of diagnostic tools that aid in instrument checkout. Projects may also provide a spacecraft simulator that can connect either to the instrument directly or to the GSE, enabling early checkout of instrument interfaces to the spacecraft. The science data processing development engineers should become involved as early as possible in the instrument design and checkout cycle. Ideally, the GSE or spacecraft simulator should generate instrument commands in flight formats, and the instrument or spacecraft simulator will produce packetized data in flight format. In this case, we can do early checkout of many science data system elements with the flight instrument hardware well before integrating the instrument payload with the spacecraft, significantly reducing the costs of that level of verification.

- **Provide Data Compression Bypass Path.** The science data processing system can verify data compression and decompression software if a separate path is provided on the spacecraft or by the instrument that bypasses the data compression routines on the vehicle. Alternatively, the GSE can inject known realistic data patterns into the instrument

data processing chain, providing an alternate mechanism to verify correct operations of data compression software on the vehicle.
- **Perform Extended Testing on the Flight Hardware Prior to Launch.** The team developing the science data processing system is probably the best group to define the operational scenarios that will represent maximum use of resources on the spacecraft, in terms of CPU, memory usage, or bandwidth challenges. That team should be involved in defining lengthy end-to-end tests on the ground data system and the flight hardware prior to launch. The test(s) should be long enough (i.e., multiple days) to represent nominal operations periods anticipated during flight operations. These lengthy tests can reveal potential problems with the instrument or spacecraft flight software when operated for significant periods of time (examples include buffer overflows, resource conflicts within the spacecraft data system, data rate mismatches between the instrument and the spacecraft triggered by specific instrument commands, etc.).

13.7 Science Data Processing System Architecture

Figure 13.3 shows the basic functional elements of a typical science data processing system for a flight project. It also indicates the major interfaces between functional elements. Note that some of these functions may be located within other elements of the project's ground data system, outside the science data processing system, as indicated in the diagram. The remainder of this section discusses each of the functional elements shown in Figure 13.3.

Science Operations Database. The heart of the science data processing system is the Science Operations Database (SODB). The SODB is the active database that stores all data required for successful operation of the science data processing system. Figure 13.3 shows transfer of data to and from the SODB by the subsystems within the science data processing system. High level data transfer paths between subsystems are shown in Figure 13.3 by the dashed lines. The SODB provides the mechanism for the transfer of critical information between the uplink and downlink systems involved in science data processing. It also includes the repository for the instrument data that has been processed through the pipeline, but has not yet been validated and transferred to the publicly accessible archives maintained by the project. The SODB must be able to accommodate instrument data reprocessing requirements if applicable.

Planning and Scheduling System and Proposal Processing and Selection Subsystem. The planning and scheduling system generates the set of sequences to be executed on board the spacecraft. It schedules events for specific periods of time during the mission. It may generate high level descriptions of the mission events that are forwarded to the mission operations system. There they are expanded into the detailed command strings to be executed on the spacecraft. Alternatively, the

13.7   Processing Data and Generating Science Data Products   513

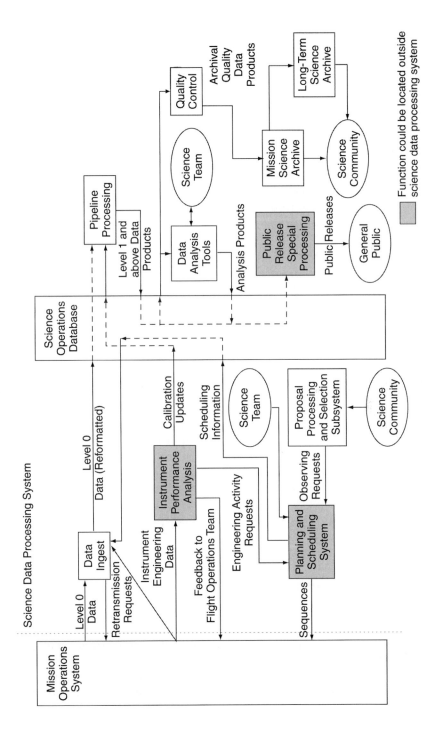

Fig. 13.3. Science Data Processing System Functional Diagram. Data processing subsystems transfer data to and from the Science Operations Database.

planning and scheduling system may generate the command strings directly, for uplink to the spacecraft. Requests for execution of activities on board the spacecraft can come from various sources. The mission may be operated in an observatory mode, where the science community submits observing proposals for competitive evaluation and selection. In that case, we need another subsystem to support proposal submittal, evaluation, and selection. Alternatively, observation requests may come from the science team associated with the particular project. For some missions, involving surveys of specific areas of the sky or a planet, or repetitive data acquisition of earth remote sensing data on some timescale, the science team may specify the priority in which observations are made. An operations sequence team may then deal with the mechanics of developing the detailed sequences that support the survey requirements. The proposal processing and selection subsystem is normally developed and operated as part of the science data processing system, while the planning and scheduling subsystem may reside in either the science processing system or elsewhere in the project's ground data system.

Instrument Performance Analysis. This subsystem primarily supports analysis of the instrument engineering data as it is received on the ground to determine health and safety of the instrument. It can also perform long term trending analysis of instrument behavior. If it detects anomalous behavior, this subsystem can generate requests for engineering sequences to help resolve the anomaly. This subsystem can also produce instrument calibration file updates required by the pipeline processing subsystem.

Data Ingest. This subsystem has an interface to the mission operations system that converts downlink telemetry into level 0 science data records. The input to the data ingest system consists of level 0 data records, along with data records containing instrument and spacecraft engineering data provided by the mission operations ground system. All of this information is merged into reformatted science data records with header information that contains selected subsets of the engineering data relating to each of the science observations. The data ingest system should also be where science data accountability occurs. We should design the uplink planning and scheduling system to provide the data ingest subsystem with information regarding the science observations planned for a particular data downlink. The data ingest system can then provide an estimate of the percentage of anticipated science data that has been received from the spacecraft for a given downlink pass. The ingest system will initiate retransmission requests to be forwarded to the mission operations system. The mission operations teams can then determine if the data transmission problems are between the ground receiving station and the mission operations system, or if the data are available on the spacecraft for transmission during a later downlink.

Pipeline Processing. This subsystem receives the reformatted level 0 data records from the data ingest subsystem and then produces Level 1 and above science data records by applying various algorithms and instrument calibration data to convert raw instrument telemetry data into physical units. The output from this subsystem consists of Level 1 and above data records in a format acceptable to the science archive.

Data Analysis Tools. Scientists will use a general purpose scientific data processing and analysis toolkit to do various types of processing on science data records. These tools often include commercially available software packages. The SODB should be able to store analysis products generated by this subsystem, or a separate data management system might be needed. We may not be able to predict the exact nature of the processing done on the data before actual flight data arrives from a given instrument. Even so, scientists associated with the project should be able to specify the types of processing, and their preferences regarding commercially available software packages, prior to flight operations to ensure that the proper toolkit is available for use during operations.

Quality Control. A set of tools enables final review of science data products before their release to the science community and a public archive. This review ensures that the data products meet the requirements established by the science teams prior to operations. Typical requirements may include specifications for spatial registration and photometric accuracy of the data. We must establish the requirements for the quality control system during the requirements definition phase of the project. The requirements should specify the degree of sampling that this subsystem must do. Generally, these systems use automated analysis algorithms that flag suspect problem data for an analyst to investigate, since it's usually not practical for an analyst to review individually each data product for missions generating significant volumes of data.

Mission Science Archive. If there is a requirement to support access to the science data products by external users during the life of the project, the project must fund and maintain a science data product archive during the life of the mission. We periodically transfer Level 1 and above science data products that pass quality control to the mission science archive. This archive is accessible to the scientific community outside the project, and in some cases (e.g., the recent NASA Mars missions) to the general public.

Long Term Science Archive. There may be a requirement to transfer science data products from a given mission to a facility that supports long term archival of the data for public access and retrieval. If so, we must identify before the end of the mission the data products to be transferred to the long term archive. In the simplest case, we transfer the entire mission science archive to the long term science archive. In other cases, we select some subset of the mission science data products as being appropriate for long term archival. USGS, NOAA and NASA have established several long term archival storage facilities for science data during or following the end of active flight missions. Public archives for science data produced by earth orbiting missions include the Earth Observing System (EOS) Data Center for land data, the NOAA Comprehensive Large Array-data Stewardship System (CLASS) for atmospheric and ocean data, and the EOS Data and Information System (EOSDIS) for data produced by various NASA funded earth observation missions. Data from NASA's planetary exploration missions is archived at the Planetary Data System (PDS). Data from NASA's astrophysics missions is stored at several different archives, based primarily on the wavelength of the data. The Infrared Processing and Analysis Center (IPAC) at Caltech maintains the Infrared Space Archive (IRSA). The Space Telescope Science Institute (STScI) maintains the Multimission Archive at STScI (MAST), with data products from the Hubble Space Telescope in the visible and near-infrared regime and other missions. The Chandra Data Archive at the Chandra X-Ray Center archives X-Ray data from the Chandra Space Telescope. The High Energy Astrophysics Science Archive (HEASARC) at Goddard Space Flight Center archives high energy astrophysics data from NASA missions.

Public Release Special Processing. Some missions funded by government agencies are required to provide periodic updates to the general public regarding mission status and science results emerging during the life of the mission. For such missions, we often produce specially enhanced image or graphics products, including animations, using the science data records in the SODB as the source. The science data processing system is often the best place to locate the tools needed to produce these special products, based mainly on ease of access to the science products and proximity to the scientists involved in selecting candidate press releases.

13.7.1 System Performance Metrics

It is important to recognize the key performance metrics for a science data system well before launch, and to ensure that suitable tools are available to produce and analyze those metrics [Doxsey, et. al., 2002]. System managers normally produce the usual data processing system statistics regarding system reliability, percentage of "uptime", significant disruptions of service to users, percentage of time spent in preventative maintenance, etc. Beyond that, we may need to produce statistics relating to top level mission science requirements. These might include statistics showing percentage of time spent performing science observations versus other spacecraft activities (the "observing efficiency"), the number of observation requests that were satisfied versus those that have not yet been scheduled or executed, the science user community if the community is allowed to submit proposals for observing time, etc.

13.8 Science Data Processing System Implementation

Science data processing system design engineers must capitalize on constantly emerging technology developments in the computer industry to minimize lifecycle system cost and maximize system performance. We base the design of the data processing system on the functional subsystems outlined in the last section. The major design tradeoffs in design of the data processing system supporting these functions are described below:

Processor Architecture. Some science data processing systems use large capacity centralized processors, with various functions sharing the resources of one or more centralized processors. Other system designers have taken advantage of the emergence of low-cost processor capabilities and based system designs on a networked approach, allocating specific processors to specific functions. Another approach involves a combination, where large capacity processors are installed in a networked topology that also includes dedicated processors for individual functions. The final system design generally trades required throughput with cost effectiveness.

The selection of processor architecture is generally most critical for the pipeline processing function, since the pipeline is usually the most computationally intensive component of the system. Pipeline calculations can often be decomposed into components that can be executed in parallel, and can thus be supported by a design that dynamically assigns the next incoming task to a specific processor. This type of design requires the development of a software executive that controls the overall pipeline process, the dynamic allocation of tasks to specific processors, and the final collection of results and production of the final data products for transfer to the SODB [Moshir, 2005].

A pipeline system that can be decomposed this way can use several different approaches to hardware implementation:

- **Blade Processors.** Some recent system implementations have used an array of blade processors to support pipeline processing. Blade processors are minimal processor configurations installed on a common backplane, accessing a common set of network and data storage resources across a shared high speed bus.
- **Networked Workstations.** An array of networked workstations, operating under control of a pipeline executive, is another option for this design approach.
- **Partitioned Large Processor Configuration.** Large capacity processors are available that use multiple processors, along with supporting software that allows partitioning of resources within the configuration. Here again, a pipeline executive allocates tasks dynamically within the configuration.

On-line Data Storage. Science data processing systems typically require large amounts of on-line data storage to support ongoing processing requirements (NOTE: archival data storage systems will be discussed in the next section). This is an area in which technology evolution continues to be explosive. The cost per unit storage continues to decline dramatically, and the floor space and power requirements supporting spinning disk storage units continue to decrease as well.

The use of Redundant Arrays of Inexpensive Disks (RAID) technology to support science data processing systems has expanded rapidly in recent years. The basic RAID concept involves distribution of data sets across arrays of inexpensive disk drives. There are various levels of redundancy available within the RAID technology, designated by the terminology RAID N (e.g., RAID 5). Each level of RAID involves replication of data across more than one disk drive and other methods designed to minimize or eliminate the impact of a disk drive failure on continuing operations. RAID can minimize the cost of large volumes of disk storage while ensuring that mission operations can proceed with no interruption of service if a single drive fails.

Communication Support—Internal. One of the most demanding areas of system design involves the topology for communication among the individual elements of the data system. The final system design must include considerations of bandwidth within the science data processing system and an analysis of potential conflicts between competing demands for system resources. Overall performance of a system designed with an array of blade processors requiring access to a RAID storage array can be less than optimal if each of the functions operating in the blade processor array has high simultaneous data input/output demands that cannot be met using the common backplane on which the blade processors reside. Performance of a system that uses a centralized multiprocessor system for major processing functions connected to a RAID array can also be less than optimal if there is only one data path between the processors and the RAID array, with insufficient bandwidth to accommodate data transfer demands.

System design should include detailed modeling of system performance under typical, and the most demanding, anticipated operations scenarios, to ensure that data communication paths within the system can handle the anticipated loading plus some design margin.

Communication Support—External. System design must consider the external demand on the science data processing system. Considerations should include the following:

- Do any users require access to data or support functions located within the science data processing system (e.g., the SODB, the analysis tools subsystem)? If so, we must provide secure access to those users.
- Is public access to the Mission Science Archive required? If so, what is the anticipated demand that network connections to the outside world must support? How do we limit external user access to the mission science archive, preventing unauthorized access to other project resources?
- Is bulk transfer required to remote science facilities involved in the project? If so, what is the transfer volume per unit time requirement, and how will we meet it?
- For projects requiring press releases, are other facilities available to support peak loading demand for public access when press releases are made?

13.9 Science Data Archive Considerations

There are two basic issues we must address when designing a science data archive—(i) the software support for user query and data retrieval and (ii) the hardware systems that store the data and transfer it to users as requested. These issues apply to both the mission science archive and the long term archive that will store the data after the end of the mission.

13.9.1 Archive Software User Interface Functionality

Software must enable users to execute queries against the archival data base and to retrieve data based on a variety of search criteria. One tool in developing a design that's responsive to user needs is the use case. A use case describes how the user expects to access and use data stored in an archive. There can be several different use cases for each data archive, and they can be defined by asking the following types of questions during the design phase:

- How does the user want to search the data—by georeferenced coordinates, by instrument, by spectral band, by the time that the data was acquired, etc.? Will typical searches involve combinations of the search parameters?
- Would the user prefer to enter coordinate values as search parameters, or is a graphical interface preferable, or is a combination of both preferable?
- How would the user like the results of a query to be displayed by the user interface?
- How large do we expect the user community to be, and where are the majority of the users located (within the project operations facility or external)? Will there be international users?
- About how much data will be retrieved under each use case? Do the users want to transfer large quantities of data to their home institutions? What response time is acceptable (as a function of data volume requested) for delivery of data to the user?
- How will we accommodate improved knowledge of instrument characteristics, and updates to instrument calibration files? Will the data be reprocessed with the latest calibration data before delivery to an external user ("reprocessing on the fly" [Swam, et. al., 2001]), or will the user simply receive the latest version of the data in the archive, which may not reflect up-to-date calibration information?
- What is the most common standardized data format for the user community that's most likely to use the data archive?

Answers to these types of question will help construct use cases that represent the best available information to guide design and implementation of the user interface. The software design should be modular and easily adaptable to changes and updates needed to reflect actual user experience and feedback with using the archive during operations [Berriman, et. al., 2000].

13.9.2 Archive Hardware Selection

Technology for storing large volumes of digital data continues to evolve rapidly, fueled primarily by exponential growth in public use of the internet. We select hardware to support archival storage and data retrieval for each project based on available technology, cost, and schedule considerations. We typically base this selection on the response time required for bulk delivery of requested data sets to end users.

Generally, the fastest response time for bulk data retrieval is by retrieval from spinning disk and delivery via high speed network. Although hardware costs for spinning disks continue to decline, this is still the most expensive data storage option compared to retrieving data from off-line storage media by either human operator or robotic device. Another on-line option that provides access to large data volumes at lower speed than spinning disk is a robotic mechanism called a "tape library" or "disk jukebox". These devices are connected to a processor via a high speed interface, and incorporate racks containing digital media (high density digital tape or disk) and appropriate media readers. When the host processor sends a data request, the appropriate physical media is transferred from its storage location into the appropriate media reader, and the data is read and transferred to the host processor. The time delay associated with these devices reflects the need for physical transfer of media into the read device. Because of the time delay in data retrieval (compared with spinning disk), these devices are often referred to as near-line storage devices. Most archives that store large volumes of science data provide a combination of on-line spinning disk and near-line and/or off-line storage solutions. The exact configuration of on-line, near-line, and off-line storage depends on the project's requirements for the speed of data access, and is also constrained by cost.

The hardware solution selected for an archive should be flexible enough to accommodate the integration of new technology as it evolves. Many archives start with initial hardware able to accommodate data from some initial time period of the project, with significant design margin. As the project continues, and the data volumes continue to increase, we can introduce new technology to expand the initial configuration while benefiting from the declining cost per unit storage as commercially available products improve.

Table 13.2 provides an overall view of archival service levels in terms of response time for data transfer to the user. In this table, "HD Tape" refers to high density digital tape in a standardized format that is compatible with a broad range of hardware platforms and systems. "Digital Disk" refers to physical media, also in a standardized format compatible with multiple platforms. Current examples of "Digital Disk" include Compact Disk—Read Only Memory (CD-ROM, with capacity on the order of 650 Mbytes) and Digital Versatile Disk (DVD, with double-sided capacity of at least 10 GBytes and beyond as the standards continue to evolve).

Table 13.2. Support for Archival Storage Service Levels.

Response Time	Immediate	Immediate to Hours	24–48 Hours	> 2 days
Delivery Approach	Network transfer from on-line spinning disk	Network transfer from on-line spinning disk and/or near-line storage	Ship data through HD tape or digital disk, or network transfer of data retrieved from near-line or off-line digital media	Ship data through HD tape or digital disk
Technology Options	Spinning disk	Spinning disk and/or near-line media library	Digital media writer if required	Digital media writer

13.10 Examples

This section contains some examples relating to topics discussed in this chapter.

13.10.1 Spitzer Space Telescope Science Operations System

The Spitzer Space Telescope is an infrared space telescope launched in August, 2003 and is expected to operate until at least 2008 when the coolant used to maintain the infrared instrument operating temperatures is exhausted. The Spitzer Science Center (SSC) at the California Institute of Technology has responsibility for science proposal solicitation, proposal selection, observatory planning and scheduling, instrument performance monitoring, production of various levels of science data products, transfer of mission data products to the long term archive (the Infrared Science Archive (IRSA) at Caltech's Infrared Processing and Analysis Center), and support of the project's education and public outreach activities.

Scientists submit requests for specific observations with the payload instruments by completing Astronomical Observing Requests (AORs) using an on-line form and electronic submittal of the completed form to the SSC. Instrument and spacecraft engineers submit Instrument Engineering Requests and Spacecraft Engineering Requests (IERs and SERs) to the mission planning and scheduling team, and that team produces a reconciled schedule of activities to be executed on board the spacecraft. The AORs, IERs, and SERs are expanded to the individual command level before transferring the mission plan to the flight operations team for final packaging and transmittal to the spacecraft.

Downlink data processing includes pipeline processing for all three payload instruments, specialized processing by instrument science team members, processing for press release and public education products, and production of various levels of science data products. The project maintains an active science mission archive, and will eventually transfer the data for long term archival to IRSA.

Figure 13.4 shows the pre-launch configuration of the Spitzer Science Operations System (SOS) (the Spitzer project was called SIRTF prior to flight operations). A comparison with Figure 13.3 will show that the Spitzer SOS supports the majority of functions shown in the earlier figure.

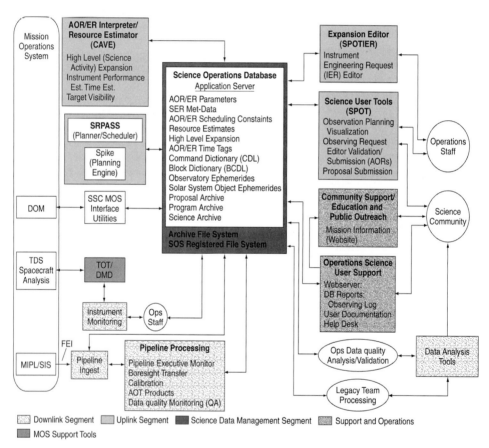

Fig. 13.4. The Spitzer Science Operations System Architecture. (SPOT is a versatile tool for users to access the Spitzer Science Operations Database; CAVE = Client AIRE Visibility Expeditor; MOS = Mission Operations System; TDS = Telemetry Delivery Subsystem; TOT/DMD = Telemetry Output Tool/Data Management Display; MIPL/SIS = Multimission Image Processing Laboratory/Science Information System; FEI = File Exchange Interface; DOM = Distributed Object Manager)

The Spitzer Science Operations System supports submittal of proposals for observing time by the scientific community. A tool that can be downloaded to a variety of platforms at user facilities enables scientists to plan an observing program and then submit it electronically to the Spitzer Science Center for competitive evaluation. The tool is called the Spitzer Planning and Observation Tool (SPOT). Figure 13.5 shows a view of the SPOT user interface. In this example, the user is interested in observing two different astronomical objects, denoted as M51 and NGC 17 with some of Spitzer's instruments. SPOT provides interfaces with existing astronomical databases, including images obtained by the 2MASS all-sky infrared survey (a ground-based survey) and the Sloan Digital Sky Survey (DSS, a ground-based survey performed at visible wavelengths). SPOT also includes an interface with the NASA Extra-galactic Database (NED), which contains references to published literature for millions of extragalactic sources. The three images in the top row of Figure 13.5 show images obtained from the DSS, 2MASS, and NED databases. The center top image contains an overlay showing the area that would be imaged by the Spitzer instruments, as an aid to the scientist planning the observation. The bottom image is a DSS image of NGC 17, and the lower right hand listing shows a variety of references obtained from the NED database. This tool enables an observer to plan observations using the latest available relevant imagery, with instant access to information obtained from prior observations of the area of interest. Eventually, the user will produce a table of AORs. which will be submitted electronically with the proposal to the Spitzer Science Center. If the proposal is accepted, SPOT interfaces with other supporting software at the Spitzer Science Center to generate the command strings required to execute each of the AORs. SPOT also forwards other information on the requested observing program to the planning and scheduling system, which attempts to schedule the requested observations as efficiently as possible.

13.10.2 Science Archive Examples

This section describes the operation of two different science archives. EOSDIS is possibly the largest scale science data processing system to date. The system supports acquisition of approximately 300 GB of data daily, and the archival data product volume increases by approximately 4.5 TB daily. Data from several active missions is processed every day, and the system is also the repository for several past missions. It archives data from over sixty different flight instruments, obtained from various NASA earth resources missions.

EOSDIS is a distributed system. NASA has established several Data Active Archive Centers (DAACs) around the country, with each DAAC focusing on data relating to a specific earth resources research area. The DAACs are responsible for archiving data relevant to their charter from past missions. They also receive instrument data from EOS missions relative to their research area, and are responsible for producing Level 1 and above data products and maintaining a public archive enabling access and retrieval.

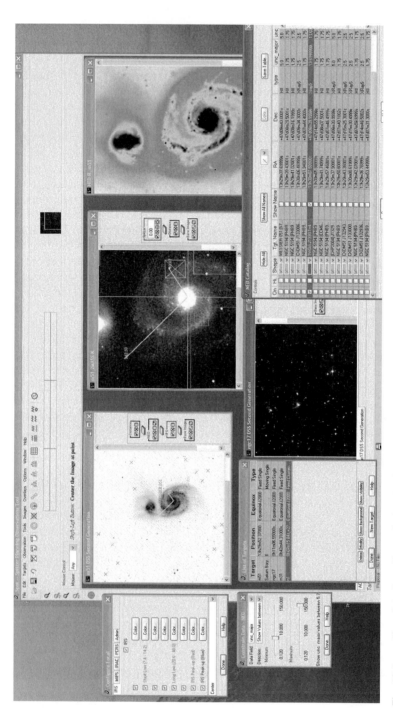

Fig. 13.5. User Interface Example for the Spitzer Planning Observation Tool (SPOT). *(Courtesy of Trey Roby, Spitzer Science Center, Caltech)*

13.10   Processing Data and Generating Science Data Products   525

Figure 13.6 shows the query interface to a DAAC located at Oak Ridge National Laboratory. This DAAC is responsible for data products related to biogeochemical dynamics, which result from interactions between biological, geological, and chemical components of the Earth's environment. A user searching the archive maintained by this DAAC uses an interface that limits the search inquiry to a specific set of missions and instruments. This DAAC is an example of a project specific scientific archive for an active flight mission.

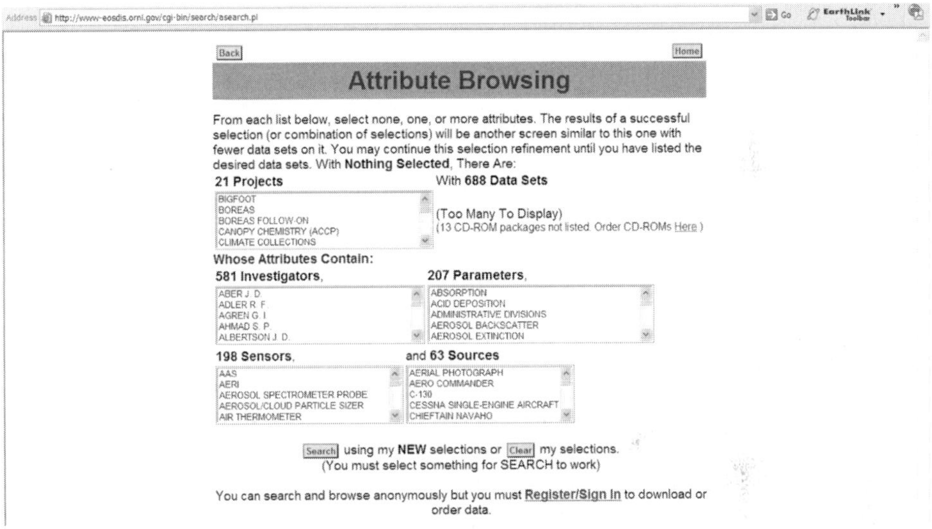

Fig. 13.6.   User Access Interface at the EOSDIS ORNL DAAC. This DAAC focuses on biogeochemical data products. *(Courtesy of ORNL DAAC)*

An example of a different type of user interface is shown in Figures 13.7 and 13.8, from the Multimission Archive at the Space Telescope Science Institute (MAST) [Padovani 2000]. This archive contains data from the Hubble Space Telescope and other NASA astrophysics missions.

Figure 13.7 is the top level user interface screen for this archive. In the example shown here, the user has requested available images in all wavelengths archived by MAST for the Messier object M4. Figure 13.8 shows the first part of the response to the query, showing datasets available from various instruments from different sources stored at MAST. The user can then request specific data sets for downloading to their home system for further analysis.

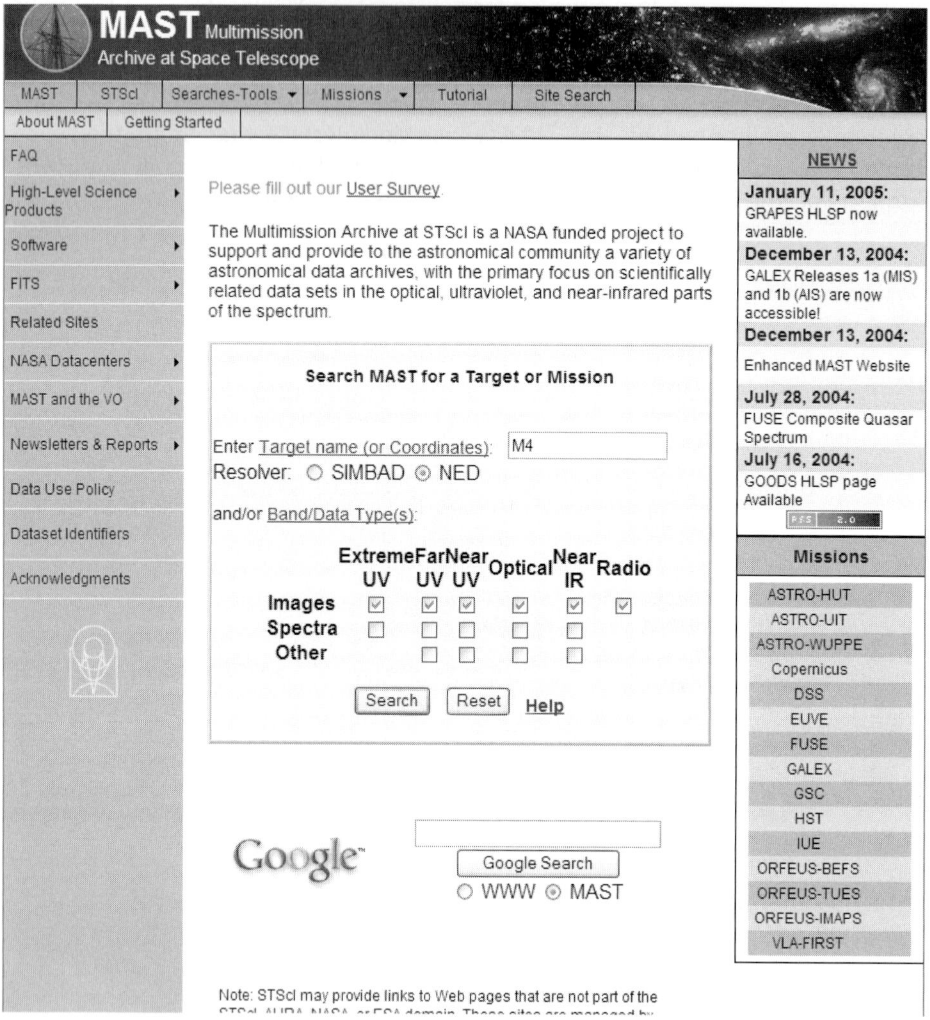

Fig. 13.7. **User Interface Screen for MAST Archive at STScI.** This archive is for astrophysics data products. *(Courtesy of STScI)*

13.10 Processing Data and Generating Science Data Products

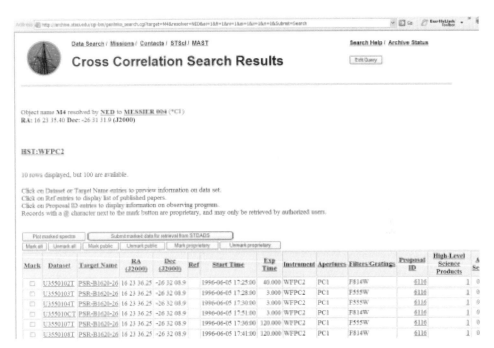

Fig. 13.8. Results of MAST Query Shown in Figure 13.7. *(Courtesy of STScI)*

References

Adler, D. S., D.K. Taylor, A.P. Patterson. 2002. *"Twelve Years of Planning and Scheduling the Hubble Space Telescope: Process Improvements and the Related Observing Efficiency Gains."* Proc. SPIE 4844, 111.

Berriman, G. B., N. Chiu, J. Good, T. Handley, A. Johnson, M. Kong, S. Monkewitz, S.W. Norton, and A. Zhang. 2000. *"Advanced Architecture for the Infrared Science Archive,"* in ASP Conf. Ser., Vol. 238, Astronomical Data Analysis Software and Systems X, eds. F.R. Harnden, Jr., F. A. Primini, and H. E. Payne (San Francisco, ASP) 36.

CODMAC (Committee on Data Management and Computation), National Research Council. 1982. Data Management and Computation, Volume 1: Issues and Recommendations, National Academy Press, Washington, D.C.

Green, W. B., and the Staff of the SIRTF Science Center. 2002. *"The SIRTF Science Operations System."* Proc. SPIE Vol. 4850, 130.

Moshir, M. 2005. *"Spitzer Space Telescope Data Processing and Algorithmic Complexity,"* in ASP Conf. Ser., Astronomical Data Analysis Software and Systems XIV, eds. P. Shopbell et. al. (San Francisco, ASP).

Spitzbart, B. D. and S.J. Wolk. *"Chandra Monitoring Trends and Response."* 2002. Proc. SPIE 4844, 476.

Swam, M. S., E. Hopkins, and D.A. Swade. 2001. *"Using OPUS to Perform HST On-The-Fly Re-Processing (OTFR),"* in ASP Conf. Ser., Vol. 238, Astronomical Data Analysis Software and Systems X, eds. F. R. Harnden, Jr., F. A. Primini, and H. E. Payne (San Francisco: ASP), 291 (ADASS 2000).

Padovani, P., F. Abney, D. Christian, T. Comeau, M. Donahue, R. Hanisch, J. Harrison, C. Imhoff, R. Kidwell, T. Kimball, K. Levay, M. Postman, J. Richon, M. Smith, and R. Thompson. 2000. *"The Multimission Archive at the Space Telescope Science Institute,"* in ASP Conf. Ser., Vol. 216, Astronomical Data Analysis Software and Systems IX, eds. N. Manset, C. Veillet, and D. Crabtree (San Francisco: ASP), 168 (ADASS 99).

Chapter 14

Assessing Payload Operations

Paul Ondrus, *NASA Goddard Space Flight Center*

14.1 Developing the Payload Operations Concept
14.2 Assessing Requirements for Payload-Data Systems
14.3 Assessing Drivers for Payload and Platform Interfaces
14.4 Characteristics of a Good Payload Operations Plan

We address this chapter to the new mission operations manager (MOM), who needs to know about important payload activities in order to have a successful mission and to control mission operations costs. People often discuss payload issues during the mission-design phase without understanding or identifying the implications for the operations phase of a mission. The tables in this chapter give a framework for identifying drivers for complexity and cost for payload operations, to enable trades among the MOM, the project managers, and subsystem engineers during mission design.

One of the MOM's main responsibilities is to ensure identification and resolution of the operational issues of a mission as early as possible in development. That's because it's cheaper to solve a problem in design than in test. A large part of this responsibility is to emphasize payload operations and its role in end-to-end mission activities. This emphasis must start early in mission definition and continue throughout development to ensure the appropriate system trades from a lifecycle perspective. Project managers too often trade the present-development budget for the future-operations budget and ignore lifecycle costs. The MOM must keep the project manager aware of lifecycle issues and implications.

In the past, spacecraft operations and payload operations have been conducted with distinct subsystems. The current trend is to blur this distinction for smaller, low-cost missions, which makes early identification of payload operations issues and payload and spacecraft cost drivers even more important. The trend to increase autonomy by automating functions and moving them on board the spacecraft further impels the need for early operations involvement.

The foundation of this chapter is a mission-engineering process, based on past experience, that will help to plan a payload's operation and identify its major operational issues. This leads to an activity plan for operations that details key activities and how people and/or systems will do them. The plan provides the framework for defining payload operations and for identifying issues that warrant the project manager's attention. The development of this plan ties in closely to the operations concept tools discussed in earlier chapters. These tools help the MOM understand the mission's objectives and clarify its program environment, so he or she can develop appropriate trades.

Spacecraft builders tend to see a mission in terms of functions. They fragment the payload and mission operations activities into sets of thermal, power, mechanical, and control requirements and resource budgets for spacecraft subsystems. The MOM and the support team need to work with systems engineers to translate the functions and requirements back into a timeline, scenarios, and process flows that show how the spacecraft operates on orbit. Our mission-engineering process also presents a structure for characterizing the mission's customer interface and seeing how the payload and spacecraft bus interact to carry out the mission's objectives over the life of the mission.

A mission operations manager entering a new project faces many ongoing, parallel activities. Our checklists indicate issues to watch for and understand. They help to avoid the tight constraints on flight resources that complicate normal payload operation, characterize the information needed for mission planning, factor in lessons learned from previous missions, and identify new technology trends that could significantly change mission operations. We may not be able to change a design, but we should address effects on operations costs to encourage proper lifecycle decisions.

The mission-engineering process guides us in identifying and resolving potential issues with payload operations (see Fig. 14.1) as we move from the mission concept to actual operations. Here, we'll consider operations planning, which interacts with support for the space-element design and ground-element definition. Mission engineering emphasizes how to define requirements for acquiring the payload data, to develop payload operations concepts and scenarios, and to critique the payload design from an operational standpoint. It helps us support system trades and identify drivers of operational costs. The main product of operations planning is the activities plan for flight operations. Results also influence design requirements for the ground system and design reviews for the spacecraft and payload.

Mission engineering is iterative. It uses analysis tools to help the operations staff visualize on-orbit operations. The operations concept is one such tool that we

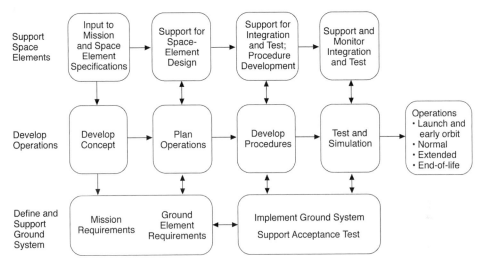

Fig. 14.1. Mission-Engineering Process. We must develop the space, ground, and mission operations elements concurrently to reduce lifecycle costs.

use to translate requirements into timelines and operational flow. The MOM must emphasize time and information flow as part of defining the mission. This process is effective for both small and large missions. The level of complexity of the mission determines the level of effort required, but the steps are the same.

As we see in Fig. 14.2, we start the concept development by understanding the mission objective and the payload's role in meeting the mission objective. This means understanding the payload's tasks, success criteria, constraints, and products its customers require. A MOM must keep an end-to-end systems perspective, recognize how elements interact, and understand the detailed workings of each element. This means working closely with spacecraft developers, payload developers, and users of the payload's data. With advances in processing capabilities, many organizations maintain test beds where various spacecraft and payload interfaces can be prototyped, modeled, and evaluated. This takes this step from an abstract to a practical process.

As part of this effort we need to identify and understand any major program interfaces. A program's interfaces determine how much flexibility we have in developing the operations plan for a mission. NASA's Small Explorer (SMEX) missions, for example, are a series of high-risk, individual scientific missions usually employing a single payload and minimal redundancy in the spacecraft subsystems. These missions have a flexible interface structure, so we can try innovative operations concepts on individual missions. A mission that is part of a large structured program has many fixed interfaces with rigid data structures, so the implementation of new concepts is limited. The Earth Observing System

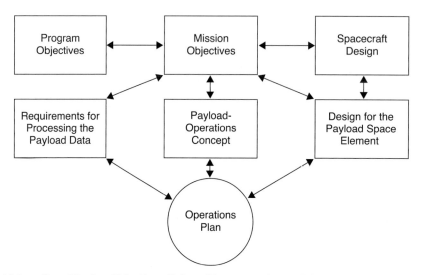

Fig. 14.2. How Mission Objectives Relate. We must understand the payload's tasks, success criteria, constraints, and required products to develop a cost-effective payload operations plan.

(EOS)—because of its large customer base and correlative research requirements—is an example of a program that limits mission design flexibility.

Our assessment of payload operations divides into three main activities that lead to the operations plan: developing the payload operations concepts, assessing requirements for payload-data systems, and assessing requirements for payload operations.

14.1 Developing the Payload Operations Concept

As discussed in Chap. 4, mission engineering continually refines and adds detail to an operations concept. In most missions the concept for payload operations extends from the concept for mission operations. But to make sure the concept is complete, we should follow four steps: (1) determine the class of operations; (2) identify drivers of operations complexity; (3) develop contingency operations; and (4) analyze launch and early-orbit scenarios.

14.1.1 Determine the Class of Operations

We classify payloads as survey, event-driven, or adaptive. The class of payload determines operational characteristics and requirements. *Survey payloads* tend to have simpler real-time operations and planning activities. An example is the SAMPEX payload, a set of space-physics instruments that operate continuously and measure the same physical events throughout each orbit. The Moderate Resolution Imaging Spectrometer (MODIS) is another example of this type of instrument because it surveys the earth continuously.

Event-driven payloads must operate at a specific time. They're usually more complex and require more sophisticated planning than survey missions because they tend to target an activity. The more accurate the time/position requirement, the more costly a payload's operation. Examples of event-driven payloads are any of the instruments on the Hubble Space Telescope (HST). These payloads operate at specific times to observe planned astronomical targets. The need for pointing accuracy to milli-arc seconds complicates payload planning significantly. An earth observing instrument example would be the Enhanced Thematic Mapper (ETM+) instrument on the Landsat 7 spacecraft, a multispectral imaging instrument, which is turned on at certain times to image geographic locations. Figure 14.3 shows how various spacecraft must be coordinated on an event basis to correlate data.

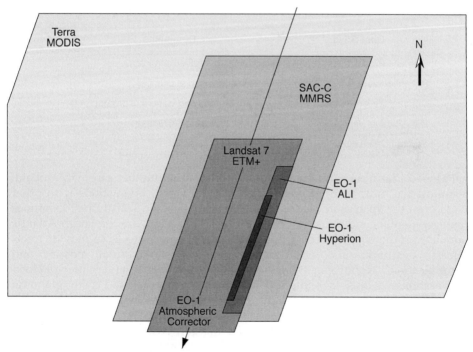

Fig. 14.3. Coordination of Spacecraft Events for Constellation Science. Overlapping footprints of constellation instruments.

Adaptive payloads transition to different states depending on unplanned events. An Earth-observing payload that automatically adapts to cloud cover is one example. If the payload operates autonomously, this type of operation will be relatively inexpensive because it has little effect on ground support and mission operations. But if the ground element needs to interact periodically with the payload concerning future operations, this type of payload can create very complex payload operations. This complexity arises because we must reconstruct payload operations and understand the payload's status in order to plan future operations.

14.1.2 Determine Complexity Drivers for Normal Operations

We must understand some important characteristics of payload operations to assess a payload's operational complexity. If we understand them early enough in the mission lifecycle, we may lower lifecycle costs by changing the payload or ground-system design. These characteristics are (1) payload planning, (2) payload reconfigurations and calibrations, (3) communications needs, (4) payload-product verification, and (5) payload orientation (see Table 14.1).

Table 14.1. **What Makes Operations Complex.** We include these cost and complexity drivers in our complexity metrics in Chap. 5.

Cost Driver	Comments
Planning Process	Number of constraints and interfaces
Reconfigurations	Number and timing of changes
Communications Needs	Onboard storage and volume of payload data
Data-Product Verification	Data structures and timelines
Payload Orientation	Maneuvers, accuracy

Payload Planning. This characteristic reflects the number of environmental constraints we must take into account for the payload, interactions with other payloads on the spacecraft platform, and internal payload constraints. Examples of environmental conditions that can complicate planning are the South Atlantic Anomaly, bright objects, occultations, and target visibility. Examples of inter-payload constraints are requirements for platform thermal, power, and communication resources. The more environmental and inter-payload considerations we must take into account, the more complex and costly planning will be. We want to encourage designs that minimize these types of constraints.

One subtle challenge to planning is the desire to over-optimize the payloads on a platform, which results in increased mission costs. The more we try to schedule all of the payload's abilities, the more complicated and costly the planning process. This is characteristic of deep-space missions, such as Voyager, which have had long cruise periods and short opportunities for full science during planetary encounters. Planning and replanning every millisecond of the encounter involves many subsystem and payload people, so it drives up complexity and cost.

A new developing consideration for payload planning is formation flying or coincidental imaging constraints on a payload. For the earth science missions there are two different constellations. The AM constellation consists of the Landsat 7 mission, the Earth Observation–1 technology mission, the Terra Spacecraft, and the SAC-C mission. All of these missions operate on the same Worldwide Reference grid within 15 minutes of each other. In this approach, payload planning may involve coordination across multiple spacecraft. The constellation of missions expands the planning interactions to a higher level of complexity (see Figure 14.4). The missions in the constellation must now maintain strict attitude and altitude control to ensure that their relative positions in the constellation are maintained. The Calipso, Cloudsat, and Aqua missions maintain relative positions seconds apart and must work with each other to maintain the required spacing. The constellation creates a virtual spacecraft made up of all the instruments. By adding and deleting missions to the constellation a program can adjust the emphasis of the measurements being taken.

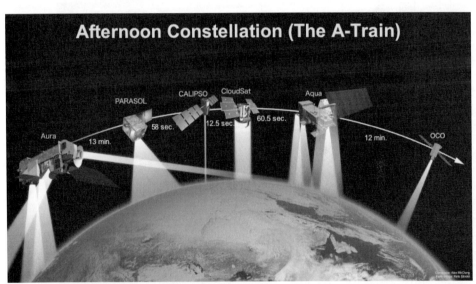

Fig. 14.4. **Coincidental Imaging of Landsat 7 and Earth Observation-1.** This type of imaging puts extra constraints on payload planning.

Number of Reconfigurations. Number of reconfigurations refers to the number of changes in a payload's or spacecraft's state required to operate the payload. This number has two parts: the number of a payload's possible modes and timing of the reconfigurations. Because nearly all payloads need reconfiguring, we should try to isolate the effects of the reconfiguration so that it doesn't ripple through the end-to-end system and increase mission operations costs unnecessarily. Examples of operational modes changes are filter changes, use of different apertures, or different sampling rates. To save money, a MOM must make sure these different modes have the least effect on the end-to-end system.

Timing of the reconfigurations is our other main concern. If, for example, we need to switch a payload instrument from a survey mode to a special mode for a target of opportunity, quick action can drive up costs. The Multi-Angle Imaging Spectro Radiometer (MISR) on the Terra spacecraft, for example, has two different imaging modes: global and local. The global mode provides continuous planet-wide observations, with most channels operating in moderate resolution. The local mode provides data at the highest resolution in all spectral bands and all cameras selected for 300km x 300km regions. These two modes have different constraints and place different demands upon the spacecraft data system elements. The impact on other payloads and the spacecraft bus needs to be understood early in the mission and managed. Figure 14.5 shows an example of a MISR image being used to observe forest fires over the western United States.

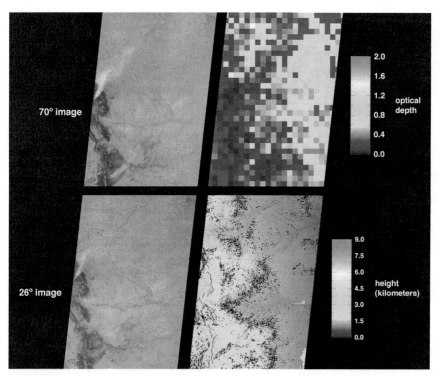

Fig. 14.5.   **Imagery to Support U.S. Forest Service.** Colorado wildfire image taken by MISR. *[Courtesy of Jet Propulsion Laboratory/NASA]*

Communications Needs. Communications are at the core of how payloads are remotely operated: they influence payload interactions directly, and strongly drive a mission's operational costs. Communications requirements depend on the payload's need for uplinks and downlinks. The amount of data a payload collects and the payload's commanding needs drive the number of contacts. The HST's payloads, for example, initially required at least three daily contacts, plus back-ups to load the daily payload commands, because the original onboard computers had limited memory. The payload's communications needs are one of the first system trades for a new mission. If the payload requires several manual reconfigurations or checks to verify operations, communications contacts become a major cost driver. A survey mission that requires only a low bandwidth for data transmission and has appropriate onboard storage can operate with one contact per day. A number of the small explorer series (simple, single, survey payloads) can operate this way.

Payload-Product Verification. One of the first payload operations trades is the timeliness required to validate payload operations. Rapid verification costs money because mission experts or systems must analyze payload products and make timely, interactive corrections to the payload operation. An orbit that allows for long contacts will help manage these costs. For example, the Solar Heliospheric Observatory (SOHO) operates at the L1 libration point, so it has continuous communications contact for up to 16 hours per day. This continuous contact allows for online reconfiguring of the payload and immediate verification of payload operations. The L-7 mission, on the other hand, records data blindly and has a three day delay between data acquisitions and verification of the payload's operation. This operation was driven by the communications cost of moving high rate data from remote arctic sites.

Payload Orientation: Does the platform have to provide unique orientations for the payload to work? Payloads that require pointing increase planning and on-orbit operation of the platform. The Cassini mission is a good example. To lower development costs, designers fixed the instruments to the spacecraft structure. The spacecraft therefore must do complex maneuvers to meet the payload's pointing needs. The more stringent the pointing requirement, the more significant the cost drivers and the greater the impact from environmental constraints.

Another factor of orientation is calibration maneuvers. Frequent calibration maneuvers to point to the moon, or deep space, interrupt the operations of other payloads on a spacecraft. This can become a cost trade between internal calibrators and on-orbit maneuvers that interrupt routine operations.

14.1.3 Develop Contingency Operations

When payload operations are normal, associated activities become routine. The challenge for a new MOM is to provide an operations scenario and staff that are optimized for normal operations, but can handle on-orbit contingencies. This means we must plan for contingencies while developing the operations scenario—planning and developing safemodes, establishing fault-isolation trees, and developing recovery operations.

Needing to respond quickly from the ground will drive up costs for payload operations because it entails almost continuous ground monitoring. If we don't have safing abilities on board, lifecycle costs for ground operations can skyrocket.

All payloads should have two safing mechanisms. The first is a safemode that the payload itself enters when it detects an anomaly; the second is a safemode that the spacecraft platform can force the payloads into when the spacecraft bus has a problem such as loss of attitude control or a significant power anomaly. *Safemode* should be a state in which the payload can't damage itself through unwanted mechanism movements and can continue indefinitely while creating minimal thermal or power concerns. Given the present state of technology, keeping a large staff waiting for anomalies is too expensive. Instead, the spacecraft or ground operations should safe the payload to a given state, which will allow ground support to take its time to find the cause and start recovery, thus buying time for a minimal staff to assess the problem and develop a course of action.

The MOM, in concert with the payload developer, needs to ensure that the payload can be placed in the correct state when a fault triggers. We must take special care in defining telemetry trip points and onboard logic sequences that define anomalous states. For example, the Aqua mission has a multilevel onboard fault detection scheme that assesses mission operations and triggers the appropriate spacecraft and instrument safemode. It has also implemented a 911 call home communications capability that notifies operations personnel immediately when the spacecraft bus enters safemode. This is an important timesaving capability since much of Aqua's time is spent outside of ground station communications contacts. It adds a level of automation to the process and minimizes the need for on-console staff.

The first step in developing contingency operations is creating fault trees for the payload. When we do this early enough, operators and developers can walk through the failure scenarios and determine if they have enough information to recognize them. The payload fault trees should augment the failure analysis for the spacecraft platform. We should also plan to test them, whenever possible, during integration and test, thermal vacuum testing, and simulations. Figure 14.6 shows a fault tree for a CERES instrument on the Terra/Aqua spacecraft [NASA, 1999]. Much of the failure analysis provides a basis for simulation exercises in which operators react to failures and verify that they have access to the appropriate information to analyze an on-orbit anomaly. The extent of this effort should match the size of the mission. If the payload is a significant instrument that will operate for many years, more effort on the failure exercises will eventually save money. If the mission is a simple, short-term instrument, we should keep the failure analysis simple.

The final consideration for contingencies is planning recovery activities, including communications needs for carrying them out on orbit. The recovery procedures exercised during payload integration and test are often based on continuous contact. But we can't do them during short-pass contacts, such as those for a low-Earth orbiter, so we must ensure the procedure can be done on orbit. We also need to control these procedures for consistency, so staff changes don't create problems caused by unfamiliarity with procedures.

To succeed in contingency operations, we must keep them simple and have the time to react properly to an anomaly. The worst danger is in responding quickly

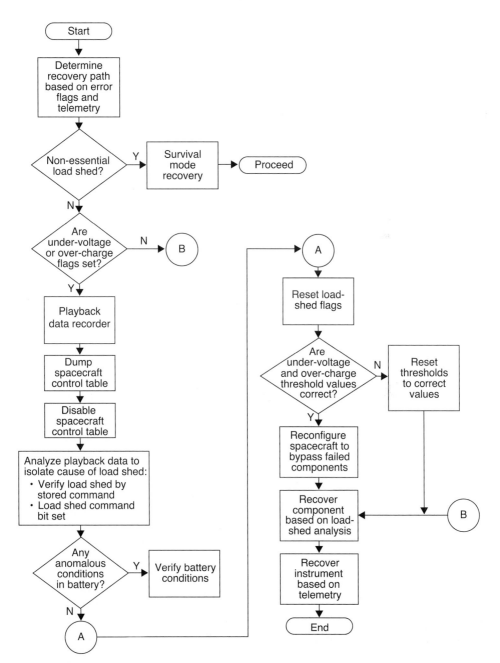

Fig. 14.6. Fault Tree for WIND Spacecraft. We would follow these steps to identify the cause of, and to recover from, an unexpected power failure on the spacecraft. [NASA, 1994]

and improperly to an anomaly, which usually creates even bigger problems, especially if we experience an anomaly that was not anticipated by pre-launch planning. Also, we'll encounter problems failure analysis didn't discover, particularly for payloads that employ significant onboard software. See Chap. 16 for more on handling anomalies.

14.1.4 Analyze Launch and Early-Orbit Scenarios

Launch and early-orbit activities are critical to all missions. Activities such as turning on the payload, out-gassing, deploying appendages, and initializing processors are significant because they usually occur only once and can damage the payload if they are not done properly. This is also the time to see how the platform systems function in a dynamic environment after experiencing launch stresses that are difficult to simulate completely on Earth.

The payload cost drivers in launch and early orbit activities are special, time-driven orbital constraints for working with the payload; special platform dependencies; appendage deployments; and calibration activities. We must understand and schedule all these activities so breakpoints in the timeline allow us to address anomalies without heavily replanning follow-on activities.

Our first concern is any special orbital constraints or restrictions on the payload. Examples of constraints are requirements to turn on the payload during darkness, to avoid operations in the South Atlantic Anomaly, or to avoid bright objects. The MOM must work with payload developers to ferret out unique requirements and to make sure the plan for launch and early orbits can support them comfortably. If we identify this activity early enough, we can help design the payload to avoid problems on-orbit with expensive replanning and extended special operations.

Our second consideration is the platform checkout and corresponding limits on attitude, power, communications, and thermal characteristics as the platform transitions to operations. We usually can't turn on the payload until the platform has reached a stable state. Payload operations must be ready to respond to any problems encountered during platform checkout, particularly if checkout is placed on hold while we deal with a platform anomaly. This is where adequate operational margins for power and temperature can avert complex activities to work around platform problems.

The third activity is planning for any special deployments of payload appendages or covers/doors. A simple survey instrument probably won't have any appendage operations, except disabling latch mechanisms. But special deployments of antennas are crucial for communications missions. For payloads on platform appendages or having instrument covers or doors, this effort takes on more significance. We should exercise these deployments during integration and test, using the same procedures we'll use on orbit. Operators must understand these deployments and practice contingency operations. It's important to schedule free time during which operations can resolve anomalies.

The final launch and early-orbit activity is turning on and calibrating the payload. Calibration should be straightforward. Requiring operators to use special external sources to calibrate a payload increases operational costs. The frequency and extent of calibration are also cost drivers. The accuracy requirements of the HST's instruments and their relationship to the platform led to a process for calibrating and validating instruments that took months. The Gamma Ray Observatory, on the other hand, was operating within weeks. Initial accuracy requirements for calibration translate directly into operations costs.

Chapter 9 provides more detail on planning launch and early-orbit activities. Here, we've discussed allowing enough time, planning the mission to have enough thermal and power margins, and providing appropriate self-calibration activities to avoid complex calibrations.

14.2 Assessing Requirements for Payload-Data Systems

To do this second and probably most significant step properly, we need to address seven considerations: (1) payload health and safety needs, (2) duty cycle, (3) data structure, (4) data volume and timeliness, (5) calibration and validation, (6) product development and data completeness, and (7) data access (see Table 14.2).

Table 14.2. Drivers for the Payload-Data Systems. On some systems, the cost of the payload-data systems represent 75% of the total operations budget.

Drivers	Comments
1. Health and Safety Requirements	Issues are autonomy, data downlinks, and payload interactions.
2. Duty Cycle	Issues are data volume, communications bandwidth, and pass duration.
3. Data Structure	Issues are data completeness, processing requirements, data quality, and user interfaces.
4. Data Volume and Timeliness	How much data must be transferred and how soon must users receive the data?
5. Calibration and Validation	Issues are frequency and complexity of calibrations.
6. Product Development and Completeness	How is data translated to meet users' requirements? Are standards used? How does loss of data affect the users?
7. Data Access	Who needs data and how do they get it?

These considerations are important because they help define operations costs and fundamental mission trades. We need to visualize how they affect operations and to flag cost drivers for the systems engineers and payload developers. We consider these cost drivers to be part of the total system. On most scientific missions, data-operations costs represent nearly 75% of the mission's operational costs, with mission operations for the spacecraft covering the other 25%. Controlling costs of data operations is crucial to controlling mission costs.

14.2.1 Health and Safety Requirements

Designs for the mission and spacecraft drive requirements for health-and-safety data. Basic drivers are the spacecraft's autonomy, data-downlink transfer, and payload interactions. A low-cost mission would have a highly autonomous payload and spacecraft—meeting their needs with few data dumps and minimal payload uplinks. The Alexis mission is one such example. It requires one uplink and one downlink per day. Observatory class missions such as the HST and Terra define the opposite end of the spectrum because they need continuous monitoring, have frequent downlink transfers, and require regular user interaction for fine pointing and instrument verification. Most other missions fall somewhere in the spectrum between Alexis and HST.

To lower costs, we try to develop an operations concept for which health and safety operations are carried out in a single shift or shared with another mission, the payload has the autonomy to do routine operations, and the payload can recognize a problem and safe itself. The concept should also include enough onboard storage and link capacity to allow file-oriented data transfers, for which the transfer protocol verifies the quality of transfers. Because of what the payload must do, most missions can't meet these simple goals. But we still need to show users and mission developers clearly how these issues affect costs.

14.2.2 Duty Cycle

Duty cycle provides a way of measuring the time margin available to handle both routine operations and on-orbit contingencies before payload data is lost. In Fig. 14.7, mission A has a low duty cycle, requiring one data transfer per day. Mission B has a high duty cycle, demanding four transfers per day. The duty cycle is a system trade that involves data storage, communications bandwidth, and ground operations.

Because all spacecraft payloads operate remotely, the mission designers must understand how the payload creates data volume and how the spacecraft and communications systems deliver this volume to the ground. The payload's duty cycle influences interactions with the spacecraft bus, onboard storage requirements, schedules for tracking and acquiring data, and mission-planning activities.

The major system trade on this requirement is the amount of space assets (on-board storage) versus ground operations. The Terra mission, for example, moves one terabyte of data every day. It has an onboard storage capacity of only 1.2 ninety minute orbits. This means data from the payload must be transferred to the ground every orbit, 7 days per week. A mission with a low duty cycle, such as the Total Ozone Measuring Systems (TOMS), collects data and dumps it once per day at a ground-station site. The onboard storage can hold one complete day's volume

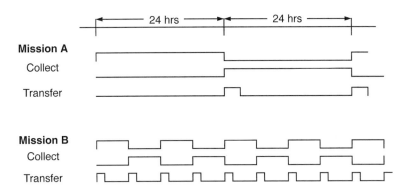

Fig. 14.7. Effect of the Payload's Duty Cycle on Data Systems. Mission A has a low duty cycle and requires only one data transfer per day. Mission B has a high duty cycle and requires four data transfers per day. Variations in collect cycles represent new cycles for the next transfer of data.

and requires only one tracking pass per day for data transfer. Proper sizing of on-orbit storage has a significant impact on mission operations. Automating routine ground activities helps to minimize the effects of a high duty cycle. The Landsat 7 mission has been very successful in automating all routine evening and weekend data acquisitions and transfers.

14.2.3 Data Structure

A major challenge for mission designers is to minimize the cost of the operational link between a payload and its users, so we must make sure the payload's data structure can handle this link at an acceptable cost. The MOM must explore the operability of the end-to-end system in terms of data completeness, data-processing flow, the end-user interface, and cost.

In considering *data completeness*, we need to decide whether the payload's data structure provides for accountability and a means of measuring data quality. An initial consideration is to pack the payload data into logical segments that match the measurement or services the payload provides. The object is to structure the data so we can contain periodic data losses from the space-to-ground link or ground processing. Doing so prevents small data losses from contaminating a much larger segment of the data.

Failure to address these needs increases the cost and complexity of data processing and staffing on the ground. Examples of logical data structures are a Landsat image or a scan of ozone data from the TOMS instrument.

Second, we need to analyze processing flow to determine whether we need special processing and ancillary information. The processing flow can be broken into a series of levels leading to stand alone data products for the end user (see Sec. 13.3). We critique these interactions to get the fewest retransmissions and reprocessing between levels and, therefore, the lowest costs for future operations. The trend is toward international packet standards or IP networks that encapsulate data logically and allow for standard interfaces on the spacecraft and the ground. Using nonstandard data structures is likely to increase the amount of unique processing software and corresponding lifecycle maintenance. On the Cosmic Background Explorer (COBE) mission, for example, higher level data (different calibration steps) became a cost driver because of the unique nature of the measurements. These processing costs were four times all other mission operations costs combined.

We must understand how our data structure will help or hinder data management during operations. Spacecraft in the past used mechanical recorders that required data reversal on replays and extensive processing to remove data redundancy, which demanded significant initial data processing. Solid-state memories and a logical structure for the payload packet allow data to be logically requested for playback, thus greatly simplifying initial data processing. Combining file-transfer protocols and solid-state memory enables data structures to lower operational costs. Chapter 13 discusses data-processing levels in detail.

Finally, we must make sure the data structure allows user access. Is the payload part of a correlative program or is it operated by a single user? Correlative programs often impose stricter data-quality standards and cost more by increasing the number of interfaces operators must deal with. They also increase pressure to have the best-quality algorithms for payload processing available early. On low-cost missions, much of this responsibility falls to the principal investigator.

14.2.4 Data Volume and Timeliness

These two issues directly influence the number of spacecraft contacts and determine whether we need multiple shifts for spacecraft operations. Data volume tends to be the major driver of complex systems for transferring data from the payload to ground. Timeliness or data latency drives the system trades among the spacecraft's storage capability, downlink bandwidth, and the availability of ground resources to accept the data.

For example, even though Fast Auroral Snapshot Explorer (FAST) is a small, single-instrument mission, it creates more data during its six-month winter campaign than the Hubble Space Telescope does during the same period [NASA, 1993]. In spacecraft design, small doesn't necessarily mean simple. The volume of FAST data drives the mission to have 12 ground contacts per day, whereas a simpler survey mission may require only one contact per day. FAST's volume necessitates ground operations around the clock, whereas SAMPEX needs only one shift. The data volume also affects ground-processing operations by requiring the complete sequence of science-processing steps to operate around the clock. The latency issue is also significant when we use mission data to update weather models, for example.

We must also consider how data volume affects storage and staging. The flow of information from the payload to the user ends up being staged on the spacecraft, at the acquisition site, and at the user's facility in order to work around various line and link failures. Staging of data at each of these facilities costs money. The more straightforward the transfer of data to the user, the lower the fundamental operations costs. This concept encourages direct downlinking of data to the end user's facility whenever possible because it avoids creating infrastructures that add marginal value to the overall mission. The direct downlink is possible only when the payload's orbit allows it to pass over the user's facility. The Terra and Aqua missions routinely direct broadcast MODIS data to a growing number of sites around the world.

Each of these staging points is a decision point for operators. Because most spacecraft can store or record only so much data, operators or automation must decide whether to replay data after a downlink problem or allow the data to be overwritten. This is a fundamental cost trade-off. The more timely the data, the more online processing, and thus, the higher the cost. The direct-downlink option helps to mitigate some of this cost, but forces around-the-clock operations at the user's building unless they are fully automated.

The marginal utility of obtaining high numbers for access to payload data also drives operational costs. If, for example, we require 99.9% of the payload data for mission success, operations will be very complex. This complexity is related to data structure and duty cycle, but it also requires quick decisions to avoid overwriting data at any of the staging points. Data completeness (see Sec. 14.2.6) also affects its quality. The data-packaging scheme can strongly reduce the effect of data loss on a payload by considering the error characteristics of the data link between the spacecraft and the ground. A number of missions are lowering the requirement for data recovery percentage to reduce the mission costs. The Small Explorer missions routinely set a requirement of 95%.

14.2.5 Calibration and Validation

Payload calibration involves two actions: calibrating the payload itself and applying algorithms to the data received to convert it into usable products. This section addresses concerns about operating the instrument. The section on data products addresses ground processing.

The frequency and complexity of calibrations drive operational costs. The first cost driver is how often we need to calibrate a payload. Depending on the accuracy requirements and the payload's construction, this activity could be per orbit, daily, or weekly.

The second cost consideration is the complexity of the calibration. A calibration that requires periodic observation of an external source or deep space will complicate operations and planning. A payload with access to an internal calibration source that has been validated externally can calibrate automatically. If we don't design calibration into the payload's operation, on-orbit operations and data processing will be more expensive. It's costly to add calibration maneuvers that we identify after spacecraft bus development has started. This is particularly true of deep space calibrations required of earth pointing missions which require leaving nadir pointing.

One final operational consideration is tracking configuration changes to the payload in response to calibration problems discovered on orbit. We have to be able to match changes in configuration with the appropriate payload to avoid significant reprocessing of data sets.

14.2.6 Product Development and Completeness

Another challenge facing mission operators is translating the payload's data into meaningful products that are easily available to customers who can use them. Proper format and completeness are the two main issues.

The various science communities are developing standards for their types of payload data so they have common tools. The astrophysics community, for example, has settled on the High Energy Astrophysics Archive (HEASARC) format, whereas the Earth-observing community has decided on the Hierarchical Data Format (HDF). The MOM must be able to work with users and understand product formats to help design a mission that will provide these products as per the standards. Particularly important is ancillary information, such as definitive attitude information, and calibration data needed to build a meaningful data product. By addressing these characteristics with users early in the mission, we can avoid unnecessary data-processing steps.

Data completeness is as important as product development. Various types of payloads have different fundamental data sets: a frame for an imagery payload, a measurement at a given time for a survey mission, or a spectrum of an object for a spectrograph. Each set of the payload data has characteristics that tie it to the fundamental activity they're measuring. The MOM must be sure the data structure is properly configured, so normal transmission errors don't propagate and cause unexpectedly high data loss. Knowing the payload and its drivers early in the mission helps to make early trades and to create appropriate margins in the end-to-end system.

We need to walk through the complete data-processing system to ensure that none of these problems exists. The actual information content should approach the level of the data collected. If data compression is used, for example, and not properly channel coded, data losses translate into burst errors and larger information loss. Although we can develop special post-processing algorithms to work around these types of data issues, it's expensive and worth avoiding. The development of packet standards has helped to improve this situation, but we still have much to learn about matching these packets to the nature of the science data.

14.2.7 Data Access

Missions in the past defined payload products and created significant infrastructures to deliver these products to users. In today's information-processing world, the less expensive approach is to enable users to obtain data as they need it. This fundamental shift to a file-server approach reduces the infrastructure and allows users to get only what they need, when they need it. It averts creating mini-archives throughout the user community.

Of course, mission operators must develop directories that allow users to determine if the data they seek exists. Thus, products must be identified and available to an archive the community can access. The Earth Observing System (EOS) is trying to meet this need by creating distributed data centers with standardized interfaces for directories and common data formats.

The requirements for monitoring data quality and the post-processing functions above initial processing fall to the data centers. This approach also allows adding data centers that can do further processing to make a higher level of data products available to a less sophisticated, but larger, data-using community. The people in these data centers also become experts on the issues with their type of data from different missions. Those at the EROS Data Center of the United States Geological Survey, for example, are experts on land data processing issues because of their history with the Landsat program.

14.3 Assessing Drivers for Payload and Platform Interfaces

With the operations concept and data systems in hand, we can begin the third phase of assessing payload operations. The steps for this phase are listed in Table 14.3 and described below.

Table 14.3. **Assessing Drivers for Payload and Platform Interface.** This is the third phase of assessing payload operations.

Step	Description	Comments
1	Define operational scenarios	Orbital constraints, products, timelines, and mission interactions
2	Characterize requirements for onboard command and data handling	Payload/platform interactions—level of autonomy
3	Define maintenance constraints on the platform	Power, maneuvers, and multiple payload interactions
4	Determine the need for ancillary information	Time and position accuracy
5	Define the need for special events	Safing, software updates, and fault isolation aids

This section continues the mission-engineering process outlined at the beginning of this chapter. Because this process is iterative, many of the steps seem redundant, but we're taking them to a new level of detail. These iterations allow the mission team, under the MOM's direction to refine mission operations continually for lower cost. The final output of these iterations is the activity plan for payload operations.

The key to cost-effective payload operations is not creating large documents or lengthy analysis but identifying the major issues. The details in a payload operations plan should be consistent with the import of the mission being developed. For a simple payload that just remains on in a monitor mode, the activity plan might be only two to three pages.

14.3.1 Define Operational Scenarios

Developing an operational scenario consists of (1) characterizing orbital constraints; (2) identifying all required operational products and actions; (3) determining when these products or actions are required; and (4) developing an integrated timeline that depicts these activities in relation to mission activities. The timeline should reflect the fundamental time unit of operations which, for many missions, is 24 hours. But for missions with highly elliptical orbits, it might be one orbit, which could be several days.

An operational scenario needs to address normal operations for this fundamental period and for any other major events, such as launch and early-orbit deployments and orbital maneuvers. It's important to identify the time relationships of any important activities and product development in this scenario. Required operational products can drive up operational staff and cost very quickly.

Figure 14.8 gives a day-in-the-life timeline for X-ray Timing Explorer (XTE). The XTE is a scientific pointing mission that uses TDRSS as its main communication link. A key to developing a scenario properly is characterizing and documenting operational constraints. Operational constraints come from two sources. The first source is system-engineering trades to work around environmental events. For example, XTE added a second TDRSS antenna as a system trade to ensure contact during slewing maneuvers. Most of these events are defined by the orbit, the capability of the launch vehicle, and the capability of the communications architecture. Examples of such phenomena for low-Earth orbiting payloads are day and night periods, South Atlantic Anomaly, and guide-star availability.

Spacecraft and payload limitations or problems encountered during mission integration and test are the second source of operational constraints. These constraints lead to a list of operational workarounds. It's important to maintain a complete list of these workarounds, so we can place them into the procedures, which is the next level of detail under the scenarios. A significant benefit of having a core team on a small mission is that they bring to operations any lessons they've learned from integration and test. One of the first questions asked during an on-orbit anomaly is, "Did we see that during integration and test?"

Another important step in developing a scenario is identifying the timing of all required products and support actions, which involves developing schedules for communications support, science plans, and data products. This step helps define operational interfaces and sets performance needs for payload and spacecraft

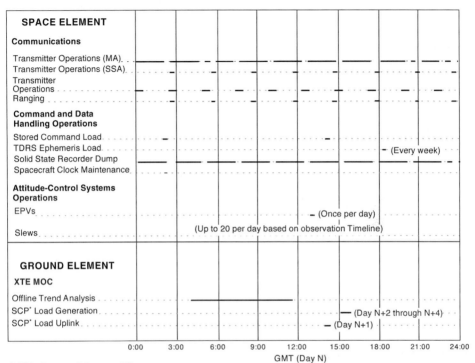

Fig. 14.8. Timeline for the X-ray Timing Explorer (XTE). XTE uses TDRS multiple access (MA) and single-subscriber access (SSA) communications links. [NASA, 1995]

interactions. Examples of actions are onboard memory management, management of data buffers, table loads for pointing accuracy, and calibration of attitude sensors.

The final step in developing a scenario is producing an integrated timeline, which documents the effects of constraints and actions to ensure the payload can accomplish its mission. The MOM must hold the timeline to ensure the mission is operable for a reasonable cost.

In summary, developing mission timelines for operational scenarios helps both MOM and the mission operations team visualize the payload's normal operations, as well as identify external interfaces and required operational products and activities. Visualizing activities helps to lay out trades for onboard autonomy versus manual operations by showing which activities parallel one another. It provides the first meaningful analysis of mission operations staffing.

14.3.2 Characterize Requirements for Onboard Command and Data Handling (C&DH)

While developing scenarios, we should check them from the perspective of command and data handling on the spacecraft bus and payload. Doing so tests the latest design.

The main concern is the bus and payload interface for command flow to, and data transfer from, the payload. This activity helps clarify the duty-cycle driver identified in Sec. 14.2.2 for given mission. We should do this trade as early as possible to see operational effects and work them with the C&DH designers. An early review also gives us the chance to walk through data-transfer activities and see if we can simplify them based on end-system needs.

Now we can identify interactions required between the payload and the spacecraft bus and verify them operationally through test scenarios. At this time, we must also establish guidelines for recovery and verification of physical constraints. An example is characterizing payload temperature during a safing or load-shedding activity and determining how long it could stay at that temperature without being damaged. This is also the time to validate all telemetry limits that match these states.

We must also characterize the level of autonomy provided by the mission and the corresponding staffing activities by iterating the day-in-the-life scenario. The key is to be able to make end-to-end system trades that accurately reflect lifecycle operations costs. As a rule of thumb, it's best to automate routine activities the spacecraft can react to rather than developing predictive planning tools in the ground system, which is usually much more costly.

14.3.3 Define Maintenance Constraints on the Platform

For normal operations, we have an interface-control document that defines the interface between the payload and the bus, but we need to add platform activities that affect payload operations. Three major bus constraints are power, attitude, and multiple payload interactions. We need to assess payload operations that result from activities required to maintain the spacecraft bus.

Let's first consider power. Most low-Earth missions are designed to operate through various shadow events with little effect on payload activities. However, for many small spacecraft, the power margins can constrain payload operations by limiting the length of data transmissions because the transmitter typically consumes a significant amount of power. The power system and its limitations are significant inputs to mission and payload operations, so we must put into the timeline scenario any unique limitations or power-system maintenance activities, such as battery management. One aspect of this consideration is cyclical yaw maneuvers to work around fixed solar arrays for the simpler missions.

We also need to understand how the platform starts shedding loads. This activity ties back into the automated fault management logic that is built into the spacecraft bus. It also ties into understanding the impact that the various spacecraft safemodes have upon the payloads.

Second, we have to consider maneuvers, including those for payload pointing and for orbit adjustment and maintenance, such as station keeping for a communications payload and drag make-up maneuvers for maintaining altitude. Most of these missions have a maneuver mode that precludes payload operations. For many Earth-observing systems, yaw maneuvers are required twice a year to correct the payload's orientation to the Sun.

Third, consider interaction with other payloads. This interaction can involve conflicts in field of view, noise interference, or contention for data-transfer resources. These fundamental issues need to be understood and resolved during the concept phase. The MOM must advocate solving these problems early to avoid many constraints after launch.

The payload's availability to do its job is one of the measures of success for a mission. One way to ensure availability is to integrate the payload to a spacecraft bus that is almost transparent to the payload. Avoid spacecraft buses that require continuous monitoring and reconfiguration. Instead, identify bus-maintenance constraints and work with the designers to minimize them because they tend to increase mission operations costs. We should move toward more autonomy by decoupling routine spacecraft-bus operations and payload operations.

14.3.4 Determine the Need for Ancillary Data

While considering end-to-end data processing, we identify the need for ancillary information, such as time and position information needed to start processing the data. With the Global Positioning System's technology, we can now do many of these functions on board and provide the ephemeris information with the data itself. This approach eliminates costly post-processing of tracking data. It's always more costly to recreate an activity than to have it measured immediately.

We also need a data-quality statement that can be appended to the data. In the past this statement was recreated on the ground by reprocessing the raw data received at tracking stations. Now, with packetized systems, the data-transfer protocol gives us a way to develop quality statistics based on the data format. Using solid-state memory and a file-transfer protocol with retransmissions would significantly change the initial phase of data processing. Configuration and calibration are other types of ancillary data to consider. It's more cost-effective to address needs for ancillary data during system definition and development than to try to recreate ancillary data during the operational phase of the mission.

14.3.5 Define the Need for Special Events

Once we understand normal operations, we must prepare for special operations, such as safing and recovery, reprogramming the payload's processor, and isolating and trouble shooting faults.

Spell out the safing activities and their corresponding recovery in a scenario as a basis for detailed procedures. The scenario gives operators the chance to visualize how the safing should take place, how long it should take, and how it would be detected on the ground. The recovery scenario should identify the sequence of steps or stages required to bring the payload to full operations in a controlled manner, particularly for payloads that use high-voltage power in their measurements.

Almost all payloads employ processors in their operations. As payloads operate, the processor's software needs maintenance. Operators need to develop a scenario that defines how they can update the program on board the payload to correct any problems or react to equipment anomalies. This scenario should determine the time required to complete an upload given the forward-link bandwidth and the throughput for C&DH. Not designing an efficient way to update the payload could necessitate complex, costly workarounds. For major payloads, it is worthwhile to maintain a maintenance copy where software changes can be simulated on the ground before being uploaded to the payload.

Finally, include fault-isolation and trouble shooting exercises to help determine if information is available to identify the problem. Doing a few fault-isolation exercises helps to validate and refine the fault trees.

Developing meaningful operations scenarios early in a program helps the MOM and the team appreciate the operational implications of various design decisions. By seeing these implications as early as possible in a mission, we can avoid unexpected staffing and cost just before launch.

14.4 Characteristics of a Good Payload Operations Plan

The payload operations plan is the end product of this phase. It's a more detailed iteration of the operations concept, with a complete timeline of events. The next level of mission engineering would be the detailed operational procedures required to carry out the plan.

We should match the efforts on areas of this plan with their effect on the mission. Pivotal events that mean mission success or failure warrant corresponding detail in the operations plan. The plan proposed here is organized around the mission's major timelines. The main sections are launch and early-orbit timelines, special operations, and normal operations. The payload operations plan can be part of a spacecraft operations plan, or at least complement it, depending on the complexity of the payload and the mission.

The plan's size depends on the payload's size and complexity. It could range from 5 to 100 pages. The object is not to create paper but to think through the operations and maintain a baseline of how the payload is to be operated. For many small payloads and missions, it incorporates the baseline of the payload's operation as the payload evolves through the mission phases.

14.4.1 Launch and Early-Orbit Timeline

Events during launch and early orbits typically focus on the spacecraft bus and turning on and checking out its subsystems. The main payload activities are usually turning it on and activating it. We have to concern ourselves with deploying appendages, such as booms attached to payloads, releasing latches, or removing covers that protect payloads during launch. We must define these activities in detail and identify rest points for contingencies. We must also be sure the payload-activation sequence allows time to handle anomalies.

Activities that require special attention are turning on high voltages, initial calibration, instrument processors, and handling special orbital-activation constraints such as lighting or a calibration target. See Chap. 9 for launch and early-orbit activities.

14.4.2 Special Operations

Special operations occur because of anomalies or unique mission activities. In many cases, the special operations used to turn on the payload are similar to those needed to recover from safe-mode due to an anomaly. By recognizing this relationship early enough, we can avoid developing unnecessary procedures.

Special operations for anomalies need to define safemode entry and recovery. They must specify the events that will cause the payload or spacecraft platform to place the payload into a safe state. Safemode should allow ground operators the time to analyze the problem and recommend how to proceed. This is particularly true during the early phases of a mission, when there is no history of the payload problems that will occur over the life of the mission. With the increasing complexity of a payload's processing ability, we need to build the payload processor so it can gather data that will enable us quickly to determine the reason for the anomaly. Soon, the payload could actually start isolating faults and recommending actions to the ground. (Of course, we shouldn't apply this advanced, expensive technology to simple instruments.)

Once we understand a problem, we need to develop a detailed sequence for recovering the instrument. We must also identify payload calibrations or special operations to return the payload to full operation. Examples of such special operations are decontaminating the wide-field planetary camera instrument on HST or the high-voltage turn-on sequence for any number of instruments.

The second aspect of special operations is unique mission activities. Many scientific payloads, for example, have encounters or campaigns centered on special scientific events. Examples of such events are the Cassini encounter with the Jovian system and the FAST mission's campaign for studying the northern lights during the winter.

Targets of opportunity demand another special operation for survey missions. For these events, we must spell out in the operations plan required replanning and interruption of the payload's normal timeline (and, many times, of the spacecraft's operation). The plan should articulate timing requirements for operations and unique communications and payload-processing support. It should also detail how we'll start and carry out these interactions with integral support areas and spacecraft operations.

14.4.3 Normal Operations

Most missions spend more than 95% of their time in normal operations. The MOM must present normal operations for a payload to reflect a day in the life of a mission. The main cost consideration is the level of daily support needed for payload operations. To save money, we must size the operations team for normal operations and allow time through safing sequences to handle anomalies. The faster the required response time, the costlier the mission operations. For example, operating from the Space Transportation System (STS) is a challenge because the short mission and public visibility require quick handling of anomalies. These pressures force all necessary support experts to be on hand, just waiting for an anomaly to occur.

To assess workload for normal operations, we need to develop a detailed mission profile for a 24-hour period. We may use a different period if it more aptly reflects the payload's cycle of operation. The activity plan should identify online and off-line activities and estimate the amount of labor required for all tasks. Laying out the work properly helps to identify peak activities and shows the amount of effort needed.

The activity plan must address health and safety monitoring, mission planning, command generation, command execution, and data-product processing. For a small mission, one or two people can do these activities; larger missions may require dedicated teams for each activity. Health and safety monitoring depends on payload design. If the payload is intelligent and has a number of safe modes, we don't need to do much monitoring. Instead, we could review parameters that affect the quality of data products, thus reducing expensive real-time operations in favor of smaller, off-line operations.

In the scenario for mission planning and scheduling, we must identify mission drivers that bound the operation. Examples of such drivers are orbital constraints, communications abilities, onboard data-storage ability, and the handling of anomalies in transporting mission data. Orbital constraints due to spacecraft or payload limitations are expensive to plan from the ground. Examples of such constraints are limits on pointing toward bright objects, operations in the South Atlantic Anomaly, or guide-star availability. It's a mission trade to obtain an orbit that minimizes these types of constraints in order to lower operations costs.

The commanding requirements of a mission depend on the spacecraft and payload designs. If we can preplan payload operations or execute them on board with script direction from the ground, we can lower the level of command interaction. If the payload is dynamic and requires continuous interaction with the ground, the commanding activities will be extensive and the operations cost will be relatively high. In the first case, we can automate many of the spacecraft operations and simply have ground controllers manage tables that the spacecraft processor was to operate. This activity is far less costly than having to uplink large numbers of single commands to a dumb spacecraft.

The volume of data that the payload creates and the spacecraft's ability to transfer it to the ground will drive pass contacts. Present network technology makes us isolate pass-contact faults manually—a costly job. The growing presence of solid-state memory makes storing large volumes of data and sending them to the ground more cost effective because it lowers the number of pass contacts.

The MOM needs to review the constraints and data requirements and create a 24-hour mission profile that shows their relationship. Figure 14.8 gave an example of a mission profile for the XTE spacecraft and its instruments. The top few lines list the orbital constraints, the next lines describe payload operations, and the final few lines describe network operations with the TDRS system.

The final activity is to control and monitor the transfer of the payload data to its destination for processing. For a simple survey payload, this monitoring can be very straightforward and easy to automate. Solid-state memory and file-transfer protocols are allowing this automation. Balance between the forward-link bandwidth and the downlink bandwidth is one of the limitations for this type of protocol. We need to build the accounting capability into the payload and spacecraft, so we won't have unique processing requirements to reconstruct the payload data on the ground. Planning the data properly can take care of such items as time tagging and ancillary position data.

The keys to low-cost mission operations are early involvement of operators to identify development/operations trades and the discipline to enforce these trades. Users of the data must participate in the systems trades because they are the only ones who can assess the trades' effect on the mission objective. We must make these trades as a team and encourage alternatives to meet the mission objectives at the lowest cost. If we consider only the mission objectives, trades will be made with only development in mind and won't consider the mission's lifecycle costs.

The mission-engineering process structures iterations on the operations concept that develop increasing levels of detail to prepare for operations. The tables in this chapter start a new MOM toward identifying cost drivers for payload operations.

References

Biroscak, Losik, and Malina. 1995. *Re-Engineering EUVE Telemetry and Monitoring Operations: A Management Perspective and Lessons Learned from a Successful Real-World Implementation.* Publication number 669. Berkley, CA: Center for EUV Astrophysics Technology Innovation Series.

National Aeronautics and Space Administration. 1995. *Flight Operations Plan for the X-Ray Timing Explorer (XTE).* GSFC-410-XTE-031. Goddard Space Flight Center, MD.

National Aeronautics and Space Administration. 1994. *GGS Mission Operations Procedures. Volume IV: Contingency Operations Data Base.* CDRL 405, NAS5-30503. Goddard Space Flight Center, MD.

National Aeronautics and Space Administration. 1993. *Fast Auroral Snapshot Explorer (FAST) Flight Operations Plan.* FAST-OPS-006. Goddard Space Flight Center, MD.

Chapter 15

Spacecraft Performance and Analysis

Rob Zmarziak, *Northrop Grumman*

Mac Morrison, *TRW (Retired)*

> 15.1 How Typical Spacecraft Subsystems Work
> 15.2 How We Carry Out Typical Spacecraft Operations

One of the mission operations manager's (MOM's) most important tasks is to ensure that spacecraft-bus operations will maintain the spacecraft's health and safety while meeting mission objectives. Before launch, the MOM (or the spacecraft planning and analysis function) monitors the bus design and develops procedures for operating the spacecraft bus. Following launch, the same person(s) analyzes spacecraft-bus performance, generates commands, maintains flight software, and plans required calibrations. This chapter describes how we do the steps in spacecraft planning and analysis, as listed in Chap. 3.

Successful space mission operations require a well designed spacecraft bus tailored to accommodate the specific flight requirements for the payload or experiments. The bus typically provides power, pointing control, maneuvering ability, communications, mounting structure, and thermal control for mission operations. Failure of the spacecraft to provide these functions because of design flaws, random failures, or operational errors can compromise mission success. Reviews should discover and eliminate faulty designs, but designs that work in breadboards, engineering models, and even during spacecraft assembly and test may be difficult to operate in the space environment. For instance, does the design include the ability to use ground-commanded overrides? Does it cross strap subsystem elements adequately? Does it include the right sensors and defined telemetry to allow us to determine the spacecraft's on-orbit state of health?

Good engineering practices that include review and testing throughout the design and development phases can minimize design flaws; careful parts selection and screening may lower the failure rate of piece parts; and thoughtful design of the spacecraft's onboard software and ground software, coupled with rigorous operations training, can minimize operational errors. But cost-effective mission operations means coupling the right spacecraft bus with the mission requirements and recognizing the operational effects of these designs early in a program. The earlier the better! The other component of cost-effective mission operations is determining what fidelity level of simulation and training will ensure mission success while minimizing operational planning, preparation, and execution costs. Finding the correct balance of "how much is enough" is difficult and falls under the purview of the MOM.

The MOM is a crucial player in determining and analyzing operations that can affect spacecraft design and reduce overall program costs. By making sure that adequate plans and procedures are developed before launch, the MOM can help reduce operational errors resulting from inadequate planning, human errors in commanding the spacecraft, misinterpretation of telemetry data, and inadequate time to complete operations.

We must evaluate carefully the requirements for operating a spacecraft to determine the concept for mission operations. These requirements include proper constellation size, the right ground-control environment, and protecting mission data.

Multiple-spacecraft constellations may develop an operations concept that allows loss of individual spacecraft as a trade between spacecraft complexity and replacement costs. Concepts for programs that feature tens to hundreds of satellites in low-Earth orbit, such as Motorola's IRIDIUM and some of the now canceled SDI programs, faced this trade.

Dedicated ground-control facilities may drive trades between onboard abilities and ground-commanded abilities. If a station is available to a mission 24 hours a day, we may control many spacecraft operations in real time by uplinked commands, thus potentially simplifying onboard subsystems. Operations using institutional ground control, such as the Air Force's Satellite Control Network, must share resources. Spacecraft operations in this environment depend a lot on scheduling. The ground-control network may not provide the required coverage for uplink and downlink, thus driving the spacecraft design to more autonomy. This is a classic trade in spacecraft operations—ground control versus onboard control. We have to trade the one-time non-recurring and the recurring costs of highly autonomous spacecraft against the recurring costs of operations manpower for spacecraft that need more ground control.

Other mission complexities such as rendezvous and/or landings require autonomous operations for complex maneuvers, since distances do not allow for man-in-the-loop control. Questions arise such as: Must we regularly interrogate the spacecraft? How immune is the spacecraft to anomalies that occur when a ground station isn't "up" on the vehicle? Does the spacecraft have built-in safe modes and can we use stored, onboard commands? What action should the spacecraft take if the descent telemetry doesn't match anticipated values?

Constellation design, specifically the amount of ground control versus autonomous operations, can result in undesirable outcomes. The potential exists for both data loss and catastrophic damage to the payload and/or spacecraft. Some of the questions that we should ask when determining operations concepts and the level of autonomy are: How important is the mission data the payload will obtain? Are there any one-time or mission critical events for which data loss is unacceptable? What delay in commanding, if any, is acceptable to meet mission requirements? The operations planning and spacecraft design must satisfy any critical mission requirements. However, comprehensive operations plans and procedures may allow for simpler spacecraft designs and a lower level of autonomy for cases where intermittent data outages and/or loss of communications during certain operations is tolerable. The MOM ensures that derived requirements of this type, identified during the definition of the operations concept, are incorporated into the system design requirements.

In some cases, wanting to provide mission flexibility runs up against economic realities—the money just is not available to do everything we want. Spacecraft complexity (and weight) drive technical decisions on the amount of onboard redundancy to include in the spacecraft design. The specified or chosen booster used to launch the spacecraft into orbit may also influence the design. An example of a more complex design is a mission launched and serviced from the Shuttle. The requirement of "fail operational—fail safe" drives spacecraft to triply redundant subsystems in some cases. All of these complexities mean we must know the operations requirements early in the program. We must also identify and analyze failure modes and recoveries.

15.1 How Typical Spacecraft Subsystems Work

Normal operations for spacecraft subsystems are real-time, ground-controlled actions or stored, onboard actions to provide and maintain power, pointing control, maneuvering ability, communications, and some thermal control for the mission operations. Design dictates how we operate—simplified, easily accomplished spacecraft operations lead to cost-effective missions.

Although we build many spacecraft, the types of spacecraft subsystems are relatively few. Power subsystems provide electrical power through solar arrays, batteries, radioisotope thermoelectric generators (RTGs), or fuel cells. For attitude determination and control, we

- Spin all or part of the spacecraft

- Use the Earth's gravitational field to point a spacecraft axis toward the Earth's center
- Use the interaction of the Earth's (or other planet's) magnetic field and onboard torquers to align a spacecraft axis toward the Earth or other planetary body the spacecraft is orbiting
- Use a combination of onboard gyroscopes or reaction wheels to provide three-axis attitude control

We maneuver spacecraft by using an onboard propulsion subsystem coupled to Sun, Earth, or star sensors or to data received from global positioning system (GPS) satellites. Communications to a ground-control station or other spacecraft depend on receivers and transmitters coupled to either fixed or maneuverable antennas. Thermal control can be passive—using insulation, second-surface mirrors, and paint; or it can be active—using thermostats, heaters, or heat pipes. Table 15.1 lists spacecraft subsystems, options, and major considerations.

These subsystems normally provide the mission with continuous power throughout eclipse periods and maneuvers, keep a spacecraft's axis pointing in the desired direction, maintain communications between the spacecraft and with the control station, and provide a benign thermal environment for all spacecraft equipment. Chapters 10 and 11 of *Space Mission Analysis and Design* [Larson and Wertz, 1999] discuss these subsystems, showing how to design them and how requirements drive their design. Here we discuss the way they operate.

Attitude determination and control for many Earth-orbiting spacecraft depends on establishing the relationship of the spacecraft body's axis to the Sun, to the center of the Earth (or the Earth limb), or to a distant stellar object. In the first case, onboard Sun sensors attached to the spacecraft body measure the angle between incident visible light and the fixed sensor, usually through a slit in the sensor housing, as shown in Fig. 15.1. Timing the occurrence of Sun pulses can establish the rotational rate. Incident light on the slit in the sensor housing provides angle data.

Earth sensors typically measure the temperature difference between the Earth's atmosphere and deep space. This difference defines the Earth's horizon and provides information to determine the geometry for spacecraft pointing. Earth sensors with oscillating mirrors view the Earth limb to limb and provide nadir pointing data to the spacecraft's attitude-control system. See Fig. 15.2.

Star sensors used on three-axis stabilized spacecraft are usually either trackers or mappers. After the tracker locates a predetermined star and tracks it, the vehicle's motion will result in an apparent movement of the star. We use this error to control the spacecraft's attitude. Star mappers use similar logic but track all stars in the sensor's field of view above a certain brightness. We usually use the data from two or more stars to establish the spacecraft's inertial attitude. We can couple magnetometers that measure the geomagnetic field with magnetic torquers to control the spacecraft's attitude. However, the attitude accuracy possible with this system is less than that achievable with the other systems. [Larson and Wertz, 1999, Chap. 11]

15.1 Spacecraft Performance and Analysis 561

Table 15.1. Operating Spacecraft Subsystems. These lists of options and considerations are incomplete but represent the issues we must address for this function.

Subsystem	Options	Key Operational Considerations	Where Discussed
Attitude control	• Gravity gradient • Spin stabilized • 3-axis stable • Sun, Earth, and star sensors	• Deploying boom • Initially acquiring Earth • Determining nadir angle • Initial spin-up • Nutation control • Controlling spin speed • Managing momentum • Acquiring attitude • Disturbance torques • Losing Earth • Number of independent fields of view • Pointing constraints	Chap. 15; Chap. 11 [Larson and Wertz [1999]
Electrical power	• Deployed arrays • Body-mounted arrays • Batteries • Radioisotope thermo-electric generators	• Deploying arrays • Eclipses and shadowing • Solar tracking • Degradation • Re-conditioning batteries • Charge/discharge cycles	Chap. 15; Chap. 11 [Larson and Wertz, 1999]
Propulsion	• Cold gas • Mono-propellant • Bi-propellant • Electric • Blow down or regulated • Thruster size	• Location of thrusters • Thrust duration and level • Maneuver complexity • Amount of redundancy	Chap. 15; Chap. 17 [Larson and Wertz, 1999]
Thermal	• Passive coatings • Multi-layer insulation • Active heat pipes	• Number and location of heaters • Thermal bathing • Temperature gradients	Chap. 11 [Larson and Wertz, 1999]
Communications	• Fixed antenna • Steerable antenna • Satellite cross-links • Operational frequency • Optical	• Encryption • Pointing requirements • Number of data modes • Real-time or playback • Data volume, data rate, and pass duration	Chap. 11 and 15; Chap. 13 [Larson and Wertz, 1999]
Command, control, and data handling	• Onboard processing • Types of recorders • Memory available • Stored command files • Real-time data only	• Memory duty-cycle • Data completeness • Autonomy • Pass duration • Data compression	Chap. 13 and 15; Chap. 11 [Larson and Wertz, 1999]
Structures and mechanisms	• Deployable booms • Articulated devices	• Coupling of articulation and attitude control • Deploying boom • Solar pressure effects	Chap. 11 [Larson and Wertz, 1999]

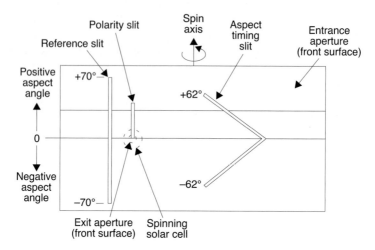

Fig. 15.1. **Geometry of a Sun Sensor.** The geometry of the slits in the sensor housing allows us to determine the spacecraft's spin speed and the Sun aspect angle.

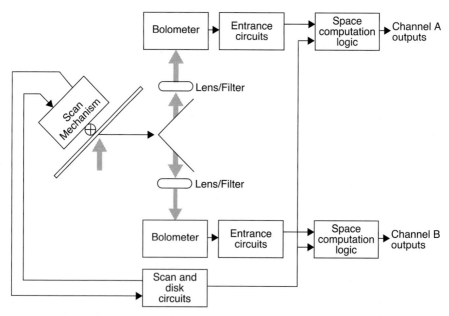

Fig. 15.2. **Block Diagram of a Typical Earth Sensor.** This Earth-sensor design has a scanning mirror/bolometer to find temperature differences between the Earth and deep space. Redundant channels avoid the Sun's or Moon's intrusion into the sensor's field-of-view.

Propulsion subsystems enable us to maneuver the spacecraft in the orbit (ΔV maneuvers) and to control spacecraft pointing. Most propulsion subsystems for normal on-orbit operations are hot-gas, catalytic-thruster systems using hydrazine as the fuel. The system operates by passing the fuel from a pressurized source over a heated catalyst bed containing alumina particles. The hydrazine decomposes into nitrogen, hydrogen, and ammonia gases, which exit through an exhaust nozzle to supply thrust. Bi-propellant hot-gas systems using N_2O_4/N_2H_4 provide higher specific impulse than hydrazine alone, but the systems are usually more complex.

If it's necessary to avoid the contaminating products from a hot-gas system—as for a payload with exposed optical surfaces—we might use a cold-gas system with pressurized helium. Table 15.2 compares the specific impulse for these systems.

Table 15.2. Specific Impulses (Isp) for Spacecraft-Propulsion Systems. Performance versus clean operation is a typical propulsion-system trade.

Type of fuel	Isp Range	Risk of Contamination
Bi-propellants	305 – 310	High
Hydrazine	220 – 240	Medium
Cold gas	30 – 70	Low

If the fuel source is in a tank that holds fuel and pressurant separated by a bladder, the thrusters will have decreased performance as the fuel is used and the pressurant expands. This is called a blow-down system and requires adjustment of thruster firings (either duration or number of firings) to maintain performance. Regulated systems have a separate pressurant tank at very high pressure; the pressurant releases into the propellant tank to provide a constant pressure over the life of the mission. Figure 15.3 illustrates these two options; Figs. 15.4 and 15.5 show their relative complexity.

The thrust range used for on-orbit operations varies from a fraction of a newton to a few newtons for attitude control and tens of newtons for maneuvering. This means that operators can typically select the thrusters. Thrusters on the spacecraft structure provide thrust in the pitch, roll, and yaw axes for three-axis attitude control and along the spacecraft's velocity vector (+ and –) for delta velocity maneuvers. Maneuvers out of the orbital plane use thrusters mounted at an angle to the orbital velocity vector or yawing of the spacecraft. Many spacecraft have thrusters canted away from the spacecraft's x, y, and z axes, so we can use components of the thrust to control in more than one axis. Contamination products and heating from the thrusters may also require us to move the thrusters off the spacecraft's orthogonal axes. Although opening or closing valves in the system's fuel lines allows us to use redundant tanks and thrusters, we don't have to alter the subsystem's configuration unless failure occurs.

Communications subsystems—sometimes referred to as telemetry, tracking, and command (TT&C) subsystems—are the links between the spacecraft and

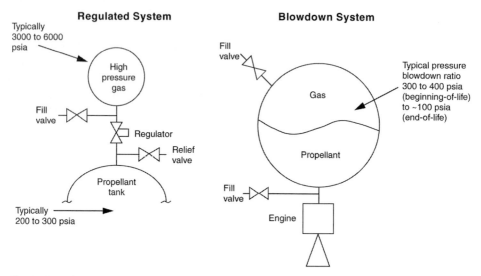

Fig. 15.3. Ways of Pressurizing Two Pressurization Subsystems. The regulated system, though more complex, maintains a steady pressure in the propellant tank and, therefore, constant performance.

ground-control stations (or, in some cases, from spacecraft to spacecraft). A concise, high-level summary of communications subsystems is in Chap. 11 of this book and in Chap. 13 of Space Mission Analysis and Design [Larson and Wertz, 1999], which we use to give you the following information. The communications subsystem allows the spacecraft to receive and track carrier signals, receive commands, and transmit telemetry data. Table 15.3 summarizes these functions. Figure 15.6 shows a typical communications subsystem.

For the communications subsystem in Fig. 15.6, the uplink RF signals are received through the antennas and pass through the diplexer to the receivers. In the case of commands, the data stream is demodulated from the carrier and subcarrier and then routed to the command detector, which validates the data and forwards it to the subsystem for command and data handling. For ranging signals, the tones or pseudo random noise (PRN) codes are demodulated in the receiver and then routed to the transmitter for conditioning and modulating onto the downlink carrier. Spacecraft telemetry data that contains configuration and state-of-health information is conditioned in the telemetry-control unit(s), modulated onto the downlink subcarriers, and routed to the transmitter for modulation onto the carrier and transmission. The transfer RF switch allows us to select a redundant transmitter and antenna. The diplexer allows a transmitter and receiver to share the same antenna. It also isolates the transmitter from the receiver port at the receiver's center frequency to keep from damaging the receiver.

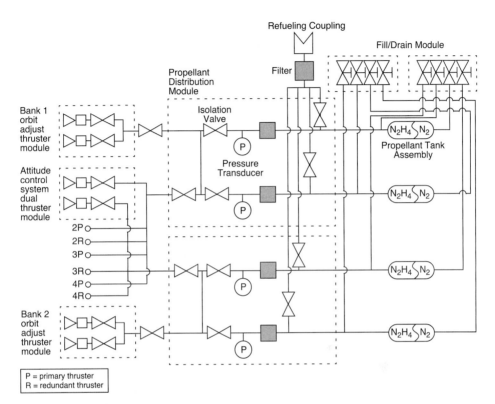

Fig. 15.4.   Schematic of Propulsion System for the Gamma Ray Observatory. Monopropellant hydrazine fuel with a blowdown fuel system going from a tank pressure of 400 to 100 psia. [Larson and Wertz, 1999]

Table 15.4 summarizes the characteristics of three S-band and two other communications subsystems, which are standard in NASA and many military programs. The systems for spacecraft-to-spacecraft communications require more operational control than S-band systems for Earth coverage because the narrow-beam antenna requires precise pointing for link lockup. Space-to-Space Laser Communications Links are becoming more common. These systems establish the link autonomously via a search mode until they locate the beam. The system then transitions from search mode to either passive or active tracking. Laser links provide the capability to transfer very large data streams without the frequency licensing constraints associated with traditional RF links. The major drawbacks of laser links are cost and added complexity.

Equipment for commanding and data handling decodes command information and routes the data to the proper subsystem for execution. Each

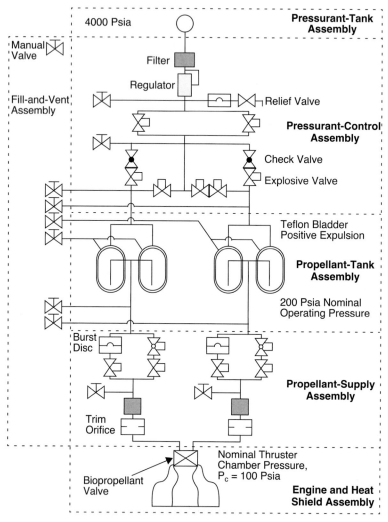

Fig. 15.5. Pressure-Fed Propulsion System Using Earth-Storable Bipropellant (N_2O_4/MMH). A bi-propellant fuel system maintains 200 psia operating pressure through regulated repressurizing of propellant tanks. [Larson and Wertz, 1999]

subsystem is interrogated for equipment status, and the data is arranged and stored in accessible formats, so the communications equipment can transmit it to the control station. Command data includes information on the action to be taken (e.g. turn off a unit, select a temperature range) and routing information. Figure 15.7 compares the command formats for two different spacecraft. Note the

Table 15.3. **What a Communications Subsystem Does.** The communications subsystem provides the ability to receive and track carrier signals, receive commands, and transmit telemetry data.

What a Communications Subsystem Does

• Carrier tracking
  • 2-way coherent communication (downlink frequency is a ratio of the uplink frequency)
  • 2-way noncoherent communication
  • 1-way communication

• Receiving and detecting commands
  • Acquire and track uplink carrier
  • Demodulate carrier and subcarrier
  • Derive bit timing and detect data bits
  • Resolve data-phase ambiguity if it exists
  • Forward command data, clock, and in-lock indicator to the subsystem for command and data handling

• Modulating and transmitting telemetry
  • Receive telemetry data streams from the subsystem for command and data handling or for data storage
  • Modulate downlink subcarrier and carrier with mission or science telemetry
  • Transmit composite signal to Earth or relay satellite

• Ranging
  • Detect and retransmit ranging pseudorandom code or ranging tone signals
  • Retransmit either phase coherently or noncoherently

• Operating subsystems
  • Receive commands from the subsystem for command and data handling (C&DH)
  • Provide health and status telemetry to the C&DH subsystem
  • Point any antenna requiring beam steering
  • Operate mission activities through command files stored in software
  • Autonomously select omni antenna when spacecraft attitude is lost
  • Autonomously detect faults and recover communications using command files stored in software

differences between real-time commands and commands to be loaded into onboard processors.

The telemetry data formatted for downlink transmission includes equipment status (on/off) and engineering parameters (voltage, current, temperatures, pointing information). An important operations consideration in the design is the frequency of sampling individual parameters for transmission. Telemetry-data formats provide for spacecraft data to be sampled at the mainframe rate or subcommutated for less frequent sampling. One bit allows us to determine data that represents equipment status (on/off), so a telemetry word can contain data on the status of several items. State-of-health data may require one or more telemetry words. Operations requirements on the type and sample frequency of data contribute heavily to the complexity and cost of spacecraft and ground-telemetry

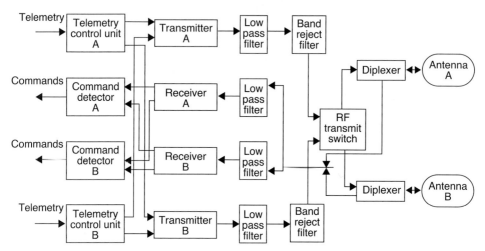

Fig. 15.6. Diagram of a Communications Subsystem. To complete a communications link, both receivers must be ON at all times. The path for the uplink signal depends on which receiver locks to it first. We may select a receiver by testing for receiver sensitivity to the uplink-signal strength and lowering the uplink transmitter's power to select the more sensitive of the two units. But because we want effective mission operations, we normally set the uplink power high enough to allow either receiver to lock up on the signal.

Table 15.4. Attributes of Some Common Communications Subsystems. Each system can support various modulation schemes (see Chap. 11). We use Earth-coverage antennas for normal operations and omnidirectional antennas for launch and contingency operations.

Application	Frequency		Modulation		Antenna Characteristics	Remarks
	U/L	D/L	U/L	D/L		
Space ground link subsystem (SGLS)	S-band 1.75 to 1.85 GHz	S-band 2.20 to 2.30 GHz	FSK AM PM	PCM PM FM	Earth coverage; omnidirectional coverage	SGLS standard
Goddard Spaceflight Tracking and Data Network (GSTDN)	S-band 2.02 to 2.12 GHz	S-band 2.20 to 2.30 GHz	PCM PSK FSK	PCM PSK PM	Earth coverage; omnidirectional coverage	GSTDN is slowly phasing out. The Deep Space Network is absorbing some of its assets.
Cross-link within constellation	W-band 60 GHz	W-band 60 GHz	Any	Any	Narrow beam 0.1 degree typical	Modulation, coding, and encryption can be customized

15.1 Spacecraft Performance and Analysis

Table 15.4. Attributes of Some Common Communications Subsystems. (Continued) Each system can support various modulation schemes (see Chap. 11). We use Earth-coverage antennas for normal operations and omnidirectional antennas for launch and contingency operations.

Appli-cation	Frequency		Modulation		Antenna Charac-teristics	Remarks
	U/L	D/L	U/L	D/L		
Cross-link to TDRSS	S-band K-band	S-band K-band	QPSK Spread spectrum	QPSK Spread spectrum	Narrow beam	TDRSS User Standard (See TDRSS Users' Guide)

Legend:
U/L = Uplink
D/L = Downlink
FSK = Frequency-shift keying
AM = Amplitude modulation
PM = Phase modulation
PCM = Pulse-code modulation
FM = Frequency modulation
PSK = Phase-shift keying
QPSK = Quadrature phase-shift keying

processing, so we must analyze them carefully early in a program to make good design trades.

Electrical-power subsystems usually use solar arrays that convert sunlight into electrical power through the photovoltaic process, combined with a source of stored energy to provide power when the spacecraft is in the Earth's or Moon's shadow. The stored energy typically comes from rechargeable batteries or fuel cells. The equipment in this subsystem also controls and distributes the electrical power throughout the vehicle. The solar arrays must face the Sun to provide power. On spin-stabilized spacecraft this usually means covering most of the spacecraft's body with solar panels to meet power requirements. Compared to spinners, arrays on articulated booms allow for a smaller total array area to meet requirements because we can point the arrays at the Sun. We operate electrical-power subsystems by pointing arrays and switching to stored power when needed.

While the basic elements of spacecraft in low-Earth and geosynchronous orbits are similar, their operations are different. For geosynchronous spacecraft, the batteries are required during the 45-day eclipse seasons around the times of the vernal and autumnal equinoxes. A major seasonal operations task is preparing the batteries for these periods. For spacecraft operating in low-Earth orbits, solar arrays go into shadow on every orbit, so operators must recondition batteries continually.

The FltSatCom spacecraft has two solar arrays that are clocked to point toward the Sun. They supply primary power through regulators to the main spacecraft bus and three nickel-cadmium batteries rated for 34 ampere-hours. The Total Ozone-Mapping Spectrometer—Earth Probe's (TOMS—EP) power subsystem uses two solar-array wings to supply power to the spacecraft and to recharge the battery so the spacecraft can operate during eclipse. By mounting solar cells on both faces of the wings, we can generate power throughout the orbit without moving the arrays. A "super" nickel-cadmium battery with 22 cells, rated for nine

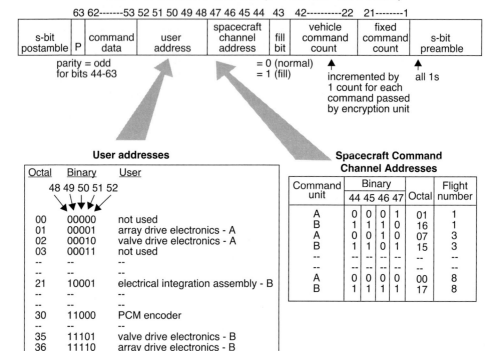

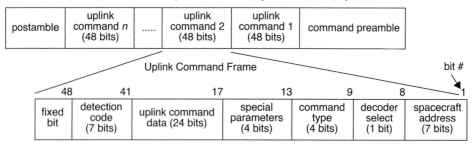

Fig. 15.7. **Command Data Formats for FltSatCom and the Total Ozone-Mapping Spectrometer—Earth Probe (TOMS-EP).** We first verify the content of command-format data during factory testing of the spacecraft. Operationally, a mnemonic or a command number usually identifies each command. [FLTSATCOM and TOMS, 1994]

ampere-hours, supplies stored energy. Solar-array regulators allow us to transfer power from the arrays to the main power bus in the power-control unit. For more on power systems and power regulation and control, see Larson and Wertz [1999].

We control the spacecraft's thermal environment passively, actively, or by combining both techniques. Passive techniques include using multilayer insulation (MLI) on the structural members, shades and baffles to protect equipment from direct sunlight, mirrored surfaces to reflect incident energy, and energy-absorbing or reflecting paint and coatings. Heat pipes use a liquid-vapor cycle to transmit heat to a radiator. For these techniques, design analysis determines the thermal control, so we don't need on-orbit control. Active thermal control uses commandable electric heaters, thermostatically controlled heaters, and in some cases, duty-cycled equipment.

The spacecraft structure is the load-carrying framework for attaching subsystem equipment. Structural mechanisms, such as booms, may deploy and fix in place throughout the mission. If we can retrieve the spacecraft for repair, refurbishment, or upgrading, we may make booms retractable and redeployable (e.g., the Gamma Ray Observatory). Other mechanisms include motors to rotate solar arrays, gimbals to move antennas, and Sun shades.

Describing how individual subsystems work doesn't give a complete picture of spacecraft operations. We must understand how maintaining power, attitude, and communication links affects all subsystems. Powering equipment on or off changes the temperature environment within the spacecraft. Operating heaters to supply heat lost when other units are off is part of many operations procedures. Attitude-sensor data that goes into the attitude processor affects how we operate subsystems for propulsion, electrical power, communications, heating, and even structures.

Suppose we're operating a three-axis-stabilized, geosynchronous spacecraft (orbits with an inclination near zero) within a small box defined by roughly a degree in longitude and a few degrees in latitude (or inclination). In this case, disturbances caused by solar torques and geomagnetic field irregularities make the spacecraft drift in longitude and latitude. When it reaches the edge of the box, we may need to activate the propulsion system to fire thrusters (in most coordinate systems, pitch or roll thrusters) that will counteract the spacecraft's motion. Usually, we may select a smaller, tighter deadband, which requires more thruster firings but gives us more accurate pointing. Interaction of Sun sensors and propulsion thrusters permits control of the spin speed and spacecraft nutation resulting from external torques. When we move spacecraft-body axes, communications coverage and link margin may be affected, and antennas mounted on articulated appendages will require repositioning. For communication subsystems with narrow beam coverage on downlink antennas, this control is essential to mission success. Table 15.5 describes some typical interactions between the communications subsystem and other subsystems.

Table 15.5. Interactions of Spacecraft Subsystems with a Communications Subsystem. This table shows interactions and their effect on normal operations.

Subsystem	Interaction	Operations Impact
Attitude Determination and Control	Spacecraft pointing and attitude knowledge for fixed antennas	Link losses may require tighter attitude control by reducing deadband
Thermal	Frequency shifts may occur on non-oven controlled equipment	Use heater control to stabilize temperatures
Electrical Power	Power required to operate communications equipment	Manage power needed by pointing arrays; manage battery conditioning; possibly duty-cycle equipment
Structures	Clear field of view for antennas	Plan for obstructions to field of view for gimbaled antennas during all contact periods
Command and Data Handling	Onboard command routing affected by link errors	Verify commands by telemetry response—load critical commands into onboard processors for later execution

15.2 How We Carry Out Typical Spacecraft Operations

Spacecraft operations has been described as long periods of boredom punctuated by moments of panic. Successful operations extend the long periods and eliminate the panics. This means all the equipment is operating normally and careful planning is in place to anticipate approaching events. In fact, spacecraft operations can be thought of as a series of tasks that prepare the vehicle to respond to planned external requirements. These requirements could arise from orbital conditions (spacecraft eclipsed by the Earth's shadow); changes in mission that require us to alter the spacecraft's position or attitude (a delta velocity maneuver to change in-plane orbital location); or a need to manage equipment because the spacecraft is aging (changing thruster-firing durations over the life of a blow-down propulsion system).

Such tasks are common to all spacecraft. The challenge is to do them within the constraints of the mission requirements and the ground-station environment. Doing routine operations tasks requires careful, detailed planning to make sure the spacecraft acts on the correct command data. Commands must certainly be valid, but if operators have to monitor and verify telemetry, they also have to plan carefully in order to know when they can halt an action or abort a procedure without harming the spacecraft. Table 15.6 lists some typical spacecraft activities and their effects on mission operations.

An example of a medium-complexity task is battery reconditioning, which places the spacecraft's stored-energy system in readiness to supply power during

15.2   Spacecraft Performance and Analysis

Table 15.6. Spacecraft Operations. The operating tasks listed here are the same ones discussed in Sec. 15.1—maintain communications, power, pointing, and thermal control. The "actions to take" column includes both routine responses to the tasks and first-order responses to anomalies. We've rated the complexity of the tasks from low to high based on normal operations and the need to respond in case of abnormal operations.

Operation Task	Operator Action	Data Collection and Trending	What to Look For	Action to Take	Complexity
Transmit commands to spacecraft	• Monitor data receipt through telemetry response	• Correct number of commands transmitted	• Correct telemetry response, if applicable, for transmitted commands	• Retransmit command • Switch to redundant equipment	Low Low
Receive telemetered data from spacecraft	• Monitor data and compare to expected values • Monitor against alarm limits	• Plot data for diurnal variations • Plot data for seasonal conditions • Compare spacecraft configuration status to expected	• Out of tolerance conditions • Onboard equipment failures • Trends in data that could lead to anomalous conditions	• Archive data • Modify telemetry limit checking for aging/seasonal conditions • Determine if onboard anomaly has occurred	Low Low Low to high
Maintain power	• Monitor output of solar array • Monitor all bus voltages, currents	• Beginning-of-life output vs. predicted end-of-life • Spacecraft configuration status	• Aging due to UV exposure • Difference in configuration • Equipment failures	• Possibly duty-cycle equipment • Command redundant equipment on-line	Medium to high Medium to high
Prepare batteries for eclipse season	• Recondition batteries	• Battery voltage, current, temperatures	• Battery discharge—recharge performance	• Command battery to discharge—recharge	Medium
Maintain power during eclipses	• Monitor battery performance	• Beginning-of-life performance vs. lifetime	• Cell failure • Cell memory	• Actively control battery	High
Maintain operating temperature	• Monitor temperature sensors on equipment or compartments	• Diurnal variations • Seasonal variations • Equipment on-off status	• Degraded passive thermal control • Failure of thermostat • Failure of equipment unit • Unpredicted change in spacecraft pointing relative to Sun angle	• Turn equipment on/off	Low to medium

Table 15.6. **Spacecraft Operations. (Continued)** The operating tasks listed here are the same ones discussed in Sec. 15.1—maintain communications, power, pointing, and thermal control. The "actions to take" column includes both routine responses to the tasks and first-order responses to anomalies. We've rated the complexity of the tasks from low to high based on normal operations and the need to respond in case of abnormal operations.

Operation Task	Operator Action	Data Collection and Trending	What to Look For	Action to Take	Complexity
Maintain pointing	• Monitor Sun, Earth, star sensor data • Monitor wheel speed	• Pointing x, y, z • Wheel speed variations • Sun's or Moon's intrusion into the Earth sensor's field-of-view	• Alarms • Excessive thruster firings	• Switch to redundant sensors • Switch to redundant wheels • Switch to scanning by redundant Earth sensor	Medium Medium Medium
Spacecraft repositioning	• Command thrusters to start and end delta-velocity maneuvers (ΔV)	• Thruster firing duration • ΔV start and stop times • Thruster performance • Fuel use	• Correct spacecraft repositioning • Proper thruster performance	• Determine position from ranging data or GPS data • Archive performance data	Low to medium Low
Load onboard computer	• Prepare command file • Transmit load commands • Verify correct load	• Old and new contents of memory	• Errors in onboard load	• Verify load commands • Retransmit required commands	Low Low to medium
Move appendages	• Reset solar arrays to Sun line • Point antennas	• Output of solar array • Strength of receiver signal • Bit error rate	• Increased output • Increased strength • Reduced rate	• Calculate Sun angle on array-reposition • Switch to redundant array drive • Recalculate and repoint	Medium Medium Medium

periods of shadowing by the Earth. For spacecraft operating at geosynchronous altitude, eclipses occur around the spring and fall equinoxes. At these times the Earth's equator lies along the ecliptic, so spacecraft operating with very low inclinations are in the Earth's shadow for part of the orbit. We determine the beginning of the eclipse period by analyzing the spacecraft's orbit. The eclipse seasons start with partially shadowed orbits (penumbral eclipses), move to fully shadowed orbits (full eclipses), and end with more penumbral eclipses.

For geosynchronous spacecraft, the two eclipse seasons last approximately 45 days each—centered on the equinox. Therefore, the battery system should be fully reconditioned 22–23 days before equinox. First, we determine the period of shadowing by analyzing the ephemeris data. Then we do detailed scheduling of spacecraft contacts required to command and monitor the spacecraft and finish by analyzing the battery's state before the eclipse season begins. Figure 15.8 shows typical battery discharge/recharge performance. Table 15.7 shows a procedure for reconditioning one of the batteries on a geosynchronous communications satellite.

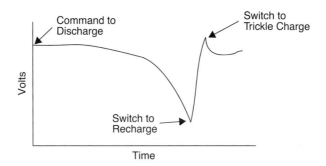

Fig. 15.8. Battery Reconditioning Performance. Voltage, current, and temperature are monitored as a normal state-of-health function. We then command the battery to DISCHARGE by switching the reconditioning circuit into the system. The battery voltage falls to a predetermined level and is switched to RECHARGE.

Battery reconditioning is just one example of the tasks people in spacecraft planning and analysis must do. Table 15.6 lists several other typical tasks, but this list is certainly not complete. The MOM must develop a complete list of spacecraft-bus operations, create a procedure for doing each task, identify the information required to complete the task, and establish a way to verify procedures. The MOM must also decide if each task will be automated on the spacecraft, automated on the groundstation, or done by operators.

Another challenge of cost effective space operations deals with managing operations staffing levels over the life of the mission. For example, in the case of a Mars scout mission, there is a considerable amount of down time (on the order of several months) while the space vehicle is traveling between the Earth and Mars. The operations staff requirements grow leading up to launch and early on-orbit checkout, but then severely decrease during the vehicle's trip between planets. Operations during the trip include monitoring telemetry and performing routine housekeeping functions. As the vehicle nears Mars, operational staffing requirements increase in a step function as the vehicle maneuvers into orbit around the planet and begins its scouting mission. The MOM is responsible for developing a training program that ensures adequate operations personnel with the proper skills are available at the correct times to support mission requirements. The MOM must also consider the unplanned departure of personnel and response to anomalies during the interplanetary transit period of the mission.

Table 15.7. **Reconditioning Procedure for the FltSatCom Battery.** The time required to recondition one battery with this procedure is over 90 hours, and this spacecraft has three batteries. This means reconditioning operations must start more than 11 days before the first eclipse (no later than mid-February for the vernal-eclipse season). [FLTSATCOM, 1986]

Command Transmitted	Operations Action
	Verify battery is on charge channel
	Configure for battery discharge
Select channel automode	Verify charge channel to automode
	Verify charge channel in trickle charge (initially charger may be in full charge for a short period before switching to trickle)
Disconnect channel battery charger	Verify channel battery charger is disconnected
Connect redundant AC source to battery	
Disconnect redundant AC source	
Start reconditioning discharge with auto stop	Begin reconditioning
	Verify the channel battery current shows discharge
	Monitor battery voltage, current, and temperature until discharge ends automatically (about 75 hours)
	Verify recondition discharge status; wait 30 minutes. (If reconditioning discharge hasn't stopped and the battery voltage is less than 16 volts, transmit STOP RECONDITION DISCHARGE and verify the telemetry parameters.)
Switch channel to trickle charge	Verify that the channel is on trickle charge
	Verify the battery-current monitor shows approximately 0.24 A charge; wait ten minutes
Select automode for channel	Verify the channel is in automode
	Verify the channel is on full charge
	Verify the battery-current monitor shows full charge—about 2.1 A
	Monitor battery voltage and charge current to make sure it has automatically switched to trickle charge—about 18 hours

References

FLTSATCOM Orbital Operations Handbook. Vols. 1 and 2.

Orbital Requirements Document Fleet Satellite Communications Program (FLTSATCOM). Aerospace Report No. TOR-0076 (6724-01)-2. Reissue B, 1*, 15 October 1986.

Larson, W. J. and J. R. Wertz. 1999. *Space Mission Analysis and Design*. Third Edition. Netherlands: Kluwer Academic Publishers.

Total Ozone Mapping Spectrometer (TOMS) Earth Probe (EP) Orbital Operations Handbook. 1994.

Chapter 16

Spacecraft Failures and Anomalies

David E. Kaslow, *Lockheed Martin Corporation*
Emery Reeves, *Aerospace Consultant*

> 16.1 Spacecraft Failures and Anomalies
> 16.2 Defining Anomalies
> 16.3 Resolving Anomalies
> 16.4 Planning for Anomalies
> 16.5 Case Studies

16.1 Spacecraft Failures and Anomalies

Understanding spacecraft failures and anomalies, that they do happen, that some can be resolved and that others cannot, gives the MOM a good understanding and background that will underscore the importance of preparing pre-flight for the detection and resolution of failures and anomalies. This chapter is divided into two sections. The first section gives a summary of the categories of failures and anomalies. It also includes examples of missions and brief summaries of the mission problems to illustrate the type of failure/anomaly under discussion. In addition we discuss factors in mission malfunctions and factors in mission success. The second section deals with anomaly resolution. The section on anomaly resolution was Chap. 16 in its entirety in the first edition. By adding the first section on spacecraft failures and anomalies, we are emphasizing the importance of viable and robust anomaly resolution processes, procedures, and training.

Designing, launching, and operating a spacecraft is a risky endeavor. Many things can go wrong, ranging from a degraded mission to a complete failure. The spacecraft could be destroyed in a launch failure. Images could be degraded because of a flaw in a mirror or the loss of pixels in a charge-coupled device.

Mission life could be shortened due to the over-use of propellant or to under-performing solar cells.

We've made many improvements in design, production, and operations over five decades of space mission operations. Software design, hardware design, and production processes have been expanded and refined. Component and system modeling techniques are more comprehensive and accurate. Materials are stronger and lighter. Standards, such as ISO-9001 and CMI Level 5 that provide for higher-quality software products, have been developed and are being implemented in the aerospace industry.

It would seem that as design and production methods increase in maturity, mission degradations and failures should gradually fade away. But more sophisticated design and production methods give rise to more complex components, subsystems, and systems. The opportunity for degradation or failure still exists and may actually be higher.

It might also be thought that as the aerospace industry matures, degradation and failures should occur less often since the knowledge of how to build a good system should be increasing. However, processes and procedures are just documentation: they are not knowledge. Concepts, requirements, and design specifications are not knowledge; they are only documentation of knowledge. The knowledge exists within the engineers, designers, developers, production personnel, testers, maintainers, and operators. The knowledge does not exist within a company; it exists within the people within the company. Consequently, it must be passed from one generation to the next through teaching and mentoring.

Table 16.1 is a listing of some of the categories of the root causes of degradations and failures. Table 16.2 shows some typical responses to launch failures. Table 16.3 lists some of the factors that contribute to component, subsystem, and system design and operations being resistant to degradations and failures.

The information in these three tables would appear to be merely common sense. We all know, generally, how rockets and spacecraft can degrade or fail. And we all should know, in general, how to build robust systems. But knowing in general is not good enough. Proposals and Program Management Plans must commit in direction, schedule, and funding to developing and operating a resilient system. Risk analysis, as well as failure modes analysis, must actively address degradation and failure at the component, subsystem, and system level. We must design and build with suitable emphasis on accommodating degradation and failure. System testing and demonstrations must prove system resiliency.

The items in Tables 16.1, 16.2, and 16.3 are intended as an initial checklist for investigating system weakness and for assessing system robustness. This initial checklist should be developed into a checklist specific to the system we are building. The checklist is first applied during proposal preparation to determine the level of resiliency required. This determination is based on customer requirements and cost-benefit trades. This checklist is not a "once and done" item. We should periodically review and refine it as the details of the system design are developed. Conversely,

16.1 Spacecraft Failures and Anomalies

Table 16.1. Categories of Root Causes of Degradations and Failures. We analyze space and ground components for points of potential degradation and failure.

Factors in Mission Malfunctions
Ground Testing Damages Component
On-Orbit Testing Damages Component
Flawed Ground Testing
Inadequate Ground Testing
Flawed Design
Flawed Design Upgrade
Improper Design Reuse
Environment
Launch—Hardware and Software
On-Orbit—Hardware and Software
Inattention to Events
Single Event Upset

Table 16.2. Typical Responses to Launch Failures. We evaluate candidate responses to potential launch failures.

Responses to Launch Failures
Allocate Mission Objective to Another Program/Spacecraft
Recover, Return, Refurbish, and Launch
Establish a Secondary Mission
Execute Recovery Orbital Maneuvers

Table 16.3. Factors that Contribute to Design and Operations that are Resilient to Degradations and Failures. We employ processes that discover and mitigate potential component degradation and failure.

Factors in Mission Success
Lessons-Learned Process
Independent Review Process
Risk Analysis
Availability Analysis
Failure Mode Analysis
Recovery Mechanisms
Anomaly Response

we should periodically review the system design against the checklist and refine it, if necessary, to provide the desired level of resiliency.

We will examine specific occurrences of degradation, failure, and recovery. They are presented to reinforce the notion that anything can go wrong in a space mission, but a robust design and ingenious work-arounds can often save it.

16.1.1 Factors in Mission Malfunctions

Ground Testing Damages Component

One of the last phases of spacecraft development includes ground testing for correct operations within specified limits and the vacuum, thermal, and vibration testing. There is always the possibility that a spacecraft component will fail in a manner that results in additional component or subsystem damage. There are also instances of test equipment malfunctions or improper operations resulting in damage to the spacecraft.

The solar arrays on the High Energy Solar Spectroscopic Imager (HESSI) satellite were damaged during vibration testing in 2000. A piece of the test equipment was stuck and then broke free, subjecting HESSI to a 20g force rather than 2g's. A careful inspection before the vibration test would have uncovered the fault in the test equipment.

On-Orbit Testing Damages Component

The last phase of spacecraft delivery before turnover to operations is the on-orbit check. This on-orbit testing itself can result in damaged components.

The AMSC-1 communication satellite experienced a power problem on one antenna in 1996 but the satellite was reconfigured to overcome the problem. The problem resulted from damage that occurred during on-orbit testing when a high power level on a signal caused a power amplifier to burn out.

Flawed Ground Testing

Sometimes the test equipment can be flawed, resulting in a design defect going unnoticed.

A well known example is the primary mirror on the Hubble Space Telescope. A flaw in the equipment used to measure the curvature of the mirror was such that it compensated for the flaw in the mirror. The incorrect curvature was not discovered until the telescope was on-orbit. Corrective optics were developed and installed on the telescope, assuring the on-going success of the telescope.

Inadequate Ground Testing

Limited funding and a tight schedule can result in a missed opportunity during ground testing to uncover a catastrophic failure mode.

The Mars Polar Lander was launched in January 1999 and arrived at Mars in December 1999. Controllers were unable to establish contact with the spacecraft after initiating the landing sequence. The mission was declared a failure after two months of searching for the Lander using the Mars Global Surveyor and listening for a signal using Deep Space Network. The failure inquiry concluded that the most likely cause of the failure was a premature landing engine shutdown due to a false signal that indicated the Lander's legs had touched ground. Experiments with the Mars 2001 Lander revealed that the force of the deployment of the legs prior to descent engine firing could prematurely set the touchdown flag. The software would have checked this flag when the Lander was 40 meters above the ground and would have shut the engines down if the flag was set. This would have resulted in a free-fall touchdown of about 15 meters per second instead of the desired 2.4 meters per second. This sequence of events existed during the testing prior to launch, but was not uncovered due to the interruption of a test to rewire a switch and insufficient time to restart the test from the beginning.

Flawed Design

Faulty solar array design has resulted in problems or disaster on several spacecraft.

The original solar arrays on the Hubble Space Telescope expanded and contracted due to heating and cooling when the telescope passed into and out of the Earth's shadow, causing a vibration in the telescope and a blurring of the images. The work-around was to cease imaging for a short time after entering and exiting the Earth's shadow to allow the vibrations to dampen. The solar arrays were replaced on the first servicing mission in 1993.

The solar arrays on Advanced Earth Observing Satellite-1 (ADEOS-1) experienced a catastrophic failure in 1997. The failure investigation concluded that the heating and expansion of the solar arrays encountered during the orbital placement weakened the solar array structures. Subsequent vibrations led to further weakening and finally failure.

The solar arrays on the EchoStar 4 communication satellite did not fully deploy after launch in May 1998. The substance used to prevent the solar arrays from rubbing together in the folded position melted, causing the arrays to stick together. The melting occurred during the sunlit orbital transfer to the geostationary position.

Flawed Design Upgrade

There is a lot of wisdom in the phrase, "The enemy of good enough is better." Upgrades to components can be well-intentioned, but ill-advised in retrospect, with flaws uncovered after many years in orbit.

Boeing added reflectors on the solar arrays to increase the power on their BSS 702 bus. However, several satellites launched from December 1999 through May 2002 experienced small and gradual power losses due to the reflectors degrading in orbit. As of late 2002 the cause of the degradation had not been discovered.

Subsequent BSS 702-based spacecraft were built without the reflectors but with additional solar arrays and other modifications to increase the power level.

Improper Design Reuse

It may be thought that design flaws are most prevalent in new components. However, a component used successfully in one mission can fail in another.

The Lewis satellite started to spin slowly a few days after its launch in 1997. The spacecraft had been out of contact with the ground for over six hours before the spin was discovered. Engineers shut down non-critical spacecraft functions but were unable to reestablish contact and recover the satellite. The spin resulted in insufficient solar illumination and a critical loss of power. The review panel found that a poorly designed control system, inadequate testing, and insufficient operator monitoring led to the Lewis failure. The control system had been reused from the Total Ozone Mapping Spectrometer - Earth Probe (TOMS-EP), but operating conditions were not the same.

The Mars Observer, launched in September 1992, experienced a catastrophic loss before arrival at Mars. The investigation board concluded that the most probable cause was a rupture in the pressurization side of the propulsion module. The board also commented that there was too much dependence on hardware, software and procedures developed for near-Earth missions.

The inaugural flight of an Ariane 5 rocket with four Cluster scientific satellites in June 1996 ended with the destruction of the rocket and satellites. There were two Inertial Reference Systems (SRIs) units operating in parallel. These units were based on the design used for the smaller Ariane 4. The primary SRI unit detected a large horizontal movement and generated an output data value that exceeded the allowable range of the parameter. This was interpreted as an error by the SRI software, triggering a shutdown. Control was then handed over to the second SRI unit. The second SRI unit encountered the same scenario and also shut down. The fundamental error was not modifying the software to accommodate the larger horizontal movements of the Ariane 5. The Cluster mission was resurrected with a combination of newly built components and components left over from development of the first Cluster.

Environment

Spacecraft components are designed to survive on-orbit radiation, solar flares, and temperatures for the specified spacecraft design life. However, design failures can occur even after many decades of experience in developing space-certified components.

The Tempo 2 communications satellite experienced a power loss in April 1997 during a period of high solar activity. The cause of the electrical shorts was related to the high voltage differences between adjacent solar cells, causing sparking from the highly charged environment during a solar storm.

The GOES-8 sounder and the GOES-9 imager experienced failed backup motors. The motors in the three remaining satellites were redesigned after ground tests showed that the motors are likely to fail under extreme temperature changes. One of the proposed modifications was to orient GOES so that the sounders and imagers are pointed away from the Sun for portions of the day. GOES would not be imaging for a short time, but the heat stress would be reduced.

Occasionally GOES-8 has an electrostatic discharge that interrupts operations for several hours. This is caused by temperature differences between components creating an electrostatic charge build-up. GOES-9 has additional insulation to prevent this problem.

Launch—Hardware and Software

Launch is an extremely risky event because there are many opportunities for problems to arise with very little time for corrective actions. All aspects of launch are equally at risk, from the booster liftoff to stage separations to payload deployment. We list several examples of launch failure to emphasize that every detail must be checked and rechecked, from hardware design, materials, assembly, and preparation to software design and initialization.

The launch of the Express A1 communications satellite on a Proton in October 1999 ended with the explosion of the second stage. The second stage engine was manufactured in 1993 following a nine month period of inactivity. An investigation determined that the accident was likely caused by asbestos fabric left in a second stage fuel line.

The third stage engine on the Soyuz launch of two Cluster scientific satellites in July 2000 ran out of fuel three seconds early due to an error in installing a fuel system component. However the upper stage compensated by changing its thrust profile.

The inaugural Delta 3 flight in August 1998 ended with the destruction of the rocket and the Galaxy 10 communication satellite. The failure was caused by the attempt to control a natural rocking motion of the booster, which used up all the hydraulic control fluid. As a result, the rocket was unable to react to a wind shear. The self-destruct mechanism was activated as the rocket broke up from high aerodynamic forces. The software was subsequently modified to ignore those natural oscillations.

The Apstar 2 communication satellite was destroyed in a January 1995 Long March launch when wind shear tore apart the first and second stages due to a failure of the payload fairing.

The April 1999 launch of the DSP-19 geostationary military early warning satellite on a Titan 4 ended with the satellite in a highly elliptical orbit due to an improper stage separation. An electrical connector between the stages failed to disengage because thermal tape wrapped around the connector disabled the separation mechanism and because of connector misalignment due to tolerance buildup. The details of the separation mechanism that could have alerted engineers to the potential danger of adding thermal wrap around the connector were not included in the electrical connector specification, dating back as far as 1978.

A Pegasus launch of High Energy Transit Experiment (HETE) and Satelite de Aplicaciones Cientificas-B (SAC-B) in November 1996 ended in failure when the two satellites failed to separate due to a malfunction in a power unit associated with the jettison pyrotechnics. It was determined that the shock of the third stage separation damaged a power transfer switch.

An Ikonos Earth imaging satellite launched on an Athena 2 rocket in April 1999 did not achieve orbit because the payload shroud did not separate properly. The extra weight resulted in insufficient velocity to achieve orbit. A review of the shroud separation design revealed that the first two ordnance charges that separated the shroud at the circumference momentarily disconnected the circuit responsible for the firing of the ordnance that separates the two clamshells.

On-Orbit—Hardware and Software

Many things can go wrong on-orbit. Some problems arise from premature equipment failures, unanticipated situations, or errors in commanding. Other problems occur when the satellite is operated beyond its design lifetime. The equipment wears out because of excessive operating cycles or degrades due to long term exposure to the space environment.

The Far Ultraviolet Spectroscopic Explorer (FUSE) was launched in June 1999. Its mission is to examine the origin and evolution of stars and galaxies. FUSE has four reaction wheels, with one being a backup, to control spacecraft attitude. Two reaction wheels failed and FUSE entered a safe mode on December 10, 2001. Procedures were developed to use the interaction of the magnetic torquer bars and the Earth's magnetic field to control the spacecraft attitude. The torquer's magnetic field was controlled by the direction and amount of current flowing through the torquer bars. Attitude control was reestablished on January 24, 2002.

Landsat 5, launched in March 1984, operated well beyond its five-year design lifetime. The bumper springs for the scanning mirror in the Thematic Mapper eventually fatigued, causing the mirror's movement to be out of synchronization with the calibration shutter. This resulted in tracks across the images. The thematic mapper was switched to its backup mode of operations, where the scanning mirror could be controlled to eliminate the tracks.

The Solar and Heliospheric Observatory (SOHO), launched on December 1995, is a joint NASA and ESA project to study the Sun from an orbit about the Lagrangian L1 point.

Contact was lost on June 24, 1998 when SOHO went into its Emergency Sun Reacquisition (ESR) mode during momentum management maneuvering. Contact was eventually re-established on August 3. Procedures were developed to gradually recharge the batteries and thaw the hydrazine fuel. It was then discovered that one of three gyros had failed due to a broken wire and another gyro was operating only intermittently. Operations were eventually recovered and all twelve instruments were recommissioned by the end of October.

However, the last of the gyroscopes failed on December 21, resulting in SOHO expending about 7 kilograms of hydrazine per week attempting to maintain a sun pointing attitude. SOHO started with 180 kilograms of hydrazine, which would have lasted 20-25 years under normal operations. Software changes for gyro-less operations were developed and SOHO successfully resumed science operations on February 2, 1999.

The investigation revealed three errors: a) a preprogrammed command sequence lacked a command that activates a gyro needed in ESR mode; b) another preprogrammed command sequence resulted in an incorrect reading from a gyroscope; and c) a command was sent that turned off a gyro in response to unexpected telemetry values. The special investigation board concluded that the loss of contact with SOHO occurred because of ground operations team errors. The board recommended that NASA and ESA conduct follow-on reviews in four areas: 1) operational change procedures; 2) flight operations and decision authorities; 3) division of responsibilities between NASA and ESA, including staffing levels; and 4) planning and verification of operating sequences, including use of the SOHO simulator.

Inattention to Events

The effects of some malfunctions are present before mission failure but are not discovered due to inattention to events.

A Titan 4-B/Centaur launch in April 1999 ended with the Milstar Block 2 satellite in an ellipsoidal orbit rather than a geostationary orbit. The roll rate constant loaded in the Centaur software was one tenth of its correct value. This resulted in incorrect attitude calculations which, in turn, resulted in multiple attitude control firings and a depletion of propellant. There were two opportunities to detect the error prior to launch. A week before launch, questions regarding a too low rate filter were not followed up. After tower roll-back, attitude rates that were not reflecting the Earth's rotation and the effects of the wind were not monitored.

The Mars Climate Orbiter was launched in December 1998 and arrived at Mars in September 1999 with a closest approach of approximately 60 km with the lowest survivable closest approach being 85 km. The spacecraft was to have a closest approach of 140–150 km. Contact could not be established after the orbital insertion burn. The thruster calibration table contained data in units of pound-foot-second instead of Newton-second. There were signs of the error even while the Climate Orbiter was enroute to Mars. The findings of the failure assessment board included:

1. The error could have been detected in ground based modeling of thruster firings.

2. The navigation team did not have detailed attitude data and did not receive an independent expert review.
3. There was inadequate training on navigation characteristics and anomaly reporting.
4. System engineering functions of tracking and checking mission performance were not robust enough.

Single Event Upset

Sometimes components are recovered after an unexplained, one time malfunction, yet continue to perform flawlessly.

The Mars Odyssey spacecraft, launched in April 2001, has a Martian Radiation Environment Experiment (Marie) payload to determine the radiation risk for humans making a journey to Mars. Marie was turned off by operators in August 2001 following a failure to execute a commanded downlink of data and was reactivated in March 2002. It was determined that the original malfunction was a single event upset and not an identifiable malfunction.

16.1.2 Responses to Failures

Allocate Mission Objectives to Another Program/Spacecraft

Some mission objectives are important enough that considerable effort and funds must be expended following the loss of a mission payload to ensure that the mission objective is still met. The mapping of ozone in the Earth's atmosphere is one such mission. Scientists want a continual set of measurements, and in addition, the overlap of old and new payloads provides for the best calibration.

The mapping of ozone began with the first Total Ozone Mapping Spectrometer (TOMS) payload, which was on Nimbus-7 launched in October 1978 and operated until May 1993. A second TOMS instrument on Meteor-3, launched in August 1991, provided data until December 1994. Advanced Earth Observing Satellite (ADEOS), launched in August 1996, had a TOMS payload but the satellite experienced a catastrophic failure in June 1997.

TOMS-Earth Probe was launched July 1997 and placed into a 500-km altitude orbit, rather than the planned 950-km orbit, to increase the payload resolution. TOMS-EP was moved to a 740-km orbit to make up for the loss of coverage of ADEOS. The higher orbit increased the extent of the coverage. It also reduced the drag, reducing the frequency of orbit maintenance thruster firings, thereby increasing the on-orbit lifetime.

Meteor 3M was to have been launched in 2000 with a TOMS instrument. However, the launch was delayed, ultimately until late 2001, due to lack of funds. The uncertainty of the Meteor 3M launch forced NASA to allocate the TOMS payload to another satellite. A contract was awarded in July 1999 for the development and launch of QuikTOMS by August 2000. Unfortunately, the launch of the Taurus rocket with QuikTOMS and Earth imaging Orbview-4 in September 2001 malfunctioned and the satellites did not achieve orbit.

Recover, Return, Refurbish, and Launch

The only means of recovering a satellite is to use the Space Shuttle, as long as the satellite orbital altitude and inclination are within shuttle range. At one time, that was an acceptable mission for the space shuttle, but no longer.

Westar VI and Palapa B2 communication satellites were launched from the space shuttle in February 1984 but were stranded in a too-low orbit due to the failure of an upper stage motor. They were retrieved during a Space Shuttle flight in November 1984. Westar VI was re-launched as AsiaSat 1 in April 1990 on a Long March rocket. Palapa B2 was re-launched as Palapa B2R in April 1990 on a Delta rocket.

Establish a Secondary Mission

Some satellites can be used even after the primary mission has ended or failed.

Engineering Test Satellite-6 (ETS-6) was launched in August 1994 with a mission to verify the design of a three-axis-stabilized geostationary bus and to test advanced communication equipment. However, the apogee kick motor failed, leaving ETS-6 in a low-Earth orbit. ETS-6 was subsequently used to demonstrate laser transmission from an observatory in California to the satellite and continues to serve in space radiation studies.

Landsat 4 Earth imaging satellite was launched in July 1982 and Landsat 5 was launched in March 1984. Landsat 4 continued mission operations well beyond its two year design lifetime, until a data transmitter failed in August 1993. It then served as a test bed for Landsat 5 software modifications.

Wide-Field Infra Red Explorer (WIRE) was launched in March 1999. The primary telescope cover was released three days early due to an error in a computer chip, resulting in sunlight falling on the cryostat, causing the hydrogen to warm up and vent into space. The entire supply of frozen hydrogen evaporated, ending WIRE's primary scientific mission. A secondary mission was defined to use the onboard star tracker to study bright stars for oscillations that reveal their inner structures and for occultations that indicate the presence of planets.

Execute Recovery Orbital Maneuvers

A rocket motor could under-perform, leaving the satellite in an incorrect orbit, but the mission can be recovered using secondary propulsion or other creative measures.

Artemis and BSAT-2b communication satellites were colaunched in July 2001 but ended up in an incorrect orbit due to a failure in the Ariane 5 upper stage. The launch failure investigation reported that improper mixing of the two propellants in the upper stage engine was one of the causes of the thrust shortfall. Onboard software modifications subsequently corrected the mixing process. Artemis was moved to a geostationary orbit using its two ion thrusters, each of which produced 15-millinewtons of thrust with a resulting speed of one kilometer per hour. BSAT-2b was declared a loss since any maneuver to a geostationary orbit would have left insufficient fuel for station-keeping.

The AsiaSat 3 communication satellite, launched in December 1999, failed to reach orbit when the fourth stage engine shut down too early due to the defective wear-resistant coating on the internal seals causing a leak. The satellite was eventually placed in a low inclination orbit by executing several lunar flybys and then renamed HGS-1.

The TDRS 9 communication satellite was launched in April 2002, but the flight from parking to geostationary orbit was placed on hold due to a pressurization problem in one of the four hydrazine propellant tanks. TDRS 9 was then moved to geostationary position after engineers devised a way to configure the connections within the propulsion system to use helium from an exhausted propellant tank to pressurize the inoperative propellant tank. NASA will not take delivery of a TDRS satellite until it is operational at full capability and with the proper fuel reserves.

16.1.3 Factors in Mission Success

Lessons-Learned Process

The preceding examples show that there are many lessons to be learned when developing a space system. We are very fortunate if we can learn from someone else's developments rather than from our own. Almost all programs at some point will go through a lessons-learned process. This process analyzes the problems that have occurred and recommends changes to methodology to mitigate or avoid those problems in future programs. Sometimes a list of lessons learned appears in a proposal to give the customer confidence (unfounded as it might be) that the proposed development will benefit from the problems of others. Sometimes the lessons-learned process is applied at the end of a program so future programs will reap the benefits. More often, the lessons-learned process is applied during a program when significant difficulties have occurred. In that case, it helps us discover why the problems occurred and to determine corrective actions, as well as recommendations, for avoidance or mitigation in future programs.

These applications of lessons learned are beneficial but are limited in effectiveness as they are often done once, usually in a reactive situation, and then set aside. A more robust application of lessons learned is to apply them proactively. One approach is to make them part of the on-going risk management process. Lessons learned should be culled from programs within the organization, and to the extent possible, from external sources. The lessons learned should be categorized as to type and program phase. We should decide how to apply and monitor that lesson learned or why that lesson learned does not apply. This approach seems very easy to implement but it could fall by the wayside during proposal writing and negotiations if the customer does not want to pay for something that does not directly produce a product. It could also be neglected during development and production where there is limited time and funding to learn from the problems that have already occurred.

The important message from this chapter is to review and use the information that other missions have learned and documented! All too often the pressures of project schedules are such that this review and understanding of applicable lessons learned is omitted. The mission operations manager is in a particularly good position to review and bring to the attention of project management applicable lessons learned during the conceptual and preliminary design phase of the mission.

Independent Review Process

All of the standard approaches to designing and developing software and hardware have a number of internal team and customer reviews. The internal team reviews have their pluses and minuses. A plus is that the internal team reviewers are well acquainted with the details of the design. A minus is that reviewers need to take time away from their own tasks in order to review the material in sufficient detail to provide meaningful comments.

The customer reviews are usually at a high level and are structured to convince the customer that the design is mature and everything is on schedule. You don't get bonus points for discussing the weaknesses or incomplete parts of the design. Additionally, the customer's technical background may not be strong enough to probe for weaknesses in the design. Some programs permit customer representatives to attend internal reviews. The benefit to this is that the customer has an opportunity to review the lower level design. The caution is that the presence of the customer may put a damper on the free flow of questions and answers.

These issues with team and customer reviews have always existed and will never go away. But a good program must minimize the factors that might degrade the effectiveness of the reviews. The review process must be built into the program development plan and must be allocated sufficient time and funds. It should be monitored to assure its effective execution.

These review processes should be supplemented with an independent review panel made up of experts in the applicable technologies. The obvious benefit is that the review panel has its own specialized knowledge and lessons learned. More importantly, they have viewpoints unencumbered by the politics of the program and can openly challenge requirements and design principles. The panel should not only assess the design, but also the process used to produce the design.

The use of an independent review team should be established at program inception to ensure that the reviews are properly funded and planned. The team should initially meet with program and technical management to establish the review methodologies, topics, materials, and schedules. In addition to the usual formal presentation of design, the review should include one-on-one discussions with the technical leadership to determine if there are areas of the design that should be explored in greater detail at another time. Each review should be concluded with a written report and action items. There are many benefits if the independent review team can be formed at the beginning of the project and be committed to reviewing the project throughout its development life cycle. It is also extremely important to ensure that there are mission operations experts on the independent review team from its inception.

Risk Analysis

A well-defined and properly executed risk management process is vital to the success of a program. The process applies the following to all program phases: requirements definition, concept development, design, production, integration, verification, deployment, operations, and maintenance. The process should even be applied during the proposal phase to make certain that all programmatic and design risks have been discovered, assessed, and eliminated or mitigated.

Saying that something is at risk is stating that the outcome may be less than desirable. That is, there is a likelihood of an undesirable consequence. Risk can exist in all aspects of the program. A system, subsystem, or component is said to be at risk if there is a chance that it will not function as desired, will not be completed on time, or will cost more than budgeted. The risk can be in the software, hardware, or operations functionality.

There is a difference between a risk and a critical issue. For example, the fact that a program is currently over a cost budget is not a cost risk. It is too late to call it a risk. It is a critical issue and must be addressed immediately. A cost risk is a judgment that a cost budget may be exceeded at some future time.

Program management is at risk if it appears as though shortcomings in funding, staffing, development methodology, or other aspects of program management will adversely impact the cost, schedule, or technical capability of a system, subsystem, or component.

All programmatic and technical aspects of a program should be analyzed with respect to risk. The senior staff should carry out the analysis, since they have the many years of experience needed to identify the risk items. However, each individual on the program should participate at some level in risk identification.

Each risk is categorized as a technical, schedule, or cost risk. A technical risk means that a product may not be satisfactory. The assessment can be as unambiguous as a specific requirement may not be satisfied or it can be as subjective as an operator procedure may be cumbersome to execute. Assessing the likelihood of satisfying a requirement can be as simple as reviewing detailed requirements and design documents or as involved as developing and evaluating models or prototypes. A set of requirements themselves can be judged to be a risk if they appear to be incomplete, imprecise, or unverifiable.

A schedule risk means that an essential milestone may not be met. Examples of essential milestones are an event on a critical path, a major design review, integration test closure, product deployment, or even a demonstration of special interest to the customer.

A cost risk means that a cost budget may be exceeded. The cost budget could be for a program phase such as design or production or could be at the system, subsystem, or component level.

An example of a cost risk is the addition of requirements and capabilities to a subsystem without the addition of funds. Program baselines are commonly and frequently modified by Requests for Changes (RFCs). The addition of requirements in any one RFC may be so slight that it is difficult to justify requesting additional funds. But the cumulative effect of many such RFCs over a period of time may indeed justify the identification of a cost risk item and even a schedule risk item. Cost risk can also be introduced during the proposal phase when program management challenges costs in order to lower the bid.

Each risk is assessed and ranked as to the likelihood of occurrence and also to the consequence if the risk is realized. Rankings of low, medium, and high are commonly used. The rankings of these two factors are combined to establish the overall severity of the risk as small, medium, or large. For example, an occurrence risk rank of small and a consequence risk rank of medium would combine for an overall risk rank of small. But if the consequence risk rank were high, then the overall risk rank would be medium. We need definitions for what constitutes a ranking of low, medium, and high for occurrence and, especially, for consequence.

The possible strategies for handling each risk are 1) accepting the risk with no further actions; 2) acknowledging the risk but just monitoring it for an increase in severity; and 3) mitigating the risk. The risk mitigation strategy reduces the likelihood of occurrence or lessens the undesirable consequence. Factors to consider when developing a strategy include available funds, timeline, and personnel needed to implement the strategy, as well as any compromise of the system, subsystem, or component capability. Additionally, each risk item needs a contingency plan that will direct further action in the event the risk is realized.

Development of a mitigation strategy can be difficult, especially with the intertwining of technical, cost, and schedule risk. A technical risk may be mitigated by modifying a design. A design modification, especially late in a program, could require additional effort, with the possibility of a cost or schedule risk. A schedule risk may be mitigated by adding staff, which could require additional funding, thereby introducing a cost risk. A cost or schedule risk may also be mitigated by simplifying a design for a smaller development effort, which could require a relaxation of requirements, thereby introducing a technical risk.

Additionally, the mitigation strategy could require customer concurrence if there is an impact to system capability, cost, or schedule. A technical risk that is be mitigated by relaxing requirements or a cost risk that is mitigated by additional funds will require customer concurrence. The funds could come from a program management reserve, from another part of the program that is under budget, or from the customer.

Availability Analysis

System specifications should include the required level of availability expressed as a number. An availability of 0.95 means that the system should be available for operations ninety five percent of the time. Space systems consist of hardware, software, and operations subsystems and components. The overall availability at the system level is allocated downward from system to subsystem to component and, eventually, to individual units. The availability number is associated with hardware. Although software does fail for a multitude of reasons, the availability of a software component cannot really be determined or measured.

MIL-STD-756B, Reliability Prediction, is frequently referenced as a methodology for determining availability. In its simplest definition, availability (A) is calculated from mean time between failures (MTBF), and mean time to restore (MTTR). That is, A = MTBF/(MTBF + MTTR). Individual hardware units can be designed to satisfy specified values of MTBF and MTTR and thus a specified availability A. The units can be designed and assembled into a hardware component with a component level of A, MTBF and MTTR calculated from the values for the individual hardware units. The calculation accommodates the level of redundancy designed into the hardware component.

This process is repeated upward in the design until the system level availability, mean time between failures, and mean time to restore have been calculated. The calculated value of system level availability is compared to the required value. If the calculated value is less than the required value, design modifications should be made to bring the value into conformance. If the calculated value is greater than the required value, design modification can be considered if the modifications result in a less expensive design with a lower availability that still meets the requirement.

Specifying just one value of availability for a system may be unwise. It may be better to specify a high level of availability for the mission-critical capabilities and a lower level for non-mission-critical capabilities. This division of availability is important for the ground portion of the system since there may be significant computer and network costs in a high availability design.

A mean time between failures (MTBF) of 400 hrs and mean time to restore (MTTR) of 2 hrs results in an availability (A) of 0.995. This would mean that over the period of a year, the system could be unavailable 16.7 times a year for a total of 43.8 hrs and still satisfy the availability requirements. Those numbers may seem too high for mission-critical capability. MTBF of 1000 hrs and MTTR of 1 hr result in an availability of 0.999, which may be more acceptable for mission-critical capabilities.

The mean time between failures is strictly a hardware reliability number. The mean time to restore depends upon getting repair personnel in place, identifying and removing the failed unit, locating and installing a spare unit, and restoring system operations. Although the MTTR number usually does not include getting personnel in place and locating a spare unit, those factors must be carefully considered, especially for mission-critical capabilities.

Failure Modes Analysis

The majority of the hardware, software, and operations components within space and ground subsystems are designed to support the normal mode of operations. But some of the components within a subsystem must be designed to detect and respond to substandard performance or failures.

Failure Modes, Effects, and Criticality Analysis (FMECA) is a mature and well defined process for identifying failure modes, as well as the effects (consequences) and the criticality of the failures. Like Availability Analysis, FMECA is generally carried out with respect to the hardware components within a system. However, FMECA should also consider failures triggered by software malfunctions and failures, as well as operator miscues. For example, software malfunctions and failures can result from invalid input data, late arriving input data, processor overload delaying the output of data, or a processor crash resulting in a failure to output data.

The results of FMECA are used to modify system design to eliminate failure mechanisms or reduce the likelihood of occurrence and also to reduce the consequences of the failures. The results of FMECA are also incorporated into the operations procedures that identify the onset of failures and prescribe corrective actions.

FMECA is carried out in a bottom-up fashion so that the cascading effects of a lower level failure can be well understood. Carrying out FMECA on spacecraft components and subsystems is vital to producing a robust spacecraft design. The ability to failover and recover spacecraft has saved missions in ways beyond the original intent of the designer. FMECA is also carried out on ground components and subsystems. This includes the support software and hardware infrastructure, the mission software, and any special purpose processing hardware.

One of best known guides to FMECA is MIL-STD-1629A, Procedures for Performing a Failure Mode and Effects Analysis. It is a two step process with failure modes and effects analysis followed by criticality analysis. The process begins with 1) a definition of the operating conditions and constraints for each component; and 2) a definition of what constitutes a failure with respect to component output falling within allowable functional and performance levels. Each level of design is analyzed with respect to the failure modes and their effects on the components, subsystem, and system. Also identified is a means of detecting the failure mode along with the design corrections and operator actions that can be taken to mitigate the failure mode.

Each probable failure mode is assigned a probability of occurrence and a level of severity. Probability categories are frequent, reasonably probable, occasional, remote and extremely unlikely. Severity categories are catastrophic, critical, marginal and minor. If quantitative failure rate data is known, the probability of occurrence is replaced by a criticality number. MIL-STD-1629A provides the methodology for constructing the criticality matrix.

The response to FMECA will depend on several factors, including the cost and schedule impact of a design modification, the mission-criticality of a capability and the intent of any associated upper tier requirements. Overall, space system capabilities have a higher level of criticality than ground systems. Within the space system, spacecraft base capabilities such as maintaining communication, attitude, and power are more critical than mission capabilities. Ground subsystem criticality ranges from the less critical planning capability to the more critical commanding capability.

Recovery Mechanisms

Effective recovery mechanisms are essential to having a high level of system availability. A recovery mechanism consists of first detecting a degraded or failed subsystem and then suspending the affected subsystem and all interacting subsystems. The next steps are isolating the degraded or failed component within a subsystem, activating a recovery component, resetting the subsystem and all interacting subsystems and finally, resuming operations.

Space systems can be broadly categorized as providing a service or collecting mission data. Communication and navigation satellites are in the service category. Military intelligence gathering satellites and science satellites are examples of the mission data collection category. The fundamental difference between the two categories is that the service satellite is a relatively simple system for data processing as compared to the mission data satellite that executes a complicated command load sequence. The command load sequence includes mission activities and spacecraft support activities. Mission activities include payload configuration, pointing, and data collection. Support activities include calibrations, alignments, and power gathering.

The response to component failure on a service satellite is generally a failover to a redundant component and a resumption of operations. The failover can be automatic or commanded by an operator. The response to a component failure on a mission data satellite is usually to place the spacecraft into a protective mode and await ground commanding, which includes a reconfiguration of components and a new command load sequence.

Fundamental options for responding to hardware failure in a ground system are 1) failover to standby hardware within the operations environment; 2) failover to alternate hardware within the primary operational site, such as hardware used in the maintenance environment; and 3) failover to an alternate operational site. All three options require saving, and then restoring, checkpoint data and reloading software. Failover to an alternate site also requires relocating personnel. We can also use options 2 and 3 when all, or a portion, of the operations environment is undergoing a scheduled maintenance or upgrade.

Additionally, we can use the following strategies to keep a mission data satellite operating through an isolated ground system malfunction or failure:

Command Uplink Malfunction. Generate and uplink, well ahead of time, a backup command load that can be executed if the active command load expires without replacement.

Command Generation Malfunction. Recover and resume command generation if there is time before the next uplink opportunity. Otherwise, uplink a backup command load that was constructed well ahead of time.

Schedule Generation Malfunction. Recover and resume schedule generation and then command generation if there is time before the upcoming uplink opportunity. Otherwise, provide to command generation an alternate schedule that was constructed well ahead of time.

Plan Generation Malfunction. Plans typically cover many days of spacecraft activities with new days added to the plan as the old days expire. Any addition to the plan should take place well enough in advance of the need date so that there is sufficient time to recover from a plan generation malfunction.

16.2 Defining Anomalies

Anomalies are extraordinary spacecraft events or occurrences. These include out-of-tolerance measurements, off-nominal telemetry points, and in a larger sense, any unexpected or abnormal behavior. Table 16.4 gives some examples.

Table 16.4.   Examples of Anomalies. Any data or performance that is not nominal or is unexpected is an anomaly.

A temperature or a group of temperatures too high or too low
An attitude-error signal larger than allowable or expected
No activity on an attitude-error signal
Received signal strength too high or too low
No indication of appendage deployment
Unexpected configuration or uncommanded configuration change

An *anomaly* is anything that is wrong, or seems wrong, or is not quite right. In classifying an occurrence as an anomaly, suspicion should be the rule. It's better to be too worried than to overlook a seemingly minor occurrence that later kills a mission. There is no such thing as a glitch. An out-of-tolerance data point is an anomaly and has a reason. The only question is how much effort we can afford to put into understanding it.

This chapter deals with Earth-orbiting spacecraft that are either under ground control or can be accessed rapidly by ground control. It assumes what experience has shown to be true: that anomalies don't occur very often even under the very general definition provided above. A well designed spacecraft may hiccup once every few months, but if it does so more often, the design is seriously flawed. We also need to distinguish between launch anomalies and anomalies that occur after the spacecraft has been placed on orbit and checked out. Launch is a particularly traumatic event. Environmental stress on the equipment and emotional stress on the launch and operations crew are very high. Many anomalies occur during launch. Fortunately, the launch crew is conditioned to react to launch anomalies, and the operations crew is usually augmented with engineering support to handle them effectively. Even though anomalies occur with greater frequency during launch, the techniques used to investigate and resolve them are no different from those used later in operational life.

This chapter assumes that the operations crew has engineering support. When an anomaly occurs, the initial actions always fall to the operations crew. However, the detailed investigation of the anomaly is best conducted by people who aren't the minute-to-minute operators of the spacecraft. Engineering support—and in some cases, outside people such as the spacecraft designers—should do the detailed analysis of the anomaly.

16.2.1 The Fundamental Rule

The fundamental rule in anomaly resolution is that an anomaly, no matter how complex, has one and only one cause.

Many times we can synthesize failure scenarios involving multiple events that explain a set of anomalous data. Unless the scenario reduces to a single event, it is wrong. Multiple failures, no matter how attractive, don't occur unless they cascade from a single root cause.

16.2.2 Categories of Anomalies

The most important way to categorize anomalies is by criticality. Table 16.5 gives standard criticality definitions. The most severe anomalies can destroy the spacecraft or cause loss of mission. Less critical anomalies cause out-of-tolerance performance, loss of functional redundancy, momentary or partial loss of function, or simple annoyance. Time is also an important consideration. By the criticality definition, an anomalous condition that takes a week to kill the spacecraft has the same criticality as one that can destroy the machine in five seconds. However the danger associated with the shorter response time (not to mention the panic level) is clearly much higher. Table 16.6 considers possible reaction times and ability of a trained ground crew to respond. Even the best of ground crews can't usually react in just a few seconds. Such reaction is possible only when we've loaded and pre-approved a canned reaction plan and the operator's finger is on the button.

16.2 Spacecraft Failures and Anomalies

Even then, in most cases, the reaction is delayed. On the other hand, a trained crew should be able to render a spacecraft safe within an hour, and several days would be "fat city."

Table 16.5. Criticality Definitions. Criticality is a key measure of an anomaly's importance and a guide to the effort we ought to expend in preparing for it.

Level	Category Description
4	Complete loss of mission. Single-point failure. Loss of life.
3	Degraded payload performance outside specified limits. Total loss of operational mode(s). Loss of channel(s). Major injury.
2	Loss of the payload's or spacecraft's functional redundancy within specified limits. No loss of modes or channels. Minor injury.
1	No mission impact. No-effect failure conditions. No loss of functional redundancy.

Table 16.6. Reaction Time. Examining required reaction time for anomalies helps define where we can best apply planning and practice.

Category	Definition	Examples	Comments
Very rapid	Response required in a few seconds or tens of minutes	• Thruster stuck in on position • Runaway chemical reaction • Runaway propellant temperature or pressure	Nearly impossible for ground crews to correct. Automatic safing of the spacecraft is the preferred approach. Propulsion and attitude control are main sources.
Rapid	One hour to a few hours	• Temperature near upper limit and climbing • Battery-cell voltage approaching lower limit • Excessive power drain	Well-trained crew with good technical data and crisp decision process should be able to safe the spacecraft and prevent mission loss.
Quick	One day	• Spacecraft thermally unbalanced with slow rise or fall in temperature • Power imbalance with failure to recharge battery fully • Spacecraft in a damaging orbit (belt flying for instance)	A day should be long enough for a trained crew to safe the spacecraft and prevent mission loss. It's not long enough to scramble a crew from scratch or to get technical data that aren't prepared.
Leisurely	One week	Similar to those for quick reaction time	A week is enough time to marshal people and find data.
Steady state	Not time critical	A spacecraft that is in safe haven and awaits further actions	No immediate action needed.

Table 16.7 lists dangerous types of anomalies. We can use this list as a framework for evaluating the seriousness of an anomaly and in designing techniques (either on board or on the ground) to protect the spacecraft. Most high-reliability spacecraft are designed to detect these dangerous anomalies (or some of them) and change to a safe operating mode—a *safe haven*. Table 16.8 presents typical safe havens.

Table 16.7.   Dangerous Anomaly Types. This table lists types of anomalies that can cause loss of mission.

Type	Description
Momentum	Anomalies that accelerate the spacecraft or spin it up fast enough to produce destructive loads
Temperature	Anomalies that cause components to overheat or get too cold
Power	Anomalies that cause excessive power or energy drain
Command	Anomalies that block out the command system or prevent acceptance of corrective commands. Telemetry blockouts might also be included.

Table 16.8.   Safe Havens. Operating modes or states that are inherently benign are called safe havens.

Type	How detected	Actions
Momentum	Attitude-error signal out of limits Thruster temperature too high	Close propulsion isolation valves
Temperature	Temperature measurement too high or too low	Turn off equipment Turn on heaters
Power	Voltage too low	Turn off loads
Command	Command not received in timed interval	Turn on command receivers Couple receivers to omni antennas

The simplest of anomalies involve only a single subsystem. Compiling data and finding appropriate people to analyze the data are straightforward and readily done. Of single-subsystem anomalies, propulsion and attitude control are the most difficult. Propulsion anomalies are difficult because they tend to result in excess momentum or explosion; attitude-control anomalies are tough because of the complexity of closed-loop operation. Anomalies that have symptoms covering multiple subsystems or diverse technical areas are harder to work. A system engineer or system manager—someone with a broad technical background—must interpret symptoms and allocate actions (detailed investigation or analysis) to subsystem people or technical specialists.

Sometimes an anomaly results from operating equipment properly; it's anomalous only because ground operators don't expect or understand it. A useful term for this type of anomaly is *pseudo anomaly*. Early in the lifecycle of a new system, many (perhaps most) anomalies occur because of improper expectations and are thus pseudo anomalies. But anomalies resulting from incorrect commanding or procedures are true anomalies, even though the anomalous data may come from good equipment operating properly. They may be as dangerous or damaging as an equipment failure.

16.3 Resolving Anomalies

Table 16.9 lists the steps for resolving anomalies. First, evaluate the danger to the spacecraft. If the operating mode or the symptoms indicate danger to the spacecraft's health, command it into a safe haven. Most of the time, spacecraft are operating in relatively benign modes, so we can work many anomalies without changing operating modes. If in doubt, however, abort the current operations and seek safe haven.

Table 16.9. Steps for Resolving Anomalies. This table suggests a common-sense approach based on experience.

Step	Action
1	Safe the spacecraft. Get it into a safe operating mode so there is time to think.
2	Get all the data, even data from unrelated subsystems and sources.
3	Establish accurate timing.
4	List possible causes. Canvass all sources for candidates.
5	Analyze the data, examine the possibilities, eliminate possibilities until we find the culprit. Validate the answer, by analysis if necessary but preferably by experiment.
6	Figure out how to fix the problem, check out the fix, and do it.

Having determined the spacecraft is safe, collect all data concerning the anomaly and establish an accurate timeline. Organize an anomaly team and start searching for explanations of the data. The search for anomaly causes is a form of organized invention (also known as a "group grope"). The anomaly captain or team leader convenes knowledgeable people and then canvasses them for possible scenarios. We analyze and investigate these scenarios in detail until we identify the root cause of the anomaly. For complex anomalies this is a lengthy process. An important part of solving anomalies is verifying the answers. This verification isn't always possible, but even if we can't do it completely, we should investigate and verify our answers as much as possible.

16.3.1 Safe the Spacecraft

Clearly immediate action is necessary when failure to act will cause loss of mission or endanger people. Any team that operates a space system should have prepared plans *(contingency plans)* for anomalies that require immediate action. The table listing categories of dangerous anomalies (Table 16.7) is a good starting point for preparing contingency plans. Given infinite resources, we'd like to list everything that can go wrong and what to do. But given finite resources, we must identify the top contenders and prepare contingency plans for these. A Failure Modes and Effects Analysis (FMEA) is also a way to prioritize contingency planning, but we need to augment the FMEA by identifying corrective action.

Anomalies occur about once for every seven months of spacecraft operation, as suggested by a recent survey of several operating space systems (See Tables 16.10 and 16.11). The data covers two years of operation of four separate systems (13 separate spacecraft). It identifies 21 anomalies, and includes two launches, one of which was unsuccessful, and one mission ending in failure.

Table 16.10. Distribution of Spacecraft Anomalies by Criticality. Criticality levels are defined in Table 16.2. This table includes 13 spacecraft and two years of operation.

Type of Anomaly	1 - No Mission Impact	2 - Loss of Redundancy	3 - Degraded Performance	4 - Loss of Mission
Number of Anomalies	4	11	4	2

Table 16.11. Spacecraft Anomalies by Subsystem or Element. The same anomalies shown in Table 16.7 are distributed by subsystem. Note lack of structural anomalies.

Location of Anomaly	Launch Vehicle	Electric Power	Data	Payload	Attitude Control	Thermal	Ground
Number of Anomalies	1	2	9	5	2	1	1

16.3.2 Get All the Data

Spacecraft-status telemetry is by far the most important source of anomaly information. Table 16.12 summarizes types of telemetry.

From an anomaly standpoint, bi-level measurements tell us the state of the spacecraft, and other measurements tell us the value of variables within the system. Although most anomaly investigations center on the behavior of the variable signals (such as voltages, temperatures, and error signal values), we must also know the system's configuration (state). Ask these sorts of questions: "Which side of redundant equipment are we using? What mode are we in? What is our source of voltage? Which command channel are we using?" The answers to these questions affect how we'll interpret the data. Sometimes, telemetry measurements depend on

Table 16.12. Types of Telemetry. This table reflects the classical separation of data into discrete (bi-level) and analog types.

Type	Description	Examples
Measurement	Value of a signal typically 8 bits. Basic signal may be analog or digital but is digitized for telemetry transmission.	Temperatures, pressures, signal strength, voltages, currents, and attitude-sensor output
Bi-level	Condition or state of a bi-level signal such as a switch	Relay state (energized/de-energized) Switch state (on/off) (closed/open) Microswitch

operating mode; that is, the designers assign different signals to a specific telemetry word depending on operating mode. If the operating mode on the spacecraft differs from the mode used in the ground station, serious confusion can result.

Some measurements tell us much about the system even though they may not be directly concerned with a particular anomaly. These include supply voltages, temperatures, and rf signal strength. Nearly all systems telemeter voltage and current for the primary power supply—sometimes in expanded scale. From these measurements, we can often determine when transient events occur, such as when a switch closes or a new load is turned on. We can also verify power drain and compare it with the equipment we believe is activated. Bus voltage and current may also indicate spacecraft attitude. Secondary voltages are also usually telemetered but secondary currents are only rarely telemetered. Secondary voltages can show equipment operating status, or if out of tolerance, can indicate where a failure is.

We can use temperatures to infer operating status of electronic equipment. Deviations from expected temperature can also indicate mechanical damage (particularly to insulation blankets). Propulsion-component temperatures can reflect leaks by abnormal cooldown rate or low overall value. On well-designed spacecraft, actual temperature profiles should be within about 5° C of projected levels. Regard deviations beyond this level as abnormal.

Rf signal strength is usually measured on the ground and in the spacecraft. The spacecraft measurement is usually telemetered. These measurements provide information about the spacecraft attitude and can reflect the antenna's state of health. We can usually measure spin rate by observing the periodic variations in signal strength. Additionally, we can infer antenna blockage by observing the variation of signal strength within a spin cycle. We can sometimes establish the exact timing of mechanical damage by fluctuations in signal strength.

Although not commonly used, direct observations of the spacecraft can provide anomaly information. Such observations are possible using ground cameras and ground radars. Optical resolution depends on range and is therefore most useful for spacecraft at low altitudes. Routine film (or video) of launches can provide information about launch anomalies.

16.3.3 Establish Accurate Timing

A complex anomaly can have various symptoms. Out-of-tolerance measurements can appear almost simultaneously in a number of subsystems or components, and the plethora of data can be overwhelming. Establishing the exact timing of occurrences will be central to sorting out cause and effect.

Telemetry measurements are usually sampled and formatted in a telemetry frame for transmission. The position of a given word in the frame tells when the measurement was sampled, and such timing can be important. Subcommutated measurements are sampled at multiples of the frame period; their position in the subcom frame also tells the sample time. The current trend toward packet telemetry tends to obscure event timing. Packets can be multiplexed with other data and transmitted with an intervening delay. We must insert a precise time standard in the raw telemetry stream if we're placing the data in packets or multiplexing it with other data streams. The Space Shuttle's telemetry system is particularly bad about telemetry timing. By the time Houston gets through handling the data, it may be several seconds old and very difficult to interpret from a timing standpoint.

One of our first steps in investigating an anomaly is to establish a time base. One person should be responsible for this time base who will correlate the various standards. All telemetry systems used today have time standards. Range telemetry at the Eastern Range provides an Inter-Range Instrumentation Group (IRIG) channel that establishes timing to Greenwich Mean Time (GMT) with microsecond accuracy. The Goddard Spaceflight Tracking and Data Network (GSTDN) and the Satellite Control Facility (SCF) have similar standards. However, if the telemetry isn't received in real time, these accurate standards may be lost. Normally the spacecraft telemetry stream gets its timing from an internal oscillator. One of our challenges in establishing event timing is correlating this internal oscillator to GMT or some other universal standard.

Some types of telemetry record events that occur between sampling intervals. Thruster firings are a prime example. Firings of thrusters for attitude control are often quite short (typically 10 to 50 ms). Telemetry sampled at 0.5 s intervals will miss most thruster-firing pulses. For this reason, thruster-firing indications usually remain in memory from the time of occurrence until telemetry reads it out. Sometimes the thruster-pulse length is quantized, so telemetry provides pulse length as well as occurrence during the sampling interval.

16.3.4 List Possible Causes

This step is essentially an inventive process. We can poll the people working the anomaly for possible causes and then tabulate and discuss their answers. If we do the polling in an interactive meeting, we'll normally raise more questions than answers. In this process, the role of anomaly team leader is to keep the discussion focused on finding the important possible causes without discarding any likely candidates. As in any group interaction, some people want to adopt favored explanations and others tend to go off on tangents. Organizational conflict or an attitude of "who's the guilty party" may lead to recriminations or other destructive interpersonal behavior that gets in the way of orderly discussion. Structuring the discussion to keep the emphasis on compiling possibilities is sometimes a daunting management challenge. We need to take detailed discussion off line and, as the investigation progresses, winnow down the list of possibilities until only the most likely are left.

16.3.5 Analyze Data and Find the Cause

The general approach to anomaly analysis is to gather the data, establish event timing, categorize observations by subsystem or equipment group, postulate failure modes or scenarios, look for confirming or contradictory evidence, discard scenarios until a base set is reached, and devise tests or experiments to further prune the scenario set if necessary. While gathering data and categorizing observations, people usually ask more questions than they answer. Typical questions are: "What was the exact mode or setting of signal-flow switches? How is the spacecraft wired? What exactly does a particular component look like? How are the wires (insulation, plumbing) routed?" Sometimes, we find answers in the telemetry data, but often we need information from the spacecraft-build chronicles.

Anomaly data includes all pertinent telemetry data, command records, system state diagrams, records of personal observation (including operator and specialist logs), and range data such as signal-strength readings, photographs, and radar data. Normally, we investigate anomalies off line with people who aren't involved in day-to-day operation of the spacecraft. This is not to say that we shouldn't use operators or draw on operator experience. But we should investigate as much as possible in parallel with continuing spacecraft operation, not within it. A separate work area is almost mandatory and we need some way to log and keep track of data. In a well-ordered operation these preparations are all in place as part of the setup, so we simply invoke them whenever we need them.

The recording of telemetry-data streams often doesn't support anomaly investigation. Typically the senior operator has an event display—a display of telemetry measurements that change in value—and selectable page displays that show measurements by subsystem or other logical grouping. Often the page displays are selectable and are sometimes supported by alert lights or switches that signal out-of-tolerance conditions and allow page call-up by pressing the indicator. Other operators have similar displays that may include time records (strip charts or computer display monitor equivalent). These displays have measurements reduced to engineering units and are quite general and powerful. However, they aren't recorded as presented and hence aren't usually available for anomaly investigation. Only the raw telemetry stream is recorded. Getting the telemetry off the primary record and into a usable form (delogging) is frequently time consuming and difficult. Clearly, capturing the raw-telemetry record is an important step in accumulating anomaly data. Methods of delogging the data and providing engineering output are also important.

Aerospace hardware is manufactured according to carefully prepared and controlled engineering data. These data include drawings, specifications, and procedures. Schematics and signal-flow diagrams are sometimes used in design but aren't usually controlled or used to construct hardware. For anomaly investigation, however, schematics and signal-flow diagrams are of utmost importance, whereas fabrication drawings are much less useful. Wire lists (connection data for intercomponent wiring) are another matter. These are typically formal engineering data used to make and test wiring harnesses. They're extremely useful in anomaly investigations.

True manufacturing data can sometimes help the anomaly investigation because it records how a component was built. Manufacturing and test-data packages typically include manufacturing travelers, inspection records, and test records. Sometimes an assembly whose manufacture is particularly sensitive is designated as a special attention item. These items include single-point-failure parts or assemblies and any areas whose performance depends on precise assembly. A special attention item typically has a control plan prepared for it which delineates extra inspections, special tests, and other precautionary measures. We can use records of these measures, normally included in the manufacturing data, as evidence of actual construction. Photographs of the component during manufacture are an even better source of as-built information. In fact, photographs are so valuable that I started the practice of including 8.5" × 11" color photographs of each assembly in the manufacturing-data package.

To find a pattern in an anomaly indication that may reveal the root cause, we sometimes construct a *failure tree*—a systematic way to list possible causes of anomalous data. Presented with a single out-of-tolerance reading, we can develop a list of failure modes that might have produced it. Because the failure modes are interrelated, a useful way to depict them is as an organizational block diagram or tree structure. Initially, the tree may be very large, but using additional observed data may allow us to eliminate or prune many of the possible causes. The idea is to list systematically everything that might have gone wrong; then, eliminate suspects until only one remains.

It would be nice if an anomaly investigation could come up with only one plausible cause of the observed data and everyone would agree that's what happened. Most of the time we can. Several keys to reaching a conclusion are: (1) Identifying a single root cause. Scenarios that require multiple events are incorrect. (2) Survival under challenge. The postulated root cause should hold up under examination or challenge by everyone involved. (3) Support consensus. The anomaly team should be able to agree on the culprit. In many cases, we can pin down the exact semiconductor that has failed. In nearly all cases, we can narrow the cause to the switchable assembly causing the difficulty. In a few cases, we don't get a unanimous opinion, so we issue minority reports.

If we can't pin down the exact root cause from the data available, can we devise an experiment to verify or refute any of the candidates? Experiments can sometimes be done on the flight vehicle, but the preferred method is to use a test article such as a qualification spacecraft or component, or a subsequent unit in the production cycle.

16.3.6 Determine the Corrective Action and Carry It Out

After we've figured out what's wrong, how do we fix it? The simplest thing is to replace the broken element. High-reliability spacecraft are almost completely redundant, so we can replace nearly any failed component. If a direct replacement isn't available, we can sometimes use a different operating mode to accomplish the mission. In either event, a series of commands to the spacecraft and in some cases, reconfiguring the ground or user elements is required. This means preparing a command plan or a reconfiguration plan and processing it through the operations segment of Ground Control.

In devising a scheme to correct an anomaly, we may need to experiment. If so, we try to experiment with test hardware or a simulator rather than the flight spacecraft. If we do need to experiment with the spacecraft, we must be careful not to foul it up by sending destructive commands. Most anomalies involve extraordinary operating modes or equipment sets, so we may not understand exactly what a command sequence will do. We have to be cautious.

In extracting the system from an anomaly, we're often going from a safe haven to a more dangerous operating mode. If we can reconfigure using successive safe modes, we should do so. We have to take our time and verify each step before taking the next step. If we must pass through a dangerous configuration, we should consider having an operator prepared to abort the sequence and safe the spacecraft. (This is called a "dead man switch" after the practice on the Manhattan Project of having a person suspend a weight over the nuclear pile. If the reaction went run-away, the dying person would drop the weight, which would disrupt the pile and prevent more serious damage.)

16.4 Planning for Anomalies

To deal with an anomaly we need a set of competent people and a good leader. A complete anomaly team has engineers responsible for each subsystem; system engineers who understand orbit mechanics, rf-link characteristics, controls analysis, thermal design, structural design, dynamics, and computer hardware and software; and design-integration engineers who know how the spacecraft is built. Table 16.13 summarizes crew composition.

Part of developing a space system is preparing an operations manual, which is the main reference for the operations crew and their engineering support. Materials for training the operations crew may supplement the operations manual. The engineering-support team will have prepared specialist manuals to help

Table 16.13. Crew Composition. The composition and organization of an anomaly team affect the team's efficiency and ability to reach conclusions quickly.

Members of Team	Description of Team Members
Team Leader or Anomaly Captain	Senior person with enough overall knowledge of the system and sufficient management ability to partition the work, assign actions, evaluate progress, and control the investigation
System Engineer(s)	Person(s) who know orbit mechanics, rf-link characteristics, controls analysis, mechanical design and dynamics, and computer resources
Design-Integration Engineers	People who know how the spacecraft is put together. Mechanical-design integration involves location, mounting, and routing. Electrical-design integration involves wiring and signal characteristics. Thermal-design integration involves temperature ranges, heat dissipation and flow paths, and thermal-control methods.
Subsystem Engineers	Specialists in the spacecraft's individual subsystems

analyze flight data. They should also have a data center that contains spacecraft drawings, schematics, and wire lists. They also need configuration-management indices that relate drawing numbers to specific components. Even relatively simple spacecraft may involve 10,000 drawings or more. Despite this mass of data, a good data center should be able to locate any drawing in the system within a few minutes. We also need telemetry and command lists (books) or a database. These lists should show how the telemetry confirms command receipt and execution. We can use telemetry-calibration plots and schematics showing the telemetry-signal conditioning to interpret the data. Table 16.14 lists documents, manuals, and engineering data useful for anomaly investigations.

We must train the engineering-support team for anomaly investigation. For launch-phase support and for first launch in particular, we can draw engineering support from people in spacecraft design and test. For continuing operations, engineering support may be by contract. Prelaunch training should involve rehearsal for the entire crew and should have both nominal and off-nominal (anomalous) sequences. Sometimes a program will have a simulator that can model spacecraft performance and rehearse realistically. If a simulator isn't available, we can use recorded or canned telemetry streams to educate the team on what to expect. A well-conducted rehearsal exercises the entire crew—all subsystems and, because multiple shifts are usually involved, all shifts.

The training of engineering-support people during continuing operations is pretty much a hit-or-miss proposition. For some systems, people rotate through sessions with a simulator in which anomalies are programmed and the team response is evaluated. However, the more normal situation is to use engineering-support people to analyze trends or do other make-work jobs while not resolving anomalies. The infrequency of anomalies tends to make even diligent support people rusty, so periodic rehearsals are a good idea.

Table 16.14. Documents for Anomaly Investigations. Complete and accurate documents must be available for an anomaly team to operate efficiently.

Document	Description
Operations Manual	A manual describing the spacecraft and its operation. Typically several thousand pages long and containing detail down to the block-diagram level. May summarize operational procedures and contingency plans.
Training Manual	Materials used to train operators
Command and Telemetry Documents	Command and telemetry lists sorted by number, subsystem, alphanumeric designator, and ground-system designator. May be augmented by expected telemetry response for each command, telemetry-calibration files, or schematics for the telemetry-conditioning electronics.
Procedures	Step-by-step procedures used by spacecraft operations
Specialist Manuals	Manuals or notebooks prepared by subsystem and system engineers to help them understand and analyze system performance
Schematics, Wire Lists, Signal Flow Diagrams	Aid in tracing signal flow through the spacecraft
Engineering Data	Drawings, specifications, and procedures. Indexed by indentured drawing lists and configuration guides
Manufacturing Data	Usually available from the spacecraft manufacturer. Data packages should include manufacturing and test data for each component and assembly, plus integration and test data for the spacecraft. May include as-built photographs.

Practically any form of operational gaming of a space system will reveal a set of anomalies or equipment failure modes that are both likely to occur and serious. Prudence requires us to prepare contingency plans for these occurrences. Some ways to identify contingency candidates are:

1. Identify what to do in case any box fails. This is relatively simplistic but it's a start. The next level below the box level is to plan for each switchable assembly.
2. Use the Failure Modes and Effects Analysis (FMEA) to identify critical or probable failure modes.
3. Use operational procedures and identified decision points as likely sources.
4. Use people's experience. If there is a corporate database, review it for propensities to failure.

The standard operator displays are event screens and system or subsystem pages. The standard medium is a computer monitor, and the displays are normally high-resolution color. Telemetry measurements are shown in engineering units and changes or out-of-tolerance conditions are sometimes annotated in color or by a blinking marker. Operators can usually call up (by push button or by keyboard) various display pages so they can scan the spacecraft's status rapidly. Back-room displays are similar to the operator displays but usually lack command capability.

Anomaly investigators also use dynamic displays—strip charts or history plots. These plots are invaluable in examining control-system performance.

A well-prepared anomaly team will have general-purpose tools available for analysis as well. These tools include computers with standard application programs and special-purpose simulation programs for on-the-spot analysis. Table 16.15 summarizes some of the useful analysis tools.

Table 16.15. **Analysis Tools.** The ability of an anomaly team to function depends on access to usable analysis tools.

Analysis Tool	Description
Orbit Mechanics	Any ground station worth the name has a powerful, general-purpose orbit program. In addition, the anomaly team should have quick-and-dirty, conic-fit programs or tools to assess orbit changes rapidly.
Controls Analysis	A method of sanity checking the attitude-control response to known transients (appendage deployment for instance) and unknown disturbances
Power Profile	Either a simulation or tabulation of power vs. equipment complement

16.5 Case Studies

16.5.1 Anomaly Case Study #1: FLTSATCOM 5

FLTSATCOM 5 was launched on August 6, 1981. The engineering-support team for the launch was deployed at the Eastern Test Range (ETR), the Satellite Control Facility (SCF) at Sunnyvale, CA, and Space Park (TRW contractor plant in Redondo Beach, CA). Almost all members of the team had worked on previous FLTSATCOM launches.

The Air Force Satellite Control Network collected telemetry data and relayed it to the SCF, where it was decommutated and displayed for the operators and the engineering-support team. Coverage was almost complete except for a small section of the parking orbit over the eastern Atlantic and Africa, when the spacecraft and the Centaur booster were out of sight of ground stations. Although telemetry from this section of the parking orbit was not relayed to the SCF, an ARIA aircraft flying over the Atlantic recorded the data. During this uncovered section of the orbit, the spacecraft and Centaur passed over the equator, and the Centaur executed its final burn, putting the spacecraft in a transfer orbit to geostationary orbit. The Centaur then oriented the spacecraft for its orbital-injection burn and released the spacecraft. After release, the spacecraft fired thrusters to spin-stabilize its attitude. A short time after spinup, the spacecraft came into view of a ground station, and telemetry started to arrive at the SCF [FLTSATCOM OOH].

For the previous launches, which had all been nominal, the arrival of telemetry at the SCF was signaled by a rise in received signal strength and increased activity on the monitors. This activity was garbled at first but rapidly cleared to the expected pattern as the decommutator and the telemetry stream synchronized. For FLTSATCOM 5, the readouts remained garbled, so anomaly action began. A canned contingency plan existed for resetting the spacecraft equipment converter. Because the converter was the main component in the spacecraft's fault-protection design, its upset was the most probable cause of the observed data. The contingency plan was executed, and the telemetry stream locked up.

Initial telemetry showed that the spacecraft was spinning and that the appropriate thrusters were warm (indicating they had fired). The spin-attitude telemetry was not nominal, raising the possibility that mechanical damage had altered the moments of inertia. Telemetry from the electric-power subsystem was seriously anomalous. Battery-charge indications were fluctuating violently, as were main bus current and voltage. Plans were immediately drafted to connect the batteries directly to the main bus, thus bypassing the onboard charge-control channels in favor of direct ground control of the batteries. After these actions, the power-subsystem's telemetry stabilized, but there were still large fluctuations in main bus and battery currents. The subsystem engineers were canvassed to explain the data. They concluded the solar arrays were damaged near the solar-cell strings used to charge the batteries. It was even possible to map the exact location of damage to the arrays using wiring data the subsystem engineer had included in his specialist notebook.

With the spacecraft in a more-or-less safe and stable condition, it was possible to review the data, organize an anomaly team, and plan future actions. The top priority was planning the spacecraft apogee-kick motor's (AKM) orbital-injection burn. The spacecraft was in a highly elliptical orbit, passing through the Van Allen belts twice on each revolution. Furthermore, the orbit was decaying with each perigee passage, so injecting into final orbit was somewhat urgent. Stability during AKM firing required an increase in the spacecraft's spin rate from the initial 30 rpm to a final value of 60 rpm [FLTSATCOM OOH]. This was a normal command sequence, so operators decided to do it. The spacecraft responded normally up to about 45 rpm; there it went into nutation and commanding was suspended. But the nutation damped out, so commanding was resumed up to 60 rpm.

Injection into the proper final orbit required knowing the spin attitude. Because the spin-attitude telemetry was anomalous and mechanical damage to the spacecraft was highly probable, the adequacy of the normal methods for measuring and controlling spin attitude was questionable. To bound the accuracy of spin-attitude knowledge, people began a dynamic analysis at the spacecraft contractor's plant. This analysis, completed in a matter of hours, showed that the spacecraft's spin axis might be misaligned from the AKM axis by as much as 15°. If this were the case, the velocity increment produced by the AKM would be grossly inadequate, and 24-hour orbit wouldn't be possible.

To avoid this possibility, the anomaly team devised an alternate targeting strategy. By aiming the spacecraft for injection at an inclination of 7° instead of the nominal 2°, the spacecraft's final velocity would be close enough to a 24-hour period for the liquid propellant on the spacecraft to make up the deficit. If, on the other hand, the AKM achieved full performance, the final velocity would exceed that required for 24-hour orbit but would still be within the liquid system's correction range. The team carried out this strategy by retargeting the spacecraft to 7° inclination. They achieved almost full efficiency of the AKM. After correction, the orbit's inclination was 5.4° and it had a 24-hour period.

After orbital injection, the spacecraft appendages were deployed. The solar arrays, despite damage, deployed successfully. The receive antenna and the transmit antenna failed to deploy despite execution of contingency plans aimed at spinning the spacecraft fast enough to dislodge them if they stuck. Test of other spacecraft equipment showed it all to be working correctly except for failure of all receive channels that used the UHF receive antenna and reduced gain of the UHF transmit antenna. Patterns of the omni antenna showed asymmetry, which was attributed to a bent mast on the UHF transmit antenna.

Failure analysis after the fact revealed that a high-energy event had occurred during booster-powered flight. Synthesized failure scenarios revealed that the inner face sheet of the aerodynamic shroud had explosively delaminated. The face sheet hit the solar arrays, the receive antenna, and the transmit antenna. The receive antenna, although broken at its base, remained attached until the spinup maneuver, when it broke free at 45 rpm.

16.5.2   Anomaly Case Study #2: FLTSATCOM 1

FLTSATCOM 1 was launched on February 9, 1978. By all indications, the launch was nominal until about 2:00 A.M. on February 10 when the TT&C subsystem engineer informed me that the telemetry was indicating a command count of zero. Because the command link was encrypted, and proper operation of the decrypter depended on knowledge of the command count, this was disturbing.

We had drafted a contingency plan for enabling the backup command channel, so we used it. The backup channel responded properly, so we knew we had command capability and were in no immediate danger. But we still didn't understand the anomalous command-count telemetry. We pulled the schematics and examined the circuitry. No explanation suggested itself. We called the engineering-support office at Space Park and asked them to find the unit engineer for the command unit (the decrypter interface box). Although we had prepared a telephone list of responsible design engineers (RDEs), the command unit's RDE had recently moved and didn't have a phone. We dispatched a runner who got the RDE to a neighbor's phone at 4:00 A.M. He couldn't explain the data.

At this point, I ordered the engineering-support office at Space Park to pull in a test crew and run a test on the qualification spacecraft. This involved bringing the test crew in (about a dozen people), writing a test procedure, and doing the test. It was complete by 8:00 A.M. and showed that when the uplink rf-signal strength dropped below a threshold (a condition that occurred each time we changed the ground station), the telemetry indication of command count reset to zero. But the command count in the decrypter didn't reset, and on the next successful command, the telemetry would reset. What we were seeing was proper operation of the equipment, so we classified this anomaly as a pseudo-anomaly.

16.5.3   Anomaly Case Study #3: FLTSATCOM 2

FLTSATCOM 2 was launched on May 4, 1979. On the launch of FLTSATCOM 1 during powered flight, we had observed a single telemetry measurement of high bus current. On FLTSATCOM 2 there was an extended period (approximately ten seconds) of high current drain (10–20 amps). It was frightening, but the spacecraft seemed to survive and checked out perfectly on orbit except for a telemetry measurement of solar-array temperature.

An off-line anomaly team was designated to analyze the data. It determined that an exposed steel deployment cable running behind the folded solar array could vibrate and strike exposed connections on the array. The spacecraft was designed to clear inadvertent short circuits by dumping battery current through the short and burning it out. This had apparently happened and vaporized the cable. Experiments duplicated both the magnitude and duration of the short. Furthermore, clearing the short would also burn out a ground wire in the solar array's slip-ring assembly, which provided the ground for the array-temperature measurement. Thus, the investigation was able to trace disparate data to a single root cause—demonstrating our fundamental rule.

Reference

FLTSATCOM Orbital Operations Handbook (OOH), Vols. 1 and 2.

Chapter 17

Interplanetary Space Mission Operations

Ray Morris and David W. Murrow,
Jet Propulsion Laboratory

17.1 Elements of Interplanetary Mission Operations
17.2 Differences Between Interplanetary and Earth-Orbiting Missions
17.3 Operations Activities Throughout the Mission Lifecycle

The mission operations manager for an interplanetary mission should understand basic elements of the interplanetary environment and how they affect mission operations. Because this book emphasizes Earth-orbiting missions, this chapter focuses on differences between them and interplanetary missions. This chapter will explain these differences and how operations change throughout an interplanetary mission. Where appropriate, "rules-of-thumb" and tables will explain how different mission types or planets affect operations.

Many exploratory missions to other planets have been launched since Mariner 2 traveled to Venus in 1962. The United States, the former Soviet Union, and the European Space Agency have combined to visit all of the planets except Pluto, as well as some comets and asteroids. Interplanetary exploration will continue, but with smaller, more focused spacecraft and lower mission-operations costs. Plans include the NASA-sponsored Discovery program for small missions to the planets and international missions to Mars, Saturn, and other targets.

Originally, each interplanetary mission—Pioneer, Mariner, Voyager—was the first chance to explore a new planet. But expanding costs now make missions appear to be the last chance to visit a planet. This is particularly true of ambitious missions to the outer reaches of the solar system, such as Galileo and Cassini. Future planned missions should be neither the first nor the last chance, but one chance in an ongoing program of exploration.

Both first-chance and last-chance missions strive hard for maximum performance. Scientific payloads tend to be comprehensive, adaptive planning is desirable, and the use of resources is optimized. Optimizing leads in turn to operating with small margins in most spacecraft resources, so managing these resources requires a lot of labor. In this chapter we'll discuss how optimizing affects operations.

First we'll discuss the major elements of interplanetary missions and detail the main differences between Earth-orbiting missions and interplanetary missions. Then, we'll look at mission operations activities throughout the mission. The reader can best use this chapter by becoming familiar with interplanetary objectives and architectures in Sec. 17.1, and then comparing the information in Sec. 17.2 to the rest of the book's material on Earth-orbiting missions. Section 17.3 is a generic plan that identifies typical activities throughout an interplanetary mission, so it can serve in planning the lifecycle of one's own mission.

17.1 Elements of Interplanetary Mission Operations

Tables 17.1 and 17.2 illustrate the main differences between interplanetary and Earth missions and show where this chapter discusses them. They match the space mission elements in Chap. 1 and mission operations functions in Chap. 3.

17.1.1 Orbit: Types of Interplanetary Missions

For this discussion, we'll divide missions into five types: flyby, mapping, tour, landers, and probes. Table 17.3 summarizes the different types and some of their main characteristics.

Flyby missions have widely different spacecraft-target ranges during science gathering times. Flyby missions were among the first missions flown to other planets and have short periods of detailed observation compared to the time for the trip to the planet. Flyby missions have been useful for discovery or initial exploration of the solar system. A flyby will likely discover something at a distance which requires plans to change for the near encounter. An interplanetary flyby results in a gravitational impulse to the spacecraft, and we can string flybys together if the geometry allows, as in the Voyager "Grand Tour" of the outer solar system. Navigation to accurate flyby conditions is important because it creates a link between the downlink part of mission operations and the planning.

Mapping missions—most like Earth missions in their repetitive nature—have been used for more detailed exploration of the solar system. Examples include Magellan at Venus, Mars Global Surveyor, and Mars Odyssey. They use orbital geometry much like that for Earth missions to view the entire planet's surface at similar resolution. This similar geometry also allows for the same kind of

Table 17.1. Differences Between Interplanetary and Earth-Orbiting Missions. Each mission element differs based on the differences in mission subjects.

Space-Mission Element	Key Differences	Where Discussed
Subject	• Lack of knowledge of target • Users represent scientific community	Sec. 17.2.3
Orbit	• Long cruise times • Very high activity periods • Interplanetary flybys and gravity assists • Different coordinate systems which change with mission phase and subject	Sec. 17.1.1 and 17.3
Space Element	• Historically small and negative resource margins • Environmental extremes • Lack of adequate solar power at outer planets	Sec. 17.2.6
Launch Element	• Highly constrained launch window • Higher launch energy required	Sec. 17.2.1 and 17.2.2
Ground Element	• Different activity levels for different mission phases • Very few capable antennas for all missions, causing scheduling conflicts	Sec. 17.3
Mission Operations	• Range and light time delay • Lack of knowledge of target • Uniqueness of opportunities • Time-critical operations	Sec. 17.2.1, 17.2.4, and 17.3.5
Command, Control, and Communications	• Range and light time delay • Very low data rates	Sec. 17.2.1

operations. Ways in which they differ from Earth-mapping missions include telecom links beyond Earth ranges and constraints imposed by solar illumination or solar conjunction.

Tour missions contain repeated gravity assists using the planet's natural satellites. The satellite flybys are used for trajectory shaping as well as for science gathering. Tour missions, such as Galileo at Jupiter and Cassini at Saturn, stay at their target for a long time, but the geometry is less repetitive than for a mapping mission. The repeated gravity assists mean that we operate the mission like several flybys in a row, but tour missions still focus on the primary body between satellite encounters. Also, the geometry for flying by a satellite changes very rapidly in a flyby mission, which makes navigation and timing accuracy important. To plan the mission, we must compromise between the objectives of a satellite flyby and our future objectives. For example, Cassini tour flybys of Titan were designed to begin at a high altitude to ensure operator control of the spacecraft encountering the atmosphere. This desire to maintain the long-term goals of our mission conflicts with the desire to acquire low-altitude, high-resolution imaging and aeronomy data as early as possible.

Table 17.2. Key Differences in Mission Operations Functions. Chapter 3 defines the differences between Earth-orbiting and interplanetary missions in each of the 13 operations functions.

Mission Operations Function	Key Differences	Where Discussed
Mission Planning	• Range and light time delay • Adaptability to changing environment or updated knowledge	Sec. 17.2.1, 17.2.3, and 17.3.8
Activity Planning and Development	• Range and light time delay	Sec. 17.2.1
Mission Control	• Critical or unique commands	Sec. 17.3.5
Data Transport and Delivery	• Range and light time delay • Small or negative margins	Sec. 17.1.3 and 17.2.1
Navigation Planning and Analysis	• Number of coordinate systems • Critical maneuvers for flybys	Sec. 17.1.1 and 17.2.1
Spacecraft Planning and Analysis	• Small or negative margins • Environmental extremes	Sec. 17.2.1 and 17.2.6
Payload Planning and Analysis	• Small or negative margins • Lack of prior knowledge of target	Sec. 17.2.7 and 17.2.2
Payload Data Processing	• Variety of users • Need for level 0 data*	Sec. 17.2.4
Archiving and Maintaining the Mission Data Base	• Need to archive level 0 data	Sec. 17.2.4
Systems Engineering, Integration, and Test	• Small or negative margins • Environmental extremes • Long cruise phases before key engineering and science events	Sec. 17.2.1, 17.2.7, and 17.3.3
Computers and Communications Support	• Range and light time delay • Software uploads during mission	Sec. 17.2.1 and 17.3.3
Developing and Maintaining Software	• Software uploads during the mission	Sec. 17.3.3
Managing Mission Operations	• Workload difference between long cruise phase and brief encounter • High turnover rate in team personnel due to long cruise phase	Sec. 17.3.3 and 17.2.4

* Project scientists usually want level 0 data.

To maintain a planned satellite tour, the spacecraft must undergo propulsive maneuvers with rapid turnaround before and after flying by the natural satellite. A maneuver before the flyby ensures the flyby is as close as possible to the prearranged science sequence. Following the flyby, targeting errors that result in velocity errors on the trajectory (because the satellite's gravity is different than we

Table 17.3. Types and Characteristics of Interplanetary Missions. Each of the five mission types has a distinct profile for science gathering and time for cruising between planets.

Mission Type	Examples	Science-Gathering Life	Cruise Duration	Primary Effects on Mission Elements
Flyby	• Mariner to Venus, Mercury, and Mars • Pioneer to Jupiter and Saturn • Voyager to Jupiter, Saturn, Uranus, and Neptune	Months per encounter, split between far and near encounters	Months to 2 years between encounters	• Mission Operations: Short burst of high activity • Space Element: Need for large data storage • Subject: Usually little prior knowledge
Mapping	• Magellan at Venus	>3 years actual	15 months	• Mission Operations, Orbit: Repetitive • Command and Control: Continuous tracking required
Tour	• Galileo • Cassini	22 months 4 years	6 years 7 years	• Mission Operations: Generally high activity, with burst of higher levels • Subject: Target changing from day to day
Lander	• Mars Pathfinder • Viking • Spirit • Opportunity	30 days >1 year 3+ months* 3+ months*	9 months 9 months 7 months 7 months	• Mission Operations: Interplanetary daily cycle vs. Earth day, low navigation requirement • Orbit: On surface during prime mission
Probe	• Galileo • Cassini/Huygens	90 minutes 3 hours	6 years 7 years	• Space Element: Multiple spacecraft

* As of this writing, NASA has extended the science lifetimes of both Spirit and Opportunity to at least 18 months.

thought) are corrected quickly to keep their cost from becoming prohibitive. The cost grows with time from the flyby for two reasons. First, a velocity error propagates into larger position errors with time. Second, the energy change that results from flyby errors is better corrected near the satellite, where the differential energy change ($\Delta E \sim V\Delta V$) is larger due to the spacecraft's higher velocity.

Lander missions, such as Spirit or Opportunity, explore surfaces while surviving for a long time after landing on a planet's surface. Their "buttoned-up" configuration, which makes cruise activities manageable, limits visibility during cruise. Usually the only required activity is a periodic checkout of the lander systems. The entry, descent, and landing of the lander require many activities, but we must preprogram them into the onboard sequence because of visibility and timing constraints. Post-landing activities are similar to post-launch activities for orbiters, except that navigation is required only once to determine the lander's location on the surface. The communications link is usually direct, with the planet's rotation modeled in the visibility algorithm.

Probe missions, such as Cassini/Huygens, require a space vehicle or instrument to hit the target body. Probe missions are similar to landers in that they enter the atmosphere, but their planned operations occur only during the atmospheric descent. The spacecraft takes scientific data through direct sampling of the atmosphere or remote sensing of the surface. The cruise is typically very long compared to the encounter. A relay spacecraft receives, records, and protects the data before playing it back to the Earth.

For operations, the cruise duration is a key difference between mission types. The mission operations manager must keep the teams engaged enough during the cruise to perform at their peak during the high-activity encounter. All interplanetary spacecraft cruise several months before operations, and some cruise up to seven years. Additionally, the proportion of time in cruise to time during encounter is very high for flyby missions and probe missions. Tour and mapping missions have more science-gathering time but still have relatively long cruises. For example, the Galileo and Cassini tour missions cruised for six and seven years, respectively. Following cruise, the prime mission lasts two to four years. Many of the mission operations functions during cruise are dedicated to planning for the encounter, but advance planning competes against the desire to retarget observations when we discover new phenomena during the encounter.

17.1.2 Subject: The Planets

The subjects of interplanetary missions are widely varied, largely unknown, and very distant from the Earth [Beatty, 1990]. These three attributes set the tone for interplanetary mission operations. First, varied targets lead to different operations strategies and mission types, which in turn reduce inheritance from one mission's operations to the next. Common ground elements can offset this effect somewhat, but the space element and the mission timelines are often unique. Second, unknown targets require adaptable operations and close coupling between the uplink and downlink functions. Finally, the distance from the Earth has many implications for operations. In Sec. 17.2, we'll talk more about these three differences as they relate to the mission-operations elements.

17.1.3 Command, Control, and Communications and Ground Elements

Command, control, and communications for interplanetary missions usually require a limited number of antennas on the Earth linked directly to each spacecraft. The number of spacecraft has so far kept any ongoing spacecraft-to-spacecraft links from being cost effective. Because interplanetary spacecraft are distant, the Earth's rotation dominates their apparent motion, so antennas can view the spacecraft from horizon to horizon without requiring large angular rates. However, because of the distance, the antenna aperture must be able to receive a weak signal. These combined effects lead to a few large antennas for commanding and controlling the interplanetary spacecraft.

NASA's Deep Space Network (DSN) collects telemetry and radiometric navigation data and radiates commands to the spacecraft. The DSN comprises three antenna complexes spaced around the Earth—at Goldstone, California; Canberra, Australia; and Madrid, Spain. A major difference between Earth-orbiting and interplanetary missions is that the DSN provides a limited number of tracking facilities. Several 26-m and 34-m antennas, plus one 70-m antenna, are at each complex, each with different capabilities and commitments. Planning to use the DSN starts at the earliest stages of a mission's development, sometimes 20 years before the tracking is required. Because of the high demand and planning based on uncertain missions, scheduling the DSN resource is a key driver for interplanetary mission operations. (See Chap. 12 for a description of DSN capabilities.)

A DSN station's performance depends on the declination of the target, which can be thought of as the latitude of the spacecraft in an Earth-centered frame. Much as the length of day increases during the summer, the duration of a spacecraft pass increases if it is at the same latitude as the station. Thus, for a spacecraft at 23° declination, a northern-hemisphere pass will be 12–13 hours, whereas a southern-hemisphere pass will be 8-9 hours. Similarly, same-hemisphere coverage will result in higher maximum elevations and therefore higher data rates.

An aggressive operations concept will count on the best station, whereas a conservative concept will plan for some off-peak performance. If the mission is at a critical point, the DSN will be firmly committed. But if the mission is in a routine phase such as mapping, we may substitute or cancel the DSN antenna's time if another spacecraft has a higher priority. If the mission return has counted on the most capable station, substituting a shorter or a less capable pass will entail some replanning.

17.1.4 Mission Operations Element

Interplanetary space missions typically have a large flight team devoted to uplink and downlink. The uplink portion of the flight team does everything required before the spacecraft executes an activity. The downlink portion does everything after execution. The work of the two teams interacts when the spacecraft's performance, characterized by the downlink, affects planning by the uplink function. For interplanetary missions this is a strong interaction, because the uplink element will typically re-plan an activity when performance estimates or navigation predictions change.

The uplink design includes incorporating the spacecraft and instrument performance into an integrated plan. We synthesize the best information about the spacecraft, the tracking configuration, and the science desires for a certain amount of time. Deciding on the boundaries of the time blocks is another, very early function of the uplink team. When the plan is at a stage where few changes are expected, and conflicting science desires have been reconciled, the team generates and simulates a spacecraft command file. After a successful simulation, the team stores the command load until just before its execution, then transmits it to the spacecraft. Uplink also includes adding so-called "real-time commands" to the spacecraft memory while a stored sequence is executing.

For downlink, we monitor the spacecraft's health and safety and track its movement. We periodically schedule subsystem calibrations and then fold detailed analytical results into future plans. When the Jet Propulsion Lab flies interplanetary missions, they focus on the timing of subsystem performance for two main reasons. First, they must understand the spacecraft and anticipate any future changes in observed parameters. Second, tiny changes in subsystem behavior can strongly affect a plan optimized for performance.

Navigation tracking is also crucial to the downlink. The navigation team determines the spacecraft's trajectory, maneuvers to optimize future geometry, and updates future plans as required.

17.1.5   Launch Element

The launch element for an interplanetary mission consists of the launch vehicle itself and any supporting services needed to deliver the spacecraft onto an interplanetary trajectory. An interplanetary trajectory requires higher launch energy to escape the Earth's gravity, as well as a specific alignment of the planets. The latter means we can launch the spacecraft only at certain times. Because the launch energy varies with the launch day, we can get several possible launch days by restricting the spacecraft's mass to the lowest allowed value (corresponding to the highest required launch energy) during the desired number of days. Sections 17.2.1 and 17.2.6 cover the launch element.

17.1.6   Space Element

The space element of an interplanetary mission is a spacecraft (or perhaps two) that is uniquely designed for the mission being flown. While there may be heritage in parts of the hardware or software, the spacecraft is usually original enough to require a dedicated, specialized flight team. In the early days of space exploration, pairs of interplanetary spacecraft typically flew to increase mission reliability. As missions became more complex, single spacecraft (such as Galileo) flew. The principal distinguishing factor in the spacecraft is its ability to survive lengthy space exposure, do large orbit-insertion maneuvers, and communicate over large ranges. For a description of the engineering systems and subsystems, see Larson and Wertz [1999].

The science portion of the space segment can be very complex and comprehensive. Multiple science objectives are often satisfied by a payload that has differing observation types and spacecraft and operations drivers. These comprehensive payloads may lead to conflicts over requirements and resources. Table 17.4 describes the typical science characteristics and their operational requirements.

17.1 Interplanetary Space Mission Operations

Table 17.4. Instrument Types and Operations Characteristics. The type of science instrument affects both the operations activities and the spacecraft design. If more than one type of instrument is flown, we must often structure operations activities to provide different spacecraft or target conditions.

Instrument Type	Characteristics
Remote Sensing 　Passive 　Active	• Resolution proportional to distance from target; requires target fixed pointing • High power due to active transmission; requires target relative pointing
In-Situ 　Particles 　Magnetic and radio wave investigations	• Sensitive to spacecraft contamination; requires orientation to acquire samples • Requires magnetically clean spacecraft; may require special calibration to remove spacecraft signature from data
Detection of High-Energy Particles	• Low time constant for three-dimensional temporal phenomena; may require rapid scanning

Remote-sensing instruments require a spacecraft to point at the target. They can be either passive, where the instrument receives energy directly from the target, or active, where energy emitted by the spacecraft is reflected by the target and received by the spacecraft. Different portions of the electromagnetic spectrum (e.g., microwave, infrared, visible, or ultraviolet) are covered by different instruments.

Optical remote-sensing instruments (cameras) are passive, receiving light from the target. The resolution is inversely proportional to distance from the target, so the mission will try to fly as close to it as possible. The target's apparent motion is directly proportional to the flyby distance. If the relative motion is too high and the exposure is long, the image may be smeared. The tradeoff between unsmeared and high-resolution images will determine the best flyby distance for an imaging instrument.

Imaging instruments have large data rates and very high public interest. For example, consider an imaging frame 1000 by 1000 pixels, with each pixel represented by 12 bits. If the framing time (time between images) of the camera is one second, the data rate is 12 Mbps. For a 20-watt, X-band radio system from an outer planet, a typical data rate is 100 kbps. Obviously, a camera with these characteristics can oversubscribe the downlink if it takes data continuously.

Active remote-sensing instruments send a signal from the spacecraft. The most common active remote-sensing instrument is a radar, which is used to image the surface of a body obscured in the optical wavelengths. Examples of such use of radar are Magellan at Venus in the early 1990's and the Cassini Radar at Saturn's moon Titan, beginning in 2004. Operationally, the most noticeable feature of a radar is its use of the spacecraft's high-gain antenna as the instrument aperture, which requires data storage because the antenna can't simultaneously view the Earth and the target.

In-situ instruments measure the fields they're in or particles that the spacecraft samples. Unfortunately, because the spacecraft contaminates its surroundings, it can't record an environment identical to the one of scientific interest. For example, the thruster plume will affect a sensitive mass spectrometer, so we must take data when the thrusters are inhibited, which may imply allowing less control than originally planned. A magnetic-field instrument is also susceptible to contamination by spacecraft fields from operating current loops. Because eliminating current loops is not practical when science activities are high, we require a magnetometer to calibrate sources of the loops on the spacecraft. In other words, we'll have to run a calibration sequence when activity is low, cycling subsystems on and off, so the magnetometer may characterize the fields. The characterization allows later removal of the fields from the science data set.

Because the particles being measured by radiowave sensors have a very short time constant and are three-dimensional, a scanning motion may be required during data gathering. If the spacecraft can't sweep the instruments independently through space, we may move the entire spacecraft, as in the case of Cassini. Operationally, we must trade this scanning motion against the target's fixed attitude required for remote-sensing measurements.

Infrared-sensing instruments operate at cool temperatures, thereby requiring less solar illumination and a thermally stable platform. Infrared spectrometers are typically the most sensitive, operating at 70–100°K. To operate at this temperature the detector connects by a cold finger to a radiator, which we then point at deep space. The plan for operating the spacecraft must ensure the radiator doesn't see a warm body such as the Sun or a planet. If it does, the viewing time is shorter. The energy input scales as the square of the solar range, so the exposure limits change during the mission.

Remote-sensing instruments typically generate a lot of data. Therefore, they drive the downlink design and, if they over-subscribe the bus, the data-management system. Objects that produce less data, such as fields, particles, and in-situ instruments, are easier to manage operationally. And radio science is easiest of all because it has no telemetry modulated on the downlink. However, if radio science demands highly stable frequencies, we must operate the spacecraft while keeping it as motionless as possible. Thruster firings that induce a translation are undesirable, and even reaction-wheel control can cause a frequency shift. This shift occurs because the antenna is offset from the spacecraft's center of mass, so a pointing offset of the spacecraft introduces a translational motion in the antenna. For an antenna offset one meter from the spacecraft's center of mass, an angular pointing correction of one microradian in one second would induce a motion of 0.000001 m/s, which corresponds to a Doppler frequency shift in X-band of 0.0001 Hz. This frequency shift is comparable to the signal our radio-science experiment will be trying to measure, so we may need to do special attitude control.

Increasing the number of instruments increases operations costs because we have to do more integration and resolve more conflicts between the instruments. If staring instruments fly with scanning instruments, we must plan the spacecraft's attitude to accommodate both. The mission plan of integrated activities results from compromises we typically develop through negotiation and advance planning. If the planning is premature, the operations team may do the planning over again. Conversely, if our planning is too close to the event, we'll have trouble finding compromises.

17.2 Differences Between Interplanetary and Earth-Orbiting Missions

The interplanetary environment, mission duration, and mission objectives cause most of the differences from Earth-orbiting missions. The spacecraft's design for interplanetary missions is also different, and mission operations must deal with these differences.

17.2.1 Orbit Differences

Range and Light Time Delay. The distance from the Earth to the target is the most dramatic difference between interplanetary and Earth missions. Earth missions vary in distance from hundreds to tens of thousands of kilometers, but interplanetary missions range over millions of kilometers. The basic distance unit used is the Astronomical Unit, or AU. The AU is defined as the semi-major axis of the Earth's orbit, equal to 149.6×10^6 km, or 92.96×10^6 miles. The range for the Earth to the spacecraft changes because of the spacecraft's trajectory and the Earth's motion around the Sun [Bate, 1971].

The most apparent difference that the increased range makes is that we have no real-time visibility into the spacecraft's operation. As shown in Table 17.5, even for a mission at the closest interplanetary range (Venus at conjunction), an event seen by the ground has already occurred 2.2 minutes ago. As a result, for any operation that requires ground interaction, we must include at least a round trip light time in the plan. The first one-way light time is used by the signal arriving from the spacecraft to the ground operations; the second half of the round trip is for the return command. A handy rule of thumb is that the speed of light $c \sim 3 \times 10^5$ km/s $\sim$ 0.12 AU/min, or light time ~8.3 min per AU. Similarly, light time is 3.33 seconds per million kilometers. For a Saturn mission, then, we need to know only that the semi-major axis of Saturn's orbit is 9.5 AU to know that one-way light times to Earth will be in the range of 70–90 minutes. Therefore, each interplanetary mission has a time constant associated with its target.

The change in Earth range changes the telecom signal strength during the mission. The cruise phase of the mission will have ranges near zero close to launch and during any Earth flybys, and the science-operations phase will have a range which itself changes by the Sun-Planet distance ±1AU, as the Earth travels around the Sun. As shown in Table 17.5, the telecom signal strength varies in strength from the range alone up to almost 17 dB as the Earth and the target planet approach and recede. For a flyby mission, the telecom signal changes less during the observation of one target because the time is short compared to one year, but the signal strength will vary from target to target.

Table 17.5. Interplanetary Distances and Resulting Environments. Sun-to-spacecraft and Earth-to-spacecraft ranges vary with the planet and the duration of a mission, affecting designs for thermal management, telecommunications, and radiation tolerance.

Planet	Sun Distance (AU*)	Earth Distance (AU)	One Way Light Time Range (min)	Telecom Space Loss Difference Min - Max Range (dB)	Solar Flux at Mean Distance (W/m^2)	Synodic Period[†] with Earth (years)
Mercury	0.31–0.47	0.53–1.47	4.4–12.2	8.79	9066	0.32
Venus	0.72–0.73	0.27–1.73	2.2–14.4	16.07	2601	1.60
Mars	1.38–1.67	0.38–2.67	3.2–22.2	16.88	586	2.13
Jupiter	4.95–5.45	3.95–6.45	32.8–53.5	4.26	50	1.09
Saturn	9.0–10.0	8.0–11.0	66.4–91.3	2.78	15	1.04
Uranus	18.2–20.3	17.2–21.3	142.8–176.8	1.80	4	1.01
Neptune	30.0–30.3	29.0–31.3	241–260	0.66	2	1.01
Pluto	29.5–50.0	28.5–51.0	237–423	5.06	1	1.00
Geostationary	1.0	35,786 km altitude	0.12 s	0 dB	1358	N/A
Moon	1.0	384,400 km	1.28 s	0 dB	1358	N/A

* 1 AU = 149.6 × 10^6 km
† Synodic period is the time required for any phase angle to repeat

Environmental Extremes. The change in Sun range, also shown in Table 17.5, changes the thermal environment that the spacecraft experiences. This thermal change leads to more environmental variation than during Earth missions. The spacecraft must operate through the cruise phase in environments that are different from that at the primary target. Earth missions may have to operate with variations in solar input because the length of occultation varies during the mission, but these durations are typically very short, and the spacecraft re-enters the sunlight before it reaches a steady-state temperature. Radiation and particulate environments also vary during the cruise phase.

The thermal environment requires us to choose the spacecraft's attitude so certain areas are in the shade, and some subsystems are powered down to prevent operations (and heat generation) during peak heating periods. Thermal variations during cruise dictate attitudes and hence maneuver and telecom profiles, as opposed to Earth missions, which have no cruise phase. The telecom profile is much lower if the spacecraft is not allowed to point its high-gain antenna toward the Earth, as is the case for a Sun-pointed spacecraft with a fixed antenna.

The Galileo and Cassini missions, which include Venus flybys to supplement propulsion and therefore allow the spacecraft to have more mass, must survive the solar environment from 0.7 to 5 AU and 0.67 to 10 AU, respectively. The effective input of solar thermal input then changes by up to a factor of 200 from Venus flyby to their final destination. As a result, the operations system has to follow flight rules that depend on the mission. In this example, when near the Sun, the spacecraft must be pointed toward the Sun to hide the electronics and instruments behind a shade. This pointing complicates the maneuver design and forces our fault protection to be able to reacquire the Sun quickly, which might not be necessary during the primary mission.

We must also take into account the solar cycle. Solar activity fluctuates over 11-year cycles. The most recent maximum of the solar cycle was in 2002, so the next solar maximum will occur in 2013. If a mission lasts past the solar maximum, the spacecraft will experience less solar radiation.

Opposition and Conjunction. The relative geometry between the Sun, spacecraft, and Earth for an interplanetary mission changes in ways Earth missions don't see. Figure 17.1 shows the geometric relationships. When the spacecraft and the Sun are aligned as seen from the Earth, the spacecraft is said to be in *conjunction*. When the spacecraft's conjunction places it on the far side of the Sun from the Earth, it is in *superior conjunction*. Superior conjunction restricts communications with the spacecraft due to interference with the radio signal as it passes close to the Sun. Mission planning will disallow or restrict activities around the time of superior conjunction. For a spacecraft whose sun range is less than 1 AU (i.e., its orbit is inside that of the Earth), the spacecraft is in *inferior conjunction* when it's aligned between the Earth and the Sun. When the Sun is on the side of the sky opposite from that of the spacecraft as seen from the Earth (this will occur for a sun range of greater than 1 AU), the spacecraft is at *opposition*.

Coordinate Systems. Having to model coordinate systems is another way an Earth-orbiting mission differs from an interplanetary mission. Understanding the different coordinate systems helps us avoid confusion and potential mistakes during operations. For an Earth mission, an Earth-centered coordinate system is the only one necessary to describe a satellite's motion. Ground stations are modeled in a local geographic coordinate system. For an interplanetary mission, however, the central body of the coordinate system changes, and the reference plane may be a planet, a planet's orbit, or the spacecraft orbit itself.

As shown in Table 17.6 and Fig. 17.2 and 17.3, several different coordinate systems are used. An Earth-equatorial coordinate system is appropriate for launch and for most applications involving tracking of the spacecraft. For interplanetary cruise, the spacecraft is usually in a plane quite close to the ecliptic plane (the plane

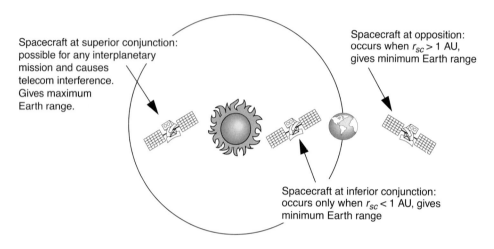

Fig. 17.1. Conjunction and Opposition. At superior conjunction, when the Sun is between spacecraft and Earth, solar activity in the signal path disrupts communications. Conservative interference regions are ±10° for S-band and ±3° for X-band.

of the Earth's orbit about the Sun), so we use an ecliptic-referenced coordinate system for the trajectory and maneuver modeling. For an interplanetary approach, and for the orbital phase of a mapping or a tour mission, the target planet's equatorial coordinate system is the logical choice because this coordinate system most easily represents the orbit dynamics. As shown in Fig. 17.2, the transformation between coordinate systems centered on a planet's equator and the ecliptic coordinate system first requires a rotation through the obliquity of the planet's rotation. The *obliquity* is the angle between a planet pole and the normal to the ecliptic plane. The obliquity of a planet's rotation causes the planet to have seasons.

For targeting to the interplanetary approach, we use the B-plane (Fig. 17.3), referenced to the target planet and the incoming spacecraft's hyperbolic orbit. The encounter conditions are described in terms of the aimpoint or the point where the asymptote crosses the B-plane. The aimpoint can be described in polar coordinates as shown (B and Q), or as a projection on the R and T axes referred to as $\hat{B} \cdot \hat{R}$ and $\hat{B} \cdot \hat{T}$, respectively. The third dimension of the target point is defined in terms of the arrival time, which is equivalent to a displacement along the S direction. The incoming asymptote, not the trajectory, defines the B-plane. Thus, the target plane itself doesn't change if we select a different aimpoint.

Differences in Orbit Mechanics—Interplanetary Escape. We must place the spacecraft on a hyperbolic trajectory to escape Earth and start its interplanetary transfer. On that trajectory, of course, it's still in orbit around the Sun in an elliptical path, which will intersect the target planet's orbit at the arrival date.

Table 17.6. Common Coordinate Systems for Interplanetary Missions. The coordinate system we choose depends on the data we're collecting, the mission phase, and the customer.

Coordinate System	Reference Plane	Reference Direction	Use
J2000 (or EME50)	Earth's equator	Vernal equinox on Jan. 1, 2000 (or 1/1/1950)	Earth-referenced position, DSN viewing geometry
EMO2000 (or 1950)	Earth's orbit (ecliptic)	Vernal equinox on Jan. 1, 2000 (or 1/1/1950)	Interplanetary trajectory
Interplanetary equator of date	Planet's equator	Planet's prime meridian of date	In-orbit geometry
B-plane	Perpendicular to incoming asymptote	Intersection of B-plane with planet's equator	Flyby and orbit insertion targeting

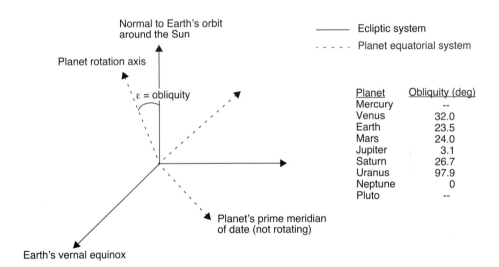

Fig. 17.2. Ecliptic and Equatorial Coordinate Systems. The interplanetary trajectory and interplanetary orbits are expressed in an ecliptic coordinate system. Once in orbit, the trajectory is described in a planet-centered equatorial coordinate system. The two coordinate systems differ by the planet's obliquity and the reference direction.

Launching to the hyperbolic escape trajectory requires the launch site to be in the plane of the hyperbolic orbit at launch. Because the launch site is rotating with the Earth, this requirement constrains the time of day for launch. A typical ascent profile will place the spacecraft into a low-Earth coasting orbit. The rocket's lower stages then fire to ensure the proper direction of the escape asymptote. Finally, an upper stage fires to provide the escape energy.

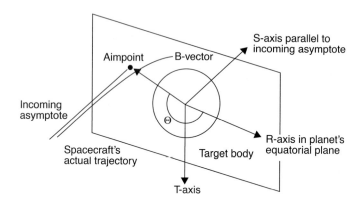

Fig. 17.3. The Planet Target Plane or B-Plane. This coordinate system describes the geometry of an interplanetary encounter by specifying the "aimpoint" in a plane perpendicular to the approach asymptote and containing the target body's center.

Gravity Assist. An interplanetary trajectory to the outer planets may involve intermediate gravity assists to increase the spacecraft's energy. The gravitational attraction of the flyby body changes the direction of the spacecraft's heliocentric trajectory and therefore can increase or decrease the energy of the spacecraft orbit. The Mariner 10 mission first used a gravity assist from Venus to target it to a subsequent Mercury flyby. For the Voyager mission, repeated gravity assists ensured the flyby of four outer planets. Later missions such as Galileo and Cassini rely on successive flybys of the inner planets to achieve the required energy for transfer to Jupiter and Saturn, respectively. Table 17.7 lists the trajectory types used in interplanetary missions.

Table 17.7. Trajectory Types for Different Interplanetary Missions. The trajectory type for any mission varies with the target, the spacecraft mass, the launch vehicle's abilities, and the allowed mission duration.

Trajectory Type	Mission Application	Characteristics
Ballistic/Direct	Inner planet, low energy trajectory	Few maneuvers, short flight time
Delta Velocity-Earth Gravity Assist	Asteroid flyby, outer planets	At least one, large, deep-space maneuver, set up for interplanetary encounter
Interplanetary gravity assist (e.g. Venus-Earth-Earth Gravity Assist)	High-energy trajectories to the outer planets	Repeated interplanetary encounters, maneuvers to avoid striking the planet, long flight time

Orbit Insertion. Orbit insertion is necessary for any mission that plans a long residence time at the target. The orbit-insertion burn mainly removes energy from the spacecraft's orbit, so we want to place it closest to the point at which the spacecraft approaches the target body. That's because an orbit's energy change is related to the applied ΔV through an amplification by the actual orbital velocity ($\Delta E \sim V \Delta V$). To meet this timing requirement, we insert the spacecraft into orbit by using a sequence stored on board—the light-time delay prevents real-time commanding. Figure 17.4 shows an orbit-plane view to the geometry of the orbit-insertion sequence. We target the incoming hyperbolic orbit to the desired inclination and periapsis altitude, and the burn results in the desired capture orbit.

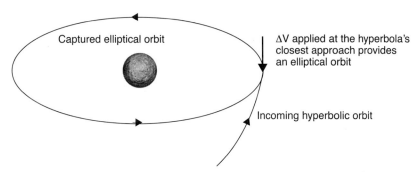

Fig. 17.4.   Orbit-Insertion Geometry. The spacecraft does a retro-maneuver to reduce energy at its closest approach to the planet and therefore goes into a planet-centered orbit.

17.2.2   Differences in Launch Element

Launch Window. For an Earth-orbiting mission, the launch window is usually constrained by geometric considerations such as Sun exposure and lighting at the launch or contingency landing sites (for manned vehicles). Also, because the desired geometry largely repeats from day to day, launch elements don't restrict the launch date. In contrast, the launch period for an interplanetary mission is constrained first by the interplanetary alignment and second by the orientation of the interplanetary transfer plane. Geometric considerations such as lighting then become constraints on the spacecraft and operations design. Section 17.2.1 describes the launch geometry, as do many references on orbital mechanics, such as those listed at the end of Chap. 10.

If we miss the defined launch period, we must usually wait months or longer before the interplanetary alignment allows another opportunity. The Voyager mission, for example, took advantage of an interplanetary alignment that won't be repeated for another 176 years.

The availability of a launch vehicle affects mission planning, as in the case of the Shuttle/Centaur loss to Galileo. We plan a mission assuming a certain capability for the launch vehicle, but program changes sometimes require us to change our assumptions. In such a case, we must replan parts of the mission before launch, often having to assume different capabilities for different phases. A longer cruise, which was required for Galileo, means that the time to wait for science gathering is either extended, if cruise science is disallowed, or shortened, if we allow science gathering early in the encounter, as Galileo did at the Venus flyby less than five months after launch.

The launch vehicle's performance affects post-launch operations mainly in the time before the first post-injection maneuver into the interplanetary transfer orbit. If the time to go before encountering a target body is large, we can change the target conditions a lot with a small ΔV. We require post-injection maneuvers because of errors in our injection maneuver. When an upper stage causes an inaccurate injection, we must correct the problem quickly so the error doesn't continue and require a larger ΔV later.

We can use low-energy launch vehicles or heavy spacecraft by doing several gravity assists during the cruise phase. This approach complicates operations by extending the cruise duration and by increasing the thermal range of exposure. For example, the Galileo trajectory used one gravity assist from Venus and two from Earth to make up for the energy shortfall of the Shuttle/Inertial Upper Stage (IUS) combination compared to the direct trajectory achievable with the Shuttle/Centaur. Galileo's trajectory had three more years of cruise, and the spacecraft had to fly at a Sun range from 0.7 to 5 AU rather than from 1 AU to 5 AU as it had been designed to do.

17.2.3 Differences in Subjects

Unknown Targets. Interplanetary missions are exploratory, so mission operations must adapt. In contrast, with Earth-science missions we know a lot about the target and can make decisions before flight. The uniqueness of interplanetary encounters and the desire to get the most information from a new discovery often require us to retarget observations. We must therefore design the observation strategy with the flexibility to re-target in mind, while being sure not to waste any of the valuable time during which the spacecraft is close to the target.

The target body's position and velocity are also uncertain. This uncertainty is especially true for first encounters with a planet, such as the early flyby missions, but even a return to a planet requires some improvement in planet ephemeris. The observation strategy in this case must allow for errors in one of two ways. As Galileo demonstrated while encountering the Gaspra asteroid, we can take enough pictures, including those of dark sky, to ensure that the target will be in at least one frame. Or, if we have enough information, we can insert timing and position changes into our sequence of observations just before the encounter.

Operations for interplanetary missions require continuing navigation, whereas most Earth missions demand only station keeping. The pre-launch plans usually include a 99% certainty that the mission will have enough propellant.

Successful navigation can save propellant and, therefore, extend the mission. We can keep the operations team in place for an extended mission, and some of the team can plan the extended mission during the primary mission. A prime example of an extended mission is the Voyager II's encounters with Uranus and Neptune, which were made possible by successful encounters with Jupiter and Saturn. In fact, the aimpoint at Saturn was chosen as a compromise between Saturn science and future science goals. The Voyager's operations team dealt successfully with this late change. [Kohlhase, 1989]

Degree of Adaptability. Some interplanetary missions must be very adaptable because the target is unique and unexplored. Other missions, particularly for mapping, can have repetitive operations. Repetition saves workforce, so we use it as much as possible through pre-tested command files. Science planning for some interplanetary missions must adapt to unexplored targets. For some observations, we purposely build in time and other margins to allow changes to the part of the target system we're viewing. New discoveries, such as the volcanoes on Io or Titan's atmosphere, lead to major changes in mission philosophy.

Many interplanetary missions have a payload designed to investigate the target comprehensively. The different objectives and observation types of the instruments create the need for a large committee, the Project Science Group, which defines priorities and resolves conflicts. Sometimes the mission objectives lend themselves to a more fixed observation design. A pair sent to the same target might be an example for which we use results from the first mission to adapt the second. Missions with either one instrument or a mapping-style objective could have very repetitive observations.

17.2.4 Differences in Command and Control Architecture and Ground Element

Continuous Spacecraft Visibility. One important consideration is whether the Earth can contact the spacecraft at any time. A spacecraft with a scan platform for instrument pointing allows us to do real-time downlink and science gathering at the same time. Simultaneously pointing the spacecraft body at the Earth and instruments at the target requires pointing planning to satisfy more constraints, and it may require more simulation. But data management is simpler if capabilities for downlink and data gathering are equal.

If the spacecraft can't communicate and gather science at the same time, the opposite is true. Fewer axes will require less pointing planning and constraint checking, but data management may be more complex.

Stored Commands Versus Real-Time Operations. Because we can't practically send commands in real time, we place most spacecraft activities in a stored file and plan them in advance. The spacecraft carries them out at the appropriate time. We have to work around the basic tension between advance planning and adaptation.

For time-critical, mission-critical events, mission planning must make sure the command file will execute with zero defects. Examples of this type of event are orbit-insertion burns, for which the time window to start the burn is just minutes. The spacecraft typically has a stored sequence of commands, including the spacecraft's response to certain potential faults. The planning, simulation, and execution of an orbit-insertion burn are among the most costly parts of mission operations.

For important, but not time-critical mission events, commands can be sent that will execute upon receipt by the spacecraft. Examples include deploying a probe, such as the Cassini Huygens probe, for which we allow a contingency release at some cost in mission performance. Another example is a ΔV maneuver that isn't time critical, for which we can abort a maneuver at some cost in propellant if anomalous conditions result.

Telecom Link. The downlink is subject to down time due to the limited number of antennas. The DSN is available to many projects, but they must compete for resources. The DSN sponsors a resource-allocation process in which all users participate, including those from ESA and Russia. If the spacecraft is undergoing a critical activity, the DSN will likely assign a high priority to the mission and tracking will be provided. If the spacecraft is doing a routine activity, its pass is more likely to be sacrificed to another mission's critical activity or anomaly. Mission planning must account for the DSN's availability and allow for the loss of some planned tracking passes.

Some missions have managed tape recorders to save data in the case of a DSN outage or to return high-priority data first. Positioning the tape recorder and managing tape travel across the heads are time consuming. Other missions use solid-state recording devices, which don't need tape management because they use software-controlled memory management.

17.2.5 Differences in Mission Operations

Engineering Health and Safety. Spacecraft monitoring relies on alarm limits set for key subsystems. These alarm limits are loaded into the flight software and interplanetary spacecraft must rely on flight software to detect many critical anomalies and safe the spacecraft. This is due to the long one-way light times. In addition, they usually require a trained flight team on station during the mission to filter out false alarms and reset limits as the subsystems become better characterized. We also characterize the spacecraft by trending, or examining the performance of critical parameters over time. Trending spots latent problems in the spacecraft's performance and keeps the team familiar with the details of the spacecraft to help resolve anomalies if necessary.

Reconstruction and Analysis. We must provide for access to preliminary and calibrated science data. The preliminary data, sometimes at level0, helps us assess the health and safety of instruments, just as the flight team assesses spacecraft health. We must also provide spacecraft and mission data to interpret the science data properly. A navigation reconstruction of the trajectory provides positions for the spacecraft, target, and Sun. Also, by analyzing downlink telemetry, we can give the science community parameters for spacecraft pointing and magnetic-field effects.

Data Release. For many interplanetary missions, it can take months to process the science data to a fully usable product. During this period, the science teams often publish their first and most important conclusions, and sometimes have to participate in public forums such as press conferences. After the data products are ready, the science data goes to the public through the planetary data system, so we've completed our task of delivering the data to the taxpayers who funded the program.

17.2.6 Differences in the Space Element

Small Margins. Nearly all resources for interplanetary missions have small operating margins, mainly because we want the best return from a unique mission opportunity. Operating with small or negative margins affects mission operations in three ways. First, there is little room for error in any spacecraft operation, which forces levels of checking and simulation unnecessary in missions with margin. Second, we use the actual performance of the spacecraft and the instruments to update models for future activities, forcing significant replanning. Third, we operate the spacecraft as a set of subsystems, rather than getting the best performance from the whole system. This approach increases workforce and cost.

Power Generation. Solar power is the usual power source for Earth missions, and it's also practical for missions to Venus and Mars. Analyzing a power system for an orbiting mission to one of these planets is similar to that for an Earth mission. The only real difference is that we must size the array to the worst-case combination of Sun range and power usage at that phase in the mission.

The large solar ranges of the outer planets preclude the use of solar power, so we typically use Radioisotope Thermoelectric Generators (RTGs). An RTG's power output degrades very slowly with time—at a rate of 3–4 watts per year. Thus, for missions to the outer planets, we don't consider solar power or reductions in power from occultation.

If no battery is flown, as has been the case with many outer-planet missions, we must manage the power peaks very carefully with time. This management results in power cycling if the payload can over-subscribe the power bus, or in software limitations on spacecraft-controlled power users, such as reaction wheels. We use fault protection to guard against overloads but manage power carefully overall to avoid safing conditions. When we have small margins to return the most science, power management uses a lot of resources.

Thermal Control. The variation in Sun range over the cruise requires us to use different rules near the Sun. For instance, the spacecraft may not be able to point towards the Earth for high-gain antenna communications without exposing its thermally sensitive areas. On those areas, we have to count integrated solar input and thermal cycles as mission consumables. We often avoid damage to science instruments by using deployable covers and special fail-safe operations.

Engineering subsystems are often coupled thermally, and varying solar input requires us to maintain proper conditions. For RTG-powered spacecraft, we sometimes use waste heat from unused electrical power to heat engineering subsystems. This approach further couples the spacecraft-engineering and science operations because the waste heat is inversely proportional to the amount of power used for the payload. An example is the use of RTG waste heat for Galileo's propulsion system. In this case, the heat input varies the operating range of the engines, so modeling of propulsion performance must include the entire spacecraft payload.

Resource Monitoring. For interplanetary missions, the cruise (non-operational) time often dominates the spacecraft's lifetime, so we use strategies to conserve resources during the cruise phase. For example, we can use a different attitude-control deadband to reduce thruster pulses, employ thrusters rather than reaction wheels to reduce wheel revolutions, or leave systems off. We must continually evaluate resource use versus cruise performance to get the best total return from the mission.

We have to model resources in advance and account for them after carrying out an activity. If we have resource margins, we simply check high-level telemetry. However, if we have small margins to acquire the most science data, we must model and count more accurately. Table 17.8 describes some of the typical resources interplanetary missions must model and shows how we treat them differently from their counterparts for Earth-orbiting missions. The resources are renewable or consumable. We budget renewable resources at different times, whereas consumable resources have only one budget for the entire mission.

17.3 Operations Activities Throughout the Mission Lifecycle

17.3.1 Pre-Launch

Designing the Operations System. The time leading up to launch goes to building the ground data system and team structures for mission operations. If mission operations gets staffing early enough in the project-development phase, we can engineer systems for the ground data and flight data at the same time. This style of project development may be the most cost effective in the long run, if operations costs are part of design trades for the spacecraft and ground system. We also may use this development style if the ground system has a lot of inheritance. For many interplanetary missions, however, the operations element is largely unstaffed early in the project. Thus, the ground-system design must react to the spacecraft data system and may not take full advantage of inheritance.

During the assembly and system test of the spacecraft, we generate spacecraft activities used to test the spacecraft and simulate important mission events. We also test critical-event activities, such as launch, maneuver, and orbit insertion.

Table 17.8. Typical Operational Activities to Best Use Spacecraft Resources. A resource-constrained interplanetary mission may increase mission return by intensifying operations activities.

Resource	Comment	How Interplanetary Missions Differ from Earth Orbiters
Renewable Resources		
Command bandwidth	Assign commands execution time within spacecraft frame time	Transmission delays prevent real-time execution
Telemetry bandwidth	Model link performance with high fidelity to ensure data capture—remove modeling if adequate margins are available	Telecom range varies less with Earth orbiters over both pass and mission duration
Command memory usage	Estimate repeatedly as planning develops—simulation verifies estimates before load is transmitted	We develop some software after launch and upload it to the spacecraft during the mission
Consumable Resources		
Fuel	Use worst case for preliminary planning; replan or extend mission if worst doesn't occur	Navigation use is usually much larger than attitude control
Thermal cycles	Plan options for both heating and cooling	Thermal environment more extreme

These tests verify that the spacecraft will behave as anticipated and that its computer can read the product of the command-generation software. The simulated activities used to test the flight team contain one or more "hiccups," which make the test more like real life. In this fashion, we develop and have pre-launch procedures. Also, the flight team, who are different from the spacecraft-development team, can learn expected spacecraft characteristics.

Mission Planning Before Launch. A major pre-launch activity is integrated activity planning for mission operations. An Integrated Mission Operations Plan (IMOP) will include activities for people who handle the spacecraft, DSN, science gathering, and ground systems. The IMOP allows us to determine the spacecraft and ground resources needed for the planned activities. Before launch, planning is at a fairly high level, so future changes won't affect it much. As in spacecraft design, it helps to have experienced operators participating in this planning, so our staffing and cost figures will be realistic. The mission plans after launch will then be more robust, and the mission will be less susceptible to cost overruns from grandiose plans or to reduced return from insufficient staffing.

17.3.2 Launch

The launch phase includes the launch event and initial characterization of the spacecraft. After lift-off, we insert the spacecraft into its parking orbit, align the parking orbit properly with the interplanetary transfer orbit, and inject the spacecraft into the transfer orbit. The first major events after injection are deployments, early pressurizing and priming of the propulsion system, calibrating attitude-control systems, and deploying of booms and antennas.

The events in the initial stages of the mission are limited to those that support the first trajectory-correction maneuver. Compared to an Earth-orbiting mission, whose main propulsive events may be complete within a few days of launch, an interplanetary mission usually runs the first of many propulsive maneuvers at 10–25 days following launch. The first correction maneuver is mainly to correct injection errors, which develop from errors in the amount of burn or pointing of the launch vehicle.

Besides initializing the spacecraft, preparations for the first maneuver include determining the initial orbit by using the DSN tracking stations. Navigation will solve for trajectories and use the final available one before the maneuver to design a correction maneuver back to the desired trajectory. Spacecraft outgassing, which produces a force like that of solar pressure on the spacecraft, complicates determination of the initial orbit. We need the spacecraft's area and reflectivity to model the effects of solar pressure, but we know them only roughly until after launch.

Launch Approval. Launch approval is different for US Air Force, non-US, and human-rated vehicles but in all cases we need to plan for contingencies or failure. For launches from the Shuttle, we must plan for interplanetary injection on more than one revolution of the parking orbit as well as for landing with the spacecraft still in the Shuttle bay.

Using RTGs to generate power complicates launch approval significantly. For an RTG-powered mission, we must take special precautions to ensure RTG safety after launch. Mainly, we have to bias the trajectory or aim the spacecraft away from the desired flyby point at the next flyby. Biasing reduces the spacecraft's probability of hitting a planet, but it requires many more larger, pre-determined maneuvers, which may be critical to the mission.

Activities During the Launch Period. The spacecraft's angular rates during injection require acquisition aids and a hand-off between 26-m and 34-m or 70-m coverage. Sometimes, Advanced Range Instrumentation Aircraft (ARIA) first acquire the spacecraft during the launch-vehicle upper stage's coast and injection. Spacecraft telemetry routes through the upper stage to provide visibility. Following the injection burn, we use the predicted state vectors for the spacecraft to provide pointing for the DSN's 26-m antennas. The 26-m antennas provide a higher angular rate than the larger 34-m antennas.

Control of the spacecraft passes between several agencies and tracking networks. Sometimes, different centers control the launch vehicle and upper stage, and a host center such as JPL controls the spacecraft.

We cover the spacecraft continuously during the critical first two to four weeks of operations, which characterizes its outgassing, trajectory performance such as solar-pressure effects, and attitude-control performance. By receiving spacecraft telemetry continuously, we can compare actual against predicted performance to understand behaviors of the spacecraft subsystems. The spacecraft's performance is largely unknown at launch, including outgassing and uncalibrated performance that we must correct.

17.3.3 Cruise

Cruise is sometimes the longest phase, typically beginning after correcting the injection and lasting until primary science observations begin. For multiple flyby missions such as Voyager, the period of up to four years between flybys is really a cruise phase as well.

Typical activities include calibrating and maintaining the spacecraft and payload, plus continued navigation. We use the cruise phase to learn how the spacecraft flies, which is usually somewhat different from the design specifications. Other engineering activities may include deployments delayed for some reason from immediately after launch. An example is the Galileo antenna, which operators tried to deploy 18 months after launch, when the thermal environment was within the design tolerance.

Cruise science includes intermediate interplanetary encounters, as well as observing fields and particles. For many missions, the cruise environment is relatively benign, but it offers a chance for the science-operations team to learn their instrument and its operational procedures.

Effect of the Launch Period on Cruise Operations. Continuous coverage must span the potential first 30 days for a 30-day launch period. We have to negotiate for the coverage from the DSN and be prepared to release coverage once we know the launch date. This may include strategies to retain coverage only around the most significant events, such as the first trajectory-change maneuver for launches that occur late in the launch period.

The coast duration from Earth orbit to interplanetary orbit varies across the launch period. We provide targeting information to the launch vehicle, typically in the form of time-varying polynomials over the launch period. For a launch from the shuttle, the performance varies depending on the deploy orbit, and not all orbits achieve the right conditions. The target polynomials, and the initial tracking and staffing plans, must account for the 90-minute difference (roughly the Shuttle's orbit period) between injection orbits. For some launch vehicles, the azimuth of the first-stage launch varies across the daily launch window, which may change our strategy for tracking or acquiring the launch vehicle.

If there are intermediate encounters, their dates will change with different launch dates, implying different coverage schedules and spacecraft sequences. For example, during Galileo's month-long launch period, the date of the Venus flyby varied by approximately one week. Most of the variation was handled between the Earth and Venus by a low-activity stored sequence, which was variable in length.

If science is planned for intermediate encounters, it must be somewhat flexible until after the launch date. This is especially true in the case of targeted observations, which will change not only in time over the launch period but in pointing direction and resolution as the flyby altitude and lighting conditions change.

Navigation. Navigation is one of the primary cruise functions; it includes calculating orbits, designing maneuvers, and propagating trajectories. Data for orbit determination contains different information for interplanetary missions than for Earth-orbiting missions. Two-way coherent Doppler data is the main type used, with two-way ranging data adding information.

The quality of the orbit determination is a function of the amount of data and the trajectory's characteristics. In particular, low-declination trajectories, or imminent interplanetary flybys, require special strategies to collect the data for orbit determination. The Doppler data has a singularity at zero declination relative to the Earth and we use additional ranging or different data types to make up for the problem. We have to plan for the additional data, and for the performance impact if critical events occur with this geometry.

The first maneuver following injection is typically large, may remove some injection bias, and is the first on-orbit use of the propulsion system. The cruise phase of a mission contains many maneuvers to shape the trajectory, correct navigation errors, and maintain biases that keep the spacecraft from hitting a planet. Compared to an Earth-orbiting mission, the propulsion system is larger and more complex, and it interacts more with future mission operations. The interaction comes about because using updated performance estimates for the mission leads to different estimates for propellant margins and then to different strategies for propulsion. A primary way to save money in the operations phase is to limit changing plans based on updated performance. This would reduce a key operations task—reconstructing the maneuver performance using data for subsystem performance and navigation tracking.

Around each targeted flyby, we must do a sequence of targeting and post-flyby cleanup maneuvers. The final targeting maneuvers are small but are required to remove most of the previous errors in carrying out maneuvers and calculating orbits while allowing little time for remaining errors to corrupt the flyby. Post-encounter maneuvers mainly correct for excesses or shortfalls in energy caused by the flyby. The maneuver occurs shortly after the flyby to regain the nominal trajectory before velocity errors propagate with time into larger position errors.

Maintaining Engineering and Science Equipment. Both science and engineering equipment must be checked out, maintained, and calibrated during the cruise phase. We don't understand clearly the physics of many mechanisms which may cause damage during the long cruise, largely because we can't test these mechanisms for the same amount of time before launch. Therefore, we do prudent but sometimes costly maintenance, such as regular motor motion.

Earth-orbiting missions rarely wait years before using important engineering capabilities. One example of such an engineering capability for an interplanetary mission is an engine used only for orbit insertion and not for regular trajectory maintenance during cruise. Magellan, for instance, used a solid rocket motor for orbit insertion after storing it while cruising for 15 months [Young, 1990]. Other than for orbit insertion, we typically use engineering equipment right away and check it out quickly.

Science instruments are mostly idle during cruise, but early checkout provides practice and often exposes problems. Routine maintenance and calibration of science instruments often cross the line into real science collection, with full demands on operators.

Maintaining and developing flight software constitute another activity during cruise. We have to maintain software because we discover errors or learn more about the spacecraft during the mission. We develop software for three reasons: planned-for capability has been deferred from before to after launch, spacecraft anomalies force a rework of algorithms, or the mission has been extended into a regime we haven't planned for. All three cases require us to know all of the spacecraft's inner workings—in effect, to become development engineers. If the original development team has disbanded, operators may face a significant challenge.

17.3.4 Intermediate Encounters

For the operations team, intermediate gravity assists mean that the spacecraft needs accurate navigation at intermediate points during the mission. For a mission such as Galileo, navigation must operate fully for a six-year cruise and a two-year tour. Also, the availability of a gravity-assist trajectory drives costs throughout the mission. Galileo's direct trajectory would have required a three-year rather than a six-year cruise.

Observing intermediate flyby planets during the flyby adds value for three reasons. First, the instruments flown on interplanetary spacecraft are often more comprehensive than those on Earth-orbiting spacecraft, allowing simultaneous observation of the target in several wavelengths of energy levels. Second, intermediate encounters provide science, which adds value to the mission and the interplanetary science data. Third, intermediate encounters give the instrument and flight teams valuable information about checkout and calibration.

17.3.5 Arrival

Arrival at the target body is crucial for practice, early observations of an unknown target, and the last science before orbit insertion. If the mission is a flyby, we must plan for science on the inbound and outbound leg of the flyby. If it's eventually an orbital mission, we'll have to trade between ensuring a successful orbit-insertion maneuver and doing pre-encounter science with a healthy and understood spacecraft. Also, unique science opportunities may occur near orbit insertion because the target planet is so close.

Targeting. Early observations of the target (optical navigation) and continued radiometric tracking combine for navigation to the target. The radiometric data determines the spacecraft's Sun-centered trajectory. Optical data for navigation, which images the target planet in the same frame as a known star, positions the spacecraft with respect to the target body. The information combines to improve the target's Sun-centered position or the planet's ephemeris. With this combination, the spacecraft and the target are mutually positioned at the insertion time with respect to each other.

With additional data, we improve our ability to hit the desired aimpoint. We use several pre-encounter maneuvers to achieve this. Early maneuvers can change the spacecraft aimpoint a lot but are therefore susceptible to large errors. Later maneuvers have less leverage but are more accurate. A typical sequence for an interplanetary encounter includes maneuvers at –60, –30, and –10 days.

Orbit Insertion. Orbit insertion is a unique opportunity because it requires the spacecraft to have a certain energy with respect to the target body and the Sun. The best point is where the heliocentric orbits of the spacecraft and the target planet are nearly aligned and are nearly the same size. That's where we need the least velocity change to orbit the planet. In a planet-centered frame, the most efficient point is at the closest approach point, where the velocity, and hence the effect on orbital-energy change, is greatest. We time the insertion burn by trading between the most efficient mission and one that allows early completion of the science, which may drive an earlier, less efficient, insertion strategy.

Due to light-time delay, this event happens in the blind, so we have to balance anxiety while no link is available against response time in case something goes wrong. We must therefore test the stored sequence for the event against any possibility of problems. Also, the spacecraft is usually in a special fail-safe mode for this critical event. This mode may include powering down science instruments to create extra power margin or powering up otherwise unused subsystems for engineering backup.

17.3.6 Encounter

Encounter lasts from hours to years for the different mission types and is the main period for science gathering. Thus, operations have historically been fairly intensive for this phase, when we take very little risk with spacecraft or science data. Reducing risk drives up the cost of operations.

Checkout on Orbit. We'll first use many instruments and observation strategies during the initial orbital phases. For some experiments, the pre-encounter time is enough to wring out problems with instrument operations. Instruments that require a target to operate successfully, such as in-situ or active remote-sensing instruments, must be checked out early in the orbital-operations phase.

Science Operations on Orbit. Sharing all of the time in a mission between different instruments gives us the best science data but uses all margins and a lot of workforce and hence dollars. Allocating time instrument by instrument is inefficient but decreases integration costs.

Strategy for Engineering Maintenance. Continuously calibrating engineering subsystems increases performance but costs time in planning, integration, and analysis of the downlink. Also, in the past, we've used continuous engineering data to get the best returns from a mission. Doing so requires more uplink and downlink activity for characterization and analysis, plus replanning of subsequent activities to take advantage of performance changes.

17.3.7 Extended Mission

An extended mission is often possible, but we must take more risks to achieve it. Interplanetary missions are often extended because we've used worst-case or 99% planning throughout the mission. If the mission ends without significant problems, the spacecraft is usually proven reliable and the consumables aren't exhausted. For example, post-landing resource margins permitted the Mars rovers Spirit and Opportunity to have extended science phases. The originally planned science gathering time was three months; as of this writing, NASA has extended this phase to 18 months.

An extended mission means two things for mission operations. First, mission planners always keep the extended mission in their thinking. An example is the Voyager flybys of Saturn and Uranus, which were constrained because of the chance of going to another target. (The Jupiter flyby was similarly constrained because of the planned Saturn flyby [Kohlhase, 1989].) Second, costs tend to exceed pre-launch estimates. But we can reduce this effect by accepting more risk and shrinking the flight team as the Magellan mission did. Extended mission costs can be reduced by an experienced operations team being able to do the same job with fewer people.

17.3.8 Ongoing Mission Planning

Mission planning is continually revisited on most interplanetary missions because of increasing accuracy of the predicted spacecraft trajectory. These updates can result in either large or small changes in the mission plan. If, for example, the spacecraft performance is markedly better (or worse) than predicted, targets may be added (or deleted). An example of this is Galileo's flyby of the asteroid Ida, which was enabled by a launch early in the month-long launch period, and also by favorable spacecraft performance. The propellant budget then allowed the "detour" from the optimal trajectory to fly by the asteroid.

Small updates to planned activities result from collecting and processing navigation data. The navigation aspect of the pointing problem is essential due to the pointing of most observations relative to the target, coupled with the unknown position of the spacecraft relative to the target. A late navigation solution is folded into the observation design by adjusting the start time of an entire block of observations.

Pre-encounter analysis may allow us to account for expected navigation errors by planning an observation strategy that includes all possible positions of the target. In this case, the mission-planning function will be in an iterative loop with the performance of the navigation system throughout the mission, in order to drive out inefficiencies with more accurate predictions.

Many interplanetary missions have used most of the spacecraft's resources at the expense of ground resources, which means that operations plans were very tightly coupled and therefore very sensitive to change. For example, a change in DSN station from one hemisphere to another will result in a shorter pass and a lower peak data rate. If we've planned the data flow to use all of the planned capability, the change in station will force us to replan activities.

Improved understanding of performance leads to changes in the telecom link, which may change the observation strategy if a planned scenario is optimized to link performance. If we don't need top performance, small changes in the link performance wouldn't result in observation changes.

We may improve the spacecraft-pointing performance by frequently calibrating elements such as gyros, but we must trade time for this activity against time for science observations.

We can adapt to some changes in the trajectory that improve science at an encounter without seriously affecting future objectives. In an orbital-tour mission, we can sometimes best correct navigation errors by reoptimizing the trajectory rather than by returning to the nominal trajectory.

References

Bate, Roger D., Donald D. Mueller, and Jerry E. White. 1971. *Fundamentals of Astrodynamics.* New York, NY: Dover Publications, Inc.

Beatty, J.K., B. O'Leary, and A. Chaikin. 1990. *The New Solar System.* Cambridge MA: Cambridge University Press.

Kohlhase, C. ed. 1989. *The Voyager Neptune Travel Guide.* JPL Publication 89-24. California Institute of Technology, Pasadena, CA.

Larson, Wiley J. and James R. Wertz. 1999. *Space Mission Analysis and Design.* Third Edition. Netherlands: Kluwer Academic Publishers.

Young, C. ed. 1990. *The Magellan Venus Explorers Guide.* JPL Publication 90-24. California Institute of Technology, Pasadena, CA.

Chapter 18
Software Engineering

Juan Ceva, Kim Chacon, Neal Gaborno, Aaron Silver, and Gary Thomas, *Raytheon Company*

- 18.1 Software Engineering Process and Software Standards—CMM® and CMMI®
- 18.2 The Eyes and Ears of Management—Project Monitoring and Software Quality Assurance
- 18.3 Sizing and Pricing Software
- 18.4 Staffing Software Projects—Organization and the Right Mix of Skills
- 18.5 Software Lifecycles
- 18.6 Focusing on What is Needed for Launch

In today's environment it is very likely that the Mission Operations Manager (MOM) will be involved in some way with the development of big software components, either for a ground or space system. If so, the MOM should be familiar with certain important aspects of software engineering. This chapter is intended to be an introduction to those aspects, a primer if you will. Section 18.1 provides an introduction to industry best practices, best practices that Raytheon and its legacy companies have helped shaped over the past decades. The rest of the chapter discusses these best practices in more detail and is intended to bring to the surface some of the most common challenges that the software manager will face. In this spirit, Sec. 18.2 describes how to stay on top of software projects. Section 18.3 continues with one of the most challenging aspects of managing software projects: its correct sizing and associated pricing. Section 18.4 discusses an effective way of organizing software projects to create efficient teams. Section 18.5 describes typical software lifecycle models and current trends in the field. Finally, Sec. 18.6 serves as a final word of advice, giving particular attention to software issues related to a satellite launch.

18.1 Software Engineering Process and Software Standards—CMM® and CMMI®

"The quality of a product is largely determined by the quality of the process that is used to develop and maintain it." [Total Quality Management (TQM) principles as established by Shewhart, Juran, Deming and Humphrey]

A MOM that oversees the production of software must be familiar with industry best practices in software engineering and see to it that his or her projects incorporate these practices (or a subset of them).

With the quote at the beginning of this section, Humphrey—the father of the formalization of software processes, together with other pioneers behind the concept of Total Quality Management (TQM)—stressed the relationship between "quality" and "process", or by extension, between "software quality" and "software processes".

Before going any further, let's define the words "quality" and "process". In the words of Humphrey [1994], "quality" is defined from the perspective of the software user[*]. Does the software work? Does the software provide the functionality the user needs in a timely manner? How easy is it to use? How compatible is it with the user's other software systems? How easy will it be to remember how to use it in the future? These questions and their answers determine the framework in establishing what "quality" means from a user-centric perspective. The Software Engineering Institute (SEI), defines "process" as: "a set of practices performed to achieve a given purpose; it may include tools, methods, materials, and/or people".

Although it is true that "process" does not always lead to quality software and successful software development projects, there is little evidence that good quality and successful software development projects are possible without it. Of course, we acknowledge that there are other aspects that contribute to the success of software development projects. We do not ignore the fact that behind good processes are good people, nor that technology is also essential to successful enterprises. In fact, traditionally, process is described as a leg of the process-people-technology triad, but we stress that it may also be considered the "glue" that unifies the other two aspects.

In summary, although process is not the only driver of software quality, it is essential in achieving it. To that end process models have been established. These

---

[*] The term "user" refers not only to persons, the traditional users, but also to non-animated systems and subsystems. In this sense for example, a spacecraft instrument can be "a user" of the spacecraft attitude subsystem.

process models are collections of elements that describe characteristics of processes that experience has proven to be effective. These models can serve as a guide for improving organizational processes. The so-called capability maturity models have emerged to provide a reference model for the maturity of the practices in a specific discipline in an organization.

18.1.1 A Brief Discussion of CMM® and CMMI®

This section discusses the SEI capability maturity models that describe a set of key practices for the development and management of quality systems and products, including software. A MOM that is just starting to become familiar with software engineering should see these capability maturity models as a set of best practices to follow rather than as a "yardstick" to measure the maturity level of his of her organization.

The SEI is a federally-funded research and development center operated by Carnegie Mellon University and sponsored by the U.S. Department of Defense. The SEI objective is to provide leadership in software engineering and in the transition of new software engineering technology into practice (see the institute website referenced at the end of this chapter). The capability maturity models (CMM® and CMMI®) establish the maturity of an organization's software processes by determining if certain best practices are being followed in the software projects and overarching organization. Simply put, the capability maturity model helps us answer the question: On a scale of 1 to 5, how mature is our software process? The CMM®/ CMMI® framework comprises five levels of process maturity, each with specific characteristics and features. Even if we do not need to know the maturity level of our organization's software practices, the key practices associated with CMM®/ CMMI® can serve as a checklist to establish well validated software development and management processes within the organization.

CMM® was established (CMM® version 1.1 was created in 1993) for software engineering. Because of the pervasiveness of software over the last decade or so, the acknowledgment that software is a part of a bigger system, and also in an attempt to bridge the artificial gap between software and system engineering, the Capability Maturity® Model Integration (CMMI®) was introduced in 2001 (the CMMI® SE-SW, Software Engineering-System Engineering version 1.1 was created in December 2001). At the level of detail that we address in this chapter, the important concept is that CMMI® integrates the disciplines of systems and software engineering into a single process improvement framework. CMMI® replaced CMM® at the end of 2003.

A description of the five levels of maturity follows* (see Fig. 18.1).

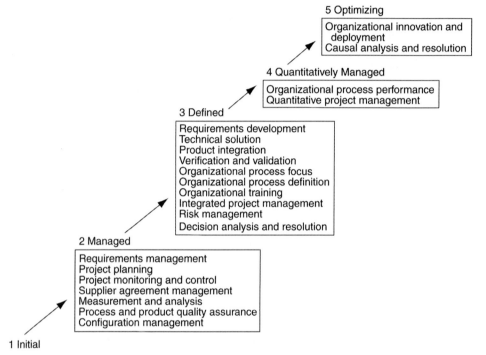

Fig. 18.1. CMMI® Levels of Maturity and Associated Practices. An organization can assess its level of maturity according to this model by checking to see if the associated practices for the different maturity levels (outlined inside the boxes) are being followed.

Level One—Initial: An organization is at Level One (the default level) when the software process, if one can call it that, is ad hoc, and even chaotic. Management panic and continual "fire fighting" by a group of heroes are the rule. Any attempts to follow existing standards and defined process are quickly abandoned during a crisis. Naturally, the cost of producing software with this modus operandi is astronomical. It pays many fold to invest in evolving to the next level of maturity.

---

* The original CMM has a "staged" representation. That is, in order to be rated at, say, level 4, we have to demonstrate that our organization meets all the criteria for Level 2 and Level 3. All organizations are Level 1 by default; in other words, there is no Level 0. CMMI® changes this; besides the "staged" representation, there is a "continuous" representation in which an organization can get recognition of maturity by demonstrating the use of certain practices of a level without having implemented all the key practices of that level or lower levels.

Level Two—Managed: Organizations at this level of maturity have established basic and consistent project management processes to track cost, schedule, and functionality. The process discipline is in place to repeat earlier successes on projects with similar applications. However, the discipline and processes, while established, vary from project to project.

Level Three—Defined: At this level of maturity, the process is uniform across the organization. The software processes for both management and engineering activities are documented, standardized, and integrated into a standard software process common to the entire organization. All projects use an approved, tailored version of the organization's standard software process for developing and maintaining software.

Level Four—Quantitatively Managed: To be at this level of maturity, the organization needs to collect detailed measurements of software processes and product quality (see Sec. 18.2). These measurements drive strategic decisions for the organization. Both the software process and the products are quantitatively understood and controlled.

Level Five—Optimizing: At the highest level of maturity, continuous process improvement is the norm. Quantitative metrics are used to seek areas of improvement. Where improvements are needed, innovative ideas and technologies are piloted. If these initiatives lead to measurable improvements, they become the new standards.

Under each of these levels of maturity, "key process areas", or activities one must have in place, are specified (see Fig. 18-1 for these key activities on each level). For detailed information, the reader should visit the SEI website.

In summary, what is important to the MOM is that certain key practices (as outlined in the CMMI® model) are essential to the success of software development projects. A MOM should not necessarily be concerned about whether the organization has a CMM® or CMMI® rating, but rather that the practices associated with CMM®/CMMI® (or something very close to them) be implemented. Certainly, those associated with Level 2 (Managed) provide the basics, and constitute a minimum. Similarly, if we have subcontracted the software to a third party, we should make sure that the activities associated with Level 2's Supplier Agreement Management are followed, whether or not the supplier has a CMM® or CMMI® rating.

18.2 The Eyes and Ears of Management—Project Monitoring and Software Quality Assurance

The purpose of Software Quality Assurance (SQA) is to provide management with appropriate insight into the status of the software project and the process that is followed. In the context of software engineering, SQA is, simply put, the eyes and ears of upper management. It is precisely because of this that SQA is important to the MOM. SQA and Project Monitoring are key practices in the basic Level 2 Capability Maturity Model (cf. Section 18.1).

Typically the Software Project Manager (SPM) will report to the MOM, and the SQA group keeps the MOM apprised of the adherence of a given software project to the defined process and its overall progress. To this end, the SQA organization—sometimes called Software Quality Engineering (SQE)—will develop a Software Quality Program Plan (SQPP), which defines the software quality plans specific to the project, including the software work products or items to be reviewed and/or evaluated. SQE also normally reviews and approves any items, such as deliverable documents and software executables, prior to delivery to a customer.

18.2.1 Management Oversight and Software Metrics

Once a project has tailored its organizational software processes to meet its specific project requirements, and the project software development plan/process documents are in place, the organizational management and MOM need regular insight into the status of the project. This can be done effectively if the SPM generates a monthly software project-reporting package, containing all project metrics and current status (as of the end of the previous financial/accounting month). The metrics should provide sufficient, yet concise, insight into the status of the project. The SPM should conduct monthly reviews for the software organization management, as well as for the MOM. The MOM should also conduct weekly status meetings and/or reviews, to ensure the effort is on track for the overall program.

SQE performs regular evaluations of the software effort, to ensure the software project team is complying with the documented software processes. For example, the SQE assigned to a program might do a regular evaluation to ensure that the SPM is compliant in implementing software metrics in the monthly reporting package. In general, SQE "grades" the SPM on his or her team's compliance with the processes. The SQE organization should be separate and independent from the software organization, so as to be an impartial judge of the software project status. SQE reports regular status on the software project to both the MOM and software organization management.

Often one hears from software project managers in less mature organizations that they are too busy to generate these metrics or status reports. In reality, they can't afford not to do it. Monthly (or sometimes even weekly) metrics reports from the SPM, together with the monthly SQE status reports, give both the MOM and software organization management sufficient insight into the status and performance of the software project. A comprehensive set of software metrics can provide a quick and simple way to obtain insight into the status of the software effort and to measure and analyze the performance relative to the plan. This is the only way to be proactive in resolving issues before they become problems. Table 18.1 shows a simple checklist for the MOM to determine if the software effort has the proper safeguards for management oversight. An affirmative answer to all the questions simply indicates that problems will be better under control. A negative answer to any of the questions should be a reason for concern.

Table 18.1. MOM Software Process Checklist. In this checklist, the MOM must understand the reason for any NO answer. Such an answer indicates increased development risks (schedule, cost and/or quality).

Item to Check	Yes/No
Does the SPM understand the software processes of the software organization?	
Have the organization software processes been tailored and documented in a Project Software Development Plan/Process document?	
Does the Software Project Manager report to the MOM?	
Does the SPM conduct monthly reviews?	
Does the MOM conduct weekly status meetings?	
Are the software metrics defined and collected?	
Are compliance reviews held for software process metric collection?	

Typical Software Metrics

Mature software organizations specify under a quantitative Software Management Plan (SMP) the practices for the organizational data collection, measurement, analysis, and dissemination of metrics. The primary focus of these practices is to evaluate the effectiveness of the software engineering processes being used, as well as to quantify the process improvement goals. Secondary objectives are: (1) "benchmarking" to determine current operational software parameters and (2) subsequent "calibration" of software process models to predict performance. Metrics provide feedback on how well a project, process or discipline is performing and allow for decisions—schedule priority, staffing resources, management emphasis, investments, process definition, etc., based on objective and quantifiable information. In a sense, we can think of metrics as providing "closed loop" rather than "open loop" control of the product.

Software data collection, measurement, analysis, and dissemination are an important field on their own. Dr. Aaron Silver has written extensively on this subject [Silver 1993, 1994], and the reader is directed to the references at the end of this chapter for a more in depth discussion. Tables 18.2 and 18.3 summarize the basic categories of software metrics typically collected. However, within these basic categories the "specific" set of metrics may be further refined. For example, Cost Performance Index (CPI) and Schedule Performance Index (SPI) as defined in the given figures are important management indicators of program "earned value". Figure 18.2 illustrates some additional management performance characteristics, including "estimate TO complete," "estimate AT complete," work efficiency, total variance, etc. Figure 18.3 gives definitions for some of the especially useful performance metrics.

Table 18.2. Typical Programmatic and Performance Metrics. These metrics and the variation of these metrics against a preplanned baseline are useful indicators of the health of a software development effort.

	Programmatic and Performance Metrics
Programmatic Metrics	Software Cost Software Schedule Software Effort (Person-Months) Software Size (Executable Lines of Code) Software Complexity (low, medium, high) based on completed projects Software Risk (Economical and Technological) Staff (Headcount)
Performance Metrics	Software Defects (Severity Levels) Software Defect Density (Defects per 1,000 source lines of code (SLOC)) Software Productivity (SLOC/Staff-Month) Software CPU and Memory (flops, MB) Software SLOC Growth Rate Software Stability and Volatility (0-1) Software Operational Availability (0-1)

Table 18.3. Typical Quality & Computer Science Metrics. The performance measurement of most of these assets is expressed in terms of the "probability" (between 0 and 1) as measured over the operational asset lifecycle. For example, "reliability" is obtained by integrating the overall system attributes over time to calculate "total system reliability". Similarly, the other assets (dependability, expandability, survivability, etc.) are measured and expressed as probabilities of success across the operational lifecycle.

	Quality and Computer Science Metrics
Quality Metrics	Software Reliability (0-1) Software Maintainability (0-1) Software Dependability (0-1) Software Inter-operability (0-1) Software Survivability (0-1) Software Expandability (0-1) Software Portability (0-1)
Computer Science Metrics	Number of Common and Unique Operators Number of Common and Unique Operands Software Cyclomatic Complexity (0-1) Software Volume Software Efficiency (0-1) OOA/OOD*—Depth of Inheritance (object-oriented software) OOA/OOD—Lack of Cohesion (object oriented software)

* OOA = Object Oriented Analysis; OOD = Object Oriented Design.

Cost Performance Index (CPI) and Schedule Performance Index (SPI) Metrics

Input Parameters:

a) BCWS: Budgeted Cost of Work Scheduled
b) BCWP: Budgeted Cost of Word Performed
c) ACWP: Actual Cost of Work Performed

Calculated Values:

Schedule Performance Index: SPI=BCWP/BCWS
Cost Performance Index: CPI=BCWP/ACWP
Schedule Variance: SV=(BCWP)–(BCWS)
Cost Variance: CV=(BCWP)–(ACWP)
Percent Schedule Variance: PCTSV=((BCWP–BCWS) (BCWS)* 100
Percent Cost Variance: PCTCV=((BBCWP–ACWP)/BCWP)* 100

Fig. 18.2. **Cost Performance Index (CPI) and Schedule Performance Index (SPI).** These metrics are part of the industry-standard Earned Value Management System (EVMS).

Performance Metrics

Input Parameters:
a) Base plan
b) Revised plan
c) Previous actuals
d) Percent complete

Calculated Values:
Cumulative actuals=(previous actuals) + (actuals)
Estimate to complete = (base plan) – (earned value)
Estimate at complete = (cumulative actuals) + (estimate to complete)
Variance = (revised plan) – (estimate at complete)
Earned value = (revised plan) * (percent complete)
Work efficiency index = (earned value) / (cumulative actuals)
Projected estimate at complete = (cumulative actuals) * (revised plan) / (earned value)
Calculated estimate to complete = (base plan – earned value) / (work effort)

Fig. 18.3. **Specific Program Management Performance Metrics.** These metrics represent other useful management metrics to track the progress of a program. The input parameters are defined as follows: The base plan is the original plan for schedule and associated cost. A modification to this plan (normally agreed to by all parties) constitutes a revised plan. Actuals refer to the "actual" cost incurred in a given (defined) period. Finally, percent complete is the percentage amount of work units completed.

Observe that the "ideal" values for both Cost and Schedule Performance Indexes are 1.0, which implies "on" cost and "on" schedule. The idea is that for each "unit" of resource expended (i.e., each dollar), one anticipates a corresponding unit of accomplishment (i.e., return). Thus, when tracking these parameters, a "bull's eye" chart of SPI (x axis) versus CPI (y axis) is commonly used to display these results. Fig. 18.4 shows the program variability associated with some sample data.

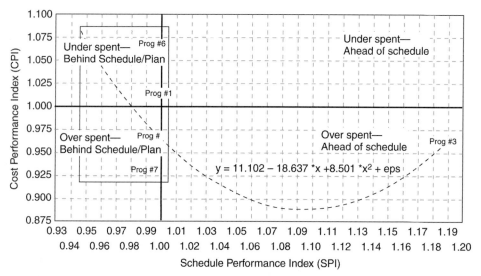

Fig. 18.4. SPI Versus CPI Performance Index. The plot is divided into four quadrants. CPI equal to 1.0 means that spending matches the planned amount of work completed. SPI equal to 1.0 means that the effort is right on schedule, or that the amount of work planned, but no more, has been completed. Counter clockwise from bottom left to top left, a project in the bottom-left quadrant [worst case] is overspent (i.e., it's spent more than planned to get the current amount of work completed) and behind schedule. A project in the bottom-right quadrant is overspent but ahead of schedule (i.e., it is completed more work than planned, which may be why more has been spent). A project in the top-right quadrant [best case] is under spent (i.e., it has spent less than planned to get the current amount of work completed) and ahead of schedule. Finally, a project in top-left quadrant is under spent, but behind schedule (which may be why the effort is under spent).

18.3 Sizing and Pricing Software

The software estimation process consists of two principal processes: estimating size and determining the amount of software effort required to develop it. A description of these processes follows.

18.3.1 Software Sizing

Over the past three decades there have been several serious attempts to formulate a definitive and objective measure of "software size." The efforts have ranged from a simple count of the number of source lines of code to the specification of "functionality" [De Marco, 1982]. More recently, with the introduction of Object Oriented Design and Object Oriented Analysis (OOD/OOA), the employment of "use case" parameters has emerged as a viable indicator of software size. In all of these methods, the primary objective is to generate a size measure that can constitute the "driver" variable for the prediction of effort and software cost, as well as to assess software productivity.

When applying sizing techniques, it is important to analyze the Request for Proposal (RFP), Statement of Work (SOW), systems specifications, Work Breakdown Structures (WBS), and above all, customer needs. From this, a "Thin Spec" or software requirement summary should be developed to support the estimation process and ensure that the estimate is focused upon expressed needs of the customer and other key stakeholders, e.g., the user communities. Furthermore, a thorough understanding of the partitioning of software and hardware is essential. Also, the potential for reuse should be addressed. Finally, systems time lines, data volume constraints, computing platforms, networks, and other important resources/drivers should be considered.

In practice, software size estimates tend to be inaccurate, usually on the low side, regardless of the method or technique used. However, to some extent, these inherent inaccuracies may be offset. For example, normal statistical variability can be addressed by using "multiple" estimates, or Wide-band Delphi techniques, or other independent size estimate techniques. Another fundamental source of error is the lifecycle stage at which the estimate is performed. The earlier in the planning phase, the greater the uncertainty or risk associated with the estimate. This is particularly important when setting senior management expectations of cost and sizing. Realism needs to prevail when estimating based upon the amount of information known about the initiative at the time. To illustrate this concept, Fig. 18.5 was developed by Dr. Barry Boehm from the USC Center for Software Engineering (USC/CSE) [Boehm, 2000]. This "funnel chart" is a graphic depiction of the sizing and cost risk and is based upon actual data from various sources.

Many experts believe that the inability to estimate the final project's product size accurately is a primary contributor to software cost underestimation. No model or algorithm calibrated with an organization's productivity factors can compensate for inaccurate size inputs. In other words, correct estimation of productivity or unit costs is often not as important as simply getting the software size right! The MOM must be aware of this common problem.

One of the most significant problems encountered is that there is no fundamental size corresponding to a particular requirement or function. That is, there is no "blue book" for software development similar to that used in the construction trades industry. So the importance of building an organizational

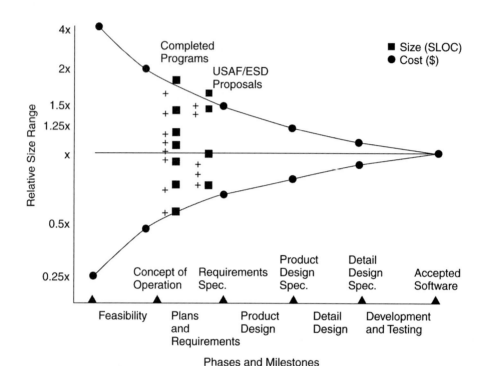

Fig. 18.5. Uncertainty in Software Size as a Function of the Project Phase. The uncertainty in the size estimate, and thus in associated cost, of a software project varies with the phase of the project. For example, in the initial Feasability phase, an estimate "x" could be off by a factor of "4x". This uncertainty diminishes as the project progresses, and is zero at the delivery, or Accepted Software stage, where the final size of the software is finally known. The lack of symmetry indicates a tendency to underestimate the software size. [Boehm, 2000]

database with the sizes of commonly performed functions cannot be overstated. There are two principal types of software sizing measures in common use: Source Lines of Code and Function Points. We will discuss each of these measures in turn.

18.3.2 Source Lines of Code

A commonly accepted definition used at Raytheon (but not the only one as discussed later) for a Source Line of Code (SLOC) is: "a complete, logical, source program statement terminated by a statement delimiter. Each line is counted only once regardless of multiple instantiations of the code during compilation or assembly. SLOC includes executable statements, macro references, data declaration statements, initialization statements, and compiler directives. SLOC excludes comments and blank lines" [Boehm, 2000].

However, the above definition includes a number of "unstated assumptions" that can cause significant variances in SLOC sizing counts. Below are only a few examples of questions that need answers:

- Should test scaffolding/support software be included or excluded?
- Should auto-generated code be included or excluded?
- Should conditionally compiled statements (e.g., "#if" in C) be included or excluded?
- Should block statements (e.g., "begin...end" in ADA) be included or excluded?
- Should Commercial Off-the-Shelf (COTS) SLOC be included or excluded?
- Should deleted code be included or excluded?
- Should compiled code be included or excluded?

Use of a standard counting tool for the project or organization helps reduce some of the ambiguities, but a SLOC counting tool can only support how the software will be counted, not what will be counted!

Some years ago, the SEI sponsored a subgroup of industry and academia experts to develop a framework to address the software source sizing issues. This group developed checklists to identify the principal attributes of a source line of code. We recommend that organizations and projects adopt this framework and the associated checklists to clarify what they mean by a SLOC.

Another definition (alas, sometimes still in use!) counts Physical Source Lines, simply a count of non-comment, non-blank source lines. All stakeholders should understand which definition (physical or logical) applies during the sizing process. Often there are 25% to 50% more physical lines than logical lines for a given component (see also the Rules of Thumb section). We also need to know whether the definition used in the current sizing effort is the same as that employed in collecting historical component sizing data that might be used for analogies.

Function Points

Allan Albrecht of IBM initially proposed the Function Point estimation process in the late 1970s. This is a methodology to estimate the amount of function provided by software as seen from a user or surface feature perspective. Software logical requirements are divided into categories and much of the current research focuses on precisely defining these categories (inputs, outputs, data files, interfaces, and inquiries). The International Function Point User's Group (IFPUG) meets periodically to address counting rules and standards. Current evidence is that this method works well for business Information System (IS) applications, assuming that the individuals counting the function points are well trained and follow the standard counting rules. The jury is still out on whether this method yields accurate estimates for highly computational-intensive, embedded applications.

One of the primary problems associated with the function point estimation method is the lack of available historical data for reference. Most historical data is recorded in SLOC, so the data first must be converted to function point data using the so-called "back-firing" method. However, one company that estimates by using function points as a matter of course (and thus has made the necessary infrastructure investment) has confirmed that the variance can be as high as 70% between the backfiring table ratios once provided in the public domain from the Software Productivity Research website and their own locally derived ratios when comparing actual Function Points to their actual SLOC. Because of this, they use their own internally developed data to make the conversions.

Many major parametric estimation models (including COCOMO II, PRICE-S™, and SEER-SEM™) ultimately use SLOC as input. Thus, after the "raw" function points (prior to the environmental adjustments mentioned below) are estimated for a project, a mapping must be made to the language(s) that is proposed for implementation. With the inherent error in the conversion of function points to language-based SLOC, the final accuracy on estimated effort suffers greatly. Dr. Boehm has stated that the variance mentioned earlier is due as much to the specific application domain as to the implementation language level. Thus, another dimension must be added to the Function Point-to-SLOC conversion table to provide more accuracy. Also at issue are the numerous so-called "environmental factors," such as "code reusability" in the Function Point estimation model, each of which can influence the final cost by no more than 5%. Some of the factors seem to have much more than a 5% degree of influence, such as the aforementioned "code reusability" factor. But perhaps the most troublesome difficulty in estimating the effort using function points (without support of a parametric model) is the fact that many shops assume a linear relationship between function points and software hours. This assumption works reasonably well if all of the projects fall within a fairly narrow size range. However, if past function point-to-effort productivity basis is on projects of 1000 function points (roughly 100,000 third generation language SLOC), it will probably grossly underestimate a project of 5000 function points because of the need for a larger staff size (and thus more personnel interfaces).

One advantage of function point sizing is that it provides early estimates independent of programming language. One of the primary drawbacks is that once implemented, the actual counting of the in-process function points cannot be automated with any degree of fidelity, which may necessitate a manually intensive effort to derive function points from the implemented code as it is under development. One reason is that function points don't necessarily map one-to-one to either functional or object-oriented entities. Time would be better spent collecting the simple core metrics espoused by the SEI: size in SLOC, software staff hours, schedule compliance, defects, and requirements volatility information. After periodic collection, metrics analysis must also be done to compare the collected data with the planned estimates and to assess the risk of values falling outside the acceptable range. In-process metrics collection without analysis and risk assessment (with predetermined triggers) does not add value to a software project. We collect in-process metrics for insight into how well the project is doing. We are buying information, giving decision makers something more tangible than "gut-feel" or promises of being "90% done."

Once the project is over, a one-time effort should be made to count the function points with a view to deriving a more accurate function point-to-SLOC table that is specific to the local organization's application domain. Personnel that do this evaluation must be well trained in function point counting techniques and should have considerable experience in doing so. It is also highly advisable to have two independent assessments of the same project to ensure accuracy. With experienced personnel counting the end of project "as implemented" function points, the results should be within 5% of each other. This could serve as a basis for the next estimation exercise, perhaps even a proposed scope increase to the current project.

Risk-Based Sizing

A manager should not accept single point size estimates. One approach is to use statistical Program Evaluation and Review Technique (PERT) inputs rather than using a single number per function. This provides a level of uncertainty to each input and thus a derivable size risk metric. The PERT technique employs a beta distribution and has been implemented in the sizing estimation portions of a number of parametric cost models. Moreover, its simplicity makes it easy to use in a spreadsheet format. The introduction of systematic, statistical methods should reduce some of the sizing error as well as provide some quantifiable indication of the uncertainty (risk) associated with an estimate.

After determining the functions that will constitute the system under bid, compiling available size data (expressed either in SLOC or Function Points) about similar functions previously developed, and identifying the differences between the similar functions and the ones to be developed, the PERT method is then applied. This method requires three numbers to be input for each sizing function/component:

> LE - Lowest possible Estimate;
>
> HE - Highest possible Estimate;
>
> ME - Most probable Estimate (not necessarily the mean of LE and HE).

From this information, the Expected Estimate (EE) and the Standard Deviation (SD) of each component are calculated using these statistical equations:

> $EE = (LE + 4 * ME + HE) / 6;$
>
> $SD = (HE - LE) / 6.$

The expected estimates for all components are summed to form the expected total lines of code bid. This is the actual number to be priced. The overall Standard

Error of Estimate (SEE) is calculated using the square root of the sum of the squares of the standard deviations of each component. The SEE provides a quantifiable level of cost uncertainty/confidence level for the software size developed.

It follows that the uncertainty (risk) in the overall software sizing is smaller when the estimates are decomposed to as low a level as possible. That is, the cost risk for a 10,000 SLOC module with an uncertainty of 10% is larger than the sum of 10 one thousand-line modules each with the same 10% level of uncertainty. This SEE provides a 3-sigma bound for the size estimates, provided that the estimator inputs values so that there is a 99% chance that the actual values lie between the lowest and the highest estimate, and the estimator has no bias towards either underestimating or overestimating. In practice though, totally unbiased estimators are hard to find! Experience indicates that the "most probable" estimates (ME's) are biased more toward the lower limit than the upper one. This supports a commonly held view that estimators tend to underestimate the size of their products. While not perfect, the PERT based approach is superior to simply providing a single number per sizing component without regard to the uncertainty associated with it. No matter how little information the estimator has at the time of the bid, he or she should feel comfortable with the range of sizing values provided to the costing process.

Software Reuse

The costs of reusing software are very often underestimated. So the manager will need to understand the amount of reused software that will be bid to determine the inherent level of risk. Reuse is handled by using various methods to normalize previously written code that will be adapted to a new project. These reused lines are expressed as "new equivalent" lines and are added to the newly developed lines to obtain an overall size for the project. This total is then input into some model or cost estimation relationship (CER) to obtain effort required. This is one area in which developers are often too optimistic in their estimates. For one thing, recent data suggest that the linear reuse model often used is seriously flawed: If we previously developed software at a given cost x, the cost of reusability of that same software is not just x, but x plus an overhead cost associated with things such as evaluating how the re-used software will interface in the new system.

Studies have shown that the reuse cost function, relating the amount of modification of the reused code to the resulting cost to reuse, is nonlinear in two significant ways: the effort required to reuse code does not start at zero—there is generally some overhead associated with assessing, selecting, and assimilating the reusable component—and further, small modifications in the reused product may generate disproportionately large costs. This is primarily due to two factors: the cost of understanding the software to be modified, and the relative cost of checking module interfaces.

Data from other supporting studies indicate that there are nonlinear effects involved in the module interface checking that occurs during the design, coding, integration, and testing of modified software. Good software structuring can reduce the penalties associated with both the software understanding and the module interface-checking. Modular, hierarchical structuring can reduce the number of interfaces that need checking, and software that is well structured, explained, and related to its mission is easier to understand. The COCOMO II model is currently the only one that reflects this in its allocation of estimated effort for modifying reusable software [Boehm, 2000].

The Raytheon software sizing methodology typically classifies each software component in terms of New, Modified, or Reused source lines of code. However, sizing expressed in function points could be substituted as well. Each classification requires the selection of Modification and/or Reuse Factors. These factors serve as the basis for calculating "Equivalent" lines of code (ELOC). In particular:

$$\text{ELOC} = \text{SLOC\_NEW} + (\text{FM} * \text{SLOC\_MOD}) + (\text{FR} * \text{SLOC\_REUSE});$$

Where:
- SLOC_NEW = newly generated code modules or existing code modules with greater than 50% modification;
- SLOC_MOD = existing code modules changed less than 50%;
- SLOC_REUSE = existing unmodified code, not edited;
- FM = factor for modified code to convert SLOC to ELOC;
- FR = factor for reused code to convert SLOC to ELOC.

Software equivalent sizing may use the following method:

The abbreviated sizing method uses Hard, Medium and Easy categories for determining the ease in adapting either Modified (you plan to change it) or Reused (you don't plan to change it) code. The factors (FM, FR) associated with these categories are supported both by internal Raytheon data and by industry data. The Modification and Reuse Factor Methods are illustrated in Table 18.4. The criteria for selecting Abbreviated Method Modification and Reuse factors (Hard, Medium, Easy) are shown in Table 18.5.

Table 18.4. Modification and Reuse Factor Methods. These factors are meant to serve as guidelines since they will vary by organization and product line.

Difficulty Adapting Modified or Reused Software Artifacts	Modified Factor (FM)	Reuse Factor (FR)
Hard	70%	40%
Medium	55%	25%
Easy	40%	10%

18.3.3 Software Effort (or Cost) Estimation

Software effort (or cost) estimation should begin soon after the sizing process begins instead of after it is completed. This is because some estimation decisions need to be made sooner rather than later. Note: software effort and software cost will be used interchangeably in this section. Typically, the best available SLOC-based estimates have been normalized to new, and all non-SLOC efforts (e.g., GUI screens, database tables, COTS integration requirements) have been quantified in hours or units convertible to hours. Any necessary effort multipliers/scalars, including productivity factors, are then applied to the estimate.

One way to generate effort estimates is to use parametric cost models. The three most commonly used are COCOMO II, SEER-SEM™, and PRICE-S™. One advantage of using parametric models is that they allow us to focus on those factors that have been shown historically to drive software costs throughout the industry. Another advantage is that they employ "diseconomy of scale" curves that Boehm mentions in his classic book "Software Engineering Economics" [Boehm, 1982]. Each of these parametric models uses a non-linear size-to-effort ratio both because historical data from across the industry supports it and because it makes intuitive sense because of the communication overhead involved in larger project staff requirements. There is a risk in using simple productivity factors as input to the effort estimation process unless the project being estimated is very similar to projects (both in estimated size and project characteristics) that are in the historical database from which the productivity factor is derived.

A commonly admitted disadvantage of using "parametrics" is that it is possible for developers to "game" the estimate unless management puts peer review and controls in place. Ideally, there should be some infrastructure to ensure that any parametric models used are calibrated (or understood) in light of the local software performance history. This involves adjusting the parametric model output to match the local effort collection system assumptions and the type of software being developed. Since SEER-SEM™ and PRICE-S™ are proprietary models, some local expertise (trained by the vendors) should be available to support local software estimation activities.

Typical examples of effort multipliers include: required reliability, multi-site coordination, process maturity, function and software complexity, number and type of interfaces, number of major subsystems/components, language type and mix, schedule constraints, team experience (languages, platforms, etc.), team capability (not to be confused with experience!), software tools, environment constraints, and requirements maturity/volatility. The MOM should ask the developers to provide their best and worst case scenarios when assessing these factors. We need to look for realism and ask questions about the assumptions. If there are external customers, they will certainly ask! Table 18.5 lists criteria for determining the modification and reuse factors for software projects.

As with software sizing, a manager should not accept single point cost estimates from developers. The typical software estimate is a "50% proposition." Fifty percent of the time it will be above that number, fifty percent of the time below. The developers should give the 25% and the 75% levels as well. Their answers should be statistically based and supportable from the inputs provided.

Table 18.5. Criteria for Selecting Modification and Reuse Factors. A number of criteria can be considered to determine if the "ease of adaption" is Easy, Medium or Hard.

East of Adaptation	Modification and Reuse Consideration Aspects for Legacy Software Adaptation							
	Engineers' Familiarity with the Code	Software Engineering Environment and Target Platform	Methodology and Standards	Product Line	Functionality	Design/Code Structure	Self Descriptiveness	Software Reliability
Easy	Completely to mostly familiar	Same environment and platform	Same methodology and development standards	Same program, new build or release	Excellent match between the code and the proposed application	High cohesion, low coupling, good information hiding	Extensive design rationale, good level of code commentary, useful documentation	Rigorous requirements testing, no outstanding issues
Medium	Somewhat familiar	Same environment, different platform	Similar methodology and development standards	Different program, same product line	Good match between the code and the proposed application	Reasonably well structured, some weak areas	Design information exists, good level of code commentary, some documentation	Requirements mostly tested, minor outstanding issues
Hard	Mostly unfamiliar	Different environment, different platform	Different methodology and development standards	Different product line	Moderate correlation between the code and the proposed application	Poor structure, moderate to low cohesion	Little design information, moderate level of code commentary, little documentation	Little requirements testing, few outstanding issues

For all formal efforts, the estimated hours are spread across the applicable work breakdown structure (WBS) elements involved in accomplishing the tasks. These hours are then spread across the proposed schedule to produce an overall staffing profile for the project. These planned profiles should be compared to actual results throughout the project. Such information provides program management with indicators of problems that may be developing so that corrective actions can be taken before a staffing shortfall occurs.

Whether using a parametric cost model or a simple set of linear productivity factors, we must understand the scope of the "software hour" that is being assumed. For example:

- Does it include only design, code, and unit test?
- Does it include software component integration?
- Does it include software management?
- Does it include configuration management?
- Does it include support for system testing and rework due to defects?

There are generally no set answers to these questions, but a manager must understand what is included in and excluded from the scope when productivities are presented. Often software productivity numbers (e.g., SLOC/hour) get tossed about with little understanding of the scope or definitions of the basic factors!

The Raytheon software cost estimation process provides three categories of elements of software effort as follows:

1) Software development "core" effort—consists of software engineering effort for design, code, unit test, and subsequent integration of the software stages;
2) Software development support costs—consists of the following six historical support activities (also called bin adders), each expressed as a percentage of the software development effort:

   - Software Management
   - Software Quality Engineering (SQE)
   - Software Configuration Management (SCM)
   - Software Requirements Analysis
   - Software Formal Qualification Test/Evaluation
   - Software Support of Systems Integration and Test

3) Software additional items costs—consists of other cost items not directly based on source lines of code, or on a percentage of software costs. This can also refer to non-labor items, such as travel and materiel or labor items that are not included within the scope of the sizing methodology being used. For example, an organization may decide to increase the staffing associated with software Commercial-Off-the-

Shelf (COTS) integration activities apart from the software development labor associated with custom-developed SLOC. In this instance, the COTS integration activities could be placed in this category. The PERT technique discussed in Sec. 18.3.2 (Risk Based Sizing) could also be applied to budgeting for these additional (and sometimes pricey) items. The only difference is that we are using hours or dollars rather than SLOC as input. This approach provides some quantifiable risk information that may be of value to decision makers.

Productivities can be expressed in this process in two ways:

- SLOC/count of software development core hours, and/or
- SLOC/(count of software development core hours plus the software development support costs).

Both productivities are useful depending upon the context. But we need to restate: understanding the scope of the hours used when discussing any productivity factors is crucial.

Finally, the MOM should ask the estimate developers for two more important artifacts:

- Key assumptions made by the software estimation team documented during the software sizing and cost estimation process. If they haven't written them down as they were developing the estimate, we should be very concerned. An after the fact compilation of assumptions is better than nothing, but not much.
- The estimation team's level of confidence in the estimate. Often a lack of confidence is based upon unclear or incomplete system requirements that may not be resolved until some time later in the development (as in the "funnel chart" example). This information should lead to a fruitful discussion of the risks that may be incurred as a result.

18.3.4 A Few Rules of Thumb for Estimation and Sizing

Here are a few rules of thumb developed by practitioners over the past few years that may be useful:

- For software schedules: Three times the cube root of the estimated development effort (in person months) provides a rough nominal schedule span in calendar months. This span should include only the period from software requirements baseline to "sell off" of these requirements—delivery acceptance by the customer. This rule of thumb should be used for large programs (>5000 lines of code) and establishes the maximum a schedule can be compressed regardless of the number of "bodies" thrown at the effort;

- For software schedules: Don't compress a software development schedule (see previous item) by more than 25% of nominal;
- For software sizing: Ratios between Physical SLOC (PSL) and Logical SLOC (LSL) range from 1.3:1 for Java or C languages, up to 1.5:1 for the C++ language. This metric may be useful if rough conversions have to be made for heritage data normalization (after the fact when it is not feasible to recount the software);
- Suppose we know the "core" software development effort (including design, code, unit test, and software-software integration). To obtain the total software effort (including software requirement engineering, formal qualification testing, software management, SCM, SQE, and system level testing support), we multiply by:
  - 1.3 for small projects (< 15 engineers)
  - 1.5 for medium projects (15-60 engineers)
  - 1.7 for large projects (61+ engineers)

18.4 Staffing Software Projects—Organization and the Right Mix of Skills

This section discusses how to organize and staff a software development effort. We base this description on Raytheon's current best practices that have been successfully applied to numerous projects in both the military and the civilian worlds.

18.4.1 Integrated Product Teams

Today's complex software development efforts cut across multiple disciplines. A typical ground system will have elements of graphical user interfaces, databases, computer communication protocols, and telecommunications, to cite a few. Individuals that are expert in all these disciplines are virtually impossible to find and in fact probably not needed. A proven approach to the multi-discipline need is to create a multi-discipline team or Integrated Product Team (IPT) [DOD, 1996]. This idea goes beyond the teaming of functional disciplines; it also integrates and applies all necessary processes (hence the name of this best practice, Integrated Product and Process Development or IPPD) to produce an effective and efficient product that satisfies customer needs. Because of this last point, customer satisfaction, it is crucial to involve customer representatives (and end users) in the IPTs. In a matrix organization, the members of the IPT are drawn from the different specialization organizations and are put under the leadership of an IPT lead. Depending on the complexity of the project this IPT lead might report to an overall program manager. This program manager, depending of the context, might report to the Mission Operations Manager (MOM), who in this case acts as a customer, or may be the MOM. Figure 18-6 illustrates such an organization.

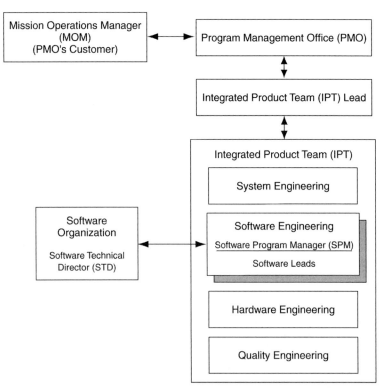

Fig. 18.6. Typical Reporting Mechanism Organizational Chart. Organizational Chart reflecting a typical reporting mechanism between the Mission Operations Manager (MOM), who acts as a customer, the Program Management Office (PMO), that sees to the delivery of the product to the MOM, as well as the interaction between the PMO and the Integrated Product Team (IPT), that actually builds the system. In matrix organizations, the IPT personnel are selected by the individual discipline organizations, e.g., the Software Organization, that maintain the processes and skilled personnel to fulfill the PMO needs.

As early as the time of the project proposal, when the initial software sizing and pricing are done, the key software leadership must be involved in the effort. If possible, the same Software Lead that is to be the eventual Software Project Manager (SPM) should head up, or at least be involved with, the software portion of the proposal effort. Then, when the contract is in place, that person would step in and take an integral role in adding the other members to his or her team.

As the program is just getting started, we must keep in mind that the overall program IPT is being formed. Identifying the software leadership is just one part of this. Leadership of the other disciplines also needs to be established. For example, in addition to selection of a Software Lead, a Systems Engineering Lead must also be chosen. The Program Management Office (PMO), including the MOM, works closely with each of the appropriate discipline organizations to establish the leads for all tiers of the IPT. The PMO must also identify all cross-product/discipline function leads, including Six Sigma (or similar process "learning/improvement" initiatives), risk management and simulation development. All leads must be formally identified on a program organization chart.

On many of Raytheon's larger software projects, such as most space mission projects, the software effort is too large to be led by one person. Often the SPM will work closely with a Software Technical Director (TD), who likely has more expertise in the specific project domain area. The Software Organization leadership plays a major role in identifying the best candidates for SPM and TD, based on their known experience levels and expertise. The SPM and TD work together as peers and as the overall leads of the software effort. They then delegate specific responsibilities to other leads, as needed.

Once the SPM and TD are in place, they must work together to identify what other leads are needed on the team, for performing the up-front work. It is usually helpful to bring a Software Requirements Manager on board, to determine the initial software requirements. If any software subcontractors were identified at the time of the proposal, a separate Subcontract Manager would also likely be needed. The idea here is that at the very beginning of the software effort, we must perform a thorough review of the requirements of the software effort, together with the software sizing and pricing from the proposal.

In line with the new spirit of CMMI®, it is important to bridge the artificial divide between Systems Engineering (SE) and Software. That is, the software leads must work closely with the SE leads not only during the program startup phase, but throughout the effort. The SE leadership normally also has expertise in the project domain, so they can be consulted, as needed, for clarification on the requirements of the system that have been allocated to the software subsystem (as opposed to the hardware subsystem).

At the time of the software proposal and software sizing and pricing, an initial software architecture is usually completed, which breaks down the software into smaller components. These Computer Software Configuration Items (CSCIs) usually provide a good way to identify the sub-teams on the Software team. The SPM and TD, together with the software organization management, play the pivotal role in identifying the leads of each CSCI team. It is usually helpful if the CSCI lead has expertise in the particular domain for that CSCI. For example, if there is a video or image processing CSCI, then the lead selected for that CSCI team would normally have expertise in that area.

In addition to having domain expertise, the CSCI leads must support performance of all of the required software processes. A good balance is needed. Finally, Table 18.6 describes "typical" skill levels for software development roles. Skill requirements for these roles are more accurately described as "typical" than "minimum", due to the wide range of experience present in each role.

Table 18.6. Typical Software Development Personnel Qualifications. For each job position, a skill level or experience and typical academic background are listed. (BS = bachelor of science; HOL = higher-order language.)

Job	Typical Skill Level	Minimum Degree Requirements
Software Project Manager (SPM) / Software Team Lead	10 to 20 years, with 5+ years management, in software development with related programs; experienced in management of software development efforts; experienced / trained in systems/software development and test, in real-time systems.	BS degree in electrical engineering, mathematics, computer science, or other scientific or engineering discipline.
CSCI Lead / Sub-Team Lead	8 to 10 years in software development; leadership skills to guide junior staff; experienced / trained in Higher Order Language (HOL), e.g., C, C++, software development and test, in real-time systems.	BS degree in electrical engineering, mathematics, computer science, or other scientific or engineering discipline.
Software Engineer	1 to 8 years in software development; experienced / trained in HOL (e.g., C, C++) software development and test, in real-time systems.	BS degree in electrical engineering, mathematics, computer science, or other scientific or engineering discipline.
Software Configuration Management (SCM) Specialist	5 to 10 years experience in software development and configuration management.	BS in computer science or mathematics or equivalent experience.
Software Quality Engineer (SQE)	2 to 8 years experience in software development and product assurance.	BS in electrical engineering, computer science or mathematics.

18.5 Software Lifecycles

A software lifecycle describes the steps that a software project takes in producing and maintaining a product. It identifies major milestones, activities, products and risk mitigation strategies for a software development effort from concept to retirement. (Software lifecycles are often confused with project software process. A software lifecycle is just an element, and a small one, of the software process followed by a project.)

A clear and recognizable software lifecycle allows proper communication and tracking of the software project. The following phases are commonly recognized and considered important in all modern, best-practice software lifecycles:

- Analysis—The concept is explored and refined; the user's requirements are elicited. The user's requirements are analyzed and presented in the form of the Software Requirement Specification.

- Planning—A Software Development Plan describing the proposed software development in detail is generated.
- Design—The specifications undergo two consecutive design processes. First comes architectural design, in which the product as a whole is broken down into components, called modules. Then each module in turn is designed; this process is termed detailed design.
- Implementation—The various components are coded and unit tested
- Integration and Test—The components of the product are combined and tested as a whole. When the developers and testers are satisfied with the product, it is released to the users.
- Maintenance—Maintenance includes all changes to the software after it has been released. Corrective maintenance consists of removing bugs while leaving the specifications unchanged. Enhancement consists of changes to the specifications and the implementation of the software.
- Retirement—The software product is removed from service.

The benefits of using a recognized lifecycle lie in standardization. A recognized lifecycle allows the statement of the basic methodology used for managing the complexity of the development effort. Further, it allows easy identification of the phases that are considered important by any modern software engineering standard, and the ability to scrutinize their effectiveness. These phases ensure that design is performed (at some level) before implementation, and that communication mechanisms are in place to support requirements management, review, and maintenance.

The following lifecycle models are generally recognized as industry best practices. They vary in the level of flexibility at milestones, and in risk analysis, but in essence they all contain the phases listed above.

18.5.1 The Waterfall Model

The waterfall lifecycle is a traditional model in which the project moves through its phases in a linear fashion (see Fig. 18-7). Typically, the phases are: requirements determination, planning, design, implementation, integration and delivery. Each phase must be completed before the next phase is begun. A phase is considered complete after the documentation for the phase is written and verified by the appropriate group (the customer, the SQA group, etc.).

Some waterfall models may include a feedback loop to correct deficiencies in an earlier stage. Any stage that is re-entered must still meet the same documentation and review criteria.

The waterfall model enforces good documentation procedures and a strict adherence to the review process. Each phase must be fully documented and reviewed or tested before the next phase is started.

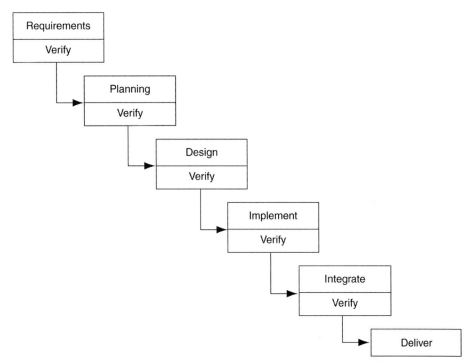

Fig. 18.7. **The Waterfall Lifecycle Model.** The software project moves systematically from phase to phase. Typically a project will spend 50% of its time during the requirements and design phase, 20% in implementing (coding) the software, and 30% in testing the product.

The main problem in using the waterfall model is that at the end of the process, the developers and the customer may agree that the product meets the requirements, but it may not be what the customer really needs. The waterfall model depends entirely on the written specifications and there is often a difference between the exact words in the documents and the concept that the customer has in mind. Various techniques such as rapid-prototyping can help to develop a more accurate requirements specification early in the design process in order to overcome this weakness.

18.5.2 The Incremental Model

In the incremental lifecycle model, the product is created as a series of increment builds. A build is defined as a piece of the complete system with stand-alone functionality. Each build is created using another lifecycle model, and then integrated into the full system when complete.

This model can offer the customer a series of products with ever increasing functionality. In the waterfall model, the customer is given the completed product only after all of the stages are complete. With the incremental model, the customer is gradually transitioned to the new system since modules of the system are being delivered throughout the lifetime of the project.

The difficulty in implementing the incremental model is that the developers must understand the complete system in order to construct a design that will allow all of the products to be integrated at some time in the future. Any problems in this high level design will be difficult to fix at later stages of the project because changes may force previously completed builds to be reworked.

18.5.3 The Spiral Model

The Spiral Model is a refinement of the traditional waterfall that accommodates and provides for development cycles. This model incorporates risk analysis into the process, and allows developers, as well as clients, to stop the process, depending on expected returns from newly discovered risks and requirements. The idea here is incremental development, using the waterfall model for each step; it's intended to help manage risks. Projects do not at first define the entire system in detail. The developers should define only the highest priority features, and implement those. This done, they should then go back to define and implement more features in smaller chunks.

The Spiral model defines four major activities within its lifecycle:

- Planning—Determining project objectives, alternatives and constraints.
- Risk Analysis—Analyzing alternatives and identifying and resolving risks.
- Engineering—Developing and testing the product.
- Customer evaluation—Assessing the results of the engineering.

A spiral divided into four quadrants, each representing one of the above activities, represents the model graphically. The Spiral Model uses iterative development, with the first iteration beginning at the center of the spiral and working outward. Successive iterations follow as more complete versions of the software are built. At the beginning of each iteration of the lifecycle, a risk analysis is undertaken. A review of the project is completed at the end of the iteration. Actions should be taken to counteract any observed risks, at any time. The Spiral Model is illustrated in Fig. 18.8.

The Spiral model is regarded as a realistic approach to software development for large-scale systems. This process model is appropriate for the development of trusted, high-performance, and mission critical systems. Because of the repetitive nature of this model, the cost of using it for lower performance, non-mission critical systems might be excessive.

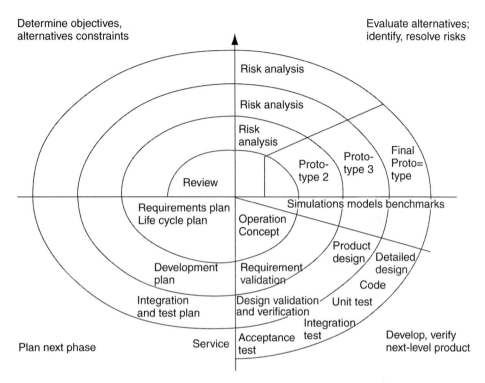

Fig. 18.8. **The Spiral Lifecycle Model.** This model stresses risk management. Although the software evolves as the traditional waterfall, prototypes are developed at each traditional stage (requirements, design, coding and testing). Risk analysis is conducted at the end of each stage before proceeding to the next stage. Early prototypes of the code aid the risk analysis.

18.5.4 The Rapid Prototyping and Evolutionary Prototyping Models

Prototyping of software has two forms: Rapid Prototyping and Evolutionary Prototyping.

Rapid Prototyping is a requirements gathering and discovery tool. Rapid prototyping is typically used at the beginning of a formal lifecycle to communicate a user interface or functional concept. The prototype is a facade program built to show a concept. In this scenario, the prototype emphasis is on communication, not design or coding technique. Prototypes can be created quickly, but construction is inherently ad hoc and design is short sighted. Therefore, *the prototype program must be discarded, not incorporated into the final product*. Too often immature projects never get out of the "prototyping mode" and the final product delivery is just a poorly documented, poorly maintainable prototype with questionable functionality.

Evolutionary software prototyping can use rapid techniques, but the emphasis is on creating a prototype in software, one that will form the basis of the final product. In a strict sense, once a satisfactory prototype has been created, the project continues on to a more 'waterfall' like method of development.

Prototyping techniques are very useful in situations where the user interface is of primary importance, such as developing software for satellite operators. However, as mentioned, there are problems with prototyping methods. At some point the prototyping has to stop, and the project continue. It is important to manage the iterations appropriately, and not continue them into actual development, where correcting mistakes is difficult and time-consuming.

18.5.5 Extreme Programming

Although strictly speaking, so-called Extreme Programming (more fashionably spelled eXtreme Programming and abbreviated XP) is more than a lifecycle model, we present it here as an example of more modern software development methodologies. Some or all of these concepts could be considered and incorporated by a software organization and/or project.

What is eXtreme Programming?

eXtreme Programming (XP) is a software development methodology founded on four values: communication, simplicity, feedback, and courage [Beck, 2000]. XP defines a set of practices to support these values. In general, these practices involve: (1) constant dialog among developers and between developers and customer, (2) keeping the design, code, and documentation simple, and (3) integrating and testing the system many times per day. Following XP practices gives developers and customers the courage and ability to respond to changes in requirements and constraints that inevitably occur during the life of a program.

An XP project is executed by planning and implementing a simple initial release and subsequent small releases based on the customer's priorities, without resorting to partial implementations of features. Every release is planned interactively, with the customer prioritizing features and developers providing estimates to implement these features. This fosters constant communication and feedback between the developers and the customer. The customer should be co-located with the developers to ensure instant, accurate feedback and to make quick adjustments to the plan.

XP developers practice "pair programming". Two developers interact while at one computer: one operates the keyboard and mouse, writing code, while the other inspects and thinks strategically. Pairs often switch roles, taking turns coding and contemplating the next move. This may sound wasteful; however, most software projects default to pair programming when efficiency is needed the most, such as during software or system integration when a deadline is only days or hours away. During a typical project's integration activity, most developers end up in one room or lab in support of the integration effort. A proper XP environment puts all the developers in one large room rather than in individual offices or cubicles, encouraging them to communicate openly and without hesitation.

XP developers practice "collective code ownership". Anybody can change any piece of code on the system when needed, and everyone has responsibility for every line of code in the whole system. By contrast, with individual ownership, only one person can change his or her piece of code at any time, which tends to limit understanding of the overall system and hamper progress. Collective ownership combined with pair programming diminishes the need for formal code reviews since the code is peer reviewed before, during, and after each change.

XP developers practice design simplicity, refactoring [Fowler, 1999], and adherence to a coding standard. At no time should the code contain redundant logic or information. It should have the fewest possible classes and methods (or functions) to satisfy the requirements and tests of the current build. It should not include functionality that might be useful in the future; instead, it should have the simplest design and code to satisfy the current need. If while a feature is being added, a developer sees a simpler way to design a portion of code, he or she must make the code simpler and/or easier to understand (refactor) before adding the new feature. All code produced must adhere to the project's agreed-upon coding standard, which includes rules and guidelines that prevent squabbles about formatting, enhance readability through simplicity and succinctness, and promote well-documented code to diminish the need for detailed external documentation.

XP developers perform continual integration and testing. After a pair finishes implementing and unit testing a feature, they integrate the changes into the current release and run the system tests, until all of the tests are passed. No change can be checked into the system if it fails any test. The result is an up-to-date system release all the time; there is no need to schedule a big integration and test effort at the end of every release cycle since the system is under constant integration and test. On the unit-test level, the unit test code should be written (and tested to fail) before one line of production code is written. As functionality is added, the pair first changes the unit test appropriately, then constantly unit tests after implementing every small but complete production code change. Unit tests should be automated so that they run in a matter of seconds.

What is NOT eXtreme Programming?

Many skeptics mistake XP for ad-hoc "cowboy" programming. In fact, because XP practices involve frequent communication, rapid feedback, constant testing, adherence to a coding standard, pair programming, and collective code ownership, it is practically impossible for an XP project to produce code that has not been reviewed or thoroughly tested by at least two people. XP takes software processes that work to an "extreme" level in order to produce high-quality software quickly. Pair programmers perform code reviews continually rather than at occasionally scheduled formal meetings. Constant testing and integration produce a reliable release more than once every day rather than once every few months. Customer reviews ensure that the development progress is aligned with the customer's needs, so XP involves the customer in a continuous, hands-on manner rather than just a few times per year. XP is an "extreme" application of process, not the abandonment of it.

Adopting eXtreme Programming

Existing projects and organizations that don't want to jump headfirst into XP could benefit from incrementally adopting XP practices, starting with the ones that parallel current software development beliefs and practices. Refactoring, automated testing and test-first design, continuous integration, small releases, and a coding standard are practices that can usually be adopted easily. Pair programming, collective ownership, an on-site customer, and setting up a one-room collaborative work environment seem to require more commitment from management and perhaps more of a change to developers' habits. An XP project does not necessarily have to commit to all XP practices. Rather, a project is considered eXtreme if it exhibits XP values; it responds readily to changes, fosters continuous communication and feedback, and strives to keep the system as simple as possible.

For projects and organizations that follow the SEI Capability Maturity Model, XP either partially or fully addresses many Key Process Areas (KPAs) at all levels [Paulk, 2001]. An XP project in a CMM environment would adhere to the organization's practices for KPAs that are not covered by XP practices.

XP is probably the most prominent of many software development methodologies categorized as "Agile". Other agile methodologies include Scrum, Dynamic Systems Development Methodology (DSDM), Agile Modeling, and Lean Development (LD). Practices of these methodologies fall into the following general categories [Highsmith, 2002]:

- Measures that assure that the project remains focused on its goals
- Short, iterative, feature-driven, time-boxed development cycles
- Constant feedback and interaction between customers and developers
- Creating and maintaining a technically excellent product (e.g., XP's refactoring)

The rapidly gaining popularity of agile methodologies such as XP is arguably due to the realization that there will always be changes during the course of any software development project. Dealing with these unknowns via the traditional over-planning (i.e., padding software bids and schedules) and over-designing (anticipating and designing for every possible feature) does not produce software as quickly and cost-effectively as possible. By promoting constant communication, continuous feedback, and simplicity so that the customer and developers are ready and willing to adapt to, rather than fear, the inevitable changes, XP can help the team reach its goal as efficiently as possible.

A good, more in depth, discussion of lifecycle models can be found in McConnell [1996].

18.6 Focusing on What is Needed for Launch

Once the MOM has all the tools and information on software development discussed in the previous sections, there are two more management tools that must be put in place early. These tools deal with requirements creep and unexpected problems related to a fixed launch date.

Managing Requirements Creep: One of the main causes of software project delays is requirements creep—the introduction of requirements by the customer and sometimes by developers without a proper control or approval mechanism. To avoid this, it is important to do two things: (1) Choose a lifecycle model (see Section 18.5) that best fits the project at hand. For projects whose requirements are expected to be volatile, perhaps because the system functionality is one of a kind, a spiral model might be more appropriate than a waterfall, where the cost of introducing requirements after the requirement phase is enormous; (2) As outlined in CMMI Level 2 (see Section 18.1 and in particular Fig. 18-1), the software project must follow the activities associated with Requirements Management and Configuration Management. This will include the creation of a Change Control Board that will evaluate the pros and cons of introducing new requirements.

Schedule Priorities and Launch: If a satellite launch is delayed, the cost of maintaining the test and integration staff and all the required logistics to support the spacecraft prior to launch is extremely high. The schedule is important; however there will be cases where there will not be enough time to get the software developed and deployed and have the operational staff trained to use the software. This will happen on nearly all satellite projects and the problem can be with the flight software, the ground software, or both. Early in the lifecycle of the development of the system, it is important for the system and software engineers to identify the software functionality that is absolutely required for launch and that which could be delayed until after launch. The Satellite Project Manager and the MOM often approve this list. This is always a difficult exercise as developers are always sure that they will meet schedule, but experience says that this is often not the case.

For example, the real-time system required for health and safety must be ready for launch. An archive system that will be used by the scientific community six months after launch can probably wait; however the system to generate the archive data should be ready at launch. The difficulty with this exercise comes in the gray areas, which usually relate to operability and quality of the final products. In most cases we can go back to the operational concept and ask, "How does the operational concept change if this specific capability is not implemented?" Then, after examining the operational concept implications, we can identify the cost delta (schedule and dollars) of eliminating the capability. The value of doing this exercise early and then updating the list periodically is that if a schedule crunch arrives, the exercise will have to be done, and it is best done when there is little schedule pressure.

References

Beck, Kent. 2000. "Extreme Programming Explained." Embrace Change, Addison-Wesley.

Boehm, Barry W. 1982. "Software Engineering Economics." Prentice-Hall.

Boehm, Barry W. 2000. "Software Cost Estimation with Cocomo II." Prentice-Hall.

DeMarco, Tom. 1982. "Controlling Software Projects: Management, Measurement & Estimation." New York, NY: Yourdon Press. 79-128.

Department of Defense. "DOD Guide to IPPD." February 5, 1996. http://www.acq.osd.mil/io/se/ippd/guide/table_of_contents.html.

Fowler, Martin. 1999. "Refactoring." Addison-Wesley.

Highsmith, Jim. "What is Agile Software Development?" Crosstalk, October 2002.

Humphrey, Watts S. 1994. "A Discipline for Software Engineering." Addison-Wesley.

McConnell, Steve. 1996. "Rapid Development." Microsoft Press.

Paulk, Mark. 2001. "Extreme Programming from a CMM Perspective." IEEE Software, November 2001.

Software Engineering Institute Website: http://www.sei.cmu.edu.

Silver, Aaron N. "Object Oriented Software Metrics Analysis and Predictability." Presented at the Operations Research Society of America (ORSA) Conference. November 1-3, 1993, Phoenix, Arizona.

Silver, Aaron N. "Dynamic Software Metrics Modeling." Presented at the Institute of Management Sciences (TIMS) Conference. June 12-15,1994, Anchorage, Alaska.

Chapter 19

Microsatellite Mission Operations

Jeffrey W. Ward, *Surrey Satellite Technology Ltd.*
Craig I. Underwood, *University of Surrey*

> 19.1 Microsatellite Teams
> 19.2 Exploring the Mission Concept
> 19.3 Detailed Development
> 19.4 Ground Element
> 19.5 Pre-Launch Operations
> 19.6 Post-Launch Operations

Although not every mission operation manager (MOM) will work on a microsatellite mission, the space industry is moving toward smaller spacecraft built quickly for inexpensive missions. This chapter discusses how such missions depend heavily on an integrated approach to mission operations planning. Managers of large missions will see here a complete example of the cost-effective techniques explained throughout this book.

Unfortunately, cost-effective doesn't necessarily mean inexpensive. Traditional space missions, even when cost-effective, are not within the financial reach of most universities and corporations or even many governments. A class of inexpensive satellites known as *microsatellites* has been developed to remove, or at least to lower, this financial barrier to participating in space missions. It's not unrealistic to plan a microsatellite mission costing only $3 to $4 million—including installation of a mission-control station and one year of mission operations.

Much of this small budget goes into getting the most from the spacecraft, and only a small fraction remains for on-orbit operations, as shown in Table 19.1. In this environment, we must take a particularly aggressive approach to cost-effective space mission operations. We've used this approach for seventeen microsatellite missions from 1981 to 2001. Many AMSAT missions have used similar procedures and have been so inexpensive that they are definitive examples of cost-effective operations. Our first-hand experience is mostly with UoSAT/SSTL and AMSAT missions, but guidelines in this chapter apply to other circumstances.

Table 19.1. **Typical Microsatellite Budget.** In a low-cost space mission less than 10% of the budget is allocated to mission operations—compared to between 20% and 50% for a typical space mission.

	US $	% of Budget
Satellite	$2,250,000	62.5%
Launch*	$750,000	20.8%
Insurance	$450,000	12.5%
Mission Control Station	$75,000	2.1%
Operations (one year)	$75,000	2.1%
Total	$3,600,000	100%

*Typical cost of a secondary payload.

Many of these guidelines apply to concept exploration and detailed development because we use the concurrent approach to design. We have to analyze all of the mission elements of a microsatellite at once because of their small size, low cost, and short timescales. Therefore, these missions have required concurrent planning for over 20 years. Cost-effective mission operations result from decisions about management structure and concept development as surely as from any isolated process for mission operations design. We show here how the thread of low-cost operations weaves through the fabric of the microsatellite mission.

We define *microsatellite missions* as those costing less than $5 million, using spacecraft between 10 and 100 kg mass, and moving from system definition to launch in 12 to 36 months. Microsatellites of this class have operated exclusively in Earth orbits, and usually in low-Earth orbits (LEO), because they ride as secondary payloads on commercial and military launches that often use LEO or geosynchronous transfer orbit (GTO). GTO presents a hostile radiation environment, suitable only for very brief missions or spacecraft that can maneuver in orbit. Larger microsatellites such as OSCAR-10 and OSCAR-13 have maneuvered from GTO to Molniya orbits, but such maneuvers are complex and expensive. The body of experience in mission operations for microsatellites derives mostly from LEO missions, so we must be careful when attempting to translate this experience to other orbits.

Section 19.3 will show that we have to be careful when selecting and designing objectives and concepts for microsatellite missions to produce cost-effective operations. Despite constraints on mass, volume, and mechanical complexity, we can still deploy microsatellites for a wide range of useful missions. The objectives and risks of such missions need to be traded carefully against cost to achieve the required outcome at the minimum cost—typically aiming to achieve 80% of the conventional mission capability at 20% of the conventional mission cost—at an acceptable level of risk.

Traditionally, microsatellite missions have focused on providing early flight opportunities for new or experimental technologies—often as an adjunct to well-established space programs. However, they have also played a major role in enabling nations to take their first steps to developing an indigenous space capability. Typical missions have often involved real-time and store-and-forward (e-mail) communications, scientific experiments with moderate data rates and attitude-control requirements, and low-to-medium resolution optical remote sensing. Indeed, through careful spacecraft design and operations planning, a single microsatellite can provide capabilities in all these areas. Thus, the key to the success of a microsatellite mission is the ability to produce a well-founded mission concept for low-cost operations, which will none the less deliver high-quality data tailored precisely to the users' needs. In this chapter, we describe the organizational and technical elements that combine to produce inexpensive mission operations that give a high return.

19.1 Microsatellite Teams

Working on short timescales within small budgets to produce a useful space mission requires teamwork. Correctly forming, training, and organizing this team form the foundation of any cost-effective space mission, but they are particularly critical for microsatellite missions. The microsatellite team must be able to cut costs radically in the space- and ground-operations elements while still developing an architecture that serves the users' goals.

Clearly, to operate at low cost, saving money in the space element can't drive up costs for ground operations. To build a practical and useful spacecraft, mission operators and the end users must be part of the mission team. A well-constituted team will be able to review continually how design decisions affect mission operations.

19.1.1 Training

The requirement for specific training in microsatellite design and operation does not just apply to neophyte space groups. Many end users, operators, and developers will enter the microsatellite arena with considerable experience in traditional space missions, so we may think they can handle the microsatellite mission with no further training. But we must train team members specifically for microsatellite missions.

Although specialists must be on the team, they must also receive general training, especially concerning mission operations. Low-cost operations are possible only if spacecraft-bus engineers, payload engineers, and operators trade ideas. The most certain way to achieve this sharing is to train each engineer in the others' disciplines.

Once we launch the microsatellite, operators take over the spacecraft bus and payloads. They have to maintain the bus and use bus services to operate the payload. The payload must return the data that end-users want, thus completing the mission's lifecycle. The communications architecture of the bus is a critical link in this chain. Both bus and payload engineers need to design for the operator as much as possible.

A development decision made by a bus engineer intending to save money or to decrease risk may end up doing the opposite when operators take over the spacecraft. To avoid such problems, bus engineers should know operations in general, as well as the specific mission operations concept selected for the mission. Initially, this investment in training increases cost. Ultimately, it results in a spacecraft that's designed to be operated, which will save effort and cost during the mission's operations and support phase. In the most extreme case, the training will help bus engineers identify operational failure-modes that they might otherwise have overlooked.

What goes for bus engineers also applies to payload engineers. The payload engineers will determine how the payload is to sense, interact, or perform. They'll do this by trading off the end user's desires against the bus's abilities and limitations, but they must also consider how their payload design affects operations. Designing microsatellite payloads is somewhat different from designing payloads for larger spacecraft. Lack of sophistication in one area doesn't mean the system as a whole lacks capability. Attitude determination may be better than attitude control. Store-and-forward collection of data may be more flexible or reliable than using data-relay satellites and proliferated ground stations. Constraints in mass, volume, and power may be offset by the availability of powerful onboard computers and the ability to integrate the payload closely with the bus. Without these insights, payload engineers may inadvertently duplicate bus services, limit the payload's operational scope, or complicate operations. Managers must give payload engineers these insights and encourage them to develop the payload accordingly.

After launch, the responsibility for mission success rests with the operations organization. This is particularly true for microsatellite missions, in which the operations organization may act alone over a wide range of tasks to save money – including looking after the bus systems. Supplied with a little high-quality technical documentation, they must define and carry out routine and emergency operations. This approach is safe and reasonable, but only if the operators are technically competent and have been trained to understand the characteristics of the microsatellite bus and payload. Rather than separating operations and support from other phases of the mission, the operators must participate in all mission phases from definition and development through testing and pre-launch operations. With such comprehensive training, they'll be able to operate the spacecraft safely, successfully, and cost-effectively.

The integrated mission team described above is an ideal organization for concurrent design of mission operations. Comprehensive cross-training is practical for microsatellite missions because we can keep the entire mission team relatively small. Indeed, if the mission is to be truly inexpensive, the mission team must be as small as possible. One way to achieve this team training is to permit team members to change roles from mission to mission. By rapidly defining and developing microsatellite missions, engineers can play several roles and see several missions to completion within only a few years. This quickly results in a team of richly trained engineers—each with experience spanning the spacecraft bus, the payload, and the mission operations—that can conduct operations that are inexpensive yet deliver high-quality data and services.

19.1.2 Organization and Management

Simply assembling a team of engineers with all-around training doesn't necessarily lead to success. Mission managers must organize the team effectively and manage it correctly. Although many management philosophies move into and out of fashion, the techniques described here have proven particularly effective for low-cost microsatellite missions. These techniques emphasize free information flow throughout the team, flexibility at all stages of the mission, and minimum bureaucracy.

Communication. To capitalize on the training that has given each engineer insight into all aspects of the microsatellite mission, all team members must be accessible to, and communicate with, all the others. Regular project meetings plus modern computer tools (electronic mail, work groups, and local-area networks) create forums and mechanisms for formal communication among team members.

Formal communication—written, electronic, or in a review meeting—carries only a small part of the information flow in a successful team. Through informal daily discussions, a well-trained team can achieve the best costs, time-scales, and mission returns. Managers must create an environment in which this informal collaboration can thrive by establishing a shallow management structure and co-locating the mission team. A shallow management has consequences which are generally favorable to the mission (see Fig. 19.1). It allows all members of the team to take responsibility and contribute to the mission, while also allowing the mission manager (who must retain overall responsibility) to detect these contributions and to incorporate them in the mission planning. It's also best to have all team members working in a single building, allowing informal communication to take place efficiently and naturally. Although geographically distributed teams can succeed, they do so at increased cost and risk, and with reduced inter-disciplinary communications. Team training gives each person an understanding of the others' domains; management and organization put this understanding to use.

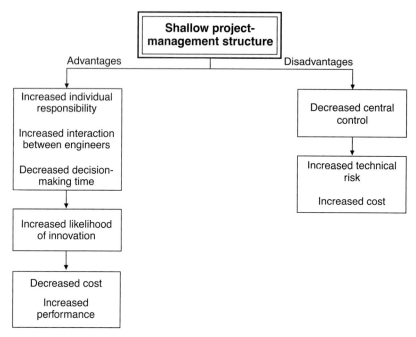

Fig. 19.1. Consequences of Shallow Management Structure. Microsatellite missions depend on innovative solutions to technical problems. Innovation thrives when teams of cross-trained engineers work within a shallow management structure. To avoid losing central control, mission managers and engineers must communicate often.

The need for close communication during mission definition and development is obvious. Less obvious is that this collaboration must continue during the operations and support phase. During post-launch operations, the bus and payload engineers will know the spacecraft's limitations and features. The operations organization will be learning about the practicalities of spacecraft operations (possible failures or unanticipated environmental constraints). The end users will be changing their mission goals based on initial results. Only a continuing team effort can best apply the available resources (including money) to meet the end users' goals.

Flexibility. Defining and developing a complete mission in as little as 12 months doesn't allow a longer mission's certainties of analysis and design. The mission team must retain flexibility for as long as possible to overcome late-breaking difficulties and exploit opportunities as they complete the analysis in parallel with development (see Fig. 19.2). Each area of the mission design has its own requirements for design and interface freezes, which will close this period of flexibility. The interfaces between the bus and payload must be frozen relatively early, especially if different teams are developing them. Hardware designs must also be frozen early, to allow time for manufacture, test, and integration. Software designs, on the other hand, can remain flexible throughout the mission lifecycle—assuming we allow properly for safely uploading software after launch. Like software, the plan for space mission operations should continue to evolve throughout the mission, in response to changing goals or capabilities. Although we must have a preliminary operations plan against which to judge mission design, we must not freeze this plan unnecessarily. By maintaining flexibility for as long as possible, we can feed the information gained through team training and communications back into the completed space-mission product, increasing returns and decreasing costs. Feedback and flexibility should continue into the operations and support phase.

Fig. 19.2. **Consequences of Late Design Freeze.** If the detailed development phase is particularly short, it's especially important to delay design freezes. Prematurely frozen designs can lead to overly conservative design and to persistent problems, both of which increase costs for mission operations.

Standards. The flexible, team-oriented approach described above can succeed if the mission's end users and developers support it. The processes and standards used in traditional aerospace projects don't promote rapid, inexpensive development of space missions. We must apply these corporate norms and industry standards only where they are necessary to ensure an appropriate level of product assurance, or where they lower costs.

One area in which we encounter standards is communication links. For example, NASA and the European Space Agency (ESA) have created standards for telemetry and telecommand links. Adherence to one of these standards allows us to use off-the-shelf "space-standard" equipment in the ground and space elements, as well as existing ground stations and data-relay satellites. But we don't know that this approach will actually reduce the cost of the ground element and post-launch operations. Accessing an existing ground element or using a data-relay satellite may cost more than a custom or semi-custom communication system specially tailored to mission requirements. There may well be a non-space-specific commercial-off-the-shelf (COTS) solution which we can use or adapt to suit our needs—often at considerably lower cost. Below, we'll see that we can build communications links with low and medium data rates by modifying ground communications equipment and assembling it into a data station dedicated to microsatellites. With such a dedicated system, conforming to industry standards isn't in itself an advantage, so we don't need to take it as a constraint during mission definition.

Similar arguments arise when considering NASA's or ESA's strict standards for parts procurement. Restricting designs to use only approved parts increases the cost of the space element and can also increase the cost of mission operations. This situation arises because approved parts are conservative, emphasizing flight heritage and thus restricting the use of new or COTS technologies. These new technologies, such as high-performance low-voltage microprocessors, high-density semiconductor memories, gate-arrays and systems-on-a-chip – usually surface-mount and plastic-packaged—are the very technologies that make low-cost, high-performance instruments possible (see Fig. 19.3). In particular, using special space-qualified microprocessors and low-capacity memories may increase the cost of software development. This, in turn, can drive up costs for post-launch operations.

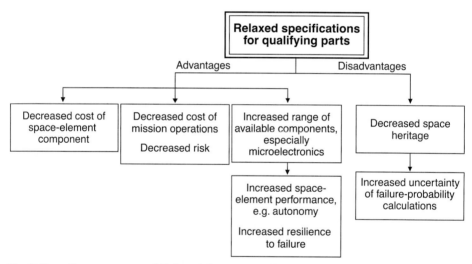

Fig. 19.3. Consequences of Relaxed Component Qualification Requirements. The use of components which have not been space qualified greatly increases the number of microelectronic components the spacecraft engineer may choose from. This increased design freedom permits rapid and inexpensive development of more complex and capable circuits.

As an alternative to a conservative parts specification, microsatellite missions have routinely employed state-of-the-art microelectronics. Relatively large solid-state data recorders (hundreds of megabits), controlled by powerful reprogrammable onboard computers, have appeared on payloads and buses. Using these systems, mission operations managers have developed cost-effective operations plans that don't rely on data-relay satellites or continuous ground-station contact. Thus, relaxed parts-procurement policies have saved money on operations. However, this approach should only be taken if the design engineers have a good understanding of the operational environment of the spacecraft (e.g., in terms of ionizing radiation, charging, thermal, etc.) and its potential effects on such components. We must also pay adequate attention to failure resilience (see Sec.19.3.2), but if costs are to be kept under control, systems should be engineered to meet the mission requirements—and no more.

Low-cost space missions must therefore substitute the initiative, knowledge, and skills of a well-trained team for rigid and comprehensive standards. We can't reject standards arbitrarily but should trade off the costs and benefits of standards just as we would any other engineering decision. The end user must accept this process, which includes risks as well as benefits.

19.2 Exploring the Mission Concept

We may strive for cost-effective operation of any mission. We may even build small spacecraft to serve a wide range of missions. Low-cost microsatellites, however, are suitable only for certain missions. Microsatellites, and the design, construction, and operation described here, have strengths and weaknesses. The success and cost-effectiveness of a microsatellite mission will depend on how well the end user's objectives correspond to the microsatellite's strengths. Concept exploration is the time to assess this correspondence. Furthermore, if the entire mission is to be inexpensive, mission-concept development must give appropriate weight to the costs of the ground element and operations. [Sweeting, 1994]

19.2.1 Mission Concepts for Microsatellites

In some cases, the space-mission budget will be tightly restricted so we can only consider a low-cost microsatellite solution. In such situations, we must identify mission concepts that we can implement within the constraints. At other times, the mission will be at least partially defined, and the question will arise, "Can a low-cost microsatellite serve this mission?" In both cases, we need to evaluate physical and operational constraints.

Microsatellites are constrained in volume, mass, and mechanical complexity. If we're launching a microsatellite as a secondary payload, we must keep its volume and mass low in order to give us a wide range of launch opportunities. Reliable, complex mechanical subsystems are usually too expensive for a microsatellite budget, so in general we don't use deployable solar panels and complex stabilizing systems. These basic constraints on the system reduce available power and pointing accuracy. We can overcome mechanical simplicity and small size to some extent by applying microelectronics creatively, but the microsatellite's basic physical limitations will rule out some missions.

If initial mission evaluation shows that a microsatellite bus can support the payload, we must determine whether or not a microsatellite can do the mission.

This is a two-part analysis: first we select critical elements of the mission architecture and then we evaluate their effect on operations. This analysis is critical to ensuring that the payload, bus, and ground element will be inexpensive and successful after launch. The main architectural elements we must select are the orbit, the data-communications system, and the attitude-control system. Once we select them, we must evaluate their demands on mission operations.

19.2.2 Orbit

Microsatellites have deployed most often in low-Earth orbits (LEO), but they've also worked in geostationary transfer orbit (GTO) and other highly-elliptical orbits. If a mission requires an intermediate or high-Earth orbit, or even an interplanetary trajectory, we shouldn't immediately assume microsatellites are out of the question. Rather, we should analyze the constraints presented by the orbit: e.g., ionizing radiation environment, illumination conditions and thermal environment, profile of visibility to ground stations, navigation requirements, and characteristics of the communications link's path. If we can define a microsatellite mission concept that meets the users' and operators' requirements and remains affordable, the mission is feasible. Our discussion in this chapter centers on LEO because many microsatellites have operated in LEO, not because LEO is the only orbit suitable for microsatellites.

19.2.3 Communications Architecture

As discussed in Sec. 11.2, the data rate between a spacecraft and ground station depends on the spacecraft's orbit, the availability of ground stations or data-relay satellites, and the mission's data requirements. In general, a low-cost microsatellite may have one or two dedicated ground stations, with communications-link rates up to ~1 Mbps—although actual link speeds are often considerably lower at 9.6 kbps, 19.2 kbps or 38.4 kbps. Higher data rates can be supported, but with a significant increase in the ground element's complexity and cost, perhaps driving the mission outside the cost-effective range. Using data-relay satellites such as TDRSS is also unlikely to be cost-effective for a low-cost mission.

A 38.4 kbps packet-switched digital downlink to a single ground-control station provides a good starting point for evaluating a typical communication architecture for a microsatellite in LEO. For example, this rate gives us approximately 18 megabits of data return per day from a microsatellite in an 850 km sun-synchronous orbit to a ground station at 50° latitude. This rate would permit continuous operation of a payload generating roughly 220 bps (assuming that suitable data storage were available on board). An experiment with a higher data rate could operate a correspondingly smaller fraction of the time. For example, a remote-sensing instrument taking 1000 × 1000 pixel, 8-bit, monochrome images could return 19 raw images per day (or ~100 with on-board data compression).

The basis of these simple analyses is a cost-effective mission operations plan involving a single ground station operating at medium data rates. If such a preliminary analysis shows clearly that the mission will work, mission development can continue. Similarly, if the data return available is several orders of magnitude too small, development should cease. Between these limits are gray areas for which we can offer cost-effective solutions. Perhaps we can add a second ground station relatively inexpensively, or we can increase the downlink rate without unacceptably increasing the ground element's cost. We must explore these options in detail, trading the cost increases for the ground and space element against the benefit of increased data return.

19.2.4 Attitude Control

We also need to examine a preliminary budget for attitude determination, control, and stabilization (ADCS), beginning with the present state-of-the art for low-cost systems. A microsatellite stabilized by a magnetic-torquer-assisted system using the Earth's gravity gradient can be expected to remain nadir pointing to within ±1° (in 850 km LEO), while attitude determination may be an order of magnitude better. Spin stabilization, achieved by magnetic torquing or cold-gas thrusters, is also reasonably inexpensive for payloads with an inertial- or Sun-pointing requirement. Recent microsatellites have used relatively simple COTS-based single momentum wheels in combination with magnetorquers and/or gravity-gradient booms to provide increased attitude and stability control. Moving away from these simple systems can increase costs significantly and may drive the mission out of the low-cost realm.

19.2.5 Operations

Having identified communications and ADCS architectures that fulfill the payload requirements, we must generate a mission operations concept. It's now particularly important to study how the mission architecture responds to phases of the mission other than normal orbital operations—especially the early-orbit phase and anomalies during orbital operations. We may have overlooked these demanding periods when we first developed the mission operations concept, which concentrates on routine orbital operations.

Microsatellites are usually released from their launch vehicles with unknown attitude dynamics (they often tumble with significant spin-rates about all axes) and all onboard systems dormant. Even if we expect some nominal attitude, the safest commissioning plan assumes little or nothing about deployment dynamics. Thus, thermal conditions, power budgets, and communication-link margins must all remain acceptable without attitude stability. The ideal spacecraft should be able to remain in this pre-commissioned state indefinitely to allow for difficulties in initial acquisition or software loading.

Once we've acquired the spacecraft and started nominal ADCS operations, thermal and link conditions should stabilize, permitting payload checkout and operation. The nominal scenario for orbital operations will then be in effect.

It's important to analyze what will happen if hardware or software failures in the space or ground elements disturb nominal orbit operations. Chapter 16 describes anomaly investigation and recovery. In a microsatellite mission, a small team operating from a single location carries out these procedures. This arrangement may affect such matters as crew composition (Table 16.10). In a microsatellite team, the anomaly captain, systems engineer, and design engineer may all be the same person.

Given this organization, we must address several issues. Can the payload and bus return to their safe pre-commissioning state? How long can the spacecraft remain "safed" before the mission objectives are completely lost? We need to ask similar questions about the communications links. How much data loss can the mission sustain? Are there periods during which data acquisition is absolutely critical? The answers to these questions will show whether the mission can employ a relaxed approach to fault detection, isolation, and recovery (FDIR). Increasing the coverage or the speed of these FDIR procedures can completely alter the mission architecture, and drive up costs.

For example, if the mission depends critically on continuous data return, both the space and ground elements must have more redundancy. The costs associated with this redundancy might overshadow savings in other areas of the design, throwing the entire mission concept into disarray. How the development team responds to such problems will determine whether or not we can achieve a reasonably low-cost mission. The team may call for increased subsystem and mission reliability (e.g., increased redundancy and autonomous fault handling) or they may call for increased subsystem and mission flexibility (e.g., accepting some data loss). If the end user's mission requirements or bureaucracy force the system design down the path of increased reliability, a low-cost mission is probably inappropriate. But if the development team can devise a mission operations plan that provides a compromise between reliability and flexibility, employing systems that fit a microsatellite's physical constraints, we may have an inexpensive, yet effective mission.

We must therefore explore mission operations concepts and choose one that neither rules out viable missions, nor admits impossible ones. In some cases, we won't be able to devise a cost-effective mission operations concept that can use a low-cost microsatellite. If so, we'll need a more expensive spacecraft or ground element. This filtering must take place early in the mission to avoid severe difficulties during detailed development. Even in successful cases, the operations concept will almost certainly identify some mission requirements that will stretch the microsatellite's capabilities. During detailed development, we'll have to give these areas more resources, to ensure we can carry out the operations concept. Thus, cost-effective mission operations turns into a design driver. Operators collaborating with developers of the ground and space elements will produce an architecture that supports cost-effective mission operations.

19.3 Detailed Development

The output of concept exploration will be a mission concept that uses a low-cost microsatellite and cost-effective operations. The mission now enters detailed development of the spacecraft bus, payloads, and ground element. For microsatellites, this phase often takes 12 months (or less).

Design decisions must be consistent with the plan for cost-effective mission operations. At all times, the team will design to cost, keeping the mission within its relatively small budget. The team must also design for operations, ensuring the space and ground elements operate cost-effectively. The following subsections discuss areas of detailed design which can strongly affect operations.

19.3.1 Mission Plan

We must refine the preliminary plan for mission operations to include detailed plans for routine operations and critical procedures during payload checkout (e.g., software loading). In keeping with the integrated philosophy, the entire team (ground-element engineers, space-element engineers, and mission operators) must help develop the detailed mission plan. This is another chance for the bus engineers to adapt to the needs of the payload engineers, as well as for both groups of space-element engineers to check how their design decisions affect the ground-element engineers and operators.

A detailed mission plan should be complete before freezing the space-element and ground-element designs, so hardware and software designs can form around the mission plan, rather than forcing the mission plan to meet rigid hardware and software constraints. This optimizing step will decrease operating costs and increase data return.

We must remain flexible while developing the mission plan, and the plan must carry that flexibility through to post-launch operations. The mission plan, developed months before launch, shouldn't be prematurely frozen, and it shouldn't dictate a static operating environment after launch.

Particularly during the checkout phase immediately after launch, we should expect the mission plan to vary daily and even from session to session. External factors, such as difficulty in getting post-launch orbital elements, or problems with the spacecraft itself may slow the pace of operations. Or everything may go smoothly, accelerating the pace of commissioning. We can't slavishly follow a plan developed and frozen before the spacecraft was fully designed.

After checkout, bus maintenance will settle into a routine involving telemetry gathering and general housekeeping. Payload operations, on the other hand, may remain fluid for months or years, depending on whether the mission is experimental or operational. A microsatellite carrying a commercial communications payload can follow a relatively fixed mission plan to maximize the owner's revenue. If the owner's business plan changes, the mission plan may also change, or, as we gain experience with the spacecraft, we may improve the operations plan to reduce operations expenses or increase mission performance. A microsatellite carrying science or technology experiments is often different. It may require a continually evolving mission plan, allowing experimenters to choose the appropriate next operation after analyzing the results of each experiment run. In both types of missions, these adjustments to the mission plan get the best return on the investment in space hardware.

Before moving from mission planning to spacecraft design, we must plan for contingencies. In low-cost missions, a mission plan that provides detailed contingency responses for all anomalies isn't cost-effective. Instead, anomaly planning concentrates on the most likely anomalies during critical periods of the mission. For example, it's valuable to have a decision tree that we can follow in case of failure when we first acquire the spacecraft. Outside of these critical periods—which primarily occur during checkout after launch—failures are unlikely, and the failure-resilient design of the bus should allow hours or even days of troubleshooting in safety. If the operations group can call upon bus and payload engineers for help with an anomaly, a comprehensive contingency plan is unnecessary. In this system, we call on the engineers' expertise as needed to fix actual problems, rather than investing it in a (perhaps unnecessary) "what if" exercise aimed at producing a comprehensive contingency document.

19.3.2 Resilience to Failures

To reduce demands on the ground element and increase the probability of mission success, we want a spacecraft intended for a low-cost mission to be failure resilient. In other words, performance should degrade gradually in case of failures, and systems to aid troubleshooting and reconfiguration should be available. Although catastrophes can still happen, we should minimize their probability.

In typical microsatellites, two systems are particularly susceptible to the possibility of temporary or permanent failure: hardware systems based on advanced microelectronics undergoing their first space use, and complex, onboard, software systems—often using many hundreds of lines of code, which, operating in a hard-real-time environment, are usually impossible to test completely adequately before uploading. Yet these systems can lead directly to cost-effective mission operations, so we must deploy them safely in microsatellite designs.

In this context, an architectural design for a microsatellite should permit the ground element to receive spacecraft telemetry and issue spacecraft telecommands without the aid of experimental hardware or complex software onboard. Furthermore, the spacecraft and its experiments should remain safe despite temporary unavailability of such high-level functions. We can achieve these goals by adopting a spacecraft architecture divided into layers. At the lowest layer are proven telemetry and telecommand systems, mostly based in hardware (perhaps including tried-and-tested firmware). In the middle layers are onboard computers and software with some flight heritage, and at the highest layer are new onboard computers and advanced software. Reliability and dependability are greatest at the bottom, but we get the best flexibility and potential mission returns at the top. (See Fig. 19.4). Working within such an architecture, engineers can manage the risk of employing the advanced hardware and software that can make operations less expensive and more effective—yet, the spacecraft doesn't depend utterly on perfection of software development or operational procedures. Through this design approach, we can handle a range of human and machine failures, fostering rapid development efforts to get the best returns.

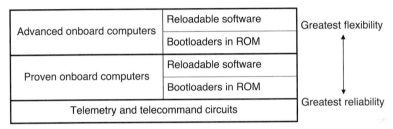

Fig. 19.4. **A Layered Spacecraft Architecture.** In this architecture, proven, reliable systems control and monitor the spacecraft and therefore back up new, unproven systems that give the mission great flexibility. Graceful and rapid transitions from one layer to another are essential (downward in the case of an anomaly and upward when full functions return).

19.3.3 Adaptability

For a layered system architecture to be successful, the data paths within the spacecraft must adapt to different modes of operation. We should extend this adaptability to allow reconnecting of the spacecraft's subsystems in any sensible configuration—not just in the configuration for nominal operations. This adaptability implies that the spacecraft must have a wide range of optional analog and digital data paths, as well as ways of making the paths operate.

We shouldn't design the spacecraft to support only a single, narrow mission plan. Instead, the bus and experiments should support several possible operating scenarios. Software should play a major role in controlling experiments and gathering data, so we can change the operating scenario after launch. The spacecraft should have a general-purpose node for handling data (the onboard computer) connected to special devices for gathering data (experiments) and actuators (bus controls). The onboard computer should use a common high-level language, so many engineers and experimenters can develop their own onboard software. The "C/C++" programming languages are most commonly used because of their ubiquity in commercial practice.

If a spacecraft and its experiments support several operating modes, we must periodically switch modes using techniques ranging from simple telecommands to reloading software tasks. We'll have the most flexibility if these changes can be routine. In particular, procedures for software reloading should be practical, safe, and reliable. To this end, the onboard computer will often run an operating system that allows several tasks to run concurrently, so that we can reload tasks without affecting the execution of the other software routines. A specialist spacecraft operating system (SCOS) has been developed for many of the microsatellites originating from AMSAT and university programs, and this remains in widespread use. However, there is increasing interest in adapting semi-commercial real-time operating systems for future missions.

Adaptable hardware and reloadable data-processing software greatly increase the spacecraft's complexity and introduce many possible spacecraft configurations, which the engineering team must understand. In return for this complexity, we get wider scope for cost-effective operations. We can adapt the spacecraft to overcome failures, to serve altered mission goals, or to respond to advances in ground-element technology.

19.3.4 Autonomy

Microsatellites must be able to operate routinely without constant contact with the ground. Onboard computers usually allow autonomous operation by executing closed-loop algorithms to control and monitor the spacecraft. Particularly for LEO spacecraft, which are in range of their control station infrequently, some autonomy is essential.

Autonomy can be cost effective if it covers enough of the spacecraft's monitoring and data-acquisition functions. Developing hardware and software for limited autonomy in controlling attitude, gathering data, and monitoring payload parameters is far less expensive than developing comprehensive autonomous systems, which deal in real time with all possible failure modes and recovery procedures. The underlying failure-resilient design for the bus should eliminate the need for comprehensive fault-detection, isolation, and recovery in real time.

The software that will make a microsatellite autonomous should be able to cope with the most likely scenarios. Upon detecting an out-of-bounds condition, the software may either return the spacecraft to a fully operational mode of gathering and generating payload data, or place it in some predefined safe state. It will do the former when the out-of-bounds condition is not completely unexpected and we have already formulated the remedy. Safe mode is reserved for truly anomalous behavior for which no automatic remedy is safe and sure. Figure 19.5 illustrates the associated program algorithm.

For instance, the HealthSat-1 microsatellite runs a program that monitors the bus battery's charge status and the transmitter's operating parameters to use transmitter power effectively. Sometimes, due to varying illumination conditions and temperature effects, the bus power can't support maximum transmitter output. The software therefore adjusts the transmitter power to match the available power. On the other hand, if the transmitter's operating parameters exceed safe values, the program places the transmitters in a safe state until operators can diagnose the fault and start the recovery manually. Thus, a combination of autonomous and operator-assisted operation produces the most cost-effective payload operations and ensures spacecraft safety.

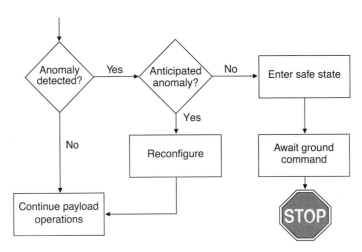

Fig. 19.5. Resolving Anomalies Onboard. Onboard software can anticipate some anomalies and use its reconfiguration routines to keep the spacecraft operating safely. Upon detecting a truly unexpected anomaly, the software places the spacecraft in a safe state and awaits ground intervention.

19.4 Ground Element

In a space-mission budget, operations costs begin with the ground element. Although the space element may include many operations-oriented features, they're not usually assigned as operations costs. The ground element for a low-cost microsatellite mission must be inexpensive in itself and must support cost-effective operations.

The key to producing a low-cost ground station is to take advantage of mass-produced consumer electronics, computers and communications equipment wherever possible. Some equipment will be directly useful, but in other cases it will require slight modifications or custom-built interfaces. Even with these changes, using mass-produced electronics as the core of a microsatellite ground station saves a lot of money.

19.4.1 Radio Frequency Equipment

We must engineer the spacecraft's communications links to allow a low-cost ground station. Unfortunately, the large Doppler shifts and high power-flux densities inherent in communicating with satellites from the ground make the adaptation of mass-produced license-free data-radio systems and mobile phone technology problematic. Alongside these COTS systems, many ground services—notably mobile and amateur radios—use the VHF (150 MHz) and UHF (400MHz) bands for voice communications. There are thus many sources of mass-produced radios for these bands. At these lower frequencies, medium-gain antennas with relatively broad beams can provide acceptable links. Broad beamwidth, in turn, permits the use of open-loop predictive tracking and low-cost antenna rotators. Selecting these lower radio frequencies eliminates the need for high-gain parabolic-dish antennas with high-speed, closed-loop tracking systems, which are ten times more expensive. Most microsatellite missions have therefore substituted VHF or UHF communications links for the microwave links used in traditional space missions.

Particularly in the Amateur Radio service, competition among radio suppliers is very fierce, resulting in low-priced radios with sophisticated features. The radios are typically computer-controlled so we can do Doppler corrections and automate ground stations. We can buy a computer-controlled radio with VHF and UHF chains for only one or two thousand dollars.

The main constraint on adapting these radios to data communications for a spacecraft is limited channel bandwidth. Frequency-modulated (FM) voice transmissions usually occupy 20 kHz, and readily-available equipment implements this standard. This type of equipment can support data rates up to 10 kbps relatively easily using frequency-shift keying (FSK) modulation. Although this data rate may be too low for some applications, data compression, onboard data filtering and even multiple ground stations can help overcome this limitation while keeping costs very low for ground equipment.

Chapter 11 shows that communications-link design is a complex subject, which can consume a lot of development effort. Standard practice for satellite communications is to use phase-shift keying (PSK) on S-Band (2.4 GHz) for telemetry, telecommand, and data recovery. Adopting FSK on VHF and UHF bands as a design baseline constitutes a different approach, one resulting in an affordable and robust system which many have adopted for microsatellites in LEO.

19.4.2 Data-Processing Equipment

Consumer electronics also play a role in data processing on the ground for low-cost missions. Personal computer (PC) technology, still increasing in performance and decreasing in cost, is a clear choice for data processing and automation in a low-cost ground station. Indeed, the performance and cost distinctions between workstations and personal computers are blurring, leaving system designers free to choose the platform more familiar to the programmers, operators, and experimenters who will use it. We thus save money not only by buying inexpensive hardware and software but also by providing ground-element developers and users with a familiar environment that doesn't require extensive retraining.

19.4.3 Delivering Data Products

The computers in the ground station should be connected as a local-area network (LAN) and equipped with communications that support access to wide-area networks (WAN) such as the Internet. The extra cost of networking equipment is offset by more operational ease and effectiveness. We can use networking to gather data, plan operations, resolve faults, and deliver data products (see Fig. 19.6).

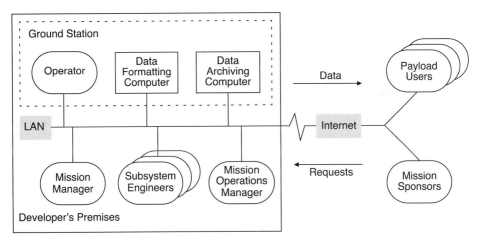

Fig. 19.6. Using Computer Networking in Microsatellite Operations. The ground station for a microsatellite mission should use local and wide-area networking to connect operators to other members of the mission team. Using suitable wide-area networking, such as the Internet, we can rapidly and inexpensively deliver operating requests from payload users and data from the payload.

If PCs control the ground station, we can easily give experimenters and engineers immediate access to all data through sharing of disk files across the network. Thus, for science or technology missions, all level 0 data should be present on the network immediately after downlinking from the spacecraft. Experimenters should be able to access the network remotely (remote users on the WAN, local users on the LAN) to collect their data. This approach has the twin advantages of reducing workload on ground-station staff and making experimental data available as rapidly as possible.

Mission planners can also use the infrastructure through which data is distributed to receive requests for specific operations and payload maintenance, and to distribute mission and activity plans to interested parties. If automatic, the scheduling software would validate and arbitrate requests; if manual, the operator would complete these tasks. In either case, it gives users immediate input to mission planning, which means operations will be most relevant to the users' present needs.

There is no reason why the network cannot also be used to allow operators in a central location to operate spacecraft from remote ground-stations. Indeed this is precisely how operators in the UK are able to command and control the PICOSat microsatellite from a remote US-based ground-station via the internet. Ultimately, with the appropriate on-board software, the satellite itself can become a node on the internet, with data transfer (and commanding) achieved by means of standard protocols such as FTP and TCP/IP, etc. Demonstrations of this concept have already been made with Surrey's satellites in collaboration with NASA.

Finally, network connectivity can improve fault diagnosis and recovery by bringing all spacecraft-development engineers to bear on a problem. If the team members aren't at the ground station, they can still retrieve data to examine the problem and contribute documents or software quickly to the ground-station staff. Effective electronic links between the development engineers and the operations group make it unnecessary for development engineers to document in detail the precise operational diagnoses and solutions for all possible spacecraft problems. These links also allow the development engineers to remain involved in the operations of one spacecraft while concentrating on the development of another.

19.4.4   Automating Ground Stations

We can easily automate a ground station equipped with powerful general-purpose computers and comprehensive networked communications, thereby decreasing operating costs while increasing data return and keeping the mission safe. A typical control center for a microsatellite mission can predict when the spacecraft will be visible, drive antenna tracking, establish a proper data-handling configuration, and gather data, all without human intervention. This automation makes it possible for a single operator to oversee the housekeeping and data gathering for several microsatellites. It also permits an unattended ground station to collect data.

Simple, inexpensive additions to this basic automated station can enhance mission safety. Software should confirm the receipt of telemetry when expected and check received telemetry for out-of-range conditions. When it detects an anomaly (no telemetry or out-of-range telemetry), automated paging equipment or Short Message Service (SMS) text messaging can notify an operator, who can further diagnose the problem. Thus, a spacecraft in LEO may be monitored on each transit over its command station, without needing second-shift operators.

Ground-station automation often pairs with onboard autonomy. The microsatellite's onboard computer may execute a stored schedule of operations that is composed weeks or months in advance. Once a mission has settled into routine operations, it may use a single schedule indefinitely. Thus, the spacecraft will continue to carry out a varied mission to the operator's plan without the operator's constant intervention. If the ground station is autonomously downlinking the generated data and monitoring the spacecraft for failure, the load on human operators drops significantly (especially if experimenters can retrieve data over the network without operator intervention). These techniques have worked routinely on LEO microsatellites that would otherwise require manning around the clock.

19.5 Pre-Launch Operations

We might assume operators and ground-element staff begin working only after we launch the spacecraft. This is not the case in a microsatellite mission. Someone must operate the spacecraft during integration, environmental testing, and pre-launch checkout. Although test engineers using special ground-support equipment could handle these operations, there are several reasons to use operators and some or all of the ground-element hardware and software we'll use after launch.

A typical team for a microsatellite mission doesn't include specialist test engineers. Instead, development engineers and spacecraft operators do subsystem-level and system-level testing. Each brings valuable information and expertise to this stage of the mission. The development engineer has detailed knowledge of the space element's hardware and software. The operators know details of the ground element and of the mission plan. This is a valuable chance for developers and operators to pass information back and forth. The presence of the specialists also increases the safety of the test period, which can be hazardous even with the most detailed documentation.

Ground-support equipment used for pre-launch operations should be a subset of the ground-station equipment for the mission, so we don't spend money on dedicated test stands and so we get the most value from pre-launch tests. Operations before launch will be the first chance to verify that the ground element and the space element are compatible. In a typical microsatellite project under rapid development, this will be an important test for new space and ground hardware and software. By conducting a single test campaign covering the space and ground elements, the team will best use all available resources.

Thermal and vacuum tests are a critical milestone in a microsatellite mission. They typically involve one week's intensive operation during which we operate the spacecraft in vacuum at various temperatures. Thermal and vacuum testing helps to identify circuit or component problems, confirms that the spacecraft meets the launch vehicle's outgassing requirements, and forces us to operate the spacecraft for a long time without physical access. In effect, it's a dress rehearsal for spacecraft on-orbit operations, and we command and monitor the spacecraft only through its radio-frequency (RF) communications links.

A well-constructed thermal and vacuum test should include periods of operational exercises alternating with periods of more intense checkout. Operational exercises confirm that we can control the spacecraft as anticipated by the operators and the mission plan, but they don't investigate the performance of every redundant path. The checkout periods test for faults that wouldn't appear during such low-coverage tests. In doing these two types of tests, we learn about the details of reconfiguring the spacecraft and ground station and confirm that the space hardware and software are operating correctly.

Just as thermal and vacuum tests are the *first* chance for full-up operational exercising of the spacecraft, the launch campaign is the final chance to verify that the space element, ground element, people, and procedures will work together. A campaign for even a low-budget launch should include at least one spacecraft operator and one spacecraft-systems engineer. This team should repeat (perhaps in shorter form) the combination of operational exercises and systematic checkout developed during the thermal and vacuum tests. These tests, done at the beginning of the campaign, verify that the spacecraft has survived shipping and is ready for mating with the launcher.

We usually operate the spacecraft through an umbilical connector to avoid radiating radio signals at the launch site. But we must still get proper clearance to run a final RF checkout, confirming all systems and procedures are ready for the initial commissioning of the spacecraft after launch.

Formal lifecycles of a space mission sometimes omit pre-launch operations. In a typical microsatellite mission, however, these operations are critical to the system developers and operators. If done well, they can decrease the cost and increase the effectiveness of integrated testing.

19.6 Post-Launch Operations

The preceding sections have discussed many parts of cost-effective microsatellite operations. Because these operations don't depend only on efficient activities after launch, we've shown how mission planning is integrated into the mission's lifecycle. But we also need to understand how the mission architecture is actually used in orbit to retrieve data or produce revenue inexpensively. So in this section, we overview post-launch operations and then show a brief practical example drawn from the S80/T microsatellite mission.

19.6.1 Overview

Post-launch operations include commissioning, checkout, exploratory operations, and mature operations. Each phase places different demands on the mission team and the space and ground elements. During commissioning and checkout, operators will work very closely with development engineers, especially if difficulties develop. During the exploratory operations, the team develops its best plan for using the end users' payload, given prevailing conditions in orbit. This phase involves end users, operators, and engineers. The exploratory phase gradually gives way to mature operations of the payload, which operators and end users can handle without the engineering team. The duration of each phase depends on the nature of the mission; the transition to mature operations can occur in days if we reuse a standard bus and payload for a new mission.

For most missions, mature operations will occupy much of the mission's lifetime. The space and ground hardware and software will remain relatively stable, and the goal will be to use the fewest people to generate the most experimental data or commercial revenue. One operator can usually sustain mature operations of a microsatellite while interacting with end users. End users request operations through the electronic-communications systems discussed earlier, and the operator produces an operating schedule, which is uploaded to the spacecraft. This efficient yet flexible operating mode is possible because we designed it into the mission architecture from the beginning.

Before describing how such a system works in practice, it's worth reiterating that microprocessors supporting operations automation and spacecraft autonomy are critical to cost-effective operation of low-cost microsatellites.

Microprocessors. More than any other single technology, the microprocessor is at the heart of the architectural design and at the forefront of post-launch operations for cost-effective microsatellites. The ground station is automated by microcomputers, linked to the spacecraft by microprocessor-controlled communications protocols, and connected to a network of experimenters, operators, and engineers through their personal computers. Onboard the spacecraft, computers in the bus and the payload execute schedules of ground commands, manage communications, maintain the spacecraft's attitude, and provide automated fault detection and recovery. These powerful, inexpensive processors and their associated software streamline every element of post-launch operations.

Automation and Spacecraft Autonomy. Microprocessors alone don't guarantee cost-effective systems. Two major decisions make this microprocessor-based architecture cost-effective. First, the decision to automate removes human operators from as many control and operations tasks as possible. In particular, operators never need to make real-time decisions under normal operating circumstances. Second, the decision to keep the ground station out of as many control loops as possible makes the spacecraft autonomous. This is a logical extension of the first decision—once we transfer routine operations to computers we may as well transfer them to computers onboard the spacecraft. Control loops implemented by onboard computers don't have the delays and uncertainties of loops involving the spacecraft's communications channels and ground-station equipment. While both rules have exceptions, they're critical parts of a cost-effective architecture for mission operations, as illustrated by S80/T.

19.6.2   S80/T—An Example of Cost-Effective Microsatellite Operations

Mission. The S80/T microsatellite mission was funded by the French space agency CNES to measure interference and propagation characteristics in the VHF frequency bands allocated to the Little LEO communications satellites. Matra

Espace was the prime contractor, and Surrey Satellite Technology, Ltd. (SSTL) supplied the microsatellite bus, ground systems, and post-launch operations. Dassault Electronique supplied the payload—an ultra-linear transponder operating in the VHF bands of interest. The mission went from concept to launch (on the third Ariane ASAP) in just 12 months. [Allery, 1994].

The orbit (dictated by the primary mission – TOPEX/Poseidon) was relatively high at 1320 km altitude, with a 66° inclination. This orbit encounters significant levels of ionizing radiation in the South Atlantic anomaly (~6 rads/day at component level), and thus the mission lifetime was expected to be relatively short, about 2–3 years. The orbit also experiences significant changes in eclipse period, posing a severe challenge to the passive thermal design, which had to maintain safe temperature limits under all illumination conditions and attitude scenarios. The varying illumination also caused significant variation in the amount of power available to support payload operations. For these reasons, at the time of launch (August 1992) S80/T, along with its partner, KITSAT-1, were the most challenging microsatellite designs Surrey had yet attempted. However, as it turned out, both S80/T and KITSAT-1 exceeded their expected design life by a considerable margin. Indeed, S80/T operated successfully for more than eight years after launch. By the end of 2000, increased current consumption by the spacecraft's on-board computer/RAMDISK (due to accumulated radiation dose damage) meant that the payload could no longer be operated under all illumination conditions. The last contact with S80/T occurred in September, 2001.

Basic Spacecraft Characteristics. S80/T was a 50-kg microsatellite approximately 600 mm tall with a 300 mm square cross section. Four body-mounted solar panels provided power in sunlight, and a 6 amp-hour NiCd battery pack provided peak power needs (e.g., during payload operations) and maintained function during eclipses. The spacecraft's nominal attitude was to have its communications antennas nadir pointing, with a slow spin about the Earth-pointing axis. A 3-axis magnetometer, two Sun sensors, and an Earth-horizon sensor provided basic attitude determination. A 6-meter, rigid, gravity-gradient boom topped with a 3-kg tip mass stabilized the spacecraft. The onboard computer used 3-axis magnetic torquers in a closed-loop control system, which removed residual libration and provided yaw-spin control. Telemetry and telecommand links operated at 9600 bps—the downlink on UHF and the uplink on VHF. The satellite's onboard system for handling data consisted of an 80C186-based primary computer for onboard control and a Z80-based secondary computer for onboard control embedded in a layered architecture. The 80C186 connected to a 16-megabyte, error-detection-and-correction-protected solid-state data recorder configured as a *RAMDISK*. The S80/T bus was based on previous SSTL microsatellites, with some changes needed to achieve the primary mission.

The S80/T payload was a 5-kg, VHF/VHF, linear repeater supported by a circularly polarized measurement antenna. Because the microsatellite structure was not large enough to support two VHF antenna systems, we used the measurement antenna for the bus telecommand uplink as well. To permit accurate interference-level measurements over a wide dynamic range, the repeater had a variable gain, which we could set using the spacecraft's telecommand system. Signal strength within the repeater passband could be measured through the spacecraft's telemetry system; hence closed-loop control of the repeater operating point was possible using the telemetry and telecommand system. Because the repeater drew 40 W from the bus (which supported only 19 W orbit average), scheduling of repeater operations and monitoring of bus performance during these operations were critical aspects of the mission operations design.

The S80/T operations concept called for two ground stations: one for mission control at SSTL and a transportable one from which CNES would make experimental measurements. SSTL provided both ground stations, using personal computers, mass-produced radio equipment, and other low-cost elements. Pre-launch plans called for the SSTL station to monitor S80/T's bus subsystems while the CNES station would conduct payload operations—keeping the ground station in the loop control while controlling the repeater through the uplink and downlink. This operations plan was based on a traditional model, using little onboard computing power. We devised a plan more appropriate to a microsatellite early in the mission, and all aspects of the mission architecture were flexible enough to support this change.

Commissioning. SSTL engineers commissioned S80/T from the SSTL Mission Operations Centre. This process started with downlink activation about six hours after launch. The operating sequence after downlink activation involves software uploading, bus checkout, and attitude-control maneuvers. Spacecraft-development engineers with detailed knowledge of the S80/T design controlled the spacecraft during these operations. Having engineers directly involved in commissioning operations paid off when S80/T entered an unforeseen mode of attitude stability. In this mode, the spacecraft's thermal controls were ineffective, and electronic systems were operating at the edge of their ability. The engineers on hand were able to make critical judgments in real time, devising a contingency plan, executing it swiftly, and restoring nominal operations before we lost communications with the spacecraft. The problem, the decisions, and the recovery took a matter of hours, exploiting all the available expertise and the streamlined decision processes typically found in a microsatellite mission team.

After establishing full onboard software operations and nominal attitude, operators concentrated on a 30-day checkout period, during which they exercised all primary and most redundant bus systems. Working with payload engineers from Dassault, SSTL operators also exercised the spacecraft's repeater, verifying that it had survived launch. They used onboard computer software to schedule the repeater operations and to monitor bus performance during these operations. Having the software end repeater operations at some specified time was especially important because the team hadn't verified positive uplink telecommand of the spacecraft during this operation. The software provided a safety net in case the ground element or command link failed.

Early-Orbit Operations. S80/T's early operations show how reloadable software and general-purpose onboard computers allow the mission team to operate the payload most effectively after launch. The initial operation plan for the S80/T payload was traditional: the spacecraft's telemetry system would measure the payload's operating parameters and transmit these measurements to the ground station for payload operations (see Fig. 19.7). Software in the ground station would decide whether or not to adjust the payload gain or operating mode, based on the telemetry readings. If changes were necessary, it would transmit them on the spacecraft's telecommand channel. Involving the ground-based computer increased uncertainties and delays in the control loop but was the natural approach taken by mission engineers unfamiliar with the power and flexibility of a microsatellite's onboard computers.

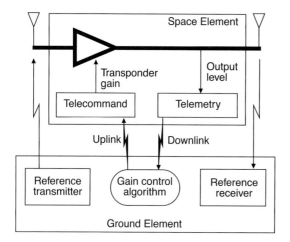

Fig. 19.7. Initial Operations Concept for the S80/T Payload. The initial operations concept for the S80/T transponder required telemetry and telecommand links during all payload operations to provide closed-loop control of the transponder gain. Telemetry downlink transmissions during payload measurements established a high noise floor for the measurement transponder.

In the final stages of pre-launch testing, we learned that interactions between the payload's RF systems and the bus's RF systems were greater than we expected. This wasn't surprising considering the proximity of the payload and bus antennas. As a result, the payload operators wanted to operate the payload without the bus's telemetry transmitter turned on. This invalidated the original mission plan, which depended on the telemetry downlink to complete the feedback loop. The solution was to use an onboard computer in the payload's gain-control loop (Fig. 19.8). We decided to use onboard control after the spacecraft had been shipped to the launch site.

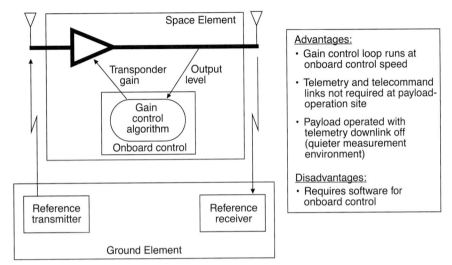

Fig. 19.8.   Actual Operations Concept for the S80/T Payload. We chose the actual operations concept for S80/T only weeks before launch. Moving the gain control loop to the onboard computer permitted operation with the telemetry downlink turned off, thus establishing a lower noise floor for payload operation.

With most members of the S80/T team occupied by the launch campaign, we hired a consultant programmer to write this critical application. This consultant couldn't access hardware models of either the payload or the S80/T's onboard computer, yet was able to develop a complete control, monitoring, and data-logging package for the payload operations. We made this development possible with earlier system-development decisions: using a familiar microprocessor, structuring onboard data handling as a general-purpose system, and using standard computer languages in a familiar programming model on the ground. The consultant was thus able to write the new software using readily available tools and high-level interfaces that completely isolated the software from S80/T's specific hardware. We shipped the software to Surrey, where people verified its basic functions (apart from closed-loop control) on a ground model of the onboard data-handling system. After initial in-orbit checkout of S80/T, we loaded the new software to the spacecraft, verified it, and placed it into the normal operations sequence.

This post-launch development and verification of a large new piece of operating software illustrates how a flexible approach to operations can greatly improve mission cost-effectiveness. Without the software, S80/T would have returned fewer results, with much more difficulty. Because the spacecraft design was general enough to support onboard control, and the mission managers were willing to accept software development late in the mission, S80/T returned a lot of data and didn't require a payload-control ground station to control the payload for every operation. This, in turn, allowed the customer to use the payload in field trials at very low cost.

It might seem that we could avoid such late changes in mission design by perfect planning during the development phase. Unfortunately, perfect planning is never possible and is especially unlikely when executing a mission with new technology on a short schedule. A successful microsatellite team will plan for change and continue to improve the mission's operating scenario during operations after launch.

Mature Operations. Once we had installed and validated the new payload-control software, S80/T entered a period of mature operations, which persisted for most of the mission lifetime. The operations plan made extensive use of the microsatellite's capabilities, so it's a good example of an effective plan for microsatellites.

The mission-control center at Surrey handled housekeeping operations and bus monitoring. The mission-control center (shared by several other microsatellites) automatically received telemetry and other housekeeping data whenever S80/T was above its horizon. A network of personal computers collected the data, which was then moved to archives on optical disks. During each transit of the spacecraft, the ground-station software compared received telemetry values with the expected high and low values. If it detected extreme or out-of-bound values, operators were automatically notified so that they could examine the data and correct any problems if necessary. The operators sent brief monthly reports of control-center activities to CNES.

Experimenters at CNES scheduled repeater operations based on the particular experiments they were running. Each operation included special settings of the control-loop parameters, a start time, and an end time. Ten days of operations were prepared as a single schedule file for transmission to Surrey. Rather than using ground-communications links, this schedule file was actually transferred between CNES and Surrey using the store-and-forward transponder onboard S80/T. Software in the CNES station automatically uploaded the message, and software in the Surrey station downloaded it. This software was essentially that used by other store-and-forward communications networks based on SSTL microsatellites, and was the same software used to recover the payload's data files.

When downloaded at the Surrey mission operations center, CNES's payload-operation schedules went through automatic consistency checks, and operators verified that the spacecraft had enough power for the proposed operations. The validated schedule was automatically converted to the binary format needed by the payload-control task in the onboard computer and was then uploaded to the spacecraft, again using the store-and-forward communications transponder. The onboard computer executed the plan autonomously for one week, and payload operations would proceed with no further operator intervention. Except for the power-budget check (a bus issue), CNES themselves controlled payload operations, and scheduling was completely automated.

As discussed previously, each payload operation consisted of a relatively brief period of repeater operation—typically one transit over a target location. During the operation, an automatic gain-control loop adjusted the payload's operating parameters by monitoring the magnitude of the repeater's input signals. The payload-control task also collected the data, which had to be correlated with ground-based measurements. Data was stored as files in the spacecraft's RAMDISK, where the store-and-forward transponder software could access it. The CNES's experimental station was programmed to download these files automatically on the transit following each operation. Once downloaded, data was immediately available in MS-DOS compatible files, time-tagged to within 10 ms, for experimenters to analyze. Thus, no one needed to download, prepare, or send data to CNES manually—an exemplary cost-effective system.

The final element of S80/T operations was fault detection. The payload repeater was a high-power device, with specific requirements for power-supply sequencing and maximum operating temperature. It placed a large current drain on the spacecraft's batteries, which also had several critical operating limitations. All of these parameters were monitored by the payload-control task executing on the onboard computer. We set thresholds for fault detection for each scheduled operation or left them at default values. If something was out of bounds, the payload operation ended and important telemetry values were logged. Certain errors would result in temporary suspension of the operating schedule, whereas others required operator intervention to resume the schedule. This fault-detection system was acceptable as long as the primary onboard computer was operating. If the onboard control crashed, a secondary computer served as a watchdog. Periodically, primary and secondary computers would exchange messages on the spacecraft's LAN. If either control computer failed to hear from the other, the spacecraft's telecommand system would end payload operations. The automatic fault detection, with redundancy, was the final link in the operations chain which allowed long periods of autonomous spacecraft operation.

While it is certainly possible to develop and operate a microsatellite in the same manner as a larger spacecraft, this type of operation won't be cost-effective and inexpensive. In this chapter, we've discussed the elements of spacecraft development which lead to cost-effective operations before and after launch, within the context of a low-cost, rapid design-to-orbit microsatellite mission. To date (2002) we've used this system to develop and operate some 17 microsatellites for various kinds of missions.

The main philosophical points we've adhered to during these missions are to keep operations in mind during all phases of development and to keep the mission plan as flexible as possible. These decisions have led to a heavy reliance on onboard microprocessors running software uploaded after launch. To make this mode safe, we develop each spacecraft to withstand temporary software failures.

A flexible mission plan supported by a reconfigurable microsatellite can provide high-quality data after launch without requiring a large operations team.

In any cost-effective project, we try to spend the available money in a way that gives the greatest return. An inexpensive and cost-effective project is particularly difficult to manage—especially in a field such as space, where the norm is for each project to be expensive and conservative. In developing low-cost microsatellites, we concentrate most of our effort on improving the return from the mission. This means putting resources into hardware or software and testing—not into bureaucracy, planning, and documentation. Of course, bureaucracy, planning, and documentation have their place, so we include the least we can for a safe mission. Reaching these minimum necessary levels is one of the keys to cost-effective mission operations in space.

References

Allery, M.N. and H. Castelbert. "Microsatellite Operations in the First Year of the S-80/T Mission." Presented at the 2nd International Conference on Small Satellite Systems and Services. Biaritz, France. June 27–July 1, 1994.

Sweeting, M.N. "Sophisticated Microsatellites for Space Exploration and Applications." Presented at the International Workshop on Digital Signal Processing Techniques Applied to Space Communications. King's College, London. September 26–28, 1994.

Chapter 20

Human Space Flight Operations

Carolyn Blacknall, *NASA Johnson Space Center*
Felix Godwin, *Teledyne Brown Engineering (Retired)*

> 20.1 Rationale for Human Missions
> 20.2 Implications of Humans in Space
> 20.3 Planning and Analyzing Space Shuttle Missions
> 20.4 International Space Station
> 20.5 Human Lunar Missions
> 20.6 Human Interplanetary Missions

This chapter will examine the unique aspects of manned space flight missions. Human space missions are a special topic because of their high public interest and other distinctive features: their mission phases, sub-phases, subsystems, trades, and program concerns are all new.

20.1 Rationale for Human Missions

Human missions are the subject of much debate. Some scientists perceive them as costly and risky efforts that displace smaller, science-oriented activities. They argue that robotic spacecraft are now so capable that expanding the human presence in space isn't necessary.

However, having humans live and work in space is a valuable goal in itself. Human crews are an important solution to working in the dynamic and challenging conditions of space. Having people in space was the primary goal of the earliest human missions, and enormous prestige came with being first in Earth orbit and first on the Moon. A thousand years from now, history books and other media will still chronicle those national triumphs.

Other areas of research in human flight soon emerged: learning what humans can do in space, how long they can work, and how far they can journey. Increased payload capabilities combined with miniature parts began to provide space crews with extensive instruments and laboratory equipment, allowing missions to take advantage of unique human adaptability, insight, and improvisation. As a result, human space activities are much more flexible and changes can be made quickly to account for unexpected situations.

This flexibility provides a key reason for having a manned space program. Astronauts can aid experiments by observing and promptly altering procedures based on their observations. They can position and activate cameras, change film, replace samples, and discuss experiments with Earthbound payload specialists. They can also exercise common sense to detect problems that a robotic spacecraft might ignore with disastrous results. These abilities add tremendous flexibility to payloads flown aboard crewed vehicles.

Astronauts can rendezvous and dock with other spacecraft, a task which can be difficult for robotic spacecraft. They can operate external devices such as the Space Shuttle's Remote Manipulator System, whose control from the ground would be hampered by time delays. Astronauts can repair and upgrade objects in space, as they did with the Hubble Space Telescope, improving its capabilities and prolonging its useful life. Astronauts can also capture a satellite and return it to Earth for laboratory examination, as was done with the Wake Shield Facility and the Long Duration Exposure Facility (LDEF).

Astronauts can perform *intravehicular activity* (IVA) in the shirt-sleeve environment of a spacecraft's pressurized and temperature-controlled module. *Extra-vehicular activity* (EVA) takes place in the hostile environment of space. This environment requires pressure suits and life support, as well as propulsive aids, such as the extravehicular maneuvering unit (EMU), for movement in micro-gravity.

Humans in space can also respond to novel and unexpected opportunities for scientific observation and are especially valuable when the payload doesn't operate as expected. Using crews also allows in-flight maintenance. The crew can change fuses, reset circuit breakers, tighten connectors, and install stored spare parts. For some payloads, we can't foresee and therefore can't automate necessary repairs. Crew members have modified hoses and creatively used spare parts to repair payloads; an example is the "fly swatter" device developed during a Shuttle flight to flip a switch on a satellite. Such unplanned activities may involve EVAs to capture, repair, and redeploy spacecraft.

Other reasons for humans in space concern politics and economics. The influence of these issues on US space operations is described by an eminent student of US space policy [Logsdon, 1986]. He explained why a space payload with no need for direct human presence may ride on a crewed vehicle, even if this method is more costly.

This approach became the US's space policy with the 1972 decision to authorize a crewed vehicle, the Shuttle, as the future principal means of accessing space. The policy was confirmed in 1982 by National Security Decision Directive (NSDD) 42 and it lasted until 1986. Then, the explosion of the final Challenger mission reminded the nation of the inherent dangers of space travel. The decision was then made to fly payloads that did not need human participation on unmanned spacecraft in the future. Incidents such as the loss of the Shuttle Columbia in February 2003 again stressed the inherent dangers in any space flight endeavor.

20.2 Implications of Humans in Space

Using crews doesn't change the 13 mission functions defined in Chap. 3, though it does change the details of how we carry them out. Mission phases of crewed flights include:

- Crew selection and training
- Ascent and abort operations
- De-orbit, landing and recovery
- Activities after landing

The Europeans of past centuries launched repeated voyages of exploration even though these voyages sometimes cost the lives of 20 per cent or more of their crews. In the 20th century, the survival rate of the first US Air Mail pilots was scarcely better. But such reckless ventures are no longer acceptable. Our value on human life now requires extensive provisions for crew survival. An example is the Shuttle's abort option in the form of suborbital diversion toward an alternative airfield. Maintaining this option reduces the Shuttle's payload on missions to high-inclination orbits because of the added fuel required to reach alternate airfields.

Thus, abort and rescue options create payload penalties and alternative branches for the mission timeline. Some of these options apply after the spacecraft is in orbit, but the mission plan must provide for unscheduled early return to Earth. This return may use a special reentry vehicle with minimum capability—functionally equal to a lifeboat.

If the mission proceeds as planned, reentry and landing occur as scheduled events that are peculiar to human missions. Like the rescue options, they create new mission subphases with their own demands on communication architecture, command and tasking structures, and the system's ground element. One effect may be to constrain the timing of earlier mission phases, due to the Earth's rotation relative to the orbit plane and the need to have a designated landing area within reach of the re-entry vehicle. Another effect is to require certain ground resources, such as rescue teams and communications relays, to be on standby, perhaps at remote locations.

Finally, crewed missions have unique phases after landing. Physiological rehabilitation of the crew may be brief or extended, depending on the length of time spent in zero gravity and the rigors of landing and recovery. A final phase is using crew members as consultants and instructors for later studies and missions.

20.2.1 Accommodations for Humans on Spacecraft

Human missions create new trades and risks, as well as much more complex functions for mission control and payload planning. Human missions must provide an appropriate atmosphere and maintain temperature, pressure, and humidity. They must have an adequate supply of food and drinking water. Provisions must be made for waste storage and containment.

Human spaceflight must have a system for environmental control and life support (ECLS). This system must provide an appropriate atmosphere and maintain temperature and pressure within safe limits. Currently fuel cells are used for electric power and these fuel cells also supply water.

The environmental system must do more than provide oxygen and water and maintain temperature and pressure within safe limits. It must guard against any biological contamination, which could quickly infect the entire crew in their small, closed environment. Certain types of illnesses, such as Legionnaires disease, flourish in air-conditioning systems, where temperature and humidity favor the growth of microorganisms.

Humans in space face two major external hazards: zero gravity and radiation. The long-term effects of zero gravity are still not fully understood, but they include loss of calcium from the bones and atrophy of cardiac and other muscles, so all long missions include physical exercise in their daily cycle of activity. In the 1950s popular media pictured space stations as rotating wheels in which centripetal force provided artificial gravity, but this rotation is uncomfortable unless the rate is very low. A typical rate would be two rpm, which for one g requires a radius of more than 200 meters and hence introduces the complicated dynamics of tethered bodies. Continuous rotation of a spacecraft would also complicate the pointing of solar panels, antennas, thermal radiators, and optical instruments. Rotation also reduces the benefits of research to study the effects of microgravity.

For near-Earth missions, we can reduce radiation hazards by orbiting below the Van Allen belts or by transiting the belts rarely and rapidly. For long missions away from Earth, the main radiation hazard is solar flares, which cause rare but abrupt increases in the solar wind (mostly protons). Shielded storm shelters for temporary occupation can counter these effects. The shelters must permit the crew to continue all activities essential to their health and operations. We can lower the storm shelter's mass by using existing masses, such as stored water, for shielding.

In addition to radiation from space, experimenters must be careful that the payloads do not expose the crew to harmful radiation. In addition, those planning spacecraft or experiments used by a human crew must be aware of the effects of loud noises, flashing lights, unusual odors, or vibration levels on astronauts.

Besides changing the way operators carry out a mission and requiring new mission phases and protective subsystems, human flight constrains payloads. If a crewed vehicle is supposed to deliver a payload to orbit, the payload would remain onboard through the landing phase if a malfunction keeps it from being released in orbit. This means we must design the payload to withstand structural loads during landing as well as ascent. Also, any payload energy sources, such as propellants and electroexplosive ordnance, must be safe through reentry and after landing.

Objects which will come in contact with the crew must be carefully checked for safety. Sharp edges must be avoided. A unique situation in manned flights is that all items taken into a manned spacecraft must be checked for offgassing, or releasing poisons or odors into the close and contained atmosphere of the crew cabin. Payloads which can operate successfully in unmanned flights may create problems if they emit even a small amount of harmful gases. As in unmanned spacecraft, fire is an extreme hazard; however, since many combustion products are toxic to humans, this also must be tested.

The crew may also affect a payload by causing undesired lighting or vibrations. Experiments must be designed so that they are not affected by possible random vibrations as the crew move about in their daily activities.

This possibility leads to operating some in-space experiments as free flyers. These experiments stay apart from the crewed vehicle, carry out their operations, and can then be recaptured and returned to ground or maintained or replenished. In 1984, the Long Duration Exposure Facility (LDEF), with a mass of 9660 kg and carrying 57 different experiments developed by 200 researchers from 8 different countries, was placed into orbit 463 km above the Earth by a Shuttle flight and retrieved over a year later. The Wake Shield Facility, a 3.7 m diameter disk-shaped platform released and then re-caught by the Shuttle's remote manipulator arm, is another example of advancements in this area.

The longer humans stay in space, the more complex and risky the mission. If we are to keep cumulative risk down, longer missions require more redundancy and provisions for contingencies, both of which increase a spacecraft's mass and cost. Missions and programs must adapt to these requirements. For example, mission planning must use human hours as efficiently as possible to keep costs down, so managers tend to micromanage crews with minute-by-minute instruction from the ground. Skylab IV had notable friction regarding tasking and workloads—a tension that could easily intensify in interplanetary missions with long durations and communication delays (see Chap. 17).

For mission managers, human flight creates new time pressure during malfunctions. If there is a malfunction in an unmanned spacecraft, ground operators may be able to place the spacecraft in a safe mode and take whatever time is needed to resolve the problem; even a month of limited operations isn't a serious loss out of a typical five- or ten-year operational life. If a human mission malfunctions, time pressure is far more severe because of risk to life and because the time available for the crew to complete tasks may be limited, such as about ten days in Spacelab. The need for more prompt response increases mission cost by requiring more elaborate provisions for contingencies.

Experimenters must be aware of the advantages and disadvantages of placing their payload on a crewed vehicle. The presence of human astronauts has specific implications for payloads on the Space Shuttle, the International Space Station, or on lunar and planetary missions.

20.3 Planning and Analyzing Space Shuttle Missions

A reusable aerospace vehicle, the Space Shuttle is launched like a rocket, orbits like a spacecraft, and lands like a glider (albeit a very heavy 68,000-kg glider). Though Shuttle flights may seem to be largely routine, each flight is unique. Each mission requires specific training of the astronauts, operations crew, launch crew, and payload operations team. Each flight requires a reconfiguration of the Shuttle computer software and information systems and each payload requires modifications to the flight operations timeline. Although reusable, after each flight the orbiter has to be carefully examined, repaired and reconstructed.

To determine whether a payload or experiment should fly on the Space Shuttle, we need to examine how people plan a Shuttle mission. Recognizing the complexity and scope of preflight and on-orbit operations permits us to understand the costs and time requirements of mission planning for the Shuttle. The following activities detail the milestones and requirements for launch of the Space Shuttle.

A necessary part of human spaceflight is planning of crew activities, which means analyzing and developing activities needed for a specific mission. First, we create and refine a flight plan that outlines specific crew activities and essential flight-support functions. This analysis results in a minute-by-minute timeline of each crew member's activities. Simulations and training then make sure the flight crew can do each operation.

Through operations-support planning, we analyze in detail the flight requirements and ground operations essential to support and control a proposed mission. Planning to support operations includes a comprehensive review of flight-controller documentation and the necessary updates as flight requirements and practices change.

Several documents detail what payload experimenters need to support flight operations: flight rules, command plans, communication and data plans, mission-control and tracking-network-support plans, procedures for operating systems, procedures for operations and maintenance, handbooks on flight-control operations, and flight-software documentation.

Payloads with specific pointing requirements must become involved with operations-support planning, which includes developing the mission's flight profile and producing, analyzing, and designing mission-planning products. These activities include assessing the flight profile and ascent performance and thoroughly analyzing the flight design to produce flight-design ground rules and constraints. Details on payload masses and centers of gravity, as well as any special requirements for payload deployment or rendezvous, are necessary to develop these mission-planning products.

We then develop crew checklists for mission-specific operations and timelines. Payload operators provide details on how much time their experiment needs to operate as well as any specific requirements such as power, cooling, attitude, and times for crew involvement.

Software is created, modified, and tested for guidance, navigation, and control of the onboard payload and laptop computers. Products to reconfigure the Mission Control Center and Shuttle Mission Simulators for operations are developed to reflect the specific flight. The simulators' configuration must reflect the characteristics of major payloads, so we can effectively train flight crews and ground controllers.

Because launch is the most critical period for a space mission's safety, a major part of preflight planning involves launch-safety planning. We plan extensively to keep the crew safe during this hazardous phase. Preflight planning also requires performance analyses—trajectory, navigation, and guidance design—for ascent, orbit, reentry, and landing operations. Performance is then analyzed in detail to support specific payloads, including payload deployment, rendezvous, proximity operations, and payload retrieval.

20.3.1 Training for Space Shuttle Flights

Astronauts are the crucial component in any crewed space flight. Before astronauts can begin work on a specific payload, they must train on Space Shuttle systems and capabilities. Payload developers need to understand this training.

Pilot astronauts serve as commanders or pilots. Commanders are responsible for crew safety, vehicle safety, and the success of the mission. Pilots are second in command, and their main duties are to control and operate the Shuttle. Commanders and pilots usually help deploy and retrieve spacecraft, using the remote manipulator arm or payload-unique equipment, and they may support payload operations.

Mission-specialist astronauts conduct experiments and monitor payload activities. They may deploy spacecraft and may move payloads with the Orbiter's remote manipulator arm. Mission specialists must know the Shuttle systems that support payload operations. They also coordinate onboard operations involving crew-activity planning and monitor the Orbiter's consumables. A payload developer will likely work with two or more payload-specialist astronauts as the mission plan progresses.

Astronaut candidates must pass one year of training before they're qualified as astronauts. After becoming astronauts, they continue to train in order to maintain proficiency and develop skills for specific flights. Following the one-year basic training for astronaut candidates, new astronauts continue with advanced courses in Shuttle systems and enter phase-related training for a specific Shuttle mission seven months to one year before the scheduled launch date. At this time, the potential payload sponsor may meet with the astronauts scheduled to operate their payload. The organization developing the payload works closely with the astronaut office to ensure that the flight crew and backup crew are well trained in operating the particular payload.

As advanced training for the specific mission begins, a special management team directs the flight crew's training. Training now involves scripts and scenarios developed uniquely for the mission, and instructors now test the individual astronauts with planned operations and staged malfunctions. The flight crews also train together on the Shuttle Mission Simulator, which simulates all mission phases. The Shuttle Mission Simulator is designed to allow installing update kits for different missions and payload configurations. The flight simulators and computers can also work with other simulators such as the European Space Agency's Spacelab Simulator.

The two simulated orbiter cockpits contain controls and displays identical to those of an orbiter and are similar to simulators for commercial airlines. A closed-circuit television display provides proper spatial views of objects from the aft window and orbiter cameras. Computer-generated sound simulations duplicate noises experienced during an actual flight, including the onboard pumps, blowers, mechanical valves, aerodynamic vibrations, thruster firings, pyrotechnic explosions, gear deployment, and runway touchdown.

20.3.2 Payload Integration

Every Shuttle payload must be integrated into the Space Shuttle vehicle and coordinated with other payloads. Power and cooling requirements of all Shuttle payloads must be integrated to stay within safe boundaries. Payload operations must also be coordinated with crew timelines to allow adequate time for set-up, stowage, and unexpected experiment activities.

20.3.3 Ground Operations and Support

The Launch Control Center at the Kennedy Space Center controls all of the Shuttle's launch activities. Because launch processing is automated, the launch countdown for a Space Shuttle takes only about 40 hours, compared with the 80 hours needed for an Apollo countdown. In addition, only 90 people now work in the control center during launch compared with the 450 needed for earlier crewed missions.

On launch day, the flight crew enters the Orbiter on the launch pad by means of its access arm, a pathway 1.5 m wide and 20 m long. Towering 45 m above the pad surface, this arm swings out from the Fixed Service Structure to the Orbiter's crew hatch. The environmentally controlled *white room*, where the flight crew undergoes final flight preparations in an ultra-clean environment, is at the end of the arm. The access arm remains in its extended position until about seven minutes before launch, so the crew can get out if needed. In case of an emergency after it has been retracted, the arm can be mechanically or manually repositioned in fifteen seconds to allow for an emergency evacuation of the crew.

People from the white room help the flight crew through the Orbiter's crew hatch. They close the hatch and check the cabin for leaks. Then they evacuate the white room, and the closeout crew leaves the launch-pad area. The pilot and commander check communications with the launch-control center and with the mission-control center at Johnson Space Center (JSC).

At T minus 9 minutes, NASA's Test Director gets a "go for launch" verification from the launch team. From this point onward, the ground-launch sequencer automatically controls the countdown. At T minus zero, the solid rocket boosters ignite and the hold-down explosive bolts and umbilical explosive bolts release. The mission's elapsed time sets to zero and its event timer starts. After the launcher clears the tower, operation of the flight transfers to the Mission Control Center at JSC. Most payloads are inactive or require no crew involvement during launch and orbit insertion.

The Mission Control Center at JSC in Houston controls all of the Shuttle's flight operations. The original Mission Control Center became operational in June 1965 for the Gemini 4 mission, and was the hub of the US's program for human spaceflight. This Mission Control Center has been updated many times to reflect the new requirements of the Space Shuttle and International Space Station programs.

A payload developer will work with the payload office of the Johnson Space Center. The Payload Operations Control Center at Mission Control is available to payload customers for use in operating their experiments. Operators can receive data and telemetry, see live and playback video of their experiments, and possibly talk to the flight crew about experiment procedures. The payload officer can receive changes in experiment timelines and send them up to the crew.

If needed, the Payload Operations Control Center at Mission Control can work closely with other payload centers. In Spacelab-1 flown on STS-9, a special Spacelab payload operations control center at JSC became operational and tied science managers at Mission Control to the Shuttle crew and to remote stations at MIT, to the European Space Agency in Bonn, Germany, and to the Goddard Space Flight Center in Maryland.

If NASA's Mission Control Center can't operate during a Space Shuttle mission, NASA's ground terminal at the White Sands Test Facility in New Mexico handles emergency mission control. Flight controllers from JSC are prepared to fly to White Sands and continue supporting the mission at a reduced level. As soon as Mission Control becomes operable again, support of the flight goes to another shift of flight controllers at Houston.

The Mission Control Center controls the Shuttle until it lands, when the Kennedy Space Center resumes control. After landing, the safety-assessment team determines when the Shuttle is safe. After the area is cleaned of any toxic gases, the recovery convoy helps the flight crew leave the Orbiter. Then another crew of astronauts enters the Orbiter and installs switch guards, removes time-critical experiments, and prepares the vehicle for ground towing. Time-critical experiments can then be taken to nearby laboratories. Refrigeration, cooling, sample storage, and data handing are available to the experimenter. Astronauts remove payloads that are less time critical from the Orbiter after it is towed from the landing area.

In the post-flight debriefing, the crew's observations and comments support payload-data analysis and may lead to future changes to the payload. Comments from the flight crew, ground controllers, and others who participated in the payload operations during the mission can be invaluable.

20.3.4 Ground Tracking

NASA's Tracking and Data Relay Satellite System (TDRSS), a set of 2500-kg position tracking satellites in geosynchronous orbits, were launched in early Shuttle missions. This capability made the old method of spacecraft tracking used for Mercury, Gemini, and Apollo obsolete. Whereas earlier space programs had to rely on being directly over a tracking station to communicate with a spacecraft, each TDRS satellite can maintain communications with spacecraft for nearly half of the globe. The TDRSS system improved the control, communications, and response between Shuttle orbiters and ground control.

20.3.5 Effects Space Can Have on Humans

People considering sending an experiment into space must be aware of the effects space can have on humans. Many astronauts are slightly nauseous the first day of launch, so this would not be a time to schedule some experiments. Some astronauts have reported that their eyes focus slightly differently in space. Regular human responses such as fatigue and eye strain must be accommodated. Experimenters must be aware of the various possible effects of space and microgravity on the astronauts that will be conducting their experiments.

Astronauts in microgravity become hypovolemic, which means they have decreased circulating fluid volume. When the astronauts are in microgravity, the fluids in their bodies migrate to their upper torsos, heads and necks. There are receptors in the carotid arteries called baroreceptors that sense this fluid in the neck and think that there is an abundance of fluid. The body's natural response is to expel what it thinks is excess liquid. To prevent the astronauts from blacking out on re-entry when the Earth's gravity begins to pull the fluids back down to their feet, they drink a combination of water and salt tablets or a hypotonic solution (like a Gatorade sports drink) so that the body will absorb the fluid quicker.

As space flight becomes more operational, the opportunities to travel in space will be open to more people. NASA studied the effects of space on older astronauts when John Glenn returned to space on Space Shuttle STS-95 in October 1998. Some day it may be commonplace for experimenters and payload operators to travel on the Space Shuttle or live on the International Space Station in order to operate their experiments.

SPACEHAB is the first company to commercially develop, own, and operate habitable modules that provide space-based laboratory facilities and logistics resupply aboard the Space Shuttles. These modules have successfully flown 13 times, supporting scientific research missions and logistics resupply missions to the Russian space station, most recently performing those roles aboard the last Shuttle visit to Mir, and as a biomedicine laboratory during the historic flight of STS-95 (Glenn, Discovery).

20.4 International Space Station

20.4.1 Space Station History

In 1923, Romanian Hermann Oberth was the first to use the term "space station" to describe a wheel-like facility that would serve as the jumping off place for human journeys to the moon and Mars. In 1952, Dr. Werner von Braun first published his space station concept.

The Soviet Union launched the world's first space station, Salyut 1, in 1971—a decade after launching the first man into space. The United States sent its first space station, the larger Skylab, into orbit in 1973 and it hosted three crews before it was abandoned in 1974. Russia continued to focus on long-duration space missions and in 1986 launched the first modules of the Mir space station.

In 1984, U.S. President Ronald Reagan announced the construction of an American space station. Later to be named Freedom, the station would house elements not only from the U.S. but from Japan, the European Space Agency and Canada. The station would have five livable modules, including two laboratories, two crew modules and a logistics module. A revised version of the station had a crew of as many as eight at one time. In 1993, a major reorganization allowed for the creation of Space Station Alpha, which was essentially Space Station Freedom with Russian assistance. This project was renamed the International Space Station (ISS) four years later. Today, a majority of Space Station Freedom's components are included in the International Space Station.

A program-level consequence of human space operations is that they are increasingly international for cost sharing. Programs such as the International Space Station must continually adjust to their partners' changing priorities. Political debate about national economic priorities has caused repeated major revisions in the Station concept, and it remains subject to change.

20.4.2 ISS Characteristics

The International Space Station is a cooperative agreement among the United States and representatives of Russia, Japan, Canada and participating countries of the European Space Agency (Belgium, Denmark, France, Germany, Italy, the Netherlands, Norway, Spain, Sweden, Switzerland, and the United Kingdom). The purpose of the ISS is to provide an Earth orbiting facility with the resources to conduct microgravity research. The objectives of the ISS program are to provide access to a world-class microgravity laboratory as soon and as easily as possible, to develop the ability for long-duration spaceflight, to develop effective international cooperation, and to provide a testbed for 21st Century technology. Over fifty flights will be needed to bring up and assemble the various modules/ elements of the ISS. Figure 20.1 illustrates the ISS at Assembly Complete.

The ISS is a large, open-truss structure supporting numerous solar panels, instruments, and modules for habitation, laboratory, and service work. The ISS represents an evolutionary approach to space science, with the Station's size, function, and abilities designed to grow over time.

The Station works like a spacecraft with unique features of on-orbit assembly, permanent habitation, and special subsystems for electrical power, thermal control, data management, guidance and navigation, attitude control, communication, and propulsion (in addition to environmental control and life support). The laboratory modules and their attached instruments yield all the advantages of direct human access as described in Sec. 20.1, so they'll support research far beyond that of any combination of unmanned low Earth orbit (LEO) spacecraft.

Routine system housekeeping is largely automated so the crew can devote their time to support research. Remote control from ground stations carries out some payload operations.

The current habitable pressurized volume on the International Space Station is equal to the habitable space in a 167-square-meter, three-bedroom house with 2.4-meter ceilings. The habitable, pressurized volume on the completed station will be 1217.6 cubic meters. That is about the volume of three average American houses, each one containing about 186 square meters with a 2.1-meter ceiling for a total of around 396.4 cubic meters. The pressurized volume will be roughly equivalent to the interior of a 747 jumbo jet.

The Soyuz is the Russian element that provides the crew emergency return ("lifeboat") capability. The Soyuz serves as an escape vehicle in the event the ISS needs to be abandoned in an emergency. As such, there is always a Soyuz docked to the International Space Station. At least every 6 months, the docked Soyuz is replaced with a "fresh" Soyuz. The Progress spacecraft is basically a cargo version of the Soyuz and is used to bring up fresh food and supplies to the ISS. Besides resupply, the Progress also reboosts the Station to a higher altitude.

The International Space Station benefits from experience gained during previous American and Soviet flights. These programs provided many answers to the question of how to sustain a long-term operation in space. (See Figure 20.1.)

20.4.3 ISS Expeditions

As an incoming crew prepares to replace the outgoing crew, there is a handover period. The current space station crewmembers communicate to the crew on Earth any unique situations they have encountered, new techniques learned, or any topic necessary for life aboard the space station. In addition to the Expedition crews, Shuttle crews arrive for periods of about a week to work on ISS assembly operations. To reduce the number of assembly tasks that the Station crewmembers must perform and therefore reduce their training time, it is usually preferable for the Shuttle crew to do as many of the assembly tasks as possible.

The effects of long-term exposure to reduced gravity on humans—weakening muscles; changes in how the heart, arteries and veins work; and the loss of bone density, among others – are being studied aboard the Station. Knowledge of these effects may lead to a better understanding of the body's systems and similar ailments on Earth. Studies of the effects of space on plants and animals conducted

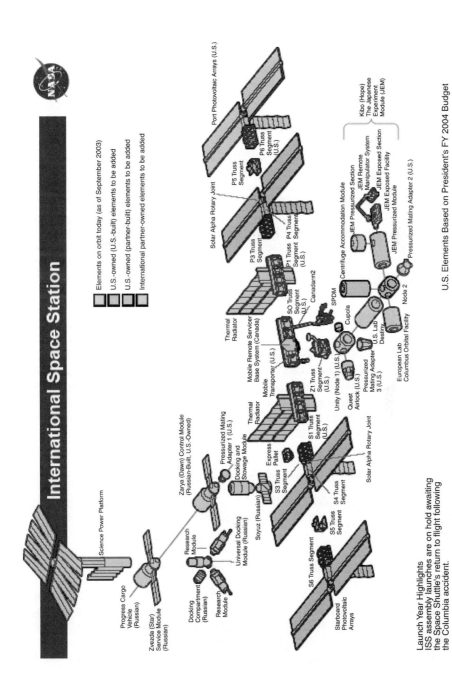

Fig. 20.1. Planned Configuration of ISS at Assembly Complete. Construction of the ISS is evolutionary, with size and capabilities growing over time. *(Courtesy of NASA)*

aboard the Station and the possible methods of counteracting these effects are needed to prepare for future long-term human exploration of the solar system.

20.4.4 International Space Station Mission Phases

The events in a simplified traffic model include the following:

- Progress docking
- Reboost
- Quiescent operation
- Shuttle docking/mated operations
- Crew handover
- Deferred assembly

The assembly sequence, altitude strategy, and crew rotation results in a traffic model. A traffic model shows what Earth-to-Orbit-Vehicles (ETOVs) arrive at the Station and when. In Figure 20.2, a crew arrives on a Soyuz, followed by a Progress resupply flight. Besides resupply, Progress also reboosts the Station to a higher altitude. The station reboost must meet two criteria it must be low enough that the Station's orbit will decay to the right altitude for the Shuttle assembly flight rendezvous but high enough in case the next Progress flight is missed.

Approximately five weeks after the reboost, the Shuttle assembly flight docks with the Station. There is no crew rotation on this flight, so the entire Shuttle uplift capability can be devoted to bringing up new Station hardware. A later Progress flight is a resupply/ reboost mission. The next Shuttle flight is another assembly flight without a crew rotation. A crew rotation occurs on the third Shuttle flight, which means that there had to be enough mass and performance margin to bring up Station hardware and three new crewmembers, who will stay on orbit approximately 156 days. Table 20.1 lists several important considerations in the selection and use of ISS crews.

20.4.5 Docking With the International Space Station

Various Earth-to-Orbit Vehicles (ETOVs) such as Shuttle, Progress, and Soyuz can rendezvous with the Station. While the Progress and Soyuz can travel directly to the ISS altitude, the Space Shuttle must perform several orbit maneuvers, or burns, to arrive. After orbital insertion at 226 km, the Shuttle performs as many as 13 burns over the next two days to raise its altitude to the ISS altitude. In a typical flight, the Shuttle performs two orbit raising burns the first day, two orbit raising burns and a planar correction burn on the second day, and three orbit raising burns and one planar correction burn on docking day. There are also four mid-course correction burns during the final orbit leading to docking to fine tune the trajectory en route to docking. The timing of these burns ensures that the Shuttle not only achieves the station altitude but achieves it in the same orbital plane and location as the ISS.

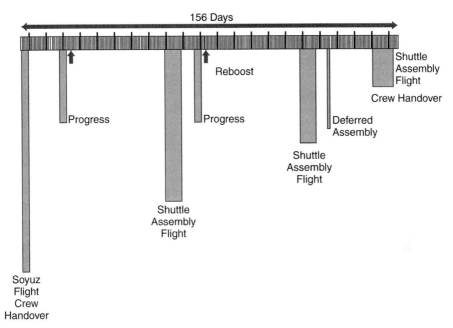

Fig. 20.2. **ISS Simplified Traffic Model.** The model lists the activities of all ETOVs during one crew's stay on the ISS.

Table 20.1. **Getting the Most from Space-Station Missions.** Continually replacing crews and payload modules will allow flexibility for learning and for improving efficiency during the years of station operations.

Crew Mix, Selection, and Training

• Train crews together to develop team cohesion
  – International program requires mix of nationalities and cultures—conflicts with usual criteria (shared background and culture) for small groups in prolonged isolation
• Decrease the training burden (allow lower fidelity training facilities)
  – Crew need not be experts on payloads
  – Familiarize them on the ground and continue on-the-job training in orbit

Ground Control Team (Payload and Systems Specialists)
• Human Space Station missions require continuous flight control monitoring: however, a highly trained Gemini team may handle emergencies until specialists can resolve the problem
Scheduling and Tasking

• Be prepared to change based on the priorities of international partners
• Exploit the flexibility humans provide
• Find the best balance between efficiency and cost
  – Don't try for 100% use of astronaut time—too costly, complex, and rigid
  – Allow margins for anomalies and benefit from the crew's judgment
• Allow for changes in crew-training criteria, scheduling, and payload recycling—be prepared to change the plan of operations (including resupply and microgravity) and maintenance cycles as space experience is gained

Ideally, one would want to launch due east from a launch site to maximize the cargo-to-orbit capability for a given launch vehicle. This is because the Earth, rotating from west to east, gives rockets a "free" head start in the right direction. Launching due east from Kennedy Space Center would place the Shuttle in a 28.5-degree inclination orbit. This inclination is the same as the latitude of KSC. Launching due east from Russia's main launch site, Baikonur, would place spacecraft in a 45.6-degree inclination orbit - the launch site latitude. However, doing so would also drop the lower stages of the boosters on China. To avoid this, the Russians crank up the minimum inclination to 51.6 degrees.

The Shuttle does trade some payload capacity for propellant needed to make up the difference between launching at 28.5 degrees vs. 51.6 degrees. An added benefit to Earth sciences of this higher inclination orbit is that the ISS flies over more of the Earth's surface and covers about 95 percent of the inhabited lands.

When the Shuttle launches, the ISS is not directly above the Shuttle, but may be on the other side of the Earth. Since objects in lower orbits take less time to go completely around the Earth than objects in higher orbits, the Shuttle must stay at a lower orbit than the ISS until it has caught up to it. How much lower depends on how far in front the ISS is when the Shuttle launches and how much time we want to take in catching up. For example, the burns on the first two days are planned to put us 74 km behind the ISS on the morning of the docking day, regardless of how far behind it we were at launch.

20.4.6 Mission Management and Operational Tasking

The Mission Control Center in Houston handles all Space Shuttle launches, oversees assembly activities, and manages overall system operations and safety. The Mission Control Center in Moscow handles Russian launches and related assembly activities. The Mission Control Center in Houston works with the Payload Operations Integration Center in Huntsville, Alabama, to integrate payload operations and carry out and coordinates all payload operations in real time.

Although one integrated plan is used by the onboard crew to ensure safety and prevent payloads from interfering with each other, each partner has mission planning responsibilities for the payloads, elements, and transportation vehicles that it provides.

Because of the potential dangers from even a simple payload malfunction, any payload commanding from the ground will only be done with the approval and monitoring of the Mission Control Center in Houston.

The Mission Control Center in Houston (MCC-H) has supported the ISS around the clock since Expedition One launched in October 2000. MCC-H personnel have grown accustomed to a stable shift schedule such that Orbit 1 works 11 p.m. until 8 a.m., Orbit 2 works 7 a.m. until 4 p.m. and Orbit 3 works 3 p.m. until midnight. This normally corresponds to the crew sleep periods. However, sometimes the station crews shift for brief periods to support major events like Soyuz/Progress docking/ undockings or station based EVAs. During these periods, the MCC-H teams maintain the standard shift schedule to avoid impacting ground personnel. The Mission Control Center in Moscow (MCC-M) has Flight Control teams that maintain a completely different shift schedule. They work a four-shift rotation of Flight Controllers who work console 24 hours in a row and are then off for three days before supporting again on the fifth day for another 24-hour shift.

The Mission Control Center in Houston is responsible for controlling and managing many Station systems, including all electrical power systems and solar array positioning. This allows the crew time to concentrate on experiments, robotic arm operations and any operations that cannot be performed by the Flight Control Team.

During early construction of the ISS, before anyone was permanently living on the Station, the modules were monitored overnight by a Station Duty Officer (SDO). The SDO would call in the appropriate personnel if any serious technical problems arose during the night.

As the ISS became operational, MCC-H soon discovered that having flight controllers rotate 8 hour shifts, including nights, weekends, and holidays, was becoming stressful. Unlike the Space Shuttle missions, the ISS mission never ended, and it was taking a toll on flight controller's time and energy.

In the spring of 2001, the Mission Operations Directorate decided to reduce the normal six person team to a two person team for the night and weekend shifts. They named this team Gemini. Each of the two Gemini operators would be responsible for three disciplines. One Gemini flight controller would be called Atmosphere Thermal Lighting Articulation Specialist (ATLAS) and the other would be Telemetry, Information Transfer and Attitude Navigation (TITAN).

20.4.7 Altitude Control and Re-boost

The International Space Station is in an elliptical orbit with an Ha (i.e. apogee/highest point of the orbit) of 407 km and an Hp (i.e., perigee/lowest point in the orbit) of 377 km. Atmospheric drag slowly reduces the Station's altitude, but propulsive reboost periodically restores it, so altitude varies cyclically.

The Station maintains an altitude margin and a reboost propellant reserve to ensure that its altitude won't become dangerously low even if resupply flights are late. Between reboosts, the freedom from propulsive disturbances allows microgravity research. The ground is heavily involved in reboost operations, since they are critical to Station survival.

20.5 Human Lunar Missions

On July 20, 1989, President George Bush asked the National Space Council to determine what was needed to return to the Moon and to continue on to Mars. The *90-Day Study on Human Exploration on the Moon and Mars* [NASA, 1989], describes the main elements of future human missions. The planned approach has three stages: the Space Station, a return to the Moon, and then a trip to Mars. The study defined the general mission objectives and key program requirements. As a result, regardless of the exact approach selected—heavy launch vehicles, space-based transportation systems, surface vehicles, underground habitats, or other concepts of support systems—returning to the Moon is a crucial step in the evolution of space exploration.

A lunar outpost is an excellent platform from which to do research in astronomy, physics, and life sciences. The crew can use rovers to explore the Moon's geology and geophysics by taking soil and rock samples for analysis in a lunar laboratory. The Moon is a unique site for synoptic viewing of the Earth's magnetosphere, which can provide long-term information about the distribution of solar energy and the consequent warming and cooling of Earth's upper atmosphere.

At just three days from Earth, the Moon is an ideal location for learning to live away from Earth with increasing self-sufficiency. The lunar outpost will serve as a test-bed for validating critical mission systems, hardware, technologies, human abilities, and self-sufficiency—all of which we can apply to interplanetary exploration. The life-support systems tested on lunar missions have essentially undergone a full-scale operations test, after which they can evolve incrementally before developing into a complete system for a Mars mission.

20.5.1 Significance of Past Missions

From 1969 to 1972, the US's Apollo spacecraft landed six times on the Moon. Pairs of astronauts spent up to three days on the Moon and explored its surface for many hours. During the last three landings, astronauts used rovers built like dune buggies to transport themselves about 30 km across the lunar surface.

These history-making successes show that human lunar missions are well within technological reach. But we should also remember that problems had to be overcome by system redundancy combined with prompt and astute action by the astronauts and the payload specialists at Mission Control. Among the six landing missions, these incidents included a computer glitch due to lightning strike during booster ascent, premature shutdown of a booster engine, difficulties in docking, computer false alarms, control-system malfunctions, a landing-radar malfunction, and a parachute damaged by contact with propellant. The explosion of an oxygen tank on Apollo 13 prevented lunar landing, and the crew survived only through prompt and ingenious measures. These events remind us of the versatility and the risk of human presence in space. Case [1993] and Wright [1993] describe typical early-1990s concepts for a return to the Moon.

20.5.2 Mission Phases

Figure 20.3 shows the phases of a human lunar mission. Translunar injection is the propulsive maneuver that gives the spacecraft enough velocity to get near the Moon. At the velocity needed for initial injection, the flight to the Moon takes about five days each way, but operators usually use slightly higher velocities, which reduces the time to three or four days. (See Chap. 10.)

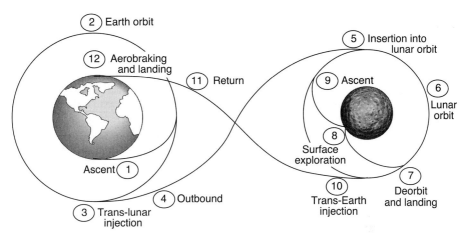

Fig. 20.3. Profile of a Lunar Mission. The drawing is not to scale—the altitudes of the orbits around the Earth and Moon are greatly exaggerated for clarity. In each of these orbits, the spacecraft may do one or more revolutions to obtain flexibility in mission timing and maneuvers.

Injection could be direct (combined with ascent from Earth in a single propulsive maneuver), but a short time in a low parking orbit routinely separates the ascent and injection. This technique doesn't penalize propulsion and it gives the flexibility of a hold period to adjust mission timing and orbit phasing, as well as to verify proper system operation or to correct problems.

These considerations also favor a parking orbit at the Moon, with the major extra advantage that propellant and structure for return (trans-Earth injection) and for landing on Earth don't have to decelerate, land on the Moon, and then be raised back to an altitude and speed for lunar orbit. Instead, only a capsule for the landed crew members and their lunar samples needs propulsion back to lunar orbit.

Without an atmosphere, the lunar orbit's lowest point or periapsis (sometimes called perilune or pericynthion) can be just high enough to clear terrain safely. The Apollo missions used altitudes as low as 17 km. Landing consists of decelerating from orbital speed and then transitioning to a descent which must reach near-zero vertical and horizontal velocity at the same time that it reaches zero altitude. This requires throttling (variable-thrust) engines and, preferably, direct human control to locate a clear and level landing area.

After landing, the crew prepares the ascent propulsive stage for return and then starts exploring the surface. Because the Moon is always about the same distance from the Earth, lunar round trips have flexible timing. The time spent on the lunar surface depends mainly on available supplies, reliability of equipment, physical endurance, and perhaps the onset of the two-week lunar night. The surface activities for Apollo included surveying the local area, documenting and collecting mineral samples, and emplacing long-lived automated instruments (Apollo Lunar Surface Experiment Packages). Thus, payload planning and data processing continued long after the astronauts returned. Apollo also showed how astronauts could work efficiently and safely in this novel environment. The purpose of future landings will be to carry out equivalent activities on a larger scale.

Like most other phases of the lunar mission, at least on the near side of the Moon, surface exploration benefits from the short distance to Earth. Voice and video links work continuously in near real time, with subsystem specialists at Mission Control available for immediate advice. However, as lunar operations mature, lunar explorers must refine ways of handling problems and not rely on a full shift of ground-control people working around the clock for their support.

After exploring the lunar surface, astronauts take the capsule up and rendezvous with the lunar-orbiting spacecraft, unless the mission uses direct trans-Earth injection. Near the Earth, the spacecraft could do an inverse of its earlier translunar injection and thereby enter an orbit around the Earth. But propellant for this maneuver would have to go to the Moon and back. Thus, an alternate approach is to take an atmospheric-reentry capsule to the Moon and back and use aerobraking for direct reentry. This reentry does require precise navigation into a reentry corridor, which must be neither too shallow to shed enough of the spacecraft's velocity nor so steep that excessive deceleration, heating rate, or impact speed destroy the spacecraft.

Future lunar missions may help human exploration of space to move into a more ambitious phase by identifying usable resources. Sources of oxygen or water would be most interesting, and processes for extracting oxygen from lunar surface materials have been described and compared, for example in Taylor [1992]. If we can get oxygen on the Moon for propulsion back to Earth, future missions could bring only fuel and not oxidizer. Thus, we could significantly increase their useful payloads to the lunar surface. In terms of propulsive effort, a low-Earth orbit is easier to reach from the Moon than from Earth—hence, spacecraft on interplanetary missions could eventually use lunar resources.

Table 20.2 is a basic checklist of issues for cost-effective lunar operations.

20.6 Human Interplanetary Missions

Beyond Earth orbit—on interplanetary missions—the rationale for crewed missions remains strong. Because unforeseen situations are sure to be encountered, interplanetary missions will benefit from the insight and flexibility of a human crew. The conquest of space by machines is not nearly as impressive as interplanetary exploration by human crews.

20.6.1 Features of Interplanetary Flight

There are basically two ways to travel from the Earth to another planet. The most direct way would involve the least time, but would require a tremendous

Table 20.2. Getting the Most from Missions to Explore the Lunar Surface. Unlike Space Station operations, a lunar mission involves only one round trip. Thus, it allows much less chance for refining operational efficiency on the job. Careful preparation best uses the limited time available on the lunar surface.

Get the Most From a Human Lunar Mission
Carefully plan the lunar mission • Assign only tasks for which human presence is essential • Logically rank and sequence these tasks
Provide the best possible resources • Crew mix, selection, and training • Ground support team (consultants/users) • Exploitation of lunar orbiters (crewed or not) • Efficient means to capture data (instruments, vehicles, communications) • Automation of routine activities
Possible Lunar Bases • Fixed stations and tele-operated rovers • Ground team to operate emplaced equipment long-term
Build future missions on the results • Data conversion and distribution—accessibility to users • Operational lessons defined and preserved to benefit later missions
Choose Most Effective Length of Time on Surface
Balance more exploration (perhaps sustained by automated resupply flights) against • Increased mass of consumables, electric power supply, tools, and spare parts • Declining human efficiency and increasing risk (crew injury, equipment failure) • Reaching limits of mission's plan and equipment (time to apply what has been learned)
Get the Most From a Series of Missions
• Choose landing sites as part of a coherent long-term plan • Search for resources and begin experiments toward using them • Build up equipment on the Moon—design for dormancy, reuse, adapting, or cannibalizing • Maintain continuity of core operational and design teams (corporate memory) • Maintain continuity of hardware lineage (retain and refine proven subsystems)

amount of fuel. Another concept involves using trajectories that use the planets' orbits and gravitational fields. The conceptual designer of an interplanetary mission will usually seek to reduce the mission's total ΔV or propulsive requirement and therefore enlarge the useful payload that will arrive on the planet. However, the designer may try to shorten the trip to decrease risks to human life.

Most likely, the first human interplanetary mission will be to Mars. Initial missions to Mars will prove the systems and equipment and further survey selected landing sites. Later missions will establish a Mars outpost with the objective of experimenting and exploring in an extraterrestrial environment.

The technology for a human Mars mission extends straightforwardly from what has already been developed for lunar missions. As early as the 1960s, NASA studies projected the first humans would land on Mars in 1984 [Manned Mars Surface Operations, 1965]. But the distances and times involved make interplanetary missions enormously more challenging than lunar missions. Going from the Earth's surface to an orbit of Mars doesn't require much more total velocity than going to the Moon, but we must also time the flight so Mars is at the correct point for arrival and return. The result is that, compared with lunar missions, interplanetary launch and return windows are sharply constrained. Also, because of the distances involved, multi-year missions are inevitable. Long Earth-Mars transit times have risks due to zero gravity and sporadic solar-flare radiation. A long stay on Mars means long exposure to an environment which may have unforeseen hazards. And finally, long total mission time means increased probability of human or mechanical failure.

Figure 20.4 shows three general types of trajectories for a Mars mission. Their times and total velocity budgets start from LEO and they use aerobraking instead of propellant for maneuver 4 (arrival at Earth). The graphs show that the trajectories with the lowest velocity requirements entail lengthy stays at Mars. A further constraint on mission timing is that the orbit of Mars is notably eccentric. This feature combines with the phasing of the movements of Earth and Mars to produce a decades-long cycle of favorable and unfavorable launch windows. For example, the years 2001 and 2018 are favorable for most types of Mars missions. See Young [1988], TRW Space Data Book [1992], and the Space Flight Handbook [1963] for details on these and other trajectory trades.

Numerous studies have proposed ways to improve this time and payload trade. One alternative is simply to wait for better technology, such as nuclear thermal rockets [Emrich, 1991]. Technologically, nuclear rockets will work. They were built and demonstrated on the ground in the NERVA program in the 1950s. A revival of research on nuclear or other non-chemical rockets and their large-scale application would depend on a degree of commitment and interest that may occur at some time in the future but is certainly not evident now.

In any case, we must minimize the total mass of environmental life support system equipment plus consumable resources so that, in interplanetary missions, the environmental processes will be almost closed-loop. The goal of the environmental system will be to operate in an almost closed loop system, using solar power to produce water and oxygen.

20.6.2 Mars Mission Phases

A Mars mission begins like a lunar mission—with a launch, a parking orbit, and trans-Mars injection. The parking orbit is assumed if the size of the spacecraft requires multiple launches with final assembly in orbit.

Upon arrival at Mars, the spacecraft has expanded mission options because an atmosphere is present. Instead of needing a major propulsive maneuver, it can use aerobraking to shed most of the arrival velocity. The atmosphere's low density makes the entry corridor very narrow (Figure 20.5), especially if preceding months

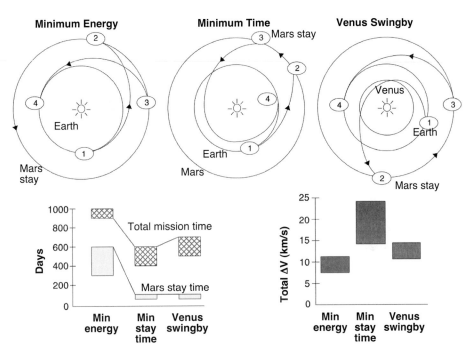

Fig. 20.4. **Trajectories for a Mars Mission.** The scale is only approximate. The basic trend is that shortening the mission time has a big price in required ΔV. The Venus Swingby obtains some free ΔV by using Venus's gravitational field to change the direction of the spacecraft's velocity vector. The phases of the mission correspond to: (1) departure from Earth; (2) arrival at Mars; (3) departure from Mars; and (4) arrival at Earth. The times and the velocities shown in the two graphs vary (shaded regions) depending on the launch date.

of zero gravity demand that the crew receive only moderate g-loads. Lynne [1992] gives entry-corridor trades in some detail with a typical corridor width of 1°, some of which we'd have to allocate to uncertainty about the actual atmospheric-density profile at that time over that region of Mars.

As with lunar missions, total payload increases if we keep the Earth-return stage in orbit around Mars as part of an orbiter module rather than bringing it to the surface and back. For this purpose, higher orbits are better. The orbiter might be continuously staffed both to ensure its readiness for return and to do remote sensing, navigation, and communications relays to support the surface exploration. If staffed, the orbiter could operate at or near one of the moons, Phobos or Deimos, to investigate their potential resources for future missions [O'Leary, 1992]. The orbit of Deimos happens to be close to the Mars equivalent of a geosynchronous orbit, where a spacecraft would appear fixed in the Martian sky.

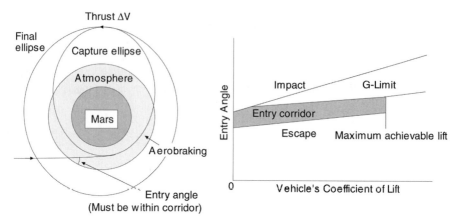

Fig. 20.5. Aerocapture from Interplanetary Trajectory. The depth of the atmosphere is exaggerated for clarity. Aerobraking sheds enough velocity to bring the spacecraft into an orbit around Mars. At apogee a small propulsive maneuver stops further aerobraking by raising perigee above the atmosphere.

If the orbiter is designed to withstand aerobraking, the lander and orbiter can aerobrake and enter parking orbit as one unit. Alternatively, the lander or Mars Excursion Module may separate early and enter the atmosphere directly from its interplanetary trajectory. Meanwhile, the orbiter expends propellant for insertion in Mars orbit but need not withstand atmospheric heating. In aerobraking any large vehicle at Mars, one concern is the physical size of the aerobraking heat shield. The mass to be landed includes an ascent vehicle as well as everything for survival and exploration on the surface, so the heat-shield diameter may need to be tens of meters, which exceeds the limits of Earth-launch vehicles. Thus, we have to extend the shield mechanically or assemble it in space, with corresponding concerns about verifying its readiness for the mission.

Surface exploration of Mars would work like exploration of the Moon. Despite advances in robotics, machine vision remains severely limited; therefore human astronauts with cameras and training in field geology will be needed as well as any robotic rover (surface explorer). As on the Moon, a high-priority objective is locating resources that could make later missions much easier. The Martian atmosphere is about 95% carbon dioxide, with the rest mostly nitrogen, and Mars is believed to hold large quantities of water as permafrost. More advanced missions could combine carbon dioxide, water, and nitrogen to manufacture propellants for rocket and surface vehicles, as well as to sustain the crew.

Godwin [1960] was among the first to point out that, if carbon dioxide is available on a planet and if we bring one ton of hydrogen from Earth, theoretically we could manufacture nearly 20 tons of rocket propellants (fuel and oxidizer) for the return journey. Recently, others have more thoroughly examined this possibility. Zubrin [1992] describes several reactions that can convert the carbon dioxide and hydrogen to methane and water, which we could then electrolyze to get oxygen. The propellants methane and oxygen will store as liquids.

Zubrin describes a very interesting mission (Figure 20.6) in which an uninhabited vehicle for propellant manufacturing and Earth return lands on Mars to produce propellants for a return flight before the crewed vehicle departs from Earth. He contends that his chosen manufacturing process is already developed and proven to a much greater extent than some of the other technology we need, so it's a reasonable part of even the first human mission. Given the enormous advantages of this propellant production, Zubrin envisages human missions to the Moon and Mars, using relatively modest masses launched from Earth. Bruckner [1993] describes a mission conceived recently by others applying Zubrin's ideas.

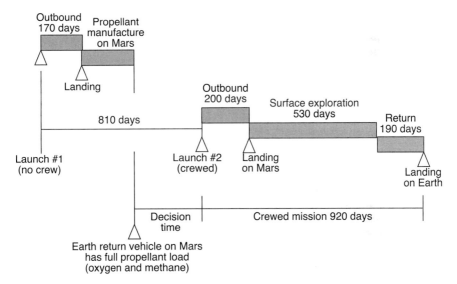

Fig. 20.6. Zubrin Mission Using Martian Resources. The first vehicle (no crew) uses nuclear power and Martian carbon dioxide to produce roughly 200 kg of propellants per day. The first vehicle accumulates enough propellant for the crewed return flight at Mars before the crewed vehicle departs from Earth.

When surface exploration is complete, the Mars lander ascends to either the parking orbit or directly to trans-Earth injection. If the orbiter has a crew, it can rendezvous in the parking orbit. Trans-Earth injection could use either a single module carrying orbit and surface crews, or two separate modules with extra capacity to provide redundancy (either one could return both crews).

Upon arrival at Earth, the vehicle may land by direct re-entry or by aerobraking to help maneuver itself into a parking orbit, which may be highly eccentric. In either case the payload for this maneuver need not include the habitation module that supports the crew during the long interplanetary flight; it needs only a capsule that can support them for the few hours until rendezvous with Earth-based vehicles or recovery teams. This approach again reduces the mass of propellant and heat shield that must go to Mars and back.

An alternative to aerobraking would be to have the approaching spacecraft met by an Earth-based recovery vehicle, which would maneuver to match interplanetary-approach velocity and allow docking with the capsule; a further maneuver would then return the vehicle to its original orbit. Or the spacecraft could land by direct entry and aerobraking.

All the items in Table 20.2 apply even more emphatically to interplanetary missions because of their long duration and other added challenges. Table 20.3 lists other issues for cost-effective operations for a Mars mission.

Table 20.3. Getting the Most from Interplanetary Missions. Add these items to the ones in Table 20.2 for lunar missions, emphasizing the challenge of enormous distances to even nearby planets.

Mission managers must allow astronauts unprecedented independence • Communication round trip time 7 to 44 minutes • Unknowns of the planetary environment
Task Mars orbiter (if used) and its crew (if onboard during Mars stay) • To best use the orbiter's advantages • To maintain the orbiter crew's morale during their stay at Mars

Continuous communication with Earth is very desirable, but difficult
• Distance and power requirements
• Other missions competing for the Deep Space Network
• Planetary rotation limits viewing between surface sites on Earth and Mars
   – A Mars orbit best for communication relay is unsuitable for other functions
   – Add a Mars-synchronous relay satellite?
   – May lose communication when Sun is between Mars and Earth

• Earth-Mars transit, typically 200 days each way, must include crew tasks
   – To get the most from the mission
   – To maintain crew morale

• Contingency of early departure from Mars surface creates special requirements
   – Ascent vehicle's propulsion increased for flexibility in launch window
   – Redundant environmental control, life support, and consumables in orbiter

Mission managers must adjust to the long communication times and the astronauts' resulting sense of isolation. Total mission time will be 500–1000 days, cost will be high, and public interest will be intense (at least for the first mission). Thus, conflict will increase between the goals of international representation (mix of crew nationalities) and team cohesion (shared background and culture).

One of the most important mission-design trades is selecting altitude and inclination for the orbiter module. Although an orbiter is valuable for relaying communications, Mars mapping, and Mars-ascent rendezvous, these functions have different orbit requirements.

Finally, Figure 20.4 shows how orbital mechanics favors long stay times. Long stay times mean more risk and more research opportunities. After a 300-day stay, would we know Mars ten times as well as after 30 days? Where is our region of diminishing returns?

As our discussion has shown, human spaceflight introduces extra mission phases and concerns into space operations. But human missions still require the functions defined in Chap. 3 and the procedural steps and trades described in other chapters.

References

Bruckner, A.P., M. Cinnamon, S. Hamling, K. Mahn, J. Phillips, and V. Westmark. *Low Cost Manned Mars Mission Based on Indigenous Propellant Production.* AIAA 93-1010, Aerospace Design Conference, Irvine, California, Feb. 16–19, 1993.

Case, Carl M. *The Lunar Campsite Mission Concept.* 30th Space Congress, Canaveral Council of Technical Societies, Cocoa Beach, Apr. 27–30 1993.

Emrich, W.J., A.C. Young, and J.A. Mulqueen. *Nuclear Thermal Rocket Propulsion Application to Mars Missions.* 28th Space Congress, Canaveral Council of Technical Societies, Cocoa Beach, Florida, April 23–26, 1991.

Godwin, Felix. 1960. *The Exploration of The Solar System.* Plenum Press. p. 81.

Logsdon, John. *The Space Shuttle Program: A Policy Failure?* Science, May 30 1986, pp. 1099–1105.

Lyne, J.E., A. Anagnost and M.E. Tauber. *Parametric Study of Manned Aerocapture.* Part II: Mars Entry, Journal of Spacecraft and Rockets, Vol. 29 No. 6 pp. 814–819, Nov.–Dec. 1992.

NASA. *International Space Station Data Book.* Space Station Program Office. NASA Johnson Space Center, 1994.

NASA. *Manned Mars Surface Operations.* Final Report RAD-TR-65-26, Contract NAS 8-11353, AVCO Corpn, Wilmington, Massachusetts for NASA Marshall Space Flight Center, 30 Sept. 1965.

NASA. *Report of 90-Day Study on Human Exploration on the Moon and Mars.* Nov. 1989.

NASA. *Space Flight Handbook. Vol. 3—Planetary Flight Handbook,* NASA SP-35, NASA Marshall Space Flight Center, 1963.

O'Leary, Brian. *International Manned Missions to Mars and the Resources of Phobos and Deimos.* Acta Astronautica, Vol. 26 No. 1 pp. 37–54, 1992.

Taylor, Lawrence. *Production of Oxygen on The Moon: Which Processes are Best and Why.* AIAA 92-1662, AIAA Space Programs and Technologies Conference, Huntsville, Alabama, Mar. 24–27, 1992.

TRW. *Space Data Book.* TRW Inc., Redondo Beach, California, 1992, pp. 4–2 and 8–21 through 8–31.

Wright, Michael J. *Conceptual Design of a Cargo Lander for the First Lunar Outpost.* 30th Space Congress, Canaveral Council of Technical Societies, Cocoa Beach, Apr. 27–30, 1993.

Young, Archie C. *Mars Mission Profile Options and Opportunities.* 25th Space Congress, Canaveral Council of Technical Societies, Cocoa Beach, Florida, April, 1988.

Zubrin, Robert. *Mars and Lunar Direct: Maximizing the Leverage of In-Situ Propellant Production.* AIAA 92-1669, AIAA Space Programs and Technologies Conference, Huntsville, Alabama, Mar 24–27, 1992.

Chapter 21

FireSat

Gael F. Squibb, *Jet Propulsion Laboratory, California Institute of Technology (Retired)*
David E. Kaslow, *Lockheed Martin*

> 21.1 Mission Information and Requirements (Inputs to an Operations Concept)
> 21.2 Clarification and Validation of Requirements
> 21.3 Operations Functions Needed and Selected Trades (Section 4.2.2)
> 21.4 Mission Operations Concept (Chapter 4)
> 21.5 Cost Estimate for 2 Satellites Versus 1 Satellite
> 21.6 Conclusions

One way to understand the main concepts in this book is to go through a specific mission as an example and to learn some of the key trades that can be made. Our example is a mission called FireSat. *Space Mission Analysis and Design III* [Larson and Wertz, 1999] describes this mission in great detail and we will use the requirements, mission concepts, and some of the design decisions from SMAD III as inputs to our operational considerations.

21.1 Mission Information and Requirements (Inputs to an Operations Concept)

Mission Statement: Because forest fires have in recent years had an increasing impact on recreation and commerce and ever higher public visibility, the United States needs a more effective system to identify and monitor forest fires. It is also desirable, but not required, to monitor forest fires for other nations; collect statistical data on fire outbreaks, spread, speed, and duration; and provide other forest management data.

The Forest Service's fire-monitoring office and rangers in the field will use the data. Data flow and formats must meet the needs of both groups without requiring specialized training and must allow them to respond promptly to changing conditions.

Primary Objectives:
- To detect, identify, and monitor forest fires throughout the United States including Alaska and Hawaii, in near real-time

Secondary Objectives:
- To demonstrate to the public that positive action is underway to contain forest fires
- To collect statistical data on the outbreak and growth of forest fires
- To monitor forest fires for other countries
- To collect other forest management data

Requirements:
- *Performance:* Four temperature levels, 30-meter resolution, and 500-meter location accuracy
- *Coverage:* Daily coverage of 750 million acres within the continental United States plus Alaska and Hawaii
- *Responsiveness:* Send registered mission data within 30 minutes to up to 50 users
- *Mission duration:* Operational for at least 10 years
- *Availability:* 98 percent, excluding weather, and 3-day maximum outage
- *Data distribution:* Up to 500 fire-monitoring offices and 2000 rangers worldwide with a maximum of 100 simultaneous users
- *Data content, form and format:* Location and extent of fire on any of 12 map bases, average temperature for each 30-square-meter grid

Orbit Description: The project office and the mission designers have chosen a low-Earth orbit, whose parameters are listed in Table 21.1.

Mission Constraints:
- The mission will be a NASA mission
- Tracking, Telemetry and Command (TT&C) will be through NOAA ground stations
- Launch will be on a Pegasus expendable rocket
- Cost will be less than $200 million plus research and development costs

Table 21.1. FireSat Orbital Parameters. A low-Earth orbit is cheaper and allows for better resolution than higher orbits.

Orbital Parameter	Value
Altitude	700 kilometers
Inclination	55 degrees
Period	98.8 minutes
Ground Track Velocity	6.76 kilometers/second
Node Spacing or Node Shift	24.8 degrees

- Initial operating capability will be five years after project start with an option for a second five year period based on the quality and usefulness of the data
- Normal operations will begin one year after launch

The sections that follow will go through a series of discussions and decisions among the MOM, the staff, and the project designers early in the project. While they clarify requirements and establish the early design, they usually identify a need for additional capabilities. The additional FireSat capabilities identified in the following discussions and decisions are summarized in Table 21.7 and formalized as derived requirements. We must document such requirements because comments like "we can do that" often become major aspects of the design. If we don't write them down, it will be difficult or impossible at a later date to trace a capability that must be incorporated and verified in the design, to a documented requirement.

End-to-End Information System (EEIS) Characteristics: The data rate from the sensor to the spacecraft data system was determined to be 8.5 megabits/s (SMAD III Table 9-15). One of the major trade studies was whether to perform the detection on board or on the ground. Two different missions have demonstrated the capability to detect hot-spots on board a satellite. Chapter 4 describes one of these missions, called BIRD. We will use one of these demonstrated approaches for onboard hot-spot detection and calculation of the average temperature for the 30-square-meter grids. During the first year after launch both the onboard processed data and the sensor raw data will be sent to the ground to support algorithm validation.

The size of the onboard computer allows it to accept and store the 8.5 megabits/s of data, simultaneously process the data, and determine fire locations and intensities. The storage device on the spacecraft is a 2 gigabyte solid state recorder. The sensor will be turned off over water, so data collection takes place about half the time. The onboard hot-spot detection and calculation of the average temperature for the 30-square-meter grids will reduce the raw data by a factor of 10,000 to one. The data is encapsulated in Consultative Committee on Space Data Standards (CCSDS) standard frames and protocols, which is a requirement of the NOAA ground network.

The ground system will be able to process the raw data and to replicate the onboard processing. If necessary, flight software changes will be developed and sent to the spacecraft. The ground system will also notify the user automatically, based on information from the spacecraft.

Although the raw data will be useful to payload designers and research scientists, FireSat is an operational system and providing the raw data would increase the costs significantly. Therefore, after the commissioning phase from launch to launch plus one year, the spacecraft will send only the processed data, except when troubleshooting onboard processing or sensor malfunctions.

Ground System Characteristics: One mission constraint is to use NOAA ground stations. The NOAA ground station characteristics are on the CCSDS web site referenced at the end of this chapter. NOAA has ground stations in Fairbanks, Alaska and Wallops Island, Virginia. Thus, there will be opportunities to contact the spacecraft from each coast of the United States.

Spacecraft (Bus and Payload) Characteristics: The EEIS section above describes the payload at a very high level. Another operational attribute is commandable sensor parameters that will be updated after launch to set the final operational capabilities of the sensor and detection algorithms.

The cross-track scanning is automatic and is not commandable. Cross-track scanning is the most inexpensive way to image. Imaging specific areas adds complexity and expense to the ground and space components.

The sensors will be turned on only over land. The transition from land to water or water to land will be detected on board automatically using GPS calculation of nadir latitude and longitude and an onboard map (table) that relates nadir location to land/water boundaries and the 30-square-meter grids. This map also identifies forest areas and habitable areas.

The spacecraft has ample margins in all of its subsystems so that simulating subsystems and allocating spacecraft resources is not necessary. There will be fault detection and protection capability to place the spacecraft and payload into a safe state that can be maintained for seven days without any adverse affects, other than the interruption of data collection. In short, it is a very operable spacecraft and payload.

End-users' Definition of Needed Data: The mission designers, the operators, and the end-users held a meeting to ensure correct understanding of the needs of the end-users. The initial system design was based on inputs from the scientists who conceived the system. Since the technology missions that have flown in the last several years have demonstrated the feasibility of detecting fires, this is to be an operational mission. The end-users that we spent a lot of time with were those that manage the operational aspects of fighting fires. The information they need is almost identical to the information generated by the MODIS instrument that is discussed in SMAD III

- Location
- Emitted energy

- Ratio of flaring to smoldering
- Area burned
- Average direction of the burn

Somewhat surprising was the fact that the fire fighting managers wanted this data in tabular form, not map form. They already had proven tools that displayed the fire in terms that they were familiar with and all they needed was the input data. We cannot emphasize enough the value of operators and designers talking often with the end-users of the system that is being developed.

21.2 Clarification and Validation of Requirements

The operator reads requirements and interprets them in a way that makes operational sense. The responsiveness requirement specifies, "Send registered mission data within 30 minutes to up to 50 end-users." This requirement does not define what registered mission data is, nor does it specify the triggering event for the start of the 30 minutes. It could be 30 minutes from fire start or 30 minutes from detection or 30 minutes from determining on the ground that there is a fire. All of these have drastically different EEIS architectures and could even affect the number of spacecraft in operation.

So the operators and the mission designers held a second meeting with the end-users. The requirement turned out to be for 30 minutes from data collection where there was a fire. However, the operations staff felt that it would be a cost driver to guarantee 30 minutes from data collection. So the end-users agreed to: "60 minutes from data collection on the spacecraft to end-user notification, with a goal of 30 minutes." The mission designers then allocated the 60 minutes to the various elements: spacecraft processing, detection, and transmission, ground reception, processing, and notification to the end-user community.

The point of this discussion is that it is okay to question requirements and request clarification to ensure that the end-user understands when a requirement is likely to become a design or cost driver. In this case the end-user was willing to make the requirement less stringent so that it would not drive the design or cost.

One of the requirements is "Coverage: Daily coverage of 750 million acres within U. S." Is this realistic and can the requirement be met? The MOM should make some rough calculations that will lead to a meaningful discussion with the mission designers and payload designers. The following is an example of a "back of the envelope" calculation as a first attempt at analyzing the requirement.

An acre is 4840 square yards or 4047 square meters. The requirement is to cover 3,035,250 square kilometers within one day. The amount of land, excluding water, in the 50 states and District of Columbia is 9,158,960 sq km [CIA Factbook]. Approximating this area as a rectangle of 4200 × 2180 km^2 validates the calculation. Therefore, covering 3,035,000 sq km per day is equivalent to covering 33 percent of the U.S. each day.

With this background the MOM should call a meeting with the mission and payload designers and ask them to show how the initial design will cover 33% of the United States on a daily basis. They would need to show ground coverage plots and discuss factors such as swath width and coverage rate. This discussion should result in an understanding of the margin that the early planners have given the operators. What happens if one pass is lost? What happens if one ground contact is lost? These questions need to be discussed early in the project so that everyone involved is aware of what is really being asked of the operations staff. Such questions often do not arise until just before launch and then the non-operational people are caught by surprise.

The MOM did call a meeting of the mission designers and the orbit experts to discuss the coverage calculations. An error in the initial design turned up at the meeting. As a result, the coverage requirement had to be reduced, the payload swath width increased, or a second satellite added. The payload could not be redesigned to increase the swath width and reducing the coverage was undesirable. The group decided to proceed with the operations design and to compare operational costs for one satellite vs. two. This enabled the customer to decide if the cost of the second satellite plus the increased operational costs justified the increased coverage. The message here is to question requirements in an operational setting and get everyone on the project to think about operations early!

21.3 Operations Functions Needed and Selected Trades (Section 4.2.2)

Let us now take what we have learned from the mission designers and determine what operational functions are needed and where there are opportunities for system trade studies. We will step through the nine primary functions established in Chapter 3 and describe briefly the operational functions associated with the FireSat Mission. The order in this section is what one usually encounters in developing the concept, not that shown in the chapter's diagram.

21.3.1 Navigation

Space Mission Analysis and Design treated the various options for navigation. Table 21.2 is the summary from that discussion.

Table 21.2. Identification of Performance Drivers for FireSat. First-order algorithms are given to allow us to estimate the performance drivers. Definition of performance drivers may change as we create more detailed definitions of the system and system algorithms. Comparison of columns two and three shows that the performance drivers may depend on the mission concept used. [Larson and Wertz, 1999]

Key Parameters	First Order Algorithm (Low-Earth Orbit)	First Order Algorithm (Geosynchronous)	Performance Drivers
Observation Frequency	(Number of spacecraft)/ 12 hr	Scan frequency	Number of spacecraft for low orbit
Time Late	Onboard storage delay + processing time	Communications + processing time	Storage delay (if applicable)
Resolution	Distance × [(wavelength/ aperture) + control error]	Distance × [(wavelength/ aperture) + control error]	Altitude, aperture, control accuracy
Observation Gap	Cloud cover interval or coverage gap	Cloud cover interval or coverage gap	None (weather dominated)

The results of these trades, summarized in Tables 21.3 and 21.4, led to the decision to place FireSat into a low-Earth orbit (LEO). The decision did not address an analysis of the operational aspects of the different orbits. The number of navigation functions and where they are performed has a cost impact in the operational phase of the mission. Although the number of required operators may be small, perhaps one person, the fact that the requirements are for a 10-year mission makes the addition of even one person significant. Assuming 3 percent inflation, a person with a cost of $100,000 a year before launch will cost $1,146,388 over the 10 years of support. The point is to involve operations staff during the early design stages of the mission to ensure that we minimize the lifecycle costs while meeting the mission requirements.

Table 21.3. FireSat Specialized Orbit Trade. The conclusion is that in low-Earth orbit we do not need a specialized orbit for FireSat. The frozen orbit can be used with any of the low-Earth orbit solutions. [Larson and Wertz, 1999]

Orbit	Advantages	Disadvantages	Good for FireSat
Geosynchronous	Continuous view of continental U.S.	High energy requirement No world-wide coverage Coverage of Alaska not good	Yes
Sun-synchronous	None	High energy requirement	No
Molniya	Good Alaska coverage Acceptable view of continental U.S.	High energy requirement Strongly varying range	No, unless Alaska is critical
Frozen Orbit	Minimizes propellant usage	None	Yes
Repeating Ground Track	Repeating viewing angle (marginal advantage)	Restricts choice of altitude Some perturbations stronger	Probably not

Table 21.4. Summary of FireSat Initial Parameter Estimates. [Larson and Wertz, 1999]

Parameter	Nominal Value
Altitude	700 km
Inclination	55 deg
Swath width	24.2 deg
Nadir angle range	±57.9 deg
Min. elevation angle	20 deg
Instrument	mid-range IR scanner
Ground resolution	30 m at nadir
Instrument resolution	4.3×10^{-5} rad
Aperture	0.26 m
Size of payload	0.4 m × 0.3 m diameter
Weight of payload	28 kg
Power of payload	32 W
Data rate of payload	85 Mbits/s
Mapping accuracy	3.5 km at elevation angle = 30° 5.5 km at elevation angle = 20°

While the decision to use GPS for navigation on FireSat removed the operational functions from the ground software and staff, there remained the cost of the flight software, and this should have been identified at the time of the decision to go to a LEO.

21.3.2 Spacecraft Planning and Analysis

The Spacecraft Planning functions for this mission are very simple. The most demanding function, in the absence of failures, is updating and maintaining the flight software. This task is especially important during the checkout and commissioning of the mission during the first year. The activity will probably be especially heavy during the first four months after launch and then slowly decrease during the rest of the first year to a maintenance level for the operational portion of the mission.

The Spacecraft Analysis functions are, after the first several weeks, limited to non-real-time analysis such as developing the trend plots and analyzing spacecraft subsystems. There are weekly requests for calibration of the GPS orbit determination and nadir pointing determination that involve sequences over ground truth points.

21.3.3 Data Transport and Delivery

One mission constraint specifies the use of NOAA ground stations. This constraint, coupled with the requirement of "notifying the end-user within 60 minutes from data collection on the spacecraft," led to a further discussion with the mission designers and the end-users. The time from data collection to notifying the end-user depends on factors that cannot be determined a priori and that are outside the design limits: the time from data collection to downlink. This time is governed by the positional geometry of the fire location, satellite and NOAA ground station. If a NOAA ground station is not accessible within 60 minutes, the requirement cannot be met.

The end-users and the project office agreed that for this first operational mission the requirement would be to "notify the end-user within 60 minutes from a successful downlink to a NOAA ground station." This is a requirement that we can design operations systems and procedures to, and one that can be effectively measured. It was also agreed that follow-on missions would consider a spacecraft trade study to determine the effectiveness, benefit and cost of using the NASA TDRSS communication satellites rather than NOAA ground stations. This would provide near-continuous contacts, allowing the mission to meet the original requirement to "notify the end-user within 60 minutes of data collection on the spacecraft." The TDRSS option was not considered for this mission because it would entail major changes to the spacecraft design.

During a brainstorming session the MOM had with his staff, the idea of sending the processed data to a commercial phone satellite came up. One member of the group had the following information:

> In September 2002 Iridium satellites relayed maintenance information from an in-flight F-16 fighter jet to a Lockheed Martin aeronautics plant.

> Rubin-2 contains an orbital telematic payload that collects satellite status parameters and transmits them back to Earth through the Orbcomm satellite network. Data is continually available since the transmission does not need to wait until passing over a Rubin ground station.

The MOM asked his telecom engineer to investigate this approach, taking into account the orbits of FireSat, and either Iridium or Orbcomm, along with the link analysis that would show feasibility. A trade study like this can be completed in about three months.

21.3.4 Payload Planning and Analysis

The payload is either off or on and the transition times are determined on board based on the transition of the satellite nadir point from land to water or water to land. Thus, there are no parameters to set to gather data.

The Payload Planning and Analysis function analyzes the payload data in non-real-time and determines if the payload parameters need to be adjusted for optimum performance. This determination is based on weekly calibration sequences in which the payload collects data over a known standard source. If the team wants to change parameters, they present and discuss the changes at a project Change Control Board (CCB) meeting. If the CCB approves the changes, all affected flight and ground parties are notified and a time is set for uplinking the changes.

21.3.5 Activity Planning and Development

This function establishes the daily sequence with associated ground processing times and uplink times, which then serve as "requirements" for the Mission Control function. This function also establishes, at the activity level, what the satellite will be doing over a 24-hour period. As described earlier, there are two primary inputs during normal operations. The Spacecraft Planning and Analysis function will be developing a weekly sequence that includes, at a minimum, a calibration of the GPS orbit determination and nadir pointing determination. The Payload Planning and Analysis function will be developing the weekly payload calibration. (Both of these functions will generate daily input if there is a problem.) The Activity Planning and Development function then generates a daily sequence based on these inputs. This sequence is checked through software and procedures for adherence to mission and flight rules. The sequence then goes to the Spacecraft Planning and Analysis, Payload Planning and Analysis, and Mission Control functions and project elements for review and approval. Once approved by a process established by the Mission Manager, the sequence goes to the Mission Control function for execution.

21.3.6 Payload Data Processing

The operational Payload Data Processing function will be performed on board the spacecraft. However, the first year has been allocated to validating the onboard algorithms. Thus, there must be a ground processing facility dedicated to processing and analyzing the raw sensor data and to developing the modifications to the onboard algorithms. The ground processing facility will be funded by the FireSat project, but built and operated by the scientists at their own location. After the first year, the project will not fund the scientists or the facility. The scientists must then find a sponsor for this activity. In discussions with the scientists that will be operating this facility, it was agreed that during the operational mission, a complete set of raw data will be down-linked one day each week. The scientists will then use this data to provide a long term validation of the onboard algorithms, and to carry out basic research that can be done only with the raw data. This is a change to the decision made in Sec. 21.1 under the EEIS discussion. Again we see the need for close interaction between scientists and engineers.

21.3.7 Mission Control

The Mission Control functions are straightforward and not complex. Pass preparation, pass execution, alarms and limits are all standard functions performed for any LEO satellite.

21.3.8 Mission Planning

During the first year, the Mission Planning function works with the Payload Planning and Analysis function to ensure fulfillment of the mission requirements and that no change in mission plan is necessary. After the first year, the Mission Planning function is needed only when there is an anomaly related to the spacecraft's ability to meet the mission requirements.

21.3.9 Archiving and Maintaining the Mission Data Base

All data is to be archived throughout the mission and maintained in on-line accessible media for five years after the end of the mission. The task of notification falls into this operations element. A system will have to be designed to automatically take fire notification messages sent by the spacecraft and route them to the appropriate end-users. This end-user notification system is to be in place and operational before launch and maintained up to date throughout the mission.

21.4 Mission Operations Concept (Chapter 4)

We stated in Chapter 4 that, "A mission operations concept describes—in the operators' and users' terms—the operational attributes of the mission's flight and ground elements...We generate a mission operations concept in ever more detailed layers."

In this section, we will develop the top layer of the mission operations concept for FireSat based on the information in the previous sections of this chapter. The seven mission operations concept products for FireSat are listed below.

21.4.1 Timelines

There are several top level timelines for FireSat. The first (Fig. 21.1) is for the period from launch to end of mission and shows the two key periods of checkout and operational certification and normal operations. The certification period timelines are shown in Figs. 21.2 through 21.5. Generating these timelines is an interactive effort. The operational scenarios and the data flow diagrams as well as the people versus hardware/software decisions are all worked on at the same time and form an integral set. Changes in one aspect of the operations scenario will affect the others.

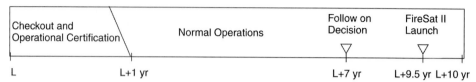

Fig. 21.1. Timeline Launch to End of Mission. This timeline shows a 1 year checkout followed by nine years of operations with a decision at L+7 years about whether or not to have a follow-on program. The timeline also indicates that the change from certification to normal operations is gradual toward the end of the first year and is not a single event. The follow-on decision of L+7 years leaves enough time for six months of dual operation of the first FireSat with the second FireSat.

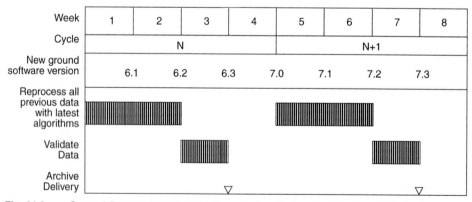

Fig. 21.2. Two Months Timeline During Certification. This shows that there is a one week sub-cycle during the certification period, with weekly uploads of new flight software, algorithms or processing changes, followed by a week of data collection. Implicit is the fact that if there are weekly flight software loads, an analysis period must stop early enough to develop and validate new flight software. There is also a one month cycle that will be explained in the ground processing timeline.

Fig. 21.3. Ground Processing Timeline Two Cycles During Certification. This timeline shows that there is a two week reprocessing cycle of all the collected data at the start of each four week cycle. At the end of the certification period the system must process 11 months of data in two weeks. This has implications on mass storage and processing speed and there will be reprocessing done in parallel with the weekly processing.

21.4 **FireSat**

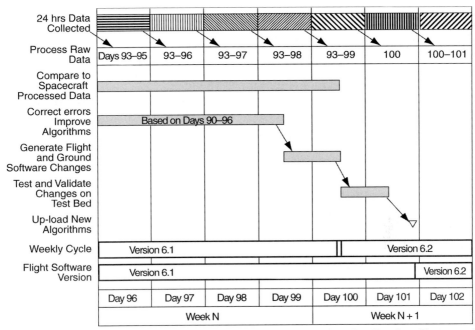

Fig. 21.4. **One Week Ground Processing Cycle Timeline During Certification.** The ground processing takes place daily. The error correction is based on data from week N and N-1. The flight software and ground processing software are out of sync for two days each week; thus no comparison with spacecraft processed data takes place on the last two days of the week cycle. The compare data for these two orbits is, however, archived. The flight software is verified on a test bed before up-linking.

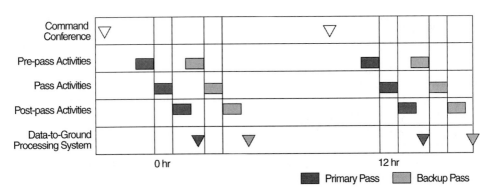

Fig. 21.5. **Daily Operations Timeline During Certification.** This shows that each day there are two primary passes and each pass has a backup pass in case there were problems on the primary pass. It implies 24 × 7 staffing or some staggered shift. Since data is transferred for ground processing, we know that there must be some type of data accountability done to ensure that all the data from the spacecraft is in the system for transfer to the ground processing system. Each of the timeline bars can be expanded into more detail and a corresponding operations scenario.

21.4.2 Data Flow Diagrams

Data flow diagrams and operational scenarios are tightly interlinked. They are usually done in parallel as they cannot be generated independently. The primary difference is that the data flow diagrams show only processing and data movement while the operational scenarios will include data processing and those functions performed by operational teams. Following is a set of data flow diagrams for FireSat that are consistent with the information discussed in previous sections of this chapter.

Figure 21.6 shows spacecraft data flow. At the start of a real time pass in contact with a NOAA ground station, the first action is to move the recording to a different solid state recorder partition. The solid state recorder has three partitions; each will hold 12 hours of data. This gives a 12 hour period to recover the stored data, if it is not downlinked per plan at the end of 12 hours. Each partition has three sub-partitions, one for payload data, one for spacecraft telemetry (health and safety) data and one for hot-spots and average temperature over a 30 square meter grid determined by onboard processing. There is a processing cycle that is always going on where the spacecraft computer is reading the payload data as it is placed in the recorder. If a hot-spot is detected, the location and intensity are determined and placed into a sub-partition within the selected primary partition.

Figure 21.7 shows the data flow and processing that take place at the tracking station and the control center during a real time pass. Shown are the basic types of data, the fact that two receiver chains are necessary (one for real-time data and one for playback data), and that tracking data is generated during the pass and sent to the control center.

Figure 21.8 shows the payload data processing flow. Here we see that each time a calibration hot-spot is detected, the calculated location and intensity are compared with the truth table of the calibration hot-spot location and intensity. This gives the scientists an indication of how good their algorithms are and if more tuning is needed. The location and intensity calculated by the payload analysis data system are also compared to the location and intensity calculated by the software on board the spacecraft. As long as the ground processing and flight processing algorithms are the same the difference should be very nearly zero. This comparison gives a good check on the validity of the onboard processing. Differences that cannot be explained need to be analyzed to see if the problems are in the flight software or ground software.

Figure 21.9 shows the data flow diagram for the end of the processing of one pass of data. It shows that the files are moved to the cycle archive, where the data is appended to that stored from previous pass processing pipelines.

The above data flow diagrams are certainly not complete but they give a top level view of the data and processing flow. Each element can be broken down into more detail as the design of the system matures.

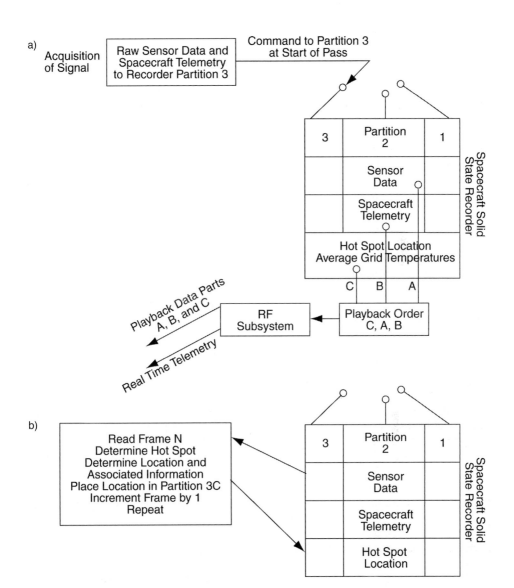

Fig. 21.6.   a) Spacecraft Data Flow During Pass. b) Spacecraft Data Flow After Loss of Signal (LOS). This figure shows the data flow on board the spacecraft during a pass.

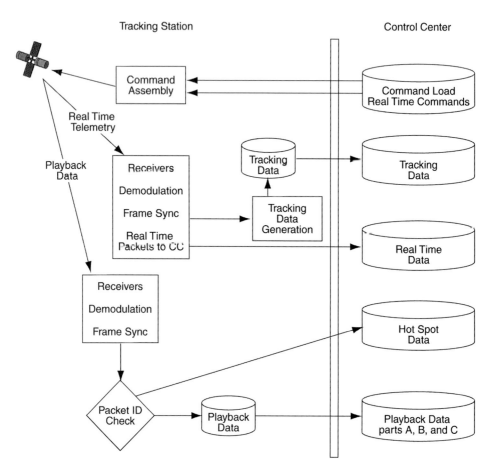

Fig. 21.7. Ground Station to Control Center Data Flow. This figure depicts the data flow between the control center and the spacecraft during a station contact.

21.4.3 Operational Scenarios

An operational scenario describes, usually step by step, the combination of the data flow events along with what the operations team must do to complete the activity. Table 21.5 shows how the FireSat MOM has chosen to document and show the FireSat operational scenario for a 12-hour period from one command conference to the next command conference. This section is somewhat detailed, but even at this top level we can identify processing steps and where team procedures and processes will have to be developed.

21.4 FireSat

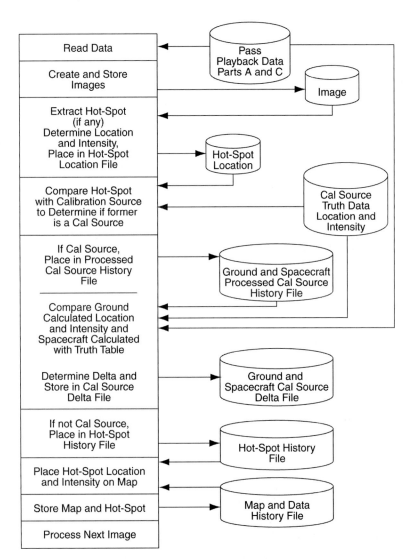

Fig. 21.8. Payload Analysis Data Flow. Processing of data for one pass.

The above operations scenario combined with the data flow diagrams gives anyone on the project—customer, user or manager—a good idea of what the mission operations tempo will be, the products to be generated, and the type of data system to be developed.

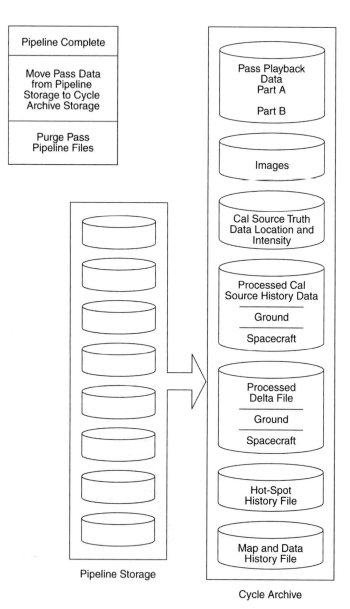

Fig. 21.9. Payload Analysis Data Flow. At the end of the pipeline processing of the data for one partition of the flight recorder, the data is moved to the archive and the pipeline storage is erased, ready for the next processing cycle.

Table 21.5. FireSat Operations Scenario 12-hour Period During Validation Phase. The following acronyms are used in the table: MOM = Mission Operations Manager; MCT = Mission Control Team; NAV = Navigation Team; SPAT = Sensor Processing and Analysis Team; SET = Spacecraft Engineering Team; DSOT = Data Systems Operations Team; SOT = Sensor Operations Team; ADT = Algorithm Development Team; STA = Ground Station Operations Team; CT = Calibration Team; SSWT = Sensor Processing Software Development Team; HST = Hot Spot Team; CC = Control Center; U/L = Up-Link; R/T = Real-Time; QQC = Quantity, Quality and Continuity of Data; SAT = Ground Station Operations Team.

Control Center (CC)	Operations <> Station	NOAA Ground Station	Station <> Spacecraft	Spacecraft
Command Conference				
SET: Verifies command load contains correct spacecraft commands SOT: Verifies command load contains correct sensor processing parameters ADT: Verifies command load contains correct sensor processing software SOT: Approves the release of data in spacecraft partition 1 as it is all accounted for MCT: Verifies that CC R/T system and ground station are operational MOM: Approves: • Command load • Release of partition 1 data				
Pre-Pass Activities				
MCT: Briefs station on upcoming pass: • Command load • 12 hours of sensor data to be downlinked • Predicted acquisition of signal • Predicted loss of signal • Confirms no anomalies since station last contacted the spacecraft • Verifies ground configuration				
MCT: Sends dummy command	> Dummy command	STA: Sends dummy command through antenna to dummy load		

Table 21.5. FireSat Operations Scenario 12-hour Period During Validation Phase. (Continued)

Control Center (CC)	Operations <> Station	NOAA Ground Station	Station <> Spacecraft	Spacecraft
MCT: Verifies receipt of dummy command confirmation MCT: Verifies that station to CC communications are ready	< Dummy command confirmation	STA: Confirms dummy command correctly processed STA: Sends dummy command confirmation		
MCT: Verifies that software load for spacecraft is ready to be transmitted				
Pass Activities				
				Passes in view of ground station
MCT: Voice notification to initiate pass activities	> Voice notification	STA: Transmits uplink radiation	> Uplink radiation	Receives radiation
		Locks onto downlink STA: Verifies telemetry lock	< Downlink	Activates telemetry downlink path
Starts receiving and storing R/T telemetry MCT: Verifies there are no out of limits alarms	< R/T telemetry	Forwards R/T telemetry Generates and stores tracking data	< R/T telemetry	Starts sending R/T telemetry
		STA: Activates command uplink path STA: Verifies command lock	> Uplink	Locks onto uplink
MCT: Sends command to switch data recording to partition 3	> Command	Transmits command	> Command	Switches data recording to partition 3
MCT: Sends command to release data in partition 1	> Command	Transmits command	> Command	Marks partition 1 available for recording

Table 21.5. FireSat Operations Scenario 12-hour Period During Validation Phase. (Continued)

Control Center (CC)	Operations <> Station	NOAA Ground Station	Station <> Spacecraft	Spacecraft
MCT: Sends command to playback partition 2 data	> Command	Transmits command	> Command	Receives command
		STA: Verifies lock on high rate channel Start recording partition 2 data	< Partition 2 data	Turns on high rate channel Starts playback of Partition 2 data
HST: Starts evaluating hot spot data HST: Starts sending hot spot notifications to FireSat users	< Hot spot data	Starts forwarding hot spot data		
MCT: Sends spacecraft software load	> Software load	Transmits software load	> Software load	Receives software load, stores in memory but does not execute
MCT: Verifies checksum is correct	< Checksum	Forwards checksum	< Checksum	Generates and sends checksum of software load
MCT: Reports playback of partition 2 data and forwarding of hot spot data are complete	< Voice notification	Detects high rate channel loss of sync Stops recording playback of partition 2 data Stops forwarding hot spot data STA: Voice notification that playback of partition 2 data and forwarding of hot spot data are complete		Completes playback of partition 2 data Turns off high rate channel
MCT: Sends command for memory dump	> Command	Transmits command	> Command	Switches R/T format to memory dump

Table 21.5. FireSat Operations Scenario 12-hour Period During Validation Phase. (Continued)

Control Center (CC)	Operations <> Station	NOAA Ground Station	Station <> Spacecraft	Spacecraft
MCT: Verifies memory dump is complete and that memory is correctly loaded	< Memory dump	Forwards memory dump	< Memory dump	Sends memory dump
MCT: Sends command to enable new software load	> Command	Transmits command	> Command	Activates new software load
Post-pass Activities				
MCT: Reports loss of signal MCT: Starts post-pass activities MCT: Generates post-pass products: • Command log • Health and safety log • Software load validation	< Voice notification	Loss of signal Stops receiving and forwarding R/T telemetry STA: Voice notification of loss of signal		Passes out of view of ground station
MCT: Voice notification to transfer playback data and tracking data	> Voice notification	STA: Initiates transfer of playback data and tracking data		
Stores playback data and tracking data	< Data	Transfers playback data and tracking data		
MCT: Voice notification to NAV that tracking data is available NAV: Initiates calculation of updated orbital parameters NAV: Sends new orbital parameters to SOT				
MCT: Initiates generation of QQC for playback data received MCT: Initiates compare of station and CC QQC reports • Determines if any data is missing • If yes, requests playback from station or prepares command to play back data on the next pass over station • If no, voice notification to SOT that all data has been received	<QQC report	STA: Initiates and sends QQC report for playback data		

Table 21.5. FireSat Operations Scenario 12-hour Period During Validation Phase. (Continued)

Control Center (CC)	Operations <> Station	NOAA Ground Station	Station <> Spacecraft	Spacecraft
MCT: Voice notification to SOT that playback data is available SOT: Initiates pipeline processing of playback data (Figure 21.8)				
Pipeline processing of playback data complete SOT: Voice notification to DSOT that pipeline processing complete				
DSOT: Initiates transfer of data from pipeline storage to archive storage (Figure 21.9)				
SOT: Voice notification to ADT that data is available for review and analysis: • Comparison of spacecraft and ground determination of calibration location/intensity source with truth				
SOT and CT: Analyze sensor data to determine if sensor parameters updates are needed SOT: Provides sensor parameter updates, if any, to MCT for inclusion in next command load				
ADT: Provides sensor processing software updates, if any, for inclusion in next command load				
SET: Provides spacecraft commands to MCT for inclusion in next command load				
MCT: Initiates generation of: • Command load • Predicted acquisition and loss of spacecraft signal at station				

Table 21.5. FireSat Operations Scenario 12-hour Period During Validation Phase. (Continued)

Control Center (CC)	Operations <> Station	NOAA Ground Station	Station <> Spacecraft	Spacecraft
Off-line Activities				
ADT: Analyzes comparison of spacecraft and ground determination of calibration location/intensity source with truth data to determine if algorithm updates are needed				
ADT: Develops algorithm updates and provides to SSWT				
SSWT: Updates flight and ground sensor processing software per algorithm updates				
Command Conference				
Where this scenario started—go to the start of this table				

21.4.4 Organizational and Team Responsibilities

The organization necessary for this mission is non-trivial. It not only has the aspects of real time command and control, but also those of science data analysis and post-launch flight and ground software and algorithm development. An organization chart for the operations phase of the mission is shown in Fig. 21.10.

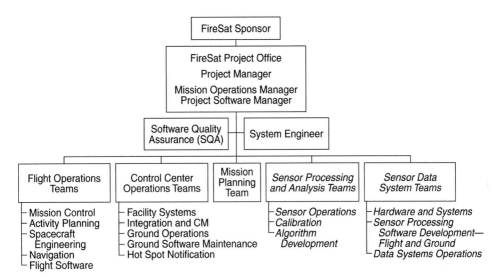

Fig. 21.10. FireSat Organizational Chart. The teams shown in Italics will not exist during the operational phase of the mission.

Looking at the organization and comparing it to the function chart (Fig. 3.1), we see that several of the functions discussed in Chap. 3 have been combined into a team. Table 21.6 shows where functions have been combined.

In some cases the functions have been combined (e.g., the Flight Operations Team) and in others the function has been distributed (e.g., Developing and Maintaining Software). As Chap. 3 points out, the 13 functions do not constitute an organization, but the organization should either cover them or the MOM should understand why we are not addressing one or more of them. We do not include the Data Transport and Delivery function because we are using the NOAA network. The interfaces with regard to scheduling and technical issues are covered by the Flight Operations Team and if necessary, the System Engineer will get involved in cross boundary technical issues.

Table 21.6. **Operation Functions vs. FireSat Organization.** Those teams listed in italics are in place only in the first year of the mission, during the payload and algorithm validation. This check ensures that the basic operations functions listed in Chap. 3 have been allocated to one or more of the teams. If there is a function that the teams are not performing, we must understand why.

Function	Team
Mission Control	Flight Operations Teams
Activity Planning and Development	
Spacecraft Planning and Analysis	
Navigation Planning and Analysis	
Payload Planning and Analysis	*Sensor Processing and Analysis Teams*
Payload Data Processing	Sensor Data System Teams
Archiving and Maintaining the Mission Data Base	Sensor Data System Teams for raw and resultant payload data products (Verification Phase) Flight Operations Teams for Spacecraft Data and Spacecraft Processed Payload data
Developing and Maintaining Software	Sensor Data System Teams Control Center Operation Teams (Ground Software) Flight Operations Teams (Flight Software)
Computers and Communication Support	Sensor Data System Teams Control Center Operations Teams
Managing Mission Operations	Project Office
System Engineering, Integration and Test	System Engineering Flight Operations Teams Sensor Data System Teams
Mission Planning	Mission Planning Teams
Data Transport and Delivery	Not in the FireSat organization

21.4.5 Requirements and Derived Requirements

The requirements were defined in Sec. 21.1. The clarification of requirements (21.2), the analysis of the operational functions needed (21.3), and the development of the operational concept (21.4) resulted in potential derived requirements. The 29 statements were compiled and analyzed and resulted in the derived requirements (Table 21.7) that are now part of the requirements baseline.

21.4.6 Technology Development Plan

We can reduce the risks in the project by investing some technology development funds in the area of algorithm development and understanding how the onboard processing will determine the location, intensity, and average temperature over a 30-square-meter grid. The German BIRD mission has demonstrated the ability to identify and determine the location of forest fires, and

Table 21.7. Derived Requirements Linked to Needed Capabilities. Analysis, development, and clarification of the original requirements lead to derived requirements.

Needed Capabilities as Determined During Requirements Clarification and Initial Design Discussions	Derived Requirements	Comment
Onboard hot-spot detection and calculation of the average temperature for the 30-square-meter grids.	Onboard hot-spot detection algorithms shall process the raw sensor data to determine the average scene temperature on a 30-square-meter grid basis.	
Onboard hot-spot detection and calculation of the average temperature for the 30-square-meter grids will reduce the raw data by a factor of 10,000 to one.	None	The amount of reduction should not be a derived requirement. The raw data is operated on to identify the hot-spots, but the raw data remains raw data – it is not reduced in quantity. The hot-spot detection data is indeed smaller in quantity than raw data, but specifying the amount of reduction is not necessary.
Required information: location, emitted energy, ratio of flaring to smoldering, area burned, average direction of the burn.	Onboard fire characterization algorithms shall process the hot-spot detection data to determine fire location, emitted energy, ratio of flaring to smoldering, area burned, average direction of the burn.	
The data is encapsulated in CCSDS standard frames and protocols, which is a requirement of the NOAA ground network.	Uplink and downlink shall be through NOAA ground stations. The downlink data shall be encapsulated in CCSDS standard frames and protocols.	
There will be opportunities to contact the spacecraft from each coast of the United States.	None	Not a derived requirement. It is a statement of fact based on the locations of the NOAA ground stations.

Table 21.7. Derived Requirements Linked to Needed Capabilities. (Continued)

Needed Capabilities as Determined During Requirements Clarification and Initial Design Discussions	Derived Requirements	Comment
The ground system will perform the user notification automatically, based on the information sent from the spacecraft.	The ground system shall, when receiving fire characterization data, automatically send fire notification messages to the end-users having jurisdiction over the fire location. The fire notification data shall contain the fire characterization data.	
The fire fighting managers want this data in tabular form, not map form.	The fire characterization data shall be provided to the end-users in tabular form.	
Notify the end-user within 60 minutes after a successful downlink pass at a NOAA ground station.	The fire notification messages shall be provided to the end-users within 60 minutes after a successful downlink.	
The user notification system is to be in place and operational before launch and maintained up to date throughout the mission.	The fire notification system shall be operational thirty days prior to launch. The fire notification system shall be maintained throughout the mission.	
60 minutes from data collection on the spacecraft to end-user notification, with a goal of 30 minutes.	The fire notification messages shall be provided to the end-users within 60 minutes from data collection on the spacecraft, with a goal of 30 minutes, when using the follow-on spacecraft.	This derived requirement was superseded by the derived requirement two rows above.
During the first year after launch, both the onboard processed data and the sensor raw data will be sent to the ground to support algorithm validation.	The commissioning phase shall last from launch to when the MOM declares the mission to be operational. The commissioning phase will nominally last one year. The sensor raw data shall be downlinked when a link is available and when the data unique downlink flag is enabled. The hot-spot detection data shall be downlinked when a link is available and when the data unique downlink flag is enabled. The fire characterization data shall be downlinked when a link is available and when the data unique downlink flag is enabled. There shall be the capability to enable each of the three data downlink flags by ground commanding.	Frees up the commissioning phase from lasting exactly one-year. Does not require all the data always to be downlinked during the commissioning phase. Provides flexibility and ground control as to when each of the three data types is to be downlinked.

Table 21.7.   Derived Requirements Linked to Needed Capabilities. (Continued)

Needed Capabilities as Determined During Requirements Clarification and Initial Design Discussions	Derived Requirements	Comment
After the commissioning phase, the spacecraft will send only the processed data, except when troubleshooting onboard processing or sensor malfunctions.	None.	Covered by an existing derived requirement.
There will be commandable sensor parameters that will be updated after launch to set the final operational capabilities of the sensor and detection algorithms.	Sensor operations shall be controlled parameters that can be updated by ground commanding to adjust the sensor for improved performance. The hot-spot detection algorithm shall use parameters that can be updated by ground commanding to update the algorithm for improved performance. The fire characterization algorithm shall use parameters that can be updated by ground commanding to update the algorithm for improved performance.	
This commandable sensor parameter is based on weekly calibration sequences in which the payload collects data over a known standard source.	The sensor shall be calibrated once a week using data collected from ground-truth sources.	
The ground system will be able to process the raw data and to replicate the onboard processing.	There shall be the capability to maintain the accuracy of the onboard hot-spot detection and fire characterization algorithms. Accuracy of the onboard algorithms shall be based on a comparison of re-processing the downlinked raw sensor data to the downlinked hot-spot identification and fire characterization data.	
If necessary, flight software changes will be developed and sent to the spacecraft.	Maintenance of the onboard algorithms shall include determining and uplinking updates.	
The sensor will be turned off over water.	The sensor shall be turned off when over the oceans.	Clarify the requirement to eliminate the need to turn off the sensor over any type of water.

Table 21.7.   Derived Requirements Linked to Needed Capabilities. (Continued)

Needed Capabilities as Determined During Requirements Clarification and Initial Design Discussions	Derived Requirements	Comment
The detection of the transition from land to water or water to land will take place on board automatically via GPS calculation of nadir latitude and longitude; an onboard map (table) relates nadir location to land/water boundaries and the 30-square-meter grids.	The determination of the transition of the spacecraft nadir point between ground and water shall be based on GPS onboard orbit position determination and an onboard map of land/water boundaries.	
This map also identifies forest areas and habitable areas.	The onboard map shall also identify habitable areas and forested areas.	
There will be a fault detection and protection capability to place the spacecraft and payload into a safe state that can be maintained for seven days without any adverse effects other than the interruption in data collection.	The spacecraft shall transition to safe mode operations and await commanding from ground operations when detecting an unrecoverable fault. The spacecraft shall operate in safe-mode operations for up to seven-days without causing any further degradation of spacecraft state-of-health.	
The requirement to cover 3,035,000 sq km per day is equivalent to covering 33 percent of the U.S each day.	None	Not a derived requirement. It is a statement of fact based on the coverage requirement.
There are weekly requests for calibration of the GPS orbit determination and nadir pointing determination that involve sequences over ground truth points.	Orbit determination and ground position determination capabilities shall be calibrated once a week based on collecting data over ground truth.	
There must be a ground processing facility dedicated to processing the raw sensor data and to developing modifications to the onboard algorithms.	There shall be a ground processing facility for processing the raw sensor data and developing modifications to the onboard algorithms for optimal performance. The FireSat project shall provide funding for the facility and one year of operations.	
During the operational mission, a complete set of raw data will be downlinked one day each week.	None	Covered by an existing derived requirement.

Table 21.7.   Derived Requirements Linked to Needed Capabilities. (Continued)

Needed Capabilities as Determined During Requirements Clarification and Initial Design Discussions	Derived Requirements	Comment
During the first year, the Mission Planning function works with the Payload Planning and Analysis function to ensure that the mission requirements will be met and that no change in mission plan is necessary.	Mission planning capabilities shall be exercised on a regular basis during the commissioning phase to ensure fulfillment of the mission requirements and that no change in mission plan is necessary.	
After the first year, the Mission Planning function is needed only when there is an anomaly that relates to the ability of the spacecraft to meet the mission requirements.	Mission planning capabilities shall be exercised during the operational phase only in response to degradation of the ability to satisfy mission requirements.	
All data is to be archived throughout the mission and maintained in on-line accessible media for 5 years after the end of the mission.	All downlinked raw sensor data, hot-spot detection data, and fire characterization data shall be archived. All sensor calibration data and onboard algorithm maintenance data shall be archived. The data archive shall be on-line and accessible to project personnel. The data archive shall be maintained for the duration of the mission plus five years.	Clarify exactly what data is to be archived.

has developed a neural network processing approach for classifying hot-spot types. (See Chapter 4 for the operations concept of the BIRD Mission). The technology development plan must describe the specific areas to be addressed, recognize that the work is to be a follow-on to the BIRD developments, and identify the point beyond which the project will not accept changes. The FireSat development staff determines when the development will have proceeded to the point where the project cannot accept new concepts and processing approaches, primarily because of risk considerations.

21.4.7   Cost Estimations

Chapter 5 discusses two different ways of getting an early estimate of the project size, and thus of the operations staff required after launch. In the next two sections we take the FireSat information discussed above, and use the tools from Chapter 5 to estimate the Full Time Equivalent Staff (FTE). Where the information was available we used it directly. Otherwise we made assumptions based on our best knowledge at this time. As more accurate information becomes available we will update the models and costs as needed.

Assessing Operations Complexity (Chapter 5.1)

The complexity model described in Sec. 5.1 gives estimated FTEs for the operations staff. It does not include post launch development, science processing and analysis or the staff that will be analyzing the data and generating the updated algorithms. These functions must be estimated and added to the FTE number that the complexity model gives. The complexity model gives an estimate of 14.6 FTEs. Referring to the organization chart in Fig. 21.10, we will add staff for the following functions:

- Sensor Processing and Analysis
- Sensor Data Systems
- Project Software Manager
- Software Quality Assurance

The size of the staff for these functions is based on the experience of the MOM making the estimates, and on inputs from various experts.

The complexity model asked for the number of shifts, which was selected as two and included in the first line of the estimate. Having only one shift and getting data only from the continental United States (as opposed to any land mass) could reduce costs, but at this time we do not need to reduce our estimates. The size of the staff for these functions is based on the experience of the MOM making the estimates, and on inputs from various experts. Table 21.8 summarizes these results.

Table 21.8. **Total Estimated Operations Team.** This total incorporates the output from the complexity model, plus additional estimates from the MOM and other project experts.

Function	Year 1 Calibration	Operations Phase (yrs 2–10)
Complexity Model Output	14.6	14.6
Sensor Processing and Analysis	6	0
Sensor Data Systems	4	0
Project Software Manager	1	0
Software Quality Assurance	2	1
Total	27.6	15.6

To get the cost per year multiply the FTEs by an average salary (the MOM may even have the salary averages for each function). If we assume an average salary of $165,517 per year (the same rate as in the cost model), the cost for the first year is $4,568,000 and for the operational years $2,582,000 per year. The total operational budget is $4,568,000 + (9 × $2,582,000) = $27,806,000 in FY2003 dollars. In making the estimate we should also add other expenses such as computer replacements every 3 years, software license fees, and reserve. During the operational phase of the mission a reserve of 5% of the total budget each year is usually added. During the first year a 10% reserve would be appropriate as software development and algorithm development is going on.

The results of the Complexity Model are given in Table 21.9.

Table 21.9. Complexity Model Results for FireSat.

Metric	Metric Value	Complexity
Operations Complexity - Mission Design and Planning *Science and Engineering Events*		
Frequency • Number per day	Fewer than 10	Low
Criticality • Science re-try opportunities • Engineering re-try opportunities	More than 2 More than 2	Low Low
Complexity • Timing accuracy • Instrument pointing events • Spacecraft pointing events	Minutes None Fixed	Medium Low Low
Data Return • Number of routine downlink data modes • Real time vs. playback	1 or 2 Playback: FIFO	Low Low
Science and Engineering Planning • Event repetitiveness • Number of consumables and margin constraints routinely tracked • Timeline duty cycle margin • Plan execution time • Plan development time • Response time to late change request	Highly repetitive None More than 50% 1 day to 1 week Less than execution time More than 50% of development time	Low Low Low Medium Low Low
Navigation, Guidance and Control		
Maneuver Frequency • Slews per day • Propulsive maneuvers per quarter	None 0 or 1	Low Low
Maneuver Criticality • Re-try opportunities	More than three	Low
Maneuver Complexity • Timing accuracy • Maneuver accuracy	Days 1 sigma or less	Low Low

Table 21.9. Complexity Model Results for FireSat. (Continued)

Metric	Metric Value	Complexity
Navigation Data Return • Number of navigation data types • Navigation data per day	1 or 2 Less than 1 hour	Low Low
Maneuver Planning • Maneuver design repetitiveness • Number of consumables and margin constraints routinely planned • Mission delta-V margin	Highly repetitive None More than 50%	Low Low Low
Ephemerides • Number of objects needing ephemerides	Fewer than 3	Low
Tracking Events		
Frequency • Number of passes per day • Hours of link coverage per week	More than 3 Not applicable	High —
Criticality • Re-try opportunities	Playback data with replay option	Low
Complexity • Station configuration changes per pass • Peak data return rate • Simultaneous, multi-station coordination	1 or 2 More than 100 kbps Not applicable	Medium High —
Tracking Facility Planning and Scheduling • Coverage repetitiveness • Track duration time margin • Schedule iterations • Toleration to late losses in scheduled coverage	Repeating weekly pattern 5 to 50% 1 or 2 More than 20% of tracks	Low Medium Low Low
Operations Complexity - Flight Systems		
Command • Commandable states • Flight rules and constraints • History-dependent commandable states • Onboard tables routinely updated • Number of instruments	Fewer than 50 Fewer than 10 None Fewer than 10 Fewer than 3	Low Low Low Low Low
Monitor • Telemetry channels • Ambiguous states	Fewer than 100 None	Low Low
Pointing • Attitude control • Pointing accuracy • Independent fields-of-view • Articulating devices • Constraints on hazard pointing of instruments	Gravity gradient Greater than 0.1 degree Fewer than 3 None None	Low Low Low Low Low

Table 21.9. Complexity Model Results for FireSat. (Continued)

Metric	Metric Value	Complexity
Automation • Unattended safing of flight system • Flight system command states requiring routine ground command for safing	2 to 7 days None	Medium Low
Flight Margins • Onboard consumables managed by operations • Onboard data storage • Onboard command-file memory • Onboard flight-software memory • Speed of the flight computer • Time to uplink full plan • Time to downlink planned data storage • Real-time downlink bandwidth (data rate) • Telecomm link margin • Power margin • Thermal margin	1 or 2 More than 3 times the downlink period More than 3 times the plan duration More than 30% margin at launch More than 30% margin at launch A fraction of a pass Approximately one full pass More than twice the data-capture rate More than 3 dB Power available more than peak load No thermal constraints on pointing or power	Low Low Low Low Low Low Medium Low Low Low Low
Operations Complexity - Operational Risk Avoidance		
Command and Control • Command errors tolerated per week • Onboard adaptive algorithms - entries tolerated per week • Activity simulation • Command reviews and approval	More than 2 More than 2 None Fewer than 2 per command file	Low Low Low Low
Data Return • Amount of lost science data tolerated • Amount of lost engineering data tolerated	More than 5% More than 5%	Low Low
Performance Analysis • Routine performance validation • Routine trend analysis and prediction • Routine performance modeling • Model calibration	Fewer than 50 parameters None Simple algorithms Fewer than 2 times per year	Low Low Low Low
Fault recovery • Tolerated timeliness in ground response to an anomaly	More than 1 day	Low

Table 21.9.   Complexity Model Results for FireSat. (Continued)

Metric	Metric Value	Complexity
Operations Complexity - Ground Systems		
Interfaces and Ground System Complexity		
• Geographical distribution of operations and ground system	1 Site	Low
• Institutions needed to operate the ground system	1 to 3	Low
• Science teams	1 or 2	Low
• Shared ground system components	1 or 2	Medium
• Users sharing instruments	1 or 2	Low
• Shared operations teams	None	Low
• Shared resource scheduling epoch	Hours	Low
• Command and control data	High-order command language	Low
Design of the Ground System		
• Ground system designed to what requirements	Designed for single user project	Low
• Ground system built, maintained, upgraded by	Combination of project and other	Medium
• Development of flight-system software after launch	More than 10%	High
Organization and Staffing		
• Staffing schedule for most operations positions	2 Shifts per day	Medium
• Number of management levels	2 Levels	Medium
• Tasking strategy	Project specialists, multi-tasked	Low
Automation - Number of separate steps requiring project operator action(s) in:		
• Alarm monitoring from decommutation through notification	1 or 2	Low
• Level 0 data capture from receipt from tracking facility to project data base	1 or 2	Low
• Real time command from entry to radiation	1 or 2	Low
• Activity generation from entered requests through command file generation	1 or 2	Low
• Scheduling of tracking coverage	1 or 2	Low
Automation - Number of transfers of non-electronic data in processes defined above		
• Alarm monitoring	1 or 2	Low
• Data capture	1 or 2	Low
• Real time command	1 or 2	Low
• Activity generation	1 or 2	Low
• Scheduling of tracking coverage	1 or 2	Low

To obtain the overall mission complexity, we take the product of all four of the above factors:

Overall Mission Complexity = 1.581 × 1.304 × 0.700 × 0.921 = 1.329.

The Mission Type Constant is 11 (Orbital).

Finally, we multiply the Overall Mission Complexity by the Mission type constant to obtain the predicted FTEs for the mission: 1.329 × 11 = 14.6 FTEs. Table 21.10 summarizes the complexity calculations.

Table 21.10. Mission Operations Complexity Summary Sheet. The complexity for each category is given as a weighted average: For example, in the Mission Design and Planning category, we have: Complexity = (27 × 1.4 + 4 × 1.6 + 2 × 4.0) / (27 + 4 + 2) = 1.581.

Range	Number of Metrics	Weight	Total = Number of Metrics × Weight
Mission Design and Planning			
Low	27	1.4	37.8
Medium	4	1.6	6.4
High	2	4.0	8.0
Total	33		52.2
		Complexity	1.581
Flight Systems			
Low	23	1.2	27.6
Medium	2	2.5	5
High	0	3.6	0
Total	25		32.6
		Complexity	1.304
Operational Risk Avoidance			
Low	11	0.7	7.7
Medium	0	1.4	0
High	0	2.0	0
Total	11		7.7
		Complexity	0.700
Ground Systems			
Low	19	0.7	13.3
Medium	4	1.7	6.8
High	1	2.0	2.0
Total	24		22.1
		Complexity	0.921

NASA's Space Operations Cost Model (SOCM) (Chapter 5.2)

The answers came out with answers similar to those of the complexity model. The questions and answers for this model are given in Table 21.11. The additional staff for the first year will be the same as discussed in the previous section.

Table 21.11. FireSat Staffing Requirements. Total staff required for operations is obtained from the Space Operations Cost Model (SOCM) model plus additional first year staff.

Function	Year 1 Calibration	Operations Phase (Years 2–10)
Cost Model Output	14.6	14.5
Sensor Processing and Analysis	6	0
Sensor Data Systems	4	0
Project Software Manager	1	0
Software Quality Assurance	2	1
Total	27.6	15.5

The SOCM cost model is very similar to the complexity model. Neither includes the sensor processing and algorithm validation during the first year. Space does not permit a discussion of why each of the parameters is the value it is, but we can play with the model and see how different assumptions change the FTE count and the cost.

Table 21.12 gives the questions and values used in the SOCM. The only values that should be explained are the Programmatic Characterizations. The values are all relatively high; 1 two and 3 three's. Although this was supposed to be a simple project, the use of NOAA ground stations, and the interfaces with the forest service and other government agencies make for complex programmatic interfaces.

Table 21.13 details the SOCM Level 1 outputs for FireSat. Table 21.14 gives the total staffing estimates and Table 21.15 gives the staffing per year estimates. Table 21.16 translates FTEs to dollar costs. Finally, Table 21.17 shows the cost estimate for adding a second spacecraft.

Table 21.12. Earth Orbiting—Level 1 Inputs. This table contains the inputs to the SOCM. The operations are 1, 2, or 3 and the value represents the input to the model. (TDRSS = Tracking and Data Relay Satellite System; DSN = Deep Space Network; LEO = Low-Earth Orbit; L1 = Libration point 1 [see Chap. 10]; NASA = National Aeronautics and Space Administration; MOC = Mission Operations Complex; SOC = Science Operations Complex; COTS = Commercial Off-the-Shelf; GOTS = Government Off-the-Shelf; PI = Principal Investigator; GDS = Ground Data System; MOS = Mission Operations System; SMEX = Small Explorer; MIDEX = Medium-Class Explorer; ESSP = Earth System Science Pathfinder.)

	Value ->	1	2	3
Mission Characterization				
Mission Type	1	Survey - Earth Science	Survey - Space Science	Targeted - Earth Science
Tracking Network	1	Ground	TDRSS	DSN
Orbit	1	LEO, circular	L1, halo	Highly Elliptical
Number of Identical Flight Systems	1	1	2	3
Nominal Mission Duration (mo)	120			
Extended Mission Duration (mo)	0			
Post-Flight Data Analysis Duration (mo)	0			
Programmatics Characterization				
Mission Risk Class	3	Technology Demo (tech > sci)	SMEX	MIDEX/ESSP
Development Schedule	2	Fast (< 2.5 yrs)	Moderate (2.5-4 yrs)	Long (> 4 yrs)
Management Mode	3	PI	NASA	--
Contract Type	3	In-House	Augmented Hybrid	Hybrid
GDS/MOS Characterization				
Operations Approach	1	Dedicated MOC	Multimission MOC	Remote MOC/SOC
Architecture Design	3	COTS	Heritage/GOTS	New/Custom
Science Team Role	1	Data Processing	Instrument Health	Sequence Planning

Table 21.12. Earth Orbiting—Level 1 Inputs. (Continued) This table contains the inputs to the SOCM. The operations are 1, 2, or 3 and the value represents the input to the model. (TDRSS = Tracking and Data Relay Satellite System; DSN = Deep Space Network; LEO = Low-Earth Orbit; L1 = Libration point 1 [see Chap. 10]; NASA = National Aeronautics and Space Administration; MOC = Mission Operations Complex; SOC = Science Operations Complex; COTS = Commercial Off-the-Shelf; GOTS = Government Off-the-Shelf; PI = Principal Investigator; GDS = Ground Data System; MOS = Mission Operations System; SMEX = Small Explorer; MIDEX = Medium-Class Explorer; ESSP = Earth System Science Pathfinder.)

	Value ->	1	2	3
Payload Characterization				
Number of Non-Imaging Instruments	0			
Number of Imaging Instruments	1			
Pointing Requirements	1	Low	Medium	High
Conflicts Among Instruments	0	Low	Medium	High
Scope of Guest Investigator Program	0	Small	Medium	Large
Number of Separate Science Investigations	1	Fewer than 2	From 2–5	2–5
Science Team Size (not all full time)	2	Fewer than 10	10–20	more than 20
Science Team Location/Distribution	1	Collocated at 1 facility	Central SOC w/ 1-2 remote	Central SOC w/ 2+ remotes
Spacecraft Design Characterization				
Spacecraft Design Implementation	1	High Heritage	Cost-Capped	Requirements-Driven
Design Complexity	1	Low (minimal number of flight rules)	Medium	High (several unique engineering requirements)
Operations Development Cost (Phase C/D)				
Ops Dev % of Total C/D		--	0.0	$M (based on total dev$)
Total Phase C/D Cost ($M)		--	--	--
Phase C/D Duration (mo)	36			

Table 21.13. Earth Orbiting—Level 1 Outputs. Level 1 cost output. Level 1 mission operations cost estimate. 2003 Constant FY $1000's

FireSat		Nominal	Extended	Post-Flight Data Analysis	Total
1.0	Mission Planning and Integration	907.0	0.0		907.0
2.0	Command/Uplink Management	2077.8	0.0		2077.8
3.0	Mission Control and Operations	2300.3	0.0		2300.3
4.0	Data Capture	1289.2	0.0		1289.2
5.0	POS/LOC Planning and Analysis	129.2	0.0		129.2
6.0	Spacecraft Planning and Analysis	263.9	0.0		263.9
7.0	Science Planning and Analysis	3172.7	0.0		3172.7
8.0	Science Data Processing	6690.3	0.0	0.0	6690.3
9.0	Long-term Archives	2415.0	0.0	0.0	2415.0
10.0	System Engineering, Integration, and Test	2587.9	0.0		2587.9
11.0	Computer and Communication Support	1218.3	0.0	0.0	1218.3
12.0	Science Investigations	1001.8	0.0	0.0	1001.8
13.0	Management	423.1	0.0		423.1
	Project Direct Total	24,476.5	000.0	000.0	24,476.5
	Phase E	24,476.5			

Table 21.14. Level 1 Mission Operations FTE Estimate. This estimate takes into account the nominal 10-year lifetime of the mission.

FireSat		Nominal	Extended	Post-Flight Data Analysis	Total
1.0	Mission Planning and Integration	4.9	0.0		4.9
2.0	Command/Uplink Management	11.3	0.0		11.3
3.0	Mission Control and Operations	19.1	0.0		19.1
4.0	Data Capture	3.7	0.0		3.7
5.0	POS/LOC Planning and Analysis	0.7	0.0		0.7
6.0	Spacecraft Planning and Analysis	2.2	0.0		2.2
7.0	Science Planning and Analysis	21.4	0.0		21.4
8.0	Science Data Processing	47.3	0.0	0.0	47.3
9.0	Long-term Archives	7.0	0.0	0.0	7.0
10.0	System Engineering, Integration, and Test	14.1	0.0		14.1
11.0	Computer and Communication Support	3.5	0.0	0.0	3.5
12.0	Science Investigations	7.1	0.0	0.0	7.1
13.0	Management	2.8	0.0	0.0	2.8
	Project Direct Total	145.1	0.0	0.0	145.1

Table 21.15. Level 1 Mission Operations FTE/yr Estimate. The total FTEs calculated by SOCM are here allocated to the individual functions. Additional staffing for the extended mission and data analysis, if approved are included but were not calculated by the SOCM.

FireSat		Nominal	Extended	Post-Flight Data Analysis
1.0	Mission Planning and Integration	0.5	0.2	
2.0	Command/Uplink Management	1.1	0.6	
3.0	Mission Control and Operations	1.9	1.0	
4.0	Data Capture	0.4	0.2	
5.0	POS/LOC Planning and Analysis	0.1	0.0	
6.0	Spacecraft Planning and Analysis	0.2	0.1	
7.0	Science Planning and Analysis	2.1	1.1	
8.0	Science Data Processing	4.7	2.4	4.7
9.0	Long-term Archives	0.7	0.3	0.7
10.0	System Engineering, Integration, and Test	1.4	0.7	
11.0	Computer and Communication Support	0.4	0.2	0.4
12.0	Science Investigations	0.7	0.4	0.7
13.0	Management	0.3	0.1	
	Project Direct Total	14.5	7.3	6.5

Table 21.16. **Level 1 Mission Operations Estimate—Operations or Phase E (see Chap. 1).** Costs are in millions of dollars and are FY2003.

FireSat	Nominal	Extended	Post-Flight Data Analysis	TOTAL
Annual FTE/$ Estimates				
Flight Ops	6.2	3.1		
Nav/Tracking Ops	0.1	0.0		
Science Ops	8.2	4.2	6.5	
Total FTEs/yr	14.5	7.3	6.5	
Annual FTE Cost	$2.4	$0.0	$0.0	$2.4
Summary				
Phase duration (mo)	120.0	0.0	0.0	120.0
Total FTE $M	$24.5	$0.0	$0.0	$24.5

Table 21.17. **Second Spacecraft Cost Estimate.** Additional resources needed to operation two spacecraft.

Function	Additions for 2 Spacecraft	New Totals for Year 1 Calibration	New Totals for Operations Phase (yrs 2–10)
Operations Teams	3	17.5	17.5
Sensor Processing and Analysis	3	6	0
Sensor Data Systems	0	4	0
Project Software Manager	0	1	0
Software Quality Assurance	0	2	1
Total	6	30.5	18.5

21.5 Cost Estimate for 2 Satellites Versus 1 Satellite

The complexity and cost models in Chap. 5 are not designed to compare costs in a mission where there is a constellation, i.e. more than one spacecraft performing the same function. Both of these estimation models were based on scientific missions, and nearly all scientific missions now use single vehicles for the scientific investigation. For FireSat the operations staff was asked to estimate the additional work that a second spacecraft would entail and the additional staff needed to support two spacecraft rather than one. Key observations were:

- Additional Mission Control Team staff are necessary to handle twice the number of passes
- Additional Activity and Analysis staff are necessary to handle the increases in uplinks
- Additional Spacecraft Engineering Team members are necessary to analyze the second spacecraft
- Navigation functions are minimal and the current estimate applies to two spacecraft
- Existing Flight Software Team can handle the second spacecraft since the algorithms will be identical
- Hot spot notification will be automated so the existing staff can handle the second spacecraft
- The Control Center Operations will be able to handle two spacecraft since the system will be the same for both
- The Sensor Processing and Analysis Team will need additional staff to analyze the data from two spacecraft and develop the algorithms for each spacecraft. The sensors may react differently on the two spacecraft and require two separate sets of algorithms
- The Sensor Data Systems Team will be able to handle both spacecraft with the original effort as the main change here is more storage

In a meeting where each team lead participated and made estimates, which were challenged and discussed by the other team chiefs, the following estimate to increased staffing requirements was agreed to:

- Mission Control Team +2 to handle a third shift
- Activity Planning and Analysis Team +1 to handle additional uplinks
- Sensor Processing and Analysis Team +3 to analyze additional data

Table 21.17 shows the additional staff required to support the operation of a second spacecraft. The increase would be 6 FTEs for the first year and 3 FTEs for the remaining nine years. Using the same cost estimate as before, $100k/per year including all fees, the total increase with 3% inflation per year would be $3.7 million. The customer, on seeing this estimate, decided to authorize the second spacecraft.

21.6 Conclusions

This chapter has gone through an example of some of the important processes contained in the first five chapters of the book. Each of the operational areas can be addressed at increasing levels of detail to understand and validate requirements and to ensure that they are written in terms that are understandable to the operator, user and developer. The idea of documenting derived requirements is very important and often overlooked. Requirements should be written in terms that can be tested and validated both pre-launch and during the mission.

There are two exercises that we recommend:

1. Go through this chapter from Sec. 21.2 through 21.4.6 and identify the changes that need to be made to accommodate the decision to add a second spacecraft. Pay particular attention to the time lines and the data flow diagrams.
2. Take a current mission and go through the same exercise that we did for FireSat. We guarantee a better understanding of the mission.

References

Larson, Wiley J., J. R. Wertz, eds. 1999. *Space Mission Analysis and Design*. Third edition. Torrance, CA: Microcosm, Inc. and Dordrecht, The Netherlands: Kluwer Academic Publishers.

CIA Factbook reference.

World Atlas reference.

Consultative Committee on Space Data Standards (CCSDS) web site: http://www.ccsds.org/documents/411x0g3.pdf, p. 3.2 22.

Appendix A

Communications Frequency Bands

Table A.1. Communication Frequency Band Designation and Frequency Range. *The International Telecommunications Union (ITU) and the World Administrative Radio Conference (WARC) regulate use of frequency band, transmission bandwidth, and power-flux density.* [Larson and Wertz, 1999]

Frequency Band	Frequency Range (GHz)
VHF	0.03 – 0.3
UHF	0.3 – 1.0
P	0.225 – 0.39
L	0.39 – 1.05
S	1.05 – 3.9
C	3.9 – 6.2
X	6.2 – 10.9
K	10.9 – 36.0
K_u	15.35 – 17.25
K_a	33.0 – 36.0
Q	36.0 – 46.0
V	46.0 – 56.0

Appendix A

Appendix B

- FOR Flight Operations Review
- FOT Flight Operations Team
- GSFC Goddard Space Flight Center
- GN Ground Network
- GND Ground
- GTAS Generic Trending and Analysis System
- H/K Housekeeping
- IPD Information Processing Division
- KSC Kennedy Space Center
- MOC Mission Operations Center
- MOM Mission Operations Manager
- NASA National Aeronautics and Space Administration
- NASCOM NASA Communications Network
- PDR Preliminary Design Review
- PI Principal Investigator
- POCC Payload Operations Control Center/Project
- POD Project Operations Director
- RUST Remote User Scheduling Terminal
- S/C Spacecraft
- SMEX Small Explorer
- SOC Science Operations Center
- SOCC Satellite Operations Control Center
- TDRS Tracking and Data Relay Satellite
- TDRSS Tracking and Data Relay Satellite System
- TLM or tlm Telemetry
- TPOCC Transportable Payload Operations Control Center
- TRK Track(ing)
- WFF Wallops Flight Facility (Virginia)
- WPS Wallops Island S-band Tracking Station

Appendix B

Mission Summaries

This appendix contains mission summaries for 14 space missions managed by the NASA Goddard Space Flight Center. The summaries include a mission statement, launch information, a description of the spacecraft bus, a description of the payload and science instruments, the orbit phases, a description of the ground system, an overview of mission operations, and a schedule. The following missions are included:

- ACE   Advanced Composition Explorer
- COBE   Cosmic Background Explorer
- ERBS   Earth Radiation Budget Satellite
- FAST   Fast Auroral Snapshot Explorer
- FUSE   Far Ultraviolet Spectroscopic Explorer
- GRO   Gamma Ray Observatory
- ICE   International Cometary Explorer
- IMP-J   Interplanetary Monitoring Platform
- IUE   International Ultraviolet Explorer
- SAMPEX   Solar, Atmospheric, Magnetic Particle Explorer
- SOHO   Solar and Heliospheric Observatory
- SWAS   Submillimeter Wave Astronomy Satellite
- TRMM   Tropical Rainfall Measuring Mission
- XTE   X-ray Timing Explorer

Acronyms and other terms that apply to only one or a few of these missons are listed along with the mission(s). The following terms apply to many or all of the missions in this appendix:

- ACS   Attitude Control System
- C&DH   Command and Data Handling
- CDR   Critical Design Review
- CMD or cmd Command
- DSN   Deep Space Network
- FDF   Flight Dynamics Facility

Mission Summaries

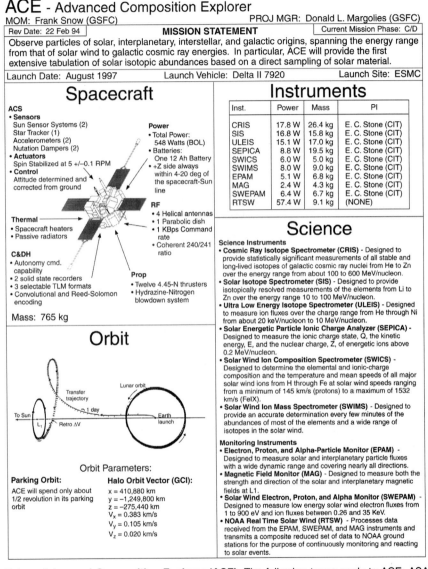

Fig. B.1. **Advanced Composition Explorer (ACE).** The following terms apply to ACE: ASARs = ACE Science Analysis Remote Sites; ASC = ACE Science Center; BOL = Beginning of Life; CIT = California Institute of Technology; CoDR = Conceptual Design Review; ESMC = Eastern Space and Missile Center; GCI = Geocentric Solar Inertial Coordinate System; GenSAA = Generic Spacecraft Analyst Assistant; Halo Orbit = So called because from Earth it resembles a halo around the Sun; L1 = Earth-Sun Libration Point; NAR = Non-Advocate Review; SOHO = Solar and Heliospheric Observatory; SOR = System Operations Review; TBD = to be determined; TBR = to be received.

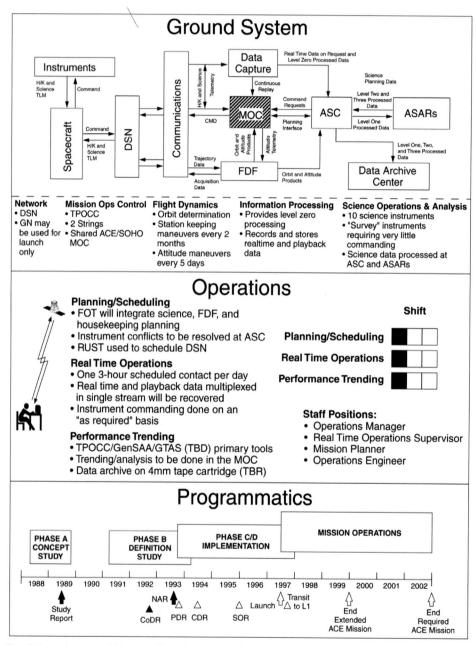

Fig. B.2.   Advanced Composition Explorer (ACE) Continued.

COBE - Cosmic Background Explorer

POD: Jim Williamson (GSFC)

Rev Date: 7 Jan 94	**MISSION STATEMENT**	Current Mission Phase: N/A
	Test the Big Bang theory of a primeval explosion 15 billion years ago. In 1992, the results from analysis of the first two years of data revealed that COBE's observations supported the Big Bang theory.	

Launch Date: November 18, 1989	Launch Vehicle: Delta 5920 ELV	Launch Site: VAFB

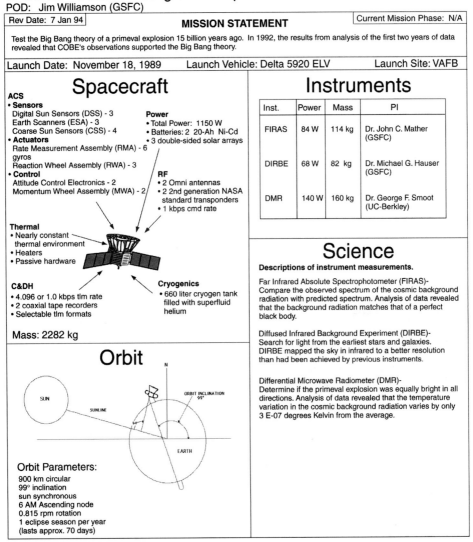

Spacecraft

ACS
- **Sensors**
  Digital Sun Sensors (DSS) - 3
  Earth Scanners (ESA) - 3
  Coarse Sun Sensors (CSS) - 4
- **Actuators**
  Rate Measurement Assembly (RMA) - 6 gyros
  Reaction Wheel Assembly (RWA) - 3
- **Control**
  Attitude Control Electronics - 2
  Momentum Wheel Assembly (MWA) - 2

Power
- Total Power: 1150 W
- Batteries: 2 20-Ah Ni-Cd
- 3 double-sided solar arrays

RF
- 2 Omni antennas
- 2 2nd generation NASA standard transponders
- 1 kbps cmd rate

Thermal
- Nearly constant thermal environment
- Heaters
- Passive hardware

C&DH
- 4.096 or 1.0 kbps tlm rate
- 2 coaxial tape recorders
- Selectable tlm formats

Cryogenics
- 660 liter cryogen tank filled with superfluid helium

Mass: 2282 kg

Orbit

Orbit Parameters:
900 km circular
99° inclination
sun synchronous
6 AM Ascending node
0.815 rpm rotation
1 eclipse season per year
(lasts approx. 70 days)

Instruments

Inst.	Power	Mass	PI
FIRAS	84 W	114 kg	Dr. John C. Mather (GSFC)
DIRBE	68 W	82 kg	Dr. Michael G. Hauser (GSFC)
DMR	140 W	160 kg	Dr. George F. Smoot (UC-Berkley)

Science

Descriptions of instrument measurements.

Far Infrared Absolute Spectrophotometer (FIRAS)-
Compare the observed spectrum of the cosmic background radiation with predicted spectrum. Analysis of data revealed that the background radiation matches that of a perfect black body.

Diffused Infrared Background Experiment (DIRBE)-
Search for light from the earliest stars and galaxies. DIRBE mapped the sky in infrared to a better resolution than had been achieved by previous instruments.

Differential Microwave Radiometer (DMR)-
Determine if the primeval explosion was equally bright in all directions. Analysis of data revealed that the temperature variation in the cosmic background radiation varies by only 3 E-07 degrees Kelvin from the average.

Fig. B.3.   Cosmic Background Explorer (COBE). The following terms apply to COBE: CMF = Command Management Facility; GSN = Gigabit Satellite Network; MSOCC = Multisatellite Operations Control Center; NCC = Network Control Center; SDPF = Sensor Data Processing Facility; SN = Space Network; TRU = Tape Recorder Unit; TSAS = TDRSS Scheduling Assistance System; VAFB = Vandenberg Air Force Base; WSGT = White Sands Ground Terminal (New Mexico).

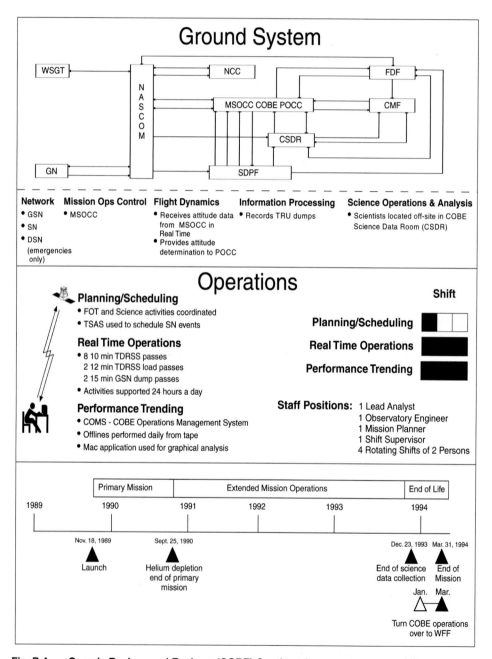

Fig. B.4. Cosmic Background Explorer (COBE) Continued.

ERBS - Earth Radiation Budget Satellite

POD: Jim Williamson (GSFC)

Rev Date: 21 Apr 94	MISSION STATEMENT	Current Mission Phase: Extended Ops

To study the Earth's atmosphere and its radiation environment.

Launch Date: October 05, 1984	Launch Vehicle: STS-41G	Launch Site: KSC

Spacecraft

ACS
- Sensors
  - Sun sensor - 2
  - Earth Horizon Scanner - 2
  - Inertial Reference Unit (IRU) - 2
- Actuators
  - Attitude Control Electronics (ACS)
  - Magnetic Control Electronics (MCS)
- Control
  - Electromagnets - 4
  - Momentum wheel - 1

Thermal
- Thermostatically controlled film heaters
- Active louvers
- Passive devices

Power
- Solar arrays: two fixed panels, 14.6 sq. meters
- Total power: 2400 W
- Batteries: two 22-Cell 50-Ah Ni-Cd

RF
- 1 high gain ESSA antenna
- 2 low gain omni antennas
- 2 NASA standard TDRS/GSTDN transponders
- 1.0 kbps command rate

C&DH
- 1.0, 1.6, and 12.8 kbps tlm rates
- 4 Tape Recorder Units (TRU)
- 4 selectable tlm formats

Prop
- Blowdown hydrazine with nitrogen pressurant
- 2 140 kg propellant tanks
- 8 2.2-N thrusters - 4 for ΔV, 4 for yaw

Mass: 2250 kg

Instruments

Instruments	Power	Mass	PI
ERBE Scanner	28 W	29.0 kg	Dr. R. Wy (LaRC)
ERBE Non-Scanner	22 W	32.0 kg	Dr. R. Wy (LaRC)
SAGE-II	14 W	33.6 kg	Dr. W. Chu (LaRC)

Science

Descriptions of instrument measurements.

Earth Radiation Budget Experiment (ERBE) Scanner/Non-Scanner:
Two-instrument package, also being flown on NOAA-9 and 10, to provide readings of the Earth's solar-absorbed radiation and emitted thermal radiation. Analysis of this data will provide:
- the solar output constant.
- the equator-to-pole energy transport gradient.
- the average monthly radiation budget on regional, zonal and global scales.
- the average diurnal variations, regionally and monthly, in the radiation budget.

Stratospheric Aerosol and Gas Experiment II (SAGE-II):
Second in a series, this instrument measures atmospheric attenuation of solar radiation during spacecraft sunrise and sunset. Analysis of this data will be used:
- to map vertical profiles of stratospheric aerosols, NO_2, water vapor, and O_3.
- to determine seasonal variations and the radiative characteristics of the stratospheric aerosols, NO_2, water vapor, and O_3.
- to identify sources and sinks of aerosols, NO_2, water vapor, and O_3 and to observe natural transient phenomena such as volcanic eruptions, tropical upwellings, and dust storms.

Orbit

Orbit inclination 57°

Orbital Parameters:
- 580 km circular
- 57° inclination
- 96 minute orbit
- 4 full-sun periods/year
- 10 180° yaw maneuvers/year to provide solar array illumination

Fig. B.5. Earth Radiation Budget Satellite (ERBS). The following terms apply to ERBS: ESSA = Electrically Steerable Spherical Array; LaRC = Langley Research Center; NCC = Network Control System; NOAA = National Oceanic and Atmospheric Administration; SDPF = Science Data Processing Facility; SN = Space Network; TDRS/GSTDN = Tracking and Data Relay Satellite/Ground Spaceflight Tracking and Data Network; WSGT = White Sands Ground Terminal (New Mexico).

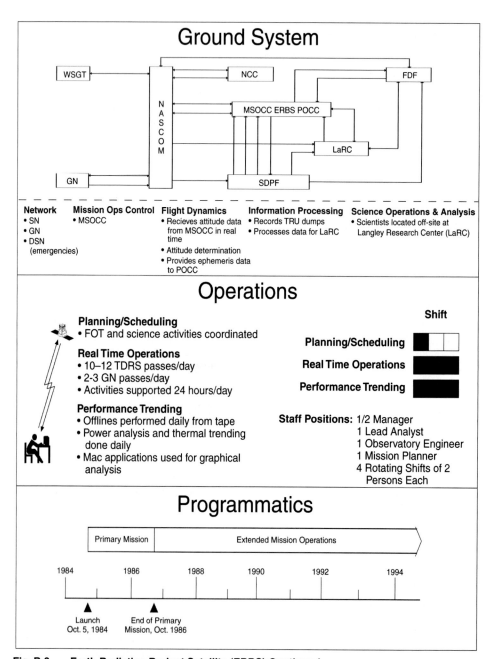

Fig. B.6. Earth Radiation Budget Satellite (ERBS) Continued.

FAST - Fast Auroral Snapshot Explorer

MOM: John Catena (GSFC)   PROJ MGR: Orlando Figueroa (GSFC)

Rev Date: 27 Jan 94	MISSION STATEMENT	Current Mission Phase: C/D

Investigate the plasma physics of auroral phenomena at extremely high time and spatial resolutions utilizing fast data sampling, a large burst memory, and triggering on various event types to collect the specific data needed to characterize different events. FAST is the 2nd mission of the Small Explorer program.

Launch Date: August 1994   Launch Vehicle: Pegasus XL   Launch Site: Vandenberg AFB

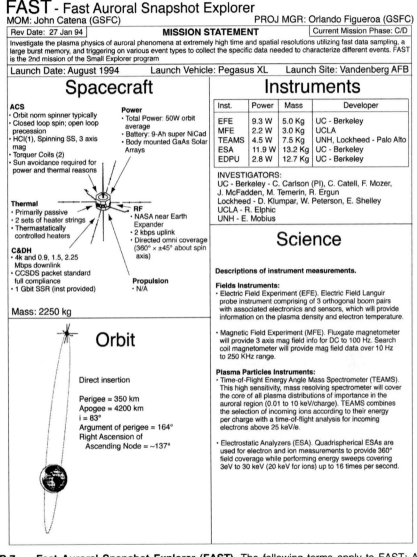

Spacecraft

ACS
- Orbit norm spinner typically
- Closed loop spin; open loop precession
- HCI(1), Spinning SS, 3 axis mag
- Torquer Coils (2)
- Sun avoidance required for power and thermal reasons

Power
- Total Power: 50W orbit average
- Battery: 9-Ah super NiCad
- Body mounted GaAs Solar Arrays

Thermal
- Primarily passive
- 2 sets of heater strings
- Thermastatically controlled heaters

C&DH
- 4k and 0.9, 1.5, 2.25 Mbps downlink
- CCSDS packet standard full compliance
- 1 Gbit SSR (inst provided)

RF
- NASA near Earth Expander
- 2 kbps uplink
- Directed omni coverage (360° × ±45° about spin axis)

Propulsion
- N/A

Mass: 2250 kg

Instruments

Inst.	Power	Mass	Developer
EFE	9.3 W	5.0 Kg	UC - Berkeley
MFE	2.2 W	3.0 Kg	UCLA
TEAMS	4.5 W	7.5 Kg	UNH, Lockheed - Palo Alto
ESA	11.9 W	13.2 Kg	UC - Berkeley
EDPU	2.8 W	12.7 Kg	UC - Berkeley

INVESTIGATORS:
UC - Berkeley - C. Carlson (PI), C. Catell, F. Mozer, J. McFadden, M. Temerin, R. Ergun
Lockheed - D. Klumpar, W. Peterson, E. Shelley
UCLA - R. Elphic
UNH - E. Mobius

Orbit

Direct insertion

Perigee = 350 km
Apogee = 4200 km
i = 83°
Argument of perigee = 164°
Right Ascension of Ascending Node = ~137°

Science

Descriptions of instrument measurements.

Fields Instruments:
- Electric Field Experiment (EFE). Electric Field Languir probe instrument comprising of 3 orthogonal boom pairs with associated electronics and sensors, which will provide information on the plasma density and electron temperature.

- Magnetic Field Experiment (MFE). Fluxgate magnetometer will provide 3 axis mag field info for DC to 100 Hz. Search coil magnetometer will provide mag field data over 10 Hz to 250 KHz range.

Plasma Particles Instruments:
- Time-of-Flight Energy Angle Mass Spectrometer (TEAMS). This high sensitivity, mass resolving spectrometer will cover the core of all plasma distributions of importance in the auroral region (0.01 to 10 keV/charge). TEAMS combines the selection of incoming ions according to their energy per charge with a time-of-flight analysis for incoming electrons above 25 keV/e.

- Electrostatic Analyzers (ESA). Quadrispherical ESAs are used for electron and ion measurements to provide 360° field coverage while performing energy sweeps covering 3eV to 30 keV (20 keV for ions) up to 16 times per second.

Fig. B.7.   Fast Auroral Snapshot Explorer (FAST). The following terms apply to FAST: AGO = Santiago, Chile Tracking Station; ATT = Attitude; CMS = Command Management System; CTV = Compatibility Test Van; DIAG = Diagnostic; ESA = European Space Agency; GSE = Ground Support Equipment; I&T = Integration and Test; KIRUNA = Kiruna, Sweden, location of the Swedish Space Physics Institute; L&EO = Launch and Early Orbit; MP = Mission Planning; PFSOC = Poker Flat Science Operations Center; Poker Flat = Poker Flat Research Range, AK; SCI = Science; SPC = Science Programme Committee; UCSOC = University of California Science Operations Center; VLSILZP = Very Large Scale Integration Level Zero Processing; WS = Workstation; WSG = Wallops Scheduling Group.

Appendix B

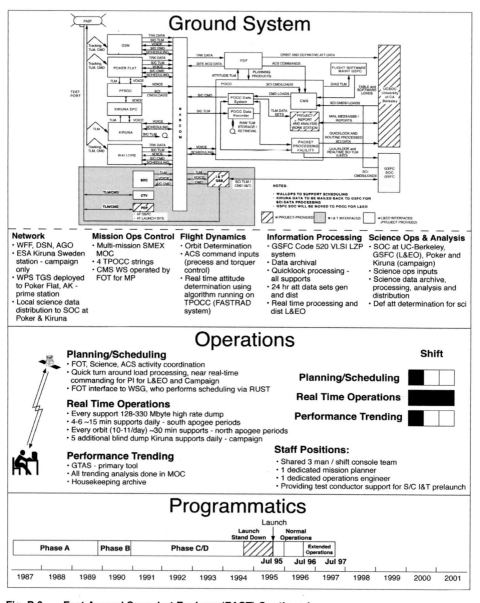

Fig. B.8. Fast Auroral Snapshot Explorer (FAST) Continued.

FUSE - Far Ultraviolet Spectroscopic Explorer

MOM: Bob Nelson (GSFC)   PROJ MGR: Frank Volpe (GSFC)

Rev Date: 22 Feb 94	MISSION STATEMENT	Current Mission Phase: B

- Study distribution and abundance of trace species in interstellar and intergalactic gas, using absorption spectroscopy of faint distant sources such as active galactic nuclei and quasars
- Study physical processes in interstellar material, stellar explosions and mass loss, active galactic nuclei, and planetary magnetospheres
- Study the mechanisms by which stars and planetary systems form and evolve

Launch Date: October 2000	Launch Vehicle: Delta II 7925	Launch Site: CCAFS

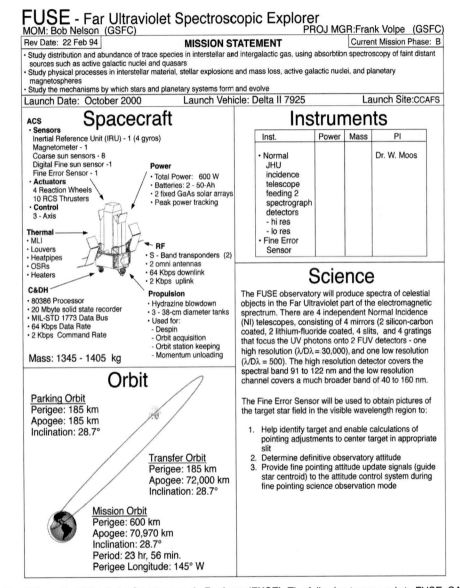

Spacecraft

ACS
- Sensors
  - Inertial Reference Unit (IRU) - 1 (4 gyros)
  - Magnetometer - 1
  - Coarse sun sensors - 8
  - Digital Fine sun sensor -1
  - Fine Error Sensor - 1
- Actuators
  - 4 Reaction Wheels
  - 10 RCS Thrusters
- Control
  - 3 - Axis

Thermal
- MLI
- Louvers
- Heatpipes
- OSRs
- Heaters

C&DH
- 80386 Processor
- 20 Mbyte solid state recorder
- MIL-STD 1773 Data Bus
- 64 Kbps Data Rate
- 2 Kbps Command Rate

Mass: 1345 - 1405 kg

Power
- Total Power: 600 W
- Batteries: 2 - 50-Ah
- 2 fixed GaAs solar arrays
- Peak power tracking

RF
- S - Band transponders (2)
- 2 omni antennas
- 64 Kbps downlink
- 2 Kbps uplink

Propulsion
- Hydrazine blowdown
- 3 - 38-cm diameter tanks
- Used for:
  - Despin
  - Orbit acquisition
  - Orbit station keeping
  - Momentum unloading

Instruments

Inst.	Power	Mass	PI
• Normal incidence telescope feeding 2 spectrograph detectors - hi res - lo res • Fine Error Sensor			Dr. W. Moos

Science

The FUSE observatory will produce spectra of celestial objects in the Far Ultraviolet part of the electromagnetic sprectrum. There are 4 independent Normal Incidence (NI) telescopes, consisting of 4 mirrors (2 silicon-carbon coated, 2 lithium-fluoride coated, 4 slits, and 4 gratings that focus the UV photons onto 2 FUV detectors - one high resolution ($\lambda/D\lambda = 30,000$), and one low resolution ($\lambda/D\lambda = 500$). The high resolution detector covers the spectral band 91 to 122 nm and the low resolution channel covers a much broader band of 40 to 160 nm.

The Fine Error Sensor will be used to obtain pictures of the target star field in the visible wavelength region to:

1. Help identify target and enable calculations of pointing adjustments to center target in appropriate slit
2. Determine definitive observatory attitude
3. Provide fine pointing attitude update signals (guide star centroid) to the attitude control system during fine pointing science observation mode

Orbit

Parking Orbit
Perigee: 185 km
Apogee: 185 km
Inclination: 28.7°

Transfer Orbit
Perigee: 185 km
Apogee: 72,000 km
Inclination: 28.7°

Mission Orbit
Perigee: 600 km
Apogee: 70,970 km
Inclination: 28.7°
Period: 23 hr, 56 min.
Perigee Longitude: 145° W

Fig. B.9. Far Ultraviolet Spectroscopic Explorer (FUSE). The following terms apply to FUSE: CAN = Canberra; CCAFS = Cape Canaveral Air Force Station; FES = Fine Error Sensors; FUV = Far Ultraviolet; GaAs = Gallium-Arsenide; JHU = John Hopkins University; MLI = Multi-Layer Insulation; NSSDC = National Space Science Data Center; OSR = Optical Solar Reflector; PC = Personal Computer; PER = Pre-Environmental Review; PSR = Pre-Shipment Review; RCS = Reaction Control System; SCR = System Concept Review; SIVVF = System Integration, Verification, and Validation File; SOR = System Operations Review.

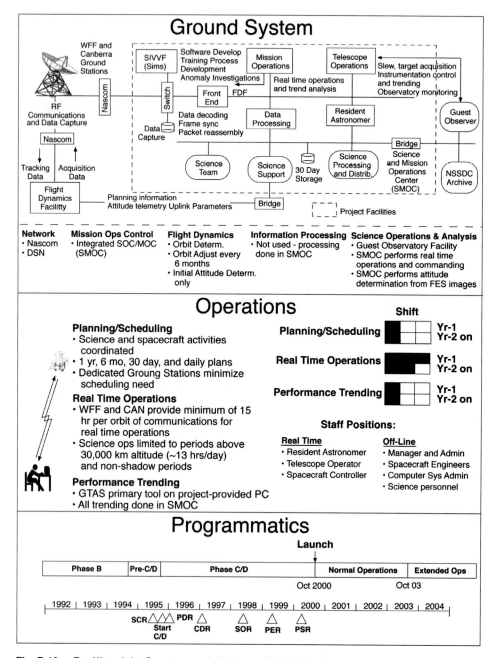

Fig. B.10. Far Ultraviolet Spectroscopic Explorer (FUSE) Continued.

Mission Summaries

GRO - Gamma Ray Observatory

POD: Robert E. Wilson (GSFC)
PROJ MGR: Paul Pashby (GSFC)

| Rev Date: 7 Jan 94 | MISSION STATEMENT | Current Mission Phase: |

Perform extensive study of Gamma Ray sources throughout the universe.

Launch Date: April 5, 1991   Launch Vehicle: STS-37 Atlantis   Launch Site: KSC

Spacecraft

ACS
- Sensors
  Fixed Head Star Trackers - 2
  Sun Sensors - 6
  Inertial Reference Unit - 1
  (contains 3 gyros)
  3 Axis Magnetometer - 3
- Actuators
  RWA - 4
  Magnetic Torquers - 3
  SA Drive Assembly - 1
  High Gain Antenna Drive - 1

Thermal
- Instruments thermally isolated from each other and the spacecraft
- Heaters
- Passive hardware

Power
- Total Power: 4304 W
- Batteries: 6 - 50-Ah Ni-Cd
- Positionable Solar Arrays

RF
- High gain antenna - 1
  1 Kbps cmd rate
- Low gain antenna - 2
  0.125 kbps cmd rate

C&DH
- 1, 32, and 512 kbps tlm rates
- Selectable tlm rates

Propulsion
- Propellant tanks - 4
- Orbit Adjust Thruster - 4
  (445 N thrust each)
- Attitude Control Thrusters - 8
  (22 N thrust each)
- Propellant distribution modules - 2

Mass: 15,909 kg

Instruments

Inst.	Power	Mass	PI
BATSE	122 W	766 kg	Dr. G. J. Fishman (MSFC)
COMPTEL	216 W	1451 kg	Dr. V. Schoenfelder (Max-Planck Institue, Germany)
EGRET	180 W	1779 kg	Dr. C. E. Fichtel (GSFC) Prof. R. Hofstadter (Stanford U., CA) Dr. K. Pinkau (Max-Planck Institute, Germany)
OSSE	161 W	1802 kg	Dr. J. D. Kurfess (Naval Research Laboratory, DC)

Science

Descriptions of instrument measurements.

Burst and Transient Source Experiment (BATSE) - Continuously monitors a large segment of the sky for the detection and measurement of bursts and other transient sources of Gamma Rays. It is able to measure time variations in such events to a fraction of a millisecond, allowing for a detailed analysis of emission mechanisms.

Imaging Compton Telescope (COMPTEL) - It has two detector arrays, one above the other, to aid in determining the direction of arrival as well as the energy of Gamma Ray photons from 1 to 30 million electron volts (MeV) of energy. It has a wide (1 steradian) field of view and the capability of rejecting background events.

Energetic Gamma Ray Experiment Telescope (EGRET) It measures the highest energy Gamma Rays, up to 30 GeV. It can measure the direction of a point source to a fraction of a degree and determine the spectrum of Gamma Ray emissions.

Oriented Scintillation Spectrometer Experiment (OSSE) It consists of four identical instruments, each mounted on a gimbal allowing it to rotate through 180°. One of a pair can measure the background while the other is observing a Gamma Ray source. Detectors are optimized for detecting Gamma Rays of up to 10 MeV.

Orbit

Orbit Parameters:
450 km circular
28.5° inclination
93 minute orbit

Fig. B.11.   **Gamma Ray Observatory (GRO).** The following terms apply to GRO: ATSC = AlliedSignal Technical Services Corporation; CMF = Command Management Facility; GGS = GSFC GRO Simulator; MSFC = Marshall Space Flight Center; MSOCC = Multisatellite Operations Control Center; NCC = Network Control Center; NRL = Naval Research Laboratory; PACOR = Packet Processor; RWA = Reaction Wheel Assembly; SA = Solar Array; WSGT = White Sands Ground Terminal.

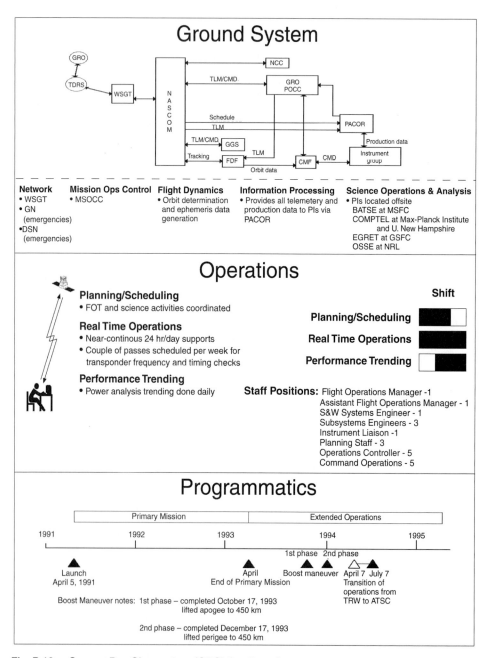

Fig. B.12. Gamma Ray Observatory (GRO) Continued.

Mission Summaries

ICE - International Cometary Explorer

POD: Gilbert Bullock (GSFC)
PROJ MGR: Paul Pashby (GSFC)

Rev Date: 22 Feb 94	MISSION STATEMENT	Current Mission Phase:

Original - Study Earth's magnetosphere and changing conditions in the solar wind.
Extended - Study geomagnetic tail, comet Giacobini-Zinner, coronal mass ejection coverage
 - Observations of the solar wind through 2014.

Launch Date: August 12, 1978   Launch Vehicle: Delta   Launch Site: KSC

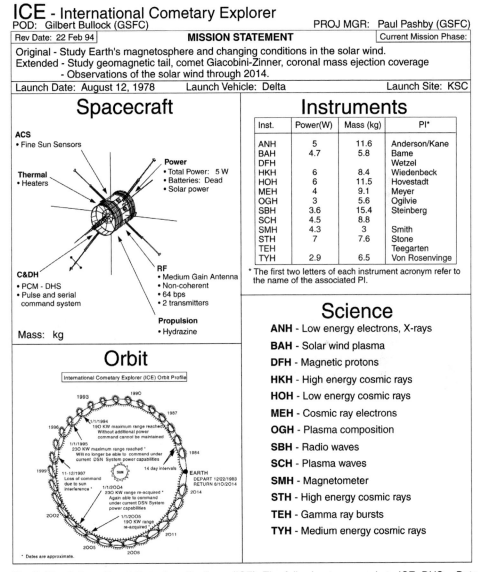

Spacecraft

ACS
- Fine Sun Sensors

Thermal
- Heaters

Power
- Total Power: 5 W
- Batteries: Dead
- Solar power

C&DH
- PCM - DHS
- Pulse and serial command system

RF
- Medium Gain Antenna
- Non-coherent
- 64 bps
- 2 transmitters

Propulsion
- Hydrazine

Mass: kg

Instruments

Inst.	Power(W)	Mass (kg)	PI*
ANH	5	11.6	Anderson/Kane
BAH	4.7	5.8	Bame
DFH			Wetzel
HKH	6	8.4	Wiedenbeck
HOH	6	11.5	Hovestadt
MEH	4	9.1	Meyer
OGH	3	5.6	Ogilvie
SBH	3.6	15.4	Steinberg
SCH	4.5	8.8	
SMH	4.3	3	Smith
STH	7	7.6	Stone
TEH			Teegarten
TYH	2.9	6.5	Von Rosenvinge

* The first two letters of each instrument acronym refer to the name of the associated PI.

Orbit

International Cometary Explorer (ICE) Orbit Profile

1993
1990
1987
1/1/1994 190 KW maximum range reached. Without additional power command cannot be maintained
1996
1/1/1995 230 KW maximum range reached. Will no longer be able to command under current DSN System power capabilities
1984
14 day intervals
1999
11-12/1997 Loss of command due to sun interference *
SUN
EARTH
DEPART 12/22/1983
RETURN 8/10/2014
1/1/2004 230 KW range re-acquired * Again able to command under current DSN System power capabilities
2014
2002
1/1/2005 190 KW range re-acquired *
2011
2005
2006

* Dates are approximate.

Science

- ANH - Low energy electrons, X-rays
- BAH - Solar wind plasma
- DFH - Magnetic protons
- HKH - High energy cosmic rays
- HOH - Low energy cosmic rays
- MEH - Cosmic ray electrons
- OGH - Plasma composition
- SBH - Radio waves
- SCH - Plasma waves
- SMH - Magnetometer
- STH - High energy cosmic rays
- TEH - Gamma ray bursts
- TYH - Medium energy cosmic rays

Fig. B.13. International Cometary Explorer (ICE). The following terms apply to ICE: DHS = Data Handling System; GEOTAIL = a joint mission between Japan and the USA to study the geomagnetic tail; GZ = Comet Giacobini-Zinner; Halo Orbit = so called because from Earth it resembles a halo around the Sun; JPL = Jet Propulsion Laboratory; MOR = Mission Operations Room; PCM = Pulse Code Modulation; Ulysses = a joint NASA/ESA (European Space Agency) probe to study the Sun.

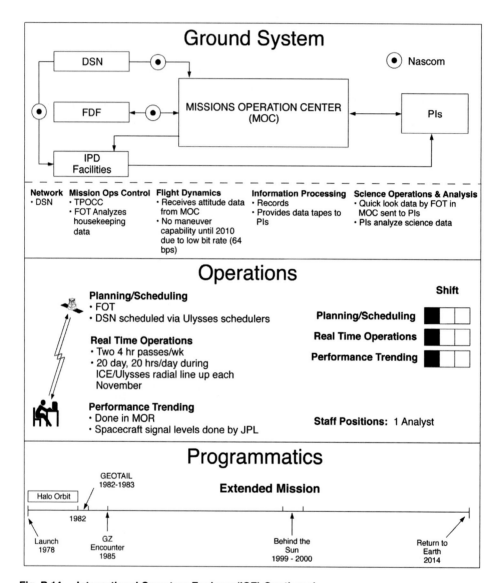

Fig. B.14. International Cometary Explorer (ICE) Continued.

IMP-J - Interplanetary Monitoring Platform

POD: (GSFC)
PROJ MGR: (GSFC)
Rev Date: 22 Feb 94
Current Mission Phase:

MISSION STATEMENT
Study solar and galactic cosmic radiation, solar plasma, solar wind, energetic particles, electromagnetic field variations, and interplanetary magnetic fields.

Launch Date: October 26, 1973   Launch Vehicle: Delta   Launch Site: KSC

Spacecraft

ACS
- Sensors Sun/Earth
- Actuators None
- Control None

Thermal

Power
- Total Power: —
- Batteries: Dead
- Solar Array

RF
- 2 Diplexers
- Hybrid circulator
- 8 Element monopole turnstile array (4 active, 4 passive)

C&DH
- Low rate tlm (3.2 kbps) + 800 bits

Propulsion
- Cold gas
- Monopropellant (Freon-14)
- Empty

Mass: 258.7 kg

Instruments

Instrument	PI
GNF	Dr. N. F. Ness
GAF	Dr. T. L. Aggson
IOF	Dr. D. A. Gurnett
GME	Dr. F. B. McDonald
CHE	Dr. J. A. Simpson
GWP	Dr. D. J. Williams
APP	Dr. S. M. Krimigis
CAI	Dr. E. C. Stone
MAE	Dr. G. Gloeckler
IOE	Dr. L. A. Frank
LAP	Dr. S. J. Bame
MAP	Dr. A. Lazarus

Orbit

35-40 Earth Radii
Transfer Orbit
30-35 Earth Radii
Final Orbit

Science

Fields
- **Magnetic Fields Experiment (GNF)** - Measure the vector magnetic field in three dynamic ranges by using three mutually orthogonal fluxgate sensors.
- **DC Electric Fields (GAF)** - Make vector measurements of the DC electric field with a sensitivity of about 0.1 mV/meter.
- **AC Electric and Magnetic Fields (IOF)** - Measure spatial and temporal characteristics of both electric and magnetic AC vector fields and their polar relationships along the IMP-J orbit.

Energetic Particles
- **Cosmic Ray Experiment (GME)** - Study solar and galactic electrons and nuclei throughout the solar cycle.
- **Cosmic Ray Experiment (CHE)** - Study the solar flare particle acceleration and particle containment in the vicinity of the sun.
- **Energetic Particles Experiment (GWP)** - Study propagation characteristics of solar cosmic rays through the interplanetary medium over selected energy ranges, electron and proton patches throughout the geomagnetic tail and near and through the flanks of the magnetopause, and entry of solar cosmic rays into the geomagnetic field.
- **Charged Particles Experiment (APP)** - Measure protons, alpha particles, $Z \geq 3$ nuclei, and X-rays in a wide energy range.
- **Electron Isotopes Experiment (CAI)** - Study local acceleration of particles solar particle acceleration processes and storage in the interplanetary medium, and interstellar propagation and solar modulation of particles in the interplanetary medium.
- **Ion and Electron Experiment (MAE)** - Determine the composition and energy spectra of low energy particles observed during solar flares and 27-day solar events.

Plasma
- **Low Energy Particles Experiment (IOE)** - Study the differential energy spectra of low energy electrons and protons measured over the geocentric radial distance of 40 earth radii.
- **Los Alamos Plasma Experiment (LAP)** - Make a comprehensive study of electrons and positive ions in the regions of space traversed by the spacecraft and coordinate them with magnetometer and other scientific data.
- **Plasma Experiment (MAP)** - Measure the properties of the plasma in the interplanetary region, transition region, and in the tail of the magnetosphere.

Fig. B.15. Interplanetary Monitoring Platform (IMP-J). The following terms apply to IMP-J: AGO = Santiago, Chile Tracking Station; B0V = a specific classification in the star spectral type classification scheme; HAW = Hawaii; IPD = Information Processing Division; MOR = Mission Operations Room; RED = Redu, Belgium; TAS = Tasmania, Australia; WMAP = Wilkinson Microwave Anisotropy Probe; WPS = Wallops Island S-Band Tracking Station.

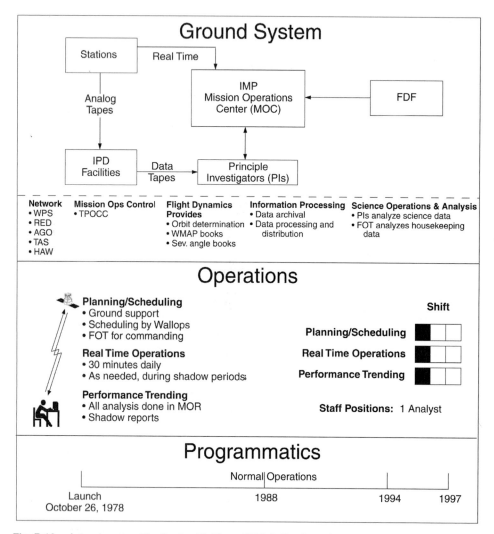

Fig. B.16. Interplanetary Monitoring Platform (IMP-J) Continued.

IUE - International Ultraviolet Explorer

POD: Gil Bullock (GSFC)
PROJ MGR: (GSFC)
Rev Date: 22 Feb 94
MISSION STATEMENT
Current Mission Phase: OPS

To serve as a general facility for observing the ultraviolet spectra of astronomical sources, leading to a better understanding of the elements, the temperatures, and the pressures within stars, and a better understanding of interstellar matter in general.

Launch Date: January 26, 1978   Launch Vehicle: DELTA 2914   Launch Site: KSC

Spacecraft

ACS
- **Sensors**
  - Digital Fine Sun Sensor (FSS)-2
  - Analog Coarse Sun Sensor (CSS)
  - Fine Error Sensor (FES)-2
  - Inertial Reference Assembly (IRA)
  - Panoramic Attitude Sensors (PAS)-2
- **Actuators**
  - Reaction Wheels - 4
  - Low Thrust Engine (LTE) - 8
  - High Thrust Engine (HTE) - 4
- **Control**
  - Onboard Computer (OBC)
  - Control Electronics Assembly (CEA)

Thermal
- HAPS Heaters
- Scientific Instrument Heaters
- Passive Hardware

C&DH
- Data Multiplexer Unit (DMU)
- 1.25, 5, 10, 20 Kbps Telemetry Rate
- Selectable Telemetry Formats
  - Fixed ROM Formats
  - Variable Programmable Formats

Mass: 700 kg at Launch

Power
- Total Power: 424W @β=67.5 (at Launch)
- Batteries: 2 - 6-Ah
- Sun must be in the (X, +Z) 1/2 Plane

RF
- VHF Transponder -2
- Turnstile VHF Antenna System
- S-band Transmitter -2
- S-band Power Amplifier and Antenna -4

Propulsion:
Hydrazine Auxillary Propulsion System (HAPS)
- Monopropellant Catalytic Hydrazine Blowdown System
- Nitrogen pressurant
- 0.89 N LTE -8
- 22 N HTE -4

Instruments

Inst.	Power	Mass	PI
LWP	20.6w	15.525kg	IUE is operated with a guest observer (GO) program. Each GO is the principle investigator for their project.
LWR	20.6w	15.525kg	
SWP	20.6w	15.525kg	
SWR	20.6w	15.525kg	
FES1	4.3w	4.3kg	
FES2	4.3w	4.3kg	

LWP - Long Wavelength Prime camera
LWR - Long Wavelength Redundant camera
SWP - Short Wavelength Prime camera
SWR - Short Wavelength Redundant camera
FES - Fine Error Sensor, #1 or #2

Science

Descriptions of instrument measurements.

Ultraviolet Spectroscopy

Mission Objectives:
- To obtain high resolution (~0.01 nm) spectra in the ultraviolet region of the spectrum from 115 nm to 320 nm of stars and planets brighter than 7th visual magnitude, for detailed analysis of stellar and planetary atmospheres in order to determine more precisely their physical characteristics.

- To obtain lower resolution (~0.6 nm) spectra over the same wavelength range for both stellar and extended objects as faint as 12th magnitude or fainter (15th magnitude desired) as a function of observing time for investigation of peculiar objects such as quasars, Seyfert galaxies, pulsars, X-ray sources, and variability phenomena to shed light on questions of cosmological significance.

Spectrographs

(1) Type: echelle
(2) Detector: Proximity focused converter and SEC vidicon camera
(3) Entrance Aperatures: 3 arcsec circular
                       10x20 arcsec elliptical

	Short l	Long l
(4) High Dispersion		
(a) Wavelength Range	119.1–192.5 nm	189.3–303.2 nm
(b) Resolving Power	0.01 nm	0.01 nm
(c) Limiting Magnitude*	7	7
(5) Low Dispersion		
(a) Wavelength Range	114.1–208.4 nm	175.0–303.2 nm
(b) Resolving Power	0.6 nm	0.6 nm
(c) Limiting Magnitude*	12	12

* Estimated for 30 minute exposure on B0V star.

Orbit
(Current)

Greenwich Meridian At Launch
Elliptical Geosynchronous Orbit
Transfer Orbit
Parking Orbit
Apogee Motor Burn
Third Stage Ejection
Perigee = 30268 km
Apogee = 41315 km
Velocity at:
  Perigee = 3.5 km/s
  Apogee = 2.7 km/s
Period = 1436 min
KSC 80.6° West Meridian At Launch
Sun January, 1978 At Launch

$a = 42170$ km $e = 0.131$ $i = 34.4°$
$\Omega = 97.4°$ $w = 31.4°$ $M = 296.2°$

Fig. B.17. International Ultraviolet Explorer (IUE). The following terms apply to IUE: AGO = Santiago, Chile; DOC = Data Operations Controller; ESA = European Space Agency; IUEOCC = IUE Operations Control Center; NSSDC = National Space Science Data Center; Sigma = series of 1980s-era computers built by SDS; VILSPA = Villafranca Satellite Tracking Station, (ESA tracking station in Spain).

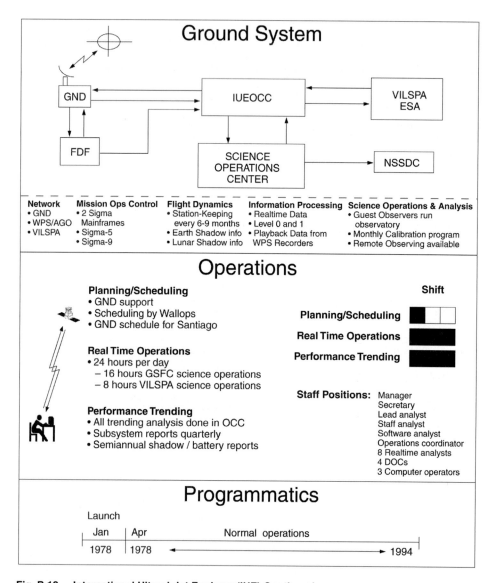

Fig. B.18. International Ultraviolet Explorer (IUE) Continued.

SAMPEX - Solar, Atmospheric, Magnetic Particle Explorer

POD: Jim Williamson / Code 513 (GSFC) PROJ MGR: Paul Pashby / Code 602 (GSFC)

Rev Date: 25 May 94	MISSION STATEMENT	Current Mission Phase: On-orbit

The primary scientific objectives of SAMPEX are to measure the elemental and isotopic composition of solar energetic particles, anomalous cosmic rays, and galactic cosmic rays over the energy range from ~1 to several hundred MeV per nucleon. SAMPEX is the 1st mission of the Small Explorer program.

Launch Date: July 3, 1992 Launch Vehicle: Scout ELV Launch Site: Vandenberg AFB

Spacecraft

ACS
- Stabilized with bias momentum using single momentum wheel aligned with sun-line.
- S/A maintained within 5 deg. of sun line during normal science modes.
- Minimize orbital debris exposure to instruments.
- Instrument zenith pointing over poles.
- 1 Fine Digital SS, 5 Coarse SS, 1 Magnetometer.
- 1 Momentum wheel, 3 Magnetic Torque Rods.
- Independent Analog Safehold Control

Thermal
- Passive in nature
- White and black paint coatings
- Silver Teflon and Kapton MLI
- Thermistors
- Thermostatically controlled heaters, survival and operational - Each set powered by its own bus, switchable off the essential bus.

Mass: 159 kg

Power
- Total power req.: 65 Watts
- Total power gen.:100 W
- Battery : Single 9-Ah Super NiCd battery.
- Four deployed non-articulating S/A panels - 4 panels of 8 strings (32 strings total).

RF
- NASA S-band transponder
- Coherent 2039/2215 MHz (Uplink/Downlink)
- Diplexer allowing the 2 identical, quadrifilar helical antennaes to both transmit and receive.
- Antenna gain : -5.0 dBi

C&DH- SEDS
- CTT/RPP/1773 Software Bus
- 26.5Mb Solid State Recorder
- 2 kbps uplink
- 900, 16 or 4 kbps downlink
- CCSDS packet standard

Propulsion
- None

Orbit

- Semi-major axis : 7008 km
- Eccentricity : 0.0099
- Inclination : 82 deg.
- Precession : 1 deg./day (with respect to vernal equinox)
  2 deg./day (with respect to sun line)
- Rotation of Argument of Perigee : -3.25 deg./day
- Orbit Period : 97 min. 19 sec.
- Perigee : 6938 km
- Apogee : 7078 km

Instruments

Inst.	Power	Mass	PI
DPU	3.4W		DAN MABRY
MAST/PET	3.4W	8.8kg	JAY CUMMINGS
LEICA	4.6W	7.4kg	GLENN MASON
HILT	5.2W	22.8kg	BERNDT KLECKER
Total	16.6W	39.0kg	

Science

Descriptions of instrument measurements.

- **Data Processing Unit (DPU)** - Provides an interface between the instruments and Small Explorer Data System (SEDS).

- **Mass Sprectrometer Telescope/Proton Electron Telescope (MAST/PET)** - Measures isotopic composition of elements from Li (Z=3) to Ni (Z=28) ranging from 10 Mev to several hundred Mev/nucleon.

- **Low Energy Ion Composition Analyzer (LEICA)** - Measures ion fluxes over the charge range from He through Ni from about 0.35 to 10 MeV/nucleon and 0.8 MeV above the mass of Ni.

- **Heavy Ion Large Telescope (HILT)** - Measures heavy ions from He to Fe in the energy range from 8 to 310 MeV/nucleon.

Fig. B.19. **Solar, Atmospheric, Magnetic Particle Explorer (SAMPEX).** The following terms apply to SAMPEX: 1773 = MIL-STD-1773, which defines a fiber optic data bus; CCSDS = Consultative Committee for Space Data Systems; CTT = Command Telemetry Terminal; DSN-RID, CAN, GDS = Deep Space Network at Madrid, Spain; Canberra, Australia; and Goldstone, CA; MLI = Multi-Layer Insulation; MP = Mission Planning; PACOR = Packet Processor; RPP = Recorder/Packetizer/Processor; SEDS = Space Electronic Detection System; SMEX = Small Explorer; SS = Sun Sensor; SSR = Solid State Recorder.

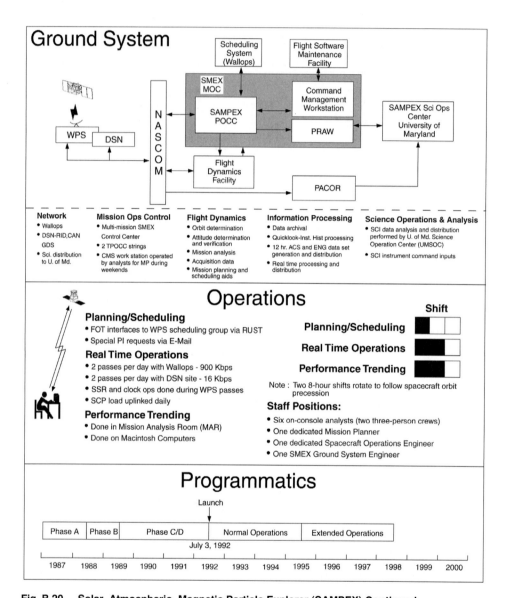

Fig. B.20. Solar, Atmospheric, Magnetic Particle Explorer (SAMPEX) Continued.

SOHO - Solar and Heliospheric Observatory

MOM: Dan Muhonen (GSFC)
PROJ MGR: Ken Sizemore (GSFC)

Rev Date: 22 Feb 93	MISSION STATEMENT	Current Mission Phase: C/D

Study solar wind, solar seismology and coronal dynamics, with emphasis on probing the interior structure of the Sun, characterizing strong and weak magnetic field regions in the chromosphere and corona, and investigating the outflow of plasma and the solar wind origin.

Launch Date: July 1995   Launch Vehicle: Atlas II AS   Launch Site: ESMC

Spacecraft

ACS
- **Sensors**
  - Fine Pointing Sun Sensor (FPSS) - 2
  - Sun Acquisition Sensor (SAS) - 3
  - Star Sensor Units (SSU) - 2
  - Inertial Reference Unit (IRU) - 2
  - Attitude Anomaly Detector - 1
- **Actuators**
  - Reaction Wheel Unit (RWU) - electronics and 4 reaction wheels
  - Control Actuation Electronics - 1

Power
- Total Power: 2240 W
- Batteries: 2 - 20-Ah
- +X-side always in full sun

Thermal
- Mode control by software
- Heaters
- Passive hardware

RF
- High Gain Antenna
- 2 Low Gain Antennas
- 2 kbps command rate
- Coherent

DHSS
- Low, medium, high rate tlm (1, 54, 246 kbps)
- Two magnetic tape recorders
- Selectable tlm formats

Propulsion
- Blowdown hydrazine
- Helium pressurant
- 8 (redundant) 4N thrusters

Mass: 1875 kg

Instruments

Inst.	Power	Mass	PI
MDI	80.0W	60.6 kg	Dr. P. Scherrer
GOLF	53.5W	27.35 kg	Dr. A. Gabriel
VIRGO	29.6W	14.6 kg	Dr. C. Fröhlich
CDS	70.0W	95 kg	Dr. R. Harrison
SUMER	80.2W	95 kg	Dr. K. Wilhelm
UVCS	60.0W	117.3 kg	Dr. J. Kohl
EIT	27.8W	14 kg	Dr. J.P. Delaboudinière
LASCO	99.2W	65.4 kg	Dr. G. Brueckner
SWAN	16.3W	13.25 kg	Dr. J.L. Bertaux
CEPAC	27.3W	20.5 kg	Dr. D. Hovestadt
CELIAS	25.0W	29.2 kg	Dr. P. Bochsler

Science

Helioseismology
- **Michelson Doppler Imager (MDI)** - Study radial stratification and latitudinal variation of pressure, density, composition, sound speed, and the internal rotation of the sun
- **Global Oscillations at Low Frequencies (GOLF)** - Study the solar eigenmodes of low degrees and the measurement of the global magnetic field
- **Variability of Solar Irradiance and Gravity Oscillations (VIRGO)** - Study the variability of solar irradiance and gravity oscillations

Solar Atmospheric Remote Sensing
- **Coronal Diagnostics Spectrometer (CDS)** - Study the heating of the corona and the acceleration of the solar wind
- **Solar Ultraviolet Measurements of Emitted Radiation (SUMER)** - Study plasma density, temperature and flow in the upper chromosphere, the transition zone, and lower corona
- **Ultraviolet Coronagraph Spectrometer (UVCS)** - Study mechanisms for accelerating the solar wind and for heating the coronal plasma and establish the plasma properties of the solar wind
- **Extreme-Ultraviolet Imaging Telescope d'Astrophysique (EIT)** - Study space-time evolution of coronal structures and coronal heating and solar wind acceleration
- **Large Angle Spectrometric Coronagraph (LASCO)** - Study coronal heating, solar wind acceleration, and coronal transients
- **Solar Wind Anisotropies d'Aeronomie (SWAN)** - Study anisotropies of solar wind and temporal variations of latitude distribution

"In Situ" Solar Wind
- **Costep-Erne Particle Analysis Collaboration (CEPAC)** - Study the physical conditions of the sun, and the processes responsible for emitting and affecting emitted particles; record interplanetary and galactic cosmic rays
- **Charge, Element and Isotope Analysis System (CELIAS)** - Study the elemental and isotopic abundances and the ionic charge state and velocity distribution of ions originating in the solar atmosphere

Orbit

Parking Orbit:
180 km circular
28 deg inclination

Transfer Trajectory:
a = 609,950 km
e = 0.989
i = 28.84°
w = 329.6°
RA = 326.4°
Perigee = 6554 km
Apogee = 1,213,400 km

Halo Orbit:
A_x (in ecliptic sun direction): 206,450 km
A_y (in ecliptic perpendicular to A_x): 666,470 km
A_z (normal to the ecliptic): 120,000 km

Fig. B.21. Solar and Heliospheric Observatory (SOHO). The following terms apply to SOHO: AGO = Santiago, Chile (Tracking station); CDHF = Central Data Handling Facility; DHSS = Data Handling Subsystem; EOF = Experimenters' Operations Facility; ESMC = Eastern Space and Missile Center; ESTEC = European Research and Technology Center; FAR = Final Acceptance Review; FOR = Flight Operations Review; FRR = Flight Readiness Review; GGS = Global Geospace Science; HDR = Hardware Design Review; HK = Housekeeping; IOS = Indian Ocean (tracking station); ISG = ISTP Scheduling Group; ISTP = International Solar-Terrestrial Physics; MDI = Michelson Doppler Imager; MOR = Mission Operations Review; ORR = Operational Readiness Review; QTR = Qualification Test Report; SDR = System Design Review; SK = Stationkeeping; SPOF = Science Planning and Operations Facility; SRR = Systems Requirements Review; SSDR = Subsystem Design Review; TR = Tape Recorder.

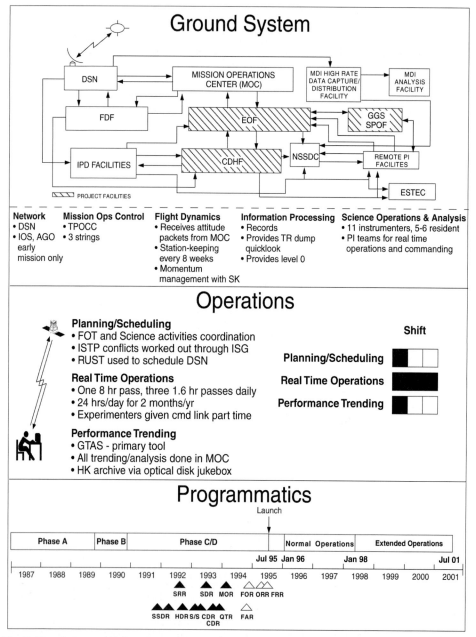

Fig. B.22. Solar and Heliospheric Observatory (SOHO) Continued.

SWAS - Submillimeter Wave Astronomy Satellite

MOM: John Catena (GSFC) PROJ MGR: Orlando Figueroa (GSFC)

Rev Date: 5 Aug 94	MISSION STATEMENT	Current Mission Phase: C/D
	The SWAS program is a pathfinding mission to study the chemical composition of interstellar galactic clouds to help determine the process of star formation.	

Launch Date: June 1995   Launch Vehicle: Pegasus - XL   Launch Site: WWF

Spacecraft

ACS
- **Sensors**
  Coarse Sun Sensors (CSS) - 6
  Digital Sun Sensor (DSS)
  Bright Object Sensor (BOS)
  Three-axis Magnetometer (TAM)
  Gyros - 4
  Star Tracker
- **Actuators**
  Reaction Wheels - 4
  Magnetic Torquer
- **Control**
  Attitude Control Electronics (ACE)

Power
- Super NiCd Battery
- GaAs Solar Arrays
- SWAS Power Electronics

RF
- S-band transponder
- Omni-directional antenna - 2

Thermal
- Survival Heaters
- Operational Heaters

C&DH
- Tlm Data Rates - 18.75k, 900k, 1.8M
- CCSDS cmd and tlm standards
- Solid State Recorder (28 hr capacity)

Mechanical System
- 3 Machined Rings
- 8 Structural Frames
- 8 Access Panels
- 8 Avionics

Propulsion
- None

Mass: 280 kg

Instruments

Inst.	Power	Mass	PI
SWAS	60.7 W	92.9 kg	Dr. Gary Melnick

Science

SWAS
Acousto-Optical Spectrometer (AOS) and receiver - Study star formation and interstellar chemistry by surveying molecular clouds within our galaxy. Spectra will be taken in the submillimeter wavelength range to measure abundances of water, isotopic water, oxygen, carbon and carbon monoxide within these clouds.

Orbit

Characteristics:
altitude = 600 km
inclination = 65 deg

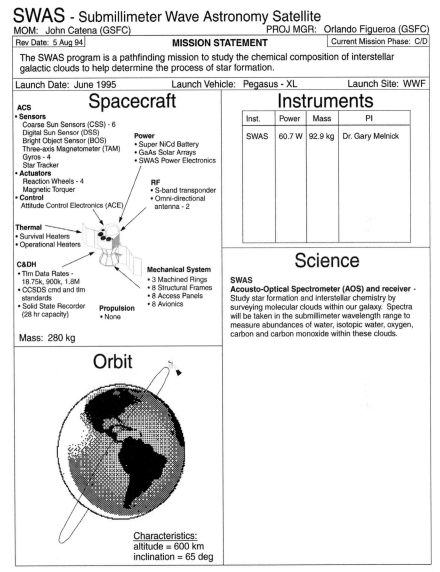

Fig. B.23. Submillimeter Wave Astronomy Satellite (SWAS). The following terms apply to SWAS: CMS = Command Management System; DDF = Data Distribution Facility; GBRS = Generic Block Recording System; GenSAA = Generic Spacecraft Analyst Assistant (Software); HUD = Heads Up Display; MAR = Mission Analysis Room; NCC = Network Control Center; PACOR = Packet Processor; PACOR II = Packet Processing Data Capture Facility; Poker = Poker Flat Research Range, AK; RTAD = Real Time Analyzer Display; SAO = Smithsonian Astrophysical Observatory; SCP = Station Communications Processor; TOTS = Transportable Orbital Tracking System (or Station); WWF = Wallops Island Flight Facility.

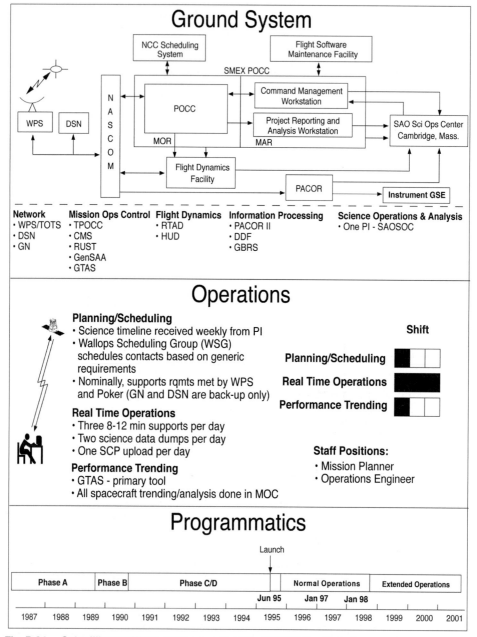

Fig. B.24. Submillimeter Wave Astronomy Satellite (SWAS) Continued.

Mission Summaries

TRMM - Tropical Rainfall Measuring Mission

MOM: Karl Schauer (GSFC) PROJ MGR: Tom LaVigna (GSFC, Code 490)
Rev Date: 13 Aug 94 **MISSION STATEMENT** Current Mission Phase: C/D

TRMM is an integral part of the NASA Mission to Planet Earth Program. As an Earth system science mission, TRMM is designed to advance our understanding of rainfall and to determine the rate of rainfall and total rainfall over the tropics and subtropics between the north and south latitudes of 35°. TRMM will study the distribution and variability of precipitation, latent heat release, Earth's radiant energy budget, and lightning over a multiyear data set.

Launch Date: August 1997 Launch Vehicle: NASDA H-II ELV Launch Site: YLC

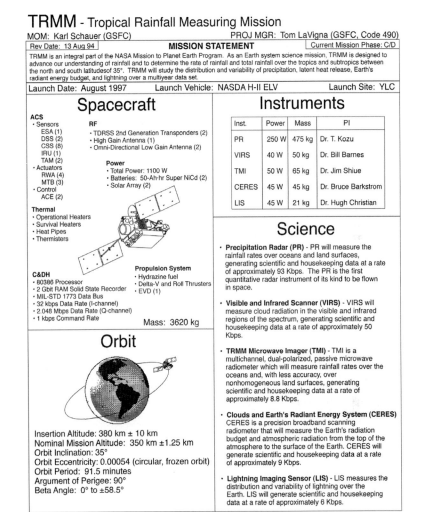

Spacecraft

ACS
- Sensors
  - ESA (1)
  - DSS (2)
  - CSS (8)
  - IRU (1)
  - TAM (2)
- Actuators
  - RWA (4)
  - MTB (3)
- Control
  - ACE (2)

RF
- TDRSS 2nd Generation Transponders (2)
- High Gain Antenna (1)
- Omni-Directional Low Gain Antenna (2)

Power
- Total Power: 1100 W
- Batteries: 50-Ah·hr Super NiCd (2)
- Solar Array (2)

Thermal
- Operational Heaters
- Survival Heaters
- Heat Pipes
- Thermisters

C&DH
- 80386 Processor
- 2 Gbit RAM Solid State Recorder
- MIL-STD 1773 Data Bus
- 32 kbps Data Rate (I-channel)
- 2.048 Mbps Data Rate (Q-channel)
- 1 kbps Command Rate

Propulsion System
- Hydrazine fuel
- Delta-V and Roll Thrusters
- EVD (1)

Mass: 3620 kg

Instruments

Inst.	Power	Mass	PI
PR	250 W	475 kg	Dr. T. Kozu
VIRS	40 W	50 kg	Dr. Bill Barnes
TMI	50 W	65 kg	Dr. Jim Shiue
CERES	45 W	45 kg	Dr. Bruce Barkstrom
LIS	45 W	21 kg	Dr. Hugh Christian

Science

- **Precipitation Radar (PR)** - PR will measure the rainfall rates over oceans and land surfaces, generating scientific and housekeeping data at a rate of approximately 93 Kbps. The PR is the first quantitative radar instrument of its kind to be flown in space.

- **Visible and Infrared Scanner (VIRS)** - VIRS will measure cloud radiation in the visible and infrared regions of the spectrum, generating scientific and housekeeping data at a rate of approximately 50 Kbps.

- **TRMM Microwave Imager (TMI)** - TMI is a multichannel, dual-polarized, passive microwave radiometer which will measure rainfall rates over the oceans and, with less accuracy, over nonhomogeneous land surfaces, generating scientific and housekeeping data at a rate of approximately 8.8 Kbps.

- **Clouds and Earth's Radiant Energy System (CERES)** CERES is a precision broadband scanning radiometer which will measure the Earth's radiation budget and atmospheric radiation from the top of the atmosphere to the surface of the Earth. CERES will generate scientific and housekeeping data at a rate of approximately 9 Kbps.

- **Lightning Imaging Sensor (LIS)** - LIS measures the distribution and variability of lightning over the Earth. LIS will generate scientific and housekeeping data at a rate of approximately 6 Kbps.

Orbit

Insertion Altitude: 380 km ± 10 km
Nominal Mission Altitude: 350 km ±1.25 km
Orbit Inclination: 35°
Orbit Eccentricity: 0.00054 (circular, frozen orbit)
Orbit Period: 91.5 minutes
Argument of Perigee: 90°
Beta Angle: 0° to ±58.5°

Fig. B.25. Tropical Rainfall Measuring Mission (TRMM). The following terms apply to TRMM: ACE = Attitude Control Electronics; AGO = Santiago, Chile (Tracking Station); BOL = Beginning of Life; CSS = Course Sun Sensor; DSS = Digital Sun Sensor; EOL = End of Life; EOSDIS = Earth Observing System Data and Information System; ESA = Earth Sensor Assembly; EVD = Engine Valve Driver; GCMR = Ground Control Message Request; I/F = Interface; I/Q = In-Phase and Quadrature; IRU = Inertial Reference Unit; LaRC = Langley Research Center; LZP = Level Zero Processing; MAR = Mission Analysis Room; MOR = Mission Operations Room; MSFC = Marshall Space Flight Center; MTB = Magnetic Torquer Bars; NASDA = National Space Development Agency (Japan); PR = Precipitation Radar; Q/L = Quicklook; RWA = Reaction Wheel Assembly; SDPF = Sensor Data Processing Facility; SN = Space Network; SSA = S-Band Single Access; TAM = Three-Axis Magnetometer; TSDIS = TRMM Science Data and Information System; UPD = User Performance Data; WSC = White Sands Complex (New Mexico); YSC = Yoshinobu Launch Complex.

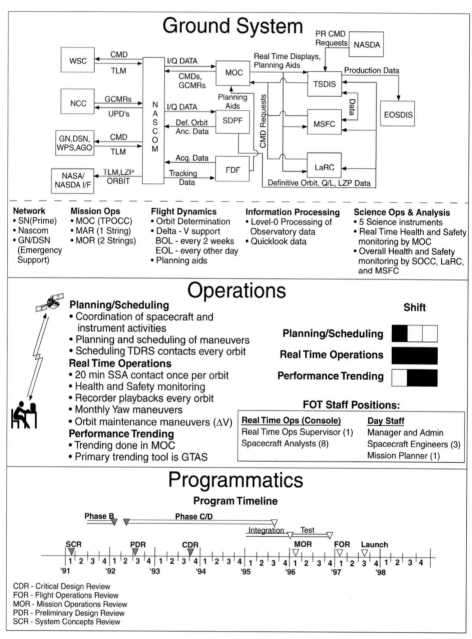

Fig. B.26. Tropical Rainfall Measuring Mission (TRMM) Continued.

XTE - X-ray Timing Explorer

MOM: J.B. Joyce (GSFC) PROJ MGR: Orlando Figueroa (GSFC, Code 410)

Rev Date: 06 Sept 95	MISSION STATEMENT	Current Mission Phase: C/D

The objective of the XTE mission is to design, develop, launch, and successfully operate an observatory capable of measuring astrophysical X-ray source characteristics with high temporal resolution over a broad energy range for a period of two years. XTE will study a variety of X-ray sources including white dwarfs, accreting neutron stars, black holes, and active galactic nuclei. Measurements will be made over a wide range of photon energies from 2 to 200 KeV. The XTE is designed to study the intensity variations and spectra of these objects over time scales as short as microseconds and as long as years.

Launch Date: October 1995   Launch Vehicle: Delta II 7920   Launch Site: KSC ER

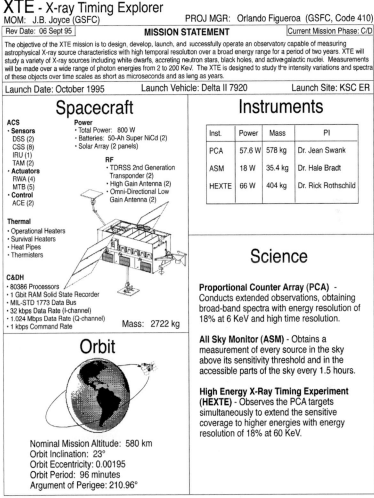

Spacecraft

ACS
- Sensors
  - DSS (2)
  - CSS (8)
  - IRU (1)
  - TAM (2)
- Actuators
  - RWA (4)
  - MTB (5)
- Control
  - ACE (2)

Power
- Total Power: 800 W
- Batteries: 50-Ah Super NiCd (2)
- Solar Array (2 panels)

RF
- TDRSS 2nd Generation Transponder (2)
- High Gain Antenna (2)
- Omni-Directional Low Gain Antenna (2)

Thermal
- Operational Heaters
- Survival Heaters
- Heat Pipes
- Thermisters

C&DH
- 80386 Processors
- 1 Gbit RAM Solid State Recorder
- MIL-STD 1773 Data Bus
- 32 kbps Data Rate (I-channel)
- 1.024 Mbps Data Rate (Q-channel)
- 1 kbps Command Rate

Mass: 2722 kg

Orbit

Nominal Mission Altitude: 580 km
Orbit Inclination: 23°
Orbit Eccentricity: 0.00195
Orbit Period: 96 minutes
Argument of Perigee: 210.96°

Instruments

Inst.	Power	Mass	PI
PCA	57.6 W	578 kg	Dr. Jean Swank
ASM	18 W	35.4 kg	Dr. Hale Bradt
HEXTE	66 W	404 kg	Dr. Rick Rothschild

Science

Proportional Counter Array (PCA) - Conducts extended observations, obtaining broad-band spectra with energy resolution of 18% at 6 KeV and high time resolution.

All Sky Monitor (ASM) - Obtains a measurement of every source in the sky above its sensitivity threshold and in the accessible parts of the sky every 1.5 hours.

High Energy X-Ray Timing Experiment (HEXTE) - Observes the PCA targets simultaneously to extend the sensitive coverage to higher energies with energy resolution of 18% at 60 KeV.

Fig. B.27. X-ray Timing Explorer (XTE). The following terms apply to XTE: ACE = Attitude Control Electronics; CSS = Coarse Sun Sensor; DSS = Digital Sun Sensor; EOR = Execute on Receipt; FOR = Flight Operations Review; GCM = Ground Control Message; GRI = Ground Reference Image; H&S = Health and Safety; HEASARC = High Energy Astrophysics Science Archive Research Center; IRU = Inertial Reference Unit; KSC ER = Kennedy Space Center Eastern Range; LZP = Level Zero Processing; MA = Multiple Access; MIT = Massachusetts Institute of Technology; MOR = Mission Operations Review; MTB = Magnetic Torquer Bars; ODM = Operational Data Message; OST = Operations Support Team; Pacor II = Packet Processing Data Capture Facility; PBK = Playback; RWA = Reaction Wheel Assembly; SCR = System Concept Review; SDPF = Sensor Data Processing Facility; SDVF = Software Development and Verification Facilities; SSA = S-Band Single Access; TAM = Three-Axis Magnetometer; TOO = Target of Opportunity; UCSD = University of California at San Diego; UPD = User Performance Data; WSC = White Sands Complex (New Mexico).

Appendix B

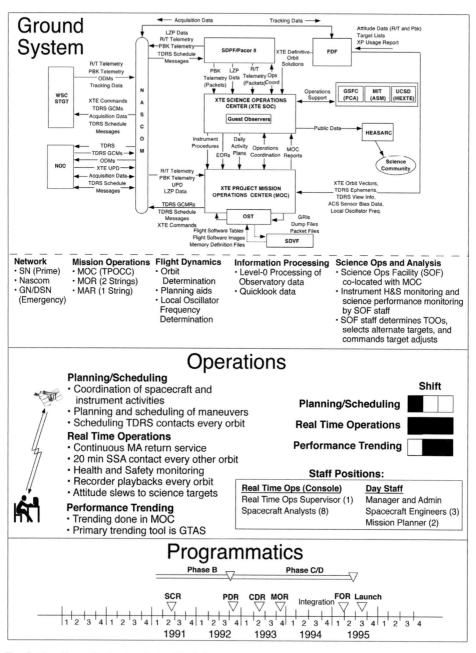

Fig. B.28. X-ray Timing Explorer (XTE) Continued.

Index

A

A/J *See* Anti-jam	
Abort and rescue options	709
Access interval	256
Access point	253
ACE *See* Advanced Composition Explorer	
Acquisition of signal	284
Active archives	76
Active ranging	291
Activities	
fixed	253
floating	253
maintenance	253, 256
mission	255
trimmable	253
Activity	41, 243, 245
Activity placement	252
Activity plan	243, 272
duration of	247
generating	249
Activity planning	243
resources	258
Activity planning and development	42
Activity requests	109
Activity-event file	277
ACTS *See* Advanced Communications and Technology Satellite	
Adaptability	631
Adaptive mission	34
Adaptive payloads	533
Adaptive schedule	23
Adaptivity	70
Advanced Communications and Technology Satellite	420
Advanced Composition Explorer	785
Advanced Range Instrumentation Aircraft	636
AFSCN *See* Air Force Satellite Control Network	
Air Force Satellite Control Network	408, 462, 479
Alarm	53
definition of	228
displays	229
levels	229
overrides	229
presentation to operator	229
AM *See* Amplitude modulation	

Ambiguous states	137
Amplitude modulation	473
AMSAT missions	678
Analysis tools	608
Ancillary data	70, 71, 74, 104, 291, 551
Anomaly	295, 595
attitude-control	598
captain	606
case study	608
categories	596
operations	28
propulsion	598
pseudo	599, 611
resolving	599
team	605
types	598
Anomaly resolution	
fundamental rule	596
Antenna	458
beamwidth	437
diameter	444
gain	436
pointing error	439
pointing loss	439
Anti-jam	399
Apollo spacecraft	724
Archival data records	73
Archive	75
Archiving and maintaining the mission database	75
Argument of latitude	350
Argument of perigee	348, 356
perturbations in	356–358
ARIA *See* Advanced Range Instrumentation Aircraft	
Articulating devices	137
Ascending node	348
Associative models	147
Astronomical Unit	623
Asynchronous transfer mode	457
ATM *See* Asynchronous transfer mode	
Attitude determination and control	559
Attitude-control anomalies	598
Attitude-control system	103
AU *See* Astronomical Unit	
Automatic planning algorithm	259
Automatic safing	597
Automation	21, 140, 144, 269

Automation *See* Ground-station automation
Autonomy *See* Spacecraft autonomy
Auto-tracking 459
Availability analysis 592

B

Backward compatibility 397
Ballistic coefficient 359
Baseline mission concept 22
Batch planning 261
Battery reconditioning 288, 572
Benign channel 402
Bent-pipe configuration 409
Bent-pipe transponders 409
BER *See* Bit-error-rate
Bi-level measurements 600
Binary files 109
Binary phase-shift keying 414–420, 473
BIRD *See* Bi-spectral Infrared Detection
Bi-spectral Infrared Detection 114
Bit synchronizer 461
Bit-error-rate 392, 473
Block encryption 395
Blow-down system 563
Boltzmann's constant 435
B-plane 626
BPSK *See* Binary phase-shift keying
Brahe, Tycho 344, 352
Broadcast communications networks 423

C

C&DH *See* Command and Data Handling
Calibration analysis 69
Capacity constraints 251
See also Constraints
Carrier frequency 436
Carrier wavelength 436
Cassegrain 459
Cassini 614, 625, 628
  mission 537
Cassini Huygens probe 632
Causal models 148

CCSDS *See* Consultative Committee for Space Data Systems
CDMA *See* Code-division multiple access
CDR *See* Critical design review
Channel 402
  characteristics 402
Chief of mission operations 85
  See also Mission operations manager
CI *See* Configuration item
Circle (Conic section) 346
Circuit-switched network 423
Class of operations 532
COBE *See* Cosmic Background Explorer
Code-division multiple access 420–422
Coherent mode 462
Coherent signal 291
Command acceptance 288
Command and Data Handling 550
Command and data processing 458
  subsystem 463
Command approval 279
Command database 287
Command dictionary 47, 64
Command errors 139, 280
Command file 110, 279
Command handbook 283
Command load 47, 240, 275
Command mnemonics 109
Command modulator 461
Command validation 280
Command verification 288
Command, control, and communications (C^3) 10
Commandable states 136
Commanding 281, 452
  non-standard 275
Command-translation software 279
Commercial, off-the-shelf equipment 396
Commercial, off-the-shelf products 260, 476
Communication bandwidth 470
Communication links 456
Communication subsystem 563, 572
Communications 560
Communications architecture 10
  optical links 427
Communications satellites 393

Communications security	395	Contact frequency	468
Communications spacecraft	19	Contact-support plan	282
Commutation	291	*See also* Pass plan	
Compatibility	471	Contamination	563
Complexity metrics	131, 145	Contingency operations	537
Computer Software Configuration Item	203	Contingency plan	600, 690
		Contingency procedures	323
Computers and communications support	80	Contingency situations	207
		Coordinate system	384, 625
COMSATs *See* Communications satellites		B-plane	626
		ecliptic	627
COMSEC *See* Communications security		EMO2000	627
		equatorial	627
Concept development	7, 94	interplanetary equator of date	627
Concept exploration	312, 685	J2000	627
needs analysis	7	planet target plane	628
Conceptual requirements	245	Copernicus, Nicolaus	344
Conceptual system requirements	212	Copper cable	456
Concurrent approach		Cosmic Background Explorer	326, 544, 787
definition of	210		
development of	210	Cost Driver Database	158, 162–170
Concurrent design	15	Cost model	148–191
Configuration control	331	COTS *See* Commercial, off-the-shelf products	
Configuration item	198		
operations concept	208	Courier (US Army communications satellite)	430
Configuration-state tables	322		
Conflict	254	Coverage area	386, 397
Conflict resolution	263	Crew	
Conflict-resolution strategies	253	selection	709
Conic section	346	training	709
Conjunction	625	Critical design review	197, 210
Connectivity constraints	251	Criticality	134
See also Constraints		Crosslink	569
Constants		Cruise operations	637
of motion	347–348	Cruise phase	637
Constraint violation	254	CSCI *See* Computer Software Configuration Item	
Constraints	46, 64, 244, 257		
capacity	251		
connectivity	251	**D**	
geometric	251		
modeling parametric	258	Data access	546
parametric	251, 258	Data code	472
satisfying	257	Data completeness	543, 546
simultaneity	251	Data flow	56
time	250	Data format	473
Consultative Committee for Space Data Systems	40, 56, 100, 473	Data rate	391, 472
		Data structure	543
Consumables	294	Data transport and delivery	54
Contact	282	Data volume	544
See also Pass			

Index

Database files 53
Data-flow diagrams 81, 92, 97, 113, 114
Data-processing plan 325
Data-pull paradigm 217
Data-push paradigm 216
Data-stream encryption 395
Decalibration 291
Decoding 413
Decommutation 291
   maps 64
Deep Space Network 462, 619
Defense Satellite Communications
   Systems 432
Defining anomalies 595–598
Definition and development phases 194
Degradations 578
Delogging 603
Demodulation 413
Demodulator 461
Deorbit 709
Deputy mission controller 286
Design
   concurrent approach 678
Design attributes
   definition of displays, alarms,
     and procedures 215
   definition of operator tasks 215
   definition of system performance 215
   optimal operations of spacecraft
     subsystems 215
   processing concepts 215
   timing and timelines 215
   verification 215
Design margins 311
Detailed development 94, 312
Developing and maintaining
   software 82
Development team for L&EO 331
Differential energy change 617
Direct downlink 545
Direct orbit 351
Direct sequence 396
Discovery program 613
Display plan 325
Displays
   definition of 228
Distributed data processing 21
Doppler ranging 291
Doppler shift 461
Dormant archives 76
Downlink 33, 110, 290
Downtime 492
Drift rate (in orbit) 369
DS *See* Direct sequence
DSCS *See* Defense Satellite
   Communications Systems
DSN *See* Deep Space Network
Duty cycle 542

E

E/E delay 442
Early-orbit operations 28
Earth Observing System 531, 547
Earth Radiation Budget Satellite 304, 789
Earth sensor 560–562
Earth-centered fixed coordinates 385
Earth-centered inertial coordinates 385
Earth-to-Orbit-Vehicles 720
East-West drift 374
Eccentric anomaly 353, 354
Eccentricity 344–348
   vector 348
ECF *See* Earth-centered fixed
ECI *See* Earth-centered inertial
Ecliptic coordinate system 627
ECLS *See* Environmental control
   and life support
ECN *See* Engineering change notice
EEIS *See* End-to-end information
   system
Effective isotropic radiated
   power 435, 444, 458
EIRP *See* Effective isotropic radiated
   power
Electrical-power subsystems 569
Element-requirements specification 203
Ellipse (Conic section) 346
Encoding 473
Encounter 640
Encrypt 464
Encryption 473
End-to-end (E/E) delay
End-to-end data processing 551
End-to-end information system 40, 737
   capabilities of 99
   characteristics of 99
Energy equation (orbits) 347, 362
Engineering change notice 295

Index

Engineering data 65, 70, 290, 607
Engineering telemetry 452
Engineering-support team 605
Enhanced Thematic Mapper 533
Environmental control and life support 710
Environmental extremes 624
EOS *See* Earth Observing System
Ephemeris
  predicted 297
  reconstructed 297
  spacecraft 297
Equatorial coordinate system 627
ERBS *See* Earth Radiation Budget Satellite
Error detection-correction 460
ESA *See* European Space Agency
ESOC *See* European Space Operations Center
ETM+ *See* Enhanced Thematic Mapper
ETOVs *See* Earth-to-Orbit-Vehicles
European Space Agency 487
European Space Agency's Network 487
European Space Operations Center 487
EUVE *See* Extreme Ultraviolet Explorer
EVA *See* Extra-vehicular activity
Event-driven payloads 533
Event-driven schedule 23
Existing networks 477
Exploratory operations 698
Extended mission 641
Extended operations 28
Extra-vehicular activity 708
Extreme Ultraviolet Explorer 270

F

Facility and voice plan 325
Failure modes analysis 593
Failure Modes and Effects Analysis 323, 600, 607
Failure scenarios 596
Failure tree 604
Failures 578
Far Ultraviolet Spectroscopic Explorer 793
Fast Auroral Snapshot Explorer 544, 791

FAST *See* Fast Auroral Snapshot Explorer
Fault detection 288
Fault protection 138
Fault trees 538
FDMA *See* Frequency-division multiple access
FECC *See* Forward-error-correction coding
FH *See* Frequency hopping
Fiber-optic cable 457
FireSat 735–780
  cost 736, 740–743, 765–779
  coverage 739
  End-to-End Information System 737
  ground system 738, 763, 770
  mission constraints 736, 738, 743
  Mission Statement 735
  objectives 736
  orbit 736, 741
  requirements 736, 739
  spacecraft characteristics 738
FITS *See* Flexible Image Transport System
Fixed activities 253
Fleet Satellite Communications System 432
Flexible Image Transport System 104
Flight consumables 137
Flight rules 40, 46, 64
Flight rules and constraints 136
Flight simulator 63, 65
Flight software 65
Flight systems 130, 134
Flight-software tables 322
Floating activities 253
FLTSAT *See* Fleet Satellite Communications System
Flyby 614
FM *See* Frequency modulation
FMEA *See* Failure Modes and Effects Analysis 600, 607
Foliage 405
Forward-error-correction coding 413, 418
Frame synchronizer 56
Framed telemetry 473
Free-space loss 438
Frequency band 474
Frequency hopping 396, 418
Frequency modulation 473

Frequency synthesizer 461
Frequency-division multiple
 access 420–422
Frequency-shift keying 414, 473, 569
 See also Modulation techniques
FSK See Frequency-shift keying
FTEs See Full-time equivalents
Fuel cells 569
Full eclipses 574
Full-time equivalents 29, 146
 definition of 151
Functional command verification 288
FUSE See Far Ultraviolet
 Spectroscopic Explorer

G

G/T 458
Galaxy satellite system 432
Galileo 614, 625, 628, 630
Gallium Arsenide, field-effect
 transistor amplifiers 426
Gamma Ray Observatory 565, 795
Gaspra asteroid 630
General perturbations 356
 See also Orbits, Orbit perturbations
GEO See Geostationary orbits
Geocentric Inertial Coordinates (GCI)
 in defining orbital elements 348
 inertial properties of 348
Geometric constraints 251
 See also Constraints
Geostationary orbit 17, 393
Geostationary spacecraft 18
Geosynchronous orbit
 period of satellites 351
Geosynchronous spacecraft 571
Geosynchronous transfer orbit 678
Global Positioning System 18, 354, 422,
 463, 560
Globalstar 412
Goddard Space Flight Center 326
Goddard Spaceflight Tracking
 and Data Network 568, 602
GPS See Global Positioning System
Gravity assist 381, 615, 628, 639
Gravity-assist trajectories 382
GRO See Gamma Ray Observatory

Ground element
 capabilities of 99
 characteristics of 99
Ground station 454
 automating 696
Ground system 130, 141, 143, 451
 availability 469
 functions 471
Ground system manager 85
 See also Mission operations
 manager
Ground systems 10
Ground tracks
 evaluation 351–353
 repeating 372–373
 satellite 351–353
Ground-data-system manager 85
 See also Mission operations
 manager
Ground-station automation 14, 26
GSFC See Goddard Space Flight
 Center
GSTDN See Goddard Spaceflight
 Tracking and Data Network
GTO See Geosynchronous transfer
 orbit

H

Hand-over briefing 274
Hardware components 209
Hardware Configuration Item 203
HDF See Hierarchical Data Format
Health-and-safety data 542
HEASARC See High Energy
 Astrophysics Archive
Heat pipes 571
Heliocentric-ecliptic coordinate
 system 376
Hierarchical Data Format 546
High Energy Astrophysics Archive 546
High-power amplifier 391
History-dependent commandable
 states 136
Hohmann transfer orbit 361–364
 as most efficient two-burn transfer 361
 total required delta V 362
 transfer ellipse 361
Hohmann, Walter 361

Index

Housekeeping data 290
  See also Engineering data
HPA *See* High-power amplifier
HST *See* Hubble Space Telescope
Hubble Space Telescope 75, 296, 304
Human interplanetary mission 726–733
Human space missions 707
HWCI *See* Hardware Configuration Item
Hyperbola (Conic section) 346
Hyperbolic arrival trajectory 380
Hyperbolic departure trajectory 379
Hyperbolic excess velocities 378

I

ICE *See* International Cometary Explorer
IM *See* Intermodulation
IMOP *See* Integrated Mission Operations Plan
IMP *See* Interplanetary Monitoring Platform
Impact-avalanche transit time diode amplifiers 426
Implementation loss 439
Inclination 348
Inertial space 354, 370
Inferior conjunction 625
Infrared Astronomical Satellite 75
Infrared-sensing instruments 622
Inlock 285
INMARSAT 432
In-situ instruments 622
Instrument data packets 70
Integrated Checkout Plan 317
Integrated console script 329
Integrated Mission Operations Plan 635
Integrated sequence of events 282
  See also Pass plan
Integrated timeline 549
Integration and test plan 80
INTELSAT 432, 433
Interface specifications 201
Interface-control document 550
Intermodulation 413
International Cometary Explorer 797
International Space Station 717–723
  characteristics 717

docking 720
history 717
mission control 722, 723
orbit 723
International Ultraviolet Explorer 801
Internet 81
Interplanetary escape 627
Interplanetary exploration 613
Interplanetary mission 613, 614, 724, 726–728
  flyby 614
  landers 614
  mapping 614
  probes 614
  tour 614
Interplanetary Monitoring Platform 799
Interplanetary spacecraft 20
Interplanetary trajectories 19
Inter-Range Instrumentation Group 602
Intravehicular activity 708
IRIDIUM 410, 423, 433, 558
IRIG *See* Inter-Range Instrumentation Group
Isp *See* Specific impulse
ISS *See* International Space Station
IUE *See* International Ultraviolet Explorer
IVA *See* Intravehicular activity

J

Jamming 399
Jet Propulsion Laboratory 487
Jupiter 628
Justified operation 113

K

Kennedy Space Center 714, 722
Kepler, Johannes 344, 352
  laws of planetary motion 344
Keplerian orbits 344–355
  See also Two-body equations of motion
Klystron 426

L

L&EO operations phase	305
L&EO *See* Launch and early-orbit operations	
LAN *See* Local-area network	
Lander	617
Launch and early-orbit	
timeline	553
Launch and early-orbit operations	301
activities	314
development team	331
environment's demands on the ground element	310
environment's demands on the operations team	309
environment's demands on the operations timeline	308
environment's demands on the space element	311
framework	313
handbook for	332
integrated timeline	318
operations management	328
planning process	312
products for	331
timeline	306, 321
tools for	332
Launch approval	636
Launch azimuth	371
Launch failures	578
Launch operations	28
Launch readiness tests	307
Launch windows	370–372
Launch-configuration table	322
Layered spacecraft architecture	691
LDEF *See* Long Duration Exposure Facility	
LEO *See* Low-Earth orbits	
Level 0 processing	72
Level 1 data	76
Level 3 data	76
Level 4 data	76
Levels of data processing	544
Levels of maintenance	491
Life cycle	
concept exploration	4
detailed development	4
operations and support	5
production and deployment	4
Light time delay	623
Link availability	394
Link margin	434
Local sidereal time	371
Local-area network	464, 694
Long Duration Exposure Facility	708, 711
Longitude of perigee	349
Long-term trending	293
Loop-back test	283
LORAN C	463
Loss of signal	289
Low-altitude, crosslinked architecture	432
Low-Earth orbits	17, 678
LST *See* Local sidereal time	371
Lunar mission	723
profile of	725

M

Magellan	641
Magnetic torquers	560
Magnetometers	560
Maintaining and developing flight software	639
Maintenance activities	253, 256
Maintenance downtime	493
Maintenance objectives	243
Major programs	7
Managing mission operations	85
Manufacturing data	607
Mapping	614
Mariner	613
Mariner 10	628
Mars Global Surveyor	614
Mars mission	723, 726–733
Mature operations	698
Maximum viewing time	386
MCC *See* Mission Control Center	
MCC-H *See* Mission Control Center, Houston	
MCC-M *See* Mission Control Center, Moscow	
Mean anomaly	352–354
Mean motion	353–354
MER *See* Message error rate	
Mercury	628
Message error rate	440

Index

MFSK *See* Multiple-frequency shift keying
Microsatellite 677
  budget 678
  communication architecture 686
  missions 678, 685
  payloads 680
  S80/T mission 699–705
  teams 679
Microwave radio 457
Minimum shift keying 414
Mission activities 255
Mission activity's access 256
Mission complexity 467
Mission concept 20, 31, 91, 97, 193, 196
Mission constraints 17
Mission control 49
Mission Control Center 455
  Houston 715, 722, 723
  Moscow 722
Mission controller 283
Mission data 290, 559
  See also Payload data
Mission database 33, 75
Mission description 98
Mission design and planning 130
Mission director 85
  See also Mission operations manager
Mission malfunctions 579–586
Mission Needs Statement 7
Mission objective 15, 243
Mission operations 10, 31
  concept 11, 15, 23, 31, 91, 114, 196, 203, 687
  cost model 29
  cost model for 131, 146
  cost of 11, 98
  element 15, 196
  functions 35, 95, 196, 616
    relative cost 36
  ground element for 193
  manager 1, 85, 91, 193, 214, 529, 737, 739, 743, 753, 759, 762, 766
    requirements 759
  scenarios 222, 235
  system 10, 32, 86, 193
  working group 331
Mission orbit 467
Mission phases 40, 107
Mission philosophies, strategies, and tactics 98
Mission plan 37, 39, 270
Mission planning 37, 271
Mission rules 40, 46
Mission sponsor 466
Mission statement 195, 245
Mission success 588–595
Mission type 467
Mission utility and complexity 98
Mission-engineering process 530, 547
Mission-phase plan 40
Mission-planning plan 327
Mission-specialist astronauts 713
Mission-type constants 145
MLI *See* Multilayer insulation
Model calibration 140
Modeling parametric constraints 258
  See also Constraints
Modulation 413
Modulation technique 473
Molniya orbits 18
  parameters 358
Molniya-orbit architecture 432
MOM *See* Mission operations manager
MOS function 106
MOS *See* Mission operations system
MSK *See* Minimum shift keying
Multilayer insulation 571
Multi-mission operations 333
Multiple frequency shift keying 414
  See also Modulation techniques

N

NASA
  Small Explorer 531
NASA Deep Space Network 462, 486
NASA Tracking and Data Relay Satellite System 462, 481
National Academy of Science 75
National Security Decision Directive 42 708
Naval Satellite Control Network 479
Navigation 638
Navigation planning and analysis 59, 297

Navigation spacecraft 19
NDI *See* Non-developmental items
Needs analysis 94
Negative margins 633
Neptune 631
Network Control Center 455
Newton, Isaac 344
Nodal vector 350
Noise jammer 399
Non return to zero-level 472
Non-coherent mode 462
Non-developmental items 396, 443
North-South drift 374
NRZ-L *See* Non return to zero-level
NSCN *See* Naval Satellite Control Network
NSDD *See* National Security Decision Directive 42

O

Oblateness of the Earth
   effect on orbits 372
Obliquity 626
Odyssey 412
Onboard processing transponder 410
One way light time 624
Operating principles 204
Operational complexity 66
Operational complexity metrics 130, 145
   flight systems 130, 134
   ground system 141
   mission design and planning 130
   risk avoidance 139
Operational constraints 548
Operational cost drivers 93
Operational demonstrations 14
Operational flight margins 138
Operational modes 535
Operational risks 138
Operational scenarios 92, 105, 107, 114, 548
Operational workarounds 548
Operations
   anomaly 28
   early-orbit 28
   extended 28
   launch 28
   routine 28

Operations and maintenance 475
Operations and support 94
Operations cases 333
Operations complexity 130
Operations concept 86
   developing an 25
Operations costs 130
Operations framework 313
Operations manual 605–607
Operations scenarios 97
Operations style 267
Operations workforce 145
Operations-management plan 327
Operations-research techniques 263
Operator
   displays 228
   loading 219
   positions 218
   tasking 218
   tasks 218
Operator role in the planning process 259
Operator tasks
   definition of 218
Opposition 625
Optical cable 456
Optical crosslinks 427
Optical tracking 461
Optimal operation of spacecraft subsystems 226
Orbit determination 354, 638
Orbit elements 348–350
   argument of perigee 348
   eccentricity 348
   eccentricity vector 348
   inclination 348
   nodal vector 350
   right ascension of ascending node 348
   semi-major axis 348
   time since perigee passage 349
   true anomaly 349
   variations
      secular and periodic 355
Orbit insertion 628, 640
Orbit maintenance 372–375
Orbit maneuvering
   coplanar orbit transfers 361–364
   one-tangent-burn 363
   spiral transfer 365

Index

Orbit period
  equations for   350
Orbit perturbations   355–360
  atmospheric drag effects   359
  due to non-spherical Earth   357–359
  due to third-body interventions   356
  general perturbations   356
  long-period variations   355
  secular variations   355
  short-period variations   355
  solar radiation effects   360
  special perturbations   356
Orbit rendezvous   368–369
Orbit transfer   361–364
  See also Hohmann transfer orbit
  plane change   365–367
Orbital constraints   540, 548
Orbital maneuvering   360–370
Orbital plan   40
Orbit-insertion burn   628
Orbit-propagation-geometry file   273
Orbits
  circular velocity   347
  direct   351
  elements   348–350
  equations of motion   344–348
  escape velocity   347
  geostationary   17
  Global Positioning System   18
  Hohmann transfer orbit   361
    See also Hohmann transfer orbit
  Keplerian orbits   344–355
  low-Earth   17
  maneuvers   360–370
  Molniya   18
  Molniya orbits   358
  retrograde   351
  semi-synchronous   17
Orbits   344–355
  See also Orbit perturbations, Sun-synchronous orbit, Molniya orbit, Geosynchronous orbit, Orbit transfer, Stationkeeping, Constellation
Orbit-trim maneuvers   298
Other-loss term   439
OTM See Orbit-trim maneuvers

P

Packet telemetry   602
Packetized telemetry   473
Packet-switched network   423
Parabola (Conic section)   346
Parabolic trajectory   347
Parametric constraints   251, 258
  See also Constraints
Partial-band noise jammer   399
Pass   282
Pass plan   52, 282
Pass report   53
Patched-conic approximation   377
Payload
  appendages   540
  calibration   534, 545
  capabilities of   100
  characteristics of   100
  complexity driver   534
  data system   541
  operational complexity   534
  operations   529, 547
  operations concept   532
  operations plan   552
  orientation   534, 537
  planning   534
  reconfiguration   534
Payload calibrations   69
Payload data   71, 290
Payload data processing   70
Payload data records   72
Payload flight software   69
Payload operability   23, 70
Payload Operations Control Center   455
Payload planning and analysis   66, 296
Payload tasking   296
Payload telemetry   452
Payload's operation   101
Payload-calibration plan   112
Payload-product verification   537
Payload-specialist astronauts   713
PCM See Pulse code modulation,
  See also Modulation techniques
PDR See Preliminary design review
Penumbral eclipses   574
Performance analysis   293
Performance models   140
Phase modulation   473
Phase-shift keying   473, 569

Phasing orbit	368
Physiological rehabilitation	709
Pioneer	613
Pipeline	510
Plan	245
Plan resources	
defining	249
Planning paradigms	260
Plan-transition time	262
Plan-visualizing tools	260
PM *See* Phase modulation	
PN *See* Pseudorandom noise code	
POCC *See* Payload Operations Control Center	
Polar equation of a conic section	345
Polar orbit	351
Positive margins	41
Post-launch activities	34
Power generation	633
Power subsystems	559
Powered-flight profile	307
Predicted ephemeris	297
See also Ephemeris	
Pre-encounter maneuvers	640
Pre-launch activities	34
Preliminary design review	197, 209
Preliminary mission operations concept	213
Preliminary mission operations concepts	245, 247
Press releases	74
Pressurization subsystem	564
Prime-focus	459
Private data links	456
PRN *See* Pseudo-random noise code	
Probe	618
Procedures	
definition of	228
Processing concepts	216
Processing delay	394
Product ciphering	395
Production and deployment	94, 312
Prograde orbit (direct orbit)	351
Program constraints	99
Program Initiation	7
Progress	718, 720
Project "vision"	78
Project Science Group	631
Propagation delay	393
Propulsion anomalies	598
Propulsion subsystem	560
Pseudo anomaly	599
Pseudo-random noise code	418
Pseudorandom noise code	396, 420
PSK *See* Phase-shift keying	
Public-key encryption	395
Pulse jammer	399
Pulse-code modulation	472, 569

Q

QPSK *See* Quadrature phase-shift keying	
Quadrature phase-shift keying	414, 473, 569
Queuing delay	394
Quick-look analysis	67, 69

R

Radar systems	461
Radiation	710
Radio frequencies	55
Radioisotope Thermoelectric Generator	633
Radiometric data	58
See also Tracking data	
Rain attenuation	404, 405
Range	461
Range controller	284
Range rate	461
Ranging data	291
Raw-telemetry record	603
RCC *See* Resource Control Center	
RDT&E spacecraft	20
Real-time command and control	322
Real-time commands	241, 275
Real-time operations	631
Receive gain	458
Rechargeable batteries	569
Reconfiguration	535
Reconstructed ephemeris	297
See also Ephemeris	
Regenerative transponder	409, 410
Regulated systems	563
Remote-sensing instruments	621, 622
Remote-sensing spacecraft	19
Repeat-back jammer	399
Replenishable consumables	137

Requirements specification 197
Requirements Validation 8
Resource Control Center 455
Resource management 250
Resource margins 21
Resource scheduling 493
Resources 244, 258
Resources controller 284
Restricted commands 287
Retrograde orbits 351
Rf signal strength 601
Right ascension of ascending node 348
Risk avoidance 130, 139
Routine operations 28, 268
RTG *See* Radioisotope Thermoelectric Generator

S

S80/T microsatellite mission 699–705
Safe haven 598
Safemode 538
Safing 137
SAMPEX mission 321
SAMPEX *See* Solar, Atmospheric, Magnetic Particle Explorer
Satellite Control Facility 602
Satellite crosslinks 410
Satellite Switched-TDMA (SS-TDMA) 420
Satellite-to-satellite crosslinks 432
Satisfying constraints 257
  See also Constraints
Saturn 628, 641
Scale height (atmosphere)
  atmospheric density 360
SCF *See* Satellite Control Facility
Schematics 604
Schematics, wire lists, signal flow diagrams 607
Science data 290
  See also Payload data
Science or engineering event 134
Scientific spacecraft 19
SDLC *See* Synchronous data-link control
Sectoral terms (geopotential) 357
Semi-major axis 348
Semi-synchronous orbits 17

Seven-layer model for satellite transmission 472
SEZ *See* Topocentric horizon
SFDU *See* Standard formatted data unit
SGLS *See* Space ground link subsystem
SGLS *See* Space-Ground Link Subsystem
Short-term trending 293
Shuttle Mission Simulator 714
Shuttle/Centaur 630
Shuttle/Inertial Upper Stage (IUS) 630
Signal suppression 410
Signal-flow diagrams 604
Signal-to-noise ratio 416, 434
Simultaneity constraints 251
  See also Constraints
Single-point failure 597
Single-point-failure parts 604
Single-shift operations 273
Single-station model 452
Skylab 711
Slant range 397
Slave tracking 459
Small Explorer 531
SMEX *See* Small Explorer
SOCC *See* Spacecraft Operations Control Center
SOCM *See* Space Operations Cost Model
Software executables 209
Software-development project 83
Software-processing architectures 207
SOHO *See* Solar Heliospheric Observatory
SOI *See* Sphere of influence
Solar Anomalous and Magnetosphere Explorer 304
Solar cycle 625
Solar flux 624
Solar Heliospheric Observatory 537, 805
Solar system 378
Solar, Atmospheric, Magnetic Particle Explorer 803
Solid-state amplifiers 426
Solid-state power amplifiers 426
SONET *See* Synchronous optical network
Source-to-destination connectivity 410
Soyuz 718, 720
Space 771

Index

Space ground link subsystem 568
Space mission
  architecture 8–9
  developer 6
  end users 5
  operators 5
  sponsor 5
Space mission elements
  command, control, and
    communications 10
  ground 10
  launch 10
  orbit 10
  payload 10
  spacecraft 10
  subject 8
Space Operations Cost Model 149–191, 772–778
  Earth orbiting mission
    example 182–189
  Level 1 149–152
  Level 2 153
  planetary mission example 173–182
  reference mission set 155
  tuning 156–159
Space Shuttle 712, 720, 722
  abort option 709
  crew training 713
  ground system 714–716
  ground tracking 716
  launch 714
  losses 708
  mission 712
  mission planning 712
  payload 710, 712, 714
Space Station *See* International Space Station
Space Telescope Science Institute 75
Space Transportation System 554
Spacecraft
  communications 19
  interplanetary 20
  navigation 19
  RDT&E 20
  remote-sensing 19
  scientific 19
  technology demonstration 20
Spacecraft acquisition scenario 236
Spacecraft analyst 286
Spacecraft autonomy 14, 21, 26, 42, 692
Spacecraft avionics 33, 106
Spacecraft bus
  capabilities of 103
  characteristics of 103
Spacecraft bus and payload
  calibrating 237
  scenario for calibrating 237
Spacecraft bus commands 64
Spacecraft command and control
  scenario 240
Spacecraft contingency plan 64
Spacecraft ephemeris 297
  See also Ephemeris
Spacecraft health and safety 50, 557
Spacecraft integration and test 50
Spacecraft loading 240
Spacecraft operability 13, 24
Spacecraft operational attributes 103
Spacecraft operations 572
Spacecraft Operations Control Center 455
Spacecraft planning and analysis 62, 292, 557
Spacecraft power
  scenario 241
Spacecraft safemode operations
  scenario 238
Spacecraft safing 66
Spacecraft scenarios
  acquisition 236
  command 236
  control 236
  payload calibrations 236
  power 236
  spacecraft bus 236
  spacecraft safemode operations 236
  state of health operations 236
Spacecraft simulators 47
Spacecraft state of health 201
Spacecraft state of health operations 239
  scenario 239
Spacecraft status and health 285
Spacecraft stored-energy system 572
Spacecraft structure 571
Spacecraft subsystems 559, 561
Spacecraft thermal environment 571
Spacecraft-bus operations 557
Spacecraft-bus performance 557
Space-Ground Link Subsystem 474
Special operations 268, 553

Special perturbations 356
See also Orbits, Orbit perturbations
Specific angular momentum 348
Specific impulse 563
Specific mechanical energy 347
Speed of light 623
Sphere of influence 376
Spread-spectrum techniques
  direct sequence 396
  frequency hopping 396
  Pseudorandom noise 396
SSA See Solid-state power amplifiers
SSTL See Surrey Satellite Technology, Ltd.
Standard commanding 275
Standard formatted data unit 57
Star sensors 560
Station pass reports 58
Station QQC reports 58
Station-allocation file 273
Stationkeeping 374
  See also Orbit maintenance
Status and announcement plan 327
Store and forward 430, 469
Store-and-forward telecommanding 56
Stored command loads 276, 322
Stored commands 631
Stressed channel 402
Strip charts 608
Structural mechanisms 571
Structures and mechanisms 561
STS See Space Transportation System
Submillimeter Wave Astronomy Satellite 807
Sufficiently inertial coordinate frame 348
Sun sensors 560
Sun-synchronous orbits
  determining inclination 358
Superior conjunction 625
Surrey Satellite Technology, Ltd 699
Survey payloads 532
SWAS See Submillimeter Wave Astronomy Satellite
Swingby trajectories 381
Switched communications network 423
Synchronizing frames 291
Synchronous data-link control 473
Synchronous optical network 457
Synodic period 624

System architecture 80
System modeling 227
System noise temperature 435
System noise-temperature term 437
System performance
  defining 223
System requirements 200
  functional approach 200
  specification 199
  topical approach 200
Systems engineering, integration, and test 78

T

Targets of opportunity 536, 553
T-carrier 457
TCMs See Trajectory-correction maneuvers
TDMA See Time-division multiple access
TDRSS See Tracking and Data Relay Satellite System
Technology demonstration spacecraft 20
Technology development plan 92, 98
Telecom space loss 624
Telecommand 56
Telemetry 452
Telemetry analysis 293
Telemetry channels 137
Telemetry dictionary 64
Telemetry, tracking, and command subsystems 452, 563
Telemetry-data streams 603
Tesseral terms (geopotential) 357
Test plans 329
Test-data packages 604
Thermal control 560, 571, 633
Thermal environment 624
Three-shift rotation 274
Thruput telecommanding 56
Time constraints 250
  See also Constraints
Time standard 602
Time-division multiple access 420–422
Time-driven schedule 23
Timeline Assessment Table 317
Timelines 46, 92, 97, 112, 114

Time-of-flight	369
in an elliptical orbit	352–354
Time-phased loading plots	319
Timing	458
Timing and timelines	220
Timing subsystem	462
TOMS *See* Total Ozone Measuring System	
TOMS-EP *See* Total Ozone-Mapping Spectrometer—Earth Probe	
Tone jammer	399
Topocentric horizon coordinates	385
Total Ozone Measuring System	542
Total Ozone-Mapping Spectrometer—Earth Probe	569
Tour	615
Tracking	452
Tracking and Data Relay Satellite System	432, 548, 569
Tracking data	58, 291
Traditional mission operations planning	2
Training	679
Training manual	607
Training plan	80
Trajectory-correction maneuver	59, 636
Transactional planning	261
TRANSEC *See* Transmission security	
Transition to normal operations	305
Translunar injection	724
Transmission security	395
Traveling-wave tube amplifier	426
Trend analysis	67, 69, 140
Trending	632
Trending analysis	293
Trimmable activities	253
TRMM *See* Tropical Rainfall Measuring Mission	
Tropical Rainfall Measuring Mission	809
True anomaly	349, 354
True longitude	350
TT&C *See* Telemetry, tracking, and command subsystems	
TT&C subsystem	460
Turnaround time	493
Turn-key systems	477
Two-body equations of motion	345–348
orbit maintenance	372–373
perturbations	357
Two-body equations of motion	345
See also Orbits	
Two-shift rotation	274
TWTA *See* Traveling-wave tube amplifier	

U

Uplink	33, 107, 282
Uplink window	282
Uplinking	281
Uranus	641
User terminals	398

V

Validate activities	46
Vehicle velocity, local horizon coordinates	385
Venus flybys	625
Verification	
methods	231
Vernal equinox	371
Very-long-baseline interferometry	291
Video-conference capabilities	81
View period	284
Viewing time	397
VLBI *See* Very-long-baseline interferometry	
Voyager	613, 628, 641
Voyager II	631
VVLH *See* Vehicle velocity, local horizon	

W

Wake Shield Facility	708, 711
WAN *See* Wide-area networks	
Waterfall approach	210
definition of	197
development	197
WBS *See* Work Breakdown Structure	
Weighting factors	145
White room	714
Wide-area networks	694
Wideband	399
Work Breakdown Structure	151, 160, 171–173

Workstations   81

X

X-ray Timing Explorer   548, 811
    timeline   549
XTE *See* **X-ray Timing Explorer**

Z

Zero gravity   710
Zero-base operations   13
Zero-defect downlink   138
Zero-defect uplink   138
Zonal coefficients   357